# PHYSIOLOGICAL BASIS OF AGING AND GERIATRICS

## THIRD EDITION

EDITED BY

# PAOLA S. TIMIRAS

# CRC PRESS

Boca Raton   London   New York   Washington, D.C.

Cover drawing of Great Basin Bristlecone Pine (*Pinus longaeva*) by Ed Monroe. According to dendrochronologists, these trees have been documented to live up to 5000 years.

## Library of Congress Cataloging-in-Publication Data

Physiological basis of aging and geriatrics / edited by Paola S. Timiras.--3rd ed.
    p.  cm.
    Includes bibliographical references and index.
    ISBN 0-8493-0948-4 (alk. paper)
    1. Aging--Physiological aspects. 2. Geriatrics. I. Timiras, Paola S.
    [DNLM: 1. Aging--physiology. 2. Geriatrics. WT 104 P578 2002]
  QP86 .P557    2002
  612.6'7—dc21                                        2002031179

**Visit the CRC Press Web site at www.crcpress.com**

© 2003 by CRC Press LLC

No claim to original U.S. Government works
International Standard Book Number 0-8493-0948-4
Library of Congress Card Number 2002031179
Printed in the United States of America    2 3 4 5 6 7 8 9 0
Printed on acid-free paper

# Preface

Given the rapidly increasing advances in the areas of gerontology and geriatrics, one of the major purposes of this third edition is to provide updated material. Never before in human history have we known so much about our bodies, and never before have so many humans lived longer, worldwide. It is predicted that by the year 2030, a large proportion of the human population will live 65 years and longer (in the United States, 18%). Presented here are clues of the physiologic, genetic, and molecular mechanisms that may determine how long humans may live. It is unlikely that a single "triggering" event is responsible for the aging and death of the human organism; rather, aging and death probably entail numerous and complex interactions at different genetic and environmental levels. The recent completion of the Human Genome Project has opened a new and exciting era to study interactions of genes with the environment. In this edition, the study of physiologic competence not only emphasizes the classic and dynamic ability of humans to adapt to changing conditions but also identifies the potential role of genetic and environmental risk factors in the etiopathology of old age.

Chapters have been grouped into three main sections: *General Perspectives* (including demography, cellular and molecular biology, theories of aging), which provides a varied and solid background in basic processes of biogerontology; *Systemic and Organismic Aging,* a survey of the aging of body systems, which focuses on maintenance of optimal function and ability to adapt to environmental demands; and *Prevention and Rehabilitation,* a synopsis of pharmacologic, nutritional, and physical exercise guidelines for a lifestyle conducive to longevity with preservation of physical and mental health.

Our anticipated goal is to provide a book that will be useful to a broad spectrum of individuals with different levels of biological and educational backgrounds. While most texts target either a professional/academic readership or a more general audience, this book targets both. It is intended to meet the needs of a diverse readership: those preparing for a career in gerontology or geriatrics, those interested in aging as a specific topic in biology, and older persons (as well as their families and caretakers) who would like to better understand the changes that are occurring with aging in their own bodies. To fulfill this goal, and using physiology as a unifying concept, the original text has been streamlined with more explicit and concise explanations. Focus has been on the established facts of physiological aging, and explanations in the text have been illustrated with numerous and comprehensive tables and graphs.

Some of the goals of the previous editions, that is, viewing aging from several perspectives, have been preserved. Information has been distilled from multiple sources and concisely integrated. According to the traditional physiological approach, aging is examined in the various organs and systems, focusing on humans, but drawing, when appropriate, on animal research.

One of the characteristics of aging is the increasing incidence of disease. Clinical correlations are included both as a practical reference for the geriatrician and as a comparison with normal aging for the gerontologist.

Aging has been viewed as an individual's "journey taking place in a community setting" from which the individual cannot be separated. Therefore, aspects of demographic, comparative, and differential aging are included with a discussion of the several theories of aging to provide a more complete profile of the aging of individuals and populations.

As more is learned of the aging process, the possibility of intervening is within reach. When aging is viewed through the eye of medicine, the emphasis is on "the elderly at risk" and on the need for assessing, managing, and reducing "risk" factors. Equally important is the strengthening of physiologic competence. The bases for future interventions to improve the quality of life are discussed with an update of those currently in use.

Generally, the meaning of aging is not perceived at its own value; rather, what is meaningful is "to reach old age." The concept of "continuity through change" is fundamental to all biological processes. As one ages, continuity of prior physiologic events may provide "a usable past" that can shape future functions. Indeed, in a previous text (*Developmental Physiology and Aging*, Timiras, 1972), identity in old age has been viewed to be a dynamic process, as it is at young ages. In this edition, reports of studies not only of physiologic but also of cellular and molecular events throughout the life span may provide a better understanding of the evolving continuum of the changes that characterize aging and their role in the physiopathology of disease and survival into old age.

# Acknowledgments

I have been greatly encouraged in preparing this third edition by the participation of several of the collaborators of the first and second editions and by the equal enthusiasm and expertise of the first-time coauthors. All have willingly accepted the task of reviewing, updating, or preparing anew their respective chapters or sections within stringent deadlines. To them, I offer my most heartfelt thanks for an extraordinarily well-accomplished performance.

Special thanks to Dr. S. Oklund for the many drawings, which effectively illustrate and integrate the complex material presented. I also gratefully acknowledge the continuing support of Dr. L. L. Rosenberg, who consistently provided constructive editorial and substantive criticism.

I also wish to thank my assistants, Marlena Chu, Mona Higuchi, and Mariel Kusano, who competently prepared the manuscripts according to the specifications of the publisher and dealt with the word processing, editing, and formulation of the many tables that effectively integrate the complex and diverse information. I would also like to thank the librarian, I. Radkey, for her assistance with the bibliography.

Finally, I would like to recognize the silent encouragement of the many hundreds of students who have taken, over the years, my class, "Physiology of the Aging Process." By their enrollment in the class, interest in the subject matter, their criticism or praise, they have inspired me to continually improve, streamline, and update the course material, a process that has resulted in the subsequent editions of this book.

**Paola S. Timiras**
*University of California, Berkeley*

# Contributors

**John K. Bielicki, Ph.D.**
Lawrence Berkeley National Laboratory
Berkeley, California

**Chester M. Brown, M.S.**
University of Illinois at Urbana-Champaign
Urbana, Illinois

**Judith Campisi, Ph.D.**
Lawrence Berkeley National Laboratory
Berkeley, California

**James R. Carey, Ph.D.**
University of California, Davis
Davis, California

**Henry F. Emerle, M.S.**
University of Illinois at Urbana-Champaign
Urbana, Illinois

**Trudy M. Forte, Ph.D.**
Lawrence Berkeley National Laboratory
Berkeley, California

**Lia Ginaldi, M.D.**
L'Aquila University, Italy
L'Aquila, Italy

**Ali Iranmanesh, M.D.**
Veterans Affairs Medical Center
Salem, Virginia

**Michael L. Johnson, M.D.**
University of Virgina School of Medicine
Charlottesville, Virginia

**Daniel Keenan, Ph.D.**
University of Virginia School of Medicine
Charlottesville, Virginia

**Joyce Leary, M.S.**
University of California, San Francisco
San Francisco, California

**Jay Luxenberg, M.D.**
Jewish Home and University of California, San Francisco
San Francisco, California

**Massimo de Martinis, M.D.**
L'Aquila University, Italy
L'Aquila, Italy

**Rolf J. Mehlhorn, Ph.D.**
Lawrence Berkeley National Laboratory
Berkeley, California

**Esmail Meisami, Ph.D.**
University of Illinois at Urbana-Champaign
Urbana, Illinois

**Franco Navazio, M.D., Ph.D.**
University of California, Berkeley
Berkeley, California

**Sally Oklund, Ph.D.**
San Jose, California

**Hal Sternberg, Ph.D.**
BioTime, Inc.
Berkeley, California

**Mary L. Timiras, M.D.**
Overlook Hospital
Summit, New Jersey
and
University of Medicine and Dentistry of New Jersey
Newark, New Jersey

**Paola S. Timiras, M.D., Ph.D.**
University of California, Berkeley
Berkeley, California

**Johannes D. Veldhuis, M.D.**
Mayo Clinic and Graduate School of Medicine
Rochester, Minnesota

**John R. Wilmoth, Ph.D.**
University of California, Berkeley
Berkeley, California

**Phyllis M. Wise, Ph.D.**
University of California, Davis
Davis, California

# Contents

## PART III  *Prevention and Rehabilitation*

# Part I

## General Perspectives

# 1 Aging as a Stage of Life, Common Terms Related to Aging, Methods Used to Study Aging

*Paola S. Timiras*
University of California, Berkeley

## CONTENTS

## I.   SIGNIFICANCE OF PHYSIOLOGY OF HUMAN AGING

Never before in human history has so much been known about our bodies and the genetic, molecular, and physiologic mechanisms that dictate their function. Never before, have so many humans lived longer, with the percentage of Americans aged 65 and older (65+) rising from 4% in 1900 to 13% in 1990, an increase that is even higher in the population of other countries (Chapter 2). Yet, little is known about why we become old. It is predicted that by the year 2030, approximately one-fifth (18%) of the population of the United States will live 65 years and longer. Such impressive and persistent increase in the elderly (65+) population underlies the pressing need for a better understanding of the aging process and the problems of the elderly. Among the challenges that greet the beginning of the 21st century is how we can maintain good health and postpone or reverse aging along with its debilitating diseases. In other words, how can we preserve youthfulness in later life?

Why study physiology of human aging? Although the pathology of old people has been studied extensively for combating specific diseases associated with this "high-risk" group, the physiology of aging has not been of primary interest, partly because of the difficulty of isolating "normal"

(the domain of physiology) from "abnormal" (the domain of pathology) aging processes. While old age may result from a "normal" process, it makes us more vulnerable to disease and injury than any other preceding age.

Aging has, thus far, defied all attempts to establish objective landmarks that would precisely signal its earlier stages. It lacks specific markers characteristic of earlier life periods such as menarche at puberty. Old age in humans is conventionally accepted as the stage of the life cycle that starts around 65 years of age and terminates with death. However, given the considerable heterogeneity of the elderly population, it is difficult to circumscribe its temporal boundaries in physiologic terms. Rather, its onset occurs at some "indeterminate" point following maturity, and its progression follows timetables that differ with each individual and vary depending on genetic and environmental factors. "Physiologic heterogeneity" is one of the consistent characteristics of the elderly population.

## II.   ORGANIZATION OF MATERIAL

The breadth of the included material reflects the accumulation of a wealth of information on aging since publication of earlier texts by this author.[1–3] Traditionally, physiology studies the normal function of biological systems. Thus, a significant part of this book is concerned with aging of the various body systems and their functions, focusing on humans but drawing, when necessary, on animal research. To provide a more comprehensive understanding of physiologic changes with aging, some clinical relationships and therapies based on physiologic concepts are also included.

Understanding aging of body systems, organs, and tissues depends on advances in our knowledge of molecular and cellular aging. Also, the various theories of aging provide a background for comprehending aging processes. All of these, together with demographic, comparative, and differential aspects of aging, are considered in the first seven chapters.

Chapters 8 to 22 describe aging of the major body systems, and Chapters 23 and 24 discuss some current interventions for prevention/delay of physiologic changes with aging and rehabilitation of the elderly with disabilities.

Our anticipated goal is to provide a book that will be useful to a broad spectrum of individuals with different levels of biologic background. It is intended to meet the needs of a diverse readership: those preparing for a career in gerontology or geriatrics, those interested in aging as a specific topic in biology, and older persons (as well as their families and caretakers) who would like to better understand the changes that are occurring with aging in their own bodies. Each chapter is organized to begin with *structure,* then to move to *function,* considered first in the adult (a brief synopsis) and then in the aged. Essential material is presented in normal text size; smaller type furnishes additional information, in-depth discussion, or less well-accepted or controversial matters.

The intent to attract a broad readership may seem too ambitious. The studies of aging and geriatrics are fast expanding, and they attract people from many disciplines who require a common understanding of the biology of aging. Current texts are directed to very specific aspects of aging or are designed for a particular group of readers. The present work, using physiology as the unifying concept, assimilates and distills information from multiple sources to produce an understandable, comprehensive text.

Any consideration of aging should not ignore the psychological and social components, particularly important for the geriatrician who deals with the individual as a whole, in its unique socioeconomic environment. Because these fields require extensive and competent presentation well beyond the scope of this book, such subject matter has not been included.

## III. THE JOURNEY OF LIFE[4]

Historically, chronological age has been used to assess the transition from one stage of life to the next. In the broadest

sense, the human life span is divided into two main periods: prenatal and postnatal. In our contemporary society, individuals, after birth, progress through a series of stages identified by a chronological timetable: "children begin school at age five, young people go to work or to college at 18, old people retire at 60 or 65….age is being taken as a criterion for sequencing…the multiple roles and responsibilities that individuals assume over a lifetime."[4] The most striking physiologic events take place in embryonic and fetal stages, but important changes continue to occur even in adult and old ages.

Many animal species are capable of an independent existence at relatively immature stages, others, including mammals, are not. The human newborn is utterly dependent on adults for food and care. Throughout infancy, childhood, and adolescence, remolding of body shape continues gradually, together with the acquisition of new functions and the perfecting of already established functions. At about 25 years, the last of these changes is completed, and the body is stabilized in the adult condition. The mature adult period lasts for approximately another 40 years and encompasses the period of maximal physiologic competence.

The view that prevailed until a few years ago was that the span from late adulthood until death was characterized essentially by a decline in normal function and an increase in pathology and disease. However, information gained during the past 20 years has allowed a reconceptualization of the aging process.[5] The substantial heterogeneity among old persons and the existence of positive trajectories of aging without disease, disability, and major physiologic decline have offered a more positive view of a possible successful (healthful) aging (Chapter 3). Such an optimistic outlook has been adopted throughout this book, while, at the same time, the text will discuss some of the depressing changes that may accompany aging.

## A. PRE- AND POSTNATAL LIFE STAGES

Both prenatal and postnatal stages may be subdivided into several periods, each distinguished by morphologic, physiologic, biochemical, and psychological features. Main divisions and the approximate time periods of the life span in humans are listed in Table 1.1.

The prenatal period includes three main stages: oval, embryonic, and fetal. The postnatal period begins with birth and continues into stages of neonatal life, infancy, childhood, adolescence, adulthood, and old age. The concept that the life span is divisible into successive stages is not new. "Ancient authors established the principal traditions of the ages of life: divisions into three, four and seven."[6] For example, Aristotle divided life into three ages—growth, stasis, and decline—modified later in Greek medicine and physiology into four ages—childhood, youth, maturity, and

**TABLE 1.1**
**Stages of the Life Span**

| Stage | Duration |
|---|---|
| Prenatal life | |
|   Ovum | Fertilization through week 1 |
|   Embryo | Weeks 2–8 |
|   Fetus | Months 3–10 |
| Birth | |
| Postnatal life | |
|   Neonatal period | Newborn; birth through week 2 |
|   Infancy | Three weeks until end of first year |
|   Childhood | |
|     Early | Years 2–6 |
|     Middle | Years 7–10 |
|     Later | Prepubertal; females 9–15; males 12–16 |
|   Adolescence | The 6 years following puberty |
|   Adulthood | Between 20 and 65 years |
|   Senescence | From 65 years on |
| Death | |

old age. The practical necessity of research to fragment the biologic study of living organisms, however, has a tendency to obscure the dynamic relationships that are obtained at many levels of organization. We should recall here that human development depends on "a program of genetic switches that turn on in a highly regulated manner, at specific places and times," and that "responses to environmental challenges fostering changes early on 'may reverberate decades later in the guise of cardiovascular diseases and diabetes."[7]

The physiologic profile of a given individual must be assessed with regard to his particular life history in all its biosocial complexity. Succeeding stages of the life span depend on the events that have occurred earlier. Indeed, as the saying goes, "growing old gracefully is the work of a lifetime!" The concept that attained growth tells of past growth and foretells growth yet to be achieved can even more appropriately be extended to the later stages of the life span. There is increasing experimental and epidemiological evidence that events in earlier ages can set the stage for disease in the adult and old ages.[7]

## B. STAGES OF MATURITY AND OLD AGE

The mature years are considered a major life stage. They are characterized by great functional stability as connoted by the attainment of optimal, integrated function of all body systems. Thus, function in adulthood is taken as a standard against which to measure any degree of physiologic or pathologic deviation. In most textbooks of human physiology, the mature 25-year-old, 70 kg, 170 cm man is taken as a model. In the present text, this period and this average man will serve mainly as a reference point for the discussion of aging and the aged.

Functional competence is multifaceted, and optimal performance may differ from age to age and from one parameter to another. In other words, it would be unsound, if not physiologically incorrect, to assume that a function is maximally efficient only during adulthood and that differences in the earlier or later years necessarily represent functional immaturity or deterioration, respectively. Rather, one must view physiologic competence as having several levels of integration, depending on the requirements of the organism at any specific age and the type and severity of the challenges to which the organism is exposed. Indeed, adequate function of several organs and systems may well persist into old age.

Studies of developmental physiology and aging provide the basis for conceptualizing physiologic competence along a continuum throughout the life span. Although the age of 65 has been accepted as the demarcation line between maturity and old age, a person of that age may be quite healthy and a long way from "retiring" from life.

If it is difficult to subdivide the adult years into physiologic stages, old age presents even more of a challenge in this regard. In the elderly, there are the following:

1. A great heterogeneity of responses among individuals of equal age
2. Changes that take place do not occur simultaneously at all levels of functional organization
3. Changes do not involve all functions to the same degree

It appears more likely that each change, whether at the molecular, cellular, tissue, or organismic level, follows its own timetable that, in turn, is differentially susceptible to specific intrinsic and extrinsic factors (Chapter 3).

Gerontology should not be restricted to the identification of pathologic factors. Old age may not be regarded simplistically as a "disease." Rather, it involves a complexity of physiologic and pathologic phenomena, all subject to numerous environmental influences.[8,9] Thus, as a life stage, aging is presented here as it affects the functioning of the whole organism as well as its specific systems; whenever appropriate, correlations will be drawn with underlying alterations at the cellular and molecular levels.

## C. DEATH VERSUS IMMORTALITY

While old age is approached gradually without any specific physiologic markers of its onset, death is the terminal event that ends life. Various types of death terminate the aging process. In broad terms, these include trauma, accidents, and disease. Trauma and accidents (e.g., high-speed vehicle crashes, dangerous occupations, drug abuse, cigarette smoking) are the major causes of death in young adulthood. Thus, death needs not to be related exclusively

**General Perspectives**

to aging: disease processes that overwhelm the defense or repair mechanisms of the body affect persons of all ages, but they are particularly life endangering in the very young and in the old. Many diseases that lead to death in the perinatal period—a period of high risk—have been conquered in developed countries, including the United States. Today, the majority of deaths from disease occur in the elderly, in whom diminished function makes the accumulation of pathology less tolerable than in the young. Indeed, some diseases occur almost exclusively in the old, and this linkage of pathology with old age justifies the argument of some investigators that aging is a disease.

Senescence is not an accepted cause of death and is not reported on the death certificate. Whether death is a natural event has never been validated scientifically. Although there may be differences of opinion over how long human beings might live, there has never been any doubt that they must die. Painless, natural (i.e., not from accidents, violence, or disease) death from "pure" old age has been predicted, but, so far, never achieved. The attitude of tacit acceptance of a debilitating old age is now being replaced by one that regards senescence as the "subversion of function," the inevitability of which is open to question. The question is being raised whether it is simply a "design fault" that we age and die.[10] Organ/cell transplantation/cloning, i.e., replacement of defective organs or cells with better functioning ones, represents, perhaps, a crude and clumsy approach to postponing death. Research for new pharmacologic and genetic interventions, for better nutrition and better habits of life (e.g., more physical exercise, no cigarette smoking) offer startling potential for the amelioration and prolongation of life. With the elucidation of the human genome, genetic research, particularly, attracts increasing support because of its potential use in diagnosis and therapy of disease.

Nostalgically, we considered immortality in the light of extending our life through our offspring; genes, in a sense, are immortal. "The genes in each of us come from our earliest humanoid ancestors and their genes from the earliest forms of life on earth."[11] However, this sense of immortality no longer seems adequate. It is frustrating that in a time when humans have gone into and returned from outer space and can manipulate DNA, they have not conquered death. Death, indeed, remains the last "sacred" enemy.

The concept of a single causative factor to account for aging and old age pervades many areas of biologic study, and gerontology is no exception. It is unlikely that a single "triggering" event is responsible for the aging and death of the human organism; rather, aging and death probably entail numerous and complex interactions at different genetic and environmental levels.[11–13]

Biologists working in the area of aging have defined two major goals: one to prolong human life and the other to significantly enhance physical vigor and viability and to prevent diseases throughout the life span. The fact that the answers to the questions on the nature of the aging process remain obscure needs not deter our search. Only by continually posing bold new questions can we hope to accomplish the spirited goals of the gerontologists, whose motto, "to add years to life and life to years," emphasizes the importance of the quality of life for all of us. New research allows a glimpse into a world in which aging, and even death, may be no longer inevitable.[10]

## IV.  COMMON TERMS RELATED TO AGING

The interpretation of aging as a physiologic process upon which pathology and disease have been superimposed has been formalized under the separate disciplines of *gerontology*, the study of aging processes, and *geriatrics*, the prevention and treatment of the disabilities and diseases associated with old age (Table 1.2). The terms aging and senescence are often used interchangeably despite some substantive differences. *Aging* refers more appropriately to the process of growing old, regardless of chronologic age; it includes all the changes—in the present context, the physiological changes—that occur with the passage of time, from fertilization of the ovum to death of the individual. *Senescence* is restricted to the stage of old age characteristic of the late years of the life span (Table 1.2). The World Health Organization classifies persons who are 60 to 75 years of age as elderly, those 76 to 90 as old, and those 90 years and older as very old. Individuals aged 80 to 85 and older are also called "old-old." *Centenarians* are persons 100 years old and older.

Other terms are listed in Table 1.2. *Life span* is the duration of time an individual remains alive under optimal conditions. It is genetically determined for each species, and, within the species, for each individual. The *average or mean life span* represents the average of individual life spans of a cohort (i.e., a group of individuals having a common demographic factor such as age) born at the same date. To determine the mean or average life span, the ages, at death, of all members of a population are added, and the sum is divided by the number of individuals in the population. This is also called *life expectancy* of the population. However, these terms represent only theoretical concepts; the life span implies the unrealistic circumstances of optimal life duration in an environment free of risk factors. Statistically, all persons in a population have the same average life span or life expectancy at the time of birth, but, over the life span, show considerable heterogeneity and variations.

*Longevity,* like life span, refers to the period of time an individual may live, given the best circumstances. It may also be interpreted to indicate the long duration of an individual's life, i.e., the condition of being long-lived. *Maximum longevity* is the age at death of the longest-lived individual(s) of the population. In humans, who are the

**TABLE 1.2**
**Brief Glossary of Aging-Related Terms**

| Term | Definition |
| --- | --- |
| Aging | Latin "aetas," age or lifetime — the condition of becoming old |
| Geriatrics | Greek "geron," old man, and "iatros," healer — a medical specialty dealing with the problems and diseases of the elderly |
| Gerontology | Greek "geron," old man, and "logos," knowledge — the study of aging and the problems of old age |
| Senescence | Latin "senex," old man — the condition of being old, used interchangeably with aging |
| Life span | The duration of the life of an individual/organism in a particular environment and/or under specific circumstances |
| Average life span | The average of individual life spans for members of a group (cohort) of the same birth date |
| Life expectancy | The average amount of time of life remaining for a population whose members all have the same birth date |
| Active life expectancy | As above, with the addition that the remaining life be free of a specific level of disability |
| Longevity | Long duration of an individual's life; the condition of being "long-lived" is also often used as a synonym for life span |
| Maximum life span | The length of life of the longest-lived individual member of a species |
| Biomarkers | Biologic indicators (morphologic, functional, and behavioral) specific to old age |

longest-lived species of mammals, the well-documented maximum longevity is that of 122 years (Chapter 2).

As indicated above, life expectancy refers to the total years of predicted survival starting from the moment (age) under study. Because the number and severity of disabilities increases with advancing age, *active life expectancy*, refers to years of survival without disability (Table 1.2). *Biomarkers* of aging refer to biologic indicators, that is, morphologic, biochemical, functional, and behavioral characteristics and signs that are exclusively specific to senescence.

Definitions of the *aging process* are numerous, indeed, there being as many definitions as theories of aging. They are better understood when presented with the theories of aging, which generated them; therefore, they are discussed in the relevant chapters (Chapters 4 through 7 and 10).

## V.   METHODS USED TO STUDY AGING

The relatively slow progress in our understanding of aging is due, partly, to the difficulty in finding adequate procedures to chart the long life of humans and in applying to humans the experimental data obtained from studies of animals and cells. Type and magnitude of changes with old age differ from one function to another. *Differential aging* of organisms, functions, and cells, is discussed in Chapter 3 and highlighted in several chapters throughout the book. In addition to the long life of humans, other critical difficulties include the lack of biomarkers of aging, and appropriate animal model systems.

Studies of human aging have utilized two main tactics based on cross-sectional and longitudinal methods. Cross-sectional methods compare characteristics among different age groups at one time. Studies with this method can be measured with relative accuracy and rapidity in a large group of people. For example, a population with a broad span of ages, perhaps birth to 90 years, is sectioned into narrow, age-defined subsets, and measurements of single or multiple functions are made in identical fashion in each group. This type of study has been used in gerontologic research for such programs as short-term testing of new drugs or regimens capable of influencing some aspects of the aging process.

Human populations live under widely different environmental conditions, and, therefore, the same measurements obtained in cohorts of individuals of the same age are subject to multifactorial influences, only one of which is aging per se. Thus, the cross-sectional study, although less time-consuming, is subject to a number of errors. For example, because of the secular trend in stature (that is, changes, in this case, an increase in stature, that occur in a population as a whole, over time), the 20-year-old in 1980 was taller than the 20-year-old in 1910. The finding that old individuals born in 1910 are shorter than those born in 1980 might suggest that there is a greater decrease in height with old age in this earlier population; rather, it may be better explained by the original shorter stature of the 1910 cohort, born when body height was lower. Similarly, differential survivorship depends on the selected survival only of individuals with a particular trait. To continue with the example of stature, tall, lean individuals may tend to live longer than the short and obese; if true, this would bias the conclusions drawn from the study.

The longitudinal study avoids some of these misinterpretations of the data and is preferred by many gerontologists. In the longitudinal study, the same individuals are examined at regular time intervals throughout the life span

**General Perspectives**

so that the process of aging can be compared in a dynamic fashion, each individual being his or her own control between two or several ages. Such investigations have been used with great success in the study of growth processes. In aging studies, however, this method has many obvious disadvantages, especially with human populations: not only is the human life span relatively long, but in industrialized countries, human populations are extremely mobile, and many members of the cohort are lost during the study.

A compromise approach combining cross-sectional and longitudinal methods involves using initial data from cross-sectional studies supplemented and corrected by data from longitudinal follow-ups. Another practical approach is to restrict the longitudinal survey to "critical" life periods (for example, 5 years before and 5 years after menopause or retirement).

Unfortunately, reliable *biomarkers of aging*, that is, biologic indicators (i.e., morphologic, functional, biochemical, behavioral signs) that may be considered exclusively specific to old age are yet to be identified. According to some gerontologists, the very paucity of reliable and valid biomarkers in humans as well as in experimental animals is one of the factors slowing progress in the field of aging. Identification of biomarkers to define the phenotype of aging is an ongoing process, and its ultimate outcome is still uncertain. Given the remarkable plasticity of human aging, which is probably due to genetic and environmental factors, it has been impossible so far to agree on a list of physical and mental measures that may exclusively and comprehensively characterize the entire group of the elderly. In humans, function, measured as *activities of daily living*, has been and continues to be used to assess physiologic competence and absence of disabilities (Chapter 3).

In rodents, nematodes, and insects, which have short lives and may be available in large cohorts, survival curves have been and are being used to identify different aging rates. Survival curves have been extensively used to study some of the characteristics of the oldest survivors and the effects of some interventions in prolonging life, such as administrations of pharmacological agents (Chapter 23) or of a caloric restricted diet (Chapter 24).[14,15] A number of different mechanisms may be responsible for the phenotypic alterations: different individuals may exhibit various patterns of aging, based on their specific inheritance and environmental experiences.[8] Indeed, one of the most common biomarkers of aging is the heterogeneity of physiologic responses in old individuals, whether humans or other animal species.

Much of the experimental work needed to elucidate physiological mechanisms of aging cannot be conducted in human beings. An alternative strategy is to choose the appropriate animal model.[16–18] Such a model should provide answers to specific questions regarding the aging

process, and these answers should be applicable to other species, usually humans. As indicated in a recent review, "The value of a specific model can only be judged in the context of a specific research agenda."[16]

The animal model is often chosen because of its availability, cost, and ease of maintenance and popularity with funding agencies, but it may not be the best model for the research in question. Criteria that are important to consider in choosing an animal model are species, sex, genetic properties, and research goals. These criteria must exhibit the following characteristics:

1. **Specificity** — the model must exhibit the trait (e.g., function) of interest.
2. **Generality** — the results observed in the chosen model must be applicable to other species.
3. **Feasibility** — the availability, cost, and convenience must be reasonable.[16]

Another currently popular approach is to create one's own animal choice. With the advent of recombinant DNA technology and the ability to genetically engineer mice, investigators may now alter a specific gene or process. The resulting transgenic mice carry a fragment of foreign DNA integrated in the genome of the organism studied or have a portion of the genome deleted or mutated.[19] In this manner, the role of individual genes may be studied in normal function (as in functional genomics) and in diseases, and its potential effects on aging may be clarified. Despite their promise in aging studies, the adoption of these transgenic models must follow guidelines similar to those that rule the choice of animal models discussed above. Particularly critical is the strain of mice chosen and the function to be studied.

Another approach to the experimental study of the aging process utilizes single cells or fragments of tissues removed from the body and cultured under *in vitro* conditions. Such studies establish standardized conditions of cell culture for cells/tissues derived from individuals of different chronological ages and compare specific parameters of cell function (e.g., replication, metabolism). For example, "cell doubling capacity" is taken as an index (among several) of cell reproductive capability (Chapter 4).[20,21] Cultured cells and tissues may serve as models for identifying specific cell functions or mimicking aging pathologies characterized by specific types of cell aging. Examples of the usefulness of the in vitro approach to the study of human aging are presented in subsequent chapters throughout this book.

## VI. HEREDITY AND ENVIRONMENT

Developmental processes and their regulation by heredity and environment have engaged the attention of biologists for many decades. The recent completion of the Human

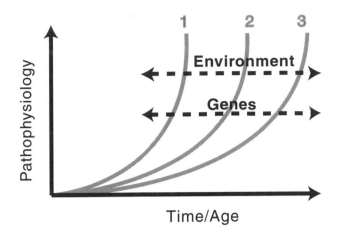

**FIGURE 1.1** Schematic diagram showing that the life span is regulated by interactions between genes and the environment. The coordinates represent age from birth to death and degree of physiopathologic changes (generic for reduced function and increased illnesses) with progressively older ages. Three different curves illustrate the considerable heterogeneity in life trajectory among old individuals. (Courtesy of Dr. J. Campisi.)

Genome Project opens a new and exciting era of studies to identify genes responsible for turning on and off those switches, mentioned above, that regulate the program for adaptation to the environment and for length and quality of life. Gene mutants for lengthening or shortening the life span have been identified in a number of animal species (Chapters 4 and 6). Genes have an important role in predisposing adult and elderly individuals to several common and complex pathologies caused by environmental risk factors (Chapter 3). At all stages of life, the directing force of heredity and the molding influences of the internal and external environment interplay in determining physiologic competence and length of the life span (Figure 1.1).

Heredity operates through internal factors, present in the fertilized egg. Chief among these are the genes, or hereditary determiners located in the chromosomes, which contain the genetic contribution of each parent. Indeed, it has been said that "the best assurance for a long life remains the thoughtful selection of long-lived parents."

It has been known for many years that many of the common disorders that affect humans have a major genetic component. The mapping of diseases with a Mendelian genetic transmission has led to major breakthroughs in our understanding of these diseases. In some animal species, one or several genes have been implicated in determining a shorter or longer life (Chapters 4 and 6). Human gene mapping in the last five years:

1. Led to a better insight, not only into the role of genes but also into the role of environmental factors on growth, development, and aging
2. Served as a comprehensive guide to a large number of common human disorders

3. May improve, by controlling for genetic susceptibility, our ability to identify and characterize additional genes, risk factors, gene–gene and gene–environment interaction
4. May provide successful therapies by relying on interventions on the genotype of the susceptible individual (Chapters 3 and 23)

The environment supplies the external/internal factors that make growth and development possible and allow inherited potentials to find expression. Environmental factors:

1. Include external factors such as temperature, humidity, atmospheric gases, drugs, infections, and radiation, as well as internal factors such as hormones, nutrition, immune responses, and nervous signals
2. May condition the appearance and modify the type of genetic characters, influence their expression, and alter their composition to make possible the creation of new inheritable characters (mutations)
3. Are operative throughout the life span
4. May provide successful therapies by modifying physical, psychological, and economic conditions (e.g., better lifestyle, hygienic habits, education, more efficient adaptability)

What is the respective influence of genetic and environmental factors in determining the phenotype? In humans, the best currently available method for estimating the involvement of genetic or environmental factors in determining or influencing the life span, is through twin and adoption studies. In a large group of identical or fraternal twins, reared together or separately, heritability was calculated by intrapair differences or similarities in the mortality rates in terms of age at death. From the data reported, genetic factors account for one-third of the variance in longevity and the environmental factors for the remaining two-thirds.[22] The evidence that genetic factors play a minor role, especially in twins dead at a young age, merits further examination.[22]

One of the main characteristics of animals is their capacity to adapt to an ever-changing environment. Adaptation is attained through a series of physiologic adjustments (regulated by neuroendocrine signals) that serve to restore the normal state once it has been disrupted by altered external conditions generating "stress" (Chapter 10). These adjustments are grouped under the term *homeostasis,* and a large part of physiology is concerned with regulatory mechanisms (e.g., negative and positive feedbacks) that maintain a constancy of the internal environment. By maturity, many such adjustments are completed. With aging, homeostasis may fail to occur or may be disrupted; repeated "stress" may lead to disease and, ultimately, death (Chapter 10).

**General Perspectives**

The new millennium, at its opening, has made available to biomedical research "the most wondrous map ever produced by humankind" with a readout of its 3 billion biochemical "letters" of human DNA.[23] Fantastic possibilities are now open to research in humans, ranging from new reproductive technologies to clones to genetic therapies to a full understanding of the aging process. However, these promises will not come to fruition unless the letters in the genome are translated into "words and sentences with functional meaning."[23] Indeed, a current type of genome analysis, so-called "functional genomics," investigates experimental approaches to assess gene function.[24-26] The early image of a stable gene is being replaced by one dependent on developmental interactions and environmental influences.[28] In addition, "chance variation," that is, variation due to neither genome nor environment, may play a significant role in development and aging.[29] For example, inbred laboratory animals exhibit a wide range of life spans despite having almost identical genes and environments. Likewise, human identical twins are not truly identical. This view that challenges the current reliance on the genes/environment hypothesis of aging is expressed in the heterogeneity of aging processes and may find its origin during development. Turning random variation into an object of study opens new research in the context of aging research.[29] The remarkable heterogeneity of the elderly population may result from the variability of genetic systems, the exposure to specific environmental conditions, and random variations.[30-31] Each should be an object of future studies as, together, they may determine, influence, predispose, and increase susceptibility to aging and disease (Chapter 3).

## REFERENCES

1. Timiras, P.S., *Developmental Physiology and Aging,* Macmillan, New York, 1972.
2. Timiras, P.S., *Physiological Basis of Aging and Geriatrics,* 1st ed., Macmillan, New York, 1988; 2nd ed., CRC Press, Boca Raton, FL, 1994.
3. Timiras, P.S. and Bittar, E.E., *Advances in Cell Aging and Gerontology,* Jay Press, Greenwich, CN, 1996.
4. Cole, T.R., *The Journey of Life,* Cambridge University Press, New York, 1992.
5. Rowe, J.R. and Kahn, R.L., *Successful Aging,* Pantheon Books, New York, 1998.
6. Burrow, J.A., *The Ages of Man,* Oxford University Press, New York, 1986.
7. Lewis, R., New light on fetal origins of adult disease, *The Scientist,* 14, 1, 2000.
8. Martin, G.M., Interactions of aging and environmental agents; the gerontological perspective, in *Environmental Toxicity and Aging Process,* Baker, S.R. and Rogul M., Eds., Alan T. Liss, New York, 1987.
9. Suzman, R.M., Willis, D.P., and Manton, K.G., Eds., *The Oldest Old,* Oxford University Press, New York, 1992.
10. Harris, J., Intimations of immortality, *Science,* 288, 59, 2000.
11. Harris, J., *Clones, Genes, and Immortality,* Oxford University Press, New York, 1998.
12. Austad, S.N., *Why We Age,* Wiley, New York, 1997.
13. Kirkwood, T.B.L. and Austad, S.N., Why do we age?, *Nature,* 408, 233, 2000.
14. Sternberg, H. and Timiras, P.S., Eds., *Studies of Aging,* Springer, New York, 1999.
15. Yu, B.P., Ed., *Methods in Aging Research,* 2nd ed., CRC Press, Boca Raton, FL, 1998.
16. Sprott, R.L., How to choose an animal model, in *Studies of Aging,* Sternberg, H. and Timiras, P.S., Eds., Springer, New York, 1999.
17. Sprott, R.L. and Austad, S.N., Animal models for aging research, in *Handbook of the Biology of Aging,* 4th ed., Schneider, E.L. and Rowe, J.W., Eds., Academic Press, New York, 1996.
18. Guarante, L. and Kenyon, C., Genetic pathways that regulate aging in model organisms, *Nature,* 408, 255, 2000.
19. Richardson, A., et al., Use of transgenic mice in aging research, *ILAR J,* 38, 124, 1997.
20. Hayflick, L. and Moorhead, P.S., The serial cultivation of human diploid cell strains, *Exper. Cell Res.,* 25, 585, 1961.
21. Hayflick, L., *How and Why We Age,* Ballantine Books, New York, 1996.
22. Ljungquist, B., et al., The effect of genetic factors for longevity: A comparison of identical and fraternal twins in the Swedish twin registry, *J. Gerontol.,* 53, M441, 1998.
23. Frank, M., EB 2001—Translating the genome, *The Physiologist,* 43, 165, 2000.
24. Finch, C.E. and Tanzi, R.E., Genetics and aging, *Science,* 278, 407, 1997.
25. Hieter, P.H. and Boguski, M., Functional genomics: It is all how you read it, *Science,* 278, 601, 1997.
26. Marcotte, E.M., et al., Detecting protein function and protein–protein interactions from genome sequences, *Science,* 285, 751, 1999.
27. Condit, C.M., *The Meanings of the Gene,* The University of Wisconsin Press, Madison, 1999.
28. Finch, C.E. and Kirkwood, T.B.L., *Chance, Development, and Aging,* Oxford University Press, New York, 2000.
29. Haines, J.L. and Pericak-Vance, M.A., Eds., *Approaches to Gene Mapping in Complex Human Diseases,* Wiley-Liss, New York, 1998.
30. Rose, M.R. and Finch, C.E., *Genetics and Evolution of Aging,* Kluwer Academic Publishers, Boston, 1994.
31. Rose, M.R. and Finch, C.E., *Genetics and Evolution of Aging,* Academic Publishers, Boston, 1994.

# 2 Human Longevity in Historical Perspective

*John R. Wilmoth*
University of California, Berkeley

## CONTENTS

## I. INTRODUCTION

Perhaps the greatest of all human achievements has been the enormous increase of human longevity that has occurred over the past few centuries. The average length of life in the early history of our species was probably in a range of 20 to 35 years (Table 2.1). By the beginning of the 20th century, this value had risen already to around 50 years in industrialized countries. One century later, the world's healthiest countries have a life expectancy at birth of around 80 years. Thus, around half of the historical increase in human life expectancy occurred during the 20th century. Of course, much of the increase in this average value has been due to the near-elimination of infant and childhood deaths. According to the available evidence, in the distant past, around a quarter of all babies died in their first year of life. Today, in the most advantaged countries, less than a half percent of infants meet a similar fate.

The increase of life expectancy at birth for one country, France, is depicted in Figure 2.1. This graph illustrates several key aspects of French demographic history over the past two centuries. First, we see the enormous increase of average human longevity over time, from a life expectancy in the high thirties during the early 19th century to values in the seventies or eighties at the end of the 20th century. Second, we witness the differential impact of the various wars on men and women. Two majors wars were fought mostly at the front and, thus, affected male life expectancy much more than female: the Napoleonic wars of the early 19th century and World War I during the 1910s. Two other conflicts involved a significant occupation of French territory by enemy forces and, thus, affected men and women in a similar fashion: the Franco-Prussian War of the early 1870s and World War II around the early 1940s. Finally, the graph illustrates the emergence of a large gap in life expectancy between men and women even during peacetime, from a difference of less than 2 years at the beginning of the interval to around 8 years at the end.

## TABLE 2.1
## Life Expectancy and Infant Mortality Throughout Human History

|  | Life Expectancy at Birth (Years) | Infant Mortality Rate (per 1000 Live Births) |
|---|---|---|
| Prehistoric | 20–35 | 200–300 |
| Sweden, 1750s | 37 | 210 |
| India, 1880s | 25 | 230 |
| United States, 1900 | 48 | 133 |
| France, 1950 | 66 | 52 |
| Japan, 1996 | 80 | 4 |

Data from Acsádi, G. and Nemeskéri, J., *History of Human Life Span and Mortality*, Budapest: Akadémiai Kiadó, 1970; Davis, K., *The Population of India and Pakistan,* Princeton, NJ. Princeton University Press, 1951; Human Mortality Database, www.mortality.org.

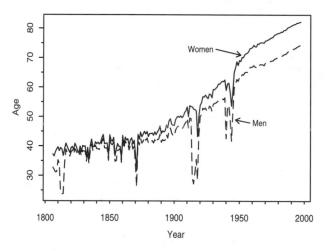

**FIGURE 2.1** Life expectancy at birth by sex, France 1806–1997. (Data from Vallin, J. and Meslé, F., Tables de mortalité francaises 1806–1997, INED, Paris, 2000.)

The rise of human life expectancy is significant for several reasons. First, it reflects the increasing material comfort of human life over this period, as well as the technological and social advances associated with modern systems of public health and medicine. However, changes also come at a certain cost. Thanks to the "longevity revolution," industrialized societies are now faced with a large and growing elderly population (see Table 2.2), which poses a significant challenge in terms of medical care and social support. The rise in the proportion of elderly is balanced to some degree by a decline in the share of young people in the population, as illustrated in Figure 2.2 for Italy. To some degree, societies must merely reorient themselves toward the care of a large dependent population at the end of life rather than at the beginning of life. Such adjustments are not without costs, however, as the needs of children and the elderly are quite different. Therefore, careful social planning is required, based on a firm understanding of historical trends.

This chapter does not provide answers about how to make the needed social and economic adjustments. Rather, it attempts to explain the driving forces behind the increase of human longevity that accounts for this momentous shift in population distribution from younger to older ages.

## II. HUMAN LONGEVITY IN THE PAST AND PRESENT

There are two important sets of questions about historical trends in mortality and health. First, how long do people live, why is longevity increasing, and how long will we live in the future? Second, given that we are living longer, are we mostly gaining healthy years of life, or are we "living longer but doing worse?" We know a lot more about the first question, partly because death is much easier to define and measure than health or functional status. In the United States, for example, there have been health interview surveys since the late 1950s and a consistent series of direct measurements of health status since the 1970s, in both cases, for a representative sample of the national population. These and similar data for other industrialized countries can be used to measure changes in health status, but it is often difficult to compare the results reliably across populations and over time.

On the other hand, we have detailed mortality data from many countries over much longer time periods. These data often include information on the attributed cause of death, although this concept, like health or functional status, is difficult to define and measure in a consistent fashion. Although there have been some attempts to measure early human longevity based on skeletal remains and other information,[1] the most useful information on historical mortality trends comes from time series of national data, collected since around 1750 in some parts of Europe. The accuracy of such data is variable, but specialists mostly know which data are reliable or potentially inaccurate. Data on cause of death must always be analyzed with great caution: although some trends are irrefutable (e.g., the historical decline of infectious disease), others appear contaminated by changes in diagnostic procedures and reporting practices (e.g., cancer trends, especially at older ages).

This section describes major trends in human longevity from the past and present. A later section of this chapter offers some guarded speculations about what the future may hold. We do not address the issue of "healthy life span," although the interested reader may refer to other sources on this topic[2,3] (see Chapter 3).

**TABLE 2.2**
**Population of Major World Regions, 1950–2025, with Percent Under Age 15 and Over Age 65**

| | 1950 Population | | | 1975 Population | | |
|---|---|---|---|---|---|---|
| Region | Total (Millions) | Percent Under Age 15 | Percent Over Age 65 | Total (Millions) | Percent Under Age 15 | Percent Over Age 65 |
| World | 2521 | 34.3 | 5.2 | 4074 | 36.9 | 5.7 |
| More developed countries | 813 | 27.3 | 7.9 | 1048 | 24.2 | 10.7 |
| Less developed countries | 1709 | 37.8 | 3.9 | 3026 | 41.3 | 3.9 |
| Africa | 221 | 42.5 | 3.2 | 406 | 45.0 | 3.1 |
| Asia | 1402 | 36.6 | 4.1 | 2406 | 39.9 | 4.2 |
| Europe | 547 | 26.2 | 8.2 | 676 | 23.7 | 11.4 |
| Latin America and Caribbean | 167 | 40.0 | 3.7 | 322 | 41.3 | 4.3 |
| North America | 172 | 27.2 | 8.2 | 243 | 25.2 | 10.3 |
| Oceania | 13 | 29.8 | 7.4 | 21 | 31.0 | 7.5 |

| | 2000 Population | | | 2025 Population | | |
|---|---|---|---|---|---|---|
| Region | Total (Millions) | Percent Under Age 15 | Percent Over Age 65 | Total (Millions) | Percent Under Age 15 | Percent Over Age 65 |
| World | 6055 | 29.7 | 6.9 | 7824 | 23.4 | 10.4 |
| More developed countries | 1188 | 18.2 | 14.4 | 1217 | 15.9 | 20.9 |
| Less developed countries | 4867 | 32.5 | 5.1 | 6609 | 24.9 | 8.5 |
| Africa | 784 | 42.5 | 3.2 | 1298 | 34.6 | 4.0 |
| Asia | 3683 | 29.9 | 5.9 | 4723 | 22.1 | 10.1 |
| Europe | 729 | 17.5 | 14.7 | 702 | 14.7 | 21.0 |
| Latin America and Caribbean | 519 | 31.6 | 5.4 | 697 | 23.7 | 9.5 |
| North America | 310 | 21.2 | 12.5 | 364 | 18.1 | 19.0 |
| Oceania | 30 | 25.2 | 9.9 | 40 | 21.3 | 14.6 |

*Note:* Figures for 2000 and 2025 are medium-variant projections.

Data from *World Population Prospects: The 1998 Revision. Volume 1: Comprehensive Tables*. New York, UN Population Division, United Nations, 2000.

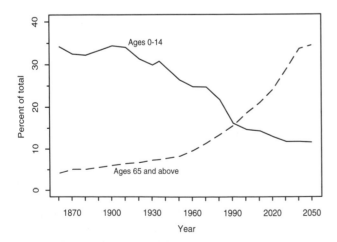

**FIGURE 2.2** Proportion of population aged 0–14 versus 65+, Italy 1861–2050. Note: Figures for 2001–2050 are projections. (Data from Italian censuses, 1861–1991; ISTAT, 2001–2050.)

## A. PREHISTORIC AND PREINDUSTRIAL ERAS

We do not know much about how long humans lived before 1750. Around that time, the first national population data were collected for Sweden and Finland. For earlier eras, we have some life tables constructed for municipal populations, members of the nobility, and other groups that were probably not representative of the national population at large.[4,5] After 1750 and even today, we have extensive and highly reliable mortality information for only a subset of national populations.

For the Middle Ages and before, mortality levels have been estimated based on data gleaned from tombstone inscriptions, genealogical records, and skeletal remains.[1] The accuracy of such estimates has been a subject of dispute.[6–9] In studies based on skeletal remains, a key issue is the attribution of age based on bone fragments. Another problem for estimation based on skeletal or tombstone data is uncertainty about the age structure of the population, which affects mortality estimates based solely on the

**General Perspectives**

distribution of ages at death. The only practical solution is to assume that the population was "stationary," implying a long-term zero growth rate and unchanging levels of fertility and mortality, and even an unchanging age pattern of mortality. Clearly, these assumptions are always violated, but the resulting estimates are useful nonetheless.

For mortality data derived from subpopulations, there is also an issue of whether the data are representative of some larger population. Who gets buried in a society, and who gets a tombstone? Which societies have regular burial practices, as opposed, say, to burning their dead? What kinds of populations have complete genealogical records from a particular time period? Thus, for many reasons, all estimates of mortality or longevity from the preindustrial period (roughly, before 1750) should be viewed with caution. Of the many sources of bias in these estimates, there are positive and negative factors, which tend to balance each other to some extent.[10] They are inaccurate or unrepresentative by amounts that cannot be well quantified.

Although these historical estimates may be too high or too low, they provide us nonetheless with a useful description of the general contours of the history of human longevity. For example, most scholars agree that life expectancy at birth (or $\overset{\circ}{e}_o$, in the notation of demographers and actuaries) was probably in the twenties for early human populations. Some very disadvantaged societies might have had life expectancies in the teens, whereas others may have been in the thirties. Because historical levels of life expectancy were in the twenties, compared to around 75 to 80 years today in wealthy countries, the average length of life has roughly tripled.

Most of this increase was due to the reduction of infant and child mortality. It used to be the case, for example, that remaining life expectancy at age 1 year was greater than at birth, because the toll of infant mortality was so high. The difference between premodern periods and today is less stark if we consider life expectancy at higher ages. Instead of the tripling of life expectancy at birth, remaining life expectancy at higher ages has roughly doubled over the course of human history. At age 10 years, for example, life expectancy (i.e., expected years after age 10) may have moved from around 30 to 33 years to almost 70 years.[11] At age 50, it may have gone from around 14 years to more than 30 years.[10]

## B. Epidemiologic Transition

The epidemiologic transition is the most important historical change affecting the level and pattern of human mortality. The transition refers to the decline of acute infectious disease and the rise of chronic degenerative disease.[12] This shift does not imply that degenerative diseases became more common for individuals of a given age. It merely means that infectious disease nearly disappeared, so something else had to take its place as the major cause

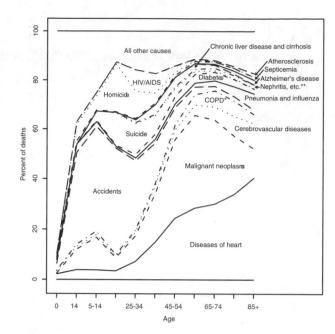

**FIGURE 2.3** Distribution of deaths by cause, United States 1997. Notes:* Chronic obstructive pulmonary disorder; ** Nephritis, nephrotic syndrome, and nephrosis. (Data from U.S. Department of Health and Human Services, 1999.)

of death. Increasingly, people survived through infancy and childhood without succumbing to infectious disease.[13] Once past these critical early years, survival to advanced ages is much more likely, and at older ages, various degenerative diseases present mortality risks even when infection is well controlled.

Thus, heart disease, cancer, and stroke became the most common causes of death in industrialized societies, as the age distribution of deaths shifted to older ages. As seen in Figure 2.3, which depicts the distribution of deaths by cause in various age groups for the United States in 1997, these three causes now account for more than 60% of all deaths at older ages. On the other hand, accidents, homicides, suicides, and HIV/AIDS are the major killers among young adults. Infants and children die mostly from accidents and "other causes" (a residual category that includes, for example, congenital anomalies and childhood diseases).

## 1. Trends in Life Expectancy

Life expectancy has been increasing not just in industrialized societies but also around the world. (During the 1990s, the two major exceptions to the worldwide increase in life expectancy were a stagnation and even reversal of earlier progress in parts of Africa, due to the AIDS epidemic, and in parts of the former Soviet bloc, especially Russia, due to social disruptions and instability.) The rise in life expectancy at birth probably began before the industrial era; thus, before national mortality statistics were first

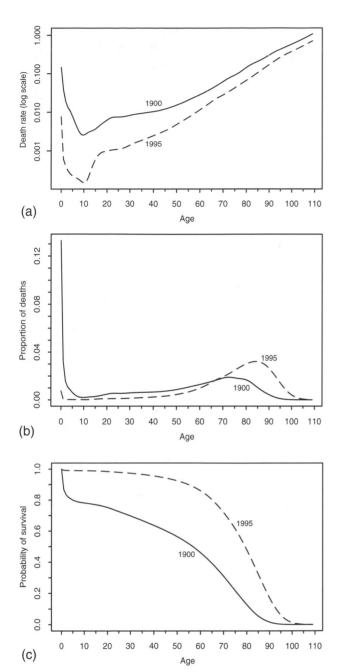

**FIGURE 2.4** Age pattern of mortality from three perspectives, United States, 1900 and 1995: a) observed death rates by age, b) distribution of deaths by age, and c) proportion surviving by age. Note: Figures 4b and 4c present the distribution of deaths and the proportion surviving in a life table for the given year. (Data from U.S. Social Security Administration, available through the Human Mortality Database: www.mortality.org.)

assembled in Sweden around 1750. As noted earlier, $\mathring{e}_0$ (life expectancy) was probably in the twenties during the Middle Ages and earlier. By 1750, Sweden (and probably other parts of northwestern Europe) had attained an $\mathring{e}_0$ of 38, so the upward trend in longevity appears to have begun before the industrial era. Over the next century or more, there was a slow and irregular increase in life expectancy.

After about 1870, however, the increase became stable and more rapid. During the first half of the 20th century, life expectancy in industrialized countries rose quite rapidly. Since 1950, the rise in life expectancy slowed somewhat, as seen in Figure 2.1 for France.

The cause of the earlier rapid rise in life expectancy and its subsequent deceleration is quite simple: the decline of juvenile mortality to historically very low levels. By around 1950, infant mortality in wealthy countries was in the range of 2 to 3% of births, compared to perhaps 20 to 30% historically. Since then, infant mortality has continued to decline and is now in the range of 0.5 to 1% of births in the healthiest parts of the world. As babies were saved from infectious disease, their chances of survival to old age improved considerably. Once juvenile mortality was reduced substantially, improvements in life expectancy due to the reduction of mortality in this age range had to slow, and further gains had to come mostly from mortality reductions at older ages.

The rise in life expectancy during the second half of the 20th century was slower than during the first half, simply because it depended on the reduction of death rates at older ages, rather than in infancy and childhood. Put simply, saving an infant or child from infectious disease, who then goes on to live to age 70, contributes more to average life span than saving a 70 year old from heart disease, who may live another 10 years. Thus, the deceleration in the historical rise of life expectancy is a product of the J-shaped age pattern of human mortality: high in infancy and childhood, low through adolescence and early adulthood, then rising almost exponentially after age 30. Gains that come from reducing juvenile mortality are quite large, whereas gains due to a reduction in old-age mortality are inevitably much smaller.

A common mistake is to assert that the deceleration in the rise of $\mathring{e}_0$ reflects a slowdown in progress against mortality. In fact, the reduction of death rates has changed its character in recent decades, but it has not slowed. At older ages, the decline of mortality has accelerated since around 1970 (as discussed below). So long as the decline of old age mortality continues, life expectancy will continue to increase, driven now by the extension of life at later ages rather than by saving juveniles from premature death.

## 2. Rectangularization or Mortality Compression

The age pattern of human mortality can be characterized in various ways. Figure 2.4 shows the American mortality levels in 1900 and 1995 from three perspectives. The first panel shows death rates by age. These death rates are used to construct a life table, which describes the experience of a hypothetical cohort subject throughout its life to the death rates of a given year. Thus, the middle and last panels show the distribution of deaths and the proportion of survivors at each age among members of such a hypothetical cohort.

**General Perspectives**

Together, these three panels illustrate some major features of the mortality decline that has taken place over this time interval. First, death rates have fallen across the age range, but they have fallen most sharply (in relative terms, since the graph has a semilogarithmic scale) at younger ages. The distribution of ages at death has shifted to the right and become much more compressed. At the same time, the survival curve has shifted to the right and become more "rectangular" in shape. This last change is often referred to as the "rectangularization" of the human survival curve.

It was once asserted that this process of rectangularization reflected the existence of biological limits affecting human longevity.[14] This notion of limits to the human life span enjoys little empirical support, as discussed below. Nevertheless, the historical process of rectangularization was both real and extremely significant. It is perhaps best thought of as a "compression of mortality," as documented in the middle panel of Figure 2.4. As the average level of longevity has increased, so has our certainty about the timing of death.

One measure of this variability is the interquartile range of deaths in the life table or the age span of the middle 50% of deaths over the life course. In the 1750s in Sweden, the life table interquartile range was about 65 years, so that deaths were spread widely across the age range. The distribution of age at death became more compressed over the next two centuries, until the life table interquartile range was around 15 years in industrialized countries by the 1950s. Since 1960, there has been little further reduction in the variability of age at death in the developed world, even though the average age at death (as reflected in life expectancy at birth) has continued to increase.[15]

Like the historic rise of life expectancy, this compression of mortality was due largely to the reduction of juvenile mortality. Once most juveniles had been saved from premature death, a pattern emerged in which deaths are concentrated in the older age ranges. As mortality falls today among the elderly, the entire distribution of ages at death is rising slowly, but its level of variability seems to have stabilized.

## C. MORTALITY DECLINE AMONG THE ELDERLY

The most significant trend now affecting longevity in industrialized societies is the decline of death rates among the elderly. Until the late 1960s, death rates at older ages had declined slowly, if at all. Traditionally, rates of mortality decline were much higher at younger than at older ages. Since about 1970, however, there has been an "aging of mortality decline," meaning that some of the most rapid declines in death rates are now occurring at older ages.[16,17] Thus, the decade of the 1960s marks a turning point, from an earlier era of longevity increase due primarily to the decline of acute infectious disease among juveniles, to a more recent era involving the decline of chronic degenerative disease among the elderly.

## 1. Cardiovascular Disease

The most significant component of the mortality decline at older ages is the reduction of death rates due to cardiovascular disease (CVD), including heart disease and stroke. In the United States, heart disease has been the leading cause of death since 1921, and stroke has been the third most common cause since 1938. From 1950 to 1996, age-adjusted death rates for these two causes declined by more than half (by 56% for heart disease and by 70% for stroke). It is estimated that 73% of the decline in total death rates over this time period was due to this reduction in cardiovascular disease mortality.[18]

The exact cause of the decline in CVD mortality is open to debate, although it is surely due to a combination of factors. For the United States, all of the following have been cited as factors contributing to this decline: a decline in cigarette smoking among adults; a decrease in mean blood pressure levels; an increase in control of hypertension through treatment; a change in diet, especially a reduction in the consumption of saturated fat and cholesterol; and, an improvement in medical care, including better diagnosis and treatment of heart disease and stroke, the development of effective medications for treatment of hypertension and hypercholesterolemia, and an increase in coronary-care units and in emergency medical services for heart disease and stroke[18] (see Chapters 16, 17).

The rapid decline in CVD mortality began around 1968 in the United States and other industrialized nations. Given the precipitous nature of this decline, it has been argued that therapeutic interventions were the most important factor, because changes in diet and lifestyle should have led to a more gradual pattern of change.[19] It is worth noting that landmark investigations, like the Framingham Heart Study, began in the late 1940s and began to provide significant breakthroughs in our scientific understanding of cardiovascular disease during the 1960s.[20]

## 2. Cancer

In most developed countries, cancer mortality has begun to decline only within the last 10 to 15 years, although in Japan, death rates from cancer began falling as early as the 1960s.[21–23] Of course, cancer takes many different forms, and trends vary greatly by site of the primary tumor. Lung cancer has become more common due to increased smoking habits, while stomach cancer has been in decline. Among women, mortality due to cervical cancer has fallen dramatically thanks to successful medical intervention (screening and early treatment), while breast cancer has been on the rise due apparently to a number of interrelated factors (lower and later fertility, changes in diet, and possibly other factors as well).

It is sometimes overlooked that some common forms of cancer may be caused by infection. For example, stomach cancer is often brought on by infection with *Helicobacter pylori*. Infection with *H. pylori*, and hence stomach cancer, was especially common in Japan prior to the widespread availability of refrigeration.[24,25] Liver cancer is related to hepatitis infection (both the B and C strains of the virus), and thus reductions in liver cancer hinge on the control of infection, as well as curbing excess drinking. A third example is infection by the human papilloma virus, which can cause cervical cancer.[26]

These three forms of cancer have tended to decline in recent decades and should decline further as the relevant infectious agents are brought under control (e.g., hepatitis B and C). On the other hand, cancers that have become more common include those strongly influenced by individual behaviors (e.g., lung and pancreatic cancer are linked to smoking, and both have tended to increase over time) and some others with causes that are mysterious or poorly understood (e.g., breast cancer and colorectal cancer, both rising but for unknown or uncertain reasons).

As noted earlier, trends in mortality among the elderly are the main factor behind the continued increase in life expectancy in developed countries. Furthermore, the main components of mortality at these ages are cardiovascular disease and cancer. These two causes have been in decline during recent decades for reasons that are complex and not entirely understood. It is clear that there are multiple causes involved in bringing down death rates due to cardiovascular disease and cancer. Medical science has played a part, but so have changes in diet and personal habits, as well as community efforts and economic changes that have reduced the spread of infectious agents. It is important to keep this complex causality in mind when speculating about future trends in human mortality.

## D. SUMMARY OF HISTORICAL TRENDS

A compact summary of major trends in human longevity in industrialized countries is presented in Table 2.3. Amidst the incredible detail available in historical mortality statistics, we cannot help but discern two major epochs: before 1960 and after 1970. The driving trend in the former period was a rapid decline of mortality due to infectious disease, which had an impact across the age range but certainly a much larger effect at younger ages. The sharp reduction in infant and child mortality led to a rapid increase in average life span and a marked reduction in the variability of age at death. It did not, however, have a major impact on maximum life span, which rose very slowly due to the more gradual improvement in death rates at older ages.

From the mid 1950s to the late 1960s, mortality trends in industrialized countries seemed to stabilize. Then, suddenly, just before 1970, death rates at older ages entered a period of unprecedented decline. Compared to the earlier era of rapid reductions in infant and child mortality, these changes yielded a slower increase in life expectancy at birth. On the other hand, the rise of maximum life span accelerated, driven by a more rapid decline in death rates at older ages. The variability of life span tended to stabilize during this period, as the entire distribution of ages at death (now concentrated at older ages) moved upward in parallel fashion. The difference between these two distinct eras is illustrated in Table 2.4 for the country of Sweden.

## III. OUTLOOKS FOR THE FUTURE

It is impossible to make a firm scientific statement about what will happen in the future. In truth, scientists can only present the details of well-specified scenarios, which serve as forecasts or projections of the future. They can also help by clearly defining the terms of the debate, for example, by discussing what is meant by the notion of "limits

## TABLE 2.3
### Summary of Major Trends in Human Longevity in Industrialized Countries

| | Before 1960 | After 1970 |
|---|---|---|
| Average life span (life expectancy at birth) | Increasing rapidly, because averted deaths are among younger people. Very rapid reduction in infant/child mortality linked mostly to effective control of infectious disease. | Increasing moderately, because averted deaths are among older people. Accelerated reduction in old-age mortality linked mostly to better management of cardiovascular disease. |
| Maximum life span (observed and verified maximum age at death) | Increasing slowly due mostly to gradual reductions in death rates at older ages. (Increases in births and improved survivorship at younger ages matter much less.) | Increasing moderately due almost entirely to accelerated reduction in death rates at older ages. |
| Variability of life span (standard deviation, interquartile range, etc.) | Decreasing rapidly due to reductions in mortality at younger ages. | Stable, because death rates at older ages are decreasing as rapidly as at younger ages. |

**TABLE 2.4**
**Change (Per Decade) in Key Mortality Indicators, Sweden**

|                                          | 1861–1960 | 1970–1999 |
|------------------------------------------|-----------|-----------|
| Average life span (life expectancy at birth) | 3.1       | 1.8       |
| Maximum life span (maximum reported age at death) | 0.4       | 1.5       |
| Interquartile range (of deaths in life table) | -5.8      | -0.3      |

*Notes:* Slopes are determined by least-squares regression.

For IQR (interquartile range), the range of the second time period is 1970–1995.

Author's calculations are based on data from Human Mortality Database, www.mortality.org.

to life span." Limits possibly affecting the increase of human longevity are the first topic of this section, followed by a discussion of extrapolative techniques of mortality projection or forecasting. Our discussion of the future of mortality concludes with a comparison of "optimistic" and "pessimistic" points of view on this topic.

## A. POSSIBLE LIMITS TO LIFE SPAN

If there are limits to the human life span, what do they look like? There are two ways to define such limits: maximum *average* life span and maximum *individual* life span.[16]

## 1. Maximum Average Life Span

Let us consider whether there might be an upper limit to the average life span that could be achieved by a large human population. Average life span, or life expectancy at birth, refers to how long people live on average in a population. In the United States, life expectancy is currently around 74 for men and 80 for women.[27] Accordingly, these numbers describe the average length of life that can be anticipated given the mortality conditions of today. For example, baby boys born this year will live an average of 74 years, assuming that age-specific death rates (as illustrated in Figure 2.4a) do not change in the future. Just as occurs today, some of these newborns will die in infancy from congenital ailments, some will be killed in car accidents as young adults, and some will succumb in old age to cancer or heart disease.

As noted earlier, death rates have been falling for several centuries. At every age, the risk or probability of dying is much lower than in the past. Thus, when we talk about life expectancy at birth, we are being conservative and asking what the average life span will be assuming that death rates do not fall any further in the

future. However, it is likely that death rates will continue to decline at least somewhat in future years, so baby boys born today in the United States will probably live longer than 74 years on average.

The question about limits to the average life span can be posed as follows: Can death rates keep falling forever, or will they hit some fixed lower bound? Perhaps biological forces impose a certain inevitable risk of mortality at every age. Thus, there might be some age-specific minimum risk of dying that could never be eliminated.[16]

Admittedly, it seems implausible that age-specific death rates could ever equal zero in any large population. However, even if the death rate at some age cannot equal zero, can it keep declining toward zero? In other words, zero might be the limit to how far death rates can drop, even if they can never attain zero. Or is there a higher limit? Perhaps there is some number, like one in a million, such that it is simply inevitable that one in a million people — say, one in a million 50-year-olds — will succumb to death over the course of a year. If true, then the death rate at age 50 can never fall below one in a million. According to this view, we have a limited capability as a society, or as a species: we cannot push the risk of death any lower than some fixed level.

If a nonzero lower limit for death rates exists, how much is it at age 50 or at any age? The answer to this question is significant, for if we knew the lower limit of death rates at every age, we could compute the maximum achievable life expectancy at birth. In this way, we would know the upper limit of the *average* human life span. However, it is quite difficult to identify a nonzero lower bound on death rates that is applicable to all human populations. Yet, if there is no lower limit to death rates except zero, then there is no upper limit to life expectancy except infinity. Nevertheless, the absence of identifiable limits does not mean that large increases in average life span are imminent. It just means that life expectancy *can* continue to increase, as death rates are pushed down further.

Why do some people think that an upper limit to life expectancy exists? In fact, there is little empirical support for such a belief. An argument frequently put forward is that the rise in life expectancy at birth slowed in the second half of the 20th century. As shown earlier, however, this deceleration resulted merely from a shift in the main source of the historical mortality decline from younger to older ages. Although the rise in life expectancy has decelerated, the decline in death rates at older ages has accelerated in recent decades.[28]

Furthermore, if death rates are approaching their lower limit, one might expect a positive correlation between the current level of mortality in a given country and the speed of mortality decline (so that those populations with the lowest level of mortality would also experience the slowest rates of mortality decline). In fact, no such correlation exists for death rates at older ages. In some cases, the fastest reduction in death rates is occurring in those countries with the lowest levels of old-age mortality, just the opposite of what we would expect if death rates were pushing against a fixed lower bound.[28]

So long as death rates at older ages keep falling, life expectancy (at birth or at any age) will continue to increase. As discussed below, current forecasts suggest that life expectancy at birth may not rise much above current levels over the next half century. Nevertheless, there is simply no demographic evidence that life expectancy is approaching a fixed upper limit. Certainly, such a limit may exist, but it is nowhere in sight at the present time.[16]

## 2. Maximum Individual Life Span

Limits to average life span, or life expectancy at birth, are one issue. When people discuss limits to the human life span, however, they often have another idea in mind: the upper limit to an individual life span. Instead of asking how long we can live on average, we might ask how long one lucky individual can hope to live. This concept is actually much easier to understand than the notion of an upper limit to life expectancy.

Who is the oldest person who has ever lived? Even if we can never have a definitive answer to this question, we can at least imagine the existence of such a person. Maybe he or she (probably she) is still alive today. Or maybe she lived hundreds of years ago but vanished without leaving a trace — no birth certificate, no census record, and not even a newspaper article about her incredible feat of longevity.

Who is the oldest person alive today? That person might or might not be the oldest person ever. However, identifying the world's oldest person is difficult even today because of the widespread practice of what demographers politely call "age misstatement."[29] Putting it less politely, some people lie about their age. Others, if asked, give the wrong age because they do not remember, because they are not numerate, or because they simply never paid attention to such matters. Such age misstatement often occurs in the absence of written records to prove or disprove the reported age.

Should we believe people who claim to be extremely old but do not have proper documentation? Certainly, we should believe them because there is no point in calling anyone a liar or questioning their memory. In terms of a scientific discussion about longevity, however, experts agree that it is best to ignore undocumented cases of extreme longevity. Thus, when we make statements about who is the oldest person alive today or in the past, we limit ourselves to cases for which solid evidence exists.[30]

To be accepted as a valid instance of extreme longevity, thorough documentation is required — not just a birth certificate, but also a series of documents and a life history

that is consistent with the written records. Ideally, if the person is still alive and mentally able, an oral history is obtained and checked against all available evidence, making sure, for example, that this person is not the son or daughter of the person in question.

Indeed, there are numerous examples of supposed extreme longevity that turned out to be cases of mistaken identity.[30] Perhaps the most notorious example was a French Canadian named Pierre Joubert, who was supposed to have died at the age of 113 years in 1814. This case was listed for many years by the *Guinness Book of World Records*.[31] When genealogical records were examined closely, however, two men named Pierre Joubert were identified — a father and his son. It was the son who died in 1814, 113 years after the birth of the father.[32] Such mistakes are not uncommon, and whether they are the result of deliberate misrepresentation or honest error is irrelevant. In either case, a complete investigation should be required before accepting such reports as factual.

The historical record is still held by a Frenchwoman, Jeanne Calment, who died at age 122 in August of 1997.[33] Madame Calment lived in Arles, a town with complete civil records (births, deaths, marriages, baptisms, etc.) going back several centuries. Fortunately, these records were not destroyed in any war, so it has been possible to carefully trace the life of Jeanne Calment. It was also possible to reconstruct her family genealogy and to document that a disproportionate number of her ancestors were long-lived as well. Of course, it is only one example, but the case of Jeanne Calment suggests that extreme longevity may have at least some hereditary component.[34]

The oldest man whose age was thoroughly verified was Christian Mortensen, who died in 1998 at the age of 115.[35,36] A Japanese man named Shigechiyo Izumi was reportedly 120 years old when he died in 1986. According to the *Guinness Book of World Records*, Izumi is still the oldest man on record.[31] However, this case has now been rejected by almost all experts who are familiar with it, including the Japanese man who originally brought it to the attention of *Guinness*.[37-39] The common belief is that Izumi was in fact "only" 105 years old at the time of his death.

It is reasonable to ask what we have learned in general from these few cases of exceptionally long-lived individuals. Admittedly, the cases of Calment and Mortensen tell us nothing about the trend in maximum longevity. Maybe these are just two cases that we have had the good fortune of documenting in recent years. Maybe there were other individuals who were just as old as Calment or Mortensen who lived years ago, and we missed them. These are valid points, so we must turn to other evidence if we want to know about trends in extreme longevity.

In order to study historical trends in extreme longevity, we need a well-defined population with reliable records over a long period of time. For that purpose, we turn to a small subset of countries that have kept reliable population statistics for many years. The longest series of such data comes from Sweden. These records are thought to be extremely reliable since 1861, even in terms of the age reporting of individuals at very old ages.[37] Vital records have a very old history in Sweden, where Lutheran priests were required to start collecting such information at the parish level in 1686. Such records were eventually brought together into a national system in 1749. In 1858, the present-day National Central Bureau of Statistics was formed, which led to further improvements in data quality. Furthermore, by the 1860s, the national system of population statistics was already more than 100 years old, so it was possible to check claims of extreme longevity against birth records from a century before. These historical developments account for the unique quality of the Swedish mortality data.

Figure 2.5 shows the trend in the maximum age at death for men and women in Sweden during 1861 to 1999. The trend is clearly upward over this time period, and it accelerates after about 1969. The rise in this trend is estimated to be 0.44 years (of age) per decade prior to 1969, and 1.1 years per decade after that date. More than two thirds of this increase can be attributed to reductions in death rates above age 70, with the rest due to mortality decline at younger ages and an increase in the size of birth cohorts.[40]

These Swedish data provide the best available evidence for the gradual extension of the maximum human life span that has occurred over this time period. Similar trends are evident for other countries as well, although patterns of age misstatement present greater problems of interpretation.[37]

## B. Extrapolation of Mortality Trends

Demographers claim some expertise in predicting future mortality levels, and their method of choice is usually a mere extrapolation of past trends. Biologists and others sometimes criticize this approach, because it seems to ignore underlying mechanisms. However, this critique is valid only insofar as such mechanisms are understood with sufficient precision to offer a legitimate alternative method of prediction. Although many components of human aging and mortality have been well described, our understanding of the complex interactions of social and biological factors that determine mortality levels is still imprecise. Furthermore, even if we understood these interactions and wanted to predict future mortality on the basis of a theoretical model, we would still need to anticipate trends in each of its components.

The extrapolative approach to prediction is particularly compelling in the case of human mortality. First, mortality decline is driven by a widespread, perhaps universal, desire for a longer, healthier life. Second, historical

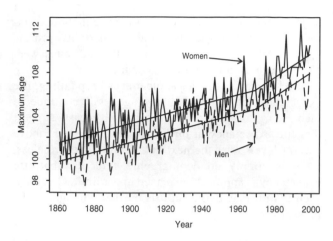

**FIGURE 2.5** Maximum reported age at death, Sweden 1861–1999. (Source: Wilmoth, J.R., Deegan, L.J., Lundström, H., and Horiuchi, S., Increase of maximum life span in Sweden, 1861–1999, *Science*, 289, 2366, 2000. With permission.)

evidence demonstrates that mortality has been falling steadily, and life span has been increasing, for more than 100 years in economically advanced societies. Third, these gains in longevity are the result of a complex array of changes (improved standards of living, public health, personal hygiene, medical care), with different factors playing major or minor roles in different time periods. Fourth, much of this decline can be attributed to the directed actions of individuals and institutions, whose conscious efforts to improve health and reduce mortality will continue in the future.

Even accepting this argument, there is still a question of what to extrapolate. Demographers tend to view death rates as the fundamental unit of analysis in the study of mortality patterns, because these rates are estimates of the underlying "force of mortality," or the risk of death at any moment in a person's lifetime. These risks change over age and time, and vary across social groupings (by sex, race, education, income, etc.). Life expectancy and the expected maximum age at death (for a cohort of a given size) can be expressed as a mathematical function of death rates by age. Thus, the usual strategy is to extrapolate age-specific death rates into the future and then to use the results of such an extrapolation to compute forecasts of life expectancy or other parameters of interest.

Predictions of future life expectancy by such methods yield values that are not too different from what is observed today. For example, recent forecasts by the U.S. Social Security Administration put life expectancy in 2050 at 77.5 years for men and 82.9 years for women, compared to 72.6 and 79.0 years in 1995.[41] These forecasts are not true extrapolations, however, because they assume a slowdown in age-specific rates of mortality decline in the future. Another study, based on a purely extrapolative technique, yielded more optimistic results — a life expectancy at birth in 2050 of around 84 years for both sexes

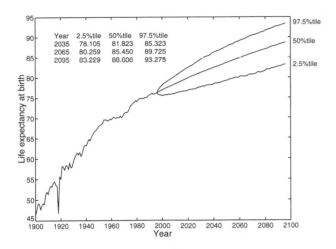

| Year | 2.5%tile | 50%tile | 97.5%tile |
|------|----------|---------|-----------|
| 2035 | 78.105 | 81.823 | 85.323 |
| 2065 | 80.259 | 85.450 | 89.725 |
| 2095 | 83.229 | 88.006 | 93.275 |

**FIGURE 2.6** Life expectancy at birth, United States, 1900–1996 (actual) and 1997–2096 (forecast). (Source: Lee, R. and Tuljapurkar, S., Population Forecasting for Fiscal Planning: Issues and Innovations, *Demographic Change and Fiscal Policy*, 1st ed., Cambridge University Press, Cambridge, Chapter 2, 2001. With permission.)

combined.[42] Plausible forecasts for Japan are only slightly higher life expectancy at birth in 2050 of 81.3 years for men and 88.7 years for women, compared to 76.4 and 82.9 years in 1995.[43]

The life expectancy forecasts of Lee and Tuljapurkar[42] are reproduced here as Figure 2.6. These projections are based on a clever extrapolative technique pioneered by Lee and Carter,[44] which has been influential in the world of mortality forecasting over the last decade. The method yields a range of estimates for each calendar year during the forecast period (in this case, from 1997 to 2096). The inherent uncertainty of future trends is represented in the graph by plotting not only the median forecast, which may be considered the "best estimate," but also by showing two extreme forecasts. The median forecast lies at the fiftieth percentile of the full distribution: half of the estimates lie below or above this value. Figure 2.6 also presents the 2.5 and 97.5 percentiles, thus showing relatively extreme trajectories of future life expectancy.

It is important to remember that these projections are mere extrapolations of the historical experience of one country during a particular time period (the United States from 1900 to 1996). The implicit assumption is that future trends will resemble past ones. This assumption is plausible given the fairly steady pace of mortality decline over the past century. Of course, extrapolation is not without its flaws. It could not, for example, have anticipated the rise of mortality in the former Soviet Union after 1990, the emergence of AIDS in certain populations during the 1980s, or the divergence of mortality trends between Eastern and Western Europe after 1960. However, such observations are less an indictment of extrapolation as a method of mortality forecasting than

a demonstration that the greatest uncertainties affecting future mortality trends derive from social and political, rather than technological, factors.

An important issue for consideration in forecasting mortality is the time frame, both the time frame of the data that form the input to an extrapolation and the time horizon of the projection. Although short-term fluctuations have been common, long-term mortality trends in industrialized countries have been remarkably stable. When mortality decline slowed temporarily during the 1950s and 1960s (in the United States and other developed countries), predictions that the rise in human life expectancy had come to an end were commonplace. Similarly, the unusually rapid decline of mortality rates after 1968 fostered expectations of unprecedented gains in longevity that would continue for decades. With the benefit of hindsight, these were both overreactions to rather short-lived episodes in the history of mortality change.

Another common error results from an undue emphasis on trends in life expectancy. Although it continues to increase, the pace of change in life expectancy at birth has slowed in recent decades relative to the first half of the 20th century (see Figure 2.1). As noted earlier, this slowdown was inevitable once juvenile mortality was reduced to historically low levels. However, it does not follow from this observation that gains against mortality in the future will be slower than in the past. Although the increase in life expectancy has slowed, the decline in death rates at older ages (where most deaths now occur) has quickened.[28] An extrapolation of current trends in death rates suggests that life expectancy will continue to increase, though not as quickly as during the first half of the 20th century. This slow but stable increase in average life span will be driven by the accelerating pace of mortality decline at older ages.

## C. Optimism Versus Pessimism

In recent years, the extrapolative approach to mortality prediction has been challenged by assertions that future changes in average human life span may come more or less quickly than in the past. The more optimistic view that life span will increase rapidly in the near future is partly a result of the acceleration in rates of mortality decline among the elderly in developed countries during the past few decades. From an historical perspective, however, this change is relatively recent and should be extrapolated into the future with caution. If the new pattern persists for several more decades, it will then constitute strong evidence that old trends have been replaced by new ones.

Another source of optimism about future mortality rates lies in the potential application of existing technologies (e.g., nutritional supplements, reductions in smoking) or the unusual longevity of certain groups, such as

Mormons and Seventh Day Adventists.[45,46] Such discussions may be a good way to improve health behaviors, but they are not so good at informing predictions, largely because this same sort of advocacy influenced past trends as well. For purposes of prediction, we need to ask whether future positive reforms in lifestyle are likely to be implemented faster or more effectively than were similar reforms in the past.

From time to time, technological breakthroughs provide another source of optimism about future mortality rates. In 1998, the manipulation of a gene that halts the shortening of telomeres during the replication of human cells *in vitro* was a source of great optimism in the popular media, provoking rather extraordinary claims about the possibility of surviving to unprecedented ages in the near future. Talk of cures for cancer and vaccines against AIDS promotes similar hopes. Such discussions should not be dismissed as mere wishful thinking but should also be seen in historical perspective.

As wondrous as they may be, recent scientific advances should be compared, for example, to Koch's isolation of the tubercle bacillus in 1882, which provided confirmation of the germ theory of disease and led to a great flourishing of public health initiatives around the turn of the century, or to Fleming's discovery of the antibacterial properties of penicillin in 1928, an event that led to the antibiotic drug therapies introduced in the 1940s. Extrapolations of past mortality trends assume, implicitly, a continuation of social and technological advance on a par with these earlier achievements.

More pessimistic scenarios of the future course of human longevity are based on notions of biological determinism or arguments about practicality, yielding the now-familiar claim that life expectancy at birth cannot exceed 85 years.[14,47] Sometimes, evolutionary arguments are invoked in support of the notion that further extension of the human life span is impossible, even though existing theories say little about whether and to what degree the level of human mortality is amenable to manipulation.[48]

Current patterns of survival indicate that death rates in later life can be altered considerably by environmental influences, and there is little conclusive evidence that further reductions are impossible. Furthermore, as noted before, trends in death rates and in maximal ages at death show no sign of approaching a finite limit. Nevertheless, although claims about fixed limits to human longevity have little scientific basis, a life expectancy at birth of 85 years, the oft-postulated theoretical limit, is within the range of values predicted by extrapolative methods through the end of the 21st century (see Figure 2.6). In contrast, more optimistic claims, a life expectancy at birth of 150 or 200 years, or even more, are much farther afield and would require a much larger deviation from past trends.

## D. LEARNING FROM HISTORY

The historical rise of human longevity is the result of a complex set of changes beginning several centuries ago. Prior to the 1930s, most of this decline was due to factors other than medical therapy,[49] and it is generally attributed to improvements in living conditions and public health. With the advent of antibacterial drugs in the 1930s and 1940s, medical treatment began to play an important role in these changes, and this role has expanded in recent decades thanks to interventions in cardiovascular disease and cancer, which have contributed to the rapid decline of old-age mortality.

It seems reasonable to expect that future mortality trends in wealthy nations will resemble past changes. Although the focus of our efforts will evolve, the net effect on death rates will probably be similar. For this reason, extrapolation is the preferred means of predicting the future of human mortality. This strategy rides the steady course of past mortality trends, whereas popular and scientific discussions of mortality often buck these historical trends, in either an optimistic or pessimistic direction. History teaches us to be cautious. Pessimism about the continuation of mortality decline is not new, and earlier arguments about an imminent end to gains in human longevity have often been overturned, sometimes quite soon after they were put forth. On the other hand, dubious claims about the road to immortality are probably as old as human culture, even though they have not influenced official mortality forecasts as much as their more pessimistic counterparts.

Although imperfect, the appeal of extrapolation lies in the long-term stability of the historical mortality decline, which can be attributed to the complex character of the underlying process. This combination of stability and complexity should discourage us from believing that singular interventions or barriers will substantially alter the course of mortality decline in the future. In this situation, the burden of proof lies with those who predict sharp deviations from past trends. Such predictions should be based on theoretical results that are firmly established and widely accepted by the scientific community. Certainly, history can be overruled by a genuine consensus within the scientific community, but not by unproven theories, intuition, or speculation.

## REFERENCES

1. Acsádi, G. and Nemeskéri, J., *History of Human Life Span and Mortality*, Akadémiai Kiadó, Budapest, 1970.
2. Robine, J.-M., Mathers, C., and Brouard, N., Trends and differentials in disability-free life expectancy, in *Health and Mortality Among Elderly Populations*, Caselli, G. and Lopez, A., Eds., Clarendon Press, Oxford, 1996, 182.

3. Crimmins, E.M., Saito, Y., and Ingengeri, D., Trends in disability-free life expectancy in the United States, 1970–90, *Popul. Dev. Rev.*, 23, 555, 1997.

4. Lee, J., Campbell, C., and Feng, W., The last emperors: An introduction to the demography of the Qing (1644–1911) imperial lineage, in *Old and New Methods in Historical Demography*, Reher, D. and Schofield, R., Eds., Clarendon Press, Oxford, 1993, 361.

5. Hollingsworth, T.H., Mortality in the British peerage families since 1600, *Population*, 32, 323, 1977.

6. Johannson, S.R. and Horowitz, S., Estimating mortality in skeletal populations: Influence of the growth rate on the interpretation of levels and trends during the transition to agriculture, *Am. J. Phys. Anthropol.*, 71, 233, 1986.

7. Paine, R.R., Model life table fitting by maximum likelihood estimation: A procedure to reconstruct paleodemographic characteristics from skeletal age distributions, *Am. J. Phys. Anthropol.*, 79, 51, 1989.

8. Sattenspiel, L. and Harpending, H., Stable populations and skeletal age, *Am. Antiquity*, 48, 489, 1983.

9. Wood, J.W., Milner, G.R., Harpending, H.C., and Weiss, K.M., The osteological paradox: Problems of inferring prehistoric health from skeletal samples, *Curr. Anthropol.*, 33, 343, 1992.

10. Wilmoth, J.R., The Earliest Centenarians: A statistical analysis, in *Exceptional Longevity: From Prehistory to the Present*, Jeune, B. and Vaupel, J.W., Eds., Odense University Press, Odense, Denmark, 1995, 125.

11. Thatcher, A.R., Life tables from the Stone Age to William Farr, Unpublished manuscript, 1980.

12. Omran, A., The epidemiologic transition: A theory of the epidemiology of population change, *Milbank Memorial Fund Quarterly*, 49, 509, 1971.

13. Preston, S.H. and Haines, M.R., *Fatal Years: Child Mortality in Late Nineteenth-Century America*, Princeton University Press, Princeton, NJ, 1991.

14. Fries, J.F., Aging, natural death, and the compression of morbidity, *N. Engl. J. Med.*, 303, 130, 1980.

15. Wilmoth, J.R. and Horiuchi, S., Rectangularization revisited: Variability of age at death within human populations, *Demography*, 36, 475, 1999.

16. Wilmoth, J.R., In search of limits, in *Between Zeus and the Salmon: The Biodemography of Longevity*, Wachter, K.W. and Finch, C.E., Eds., National Academy Press, Washington, D.C., 1997, 38.

17. Horiuchi, S. and Wilmoth, J.R., The Aging of Mortality Decline, Annual Meetings of the Population Association of America, San Francisco, April 6–8, and the Gerontological Society of America, November 15–19, 1995.

18. Centers for Disease Control, Decline in deaths from heart disease and stroke — United States, 1900–1999, *Morbidity and Mortality Weekly Report*, 48, 649, 1999.

19. Crimmins, E.M., The changing pattern of American mortality decline, 1940–77, and its implications for the future, *Popul. Dev. Rev.*, 7, 229, 1981.

20. National Heart, Lung, and Blood Institute, Research Milestones: Framingham Heart Study, http://www.nhlbi.nih.gov/about/framingham/timeline.html, 2000.

21. Cole, P. and Rodu, B., Declining cancer mortality in the United States, *Cancer*, 78, 2045, 1996.

22. Levi, F. et al., Declining cancer mortality in European Union, *Lancet*, 349, 508, 1997.

23. Gersten, O. and Wilmoth, J.R., The cancer transition in Japan since 1951, *Demographic Research*, forthcoming, 2002. Available http://www.demographic-research.org.

24. Asaka, M. et al., What role does *Helicobacter pylori* play in gastric cancer? *Gastroenterology*, 113, S56, 1997.

25. Replogle, M.L. et al., Increased risk of *Helicobacter pylori* associated with birth in wartime and post-war Japan, *Int. J. Epidemiol.*, 25, 210, 1996.

26. World Health Organization, The World Health Report, 1996: Fighting Disease, Fostering Development, WHO, Geneva, 1996.

27. Population Reference Bureau, World Population Data Sheet, P.R.B., Washington, D.C., 2001.

28. Kannisto, V. et al., Reduction in mortality at advanced ages, *Popul. Dev. Rev.*, 20, 793, 1994.

29. Myers, G.C. and Manton, K.G., Accuracy of death certification, in *Proceedings of the Social Statistics Section*, American Statistical Association, 1983, 321.

30. Jeune, B. and Vaupel, J.W., Eds., *Validation of Exceptional Longevity*, Odense University Press, Odense, Denmark, 1999.

31. McWhirter, N., *Guinness Book of World Records*, Guinness Publications, Barcelona, Spain, 1995.

32. Charbonneau, H., Pierre Joubert a-t-il vécu 113 ans? *Mémoires de la Société Généalogique Canadienne-Française*, 41, 45, 1990.

33. Robine, J.-M., and Allard, M., Validation of the exceptional longevity case of a 120 year old woman, *Facts and Research in Gerontology*, 1995, 363. Robine, J.-M., and Allard, M., Jeanne Calment: Validation of the duration of her life, in *Validation of Exceptional Longevity*, Jeune, B. and Vaupel, J.W., Eds., Odense University Press, Odense, Denmark, 1999, 145.

34. Robine, J.-M. and Allard, M., Letter to the editor, *Science*, 279, 1831, 1998.

35. Wilmoth, J.R. et al., The oldest man ever? A case study of exceptional longevity, *The Gerontologist*, 36, 783, 1996.

36. Skythe, A., Jeune, B., and Wilmoth, J.R., Age validation of the oldest man, in *Validation of Exceptional Longevity*, Jeune, B. and Vaupel, J.W., Eds., Odense University Press, Odense, Denmark, 1999, 173.

37. Wilmoth, J.R. and Lundström, H., Extreme longevity in five countries: Presentation of trends with special attention to issues of data quality, *Eur. J. Popul.*, 12, 63, 1996.

38. Matsuzaki, T., Examination of centenarians and factors affecting longevity in Japan, in *Why do the Japanese live long?*, Hishinuma, S., Ed., Dobun (in Japanese), Tokyo, 1988, 120.

39. Kannisto, V. and Thatcher, A.R., The plausibility of certain reported cases of extreme longevity, Paper presented at the Research Workshop on the Oldest-Old, Duke University, March, 1993.

40. Wilmoth, J.R. et al., Increase in maximum life-span in Sweden, 1861–1999, *Science*, 289, 2366, 2000.

**General Perspectives**

41. Bell, F.C., Social Security Area Population Projections: 1997. Washington, D.C.: Social Security Administration, Office of the Chief Actuary, Actuarial Study No. 112, SSA Pub. No. 11–11553, 1997.

42. Lee, R. and Tuljapurkar, S., Population forecasting for fiscal planning: Issues and innovations, in *Demographic Change and Fiscal Policy,* 1st ed., Auerbach, A.J. and Lee, R.D., Eds., Cambridge University Press, Cambridge, 2001, Chapter 2.

43. Wilmoth, J.R., Mortality projections for Japan: A comparison of four methods, in *Health and Mortality Among Elderly Populations*, Caselli, G. and Lopez, A., Eds., Clarendon Press, Oxford, 1996, 266.

44. Lee, R.D. and Carter, L.R., Modeling and forecasting U.S. mortality. *J. Am. Stat. Assoc.*, 87, 659, 1992.

45. Ames, B.N., Shigenaga, M.K., and Hagen, T.M., Oxidants, antioxidants and the degenerative diseases of aging, *Proc. Natl. Acad. Sci.*, 90, 7915, 1993.

46. Manton, K.G., Stallard, E., and Tolley, H.D., Limits to human life expectancy: Evidence, prospects, and implications, *Popul. Dev. Rev.*, 17, 603, 1991.

47. Olshansky, S.J., Carnes, B.A., and Cassel, C., In search of Methuselah: Estimating the upper limits to human longevity, *Science*, 250, 634, 1990.

48. Partridge, L., Evolutionary biology and age-related mortality, in *Between Zeus and the Salmon: The Biodemography of Longevity*, Wachter, K.W. and Finch, C.E., Eds., National Academy Press, Washington, D.C., 1997, 78.

49. McKeown, T., *The Role of Medicine: Dream, Mirage, or Nemesis?*, Basil Blackwell, Oxford, UK, 1979.

# 3 Comparative and Differential Aging, Geriatric Functional Assessment, Aging and Disease*

*Paola S. Timiras*
University of California, Berkeley

## CONTENTS

## I.  COMPARATIVE PHYSIOLOGY OF AGING

Although the interest of gerontologists in investigating the life span of a variety of organisms has increased considerably in the last few decades, much of the knowledge of the nature of aging in animal and plant species is still incomplete. Senescence was assumed to occur in all vertebrates and to lead to death, hence, its generalized qualification as being *universal, progressive, deleterious, and irreversible.*[1] However, in certain plants and invertebrates, aging may not occur at all.

Senescence has been viewed in evolutionary terms as an inevitable product of natural selection. The survival of the species would demand the aging and death of its members, once reproductive vigor is no longer optimal. "Evolution coddles you when young and forsakes you when old."[2] Survival beyond the period of reproductive activity would represent a luxury that few species are able to afford. Stimulated by these evolutionary ideas, a stock of flies has been developed that lives twice as long as normal (Chapter 6). Indeed, data of mortality, longevity, aging, and life span of flies suggest new possibilities for the use of model insect systems to study aging and mortality dynamics applicable to other species.[3]

Theoretically, in many species, and certainly in humans, longevity could be subject to positive selection based on criteria of fitness other than reproductive capac-

---

* Illustrations by Dr. S. Oklund.

ity. In humans, increases in life expectancy in the past century have been ascribed overwhelmingly to reductions in environmental causes of mortality that are extrinsic to the aging process. In this section, life spans of several species are compared (selective longevity), and some relationships are drawn between the life span and selected physiologic characteristics (physiological correlates of longevity).

## A. SELECTIVE LONGEVITY

There are significant differences among animal and plant species in the duration of life (longevity), the onset of aging, and the rate of mortality.

In eukaryotes, the life span varies from a two-day duration in yeast cells (*Saccharomyces*) to a duration of 5000 years in the California bristlecone pine (*Pinus longaeva*). In vertebrates, aging and death, assumed to occur in all species, were originally attributed to species-specific genetic programs. However, the early view of a rigidly, genetically preprogrammed life span is gradually being replaced by one of greater plasticity in response to environmental modifications (Chapter 1). It is now possible to envision that aging could be retarded by appropriate genetic and environmental modifications.

According to Finch,[4-7] life spans of sexually reproducing species may be categorized according to rapid, gradual, and negligible rates of senescence (Figure 3.1). *Rapid senescence* is usually observed in organisms that have a short adult life. In addition to the ones listed in Figure 3.1, the Pacific salmon (*Oncorhynkus*) represents a prime example of an organism characteristic of this group. It undergoes severe stress (with high blood cortisol levels) at the time of spawning and dies shortly thereafter at about 1 year of age.[8] However, if reproduction and the associated stress are blocked, the salmon will continue to live for several (6 to 7) years.[8] Other species fitting into this category, and listed in Figure 3.1, include yeast, some nematodes (*Caenorhabditis*), the housefly (*Drosophila*), and some longer-lived (10 to 100 years) species that die after a very short senescence period, such as the tarantula (*Eurypelma californica*) and some species of bamboo (in the genus *Phillostachys*).

*Gradual senescence* characterizes the majority of animals and plants with life spans ranging from 1 year to more than 100 years. Humans (and many animals utilized in the laboratory for the study of aging) fall into this category. In these animals, midlife is marked by a progressive decline in function and a progressive increase in the number of diseases affecting the same individual simultaneously (comorbidity).

The category of *negligible senescence* includes species that, at older ages, do not show evidence of physiologic dysfunctions, acceleration of mortality, or limit to the life span. Examples of species in this category include some

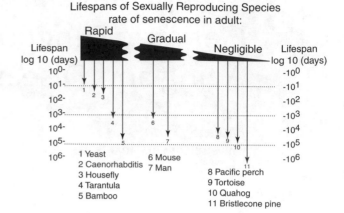

**FIGURE 3.1** Comparison of maximum life span in different animal species. Species differ widely in longevity (total life span from zygote to oldest adult in years) and in the rate of senescence in adult (semiquantitative scale ranging from rapid to gradual to negligible senescence). Data on maximum life spans depend on husbandry conditions (temperature, nutrition). Sources of data are found in Finch[4] (adapted with permission from Ref. 5). **Rapid senescence**. Yeast (*Saccharomyces cerevisae*) 2 to 4 days during asexual budding. Nematode (*Caenorhabditis elegans*) 30 days. Housefly (*Drosophila melanogaster*) 60 days. Pacific salmon (*Onchorynchus*) 3 to 6 years. Tarantula (male) 10 years. Vascular plants: thick-stemmed bamboo (*Phyllostachys bambusoides*, *P. henonis*) 120 years; Puya raimondii (related to the pineapple) 150 years. **Gradual senescence**. Mouse (*Mus musculus*) 4.2 years. Human (Jeanne Calment) 122.4 years. **Negligible senescence**. Fish (Pacific perch), rockfish (*Sebastes aleutianus*) 140 years; orange roughy (*Hoplostethus atlanticus*) 140 years; warty oreo (*Allocyttus verrucosus*) >130 years; sturgeon (*Acipenser fulvescens*) 152 years. Tortoise (*Geochelone gigantea*) 150 years. Bivalve mollusc: ocean quahog (*Artica islandica*) 220 years. Great Basin bristlecone pine (*Pinus longaeva*) 4862 years. Ring-dating often underestimates, and the true ages are probably greater than 5000 years. Inclusion of clonal, asexual reproducing species, e.g., clones of the crosote bush, an asexually reproducing species, would extend the upper range of postzygotic life spans to >10,000 years.

fishes (e.g., rockfish, orange roughy), turtles (e.g., tortoises such as the *Testudo sumerii* that may live to 150 years), and some trees (e.g., bristlecone pine[9]). Other life-forms — particularly plants and invertebrates — in which aging may not occur at all have been considered immortal. Examples of *selective immortality* include some protozoa (e.g., *Paramecium)* and metazoa (e.g., sea anemones such as *Cereas pedunculatus* in which the individual "jellies" have a fixed life span, but the larval "stub" produces a constant supply of new blanks as old ones are removed, and the specimens remain vigorous indefinitely).[10] The absence of signs of aging may not be equated with immortality; rather, these organisms may better fit the negligible senescence category; they still may die but from a cause that might kill them at any age.

**TABLE 3.1**
**Physiologic Correlates with Longevity**

| Index Studied | Correlation |
|---|---|
| Body weight | Direct |
| Brain/body weight | Direct |
| Basal metabolic rate | Inverse |
| Stress | Inverse |
| Reproductive function/Fecundity | Inverse |
| Length of growth period | Direct |
| Evolution | Uncertain |

## B. PHYSIOLOGICAL CORRELATES OF LONGEVITY

Data from several orders of placental mammals show a highly significant relationship between life span and body weight: the bigger the animal, the longer the life span (Table 3.1). For example, the elephant may reach or exceed 70 years in captivity, whereas the rat seldom lives more than 3 years. There are, however, many exceptions to this generalization. Humans may reach 122 years of age, whereas, other larger mammals show a shorter potential longevity (horse: approximately 60 years; hippopotamus and rhinoceros: approximately 50 years; bear: 30 years; camel: 25 years). Among domestic carnivores, cats, although generally smaller in size than dogs, live longer.

The same data show an even more significant relationship between life span and brain weight (Figure 3.2). For example, insectivores with a smaller, simpler brain have a shorter life span than ungulates, and the latter have a simpler brain and shorter life span than humans (Fig-

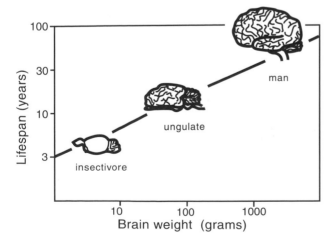

**FIGURE 3.2** *Comparison of the relationship of brain weight to life span in vertebrates.* Humans have the heaviest brain in relation to body weight among all vertebrates, and also the longest life span.

ure 3.2). Humans with the longest life span have the heaviest brain in relation to body weight and also the most structurally complex.[11] Among the major brain regions, the neocortex (the largest and, ontogenetically, most recently developed part of the cerebral cortex) shows the strongest correlation with the life span, with potentially important physiologic implications.[12] The most obvious is that the higher the brain weight to body weight ratio is, and particularly the greater the degree of cerebral cortical expansion (i.e., the process of encephalization), the more precise would be the physiologic regulations; hence, the greater the chance for longer survival. Such an interpretation seems well justified in view of the essential role the nervous system plays in regulating vital physiologic adjustments, especially responses to environmental demands. Maintenance of the physiologic optimum and reduction of the magnitude of the fluctuations occurring over time diminish the probability of irreversible changes per unit time and, thus, slow the rate of aging and reduce the incidence of death.

Appealing as this interpretation may be to a physiologist, some caution is suggested when formulating direct correlation of body and brain size with life span. The correlations are not always appropriate when the terms of comparison represent different entities, one essentially anatomical (stature, brain weight) and the other essentially evolutionary, functional, and biochemical (duration of life). Another limitation is the paucity of accurate data on the maximum life span of most animal species. While the brain is functionally a most important organ, a similarly positive relationship exists between size of several organs (e.g., adrenal, liver, spleen) and life span.[4]

Another criterion of the relationship between a physiologic parameter and the length of the life span is *basal metabolic rate* (BMR): the higher the metabolic rate, the shorter the life span[13] (Table 3.1). BMR represents the amount of energy liberated per unit time by the catabolism of food and physiologic processes in the body. It is measured with the subject at rest, in a comfortable ambient temperature, and at least 12 h after the last meal. BMR is expressed in kilocalories per unit of body surface area and can be compared among different individuals. The energy thus liberated appears as external work, heat, and energy storage.

Some have argued that the life span of a species is limited by a fixed total metabolic potential that is consumed over a lifetime. Comparison of shrews and bats shows that shrews, with the highest metabolic rate of all mammals, have a life span of 1 year compared to bats, with a much lower metabolism and a life span of over 15 years.[14] The higher metabolic rate could accelerate the accumulation of nuclear errors (DNA damage) or of cytoplasmic damage (accumulation of free radicals) and thereby shorten the life span. An example of environmental manipulation that reliably extends the life span, at least

**General Perspectives**

in rodents, is caloric restriction that would slow the oxidative damage that may underlie aging[15] (Chapters 5 and 24). Another example, in fruit flies, shows that mutation of the gene that encodes a protein necessary for transport and recycling of metabolic by-products would double the life span.[16] In fishes, exposure to temperatures lower than those optimal for the species may increase longevity by decreasing immune responsiveness and, thereby, prevent autoimmune disorders. In *C. elegans*, exposure to low temperature induces mutations of several temperature-sensitive genes responsible for the formation of larvae, which are both developmentally arrested and long-lived; when the so-called "dauer" larvae are allowed to grow, the adult *C. elegans* lives twice as long as the wild type.[17,18] Cumulative damage deriving from metabolism may result in chromosomal damage,[19] such as mutation of a DNA putative helicase that occurs in the human premature aging disease, Werner's syndrome.[20]

*Fecundity*, as an expression of reproductive function and measured by the number of young born per year of mature life, appears also to be inversely related to longevity (Table 3.1). Shrews, with a short life span, have a large litter size and produce two litters per year, whereas the longer-lived bats have only one young per year. Duration of growth (during the developmental period preceding adulthood) has also been related to the life span. For example, comparison of the chimpanzee to humans shows that growth periods last approximately 10 years in the chimpanzee and 20 years in humans, and their respective average life spans are approximately 40 and 80 years.

Duration of the growth period can be prolonged in experimental animals; the onset of developmental maturation can be delayed by restricting food intake in terms of total calories or of some specific dietary components.[21] With these dietary manipulations, not only is the life span prolonged, but some specific functions, such as reproduction and thermoregulation, are maintained until advanced age; also, the onset of aging-related pathology is delayed, and its severity is reduced (Chapter 24).

The factors described above, including body and brain size, metabolic rate, fecundity, duration of growth, as well as stress and conditions of life, superimposed on the genetic makeup, undoubtedly contribute to the difference in longevity among species. As discussed already, natural selection requires that the individual member of a group must survive through the reproductive period to ensure continuation of the species; thereafter, survival of the post-reproductive individual becomes indifferent or even detrimental (e.g., food competition) to the group. In this sense, a gene that would act to ensure a maximum number of offspring in youth but produce disease at later ages might be positively selected. Currently, not only is life expectancy increased, but humans live well beyond the reproductive years, women some 40 years beyond the child-bearing period. Longevity of a species beyond the reproductive years must be incidental to some earlier events, or the infertile individual must confer some advantage to the fertile. In humans, this must be the case, for older members contribute to the maintenance of the entire population structure and to the development and progress of our society.

## II. DIFFERENTIAL AGING IN HUMANS

### A. "SUCCESSFUL" AGING: FUNCTIONAL PLASTICITY PERSISTS IN OLD AGE

*Chronologic age* (age in number of years) and *physiologic age* (or age in terms of functional capacity) do not always coincide, and physical appearance and health status often belie chronological age. Although specialized knowledge is not required to estimate age, in many cases, people may look younger or older than their chronological age. From a physiologic standpoint, such disparities in the timetable of aging may occur among individuals or among selected populations; they result from complex interactions between genetic and environmental factors that operate on the individual as a child and also as a member of a societal group. Early attempts were made to standardize functional profiles of old persons similar to the well-established diagnostic charts of growing children. Yet, one of the characteristics of the aging human population is its substantive heterogeneity. That is, some individuals "age" at a much slower or faster rate than others, and variability among individuals of the same age in response to a variety of physiologic and psychological tests increases with old age (Figure 3.3).

As already mentioned (Chapter 1), in humans, the physiologic "norm" is represented by the sum of all functions in a 25-year-old man free of any disease, with a weight of 70 kg, and a height of 170 cm. Comparison of old individuals with this "ideal man" inevitably discloses a range of functional decrements with advancing age. Early studies were conducted in selected samples of "representative" elderly.[22] Because the prevalence of chronic disease increases in old age, a large part of the data of functional loss with aging in these studies may have been due to the effects of disease rather than to aging. In those earlier studies, comparison of several functions from young to old age revealed a gradation of decrements with old age. Another, more recent approach has challenged the inevitability of functional impairment with aging by grouping aging processes into three possible trajectories:

1. *Aging*, with disease and disability
2. *Usual aging*, with absence of overt pathology but presence of some declines in function
3. *Successful (or healthy) aging*, with no pathology and little or no functional loss

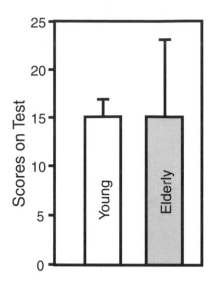

**FIGURE 3.3** *The heterogeneity of the elderly population as illustrated by scores in a hypothetical test.* Results from a large number of tests show that the mean of two scores (as represented by the barograms) is the same for both young and old individuals. However, the much greater standard error of the mean in older than in younger individuals indicates a greater variability among individuals from the old population.

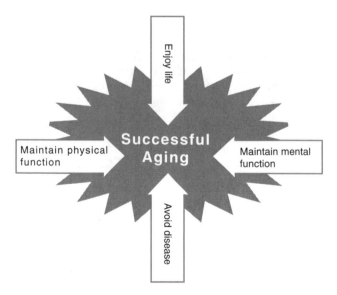

**FIGURE 3.4** *Successful Aging.* Successful aging is based on maintenance of mental and physical function, avoidance of diseases, and enjoyment of life.

Such a grouping[23,24] of aging processes:

1. De-emphasizes the view that aging is exclusively characterized by declines in functional competence and health
2. Refocuses on the substantial heterogeneity among old persons

3. Underscores the existence of positive trajectories (i.e., without disability, disease, and major physiological decline)
4. Highlights the possible avoidance of many, if not all, the diseases and disabilities usually associated with old age

Examples of successful aging include elderly individuals in whom blood pressure, serum cholesterol, body mass index, cognitive function, and responses to exercise are similar to those of much younger subjects. Conditions for successful old age include interactions among several factors, such as higher level of formal education and financial income, strong social support, and consistent participation in moderate exercise, good diet, and hygienic habits (no drugs, alcohol, or smoking) (Figure 3.4).

The distinction between usual (average aging) and successful aging (better than average aging) assumes the persistence into old age of normal function and the ability of compensatory restoration of normal function (once it has been disrupted). *Functional plasticity*, that is, the capacity of the individual to adapt to environmental demands without loss of physiologic competence, is most effective at young developmental ages; however, as discussed throughout this book, it persists into old age, albeit at a less efficient level. As we consider the many cross-cultural differences that greatly influence aging, factors such as diet, exercise, drugs, and psychosocial environment should not be underestimated as potential moderators of the aging process. Taking these elements into account, the prospects for avoidance, or eventual reversal, of functional loss with age are vastly improved, and the risks of disability and disease are reduced.

## B. HETEROGENEITY OF PHYSIOLOGICAL AGING

Changes with aging lack uniformity, not only among individuals of the same species, but also within the same individual: onset, rate, and magnitude of the changes vary depending on the cell, tissue, organ, system, or laboratory value considered.[22] An example of the heterogeneity of aging changes is illustrated by a number of laboratory values, many of which remain unchanged with aging, while a few decrease or increase (Table 3.2). Laboratory values are always given within a normal range, and many of the age-related levels do not change in absolute terms but only alter value distribution within this normal range.[25]

Successful aging is a demonstration that aging may occur with little loss of function. Thus, in certain functions, regulation remains quite efficient until advanced age; in others, it may decline at an early age. Examples of this type of differential aging include the *basal metabolic rate* (BMR) that declines continuously, in humans, throughout the life span, with a faster decline in childhood and adolescence and a consistent but slow

**TABLE 3.2**
**Laboratory Values in Old Age**

| Unchanged | Decreased | Increased |
|---|---|---|
| Hepatic function tests | Serum albumin | Alkaline phosphatase |
|   Serum bilirubin | HDL cholesterol (women) | Uric acid |
|   Aspartate aminotransferase (AST) | Serum $B_{12}$ | Total cholesterol |
|   Alanine aminotransferase (ALT) | Serum Magnesium | HDL cholesterol (men) |
|   γ-glutamyltransferase(GGTP) | $PO_2$ | Triglycerides |
| Coagulation tests | Creatinine clearance | Plasma TSH (?) |
| Biochemical tests | Plasma $T_3$ (?) |   Plasma glucose tolerance tests |
|   Serum electrolytes | White blood count | Fasting blood sugar |
|   Total protein | |   (maybe within normal range) |
|   Calcium | | Postprandial blood sugar |
|   Phosophorus | | |
|   Serum folate | | |
| Arterial blood tests | | |
|   pH | | |
|   $PCO_2$ | | |
| Renal function tests | | |
|   Serum creatinine | | |
| Thyroid function tests | | |
|   Plasma T4 | | |
| Complete blood count | | |
|   Hematocrit | | |
|   Hemoglobin | | |
|   Red blood cells | | |
|   Platelets | | |

decline in old ages.[4] If, as already suggested above, a decline in BMR may result from the accumulation of metabolic damage, the rate of such accumulation may differ in different organisms and may be responsible for differences in life span; in yeast and some worms, metabolism may be the determinant of the life span,[26,27] and in others, damage to chromosomes perhaps deriving from altered metabolism may be the cause of premature aging (as in Werner's syndrome).[20]

One classic example of a unique timetable involving an organ that develops and ages during a specific period of the life span is the ovary; it begins to function at adolescence (in humans, approximately 10 to 12 years) and ceases to function at menopause (in humans, as long as 40 years before death) (Chapter 11). Other classic examples include embryonic structures, such as the placenta, which develop and age within a relatively short period of prenatal time as compared to the total length of the life span. Whichever the organ or tissue considered, timetables of aging represent an approximation, for the onset of aging cannot be pinpointed precisely by a specific physiologic sign.

Because aging is often a slow and continuous process, some of its effects can be observed only when they have progressed sufficiently to induce alterations that can be identified and validated by available testing methods. An illustrative example is atherosclerosis. Atherosclerotic lesions begin in childhood, when they remain functionally silent. They become manifest in middle and old age, when the lesions have progressed sufficiently to induce pathological consequences affecting the entire cardiovascular function (Chapters 16 and 17).

Fasting blood sugar values are minimally affected by aging, with exception to late onset diabetes (insulin-resistant diabetes), where blood sugar levels are elevated (Chapter 14). However, even in the nondiabetic elderly, when blood sugar levels are tested under increased physiologic demand (e.g., a sugar load, as in the sugar tolerance test), the efficiency with which the organism is capable of maintaining levels within normal limits and the rapidity with which these levels return to normal are significantly reduced in old as compared to adult individuals (Chapter 14). Likewise, conduction velocity in nerves, cardiac index (cardiac output/minute/square meter of body), renal function (filtration rate, blood flow), and respiratory function (vital capacity, maximum breathing capacity) are less capable of withstanding stress in old than in young individuals. Imposition of stress reveals age differences not otherwise detectable under steady state conditions; it clearly demonstrates the declining ability of the old organism to withstand or respond adequately to stress (Chapter 10).

Other body functions begin to age relatively early in adult life and fall to a minimum before the age of 65, officially heralding the stage of senescence. For example, in the eye, accommodation begins to decline in the teens and regresses to a minimum in the mid-fifties (Chapter 9). Hearing deterioration begins at adolescence and continues steadily thereafter, culminating around 50 years of age. Auditory deterioration may also be hastened by the environmental noise to which, in our civilization, individuals are continuously exposed from young age. Some comparative studies in isolated populations living in a quiet environment (e.g., forest-dwelling African tribes) and maintaining good auditory function into old age seem to support the view that continuous exposure to noise may be harmful to hearing (Chapter 9).

Additional examples of differential aging include prominent changes that occur in stature (or standing height), sitting height, breadth of shoulder, and depth of chest — all of which show a progressive reduction with aging in contrast to head diameters, which remain practically unchanged. The considerable and progressive diminution in stature that appears to occur with aging in all humans may be ascribed, at least in part, to alterations in bone structure (e.g., osteoporosis) (Chapters 11 and 21).

Body weight usually increases at the initial stages of senescence as a consequence of increased fat deposition (especially in the subcutaneous layer), which reaches a peak at approximately 50 years in men and 60 years in women.[28] Both body weight and fat deposition progressively decrease at later ages (between the fifth and seventh decades) when morbidity and mortality increase.[29]

In view of the high incidence of pathology with aging, sudden or progressive changes in body weight, abnormal in severity and timing, may also be indicative of onset and course of disease. In this manner, body weight represents a relatively simple measure of the physiologic and, eventually, pathologic assessment of the health status of the elderly (as it is also an indicator of the growth progress during development).

A more complete discussion of changes with aging in various functions is presented in the corresponding chapters of this book.

## III. FUNCTIONAL ASSESSMENT IN THE ELDERLY

Assessment of physiologic competence is critical at all ages to determine the health status of the individual and to serve as a basis for the diagnosis and prognosis of disease. In this section, parameters of physiologic status are evaluated first and then incorporated into geriatric assessment.

## A. ASSESSMENT OF PHYSIOLOGICAL AGE IN HUMANS

*Assessment of physiologic competence and health status in humans, at any age, is a multifactorial process* (Figure 3.5). It requires quantitative measurements of numerous parameters selected as indices of physical, neurological, and behavioral competence at progressive ages.[30–33] To establish an accurate profile that will reflect the different age-related timetables for body systems and combine them to represent the health status of the individual, a number of criteria must be satisfied. These criteria should account for many variables, among which some of the prerequisites are as follows:

- The variables must be indicative of a function important to the competence or general health of the individual and capable of influencing the rate of aging.
- They must correlate with chronological age.
- They must change sufficiently and with discernible regularity over time to reveal significant differences over a 3- to 5-year interval between tests.
- They must be practically measurable in an individual or cohort of individuals without hazard, discomfort, or expense to the participant or excessive labor or expense for the investigator.

The validity of any assessment lies in the choice of the proper test or battery of tests best qualified to provide an overall picture of current health and to eventually serve as a basis for prediction of future health and length of the life span. Such a choice is complicated by the need to also take into account the financial feasibility and the facilities available for the testing. The relatively large number of tests for assessing physiologic competence and health status in the elderly reflects the current failure to reach a consensus on the best checklist. A global measure of physiologic status may be derived from many different combinations of tests. Selection will depend on the purpose of the assessment and who will use these data. With respect to purpose of assessment, measurements may be expected:

- To describe the physiologic status of an individual at progressive chronological ages
- To screen a selected population for assessment of overall physiologic competence or competence of specific functions using either cross-sectional or longitudinal sampling methods (Chapter 1)
- To monitor the efficacy of specific treatments, drugs, exercise, and diet
- To predict persistence or loss of physiologic competence, to determine incidence of disease, and to evaluate life expectancy

**General Perspectives**

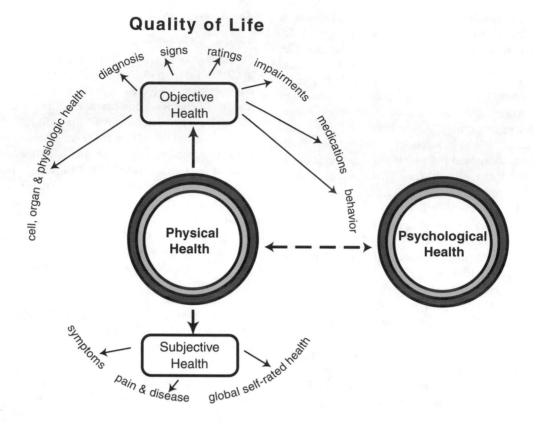

**FIGURE 3.5** *Assessment of the health status and quality of life in humans depends on multiple measures.* These include several parameters of objective and subjective physical health together with parameters of psychological health.

Approximately 10% of nondisabled community-dwelling adults aged 75 and older lose independence each year. A number of simple qualitative and timed performance tests may be useful in identifying subgroups of older persons that are at increased risk for functional dependence.[34]

The choice of tests will depend on whether the user is a health provider, specialist, researcher, or case manager. Even with a precise identification of the nature of the assessment needed and of the user who needs it, it still remains difficult to choose the most significant and feasible tests.

Some tests, although relatively innocuous in young and healthy individuals, may be troublesome for the elderly. Yet, it may be contrary to the interest of the elderly to assert that they represent a vulnerable group needing special protection. Rather, benefits may accrue for the elderly from their participation in medical and psychosocial survey research. Not only may such tests lead to the discovery of an unsuspected illness and its eventual treatment, but they may also provide altruistic commitment and mental satisfaction.

## B. GERIATRIC ASSESSMENT

Geriatric assessment involves a multidimensional diagnostic process (including physiologic assessment) designed to qualify an elderly individual not only in terms of functional capabilities and disabilities but also of medical and psychosocial characteristics (Table 3.3, Figure 3.5). It is usually conducted by a multidisciplinary team with the intent of formulating a comprehensive plan for therapy and long-term follow-up for either ambulatory noninstitutionalized individuals or institutionalized geriatric patients (including centenerians[35]).[36–40] Major purposes are as follows:

1. To improve diagnosis (medical and psychosocial)
2. To plan appropriate rehabilitation and other therapy
3. To determine optimal living conditions, arrange for high-quality follow-up care and case management
4. To establish baseline information useful for future comparison

Most of these assessment programs include several tests, which have been grouped into three categories:

- Tests that examine general physical health as represented by physiologic competence and the absence of disease
- Tests that measure the ability to perform basic self-care activities, the so-called *activities of*

**TABLE 3.3**
**"Simple" Functional Assessment of Ambulatory Elderly**

History
Physical examination
    Including: neurologic and musculoskeletal evaluation of arm and leg
    Evaluation of vision, hearing, and speech
Urinary incontinence (eventually fecal incontinence)
    Presence and degree of severity
Nutrition
    Dental evaluation
    Body weight
    Laboratory tests depending on nutritional status and diet
Mental Status
    Folstein Mini-mental Status Score
    If score <24, search for causes of cognitive impairment
Depression
    If Geriatric Depression scale is positive:
        Check for adverse medications
        Initiate appropriate treatment
ADL and IADL (see Table 3.4)
Home environment and social support
 Evaluation of home safety and family and community resources

Adapted from Lachs, M.S., et al., *Ann. Int. Med.*, 112, 699, 1990. With Permission.

*daily living (ADL)* (Table 3.4), which reflect the ability to manage personal care
- Tests that measure, in addition to basic activities, the ability to perform more complex *instrumental activities of daily living (IADL)* (Table 3.4). ADL and IADL or the lack of them are taken as reflecting the ability to live independently in the community. However, despite their current wide use in geriatric assessments for identification of health status and eventual disabilities, some cautions must be recognized in determining their validity. One such caution, among others, is that evaluations of the tests are carried out by health professionals rather than by the patients, and the opinions of the two do not always coincide.[41]

Quality of life in later years may be diminished if illness, chronic conditions, or injuries limit the ability of some to care for themselves without assistance. ADL and IADL are widely utilized as representative measures useful in home-dwelling populations and as representative of capability for independent living or, vice versa, as indicators of disability (Figure 3.6). *Disability is the inability to perform a specific function because of health or age and results from impaired functional performance.* In most testing, the degree of wellness, that is, the absence of disability or disease, is recorded and, reciprocally, the presence and severity of disability or disease is

**TABLE 3.4**
**Categories of Physical Health Index Measuring Physical Competence**

| Physical Health | Activities of Daily Living | Instrumental Activities of Daily Living |
|---|---|---|
| Bed days | Feeding | Cooking |
| Restricted-activity days | Bathing | Cleaning |
| Hospitalization | Toileting | Using telephone |
| Physician visits | Dressing | Writing |
| Pain and discomfort | Ambulation | Reading |
| Symptoms | Transfer from bed | Shopping |
| Signs on physical exam | Transfer from toilet | Laundry |
| Physiologic indicators (e.g., lab tests, x-rays,pulmonary and cardiac functions) | Bowel and bladder control | Managing medications |
| Permanent Impairments (e.g., vision, hearing, speech, paralysis, amputations, dental) | Grooming | Using public transportation |
| Diseases/diagnosis | Communication | Walking outdoors |
| Self-rating of health | Visual acuity | Climbing stairs |
| Physician's ratings of health | Upper extremities (e.g., grasping and picking up objects) | Outside work (e.g., gardening, snow shoveling) |
| | Range of motion of limbs | Ability to perform in paid employment |
| | | Managing money |
| | | Traveling out of town |

Many of the items presented are components of several measures of physical and functional health as discussed in a number of geriatric screening and assessment programs.

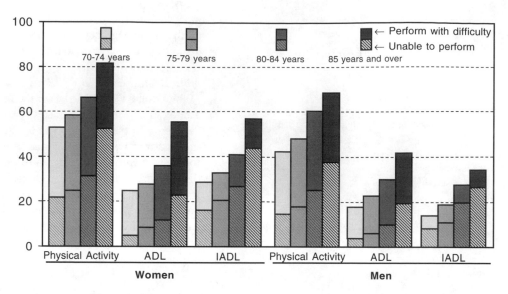

**FIGURE 3.6** Percentage of persons 70 years of age and older who have difficulty or inability to perform physical activity, activities of daily living (ADL), and instrumental activities of daily living (IADL). U.S. (1994) National Health Interview Survey, Second Supplement on Aging. Examples of activities include: for physical activities, walking for a quarter mile; for ADL, bathing, dressing, getting in and out of bed; for IADLs, preparation of one's own meals, shopping, heavy housework. From Kramarow, E.,et al., *Health and Aging Chartbook*, Health, United States, 1999.

recorded.[42–44] The severity of the disability may be measured in terms of whether a person:

1. Does not perform the activity at all
2. Can only perform the activity with the help of another person or if a person is available (but does not actually give aid)
3. Can perform the activity with the help of special equipment

Disability is coded according to five degrees of severity: (1) no disability, (2) at least one IADL disability but no ADL disability, (3) one or two ADL disabilities, (4) three to four ADL disabilities, and (5) five to six ADL disabilities.[45] For example, in 1995, among noninstitutionalized persons 70 years of age and older, 32% had difficulty performing, and 25% were unable to perform at least one of nine physical activities.

With advancing age, disability intensity increases in men and women, with the highest disability at 85 years and older (85+) (Figure 3.6). Disability is usually associated with the incidence of chronic conditions and diseases. These include in decreasing frequency: arthritis (especially in women), heart disease, stroke, respiratory disease, and diabetes (Table 3.5). It is to be noted that the greater intensity of disability of women than men becomes manifest at the later ages of 74 to 85+ years. Thus, females with a longer average life span than men (Chapter 2) live longer with disability. The cause of this sex difference is unknown. Probably, females are at higher risk for a number of chronic degenerative conditions (e.g., osteoporosis, diabetes, arthritis) that interfere

**TABLE 3.5**
**Percent of Persons 70 Years of Age and Older Who Report Specific Conditions as a Cause of Limitation in Activities of Daily Living: United States, 1995**

| Type of Condition | Percent |
|---|---|
| Arthritis | 10.6 |
| Heart disease | 4.0 |
| Stroke | 2.6 |
| Respiratory | 2.5 |
| Diabetes | 1.5 |

From the *1994 National Health Interview Survey, Second Supplement on Aging*, Centers for Disease Control and Prevention, National Center for Health Statistics.

with those functions (e.g., walking, doing housework) necessary for independent living.

The passage from independence to disability may be gradual or abrupt. It may be accelerated by the onset or worsening of disease,[46–48] or it may be delayed by a high level of education associated with high income and good lifestyle habits.[49–52] However, disease prevention and improvement of environmental conditions cannot forestall the progressive loss of function and corresponding increase of disability that occur with advancing age.[53,54] Indeed, in some populations of socioeconomic-advantaged elderly, while death has been postponed, the prevalence of disease and disability has not, especially in women.[55]

In parallel with differences in socioeconomic, educational, and hygienic habits, there are also differences in physical functioning and disability by race among the older populations. Afro-Americans report higher levels of disability than white persons. In 1995, among noninstitutionalized persons 70 years of age and older, Afro-Americans were 1.3 times as likely as white persons to be unable to do certain physical activities and 1.5 times as likely as white persons to be unable to perform one or more ADL. Similar conclusions may be drawn for some other ethnic (e.g., Hispanic) groups, although additional research is needed to further define the social and health factors that contribute to these ethnic differences.[46,47,56]

In the last 20 years, the proportion of noninstitutionalized older women and men unable to do one or more ADL or IADL has declined, the decline being particularly marked in women. This trend may be viewed as evidence of progressively healthier (and even successful) aging than in the previous decades.[57–59] However, the overall (including persons in nursing homes and other similar institutions) proportion of elderly persons unable to perform ADL continues to increase with the lengthening of the life span, even though the level remains quite low.[60]

## C. DISUSE AND AGING

Many of the changes that accompany aging coincide with those associated with physical inactivity; they are generally grouped under the term *disuse*. Changes induced by bed rest as a consequence of disease may be superimposed on aging changes and further accelerate the aging process. Studies of long-term space travel have revealed that weightlessness in space also induces changes resembling those of aging and physical inactivity. The relationship between disuse (due to bed rest, insufficient exercise, or lack of gravity) and aging has some definite practical implications inasmuch as it permits us to envision that prevention and rehabilitation of the disuse phenomena may also ameliorate some of the deficits of old age.

This association of disuse and aging is apparent at all functional levels. A brief list of the functions affected is presented in Table 3.6. At the top of the list is a decrease in maximum oxygen consumption ($V_{O2}$), which measures indirectly the ability of the organism to transport oxygen from the atmosphere to the tissues. This transport is significantly reduced with aging at the rate of about 1% per year and with bed rest; in both cases, a program of physical activity slows the decline. $V_{O2}$ max depends on cardiac output, which also decreases with age and bed rest, as a result of decreased stroke volume (Chapter 21). Simultaneously, blood pressure increases with age and weightlessness, probably due to increased peripheral circulatory resistance.

In all cases, both younger and older subjects are able to benefit from exercise. Studies of exercise and movement programs for elders (including nonagenarians) have

**TABLE 3.6**
**Physiologic Parameters in Aging, Physical Inactivity and Weightlessness (In Space)[a]**

| Reduced | Increased |
|---|---|
| Maximum oxygen consumption ($VO_2$ max) | Systolic blood pressure and peripheral resistance |
| Resting and maximum cardiac output | Vestibular sensitivity |
| Stroke volume | Serum total cholesterol |
| Sense of balance | Urinary nitrogen and creatinine |
| Body water and sodium | Urinary calcium |
| Blood cell mass | |
| Lean body mass | |
| Glucose tolerance test | **Variable** |
| Sympathetic activity and neurotransmission | Endocrine changes |
| | Altered EEG |
| Thermoregulation | Altered sleep |
| Immune responses | Changes in specific senses |

[a] Possibly responsive to physical activity.

shown significant increase in lean body (muscle) mass, improvement in several joint movements, and subjective perception of improved mobility and well-being. The capacity for improvement in muscle mass and power upon exercise is quite striking in humans as well as in experimental animals. At all ages, "practice makes perfect." However, for the elderly more than for other age groups, activity programs must be tailored to the individual for optimal benefits (Chapter 24).

## IV. AGING AND DISEASE

Aging is associated with increased incidence and severity of diseases, accidents, and stress. Deleterious factors, not themselves lethal, may, from an early age, gradually predispose the individual to functional losses or to specific diseases in later life. The study of the *epidemiology of aging,* the branch of medical sciences that deals with the incidence, distribution, and control of disease in a population, is important not only to understand the succession of events that culminate in functional failure and disease, but also to uncover valuable clues concerning the aging process.

It is difficult to isolate the effects of aging from those consequent to disease or gradual degenerative changes that develop fully with the passage of time. Any demarcation among these effects can be only tentatively drawn at the present state of our knowledge. For example, it is reasonable to question whether the *atheroma* — the characteristic lesion of atherosclerosis — represents a degenerative process or a disease; in other words, does it result from age-related cellular and molecular changes in one or sev-

**TABLE 3.7**
**Holistic View of the Elderly**

In geriatrics, it is necessary
- To differentiate the aging process from disease
- To correlate physical state with psychosocial environment

eral of the arterial wall constituents or from mechanical injury or infection of the vascular wall (Chapter 16). *Dementia* (contrary to the opinion of some pessimists!) is not a normal unavoidable consequence of aging and should be investigated as any other disease process (Chapter 8). Likewise, *anemia* is not a normal correlate of aging, and, when present, a cause for this condition should be investigated (Chapter 18). Indeed, one of the tasks of the geriatrician is to be able to differentiate aging from disease and to treat both as independent but related entities. One of the challenges of geriatrics arises from the multiplicity of problems confronting the elderly.[61] One cannot adequately treat disease without considering the psychological, economic, and social situation of each individual (Table 3.7). This global, "holistic" view of the individual should be taken for all ages, of course, but it becomes crucial for the elderly, for whom loneliness, social instability, and often, financial hardship have enormous impacts on health and well-being.

The close association between aging and disease affects the practical orientation of aging research. One essentially pragmatic approach focuses on traditional research aimed at specific functional and clinical entities rather than at aging as a whole. Clarification of the cause and pathology and eventual treatment of a specific disease spotlight the search for the immediate improvement of health and quality of life of the elderly.

Another more fundamental goal is to study the aging process per se. This study responds to the reality that progress in biomedical research and conquest over many diseases represent only two of many factors responsible for longer life span and an increasing percentage of the elderly in the human population today. While the study of a specific disease projects a much more concrete image and a task achievable within a limited time, both clinical and basic research orientations have merits and should be pursued simultaneously.

## A. COMPLEXITY OF PATHOLOGY OF AGING

It has long been accepted, and with reason, that "old age is almost always combined with, or masked by, morbid processes." Certainly, late life is a period of increasing and multiple pathology and morbidity (incidence of disease is also referred to as morbidity) during which it becomes progressively more difficult to distinguish between "normal" aging and specific diseases that affect

old people. As indicated at the beginning of this chapter, aging has been qualified as universal (although there are exceptions[3–5]), intrinsic, progressive, deleterious, and irreversible.[1] Disease may be viewed as a process that is:

- *Selective* (i.e., varies with the species, tissue, organ, cell, and molecule)
- *Intrinsic and extrinsic* (i.e., may depend on environmental and genetic factors)
- *Discontinuous* (may progress, regress, or be arrested)
- *Occasionally deleterious* (damage is often variable, reversible)
- *Often treatable* (with known etiopathology, cure may be available)

In humans, the probability of dying from disease or injury increases with the passage of time. In every organ, tissue, and in many cells, a time-dependent loss of structure, function, and chemistry proceeds slowly as the consequence of accumulating injuries. For example, deterioration of skin elastin may result from bombardment by ultraviolet photons (Chapter 22); degeneration of articular cartilage, from countless mechanical insults (Chapter 21); and opacification of the crystalline lens, from molecular injuries (Chapter 9). Indeed, it is when these injuries act at the molecular level that most significant consequences for aging may occur, such as failure of DNA repair or oxidative damage or progressive cross-linking of collagen, the major structural protein of the body. The greater susceptibility to diseases and disabilities in old persons compared to adults has raised the question of whether a higher proportion of older people in the population will result in an increase or a decrease in the prevalence of chronic disease and disability. Such a question has not only biomedical but also socioeconomic implications (Box 3.1).

As mentioned previously, one of the main characteristics of old age pathology is *comorbidity*, that is, *the multiplicity of diseases simultaneously affecting the same individual.*[62–64] The prevalence of two or more diseases, each capable of limiting ADL (Table 3.5), increases with age for men and, with a greater severity, for women (Figure 3.6). Autopsy often reveals numerous lesions involving so many organs of the body that it is difficult to know which one was responsible for death. The pathologist is often perplexed as to how the patient managed to live so long with such a "load" of diseases. This pathologic multiplicity is particularly evident in the 70- to 90-year-old age group in contrast to the 60 to 70 year olds who reveal a limited number of lesions at autopsy.

The multiple pathology of the elderly poses many problems in diagnosis and treatment. As children are not just "little or young adults," the elderly are not just "old adults." Of course, the wide range of individuality and variability makes many generalizations inappropriate; however, certain

---

**Box 3.1**
**Expansion or Compression of Morbidity?**

While observations of increasing pathology in the elderly would point to an "expansion" of morbidity, several investigators have proposed a more optimistic view of "compression" of morbidity. The latter is the term used by epidemiologists to indicate a shortening in the length of time between onset of disease and death. Delay of the onset of functional limitation of ADL and IADL and of disease, even without further improvements in life expectancy, would greatly improve the quality of the later years of life.

Compression of morbidity (resulting from changes in diet, exercise, and daily routines) would postpone (and, in fact, may have already postponed) the age of onset of the major fatal/degenerative diseases (heart disease, cancer, and stroke; Alzheimer's disease, osteoporosis, and sensory deficits).[65] It would lead to "rectangularization" of the survivorship curve (Chapter 2) and to an healthier old age. Declining function and increasing pathology would be banished to the very last months or weeks of life.

Many epidemiologists vigorously challenge this hypothesis. They emphasize that the lengthening of the life span does and will continue to result in expansion (rather than compression) of morbidity.[66] They argue that the lengthening of the life span has not resulted from a reduction in the incidence of chronic disease and concurrent functional limitations but rather from a reduction in the "lethal consequences" associated with those conditions. Modifications of the environmental and behavioral factors known to reduce the mortality risks from fatal diseases, do not change the onset and progression of most of the debilitating diseases associated with old age. Current data illustrated in Figure 3.6 indicate that the time of disability and disease is lengthened when old age mortality is postponed, with a trade-off of reduced mortality in the middle and older age for an expansion of morbidity (Chapter 2).

---

**TABLE 3.8**
**General Characteristics of Disease in the Elderly**

Symptoms
  Vague and subtle
  Atypical
  Unreported
Chronic versus acute
Multisystem disease
Altered response to treatment
Increased danger of iatrogenicity
  (medically induced morbidity and/or mortality)

---

principles should be borne in mind when considering the clinical manifestations of disease in the elderly (Table 3.8). Disease presentation is often atypical. Sepsis without fever is common. Myocardial infarction may occur without chest pain and present either without symptoms or with fainting or congestive heart failure. Even for such well-characterized diseases as thyrotoxicosis, appendicitis, peptic ulcer, and pneumonia, the senile patient often has atypical symptomatology (i.e., the combined symptoms of a disease). For example, pneumonia may present with a chief complaint of confusion but lack any history of cough.

Another characteristic of the elderly is that *diseases tend to be chronic and debilitating rather than acute and self-limiting;* symptomatology tends to be more subtle and vague. Thus, recognition and diagnosis of disease in the elderly require a high degree of alertness on the part of health care providers.

A consequence of multiple pathology is the need for *multiple therapy.* This is, in turn, associated with the danger of "polypharmacy" (i.e., the administration of too

many drugs and medications together) (Chapter 23). Due to their impaired homeostatic mechanisms, the elderly do not tolerate therapeutic mistakes as well as younger patients. Considerations of risk versus benefit, while important in all medical decisions, become crucial when dealing with elderly patients. Altered drug reactions and interactions to commonly used drugs should be considered potential dangers lest they lead, together with the decreased physiologic competence and the multiple pathology, to the most serious of complications of medical treatment of the elderly, *iatrogenic disease,* that is, medically induced disease (Chapter 23).

## B. DISEASES OF OLD AGE

Of the many diseases that afflict the elderly, certain medical problems are clearly relegated to the older population, while some overlap with those found in younger adults (Table 3.9). Diseases that are primarily limited to the elderly include osteoporosis, osteoarthritis, adenocarcinoma of the prostate, temporal arteritis, and polymyalgia rheumatica. Diseases associated with aging include adult-onset Type 2 diabetes, Alzheimer's and Parkinson's diseases, neoplasms, emphysema, and hypertension. Examples of diseases associated with aging and with more clear-cut etiologies include septicemia, pneumonia, cirrhosis, nephritis, and diseases due to the advanced stage of atherosclerosis.

A list of common diseases responsible for death, compiled from hospital records in the United States (Figure 3.7) and other developed countries, shows atherosclerosis (including hypertension and myocardial and cerebral vascular accidents) and cancer to be the diseases most related to

## TABLE 3.9
### Diseases of the Elderly

**Limited to aging**

Osteoporosis
Osteoarthritis
Prostatic adenocarcinoma
Polymyalgia rheumatica
Temporal arteritis

**Associated with aging**

*Known Etiology*

Septicemia
Pneumonia
Cirrhosis
Nephritis
Cerebrovascular disease
Myocardial infarction

*Unknown Etiology*

Adult-onset, Type 2 diabetes
Neoplasm
Hypertension
Alzheimer's disease
Parkinson's disease
Emphysema

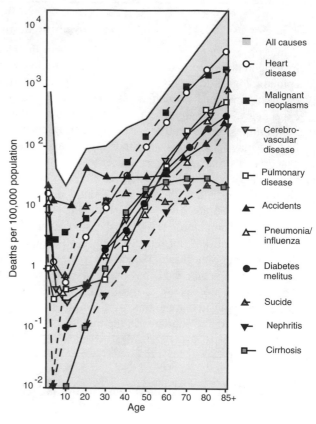

**FIGURE 3.7** *Common causes of death by age in the United States (1998).* (From Martin, J.A. et al., *Births and deaths: Preliminary data for 1998*, National vital statistics reports, Vol. 47, No. 25, Hyattsville, Maryland: National Center for Health Statistics, 1999.)

old age (Table 3.10). It should be noted that when the information is secured not from general hospitals but specifically from geriatric units, the distribution of diseases is somewhat different, with atherosclerosis and dementia being the major causes of hospitalization and deaths, while cancer is a lesser cause, especially after 80 years of age (Table 3.10).

Together with the higher morbidity, after 60 years of age, hospitalization days per person per year increase dramatically from 2 days to 2 weeks (Figure 3.8). Such an increase has a significant impact on the cost of medical care.

### SELF-AWARENESS OF DISEASE IN THE ELDERLY

It has been claimed that the elderly complain excessively about poor health, aches and pains, and a variety of symptoms. In fact, considering their multiple pathology, they do not complain enough. Overall, they view their health status positively, as is the case of 75% of the 55 to 64 years of age group; it is only in 13% of the older age group (85+) that this optimistic view is replaced by a "poor" health outlook (Table 3.11).

In addition to the possibility of self-overevaluation, the reticence of old people to seek help from health care providers may be ascribed to several factors: (1) the often mistaken assumption that nothing can be done about the problem; (2) the fear of a dreaded disease, that is, cancer, dementia; (3) the lack of knowledge of what is normal aging; and, therefore, (4) the failure to discriminate among a variety of deficits and disturbances, those due to "old age" from those due to disease.

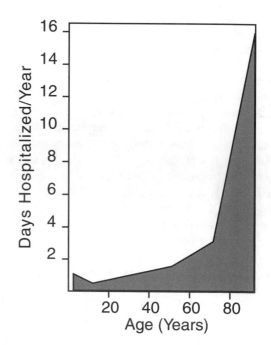

**FIGURE 3.8** *Days spent in hospital according to age.* Number of days hospitalized per year according to age increases after 55 years of age and increases precipitously after age 70.

**TABLE 3.10**
**Common Fatal Diseases in Old Age**

| | General Hospital (% Affected) | | | | | Geriatric Unit (% Affected) | | |
| | Age | | | | | | Age | |
| Disease | 65–69 | 70–74 | 75–79 | 80+ | | 80–89 | 90+ |
|---|---|---|---|---|---|---|---|---|
| Cancer | 29 | 27 | 27 | 24 | Atherosclerosis | 21 | 30 |
| Cardiovascular | 25 | 25 | 32 | 36 | Myocardial infarct | 19 | 10 |
| Respiratory | 14 | 12 | 13 | 10 | Bronchopneumonia | 17 | 25 |
| Digestive | 12 | 9 | 13 | 16 | Cancer | 10 | 7 |
| Nervous system | 11 | 9 | 8 | 6 | Cerebral thrombosis | 9 | — |
| Renal tract | 4 | 7 | 5 | 3 | Chronic bronchitis | 7 | — |
| Other | 5 | 1 | 2 | 5 | Other[a] | 16 | 28 |

[a] Senile dementia frequent in old age. Due to chronicity, patients are placed in long-term care facilities.

**TABLE 3.11**
**Self-Assessed Health Status by Selected Characteristics**

| | Self-Assessed Health Status | | | |
| | Excellent or Very Good | Good | Fair | Poor |
|---|---|---|---|---|
| Age | | | | |
| 55–64 | 44.1% | 30.8% | 16.6% | 8.4% |
| 65–74 | 36.5 | 32 | 21.1 | 10.3 |
| 75–84 | 36 | 31.1 | 20.7 | 12.2 |
| 85+ | 35 | 28.6 | 23.2 | 13.2 |
| Gender | | | | |
| Male | 39.4 | 30.3 | 19.1 | 11.2 |
| Female | 37.9 | 32 | 20.4 | 9.7 |
| Family income | | | | |
| Under $15,000 | 30.2 | 30.5 | 24.7 | 14.6 |
| $15,000 or more | 46.9 | 31.8 | 15.2 | 6.1 |
| Race | | | | |
| White | 39.5 | 31.8 | 19.1 | 9.6 |
| Nonwhite | 28.7 | 25.6 | 27.1 | 18.5 |
| Residence | | | | |
| Central city | 38 | 31.1 | 20.6 | 10.3 |
| Suburban | 42 | 32.3 | 17.5 | 8.2 |
| Rural | 35.3 | 30.3 | 21.7 | 12.7 |

*Source:* 1984 Supplement on Aging (SOA) to the 1984 National Health Interview Survey (NHIS). In addition to age, self-evaluation of health status depends on several variables, including income, race, and geographic residence (affluent, white, suburbanites have a more positive evaluation of health than low income, black residents of central cities or rural areas). Gender differences are not evident despite the greater degree of disability in old women.

To remedy some of these misconceptions, a vigorous educational campaign, geared to the level of the particular audience, should provide information on normal changes with age in various body functions and should be accompanied by brief descriptions of the diseases associated with aging. It should stimulate awareness of the many interventions available to support an healthful life span. It should not be limited to geriatric centers — education of aging should start in secondary schools in parallel with other programs (e.g., sex education) related to health and social issues.

As the characteristics of disease are different in the elderly, likewise, the goal of treatment is modified compared to treatment at young and adult ages. Frequently, a cure is not the main objective, rather efforts should be shifted toward prevention and relief (Chapter 23). Often, the priority is to maximize the ability of the elderly to function. When cure is not possible, rehabilitation can help in some cases; in other cases, the provision of proper care can assist in preventing the development of further complicating illnesses. Quality of life versus prolongation of life becomes a significant issue that creates medico-legal and ethical dilemmas that are heatedly debated but remain largely unresolved (Box 3.2).

## C. DISEASES OF OLD AGE — CAUSES OF DEATH?

The relationship between the aging process and disease is also supported by demographic studies of diseases of old age and causes of death. Deaths from major causes in the U.S. population are obtained from Vital Statistics. Patterns of death have varied considerably from the first half of the 20th century to now, but values for 1955 are not too different from those of today. In 1900, and the early years of this century, the three major causes of death in this country were tuberculosis, pneumonia, and diarrhea-enteritis. By 1950, and persisting today, the three major causes of death are cardiovascular diseases, cancer (neoplasms), and accidents (including those consequent to violence) (Figure 3.7). The death rate due to accidents increases sharply in late adolescence and young adulthood, and then remains constant until about 70 years of age, when it starts to rise again. Aging-related diseases such as those related to cardiovascular pathol-

**General Perspectives**

---

**Box 3.2**

**Setting or Not Setting Limits of Health Care for the Elderly, and the Quest for a Good Death**

Quality of life in later years and cost to be paid to achieve it continues to elicit heated debate and controversy. In the 1980s, the concept was first publicly formulated that health care should be denied or "rationed" to older persons in the United States.[67] It was argued that "In the name of medical progress, we have carried out a relentless war against death and [health] decline, failing to ask...if that will give us a better society. Neither a longer lifetime nor more life-extending technology is the way to that goal."[68,69]

This and similar statements represent a backlash against a stereotyped group of old individuals who are in need of assistance and ready to reap the benefits of a "welfare state for the elderly."[70] This negative view was compounded by the rapidly increasing costs of medical care, by the tremendous growth of federal (e.g., Medicare) spending for the elderly, by the fear of bankruptcy of the Medicare programs, and by the realization that not all elderly are poor and in need of public aid but, on the contrary, many represent a well-off elite.

Arguments against the "setting of limits" based exclusively on age, include: (1) the heterogeneity of the elderly, many of whom age "successfully" and require little or no public support; (2) the realization that denying access to high technology medicine to the elderly may not substantially reduce overall health costs (primarily incurred in neonatal care) and save money; and (3) the difficulty of managing ethical choices and legal consequences. Setting limits would: "...burden the elderly, undermine our (U.S.) democratic freedoms and would not guarantee any significant reductions in expenditures." [69,70]

Within this debate of the relationship between advances in technology and care of the elderly, we must include also "the quest for a good death."[71] The best decision to achieve it may be to decline the use of the most advanced tools for prolonging life of very old and frail individuals (for a few days or sometimes a few hours) in favor of the choice "to die in one's own bed with a minimum of intrusive therapy." The essential feature of this cultural discourse is the notion of patient- and family–centered care and respect, in conflict with "inhumane life-prolonging treatments" for those who are dying."[72] The choice between "heroic interventions" and "humanistic medical care" for the dying elderly should be based not only on medical decisions but also on a number of factors, including the responsibility of the dying to make their own choice and continuing communication with their family.

The increase in the proportion of the elderly in the population and, within this group, that of the very old, dictates that policies for health and medical care (including care at the time of death) be reorganized to meet the changing demographic demands.[73,74] Their reorganization should follow those guidelines of medicine, law, ethics, public policy, religion, and economics that are acceptable to a civilized and caring society. Americans should decide for themselves just what is a "natural life course" and a "natural death" and whether any of us is "too old" for health care. "Be careful! Your decisions about someone else's life might affect you sooner than you think!" [75]

---

ogy or cancer do not equally affect all individuals and, hence, the heterogeneity of these diseases. With increasing age, functional disabilities and comorbidity have adverse effects, or at least complicate diagnosis, treatment, and prognosis of many of the diseases of the elderly. Consequently, death may be due to a multiplicity of causes rather than a single one.

Today, the major causes of death by age in the United States show a close relationship to cardiovascular pathology, primarily atherosclerosis (responsible for heart, kidney, and brain pathology) and hypertension. In 1996, the leading causes of death were distributed as shown in Table 3.12, with heart disease representing 32% of all causes of death; cancer, 23%; and stroke, also another consequence of vascular disease, 7%. However, even mortality from respiratory diseases, although considerably reduced now compared to 1900, remains a significant cause of death, with chronic obstructive pulmonary disease representing 5% and pneumonia/influenza 4%. Pulmonary tuberculosis, which almost disappeared in the last decade, is now recurring, especially in the old and those affected with AIDS.

**TABLE 3.12**
**Leading Cause of Death, United States, 1996**

1.  Heart disease (32%)
2.  Cancer (23%)
3.  Stroke (7%)
4.  Chronic obstructive pulmonary disease (5%)
5.  Accidents (4%)
6.  Pneumonia/influenza (4%)
7.  Diabetes (3%)
8.  HIV/AIDS (1%)
9.  Chronic liver disease and cirrhosis (1%)
10. All others (19%)

It has been claimed that the life expectancy of adults would be extended by 1 to 3 years if malignant neoplasms were cured, by 5 to 7 years if atherosclerosis were prevented, and by approximately 10 years if both diseases were abolished. It is true that the spectacular advances in medicine in the last 100 years have been responsible in part for the dramatic increase in the average life span and, especially, the reduction of mortality due to cardiovascular disease (Chapter 2), but further lengthening on this basis alone appears unlikely. Efforts, therefore, should be expended to elucidate the basic mechanisms, including the physiology of the aging process per se. An appreciation of these mechanisms is indispensable for a better understanding of the aging process. Without such an understanding, the etiopathology of the diseases of old age and their rational treatment cannot be achieved and the life span neither further improved nor prolonged.

## D. Genetic Epidemiology

Many of the common diseases that affect humans have been known for many years to have a genetic component. More than 7000 human rare diseases are inherited in a Mendelian fashion (one gene–one disease), and many of the genes responsible for the phenotype have now been mapped, leading to major breakthroughs in our understanding of these conditions.[76] It is only in recent years that a comprehensive guide to human disease gene mapping in complex diseases has become available. In these diseases, genetic variation interacts with environmental influences to modify the risk of disease.[77,78] Today, genetics is at the core of research on cancers, coronary heart disease, high blood pressure, neurological and psychiatric disorders, and a host of other clinical conditions.[79–81]

Advances in genetics, molecular biology, and biotechnology have facilitated the discovery of new biomarkers of susceptibility to specific diseases and to environmental exposure that are being evaluated as part of molecular epidemiological research. A 1987 report of the National Research Council noted the dynamic nature of the various types of biomarkers underlining the continuum of cause-to-effect relationships (Figure 3.9). Precise measures of potential etiologic determinants and phenotypes have become available to epidemiological research and are providing new links between genetics, environment, and disease. A few aspects of these interrelations are briefly presented, and more information may be found in some comprehensive textbooks.[77,78]

It is apparent that monogenic (one gene) or polygenic (multiple genes) alterations may increase or decrease the risk of developing a trait. For example, the "risk" for late onset Alzheimer disease (AD), but not the disease phenotype, is increased (or decreased) depending on the type of lipoprotein mutation (Chapter 8). Lipoproteins are specific lipid-containing macromolecules in the plasma that have an important (predisposing or protective) role in the development of the atherosclerotic lesion (Chapter 17). The risk of AD is increased up to 12-fold for individuals carrying two copies of the APOE-4 allele, while the risk may be halved for those carrying one copy of

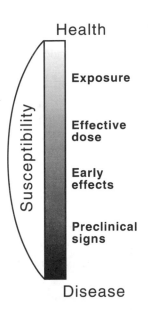

**FIGURE 3.9** *Dynamic nature of various types of biomarkers.* Note the continuum of cause-to-effect relationships from health to disease, and, superimposed a continuum of genetic effects with a corresponding increased susceptibility to environmental factors.

the APOE-2 allele. The frequency of the APOE-4 allele in the general population is 16%, while that of the APOE-2 is 7%.[82–84] As with other complex genetic traits, there is an "incomplete" correlation between the APOE-4 genotype and the AD phenotype: APOE-4 may contribute to the risk of developing AD but may not, by its presence alone, cause AD.[82–84]

In examining the role of genes in the etiology of diseases, we must distinguish: (a) *causal genes*, single gene mutations leading directly to a disease phenotype (as in Mendelian disorders, e.g., Huntington disease), from (b) *susceptibility genes* that are associated with a disease but by themselves, are not sufficient to cause the disease.[85] This would be the case of the APOE-4 mutation mentioned above with reference to Alzheimer's disease.

Determination of the genetic component of a disease depends on three major and sequential steps:

1. Determination of familial aggregation (e.g., capturing information about disease in specific relative sets, siblings, cousins, etc., with twins being a special case, or entire pedigrees, i.e., nuclear families or extended kindred)
2. Determination of evidence of familial aggregation and, further, discrimination among environmental, cultural, and genetic factors that may contribute to the mutation clustering
3. Determination of evidence of genetic factors and, further, their identification

**General Perspectives**

Complex disease genes express traits:

1. That show no clear Mendelian inheritance (one gene/one phenotype)
2. That have moderate to high evidence of genetic inheritance
3. That exhibit familial aggregation cases
4. That are polygenic (involve multiple genes) or are multifactorial (involve multiple genes interacting with the environment)

Thus, there are several ways in which genetic susceptibility may influence a disease:

1. By itself
2. By making the carrier more susceptible to the disease
3. By exacerbating the expression of a risk factor, or the risk factor may exacerbate the genetic effects

### LIABILITY OF GENETIC EPIDEMIOLOGY

As already mentioned in Chapter 1, mapping genes involved in diseases helps us to better understand the respective role of genetic and environmental factors in the etiology of human diseases. It provides new practical screening protocols for those who are genetically predisposed. It also improves our ability to find therapeutic interventions for prevention or cure of the disease, once genetic susceptibility to specific environmental conditions is identified. However, findings of molecular medical genetics, when incorporated into public health initiatives, may lead to long-term and serious social consequences. Therefore, educational programs to target the public at large as well as the health agencies must be carefully constructed and integrated to avoid stigmatizing persons susceptible to specific diseases.[86–89]

## V. DISEASE AS A TOOL FOR THE STUDY OF AGING

Sporadic cases of syndromes having multiple characteristics of premature (early onset) or accelerated (rapid progression) aging occur in humans.[90,91] It is unclear how far any of these syndromes may be regarded as a genuine acceleration of timing mechanisms that determine senescence. They are apparently *pleoitropic genetic defects*, and when one of the major features is accelerated aging, they are designated as *progeria*. One example is Werner's syndrome. When accelerated aging is associated with other prevalent defects, they are called *progeria-like* or *progeroid* or *segmental syndromes*. One example is Down's syndrome.[90]

Progeria syndromes do not quite duplicate all the pathophysiology of aging; each syndrome presents an acceleration of only some of the characteristics associated with normal aging. Progeria is described as being of two main types, infantile and adult. The infantile form (Hutchinson–Gilford syndrome) becomes apparent at a very early age and is associated with stunted growth, failure of sex maturation, and signs of aging, such as skin atrophy, hypertension, and severe atherosclerosis; death occurs in the twenties, usually consequent to coronary heart disease. The adult form (Werner's syndrome) resembles more closely the changes associated with aging, with respect to both the affected individual's physical appearance and the disease pattern. The onset of this premature aging syndrome occurs between the ages of 20 and 30 years, and death ensues a few years from the onset, usually due to cardiovascular disease. Tissue culture studies of fibroblasts, in infantile and adult syndromes, reveal a shortening of the cell-replicating ability that has been interpreted as being supportive of accelerated aging.

### A. SIMILARITIES AND DIFFERENCES BETWEEN WERNER'S SYNDROME AND AGING

Werner's syndrome (WS) is an inherited disease with clinical symptoms (forming a syndrome) that resemble premature aging.[92] Early susceptibility to a number of major age-related diseases is a key feature of this disorder. Principal WS features include shortness of stature; senile appearance; cataracts and graying of the hair beginning at 20 to 30 years; skin changes (i.e., tautness, atrophy or thickening, ulceration) designated as scleroderma; joint deformities, soft tissue calcifications, and osteoporosis; atrophy of muscles and connective tissue; early cessation of menstruation; and increased incidence of neoplasms. Most of these features occur in aging as well, but at a later age and in different degrees.[90]

The gene responsible for WS (known as WRN) is located on the short arm of chromosome 8. The predicted protein (1432 amino acids in length) shows significant similarities to DNA helicases (the enzymes capable of unwinding the DNA double helix). Four mutations in WS patients have been identified, one of the four was found in the homozygous state in 60% of Japanese WS patients examined. The identification of a mutated putative helicase as the gene product of the WS gene suggests that defective DNA metabolism is involved in the precociously aging WS patients.[20]

Among some of the major differences between WS and aging are the type of inheritance — universal, multifactorial in aging and autosomal, recessive in Werner's; the high incidence of hypertension in aging but not in WS; the presence of dementia and other degenerative disorders of the central nervous system in aging but not in WS; and the occurrence of soft tissue calcifications, uncommon in aging but common in WS.

These differences are sufficient to justify the statement that WS is not merely a process of premature or accelerated aging. Rather, it may be viewed as a "caricature" of aging. Both WS and aging may represent the result of generalized metabolic processes or aberrations thereof. Indeed, the overlap between the two entities is not surprising inasmuch as the various tissues of the human organisms have only a limited repertoire of reactions to genetic abnormalities and environmental insults. Irrespective of similarities or differences, a study of the features of WS and aging will conceivably be useful in achieving an understanding of both.

The etiology of the syndrome remains obscure, but among the several causes proposed, neuroendocrinologic dysfunction is supported by the stunted growth, failure of gonadal maturation, and diabetes, either singly or in combination. However, the phenotypes observed in the affected individuals may result from mechanisms relating to aging processes. If this were so, it could be inferred that cell autonomous functions dictate the pace of aging, at least in some organs and tissues. Further investigations of the WRN protein may reveal why particular systems and organs are differentially affected with aging.[92]

## B. Down's Syndrome

Down's syndrome (mongolism) is another example characterized by several symptoms, including accelerated aging and premature death, and is due to trisomy at chromosome 21. The incidence of the syndrome is greatest among children born from mothers 40 years of age and older, and the genetic abnormality has, therefore, been related to aging processes involving the oocytes (Chapter 11). Although in 20 to 30% of cases, the extra chromosome is contributed by the father, paternal age does not seem to have any significant effect on the incidence of the syndrome.

Individuals affected by Down's syndrome may present somatic malformations, but the major deficit is represented by severe mental disability. Affected subjects, who live to reach 30 years of age and longer, present many signs of accelerated aging, including senile dementia of the Alzheimer type (AD) superimposed on the mental retardation[93] (Chapter 8). Animal models have also been proposed for the study of Down's syndrome. The mouse is the animal of choice, for it is possible to introduce trisomy of chromosome 16 and produce individuals with some phenotypic characteristics of Down's syndrome, such as cardiovascular defects, neurologic alterations, and retardation of brain maturation as well as early aging.[94] Proposed as a good *in vitro* model to experimentally induce the premature aging of neurons is the transfer of the trisomy 21 from the fibroblast donor cells to neuroblastoma cells.

## C. Experimentally Induced Aging and Disease in Animals

The use of animals, isolated organs, tissues, and cells, has led and continues to lead to a better understanding of the aging process. This use is regulated by specific rules governing the choice of the most appropriate model to answer each specific question (Chapter 1). Within this context, cause–effect relationships between disease and aging have been variously explored, depending on the hypotheses entertained by different investigators. Most often, disease has been used to accelerate the onset and course of aging processes to test therapeutic measures that might prevent or slow disease and aging. Some attempts have focused on "segmental" aging, that is, induction of aging in a selected organ, tissue, or cell type to mimic specific aspects of aging. Examples of these experimental approaches include genetic manipulation, increased free radical accumulation, inoculation of slow viruses, interference with nervous and endocrine functions, induction of wear-and-tear and stress, administration of mutagens/carcinogens, and others. Principles and methodology for the choice of animal models as "biomarkers" of aging and for the *in vivo* and *in vitro* induction of accelerated or delayed aging and aging-associated disease are reviewed in specialized textbooks.[95–97] A few examples are briefly indicated below.

### Examples of Induction of Pathology as a Tool to Study Aging

Premature aging has been induced in a number of laboratory animals as well as in captive wild animals by various methods: one of the favorite methods is to expose the animal to stress, that is, to excessive environmental (including physical, emotional, and social) demands. Stress will activate or interfere with all major regulatory systems, that is, it will disturb neuroendocrine balance (Chapter 10), alter immunologic competence (Chapter 15), and increase the production of free radicals (Chapter 5).

Stress or injection of cortisol to mimic adrenocortical stimulation, in association with a high-lipid or high-cholesterol diet, accelerates atheroma formation in rabbits (Chapter 16). High doses of vitamin D given along with calcium to rats precipitate calcification of the skin, heart, blood vessels, and other tissues usually susceptible to calcium deposits in old age (Chapters 21 and 22). Still other interventions utilize inoculation of animals with viruses or viral particles (prions) to mimic some of the degenerative diseases of old age. Slow viruses or prions induce "scrapie" in sheep and "mad cow disease" in cows, both conditions that resemble symptomatically and pathologically (e.g., accumulation of amyloid) human Alzheimer's dementia (AD). Hippocampal slices from transgenic mice or infusion of amyloid beta protein intraventricularly in the brain of rats induce amyloid plaque deposits (Chapter 8).

Despite the availability of the currently promising animal models, none exhibits all the characteristics and symptoms of old age in humans. The paucity of adequate models and the shortcomings of the current ones have been responsible, in part, for the relatively slow progress in our understanding of the aging process. Current advances, especially in molecular biology and genetics, promise faster advances in our knowledge of this area.

**General Perspectives**

# REFERENCES

1. Strehler, B.L., *Time, Cells and Aging*, 2nd ed., Academic Press, New York, 1977.
2. Wachter, K.W., Between Zeus and the salmon: Introduction, in *Between Zeus and the Salmon*, National Academy Press, Washington, D.C., 1997.
3. Carey, J.R., Insect biodemography, *Annu. Rev. Entomol.*, 46, 79, 2001.
4. Finch, C.E., *Longevity, Senescence, and the Genome*, University of Chicago Press, Chicago, 1990.
5. Finch, C.E., Comparative perspectives on plasticity in human aging and life spans, in *Between Zeus and the Salmon*, National Academy Press, Washington, D.C., 1997.
6. Finch, C.E. and Tanzi, R.E., Genetics of aging, *Science*, 278, 407, 1997.
7. Finch, C.E. and Austad, S.N., History and prospects: Symposium on organisms with slow aging, *Exp. Gerontol.*, 36, 593, 2001.
8. Robertson, O.H. and Wexler, B.C., Prolongation of the life span of Kokanec salmon (*Oncorhynkus nerka Kennerlyi*) by castration before beginning of gonad development, *Proc. Natl. Acad. Sci. USA*, 47, 609, 1961.
9. Nooden, L.D., Whole plant senescence, in *Senescence and Aging in Plants*, Nooden, l.D. and Leopold, A.C., Eds., Academic Press, San Diego, 1988.
10. Comfort, A., *The Biology of Senescence*, Churchill Livingstone, Edinburgh and London, 1979.
11. Sacher, G.A., Relation of lifespan to brain weight and body weight in mammals, in *The Lifespan of Animals*, Wolstenholme, C.E.W. and O'Connor, M., Eds., Little, Brown, & Co., Boston, 1959.
12. Hofman, M.A., Energy metabolism, brain size and longevity in mammals, *Q. Rev. Biol.*, 58, 495, 1983.
13. Sacher, G.A., Life table modification and life prolongation, in *Handbook of the Biology of Aging*, Finch, C.E. and Hayflick, L., Eds., Van Nostrand Reinhold, New York, 1977.
14. Kirkwood, T.B.L., Comparative and evolutionary aspects of longevity, in *Handbook of the Biology of Aging*, 2nd ed., Finch, C.E. and Schneider, E.L., Eds., Van Nostrand Reinhold, New York, 1985.
15. Sohal, R.S. and Weindruch, R., Oxidative stress, caloric restriction, and aging, *Science*, 273, 59, 1996.
16. Rogina, B. et al., Extended life-span conferred by cotransporter gene mutations in *Drosophila*, *Science*, 290, 2137, 2000.
17. Guarente, L., Do changes in chromosomes cause aging? *Cell*, 86, 9, 1996.
18. Guarente, L. and Kenyon, C., Genetic pathways that regulate ageing in model organisms, *Nature*, 408, 255, 2000.
19. Kenyon, C., et al., A *C. elegans* mutant that lives twice as long as wild type, *Nature*, 366, 461, 1993.
20. Yu, C.E. et al., Positional cloning of the Werner's syndrome gene, *Science*, 272, 258, 1996.
21. Segall, P.E., Timiras, P.S., and Walton, J.R., Low tryptophan diets delay reproductive ageing, *Mech. Ageing Dev.*, 24, 245, 1983.
22. Shock, N.W., Age changes in physiological functions in the total animal: The role of tissue loss, in *The Biology of Aging: A Symposium*, Strehler, B.L., Ebert, J.D., and Shock, N.W., Eds., *Am. Inst. Biol. Sci.*, Washington, D.C., 1960.
23. Rowe, J.W. and Kahn, R.L., Human aging: Usual and successful, *Science*, 237, 143, 1987.
24. Rowe, J.W. and Kahn, R.L., *Successful Aging*, Pantheon Books, New York, 1998.
25. Cavalieri, T.A., Chopra, A., and Bryman, P.N., When outside the norm is normal: Interpreting lab. data in the aged, *Geriatrics*, 47, 66, 1992.
26. Ewbank, J.J. et al., Structural and functional conservation of the *Caenorhabditis elegans* timing gene clk-1, *Science*, 275, 980, 1997.
27. Proft, M. et al., CAT5, A new gene necessary for derepression of gluconeogenic enzymes in *Saccharomyce cerevisiae*, *Eur. Mol. Bio. Org. J.*, 14, 6116, 1995.
28. Silver, A.J. et al., Effect of aging on body fat, *J. Am. Geriatr. Soc.*, 41, 211, 1993.
29. Steen, B., Body composition and aging, *Nutr. Rev.*, 46, 45, 1988.
30. Gallo, J.J., Reichel, W., and Anderson, L.M., Eds., *Handbook of Geriatric Assessment*, 2nd ed., Aspen, Gaithersburg, MD, 1995.
31. Rubenstein, L.Z., Wieland, D., and Bernabei, R., Eds., *Geriatric Assessment Technology: The State of the Art*, Editrice Kurtis, Milan, 1995.
32. Kane, R.A. and Rubenstein, L.Z., Assessment of functional status, in *Principles and Practice of Geriatric Medicine*, 3rd ed., Pathy, J.M.S., Ed., Wiley, New York, 1998.
33. Lawton, M.P. and Teresi, J.A., *Annu. Rev. Gerontol. Geriatrics*, 14, Springer, New York, 1994.
34. Gill, T.M., Assessing risk for the onset of functional dependence among older adults: The role of physical performance, *J. Am. Geriatr. Soc.*, 43, 603, 1995.
35. Ravaglia, G. et al., Determinants of functional status in healthy Italian nonagenarians and centenarians: A comprehensive functional assessment by the instruments of geriatric practice, *J. Am. Geriatr. Soc.*, 45, 1196, 1997.
36. Thomas, D.R. and Ritchie, C.S., Preoperative assessment of older adults, *J. Am. Geriatr. Soc.*, 43, 811, 1995.
37. Silverman, M. et al., Evaluation of outpatient geriatric assessment: A randomized multi-site trial, *J. Am. Geriatr. Soc.*, 43, 733, 1995.
38. Toseland, R.W. et al., Outpatient geriatric evaluation and management: Is there an investment effect? *Gerontologist*, 37, 324, 1997.
39. Simonsick, E.M. et al., Methodology and feasibility of a home-based examination in disabled older women: The Women's Health and Aging Study, *J. Gerontol.*, 52, M264, 1997.
40. Reuben, D.B. et al., Looking inside the black box of comprehensive geriatric assessment: A classification system for problems, recommendations, and implementation strategies, *J. Am. Geriatr. Soc.*, 44, 835, 1996.
41. Reuben, D.B., What's wrong with ADLs?, *J. Am. Geriatr. Soc.*, 43, 936, 1995.

42. Fried, L.P. et al., Functional decline in older adults: Expanding methods of ascertainment, *J. Gerontol.*, 51, M206, 1996.

43. Langlois, J.A. et al., Self-report of difficulty in performing functional activities identifies a broad range of disability in old age, *J. Am. Geriatr. Soc.*, 44, 1421, 1996.

44. Ostchega, Y. et al., The prevalence of functional limitations and disability in older persons in the U.S.: data from National Health and Nutrition Examination Survey III, *J. Am. Geriatr. Soc.*, 48, 1132, 2000.

45. Guralnik, J. M. and Lacroix, A.Z., Assessing physical function in older populations, in *The Epidemiologic Study of the Elderly*, Wallace, R.B. and Woolson, R.F., Eds., Oxford University Press, New York, 1992, 159.

46. Fuchs, Z. et al., Morbidity, comorbidity, and their association with disability among community-dwelling oldest-old in Israel, *J. Gerontol.*, 53, M447, 1998.

47. Woo, J. et al., Impact of chronic diseases on functional limitations in elderly Chinese aged 70 years and over: A cross-sectional and longitudinal survey, *J. Gerontol.* (Biol. Med. Sci.), 53, M102, 1998.

48. Hogan, D.B., Ebly, E.M., and Fung, T.S., Disease, disability, and age in cognitively intact seniors: Results from the Canadian Study of Health and Aging, *J. Gerontol.*, 54, M77, 1999.

49. Amaducci, L. et al., Education and the risk of physical disability and mortality among men and women aged 65 to 84: The Italian Longitudinal Study on Aging, *J. Gerontol.*, 53, M484, 1998.

50. Pappas, G. et al., The increasing disparity in mortality between socioeconomic groups in the United States, 1960 and 1986, *N. Engl. J. Med.*, 329, 103, 1993.

51. Guralnik, J.M. et al., Educational status and active life expectancy among older blacks and whites, *N. Engl.J. Med.*, 329, 110, 1993.

52. Timiras, P.S., Education, Homeostasis, and Longevity, *Exp. Gerontol.*, 30, 189, 1995.

53. Fried, L.P. and Guralnik, J. M., Disability in older adults: Evidence regarding significance, etiology and risk, *J. Am. Geriatr. Soc.*, 45, 92, 1997.

54. Hoeymans, N. et al., The contribution of chronic conditions and disabilities to poor self-rated health in elderly men, *J. Gerontol.*, 54, M501, 1999.

55. Reed, D. et al., Health and functioning among the elderly of Marin County, California: A glimpse of the future, *J. Gerontol.*, 50, M61, 1995.

56. Tucker, K.L. et al., Self-reported prevalence and health correlates of functional limitation among Massachusetts elderly Puerto Ricans, Dominicans, and non-Hispanic white neighborhood comparison group, *J. Gerontol.*, 55, M90, 2000.

57. Crimmins, E., Saito, Y., and Reynolds, S., Further evidence on recent trends in the prevalence and incidence of disability among older Americans from two sources: The LSOA and the NHIS, *J. Gerontol.*, 52, S59, 1997.

58. Manton, K., Corder, L., and Stallard, E., Chronic disability trends in elderly United States populations: 1982–1994, *Proc. Natl. Acad. Sci.: Med. Sci., USA*, 94, 2593, 1997.

59. Freedman, V. and Martin, L., Changing patterns of functional limitation among the older American population, *Am. J. Public Health*, 88, 1457, 1998.

60. Lentzner, H.R., Weeks, J.D., and Feldman, J.J., Changes in Disability in the Elderly Population: Preliminary Results from the Second Supplement on Aging. Paper presented at the annual meetings of the Population Association of America, Chicago, 1998.

61. Fried, L.P. and Wallace, R.B., The complexity of chronic illness in the elderly: From clinic to community, in *The Epidemiologic Study of the Elderly*, Wallace, R.B. and Woolson, R.F., Eds., Oxford University Press, New York, 1992.

62. Guralnik, J.M., LaCroix, A.Z., Everett, D.F., and Kovar, M.G., Aging in the eighties: the prevalence of comorbidity and its association with disability, *National Center for Health Statistics, Advance Data from Vital and Health Sciences*, Hyattsville, MD, No. 170, 1989.

63. Verbrugge, L.M., Lepkowski, J., and Imanaka, Y., Comorbidity and its impact on disability, *Milbank Q.*, 67, 450, 1989.

64. Satariano, W.A., Comorbidity and functional status in older women with breast cancer: Implications for screening, treatment, and prognosis, *J. Gerontol.*, 47, 24, 1992.

65. Fries, J.F. and Caprol, L.M., *Vitality and Aging*, W.H. Freeman and Company, San Francisco, CA, 1981.

66. Kaplan, G.A., Epidemiologic observations on the compression of morbidity; Evidence from the Alameda County study, *J. Aging Health*, 3, 155, 1991.

67. Smeeding, T.M. et al., Eds., *Should Medical Care Be Rationed by Age?* Rowman and Littlefield, Totowa, NJ, 1987.

68. Callahan, D., *What Kind of Life: The Limits of Medical Progress*, Simon & Schuster, New York, 1990.

69. Barry, R.L. and Bradley, G.V., *Set No Limits*: A rebuttal to Daniel Callahan's *Proposal to Limit Health Care for the Elderly*, The University of Illinois Press, Urbana, 1991.

70. Myles, J.F., Old Age in the Welfare State: The Political Economy of Public Pensions. University Press of Kansas, Lawrence, 1989.

71. Kaufman, S.R., Senescence, decline, and the quest for a good death: Contemporary dilemmas and historical antecedents, *J. Aging Studies*, 14, 1, 2000.

72. Kaufman, S.R., Intensive care, old age, and the problem of death in America, *Gerontologist*, 38, 715, 1998.

73. Homer, P. and Holstein, M., *A Good Old Age? The Paradox of Setting Limits,* Siomon & Schuster, New York, l990.

74. Evans, R.W., Advanced medical technology and elderly people, in *Too Old for Health Care?* Binkstock, R.H. and Post, S.G., Eds., The Johns Hopkins University Press, Baltimore, 1991.

75. Koop, C.E., Foreword, in *Too Old for Health Care?* Binkstock, R.H. and Post, S.G., Eds., The Johns Hopkins University Press, Baltimore, 1991.

76. McKusick, V.A., *Mendelian Inheritance in Man*, 11th ed., Johns Hopkins University Press, Baltimore, 1994.

**General Perspectives**

77. Haines, J.L. and Pericak-Vance, M.A., *Approaches to Gene Mapping in Complex Human Diseases*, Wiley-Liss, New York, 1998.

78. Khoury, M.J., Burke, W., and Thomson, E.J., *Genetics and Public Health in the 21st Century*, Oxford University Press, New York, 2000.

79. Lander, E.S. and Schork, N.J., Genetic dissection of complex traits, *Science*, 265, 2037, 1994.

80. Sing, C., Haviland, M.B., and Reilly, S.L., Genetic architecture of common multifactorial diseases, in *Ciba Foundation symposium 197: Variation in the human genome*, Wiley, Chichester, UK, 1996, 211.

81. Gelehrter, T.D., Collins, F.S., and Ginsburg, D., *Principles of Medical Genetics,* 2nd ed., Williams and Wilkins, Baltimore, 1998.

82. Corder, E.H. et al., Apolipoprotein E4 gene dose and the risk of Alzheimer disease in late onset families, *Science*, 261, 921, 1993.

83. Corder, E.H. et al., Apolipoprotein E type 2 allele decreases the risk of late onset Alzheimer disease, *Nat. Genet.*, 17, 180, 1994.

84. Saunders, A.M. et al., Association of apolipoprotein E allele 4 with late-onset familial and sporadic Alzheimer's disease, *Neurology*, 43, 1467, 1993.

85. King, R.A., Rotter J.I., and Motolsky, A.G., *The Genetic Basis of Common Diseases*, Oxford University Press, New York, 1992.

86. Bartels, D.N., LeRoy, B.S., and Caplan, A.L., Eds.*, Prescribing Our Future: Ethical Challenges in Genetic Counseling*, Aldine de Gruyter, New York, 1992.

87. Rosner, M. and Jonnson, T.R., Telling stories: Metaphors of the Human Genome Project, *Hypatia*, 10, 104, 1995.

88. Peters, T., *Playing God: Genetic Determinism and Human Freedom*, Routledge, New York, 1997.

89. Condit, C.M., *The Meaning of the Gene*, University of Wisconsin Press, Madison, WI, 1999.

90. Dyer, C.A. and Sinclair, A.J., The premature ageing syndromes: insights into the ageing process, *Age and Ageing*, 27, 73, 1998.

91. Martin, G.M. and Oshima, J., Lessons from human progeroid syndromes, *Nature*, 408, 263, 2000.

92. Lombard, D.B. and Guarente, L., Cloning the gene for Werner syndrome: a disease with many symptoms of premature aging, *TIG,* 12, 283, 1996.

93. Heston, L.L., Alzheimer's disease and Down's syndrome: Genetic evidence suggesting association, *Ann. N. Y. Acad. Sci. USA.*, 396, 29, 1982.

94. Epstein C.J., Cox, D.R., and Epstein, L.B., Mouse trisomy 16: An animal model of human trisomy 21 (Down syndrome), *Ann. N. Y. Acad. Sci. USA.*, 450, 157, 1985.

95. Cole, G.M. and Timiras, P.S., Aging-related pathology in human neuroblastoma and teratocarcinoma cell lines, in *Model Systems of Development and Aging of the Nervous System*, Vernadakis, A., Privat, A., Lauder, J.M., Timiras, P.S., and Giacobini, E., Eds., Marlinus Nijhoff Publishing, Boston, 1987.

96. Yu, B.P., Ed., *Methods in Aging Research*, 2nd ed., CRC Press, Boca Raton, FL, 1998.

97. Sternberg, H. and Timiras, P.S., Eds., *Studies of Aging*, Springer, New York, 1999.

# 4 Cellular Senescence and Cell Death

*Judith Campisi*
Lawrence Berkeley National Laboratory

## CONTENTS

## I. INTRODUCTION

Cellular senescence and apoptotic cell death refer to two processes that occur throughout the life span of complex organisms such as mammals. Both processes have been conserved throughout metazoan evolution, and they most likely evolved to facilitate embryogenesis. In complex organisms, both processes are also important for preventing the development of cancer. Cellular senescence and apoptosis arrest the growth or eliminate, respectively, damaged, dysfunctional, or unneeded cells. Thus, during embryonic development and early in the life span, these processes help maintain the integrity and function of tissues. Later in life, however, both processes may contribute to phenotypes and pathologies that are associated with aging. This chapter will describe the characteristics and causes of cellular senescence and cell death. It will then discuss the role that both processes play in suppressing

tumorigenesis, and the potential roles these processes may play in contributing to aging.

## II. CELLULAR SENESCENCE — CAUSES AND CHARACTERISTICS

Cellular senescence refers to the response of mitotically competent cells to a variety of stimuli, ranging from chromosomal damage to supraphysiological mitogenic signals. Mitotically competent cells are capable of proliferation (used here interchangeably with growth), as opposed to postmitotic cells, which have irreversibly lost the ability to proliferate. Examples of mitotically competent cells are the basal keratinocytes in the skin, epithelial cells that comprise much of the gastrointestinal tract, liver, and other epithelial organs, endothelial and smooth muscle cells, T lymphocytes, and stromal fibroblasts. Examples of postmitotic cells are mature neurons and differentiated skeletal and heart muscle cells. One fundamental difference between mitotically competent and postmitotic cells is their potential to give rise to cancer. Cell proliferation is essential for the development of cancer.[1] Thus, cancers arise from mitotically competent cells and never from postmitotic cells. Cellular senescence entails an irreversible withdrawal from the cell cycle, thereby converting a mitotically competent cell into a postmitotic cell. In complex organisms, the senescence response suppresses the development of cancer. In addition, the accumulation of senescent cells may contribute to age-related pathologies.

## A.  REPLICATIVE SENESCENCE

Cellular senescence was first recognized as the process that limits the proliferative potential of mitotically competent cells.[2] Hayflick and colleagues, working with normal human fibroblasts in culture, showed that the cells gradually lost the ability to proliferate. After 50 or so population doublings, all cell division had ceased. Despite their inability to proliferate, the cells remained intact and viable. They continued to metabolize RNA and protein, but they arrested growth with a G1 DNA content and were unable to initiate DNA replication in response to physiological mitogens.[reviewed in 3–5] This phenomenon was termed replicative senescence, and cells that underwent replicative senescence were said to have a finite replicative life span. We now know that replicative senescence is a specific example of a more widespread response, termed cellular senescence (discussed below). Thus, replicative senescence refers to the cellular response to repeated cell division, whereas cellular senescence refers to response to any of many stimuli, of which cell division is but one.

Many cell types from many species have been shown to undergo replicative senescence.[reviewed in 3–5] However, the mechanisms responsible for replicative senescence may vary, depending on the cell type and species.[6–8] Except where noted otherwise, this chapter will focus on mammalian cells, and principally, on rodent and human cells. Nonetheless, a finite replicative life span may be an ancestral phenotype. Even simple organisms, such as the single-celled yeast *Saccharomyces cerevisiae*, have been shown to have only a finite capacity for cell division.[9] Likewise, there are only a few proliferative somatic cells in the largely postmitotic model organism *Drosophila melanogaster*, but those cells (ovarian epithelial stem cells) also appear to have a finite replicative life span.[10] Although there is little doubt that replicative (and cellular) senescence is important for suppressing tumorigenesis in mammals,[11,12] it may have originated for other purposes, such as preventing the proliferation of defective germ cells or contributing to embryonic morphogenesis.[13]

Most studies of replicative senescence have been done using cells in culture. These studies have led to several general conclusions about the characteristics and control of replicative senescence in mammalian cells.[reviewed in 3–5, 11, 12, 14]

1. Normal somatic cells, whether fetal or adult in origin, do not divide indefinitely.
2. Germ line precursor cells and early embryonic cells, including embryonic stem cells, have an unlimited cell division potential (replicative immortality).
3. Most cancer cells have an unlimited cell division potential.
4. Replicative (and cellular) senescence is controlled by several well-recognized tumor suppressor genes, which are inactivated by mutational or epigenetic events in most cancer cells.
5. Rodent cells undergo replicative senescence after substantially fewer doublings than human cells.
6. Replicative senescence is exceedingly stringent in human cells, which rarely, if ever, spontaneously immortalize.
7. Replicative senescence is less stringent in rodent cells, which, after a period of genomic instability termed crisis, immortalize at a low but finite frequency.
8. Immortalization greatly increases a cell's risk for undergoing neoplastic transformation. The fact that rodent cells immortalize much more readily than human cells may help explain why cancer develops at a much faster rate in mice.

The salient features of replicative senescence and spontaneous immortalization are illustrated by Figure 4.1.

### 1.  Telomeres Shortening

A prime cause of replicative senescence in human cells is progressive telomere shortening. Telomeres are the repetitive DNA sequences and specialized proteins that cap the tips of linear chromosome.[15] Telomeres are essential structures. Without them, chromosome ends would be indistinguishable from a DNA double-strand break, and the chromosomes would be subject to degradation and fusion by the cellular DNA repair machinery. As a result, genetic information would be lost and the genome would become unstable. This, in turn, could lead to cell death or neoplastic transformation.

The biochemistry of DNA replication is such that, with each round of DNA replication, 50–200 bp of DNA at each chromosome 3' end cannot be replicated.[16] This has been termed the end-replication problem. As a result

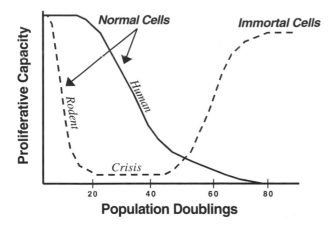

**FIGURE 4.1** *Replicative senescence in human and rodent cells.* Normal human and rodent cells gradually lose proliferative capacity as they undergo an increasing number of population doublings. This process is known as replicative senescence and is illustrated for typical cultures of fetal mouse and human fibroblasts. At the end of the replicative life span, all cell proliferation ceases, but the cells remain viable. Rodent cells undergo replicative senescence after fewer doublings than human cells. In addition, normal human cells rarely, if ever, spontaneously escape senescence and develop an indefinite or immortal replicative life span, whereas rodent cells routinely spontaneously immortalize following a period characterized by genomic instability (crisis). Immortalization greatly increases the risk for developing cancer.

of the end-replication problem, telomeres shorten progressively with each cell division. When the telomeres reach a critically short length — generally 4–6 kb, or well before the protective telomeric cap is completely eroded — cells respond by undergoing a senescence arrest.[17] Human cells proliferate until their telomeres become sufficiently short to elicit a senescence response.[18,19] Recent findings suggest that shortening actually disrupts the telomeric structure, and that cells respond to a dysfunctional telomeric structure, rather than telomere length per se.[20,21] Whatever the case, the senescence response prevents further cell division, thereby avoiding the risk of losing genetic information and developing genomic instability.

Telomeres have been shown to shorten in several cell types as the number of population doublings increase in culture.[17,22–24] Telomeres also shorten *in vivo* as a function of donor age and proliferative history of the tissue.[22–25] These findings have led to the hypothesis that telomeres are determinants of longevity. However, there is no correlation between longevity and telomere length per se. For example, laboratory mice have very long telomeres — 30 to >50 kb, compared to the 15–20 kb telomeres characteristic of human cells.[26] Nonetheless, mouse cells have a shorter replicative life span in culture, and, of course, mice are shorter lived than humans. Mouse cells clearly undergo replicative senescence in culture owing to mechanisms other than telomere shortening. These mechanisms are not

well understood, but they include damage due to the culture conditions or the relatively high oxygen level in which mammalian cells are typically cultured.[6,8] Bc that as it may, as discussed below, cellular senescence may contribute to aging, and the senescence response can be triggered in part by short or dysfunctional telomeres.

## 2. Germ Line and Stem Cells

How do the germ line precursors and early embryonic cells avoid the end-replication problem? These cells express the enzyme telomerase. Telomerase is a specialized reverse transcriptase that can add *de novo* the repetitive telomeric sequence (TTAGGG, in vertebrates) to the chromosome ends.[27,28] In the presence of optimal telomerase activity, telomere lengths remain stable despite cell proliferation.[28,29] Thus, the germ line and early embryonic cells can proliferate indefinitely, without telomere loss and without triggering the senescence response (Figure 4.2).

Telomerase expression is maintained in the testes, where sperm are continually produced, but in most cells of most other tissues, the enzyme is repressed by midgestation.[26,29,30] Nonetheless, some adult tissues express low levels of telomerase. This is more common, and expression is more robust, in mouse tissues compared to in human tissues.[26,30] This low level of telomerase expression is thought to be due to the presence of telomerase-positive, adult stem cells.[31]

Stem cells are cells that, upon division, produce daughter cells with different characteristics. One daughter retains the properties of the mother cell and thus remains a stem cell. The other daughter becomes a more differentiated cell type. Stem cells can be totipotent — that is, capable of producing all the differentiated cells in the adult organism. Embryonic stem cells, which derive from the inner cell mass of the very early embryo, are totipotent.[32] Stem cells can also be pluripotent or multipotential — capable of giving rise to only a limited differentiated lineage.[33] Hematopoietic stem cells, which can reconstitute the immune system, are an example of pluripotent stem cells.[34] In recent years, it has become apparent that there are low numbers of pluripotent stem cells in a number of adult tissues, particularly the bone marrow.[33,35] These adult stem cells can often differentiate into multiple lineages — for example, bone-marrow-derived stem cells can differentiate into cells with characteristics of neurons and muscle. Several lines of evidence suggest that the pool and/or differentiating capacity of adult stem cells declines with age. Moreover, this decline is thought to contribute to the loss of function and repair capacity that is seen in many aging tissues.[36,37]

Aside from embryonic stem cells, which express high levels of telomerase and proliferate indefinitely, little is known about the replicative life span or telomerase expression in adult stem cells. Adult stem cells may account for

**General Perspectives**

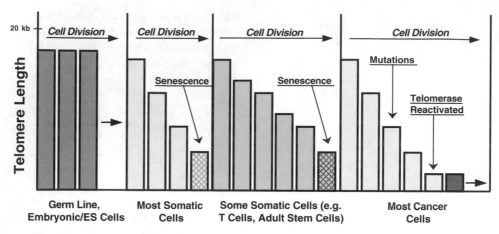

**FIGURE 4.2** *Telomere length, telomerase expression, and replicative senescence in various human cell types.* The germ line, early embryonic cells, and embryonic stem cells (ES cells) express telomerase, and telomeres are maintained at a stable length as these cells divide. By contrast, most somatic cells do not express telomerase, and the telomeres shorten with cell division. When the telomeres reach 4 to 6 kb, the cells undergo a senescence arrest (replicative senescence). Some somatic cells, including T cells and possibly adult stem cells, express telomerase, but the activity is insufficient to completely prevent telomere shortening. In these cells, telomeres still shorten, although more gradually than in other cells, and the cells still undergo a senescence arrest when they reach 4 to 6 kb. The majority of cancer cells suffer mutations that cause cells to continue to proliferate, even after the telomeres reach 4 to 6 kb. When the telomeres become very short, there is strong selection for mutations that reactivate telomerase. Only after telomerase is reactivated (or the alternative recombinational mechanism is reactivated) can cancer cells divide without limit. Reactivation of telomerase is the most common mechanism by which cancer cells overcome the end-replication problem.

the low level of telomerase activity seen in some adult tissues.[31,38] If so, why does telomerase not permit the unlimited renewal of stem cells in adults, thereby preventing the decline in stem cells with age? First, telomerase may indicate the presence of more differentiated cells that undergo limited expansion *in vivo* and express suboptimal levels of telomerase. This situation is seen in human T cells.[25,39,40] Human T cells transiently express telomerase when they are activated by antigen. However, telomerase activity is apparently suboptimal or partially inhibited in these cells, such that telomere shortening is retarded but not completely avoided. Thus, telomeres shorten in activated T cells, albeit somewhat more slowly than in other cell types, and the cells eventually undergo replicative senescence (Figure 4.2). It is also possible that the low levels of telomerase that are detected in some adult human tissues are caused by the presence of transformed or preneoplastic cells.[38,41] As discussed below, cancer cells frequently express telomerase.

### 3. Cancer Cells

The most common mechanism by which cancer cells overcome the end-replication problem is by reactivating the enzyme telomerase.[42] Although some cancer cells overcome the end-replication problem by an alternative, recombination-based mechanism,[43] the majority of malignant tumors are telomerase-positive. Many tumor cells have suffered mutations that inactivate genes that are important for the senescence response.[12,14,44] Thus, many tumor cells proliferate despite very short telomeres, eventually turning

on telomerase to stabilize them (Figure 4.2). If telomerase is inhibited, either by drugs or genetic manipulations, many cancer cells lose telomeric DNA as they divide, and eventually undergo a senescence arrest or, more often, die.[45] Thus, long-lived organisms with renewable tissues may repress telomerase as a strategy for preventing or retarding the development of cancer. As discussed at the end of this chapter, this strategy may be an evolutionary trade-off (see also Chapter 6). Although likely beneficial in young organisms, it may eventually lead to a depletion of stem cells and other mitotically competent cells, thereby causing or contributing to aging phenotypes later in life.

### B. Cellular Senescence

Short, dysfunctional telomeres are one of a growing number of signals that are now known to induce a senescence response. These signals can be broadly categorized into those that damage DNA, perturb chromatin, or deliver strong (supraphysiological) mitogenic signals, as shown in Figure 4.3.

Cells arrest growth with a senescent phenotype when they experience moderately high levels of DNA damage, such as those induced by ionizing radiation or oxidants.[46–48] Likewise, they undergo a senescence arrest when they are treated with agents or suffer mutations that disturb the organization of DNA into heterochromatin.[49,50] Loss of heterochromatin is known to cause inappropriate expression of silenced genes and is frequently seen in cancer.[51,52] Finally, normal cells undergo senescence when they experience strong mito-

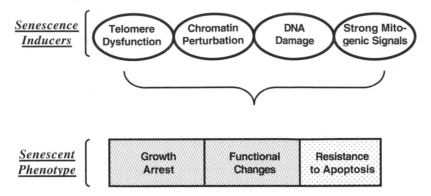

**FIGURE 4.3** *Inducers and phenotypic consequences of cellular senescence.* A variety of stimuli, all of which have the potential to cause or contribute to cancer, can induce cellular senescence. These inducers fall into the broad categories of disrupters of telomere function, disrupters of chromatin structure, DNA damaging agents, and agents that deliver supraphysiological mitogenic signals. The resulting senescent phenotype has two invariant features: an essentially irreversible growth arrest and changes in differentiated cell functions. In addition, some cells acquire resistance to signals that cause apoptotic cell death.

genic signals, such as those delivered by certain oncogenes or overexpression of certain signal-transducing proteins. Examples include activated forms of RAS, RAF, or E2F — proteins that normally act to stimulate cell proliferation.[53–56]

What do all these inducers of cellular senescence have in common? All have the potential to cause or facilitate the development of cancer. Thus, DNA damage, loss of heterochromatin, and the expression of oncogenes do not transform normal cells; rather, they cause normal cells to arrest growth with a senescent phenotype. These agents can, of course, transform cells, but do so only after cells have acquired mutations that abrogate the senescence response. As discussed below, such mutations tend to inactivate the function of p53, pRB, or both — the genes that lie at the heart of the two tumor suppressor pathways that are critical for establishing and maintaining the senescent phenotype.[12,14,44] These findings suggest that cellular senescence likely evolved as a fail-safe mechanism to ensure that potentially oncogenic cells irreversibly withdraw from the cell cycle and thus become incapable of tumorigenesis.

## C. The Senescent Phenotype

What happens when cells become senescent? Cellular senescence entails many changes in gene expression, only some of which are important for the growth arrest. In general, senescent cells appear to acquire two, or in some cases three, characteristics that, together, define the senescent phenotype:

1. An irreversible arrest of cell proliferation
2. In some cell types, resistance to signals that cause apoptotic cell death
3. Changes in differentiated cell functions

These aspects of the senescent phenotype are illustrated in Figure 4.3.

## 1. Growth Arrest

The most striking feature of senescent cells is their essentially permanent withdrawal from the cell cycle. This growth arrest is due to the repression of a set of genes that are essential for cell cycle progression, and the overexpression of a set of genes that inhibit cell cycle progression. The latter (overexpression of cell cycle inhibitors) is responsible for the fact that the senescence growth arrest is dominant — that is, when a senescent cell is fused to a proliferating cell, the proliferating cell arrests growth.[reviewed in 57] The precise mechanisms responsible for the senescence-associated growth arrest are, as yet, incompletely understood. However, it is clear that the overexpression of at least two cell cycle inhibitors — the p16 and p21 proteins — are critical for the arrest. Moreover, it is also now clear that p16 and p21 are critical components of the pRB and p53 tumor suppressor pathways, both of which are essential for cellular senescence.[reviewed in 12]

## 2. Resistance to Apoptosis

Some cell types, for example, human fibroblasts, T cells, and mammary epithelial cells, become resistant to signals that trigger apoptosis or programmed cell death when induced to undergo cellular senescence.[58,59] As discussed below, apoptosis is a cellular response to physiological signals and damage that eliminates superfluous, dysfunctional, or damaged cells in a controlled manner. Thus, some senescent cells become resistant to physiological stimuli that would normally eliminate them. This attribute of senescent cells may explain why they persist *in vivo* and accumulate with age.[60–62]

**General Perspectives**

### 3. Functional Changes

Finally, cellular senescence entails selected changes in differentiated functions. Some of the senescence-associated functional changes are common to most, if not all, cell types. These changes include an enlarged cell size, increased lysosome biogenesis, and decreased rates of protein synthesis and degradation.[reviewed in 3–5] In addition, most senescent cells express a neutral β-galactosidase, termed the senescence-associated β-galactosidase.[60] The function of this enzyme is unknown, but it can be detected by a simple histochemical reaction, and it is a useful marker for the senescent phenotype.

Senescent cells also display changes in the expression or regulation of cell type-specific genes. For example, replicative senescence of human adrenocortical epithelial cells causes a selective loss in the ability to induce 17α–hydroxylase, a key enzyme in cortisol biosynthesis.[63] Senescent human dermal fibroblasts, in contrast, show a marked increase in the expression of collagenase and stromelysin, metalloproteinases that degrade extracellular matrix proteins.[64,65] On the other hand, senescent human endothelial cells upregulate expression of interleukin-1α, a proinflammatory cytokine,[66] and downregulate expression of thymosin-β-10.[67] Interestingly, endothelial cells that simultaneously express the senescence-associated β-galactosidase and lack thymosin-β-10 (presumptive senescent endothelial cells) have been identified at the sites of atherosclerotic lesions in tissue samples from adult human donors.[67] Because many of the genes that are overexpressed by senescent cells encode secreted factors, it has been proposed that senescent cells can, in principle, disrupt local tissue integrity and thus contribute to age-related pathology.[68]

## III. CELL DEATH — CAUSES AND CHARACTERISTICS

Cell death plays an important role during the embryonic development of metazoans, in maintaining tissue homeostasis in the adults of complex organisms such as mammals, and in responding to cellular damage, tissue injury, and disease. In general, cells can die by two fundamentally different mechanisms. One is apoptosis or programmed cell death, which entails a series of controlled biochemical reactions that prevent the dispersion of the cellular contents into the surrounding tissue. The other is necrosis, which is generally less controlled (or uncontrolled), and entails cell lysis and the extrusion of cellular components into the surrounding tissue. Both modes of cell death may contribute to age-related pathologies.

### A. Apoptosis

Apoptosis is also referred to as programmed cell death, because it is, in fact, a highly orchestrated, genetically programmed process that allows cells to die in a controlled fashion.[69] All cells have the intracellular machinery necessary for programmed cell death. Whether or not they use that machinery depends on the cell type, its tissue context, the presence or absence of physiological death-promoting signals, and the extent to which a cell is damaged or dysfunctional.

Apoptosis is critically important during the embryonic development of all metazoans.[70–72] During embryogenesis, apoptosis serves to eliminate damaged or dysfunctional germ cells. It also eliminates superficial or unneeded cells during tissue morphogenesis and cells that fail to make the proper functional interactions with neighboring cells or target cells (for example, in the case of neurons, cells that fail to form proper functional connections to other neurons or specific muscle cells). The key features of apoptosis, and its major regulatory and effector molecules, have been highly conserved throughout metazoan evolution. However, during the evolution of vertebrates, the number and complexity of the proteins and reactions that regulate and execute apoptosis have increased greatly.[73] In complex organisms with renewable tissues, apoptosis is not only crucial during embryogenesis, it is also critical for maintaining tissue homeostasis in adults and eliminating dysfunctional, damaged, and potentially cancerous cells throughout life.[74,75]

As discussed below, apoptosis can be initiated by physiological signals acting through specific membrane-bound receptors or by damage to the nucleus, mitochondria, or other organelles. Once initiated, apoptosis entails an interacting series of biochemical steps that culminate in the controlled destruction of nuclear DNA and cross-linking of the cellular contents.[reviewed in 76–78] These steps are thought to proceed in three distinct phases: activation, commitment, and execution. Once cells have entered the commitment phase, the execution phase and demise of the cell is inevitable. A prominent feature of all phases of apoptosis is a cascade of enzymatic proteolysis (cleavage of proteins) in which a series of proteases, termed caspases, are sequentially activated by site-specific proteolysis.[76,78] The caspases that are activated in the last steps of the cascade, the so-called "executioner" caspases, then specifically cleave a number of cellular proteins. Most of the cellular proteins that are cleaved by caspases are inactivated by the cleavage. Some proteins, however, are activated by caspase cleavage. The net result of these cleavages includes the activation of the endonuclease that cleaves the nuclear DNA, an inhibition of DNA repair processes, and the activation of a transglutaminase that cross-links cellular proteins. Another prominent feature of apoptosis is the intimate involvement of mitochondria.[79] Mitochondria sense apoptotic signals, by mechanisms that are only partially understood, and respond by releasing calcium and cytochrome C into the cytosol. These mitochondrially derived molecules then can initiate or rein-

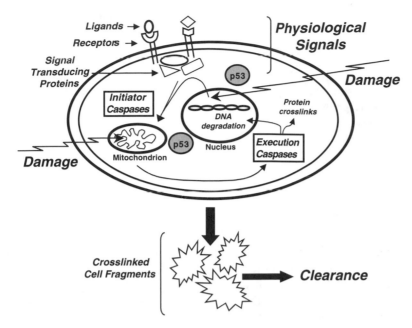

**FIGURE 4.4** *Major pathways and steps in apoptotic cell death.* Apoptosis can be initiated by physiological signals, such as the binding of ligands (tumor necrosis factor, FAS ligand, etc.) to their death-promoting receptors, or by endogenously or exogenously induced damage to the nucleus, mitochondria, or other cellular components. The initial stages of apoptosis entail the sequential activation of initiator caspases, which results in the release of mitochondrial signals. Mitochondrial signals and upstream caspases activate the executioner caspases, which cleave a variety of cellular proteins. Some cellular proteins are activated, whereas others are inhibited, as a result of caspase cleavage. The result is activation of the endonuclease that degrades the nuclear DNA, an inhibition of DNA repair processes, and activation of the transglutaminase that cross-links cellular proteins. Ultimately, the cell disintegrates into small cross-linked fragments. These fragments are then cleared from the tissues by macrophages or neighboring cells, which engulf and degrade them.

force the signals that activate the caspase cascades. Finally, during the last stages of apoptosis, the cells shrink and eventually disintegrate into small, cross-linked fragments. These fragments are then engulfed and degraded by macrophages or neighboring cells.

Apoptosis is controlled by more than 100 proteins, which act in highly complex and interrelated pathways that are linked to other basic cellular process, such as cell proliferation and differentiation. It is beyond the scope of this chapter to provide a comprehensive description of the regulation and execution of apoptosis. However, a simple description of its salient features is shown in Figure 4.4.

There is one particularly interesting molecular aspect of apoptosis that links it to cellular senescence: both processes are controlled by the tumor suppressor protein, p53.[80] This is consistent with p53 being (arguably) the most powerful tumor suppressor encoded by the mammalian genome, and the important roles that both cellular senescence and apoptosis play in preventing the development of cancer. The precise role that p53 plays in apoptosis is not yet completely understood. However, p53 is known to induce the transcription of genes that promote or facilitate apoptosis — for example, nuclear genes that encode proteins that regulate the release of calcium and cytochrome C from mitochondria.[74,75,81] In addition, p53 has been proposed to play a more direct role in regulating the

mitochondrial response to apoptotic signals.[80] Whatever the case, p53-dependent apoptosis is not crucial for embryonic development, because genetically engineered mice that completely lack p53 function develop normally.[82] However, these mice die prematurely of cancer. Thus, p53 function is more important in adult compared to developing mammals. One of its most important functions is to promote apoptosis in adult tissues, which suppresses tumorigenesis.[81,83] And, p53 may be more important for damage-induced apoptosis, as compared with apoptosis induced by physiological signals in embryos and adult tissues. Consistent with this idea, it is well-established that cells can also undergo apoptosis by p53-independent mechanisms.[73,77,84]

One very important function of apoptosis in adult organisms is to provide a mechanism for eliminating unwanted cells without cell lysis. This is important because cell lysis frequently results in local tissue destruction, owing to the release of degradative enzymes from the lysed cells and the inflammation reaction that it frequently elicits. Thus, apoptosis allows organisms to remove damaged or dysfunctional cells with minimal collateral damage to the tissue. In postmitotic tissues, where cell proliferation cannot replace lost cells, apoptosis minimizes the cell loss. In renewable tissues, where cell proliferation can replace cells that are lost, apoptosis is an

**General Perspectives**

important mechanism for maintaining the size, integrity, and health of the tissue.

## 1. Physiological Signals

Apoptosis can be induced by a variety of physiological (as opposed to damage-induced) signals. In healthy tissues and organisms, these signals serve to remove unnecessary or dysfunctional cells and to maintain the size of a tissue or organ. They also serve to attenuate or reverse proliferative response, such as the activation-induced cell death of T lymphocytes, which switches off an immune response in order to prevent autoimmune disease.

Some physiological apoptosis-inducing signals are initiated by ligand-receptor interactions at the plasma membrane and are transduced via specialized signal transduction proteins that reside on the cytoplasmic surface of the plasma membrane (Figure 4.4). Many of these ligands are active in the immune system, and include FAS ligand, tumor necrosis factor, and gamma interferon.[74,76,77] In reality, many ligands, hormones, cytokines, and other physiological regulators, can induce apoptosis, depending on the cell type, tissue, and physiological context. For example, transforming growth factor-beta, a multifunctional cytokine, can induce growth inhibition, extracellular matrix production, or apoptosis, depending on the cell type and its history and microenvironment.[85]

In addition to apoptosis-inducing signals, some cytokines, hormones, and growth factors deliver antiapoptotic signals and are required to prevent cell death.[reviewed in 77] For example, a deficiency in nerve growth factor causes certain neurons to die by apoptosis. Similarly, if T lymphocytes are deprived of certain interleukins (cytokines that act in the immune system), they die by apoptosis. Thus, some factors (such as nerve growth factor and specific interleukins) suppress apoptosis and, thus, are survival factors for certain cell types.

Finally, physiological mediators of apoptosis are not limited to factors that interact with receptors at the plasma membrane. For example, some hormones that act via nuclear receptors can stimulate or protect cells from apoptosis.[86] Examples include the action of testosterone in protecting germ cells in the testis from undergoing apoptosis,[87] the action of estrogen in protecting certain neurons from dying by apoptosis,[88] and the action of glucocorticoids in promoting apoptosis in lymphocytes.[89]

Aging entails many changes in the levels or availability of numerous hormones, cytokines, and growth factors — physiological mediators of apoptosis in adult tissues. As discussed below, aging also entails changes in the regulation of apoptosis in various tissues. Although it is not yet clear whether or to what extent age-related changes in the physiological mediators are responsible for the age-related changes in apoptosis, it is quite plausible that these two events are causally linked.

## 2. Damage Signals

Apoptosis can also be induced by cellular damage, particularly, damage to DNA. Apoptosis-inducing damage can originate from endogenous sources (for example, reactive oxygen species produced by defective mitochondria), or from exogenous sources (for example, ionizing radiation or chemicals). Very little is known about exactly how cells actually sense the damage. What is known is that, upon damage, cells mount an appropriate damage response. In doing so, cells have four basic options:

1. They can transiently arrest proliferation (generally in the G1 or G2 phase of the cell cycle) in order to repair (or attempt to repair) the damage.
2. They can undergo cellular senescence (i.e., permanently arrest growth with a senescent phenotype, as discussed above).
3. They can die by apoptosis.
4. They can die by necrosis (discussed below).

What determines which of these options a cell will choose? The important considerations are the cell type and physiological context, as well as the level and type of damage. Thus, a given dose of ionizing radiation may cause a senescence arrest in human fibroblasts, but apoptotic death in T lymphocytes. Likewise, a higher dose of radiation may cause human fibroblasts to undergo apoptosis, and an even higher dose may cause the same cells to die by necrosis.

p53 is extremely important for three of the four damage-response options: the transient growth arrest, the senescent growth arrest, and the apoptotic response.[74,81,90] As discussed above, p53-deficient mice do not exhibit any striking developmental abnormalities, but after birth, they are highly cancer-prone and die at an early age of multiple malignant tumors.[82] The same is true for humans with Li–Fraumeni syndrome, which is due to hereditary mutations in the gene that codes for p53.[91] Moreover, cancer is the primary cause of death in mice in which cells are defective in undergoing apoptosis (owing, for example, to the inactivation of genes encoding antiapoptosis proteins such as BCL2 which regulates the mitochondrial response to apoptotic signals).[74,92] Taken together, these findings suggest that the apoptotic response to damage may be particularly important for preventing the development of cancer in adult organisms.

## B. Necrosis

Necrosis is an alternate mode of cell death, which differs from apoptosis in a number of significant ways. Necrosis is characterized by an irreversible swelling of the cell and its organelles, including the mitochondria. Subsequently, the cellular membranes lose their integrity, and cellular constituents, including many degradative enzymes, leak into the surrounding tissue. This leakage often damages contiguous cells and causes exudative inflammation in the local tissue area. Thus, unlike apoptotic cell death, necrotic cell death is chaotic and is generally detrimental to the tissues in which it occurs.[reviewed in 77,79]

Necrosis is a pathological or accidental mode of cell death. It is induced by noxious stimuli, such as cytotoxins, hyperthermia, hypoxia, metabolic poisons, and direct cell or tissue trauma. In some pathological conditions — ischemic injury, for example — both apoptotic and necrotic cell death may occur. Cells are thought to die by necrosis, rather than apoptosis, when they experience more severe noxious insults. In addition, cells may die by necrosis when one or more component of an apoptotic pathway is inactivated, which can occur as a consequence of mutation or physiological processes. For example, recent evidence indicates that senescent human fibroblasts become resistant to apoptosis-inducing signals because they cannot stabilize p53 in response to damage.[59] As a result, senescent fibroblasts die by necrosis, rather than apoptosis. Thus, necrosis, and the accompanying inflammation reactions, may increase during aging owing to age-related defects in the cellular machinery that controls and carries out apoptosis.

## IV. CONSEQUENCES OF CELLULAR SENESCENCE AND CELL DEATH

Cellular senescence and apoptosis may play a role in aging and age-related pathology by two distinct means. On the one hand, as discussed above, both processes are important for suppressing the development of cancer. Because they suppress tumorigenesis, both apoptosis and cellular senescence can be considered longevity-assurance mechanisms — mechanisms that postpone the onset of aging phenotypes or age-related disease. Thus, both processes help ensure the fitness of organisms with renewable tissues (by preventing cancer), at least early in life. On the other hand, as senescent cells accumulate in tissues and apoptosis depletes cells from postmitotic tissues, both cellular senescence and apoptosis may contribute to aging phenotypes late in life. As discussed at the end of this chapter, this apparent paradox — the ability to be beneficial to the organism early in life but to be detrimental to the organism late in life — is consistent with the evolutionary hypothesis of antagonistic pleiotropy.

### A. SUPPRESSION OF CANCER

There is persuasive evidence that cellular senescence and apoptosis are important for preventing the development of cancer,[12,75,77] a major age-related disease.[93] This evidence can be summarized as follows:

1. Most naturally occurring and experimentally induced mammalian cancers harbor mutations that blunt or inactivate the senescence response to dysfunctional telomeres and other stimuli, and many cancers also harbor mutations that make them resistant to apoptotic cell death.
2. A number of well-recognized oncogenes are now known to act primarily by blocking the signals or cellular machineries that trigger or implement senescence response and apoptosis.
3. Conversely, cellular senescence and apoptosis are controlled by well-recognized tumor suppressors, of which p53 is perhaps the most prominent. These tumor suppressors, or one of their regulators, are lost or inactivated in the vast majority of cancers.
4. Several genetically engineered mouse models have been developed in which cells from these mice either fail to undergo cellular senescence or fail to undergo apoptosis in response to specific stimuli.[83,92,94,95] For the most part, these mice develop tumors at an early age and generally die of cancer.

Taken together, these findings strongly suggest that cellular senescence and apoptosis are powerful tumor-suppressive mechanisms. Nonetheless, mammalian organisms develop cancer, although generally not until around the midpoint of the observed maximum life span. Thus, cellular senescence and apoptosis are effective in preventing cancer in relatively young organisms but are less effective at doing so in older organisms. Why might this be the case?

First, mutations are required for the development of cancer,[1] and it takes time for cells to accumulate mutations that allow them to avoid cellular senescence and apoptosis. Mutations, including potentially oncogenic mutations, accumulate with age, although it is unlikely that mutation accumulation alone can account for the exponential rise in cancer that occurs during the last half of the mammalian life span.[reviewed in 8,12,96] Second, cancers require a permissive tissue structure in which to develop; in many cases, a healthy tissue environment can suppress the development of cancer from even highly mutated cells.[83,97–99] As discussed below and in other chapters, aging entails a number of changes in tissue structure and integrity, which may create a permissive environment for the growth and malignant transformation of mutant cells.

### B. CELLULAR SENESCENCE AND AGING

How might cellular senescence contribute to aging? As discussed above, senescent cells acquire altered functions. Among these altered functions is an increase in the secretion of molecules that can erode the structure and integrity of tissues if they are chronically present. In the case of senescent stromal fibroblasts, for example, these secreted molecules include proteases, inflammatory cytokines, and growth factors.[reviewed in 5, 96] Thus, senescent cells secrete molecules that can directly (by protease action) and indirectly (by stimulating inflammation) disrupt tissue structure. They also secrete growth factors that can disrupt the proliferative homeostasis of tissues and even stimulate the growth of mutant preneoplastic or neoplastic cells.[100]

There is some, albeit indirect, evidence for this scenario. Cells that express a marker of the senescent phenotype have been shown to increase with age in some mammalian tissues (for example, the skin, retina, and liver).[60–62,101] Moreover, such cells have been found at sites of age-related tissue pathology, such as the enlarged prostates of men with benign prostatic hyperplasia[102] or sites of tubulointerstitial fibrosis in the kidneys of aged rats.[103] Such cells have also been found in rabbit arteries after denudation by balloon catheter and at sites of atherosclerosis in human arteries.[67] It is possible, then, that the degradative enzymes, inflammatory cytokines, and growth factors produced by senescent cells can cause or contribute to such age-related pathologies as thinning and wrinkling of the skin, retinal degeneration, benign prostatic epithelial cell growth, kidney fibrosis, and atherosclerosis. Because senescent cells not only disrupt the tissue architecture but also secrete growth factors, it has also been proposed that they can contribute to the development of cancer in the skin, liver, and other tissues.[68,100,104]

Thus, although cellular senescence acts to retard the development of cancer early in life, later in life, it may contribute to age-related pathology, including cancer.

## C. Cell Death and Aging

How might apoptosis or necrosis contribute to aging? The most obvious potentially detrimental effect of cell death is in postmitotic tissues, where cells cannot be readily replaced. Although some postmitotic tissues such as skeletal muscle can be replenished by the recruitment of precursor cells (satellite cells, in the case of skeletal muscle), these cells have a finite replicative life span that eventually is exhausted.[105] Moreover, although stems cells that assume some features of neurons or cardiomyocytes have been identified, postmitotic neurons and cardiomyocytes do not appear to be efficiently replaced, if they are replaced at all, in many regions of the adult brain or heart, respectively.[33,36,106] Thus, eventually, cell death, whether apoptotic or necrotic, depletes postmitotic tissues of cells. Apoptosis and necrosis have been observed in a number of age-related pathologies, such as muscle atrophy and Alzheimer's disease.[77,107]

Apoptosis and necrosis may increase with age owing to increased levels of oxidative stress and ischemic injury.[107] Similar to the proposed role of cellular senescence in aging, apoptosis may benefit young organisms by efficiently eliminating dysfunctional or damaged cells. As organisms age, however, cellular reserves and functional redundancy may become exhausted, such that the loss of cells leads to tissue atrophy or degeneration.

In contrast to what is seen in postmitotic tissues, there is some evidence that apoptotic mechanisms may become less efficient with age in some mitotically competent tissues.[108] The mechanism responsible for this loss of apoptotic response is unknown. It may be related to the apoptosis resistance that occurs during cellular senescence, which appears to be due to the failure to stabilize p53 in response to damage, resulting in a loss of p53-dependent apoptosis.[59] Interestingly, caloric restriction, which extends the life span of many species, appears to increase the basal rate of apoptosis in some tissues.[reviewed in 108] This increase in apoptosis has been proposed to explain the ability of caloric restriction to delay all major age-related pathologies, including cancer.[108]

## V. EVOLUTIONARY PERSPECTIVE ON CELLULAR SENESCENCE AND CELL DEATH

As discussed in Chapter 6, aging is thought to be a consequence of the fact that the force of natural selection declines with age. One consequence of this decline is that some processes that were selected to maintain the health and fitness of young organisms can have *unselected*, deleterious effects in old organisms.[reviewed in 109] This idea has been termed antagonistic pleiotropy. It is possible that cellular senescence and apoptosis are antagonistically pleiotropic. Both processes appear to have been used by evolution to prevent the development of cancer in organisms with renewable tissues. Indeed, for the most part, such organisms are protected from cancer for a large portion of their life span, certainly for life spans equal to or exceeding those that were extant in the wild during most of evolution. However, in organisms that survive beyond those early life spans — most humans living today in developed countries, or mice housed under laboratory conditions — deleterious effects of cellular senescence and apoptosis may become apparent. The result may be a number of pathological conditions that are recognized as aging phenotypes or age-related disease.

## ACKNOWLEDGMENTS

Many thanks to present and past members of my laboratory for their hard work and interesting ideas, many colleagues who provided advice, reagents, and stimulating discussions, and the National Institute on Aging, Ellison Medical Foundation, California Breast Cancer Research Program, and Department of Energy for support.

# REFERENCES

1. Hanahan, D. and Weinberg, R.A., The hallmarks of cancer, *Cell*, 100, 57, 2000.

2. Hayflick, L., The limited *in vitro* lifetime of human diploid cell strains, *Exp. Cell Res.*, 37, 614, 1965.

3. Stanulis-Praeger, B., Cellular senescence revisited: A review, *Mech. Ageing Dev.*, 38, 1, 1987.

4. Cristofalo, V.J. and Pignolo, R.J., Replicative senescence of human fibroblast-like cells in culture, *Physiol. Rev.*, 73, 617, 1993.

5. Campisi, J. et al., Control of replicative senescence, in *Handbook of the Biology of Aging*, Schneider, E. and Rowe, J., Eds., Academic Press, New York, 1996.

6. Sherr, C.J. and DePinho, R.A., Cellular senescence: Mitotic clock or culture shock? *Cell*, 102, 407, 2000.

7. Wright, W.E. and Shay, J.W., Telomere dynamics in cancer progression and prevention: Fundamental differences in human and mouse telomere biology, *Nature Med.*, 6, 849, 2000.

8. Campisi, J., From cells to organisms: Can we learn about aging from cells in culture? *Exp. Gerontol.*, 36, 607, 2001.

9. Jazwinski, S.M., Longevity, genes and aging, *Science*, 273, 54, 1996.

10. Margolis, J. and Spradling, A., Identification and behavior of epithelial stem cells in the Drosophila ovary, *Development*, 121, 3797, 1995.

11. Sager, R., Senescence as a mode of tumor suppression, *Environ. Health Persp.*, 93, 59, 1991.

12. Campisi, J., Cellular senescence as a tumor-suppressor mechanism, *Trends Cell Biol.*, 11, 27, 2001.

13. Martin, G.M., Clonal attenuation: Causes and consequences, *J. Gerontol.*, 48, 171, 1993.

14. Bringold, F. and Serrano, M., Tumor suppressors and oncogenes in cellular senescence, *Exp. Gerontol.*, 35, 317, 2000.

15. Blackburn, E.H., Structure and function of telomeres, *Nature*, 350, 569, 1991.

16. Levy, M.Z. et al., Telomere end-replication problem and cell aging, *J. Mol. Biol.*, 225, 951, 1992.

17. Harley, C.B. et al., Telomeres shorten during aging of human fibroblasts, *Nature*, 345, 458, 1990.

18. Campisi, J. et al., Cellular senescence, cancer and aging: The telomere connection, *Exp. Gerontol.*, 36, 1619, 2001.

19. Wright, W.E. and Shay, J.W., Cellular senescence as a tumor-protection mechanism: The essential role of counting, *Curr. Opin. Genet. Dev.*, 11, 98, 2001.

20. Blackburn, E.H., Telomere states and cell fates, *Nature*, 408, 53, 2000.

21. de Lange, T., Telomere capping — one strand fits all, *Science*, 292, 1075, 2001.

22. Allsopp, R.C. et al., Telomere shortening is associated with cell division *in vitro* and *in vivo*, *Exp. Cell Res.*, 220, 194, 1995.

23. Chang, E. and Harley, C.B., Telomere length and replicative aging in human vascular tissues, *Proc. Natl. Acad. Sci. USA*, 92, 11190, 1995.

24. Yang, L. et al., Telomere shortening and decline in replicative potential as a function of donor age in human adrenalcortical cells, *Mech. Ageing & Dev.*, 122, 1685, 2001.

25. Effros, R.B., Replicative senescence in the immune system: Impact of the Hayflick limit on T-cell function in the elderly, *Am. J. Hum. Genet.*, 62, 1003, 1998.

26. Prowse, K.R. and Greider, C.W., Developmental and tissue-specific regulation of mouse telomerase and telomere length, *Proc. Natl. Acad. Sci. USA*, 92, 4818, 1995.

27. Blackburn, E.H., Telomerases, *Annu. Rev. Biochem.*, 61, 113, 1992.

28. Lingner, J. and Cech, T.R., Telomerase and chromosome end maintenance, *Curr. Opin. Genet. Dev.*, 8, 226, 1998.

29. Weng, N.P. and Hodes, R.J., The role of telomerase expression and telomere length maintenance in human and mouse, *J. Clin. Immunol.*, 20, 257, 2000.

30. Wright, W.E. et al., Telomerase activity in human germline and embryonic tissues and cells, *Dev. Genet.*, 18, 173, 1996.

31. Yasumoto, S. et al., Telomerase activity in normal human epithelial cells, *Oncogene*, 13, 433, 1996.

32. Smith, A.G., Embryo-derived stem cells: Of Mice and Men, *Annu. Rev. Cell Dev. Biol.*, 17, 435, 2001.

33. Weissman, I.L. et al., Stem and progenitor cells: Origins, phenotypes, lineage commitments, and transdifferentiations, *Annu. Rev. Cell Dev. Biol.*, 17, 387, 2001.

34. Gunsilius, E. et al., Hematopoietic stem cells, *Biomed. Pharmacother.*, 55, 184, 2001.

35. Almeida-Porada, G. et al., Adult stem cell plasticity and methods of detection, *Rev. Clin. Exp. Hematol.*, 5, 26, 2001.

36. Rao, M.S. and Mattson, M.P., Stem cells and aging: Expanding the possibilities, *Mech. Ageing Dev.*, 122, 713, 2001.

37. Schlessinger, D. and Van Zant, G., Does functional depletion of stem cells drive aging? *Mech. Ageing Dev.*, 122, 1537, 2001.

38. Harle-Bachor, C. and Boukamp, P., Telomerase activity in the regenerative basal layer of the epidermis in human skin and in immortal and carcinoma-derived skin keratinocytes, *Proc. Natl. Acad. Sci. USA*, 93, 6476, 1996.

39. Buchkovich, K.J. and Greider, C.W., Telomerase regulation during entry into the cell cycle in normal human T cells, *Mol. Biol. Cell*, 7, 1443, 1996.

40. Weng, N.P. et al., Regulated expression of telomerase activity in human T lymphocyte development and activation, *J. Exp. Med.*, 183, 2471, 1996.

41. Bickenbach, J.R. et al., Telomerase is not an epidermal stem cell marker and is downregulated by calcium, *J. Invest. Dermatol.*, 111, 1045, 1998.

42. Kim, N.W. et al., Specific association of human telomerase activity with immortal cells and cancer, *Science*, 266, 2011, 1994.

43. Bryan, T.M. et al., Evidence for an alternative mechanism for maintaining telomere length in human tumors and tumor-derived cell lines, *Nature Med.*, 3, 1271, 1997.

**General Perspectives**

44. Lundberg, A.S. et al., Genes involved in senescence and immortalization, *Curr. Opin. Cell Biol.*, 12, 705, 2000.

45. Hahn, W.C. et al., Inhibition of telomerase limits the growth of human cancer cells, *Nature Med.*, 5, 1164, 1999.

46. DiLeonardo, A. et al., DNA damage triggers a prolonged p53-dependent G1 arrest and long-term induction of Cip1 in normal human fibroblasts, *Genes Dev.*, 8, 2540, 1994.

47. Chen, Q. et al., Oxidative DNA damage and senescence of human diploid fibroblast cells, *Proc. Natl Acad. Sci. USA*, 92, 4337, 1995.

48. Robles, S.J. and Adami, G.R., Agents that cause DNA double strand breaks lead to p16INK4a enrichment and the premature senescence of normal fibroblasts, *Oncogene*, 16, 1113, 1998.

49. Ogryzko, V.V. et al., Human fibroblast commitment to a senescence-like state in response to histone deacetylase inhibitors is cell cycle dependent, *Mol. Cell. Biol.*, 16, 5210, 1996.

50. Bertram, M.J. et al., Identification of a gene that reverses the immortal phenotype of a subset of cells and is a member of a novel family of transcription factor-like genes, *Mol. Cell. Biol.*, 19, 1479, 1999.

51. Cairns, B., Emerging roles for chromatin remodeling in cancer biology, *Trends Cell Biol.*, 11, 15, 2001.

52. Wade, P.A., Transcriptional control at regulatory checkpoints by histone deacetylases: Molecular connections between cancer and chromatin, *Hum. Mol. Genet.*, 18, 693, 2001.

53. Serrano, M. et al., Oncogenic ras provokes premature cell senescence associated with accumulation of p53 and p16INK4a, *Cell*, 88, 593, 1997.

54. Sewing, A. et al., High-intensity Raf signal causes cell cycle arrest mediated by p21/Cip1, *Mol. Cell. Biol.*, 17, 5588, 1997.

55. Zhu, J. et al., Senescence of human fibroblasts induced by oncogenic raf, *Genes Dev.*, 12, 2997, 1998.

56. Dimri, G.P. et al., Regulation of a senescence checkpoint response by the E2F1 transcription factor and p14/ARF tumor suppressor, *Mol. Cell. Biol.*, 20, 273, 2000.

57. Smith, J.R. and Pereira-Smith, O.M., Replicative senescence: Implications for *in vivo* aging and tumor suppression, *Science*, 273, 63, 1996.

58. Wang, E. et al., Control of fibroblast senescence and activation of programmed cell death, *J. Cell. Biochem.*, 54, 432, 1994.

59. Seluanov, A. et al., Change of the death pathway in senescent human fibroblasts in response to DNA damage is caused by an inability to stabilize p53, *Mol. Cell. Biol.*, 21, 1552, 2001.

60. Dimri, G.P. et al., A novel biomarker identifies senescent human cells in culture and in aging skin *in vivo*, *Proc. Natl. Acad. Sci. USA*, 92, 9363, 1995.

61. Hjelmeland, L.M. et al., Senescence of the retinal pigment epithelium, *Mol. Vis.*, 5, 33, 1999.

62. Paradis, V. et al., Replicative senescence in normal liver, chronic hepatitis C, and hepatocellular carcinomas, *Hum. Pathol.*, 32, 327, 2001.

63. Hornsby, P.J. et al., Loss of expression of a differentiated function gene, steroid 17a-hydroxylase, as adrenocortical cells senesce in culture, *Proc. Natl. Acad. Sci. USA*, 84, 1580, 1987.

64. West, M.D. et al., Replicative senescence of human skin fibroblasts correlates with a loss of regulation and over-expression of collagenase activity, *Exp. Cell Res.*, 184, 138, 1989.

65. Millis, A.J.T. et al., Differential expression of metalloproteinase and tissue inhibitor of metalloproteinase genes in diploid human fibroblasts, *Exp. Cell Res.*, 201, 373, 1992.

66. Maier, J.A.M. et al., Extension of the life-span of human endothelial cells by an interleukin-1a antisense oligomer, *Science*, 249, 1570, 1990.

67. Vasile, E. et al., Differential expression of thymosin beta-10 by early passage and senescent vascular endothelium is modulated by VPF/VEGF: Evidence for senescent endothelial cells *in vivo* at sites of atherosclerosis, *FASEB J.*, 15, 458, 2001.

68. Campisi, J., Aging and cancer: The double-edged sword of replicative senescence, *J. Am. Geriatr. Soc.*, 45, 1, 1997.

69. Ellis, R.E. et al., Mechanisms and functions of cell death, *Annu. Rev. Cell Biol.*, 7, 663, 1991.

70. Jacobson, M.D. et al., Programmed cell death in animal development, *Cell*, 88, 347, 1997.

71. Vaux, D.L. and Korsmeyer, S.J., Cell death in development, *Cell*, 96, 245, 1999.

72. Meier, P. et al., Apoptosis in development, *Nature*, 407, 796, 2000.

73. Aravind, L. et al., Apoptotic molecular machinery: Vastly increased complexity in vertebrates revealed by genome comparisons, *Science*, 291, 1279, 2001.

74. Thompson, C.B., Apoptosis in the pathogenesis and treatment of disease, *Science*, 267, 1456, 1995.

75. Reed, J.C., Mechanisms of apoptosis in avoidance of cancer, *Curr. Opin. Oncol.*, 11, 68, 1999.

76. Budiharjo, I. et al., Biochemical pathways of caspase activation during apoptosis, *Annu. Rev. Cell Dev. Biol.*, 15, 269, 1999.

77. Fadeel, B. et al., Apoptosis in human disease: A new skin for an old ceremony? *Biochem. Biophys. Res. Comm.*, 266, 699, 1999.

78. Earnshaw, W.C. et al., Mammalian caspases: Structure, activation, substrates and functions during apoptosis, *Annu. Rev. Biochem.*, 68, 383, 1999.

79. Kroemer, G. et al., The mitochondrial death/life regulator in apoptosis and necrosis, *Annu. Rev. Physiol.*, 60, 619, 1998.

80. Moll, U.M. and Zaika, A., Nuclear and mitochondrial apoptotic pathways of p53, *FEBS Lett.*, 493, 65, 2001.

81. Ryan, K.M. et al., Regulation and function of the p53 tumor suppressor protein, *Curr. Opin. Cell Biol.*, 13, 332, 2001.

82. Donehower, L.A. et al., Mice deficient for p53 are developmentally normal but susceptible to spontaneous tumors, *Nature*, 356, 215, 1992.

83. DePinho, R.A., The age of cancer, *Nature*, 408, 248, 2000.

84. Joaquin, A.M. and Gollapudi, S., Functional decline in aging and disease: A role for apoptosis, *J. Am. Geriatr. Soc.*, 49, 1234, 2001.

85. Akhurst, R. and Derynck, R., TGF- signaling in cancer: A double-edged sword, *Trends Cell Biol*, 11, 44, 2001.

86. Medh, R.D. and Thompson, E.B., Hormonal regulation of physiological cell turnover and apoptosis, *Cell Tissue Res.*, 301, 101, 2000.

87. Sinha Hakim, A.P. and Swerdloff, R.S., Hormonal and genetic control of germ cell apoptosis in the testis, *Rev. Reprod.*, 4, 38, 1999.

88. Belcredito, S. et al., Estrogens, apoptosis and cells of neural origin, *J. Neurocytol.*, 29, 359, 2000.

89. Planey, S.L. and Litwack, G., Glucocorticoid-induced apoptosis in lymphocytes, *Biochem. Biophys. Res. Commun.*, 279, 307, 2000.

90. Itahana, K. et al., Regulation of cellular senescence by p53, *Eur. J. Biochem.*, 268, 2784, 2001.

91. Akashi, M. and Koeffler, H.P., Li–Fraumeni syndrome and the role of the p53 tumor suppressor gene in cancer susceptibility, *Clin. Obstet. Gynecol.*, 41, 172, 1998.

92. Wu, X. and Pandolfi, P., Mouse models for multistep tumorigenesis, *Trends Cell Biol.*, 11, 2, 2001.

93. Balducci, L. and Beghe, C., Cancer and age in the USA, *Crit. Rev. Oncol./Hematol.*, 37, 137, 2001.

94. Ghebranious, N. and Donehower, L.A., Mouse models in tumor suppression, *Oncogene*, 17, 3385, 1998.

95. de Boer, J. and Hoeijmakers, J., Cancer from the outside, aging from the inside: Mouse models to study the consequences of defective nucleotide excision repair, *Biochimie*, 81, 127, 1999.

96. Campisi, J., Cancer, aging and cellular senescence, *In Vivo*, 14, 183, 2000.

97. Boudreau, N. et al., The embryonic environment and extracellular matrix suppress oncogenic transformation by Rous sarcoma virus in chick embryos, *Mol. Cell. Diff.*, 3, 261, 1995.

98. van den Hoof, A., Stromal involvement in malignant growth, *Adv. Canc. Res.*, 50, 159, 1998.

99. Park, C.C. et al., The influence of the microenvironment on the malignant phenotype, *Molec. Med. Today*, 6, 324, 2000.

100. Krtolica, A. et al., Senescent fibroblasts promote epithelial cell growth and tumorigenesis: A link between cancer and aging, *Proc. Natl. Acad. Sci. USA*, 98, 12072, 2001.

101. Pendergrass, W.R. et al., Cellular proliferation potential during aging and caloric restriction in rhesus monkeys (Macaca mulatta), *J. Cell. Physiol.*, 180, 123, 1999.

102. Choi, J. et al., Expression of senescence-associated beta-galactosidase in enlarged prostates from men with benign prostatic hyperplasia, *Urology*, 56, 160, 2000.

103. Ding, G. et al., Tubular cell senescence and expression of TGF-beta1 and p21(WAF1/CIP1) in tubulointerstitial fibrosis of aging rats, *Exp. Mol. Path.*, 70, 43, 2001.

104. Rinehart, C.A. and Torti, V.R., Aging and cancer: The role of stromal interactions with epithelial cells, *Mol. Carcinogen.*, 18, 187, 1997.

105. Renault, V. et al., Skeletal muscle regeneration and the mitotic clock, *Exp. Gerontol.*, 35, 711, 2000.

106. Blau, H.M. et al., The evolving concept of a stem cell: Entity or function? *Cell*, 105, 829, 2001.

107. Martin, L.J., Neuronal cell death in nervous system development, disease, and injury, *Int. J. Mol. Med.*, 7, 455, 2001.

108. Warner, H.R. et al., A unifying hypothesis to explain the retardation of aging and tumorigenesis by caloric restriction, *J. Gerontol.*, 50, 107, 1995.

109. Kirkwood, T.B. and Austad, S.N., Why do we age? *Nature*, 408, 233, 2000.

**General Perspectives**

# 5  Oxidants and Antioxidants in Aging

*Rolf J. Mehlhorn*
Lawrence Berkeley National Laboratory

## CONTENTS

## I.  OVERVIEW

Considerations of natural selection suggest that life span maximization is not a significant evolutionary priority.[1] Animals in the wild must survive at least long enough to reproduce in a highly competitive environment. Survival hinges on the ability to fight or escape from predators, recover from physical trauma, tolerate toxins in food, subdue infectious microbes, and optimize energy storage and expenditure to endure long periods of starvation. The need to allocate energy resources to these various challenges diverts at least some resources from maintenance functions. Among important maintenance functions is the prevention of, and recovery from, chronic, inevitable damage. Such damage can arise from a variety of nonphysiologic reactions in tissues, including reactions of macromolecules with reducing sugars (glycation) and other aldehydes, oxidants, alkylation by methylating agents, and spontaneous hydrolytic processes. Even if some of the products of these reactions are transient, they can exert adverse effects on organisms. Incomplete repair of damaged macromolecules could well lead to the cumulative effects that characterize aging. No single damage mechanism like oxidative damage is likely to fully explain the aging phenomenon — an "understanding" of aging may prove to comprise a ranking of the various damage (and incomplete repair) processes that characterize life.

Animals are certainly subject to considerable oxidative stress. Animals suffer lung and neurological damage within a few hours and die within a few days of exposure to pure oxygen.[2] However, they can survive longer oxygen exposures if gradually adapted to increasing oxygen concentrations. Humans exposed to pressurized, i.e., hyperbaric, oxygen (HBO) suffer DNA damage within a few hours of exposure, but protection is afforded by a brief preexposure to HBO.[3] The discovery and characterization of the "antioxidant" enzyme superoxide dismutase (SOD, whose only known function is to scavenge the oxygen-derived reactive molecule superoxide, an example of a free radical molecule — see later sections of this chapter), and the observation that virtually no air-respiring animal can survive without this enzyme, have established that hazardous oxygen products are formed *in vivo*. Adaptation to elevated oxygen concentrations is consistent with the view that aerobic organisms maintain antioxidant defenses that are just sufficient for protection against hazardous oxygen derivatives under normal physiologic conditions but that can be enhanced in response to gradual increases in oxidative stress. A general theme of biologic stress responses is that low doses of some specific toxic agent elicit an adaptive response that confers protection against a relatively broad range of subsequent toxic challenges. In principle, this represents an amplifying mechanism for antioxidant protection and suggests that certain kinds of chronic stress can exert long-term net benefits. Consistent with this idea is a recent report that chronic exposure of mice to low levels of ionizing radiation, a process that produces free radicals (see subsequent sections), increased life spans by about 22%.[4]

Although destructive oxygen-derived products like free radicals may escape scavenging by antioxidants like SOD and react with vital cellular targets, organisms have the capacity to repair damage. For example, oxidative modifications of macromolecules can be (a) partially reversed by reduction, (b) repaired by excision and replacement of damaged parts, or (c) counteracted by complete turnover. Even if toxic agents trigger the death of individual cells, multicellular organisms have some capacity to compensate for cell loss by replacing irreversibly damaged cells with new ones. Hence, oxidative damage theories and, indeed, all *cumulative damage theories must contend with the question of why repair is incomplete or why compensatory processes like cell renewal fail.* One important consideration is that no repair system is likely to be absolutely efficient if the damage is so random that macromolecules are altered in virtually limitless ways. An oxidative damage wear-and-tear theory, which embodies the characteristic of unpredictable molecular alterations, is the free radical theory of aging. It was first proposed by Harman in 1955, essentially as follows: *aging results from the deleterious effects of free radicals produced in the course of cellular metabolism.*[5,6] Harman supported his theory by citing the purported detection of free radicals in biologic systems with the newly discovered technique of electron paramagnetic resonance (EPR).[7] Of historical interest is that the initial reports of free radical signals in biologic tissues were incorrect — the observed EPR signals were an artifact of the freeze-drying methods used in sample preparation and were not due to metabolic activity as postulated in the free radical theory.[8] Subsequently, the existence of tissue free radicals was confirmed, but concentrations were much lower than suggested by the initial reports.

The stochastic nature of free radical reactions implies that cellular surveillance systems cannot possibly recognize all of the structural or functional changes that can occur. Even with perfect surveillance of exposed damage-sensitive sites, the need to repair "hidden" damage, e.g., oxidized amino acids buried within proteins, requires a periodic turnover of macromolecules, probably with energy-wasting turnover of undamaged macromolecules. These considerations suggest that the pursuit of perfect maintenance necessarily entails substantial energy dissipation in dealing with the possibility rather than certainty of damage. From an energy optimization perspective, survival in a highly competitive natural world is simply not consistent with the high level of maintenance that would ensure maximum life spans.

The free radical theory of aging continues to generate great interest as indicated by a sampling of recent publications.[9–12] Considerable evidence supports an involvement of oxidative damage in major life-shortening diseases and, hence, average life span. The evidence for a pivotal role of oxidants in the antimicrobial activity of immune cells, which suggests a life-shortening effect of immune system activity in certain diseases and chronic infections, is especially compelling. However, the question of whether oxidants, or indeed any damage process, limit maximum life span, i.e., control the aging process *per se*, remains unanswered. Efforts to use predictions of the free radical theory to achieve increases in maximum life span extension, e.g., through nutritional supplementation with natural antioxidants, have failed.[13]

While there is no question that free radicals can be highly destructive, it is becoming increasingly evident that at least some free radicals and nonradical oxidants play vital physiologic roles (in addition to their involvement in immune function). The free radicals nitric oxide and superoxide and the nonradical hydrogen peroxide serve as second messengers in regulating cell growth and many other physiologic processes. Of particular interest is the role of the free radical nitric oxide in controlling mitochondrial metabolism. A growing body of evidence argues for the likelihood that many of the adverse effects of at least some oxidants may be attributable to their disruptive effects on cell regulation rather than to overt damage.[10]

## II. OXIDANTS

Oxidants of biologic importance are small molecules containing oxygen, referred to as reactive oxygen species (ROS). Some oxidants also contain nitrogen and are referred to as reactive nitrogen species (RNS). ROS readily oxidize cellular constituents like DNA, proteins, or lipids in the most basic chemical sense, i.e., in reactions that involve one or more electron transfers from an organic target molecule to the oxidant. However, some biologic oxidants, notably certain free radicals, also alter organic molecules by combining with them to form "adducts." ROS comprise a group of molecules that span a broad range of reactivities. Some ROS, like the superoxide radical, are relatively weak and are likely to collide with many other molecules before a reaction occurs. Others, like the hydroxyl radical, are extremely reactive and will react with virtually any organic molecule they encounter. Important biologic oxidants are listed in Table 5.1.

### A. What are Free Radicals?

Free radicals are molecular fragments containing "unpaired" electrons. They are produced when chemical bonds are broken. Considerable energy is required to break most chemical bonds, so free radicals rarely occur in nature. Once formed, free radical fragments will usually react with nonradical molecules to form new free radical products or combine with other free radicals to form stable products. Even if sufficient energy is available to break bonds, the proportion of free radicals among nonradicals is usually very low. Examples of free radicals include:

- The very reactive hydroxyl radical, OH· (the dot refers to an unpaired electron)
- The moderately reactive thiyl radical, RS·
- The weakly reactive nitric oxide radical, NO·

Reactive free radicals can pose a considerable hazard to biological systems because of their unique chemistry, which distinguishes free radicals from other toxic agents.

---

**TABLE 5.1**
**Biologic Oxidants and Oxidation Catalysts**

| Name | Structure | Sources |
|---|---|---|
| Superoxide radical | $O_2^-$ | Mitochondria, enzymes, some hydroquinones |
| Hydrogen peroxide | $H_2O_2$ | Enzymes, superoxide |
| Hydroxyl radical | OH· | Hydrogen peroxide + reduced transition metal ions, peroxynitrite, superoxide + hypochlorite |
| Singlet oxygen | $^1O_2$ | Aerobic free radical chain reactions |
| Nitric oxide | NO· | Nitric oxide synthase enzymes |
| Peroxynitrite | ONOO$^-$ | Superoxide + nitric oxide |
| Hypochlorite | HOCl | Myeloperoxidase |
| Transition metal ions | $Fe^{+m}$, $Cu^{+n}$ | Metalloprotein degradation, dietary overload |

The most damaging free radicals exhibit some or all of the following reaction patterns:

- Attack other molecules indiscriminately
- Initiate oxygen-consuming chain reactions, such that a single free radical effectively damages a large number of other molecules
- Cause fragmentation or random cross-linking of molecules, including vital macromolecules like DNA, enzymes, and structural proteins

## B. Oxygen

Ordinary oxygen has two unpaired electrons, i.e., $O_2$ can be represented as $\cdot O–O \cdot$. Oxygen is an example of a free radical that is normally quite stable, i.e., oxygen radicals do not combine with each other, and they do not react spontaneously with combustible organic molecules. However, oxygen can be quite reactive in the presence of other free radicals. In a free radical chain reaction, this property of oxygen is expressed as an incorporation of oxygen molecules into organic free radicals ($R\cdot$, $R'\cdot$), converting them to hydroperoxides as follows:

$$R\cdot + O_2 \rightarrow ROO\cdot$$

$$ROO\cdot + RH' \rightarrow ROOH + R'\cdot$$

$$R'\cdot + O_2 \rightarrow R'OO\cdot$$

In this sequence of reactions, oxygen combines with an organic free radical ($R\cdot$) to form a peroxyl radical ($ROO\cdot$), which in turn, oxidizes an organic molecule $R'H$. Such chain reactions exert their adverse effects by chemically modifying target molecules and by consuming oxygen required for normal metabolic processes. Another highly reactive oxygen species is singlet oxygen, which can arise during free radical chain reactions. When two peroxyl radicals react with each other, they do not form a stable product but decompose into a variety of fragments, including singlet oxygen.

Oxygen can be "activated" by the sequential incorporation of electrons ($e^-$) and protons as follows:

$$O_2 + e^- \rightarrow O_2 \cdot^-$$
(superoxide radical, weakly reactive)

$$O_2 \cdot^- + e^- + 2\,H^+ \rightarrow H_2O_2$$
(hydrogen peroxide, weakly reactive)

$$H_2O_2 + e^- \rightarrow OH^- + OH\cdot$$
(hydroxyl radical, strongly reactive)

$$OH\cdot + e^- + H^+ \rightarrow H_2O$$

A major fate of the superoxide radical is "dismutation," a reaction of two superoxide radicals with each other and with two protons as follows: $2\,O_2\cdot^- + 2\,H^+ \rightarrow H_2O_2 + O_2$. However, superoxide is not an effective electron donor to hydrogen peroxide, so reducing agents (electron donors) that produce superoxide radicals generally lead to an accumulation of hydrogen peroxide but do not produce hydroxyl radicals. Importantly, superoxide radicals can activate hydrogen peroxide to hydroxyl radicals in the presence of certain metal ions. Either iron or copper ions, loosely bound by low molecular compounds or macromolecules, catalyze this activation of hydrogen peroxide in the "Fenton reaction."

## C. Nitric Oxide

Nitric oxide ($NO\cdot$) is a free radical, which like oxygen, is essentially inert toward nonradical molecules and quite reactive with many other free radicals. However, in contrast to oxygen, when $NO\cdot$ reacts with an organic free radical, the product of the reaction is a stable nonradical product, i.e., $R\cdot + NO\cdot \rightarrow RNO$. Therefore, $NO\cdot$ can terminate free radical chain reactions. From a biologic perspective, one of the most important properties of nitric oxide is its reactivity with the superoxide radical to produce the nonradical peroxynitrite: $O_2\cdot^- + NO\cdot \rightarrow ONOO^-$. Peroxynitrite is a strong oxidizing agent that is capable of nitrating aromatic compounds, e.g., converting the amino acid tyrosine to nitrotyrosine. Peroxynitrite is weakly acidic, such that most of this species is negatively charged at physiological pH. However, even at physiological pH, a significant fraction of peroxynitrite is in the protonated uncharged form, which can readily diffuse into and through biologic membranes. Peroxynitrite can decompose into two new free radicals — nitrogen dioxide and the hydroxyl radical. This represents an alternative mechanism to iron or copper catalysis for generating the extremely reactive hydroxyl radical.

## D. Enzyme Activation of Oxygen

Many enzymes involved in energy metabolism have iron or other reduction/oxidation-active (redox-active) metals at their active sites. In their reduced state, some of these metalloenzymes have the potential to transfer electrons to molecular oxygen, producing superoxide radicals and hydrogen peroxide. Enzymes capable of reducing molecular oxygen to superoxide and hydrogen peroxide also include certain flavoenzymes. Generally, an electron transfer from an enzyme to molecular oxygen would be an "accidental" electron transfer that the enzyme was not designed to catalyze. In at least one case, however, an enzyme has the specific function of producing superoxide radicals. That enzyme is an NADPH oxidase, which was originally found in cells of the immune system but has more recently been detected in many other cells. Some metalloenzymes may form complexes with superoxide, resulting in altered enzyme activities. In the presence of low superoxide levels, enzymes with high superxoxide affinities may function as second messengers. The interaction of superoxide with some metalloenzymes causes a release of the metal from the enzyme. Aconitase, an important mitochondrial enzyme, loses iron upon reacting with superoxide, causing not only its inactivation, but leading to the formation of hydroxyl radicals by the released iron (Fenton reaction).[14]

## E. ENZYME ACTIVATION OF HYDROGEN PEROXIDE

Certain heme proteins are converted to highly oxidizing or "ferryl" species by hydrogen peroxide. These include a variety of peroxidases that are designed to exploit the oxidizing potential of hydrogen peroxide for specific molecular transformations. They also include myeloperoxidase, which is found in cells of the immune system. Myeloperoxidase activation of hydrogen peroxide leads to the formation of hypochlorous acid at physiologic chloride concentrations. Hypochlorous acid (the active ingredient of household bleach) diffuses from the enzyme and readily oxidizes a variety of molecules, including sulfhydryl groups on proteins and low molecular weight thiols. Hypochlorous acid reacts with superoxide radicals to form hydroxyl radicals by a mechanism that does not require transition metal ions. Free amino acids are oxidized to aldehydes by hypochlorous acid. One of these reaction products is the highly toxic unsaturated aldehyde acrolein, which is formed from the oxidation of threonine. Acrolein is a highly cytotoxic molecule, which is found in tobacco smoke and arises during the decomposition of cyclophosphamide, a compound used to kill tumor cells in chemotherapy.

An important aspect of enzyme-mediated oxidant production is that of enzyme dysfunction. Enzymes that have been oxidatively modified may be inactivated or may be more likely to promote the activation of oxygen or of hydrogen peroxide. For example, an enzyme of purine catabolism is xanthine dehydrogenase, which conserves the bond energy in xanthine to produce NADH. Oxidation of thiol groups in xanthine dehydrogenase converts it to xanthine oxidase, which produces superoxide and hydrogen peroxide without conserving energy. A peroxidase may lose its substrate specificity after oxidative modification and exert its powerful oxidizing potential on molecules it was not designed to oxidize. Oxidation of hemoglobin and myoglobin converts these ferrous oxygen-binding proteins to ferric products with peroxidase activity and no significant oxygen-binding capacity. A vicious cycle of oxidative enzyme modification and increased free radical production could lead to substantial damage amplification under conditions of oxidant stress.

## F. PRO-OXIDANTS IN FOODS AND OXIDANT-DAMAGED TISSUES

The world of plants survives by waging chemical warfare against predators. Many of the toxic agents that protect plants act by promoting oxygen activation or by reacting directly with important cellular targets like DNA. Oxygen activation in animals can be catalyzed by plant quinones that interact with reducing enzymes and divert electrons directly to oxygen. A number of plant toxins are activated by P-450 enzymes, mammalian enzymes normally involved in detoxification of a broad range of hazardous agents, to mutagenic products or interact with the P-450 enzyme system to promote superoxide formation. Other compounds, designated as quinonoids, can catalyze a non-enzymic one-electron reduction of oxygen by low molecular weight compounds in cells. An example of such a catalytic agent is divicine, which is derived from fava beans. This agent can be reduced by simple sulfhydryl compounds in the aqueous phase of cells and subsequently reoxidized by oxygen in a reaction that produces superoxide radicals. Pro-oxidants can also form in oxidant-stressed animal tissues. Among the oxidation products of uric acid, a product of purine metabolism, is alloxan, which exhibits a superoxide-generating activity similar to that of divicine. In the presence of alloxan, vitamin C becomes a source of superoxide radicals.[15]

## G. ANTIOXIDANTS

Antioxidants can greatly reduce the adverse impact of oxidants by: (a) intercepting oxidants before they react with vital biologic targets, (b) preventing chain reactions, or (c) preventing the activation of oxygen to highly reactive products. Low molecular weight antioxidants can be maintained at high intracellular concentrations without occupying excessive cell volume and are, therefore, good candidates for intercepting free radicals that would otherwise target vulnerable macromolecules. To be effective, such antioxidants must readily react with free radicals to form relatively harmless products. Probably the most efficacious low molecular weight antioxidant is ascorbic acid (vitamin C), which scavenges virtually all free radicals that come into contact with it. It is water-soluble and reacts with oxidizing free radicals by donating electrons or hydrogen atoms to them. An important example of such a reaction is the reaction of an antioxidant with a free radical (one-electron oxidation) product of a macromolecule like DNA or a protein. If present at sufficiently high concentrations, ascorbic acid can donate a hydrogen atom to the unpaired electron of the macromolecule before oxygen combines with it, thereby restoring its original structure. In the case of an oxidized membrane, lipid ascorbic acid, in conjunction with vitamin E (see below), can either restore its structure or stabilize its oxidation product as a

**General Perspectives**

**TABLE 5.2**
**Agents Involved in Protection Against Oxidants**

| Name | Mechanism(s) | Source (humans) |
|------|-------------|-----------------|
| Ascorbic acid (Vitamin C) | Free radical (FR) scavenger | Dietary (see text) |
| Glutathione | FR scavenger, enzyme cofactor | Endogenous agent (see text) |
| Tocopherol (Vitamin E) | FR scavenger | Dietary (see text) |
| Carotenoids | Singlet oxygen scavengers | Dietary |
| Bilirubin | FR scavenger | Heme catabolism |
| Lipoic acid | FR scavenger | Endogenous agent |
| Uric acid | FR scavenger | Purine catabolism |
| Enzyme mimetic agents | Superoxide, hydrogen peroxide scavengers | Potential antiaging drugs |
| PBN (phenyl N-tert-butylnitrone) | FR scavenger, cell signaling effects | Potential antiaging drug |
| (-) deprenyl (selegiline) | Monoamine oxidase B inhibitor | Putative neuroprotective drug |
| Protein methionine groups | FR scavengers | Endogenous agents |
| Superoxide dismutases | Superoxide scavengers | Endogenous enzymes |
| Catalase | Hydrogen peroxide scavenger | Endogenous enzyme |
| Selenium peroxidases | Hydroperoxide scavengers | Endogenous enzymes |
| Heme oxygenases | Heme decomposition | Endogenous enzymes |
| Ferritin | Iron sequestration | Endogenous enzyme |
| Quinone reductase | Quinone scavenging | Endogenous enzyme |
| Glutathione-S-transferases | Detoxification of xenobiotics, endogenous toxins,  hydroperoxides | Endogenous enzymes |
| Peroxiredoxins | Cell signaling effects (?) | Endogenous enzymes |
| Metallothionine | Transition metal sequestration | Endogenous enzyme |

relatively benign *trans*-unsaturated fatty acid. An antioxidant can also affect the reaction product of a free radical addition reaction (adduct) by donating an electron to a reactive free radical intermediate that might otherwise reduce an oxygen molecule to a superoxide radical. Important antioxidants and related protective agents are listed in Table 5.2.

## 1.  Ascorbic Acid

This widely consumed nutritional supplement is converted to the virtually inert ascorbyl radical when it reacts with free radicals. Its potency as an antioxidant is illustrated by its rapid reduction of nitroxide free radicals, synthetic organic radicals that are so unreactive as to be stable in aqueous solution. Ascorbic acid also plays a direct role in a number of enzyme reactions, serving a vital function in collagen synthesis and, hence, in the maintenance of tissue structure. The ascorbyl radical does not reduce oxygen to superoxide. It is one of the few free radicals that can be observed directly in some tissue preparations with magnetic resonance techniques, which testifies to its lack of reactivity. Most animals synthesize ascorbic acid; humans and higher primates do not. Ascorbate is found in fruits and vegetables, which normally provide an adequate supply of this vitamin. A lack of dietary ascorbate can lead to scurvy, which had once been a major cause of mortality among sailors during long sea voyages. The symptoms of scurvy can be attributed to a failure to synthesize collagen without invoking a role for

oxidant damage. In some tissues, the ascorbyl radical can be reduced back to ascorbate by specific enzymes. Alternatively, two ascorbyl radicals will dismutate, leading to the formation of dehydroascorbic acid. Dehydroascorbic acid is not stable in the body of an animal and has toxic properties. It can react with and chemically modify biomolecules and has been used as a diabetogenic agent in animals. It decomposes with a half life of 6.5 min at 37°C if it is not reduced to ascorbic acid enzymatically or by a nonenzymatic reaction with glutathione (see below) or other thiols. Ascorbate may play a role in protecting membranes by donating a hydrogen atom to free radicals at the membrane-water interface. Ascorbate can exert adverse effects by acting as a pro-oxidant, e.g., in the presence of certain metal ions. It is capable of reducing loosely bound iron or copper and can thus participate in the activation of hydrogen peroxide to hydroxyl radicals. A surprising recent discovery was that ascorbate can interact with lipid hydroperoxides to produce genotoxic aldehydes by a mechanism that was previously thought to be possible only in the presence of transition metal ions.[16] This important discovery may explain some of the disappointing failures of ascorbic acid therapy in treating diseases (see later section on oxidants in disease).

An important characteristic of reactive free radicals is the tendency for reaction products to be weaker radicals than reaction precursors. A typical reaction cascade would begin with a hydroxyl radical forming less reactive carbon-centered radicals (the free electron resides mostly on a carbon atom of an organic molecule), which in turn, would form even less reactive peroxyl radicals. The latter would be relatively persistent and eventually react with glutathione or ascorbate.

## 2. Glutathione

Thiols, and glutathione in particular, are involved in a variety of protective cell functions. Glutathione is a cysteine-containing tripeptide that is found in millimolar concentrations in most cells. The radical that is formed from glutathione (or most other thiols), referred to as a thiyl radical, is considerably more reactive than the ascorbyl radical but will generally react with another thiol like glutathione, producing a disulfide product and a superoxide radical. The latter is generally dismutated by superoxide dismutase, so no damage results from this reaction pathway. However, thiyl radicals react with phenolic molecules like tyrosine and, therefore, have significant destructive potential. A specific enzyme, glutathione reductase, regenerates glutathione from disulfide products, using NADPH generated by the hexose monophosphate pathway or by isocitrate dehydrogenases as the reductant.

## 3. Vitamin E

Vitamin E is another widely consumed vitamin. The E vitamins are a group of hindered phenolic, hydrophobic antioxidants that can donate hydrogen atoms to lipid radicals in membranes. They are also known as tocopherols, comprising a group of phenols with different substituents on a phenolic ring attached to a long hydrophobic chain that intercalates into membrane lipid domains and binds to certain proteins. After donating a hydrogen atom to a free radical, the vitamin E radical, designated as the tocopheroxyl radical, is too weak to react even with a highly unsaturated fatty acid. Moreover, the α-tocopheroxyl radical differs from simple phenolic radicals in not forming radical-radical products (dimers). Thus, vitamin E is a highly effective terminator of chain reactions in membranes. A number of reducing agents, including vitamin C, can react with the tocopheroxyl radical to regenerate vitamin E. Therefore, vitamin E usually acts as a catalytic agent that transfers reducing power derived from cellular metabolism to membrane free radicals.

The most common form of vitamin E in nutritional supplements is α-tocopherol. As noted above, it is the most effective antioxidant of the tocopherols, because the α-tocopheroxyl radical is highly resistant to reactions other than hydrogen–atom oxidation. Interestingly, γ-tocopherol may be more potent in protecting humans from degenerative diseases. This effect appears to be related to its inhibition of an enzyme (cyclooxygenase 2) involved in inflammation rather than to a free radical scavenging effect.[17] As such, the vitamin is acting in much the same manner as aspirin in reducing risks of cardiovascular disease. Major

commercial vitamin formulations consist of α-tocopherol. On the other hand, food sources contain mostly γ-tocopherol. Of concern is that α–tocopherol can exert an inhibitory effect on γ-tocopherol activity.

## III. OXIDANTS AND ANTIOXIDANTS IN CELLS AND ORGANELLES

### A. METABOLISM

A key element of the free radical theory of aging is the notion that metabolism spawns destructive free radicals. The long-known observation that animal life spans correlate roughly with inverse specific metabolic rate led to the suggestion that free radicals were responsible for aging, because free radical production was assumed to be a direct function of metabolic rate. However, our current understanding of oxidative metabolism argues against such a simple view. About 90% of the oxygen consumed by animals is utilized by mitochondria in a harmless process that directly yields water.[18] It is not known what fraction of normal mitochondrial oxygen consumption in vivo comprises ROS production, but the fraction is likely to be much less than the 2% observed in isolated organelles and cells in air-saturated solution. Superoxide production is a function of the state of reduction of the respiratory chain, not of metabolic rate. Therefore, mitochondrial oxidant production in vivo will vary depending on tissue oxygen concentrations, concentrations of reducing substrates, ATP utilization, and the presence of electron transport inhibitors (see next section), but not necessarily on the rate of oxygen consumption. Of interest is the potential effect of the general nutritional status on the mitochondrial redox potential. Recent studies on the effects of high glucose concentrations on cultured epithelial cells have shown that large increases in glucose concentration elicit a substantial increase in oxygen consumption, hydrogen peroxide production, advanced glycation end (AGE) product formation, and accumulation of the glucose reduction product sorbitol.[19] Conceivably, large fluctuations in blood glucose levels, rather than mean energy intake, are key determinants of oxidative damage.

Several considerations other than mechanisms of oxidant production argue against the idea that aging is governed by metabolic rate. These include the beneficial effects of exercise, analyses of the aging rates of organs exhibiting widely differing metabolic rates (e.g., brain vs. muscle) and animal species comparisons (substantial difference in life spans between bats and mice having the same specific metabolic rates).

### B. HOMEOSTASIS

Homeostasis, e.g., the maintenance of a well-controlled redox environment, is necessary for the proper functioning of all organisms. Animals can tolerate a range of tissue oxygen concentrations, as exemplified by a range of well over an order of magnitude between intensive exercise and rest. This range is limited, however, as shown by the toxicity of pure and pressurized oxygen (in the absence of gradual adaptation). At the other extreme, complete lack of oxygen, as in ischemia, leads to tissue damage when oxygen becomes available again, as documented in numerous publications on "reperfusion injury." Intolerance to extremes of oxygen concentrations can be reconciled with a straightforward model of oxidant production. Assuming that superoxide radicals are produced by random collisions of "autoxidizing" reducing agents with molecular oxygen, the rate of superoxide production would be directly proportional to the oxygen concentration and to the concentration of the autoxidizable reducing agents (e.g., certain quinols and semiquinones). Hence, it would be expected that exposure to pure oxygen would be associated with high superoxide burdens. On the other hand, in the absence of oxygen, the concentration of autoxidizable reducing agents would accumulate. Upon reoxygenation of the tissue, there would be a "burst" of superoxide radicals and superoxide-derived oxidants.

Although free radicals are clearly produced during normal oxidative metabolism, the fact that animals survive for years indicates that the elaborate antioxidant defensive systems and repair processes substantially neutralize the threat. Homeostasis implies that cells adapt to increases in oxidant levels by augmenting their responses to these threats. Antioxidants alone cannot confer complete protection, as indicated by a multiplicity of repair and turnover processes. Repair systems include specific enzymes for regenerating oxidized proteins with metabolic reducing power, a variety of proteases that break down oxidatively damaged proteins to amino acids that can be used for synthesizing new proteins, and a host of DNA repair enzymes, including enzymes that repair specific hydroxyl radical adducts of DNA bases.

## C. MITOCHONDRIA, PEROXISOMES, AND ENZYMES

Although it is known that the mitochondrial respiratory chain operates as an array of enzymes that perform electron transfer reactions similar to those occurring in simple chemical free radical reactions, the structure of these enzymes ensures that few or no untoward chemical reactions occur. In particular, no release of active oxygen molecules occurs from the site of oxygen binding cytochrome oxidase. However, despite the tightly controlled flow of reducing equivalents through the respiratory chain, some leakage of reductants from mitochondria appears to occur at sites other than cytochrome oxidase. This can be seen in a release of superoxide radicals or hydrogen peroxide in studies with isolated mitochondria, submitochondrial particles, and cell suspensions. Blockage of electron

flow by a variety of specific electron transport inhibitors causes part of the mitochondrial electron transport chain to become highly reducing, thereby increasing superoxide production. Electron flow is "coupled" to ion gradients (expressed as the electrochemical potential) and the ATP/ADP ratio. Because of this coupling, an increase in the magnitude of the electrochemical potential or the ATP/ADP ratio also increases superoxide production. A "safety valve" that can prevent an excessive mitochondrial electrochemical potential comprises uncoupling proteins (UCPs) that induce proton leaks in conjunction with organic anions, e.g., free fatty acids. Evidence for this concept is the report that ROS production is elevated in knockout mice lacking the mitochondrial uncoupling protein UCP3.[20] Another safety valve may be the nonspecific permeability pore of the inner mitochondrial membrane. Opening of this pore is reversibly induced by increases in ROS; when ROS levels decrease, the pores close. Failure to control the production of ROS by uncoupling proteins or pores has been proposed to cause selective degradation of vulnerable subpopulations of mitochondria, which has been referred to as "mitoptosis."[21] A release of intramitochondrial proteins as a result of extensive formation of nonspecific pores is known to orchestrate apoptosis. Mitochondria have been called the "poison cupboard" of the cell, because they contain many of the agents involved in apoptosis.[22] Knockout mice lacking the means to export ATP from mitochondria and, thus, not able to maintain appropriate ATP/ADP ratios have been reported to produce high levels of ROS and to exhibit an accelerated accumulation of mitochondrial DNA damage.[23]

It has recently been reported that mitochondrial respiration may be regulated by nitric oxide.[24] Nitric oxide binds reversibly to cytochrome oxidase, thereby inhibiting oxygen binding and, hence, respiration. At the oxygen concentrations found in tissues, respiration is substantially inhibited by nitric oxide in the submicromolar concentration range. Mitochondria contain nitric oxide synthase. Partial inhibition of respiration by nitric oxide produced by this enzyme could exert beneficial effects, e.g., by preventing steep oxygen concentration gradients across cells.[25] On the other hand, substantial blockage of electron flow by nitric oxide would render components of the electron transport chain highly reducing, thus stimulating superoxide production. This suggests that exposure of mitochondria to high nitric oxide levels would lead to significant peroxynitrite-mediated damage compounded by the adverse effects of an inhibition of oxidative phosphorylation.

Other organelles and some soluble enzymes can cause univalent and divalent oxygen reduction. Of possible significance as a risk factor in fat consumption is beta oxidation of fatty acids in peroxisomes. Estimates of hydrogen peroxide production by isolated subcellular fractions in pioneering experiments conducted in the laboratory of

**TABLE 5.3**
**Production of Hydrogen Peroxide by Isolated Cell Fractions**

| Organelle | Oxidant Source | $H_2O_2$ Production (% of total) |
|---|---|---|
| Mitochondria | Electron transport leak | 15 |
| Peroxisomes | Product formation | 35 |
| Endoplasmic Reticulum | Mixed function oxidations | 45 |
| Cytosol | Xanthine oxidation | 5 |

Britton Chance are set forth in Table 5.3.[26] These classic experiments demonstrated that a variety of potential sources of active oxygen exist in most cells. However, the relevance of these and many subsequent studies to *in vivo* oxidant production remains obscure, because isolated organelles are generally damaged, and because interpretations of *in vitro* studies have usually failed to take into account that tissue oxygen concentrations are generally much lower than those found in laboratory suspensions of cells or subcellular fractions. Nonperoxisomal enzymes that generate ROS as normal metabolic products include monoamine oxidase and some P-450 isoenzymes. Monoamine oxidase is located in the outer mitochondrial membrane and may be a major source of hydrogen peroxide in some tissues.[27]

## D. Cells of the Immune System

Cells of the immune system destroy invasive microorganisms by mechanisms that include reactive oxygen molecules. The destruction of invasive microorganisms occurs in part by the action of the potent membrane-permeable oxidants hypochlorous acid and protonated (i.e., uncharged) peroxynitrite. Stimulation of neutrophils and macrophages, ultimately triggered by cell surface antigens on invasive microbes or on oxidatively damaged macromolecules at the surfaces of host cells, leads to an "oxidative burst." This burst reduces oxygen to superoxide by a membrane-bound NADPH-dependent enzyme and production of nitric oxide by an inducible nitric oxide synthase. As noted previously, superoxide and nitric oxide combine spontaneously to form peroxynitrite. Hydrogen peroxide that results from superoxide dismutation combines with chloride ions within the immune cells in the myeloperoxidase reaction to yield hypochlorous acid. The powerful oxidants released by the immune cells oxidize a variety of intracellular target molecules in bacteria or other invasive microbes to exert their cytotoxic effects. However, the cells of the host organism, including tissue and immune cells, are not spared from the destructive effects of the oxidants. It must be considered that immune activity can be a major source of oxidant damage to animals. Oxidative damage, particularly uncontrolled lipid peroxidation, generates chemotactic factors that

attract leukocytes to areas of high free radical activity, compounding the effects of the initial oxidative damage. Such damage-amplifying processes are probably important factors in pulmonary damage associated with emphysema and pollutant exposure and in atherosclerosis. Oxidants generated by cells of the immune system are likely to play a role in the impact of many diseases and are, thus, likely to contribute to a shortening of mean life spans (Chapter 15).

A possible involvement of free radicals in aging is outlined schematically in Figure 5.1. This scheme identifies the major sources of superoxide radicals and other oxidants, protective systems that remove these oxidants, and their major cellular targets. The fundamental idea of all wear-and-tear theories of aging, including the free radical theory, is that some damage is not repaired, and that damage gradually accumulates, culminating in death. While oxidized lipids can be replaced, and most protein and DNA damage can be repaired, some damage is not repaired and accumulates with age.

## E. What is the Oxidant Burden?

An effect of oxidants on the regulation of cell growth is seen at very low concentrations compared to overt oxidative damage effects, such as those seen in necrosis (Table 5.4).[28] Tissue oxygen concentrations are low (typically about 40 μM), most of the oxygen is normally consumed by mitochondria, and antioxidant systems rapidly scavenge reactive oxygen species. Background levels of ROS under normal physiologic conditions must be low enough to allow cell signaling to function properly, i.e., the communication network involved in cell regulation cannot be distorted by the "noise" of random free radical reactions. These considerations imply that ROS levels must be very low under normal physiologic conditions and that elevated levels of ROS are likely to exert their adverse effects primarily on cell signaling. Also, to the extent that mitochondrial respiration is inhibited by nitric oxide, the background levels of this radical must be low enough to allow for an adequate maintenance of ion gradients and the generation of ATP.

It seems plausible that the highest oxidant burdens would arise from the effects of activated cells of the

**General Perspectives**

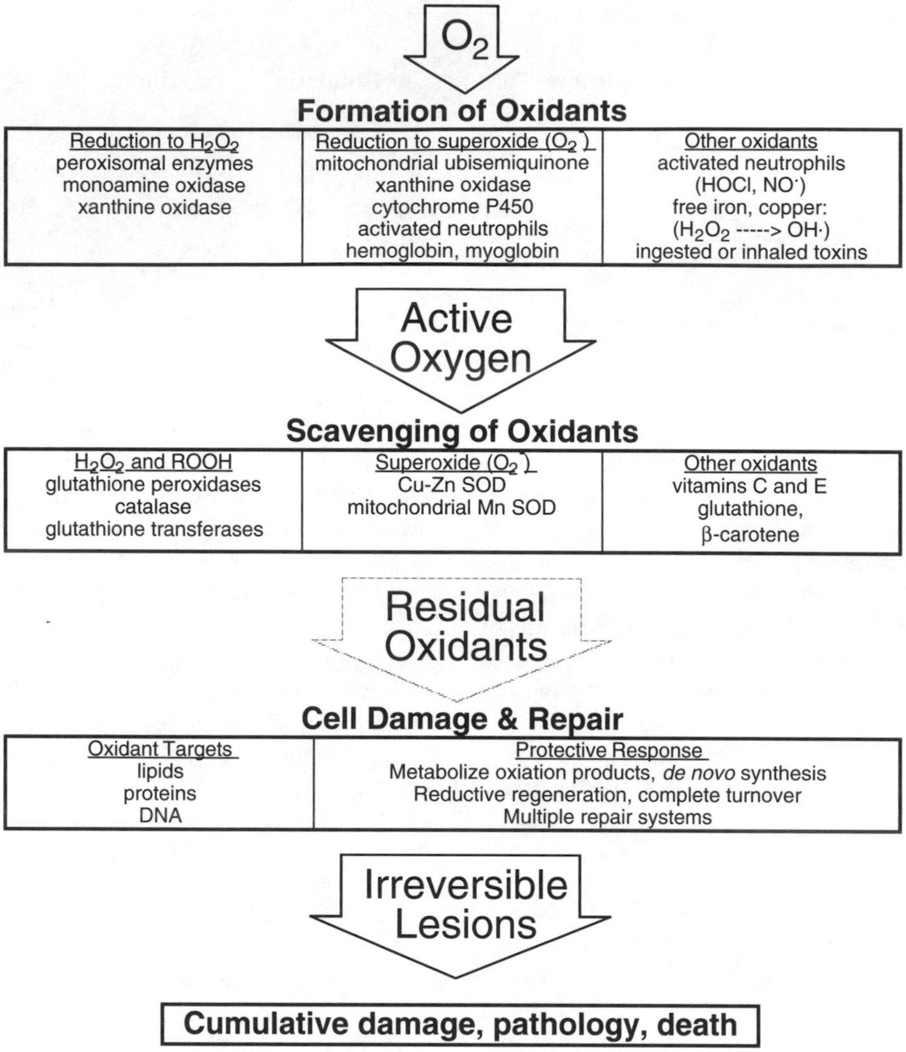

O2
Formation of Oxidants
Reduction to H2O2
peroxisomal enzymes
monoamine oxidase
xanthine oxidase
Reduction to superoxide (O2˙)
mitochondrial ubisemiquinone
xanthine oxidase
cytochrome P450
activated neutrophils
hemoglobin, myoglobin
Other oxidants
activated neutrophils
(HOCl, NO˙)
free iron, copper:
(H2O2 -----> OH·)
ingested or inhaled toxins
Active Oxygen
Scavenging of Oxidants
H2O2 and ROOH
glutathione peroxidases
catalase
glutathione transferases
Superoxide (O2˙)
Cu-Zn SOD
mitochondrial Mn SOD
Other oxidants
vitamins C and E
glutathione,
β-carotene
Residual Oxidants
Cell Damage & Repair
Oxidant Targets
lipids
proteins
DNA
Protective Response
Metabolize oxiation products, de novo synthesis
Reductive regeneration, complete turnover
Multiple repair systems
Irreversible Lesions
Cumulative damage, pathology, death

**FIGURE 5.1**

**TABLE 5.4**
**Effects of Hydrogen Peroxide on Cells**

| $H_2O_2$ Dose | Effect on HA-1 Cells |
|---|---|
| 3 to 5 μM | 25 to 45% growth stimulation |
| 120 to 150 μM | Temporary growth arrest, adaptation to $H_2O_2$ |
| 250 to 400 μM | Permanent growth arrest without loss of function |
| 0.5 to 1 mM | Apoptosis (nuclear condensation, DNA "laddering") |
| 5 to 10 mM | Necrosis (membrane disruption, protein denaturation) |

immune system on host tissues. Conceivably, the destructive effects of macrophage oxidants would be compounded by oxygen depletion during the oxidative burst. Membranes of host tissue cells in close proximity to macrophages would be especially vulnerable to oxidative damage.

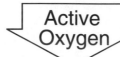

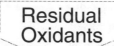

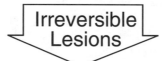

## F. Targets of Free Radical and Nonradical Oxidants

### 1. DNA

Probably the most important free radical targets are nuclear and mitochondrial DNA; debate continues on the relative vulnerabilities of these two targets. Addition of the hydroxyl radical to a double bond in a DNA base can lead to a large variety of hydroxylated products.[29] Reactive nitrogen species can nitrate or deaminate DNA bases. Free radical attack on the DNA backbone or chemical modification of bases can produce strand scission. Radical–radical reactions can lead to cross-linking among segments of DNA or between DNA and proteins.

The production of the hydroxylated nucleoside 8-oxo-8 doxyguanosine (8-oxo-8 dG) has been utilized as an index of DNA damage in a variety of cells and tissues. Unfortunately, methodology artifacts have led to overestimation of 8-oxo-8 dG and of free radical DNA damage.[30] Recent technical improvements have disclosed lower levels of DNA damage[31] than previously estimated[31,32] and suggest further assessment of this methodology. Other products that attack DNA include nonradical oxidants, adduct-forming agents, reactive species (e.g., peroxynitrite) and others.[33]

Base methylation plays an important regulatory role in gene expression and in the differentiation that characterizes multicellular organisms. Aging is associated with a change in the pattern of base methylation and dedifferentiation. The resulting organismal dysfunctions may explain some of the phenomenology of the aging process.[34] The repair of oxidant-modified methylated bases may lead to a loss of normally methylated bases,[35] suggesting one mechanism whereby free radicals could exert adverse epigenetic effects.

### 2. Proteins

As already noted, some amino acids are readily oxidized. Cysteine reacts with many free radicals and with hydroperoxides to form products that react with thiols to produce disulfides. Disulfides can subsequently be reduced again to repair the lesion. As disulfides are resistant to further oxidation or chemical modification by aldehydes, the formation of protein-mixed disulfides with glutathione may serve a protective function under oxidative stress. Many amino acids may be irreversibly oxidized by free radicals that are much less reactive than the hydroxyl radical. The reactive amino acids include histidine, tryptophan, and tyrosine. Tyrosine oxidation by free radicals can lead to bityrosine, a fluorescent free radical damage marker that is implicated in irreversible protein cross-linking.

As noted earlier, heme proteins, including hemoglobin and myoglobin, react with hydroperoxides to form highly oxidizing ferryl species that can produce amino acid radicals within the heme protein or oxidize nearby molecules, e.g., initiate free radical chain reactions in membranes. Activation of hydroperoxides by heme proteins can also exert its damaging effects more indirectly, by inducing a release of iron from oxidatively cleaved porphyrins. The released iron can subsequently catalyze hydroxyl radical production or other reactions at iron binding sites. Among the site-specific reactions of protein-bound iron, the oxidations of histidine, proline, and arginine side chains have been cited.[36-38]

Although most oxidized proteins are either repaired or degraded and replaced by new ones, these processes are not perfect, and some altered proteins may persist for considerable periods, particularly in energy-deficient cells. A dramatic example of a persistent altered protein is seen in brain myelin protein. Racemized amino acids, i.e., amino acids that have undergone an inversion from the normal L form to the mirror-image D form, accumulate in this protein throughout life (however, amino acid racemization is not thought to be a free radical process), indicating that some altered proteins are not turned over significantly during the course of an animal's lifetime.

Oxidatively modified individual proteins are selectively degraded in mammalian cells by proteasomes.[39] However, when oxidative attack causes protein cross-linking, the aggregated proteins are resistant to proteasome degradation[40] and, indeed, appear to bind irreversibly to proteasomes, thereby inhibiting them.[41] A decline in proteasome activity with age has been invoked to explain the accumulation of various polymerized age pigments. In some cells, this accumulation can be quite substantial, e.g., 75% of the volume of large motor neurons in human centenarians may be occupied by fluorescent polymeric deposits.[42] It is likely that nonoxidative cross-linking reactions including glycation-mediated reactions contribute to the accumulation of age pigments.[43]

The catalytic activity per unit of protein antigen decreases with age. An accumulation of oxidatively modified nonfunctional or dysfunctional proteins, which are not recognized or are turned over at an inadequate rate, appears to play an important role in aging.[36]

### 3. Membranes

Most cell membranes contain polyunsaturated fatty acids that are highly susceptible to free radical oxidation. As long as vitamin E is present, the extent of free radical oxidation is controlled, because the vitamin prevents or limits the occurrence of chain reactions. Depletion of vitamin E can lead to considerable oxidation of membrane polyunsaturated lipids, with adverse consequences that can include major alterations of membrane structure, release of lipid oxidation products, including cytotoxic molecules like malondialdehyde, hydroxynonenal, and

**General Perspectives**

acrolein, and an increase in permeability to ions. Lipid radicals react avidly with oxygen and membrane free radical chain reactions can potentially create an oxygen deficit with a concomitant loss of oxidative phosphorylation. The reduction of hydroperoxides by glutathione peroxidase consumes cellular reducing power, which could also contribute to energy depletion.

## IV. SUSCEPTIBLE TISSUES AND ORGANELLES

Because the cell populations in different tissues are likely to vary in their redundancy, regenerative capacity, free radical production rates, antioxidant defenses, and DNA repair rates, some tissues may be particularly sensitive to free radical damage, and these could determine aging rates in the organism.

### A. Neuronal Tissue

Cells like the neurons could well be the most susceptible targets of oxidative damage. Indeed, there is a loss of neurons with age, but such loss appears to be species-specific and to be a function of the tissue location. The surprising durability of neuronal tissue, despite its large oxygen utilization, appears, at first sight, to be incompatible with free radical involvement in age-related functional declines. However, a large portion of the oxygen utilization is in axons and terminals that are physically removed from sensitive and nonrenewable targets like the nuclear genome. Moreover, neuronal mitochondria do not appear to produce superoxide radicals at a site that is a major source of these radicals in other cells. Finally, considerable cell redundancy may prevent functional deficits from becoming clinically significant. As a rule, more than 40% cell loss is required in a neuronal system before functional loss in the central nervous system is apparent (see Chapters 7 and 8).

### B. Radiosensitive Tissues

Clinical experience with tumor irradiation has identified a number of tissues that are highly sensitive to free radical attack. The majority of these radiosensitive tissues have rapidly dividing cells with DNA that would be expected to be more vulnerable to oxidative attack than that of nondividing cells. Among these rapidly dividing cells are those of the hemopoietic system, gastrointestinal tract, and skin. However, some nonproliferating cells also exhibit high radiosensitivity, e.g., cells of the salivary gland. Interestingly, a recent analysis of intestinal radiation damage in mice has shown that radiation-induced apoptosis is responsible for the adverse physiologic consequences of irradiation in this tissue. Inhibition of apoptosis by pharmacological agents or genetic manipulations greatly reduced tissue damage, demonstrating that the direct effects of free radical reactions did not play a major role in the damage caused by the irradiation.[44]

### C. Mitochondria

Mitochondrial dysfunction and depletion occurs in aging postmitotic cells and may prove to be a major factor in aging. Miquel and his colleagues have postulated that mitochondria are the "Achilles heel" of postmitotic cells. Mitochondrial DNA is not protected by a coat of histone proteins, is in close proximity to sites of oxygen radical production and lipid peroxidation, and may have an inadequate DNA repair system.[45] However, because many mitochondrial proteins are encoded by the nuclear genome and are sensitive to the cellular environment (ions, substrates) and even hormonal stimulation (viz. thyroid hormone effects on mitochondrial replication), it is difficult to unambiguously dissociate intrinsic mitochondrial damage from extrinsic factors.

## V. SYSTEMIC EFFECTS

One of the most obvious characteristics of aging is the gradual decline of physiologic integrity, including the progressive impairment of neurological, immunological, metabolic and other functions. Evidence for free radical involvement in all of these debilitating processes is growing. However, organelle and cellular redundancy and renewal may be able to sufficiently compensate for the damage in order to sustain function. Thus, the epithelia lining the digestive tract and skin, the blood cells, and the liver may be extensively damaged and yet regenerate on a regular basis. In these renewing cell populations, the principal danger of damage appears to be neoplastic transformation.

Because aging is characterized by an increased incidence of infectious diseases, one might expect immunosenescence to play a crucial role in physiologic aging (Chapter 15). Declines in immune function begin relatively early and appear to be heavily dependent on thymic involution and selective T cell aging. Immunological declines can be, at least partially, reversed by thymic grafts or thymic hormone therapy, but complete restoration requires an additional young bone marrow graft.

Evidence for free radical involvement in immunosenescence includes a selective vulnerability of the immune system to radiation and other free-radical-generating agents. T cells, which age more rapidly than B cells, are reported to be more vulnerable to oxygen radicals and to accumulate more lipofuscin; treatment of aging mice with the antioxidant 2-mercaptoethanol delays the accumulation of T cell lipofuscin and the decline of immune function with age, and it increases the mean life span.

The thymus may also be selectively vulnerable to free radical damage. The first age-related loss in size (thymic involution) can be ascribed to the loss of the most radiosensitive (cortical) lymphocytes. The medullary epithelial cells that secrete thymic hormones are heterogenous, and the early loss is again that of the most radio-sensitive cells, which are active metabolically, require vitamin C for their secretory activity, and appear to accumulate intrinsically autofluorescent substances (age pigments). Although there are probably developmentally programmed cellular and hormonal controls of thymic involution, significant oxidative damage seems to be involved as well (Chapter 15).

Many antioxidants are immune stimulants and will enhance immune function both *in vivo* and *in vitro* with both young and old cells. Part of this effect appears to be due to the maintenance of reduced sulfhydryl groups. Although antioxidant treatment can boost immune function at any age, it does not markedly reduce the rate of basic aging of the immune system.

## VI. MODULATION OF LIFE SPANS: ROLE OF OXIDANTS

### A. Ionizing Radiation

Considerations of parallels between ionizing radiation and metabolic processes led to the articulation of the free radical theory of aging. However, experimental studies on the effects of irradiation on animals have failed to demonstrate a life-shortening effect that could be resolved from disease. Considerable effort has been devoted to the study of the health effects of ionizing radiation, prompted largely by the advent of nuclear weapons, the occurrence of fatal accidents in nuclear industries, and the utility of treating tumors with irradiation. Because of these motivations, many animal studies have focused on relatively high dose exposures or the on effects of inhaled or ingested radionuclides. It was found that high-dose irradiation causes leukemia and other cancers, and that susceptibility to radiation-induced disease varies substantially among animals, e.g., susceptibility is in the order: mice > beagle dogs > burros.[46] An examination of postirradiation survival data argues against any impact of irradiation on maximum life spans, consistent with the view that the irradiation treatments increased diseases other than cancer but did not alter aging rates *per se*. Importantly, the bulk of the early literature on life-span effects of irradiation provides no information on the possible protective effects of gradual adaptation. Moreover, the typical irradiation doses were large, comprising exposures delivered during short periods that were comparable to or greater than the background levels that the animals would accumulate during their entire lifetimes. Relatively recently, studies of the effects of very low-level chronic irradiation have been reported. These studies suggest a completely different effect of low-level irradiation than would be inferred by simple extrapolations from high-level data.[4] In seeming contradiction to the predictions of the free radical theory of aging, at least seven publications have reported an increase in life spans of mice exposed to chronic free radical stress induced by low-level ionizing radiation (references cited by Caratera et al.[4]).

It can be argued that ionizing radiation is not an appropriate model for testing the free radical theory of aging. Most importantly, it is not possible to accurately mimic endogenous oxidative damage with ionizing radiation. The energy of a given photon or particle of ionizing radiation (tens of thousands to millions of times greater than the energy of the O–H bond in water) greatly exceeds the energy of any metabolic process. Rapid dissipation of this energy gives rise to concentrated clusters of highly reactive free radicals that can produce multiple reactions in a macromolecule like DNA. Multiple free radical lesions in DNA create the potential of cross-linking and multiple strand scissions, which are much more likely to be irreparable than would be single lesions. Such multiple lesions would also be more likely to lead to cancer than would the relatively rare events produced by metabolic oxidants. Lowering the dose of ionizing radiation only changes the frequency of occurrence and not the clustering of free radicals. Although the radicals produced by radiation undoubtedly can damage DNA and other important biologic targets, chronic low-level irradiation may trigger an adaptive response, whereby protective and repair processes are increased. However, such an adaptive response would have to elicit a broadly protective stress response to explain its purported life span extending effects.

### B. Caloric Intake

The most dramatic extensions of life spans in animals have been achieved by dietary manipulations, which have achieved increases in maximum life spans of up to 50% by the imposition of "caloric restriction." The efficacy of this regime in prolonging life spans has led to the suggestion that a decrease in metabolic rate by lowering calorie intake lowered the rate of oxidant damage. However, caloric restriction leads to lowered weight, and this must be taken into account. Conflicting results have been reported when the antiaging effects of food restriction were analyzed in terms of specific metabolic rates (Chapter 24).[11]

The term caloric restriction can be misleading. *Ad libitum* fed rats should be viewed as obese,[11] and much of the literature on caloric restriction should perhaps be viewed as literature on "body mass vs. health and longevity." As is well known for humans, high calorie intake correlates with obesity and predisposes to disease and premature death. Risk of premature death in humans is

**General Perspectives**

not confined to the grossly overweight but exists over a broad range of body weights.[47] Body weights of rodents are a direct function of caloric intake.[11] Human mortality risk (inverse of mean life span) and the inverse of rodent life spans (mean and maximum) increase as smooth, non-linear functions of body mass. In the context of an oxidant damage mechanism of aging, the correlation of weight with life span could be explained if one could establish a link between calorie consumption and oxidant damage. One such link has recently been suggested by studies of tissue-cultured bovine aortic endothelial cells. High glucose in the culture medium correlated with high oxidant production and protein glycation, while also leading to increases in several manifestations of diabetic disease.[18]

## C. MUTATIONS IN RODENTS AND OTHER AGING MODELS

Aging model studies have identified a number of mutated species that outlive wild-type organisms (Chapters 1 and 3), consistent with the view that natural selection has not prioritized maximum life spans (Table 5.5). Most of these studies lend support to a significant role of oxidants in determining the rate of aging. Probably most relevant to human aging are studies of mutant mice that showed a strong correlation between increased resistance to oxidative stress and longevity.[52] Of interest is the observation that the mutated mice achieved increased life spans in the absence of caloric restriction effects.

Genetic manipulations of antioxidant systems have made important contributions to an understanding of oxidative damage in aging and disease (Table 5.6). The importance of MnSOD is underscored by the failure of an overexpression of CuSOD to compensate for its deletion. A diversity of hydrogen peroxide scavenging systems is suggested by the minimal impact of a deletion of the major glutathione peroxidase enzyme GSH-Px1. Similarly, a multiplicity of systems of controlling adverse effects of transition metal release are suggested by the minimal effects of metallothionein deletion or overexpression. The dramatic impact of deleting heme oxygenase HO-1 argues for a potent antioxidant role of this enzyme.

## D. GENETIC DISORDERS

In principle, an analysis of genetic disorders, involving defects in antioxidant defenses, could resolve whether the free radical theory of aging is valid. Unfortunately, although several examples of human mutations in one or more of the protective enzymes or antioxidant vitamins are known, including individuals with reduced levels of catalase, glutathione (GSH), or GSH peroxidase, or defective vitamin E absorption, none of these show signs of accelerated aging that can be clearly distinguished from pathology.[13]

Conversely, several genetic syndromes that exhibit some features of accelerated aging, the so-called "segmental progeroid syndromes," exhibit damage that could be consistent with increased free radical damage. These include Down's Syndrome, Ataxia Telangiectasia, Cockayne's syndrome, and, possibly, Werner's syndrome, which exhibits genetic instability, yet has normal levels of SOD, GSH peroxidase, and radiation-induced repair. The molecular basis of these "geromimetic" diseases may involve accelerated rates of chromosome breakage, and the diseases frequently exhibit radiation sensitivity that suggests a high susceptibility to free radical damage.

Fanconi's anemia also exhibits characteristics of oxygen radical damage, but the pathology is so severe that death occurs in infancy or early childhood, and the only sign of an age-related effect is increased malignancy. Devastating genetic deficiencies of this type do not permit normal development, and, frequently, disease-linked premature death precludes distinguishing symptoms of accelerated aging from pathology.

Down's syndrome is characterized by a 50% elevation of the copper–zinc SOD (Cu/Zn-SOD) above normal levels of this enzyme. This increased protection against

---

**TABLE 5.5**
**Significant Extensions of Life Spans in Aging Models (Excluding Caloric Restriction)**

| Model | Intervention (Life Span Increase vs. Wild Type) | Relevance to Oxidant Theory of Aging |
|---|---|---|
| *Caenorhabditis elegans*[48] | Nutritional SOD/catalase mimetics (44%) | Directly supportive |
| *Caenorhabditis elegans*[49] | Genetic (more than double) | Insulin response effect may be supportive |
| *Drosophila melanogaster*[50] | Genetic (Indy gene) (87–89%) | Caloric restriction analogies may provide indirect support |
| *Saccharomyces cerevisiae*[51] | Genetic (glucose mutants) (40%) | Caloric restriction analogies may provide indirect support |
| Mouse[52] | Genetic (deleted signal tranducer) (30%) | Supportive: lessens some $H_2O_2$ effects, e.g., apoptosis induction |
| Mouse[3] | Chronic exposure to ionizing radiation (20%) | Not supportive |

## TABLE 5.6
## Effects of Altered Expression of Antioxidant-Related Genes in Mice

| Genetic Manipulation | Observed Effect(s) | Implications for Aging |
|---|---|---|
| MnSOD deletion | Myocardial injury, neurodegeneration, anemia, fatty liver, severe mitochondrial dysfunction, neonatal death 1–20 days after birth[53] | Supports major role of superoxide in shortening at least mean life spans |
| MnSOD: decreased expression in SOD2-/+ heterozygotes | Mitochondrial dysfunction, no compensatory upregulation of other antioxidant enzymes[54] | Supports role of superoxide in mitochondrial damage |
| CuSOD overexpression | Cannot compensate for lethality of MnSOD loss[55] | Strongly suggests oxidative damage, occurs primarily in mitochondria |
| GSH-Px1 deletion | Conflicting reports,[56,57] including a surprisingly consistent (but statistically insignificant) increase in life expectancy[58] | Unclear |
| Metallothionine overexpression or deletion | Conflicting reports[59,60] | Unclear |
| Heme oxygenase 1 deletion | No viable offspring, progressive chronic inflammation[61] | HO-1 is likely to modulate at least mean life spans |

superoxide radicals fails to confer antiaging benefits; indeed, some aspects of age-related pathology, notably senile dementia, are accelerated. There are conflicting reports concerning altered brain lipofuscin accumulation, an *in vivo* index of lipid peroxidation. *In vitro* lipid peroxidation appears to be accelerated. In patients with trisomy 21, red blood cells are abnormally sensitive to lysis in the presence of paraquat, a molecule that causes an increase in cellular superoxide production. Fibroblasts from trisomy 21 sufferers exhibit enhanced lipid peroxidation. The apparent increase in free radical damage under conditions of elevated Cu/Zn-SOD levels can be explained by a concurrent decrease in manganese SOD (Mn-SOD) levels in some tissues. In patients with monosomy 21 ("21q- or anti-Down's syndrome"), cells possess only 50% of normal Cu/Zn-SOD but normal Mn-SOD. While these patients suffer from developmental abnormalities and poor survival, there are no obvious signs of accelerated aging.

The difficulty of dissociating disease from aging has been a major obstacle in exploiting genetic analyses to resolve aging mechanisms. Nevertheless, the available genetic data suggest that biological concentrations of catalase, vitamin E, glutathione, and glutathione peroxidase are probably not critical determinants of rates of aging, while levels of superoxide dismutase and the overall oxygen radical defensive capacity (viz., whole body radiation) may be important factors.

### E. SPECIES COMPARISONS

It has often been argued that metabolic rate determines life span, based largely on the observation that the specific metabolic rate (rate of calorie consumption per unit weight) of groups of animal species is inversely propor-

## TABLE 5.7
## Metabolic Potentials of Animals (Lifetime Energy Expenditure in Kilocalories per Gram of Body Weight)

| Species | Metabolic Potential |
|---|---|
| Dipteran flies | ~25 |
| Most nonprimate mammals | ~200 |
| Humans | ~800 |
| Birds, bats | ~1200 |

tional to the maximum life span within different groups of animals (Table 5.7).[8] This correlation can also be expressed as follows: the lifetime oxygen (or energy) allocation per unit weight is fixed for any animal in that group. Alternatively, it can be argued that oxygen is inherently toxic, and the cumulative effects of oxygen toxicity correlate directly with the total amount of oxygen that an animal is exposed to during its lifetime. The distinction between these considerations may be important in that metabolic rate is determined largely by normal mitochondrial metabolism, whereas oxygen toxicity is likely determined by abnormal metabolism.

Studies of free radical production by isolated mitochondria suggest that altered free radical production may be a more significant factor in aging than impaired antioxidant protection. Sohal and his collaborators have shown that hydrogen peroxide production by mitochondria correlates with age and species life span, and that antioxidant enzymes do not exhibit consistent correlations. They suggest that altered free radical production is a key determinant of maximum attainable life span, and that variations in antioxidant enzymes exert relatively little influence.[62-64]

## F. CIGARETTE SMOKING

Statistics on cigarette smoking may be regarded as the largest database on a human carcinogen. Several carcinogenic agents, including tobacco-specific carcinogens, have been identified in cigarette smoke.[65] Smoking is associated with numerous other human pathologies including cardiovascular and respiratory diseases (Chapters 16, 17, and 18).[66] Relatively little is known about mechanisms responsible for the association of cardiovascular disease with smoking. Cigarette smoke is known to contain free radicals, including nitric oxide in the gas phase and relatively persistent polymeric free radicals in the tar phase. The possible involvement of free radicals in the adverse health effects of cigarette smoke has long intrigued investigators and has been cited to explain chemical modifications of biomolecules (Chapter 18).[67] The positive correlation of smoking with major life-shortening diseases implies that cigarette smoking reduces mean life expectancy but not necessarily maximum life span.

One of the factors complicating an assessment of the involvement of free radicals in the damaging effects of tobacco smoke is antioxidant depletion.[68,69] Tobacco smoke contains acrolein, which reacts with thiols like glutathione. It inactivates enzymes with a function that involves thiol groups.[70] A role of thiol destruction in tobacco smoke carcinogenesis has been proposed in the "thiol defense hypothesis."[71] Depletion of free radical scavengers may not be a radical-induced process, but it could exacerbate free radical damage. Thus, antioxidant depletion could be sufficient to explain an accelerated formation of free radical damage markers.

## G. DIETARY ANTIOXIDANTS

Dietary supplementation with antioxidants in a variety of animal aging models has shown that, while the mean life span is increased, there is no significant increase in the maximum life span.[13] Side effects of antioxidants must also be considered, e.g., possible caloric restriction effects associated with foul-smelling food additives like 2-mercaptoethylamine. Another potential problem with dietary antioxidants in whole animals is a possible lack of specificity. For example, antioxidants are claimed to enhance immune function, and some phenolic antioxidants like 2(3)-tert-butyl-4-hydroxy-anisole (BHA) induce quinone reductase and UDP-glucuronyl transferase systems and, hence, may accelerate the removal of potentially mutagenic chemicals that could affect aging by nonradical mechanisms. Two other synthetic antioxidants, BHT and ethoxyquin, have been shown to induce hepatic enzymes. Thus, increased life span that correlates with

antioxidant feeding may be due to effects other than free radical quenching.

The accumulation of lipofuscin is increased with vitamin E deficiency and decreased with dietary antioxidants. However, there is no convincing evidence that lipofuscin accumulation correlates with age-related cell loss. On the contrary, among the brain stem nuclei, the inferior olive has high lipofuscin levels and shows no age-related cell loss. Moreover, there are examples of cells that accumulate lipofuscin but do not exhibit a decline in function. This was illustrated by cell sorting experiments with cultured fibroblasts, which separated populations of cells on the basis of fluorescence intensity, and which showed that fluorescent cells suffered no loss of proliferative potential.

Thus, the evidence for antioxidant-mediated life span extension in vertebrates does not argue persuasively for a causal role of free radicals in aging. This conclusion is generally acknowledged by advocates of the free radical theory, and they offer two general explanations that could sustain the basic concept: (1) the principal site of damage (e.g., mitochondrial DNA) is not protected by exogenous antioxidants, or (2) the endogenous defense system is regulated to maintain a fixed overall level of protection so that simple attempts to increase protection by some dietary manipulation of one or a few antioxidants fail because of compensatory decreases in endogenous defenses ("compensatory downregulation"). However, the fact that lipofuscin accumulation can be reduced by feeding antioxidants in the absence of life span extension, and that tumor incidence can be reduced by feeding exogenous antioxidants, suggests that antioxidants can effectively reduce the rate of free radical damage in some tissues without undue compensatory downregulation. While compensatory downregulation would seem consistent with concepts of homeostasis, at least one study of heterozygous knockout mice with decreased MnSOD activities has shown no compensatory upregulation of other antioxidant enzymes (Table 5.6).

## H. CELLULAR SENESCENCE

The finite doubling potential or "replicative life span" of cultured mammalian cells has been considered by some investigators to be a meaningful aging model. In this model, normal cells usually cease to grow after about 50 population doublings, regardless of how long they are maintained in culture, a phenomenon referred to as "cellular senescence." After growth ceases, the cells remain metabolically active. Because the replicative life span of these cells is only dependent on the number of cell doublings, and not of time in culture, it cannot be a simple function of oxidants associated with metabolism. Most studies of replicative senescence have utilized fibroblasts, and reports from several laboratories have suggested a correlation of the doubling potential with the age of the cell donor.[72] However, a recent analysis has concluded that the replicative life span of cultured human fibroblasts does not correlate with donor age.[73] Despite questions about its relevance to animal aging, the cultured fibroblast model has provided some useful information about cellular effects of oxidative stress. Of particular interest are studies of the effects of oxygen tension on cell growth. Although there were conflicting initial reports,[74–76] more recent studies have found an increase in the maximum number of doublings by lowering the oxygen concentration below ambient levels.[77,78] Elevated oxygen concentrations curtail both growth rates and the doubling limit, an effect that is not prevented by vitamin E supplementation.[76]

## 1. Exercise

Exercise can exert substantial effects on metabolic rate and tissue oxygen concentrations. The analysis of exercise effects requires that the intensity and duration of exercise as well as prior training be taken into account. Increased free radical production appears to occur with exhaustive exercise in untrained animals, and positive adaptations, conferring increased resistance to oxidants, are observed in endurance-trained animals. Strenuous exercise in untrained animals is associated with mechanical damage associated with muscle lengthening contractions.[79] Membrane damage under these conditions is inferred from a release of intracellular enzymes into the bloodstream. The muscle damage elicits phagocytic activity with attendant oxidant production. The phagocytic activity increases for several days before returning to resting-state levels.[79]

There is also an immediate oxidative stress associated with exhaustive exercise, which appears to be associated primarily with the consumption of ATP. An increase in the ADP/ATP ratio, followed by AMP formation and its degradation to hypoxanthine, culminates in superoxide and hydrogen peroxide production by xanthine oxidase. The importance of this pathway is demonstrated by the pronounced suppression of oxidative damage markers elicited by treating animals with allopurinol, an inhibitor of xanthine oxidase.[80] In contrast to the marked effects of tissue injury and xanthine oxidase activation, mitochondrial superoxide formation seems to be a relatively insignificant component of the damage induced by strenuous exercise in untrained animals.[80] A minimal role of mitochondrial oxidants is entirely consistent with an analysis of the mechanism of superoxide production (see sections on mitochondria and metabolism).

Evidence for an association of oxidants with exhaustive exercise has been inferred primarily from studies of postexercise tissue samples. For example, a widely cited study of untrained rats running to exhaustion on a treadmill showed increased lipid peroxidation products in liver and skeletal muscle homogenates and in mitochondrial fractions.[81] The same study showed that plasma glutathione disulfide levels increase significantly with exhaustive exercise and that vitamin E deficiency appears to be associated with a marked decrease in endurance capacity, prompting the authors to suggest that lipid peroxidation may play a role in muscle fatigue upon exhaustion. Human studies have also suggested that oxidative stress can arise under some exercise conditions. "Extreme exercise" in healthy young men, comprising a 30-day period of 8 to 11 h of intense exercise per day for six days per week, was reported to correlate with an increase in urinary excretion of oxidatively damaged DNA bases[82] (however, recall that this damage marker is difficult to interpret; see section on DNA damage).

The numerous reports suggesting adverse effects of exercise-induced oxidant damage have aroused skepticism, given the dramatic epidemiologic evidence of exercise benefits in humans (Chapter 24) and experimental evidence of benefits in chronically exercised animals. Female rats that have exercised throughout life have longer average life spans than their sedentary counterparts (although they do not have longer maximum life spans).[83] The mechanism of this average life-span extension is unresolved but does not seem to involve caloric restriction, because exercising animals increased their food intake to maintain peak body weights similar to those of sedentary animals. One possible factor in the beneficial effects of exercise training could be an increase the activity of antioxidant enzymes, which has been reported to occur in some animal models as well as in humans.[84] However, other work has shown no induction of SODs, glutathione peroxidase, or change in glutathione status in human muscle tissues by exercise training.[85] Arguably, the detection of certain oxidative damage markers in tissue samples does not necessarily imply adverse health effects.

## VII. DISEASE

Many diseases that shorten mean life spans are aggravated by free radical processes (see, for example, Pryor[86]). Some of the support for free radical involvement in major diseases is outlined in the following sections.

### A. CANCER

The process whereby a normal cell assumes the uncontrolled growth characteristics of a cancer cell clearly involves altered genes. As summarized earlier, oxidants damage genes and, therefore, play a role in carcinogenesis. Genes can also be altered by nonoxidants, including alkylating and glycating agents. The literature abounds with reports of exogenous chemical and biological carcinogens. Epidemiology has established an increased risk of cancer for smoking, excessive consumption of alcohol, high levels of ionizing radiation, exposure to asbestos particles, and many other exogenous factors. Of these agents, ionizing radiation (a complete carcinogen) most directly demonstrates the carcinogenic power of free radicals. An important distinction between cancer and most other diseases or aging is that cancer generally begins with a single transformed cell. Considering the number of mitotic cells in higher animals, neoplasia is an extremely rare event, testifying to the effectiveness of the various systems that protect the genome.

Some exogenous carcinogens like vinyl chloride can be categorized by "fingerprinting" damaged DNA in terms of specific patterns of base lesions. When this fingerprinting technique was originally developed, studies of unexposed DNA showed significant background levels of base lesions expected for exogenous chemicals. It thus became apparent that analogous, or perhaps identical, chemicals were being generated endogenously.[87] Subsequent studies have shown that a host of mutagenic metabolites and molecular decomposition products arise spontaneously *in vivo*. These endogenous mutagens include not only ROS, but also aldehydes, alkylating agents, glycoxidation products, estrogen metabolites, RNS, chlorinating reagents, and δ-aminolevulinic acid (the latter is involved in heme synthesis).[87] Oxidative damage mechanisms as well as direct effects on DNA are involved in the mutagenic activity of most of these agents.

While there is abundant information about risk factors for increased cancer risk, relatively little is known about effective strategies for reducing cancer susceptibility. However, epidemiology has shown that a diet rich in fruits and vegetables correlates with lowered cancer incidence. For many years, it was widely assumed that the beneficial effects of fruits and vegetables could be attributed to their content of antioxidants like vitamins C and E and carotenoids. However, it has become evident that the most potent anticancer agents are not simple antioxidants. By fractionating the highly effective cancer-preventing cruciferous vegetables, certain chemicals (e.g., isothiocyanates) have been found to have the most potent cancer-fighting activities. Acting directly, some of these chemicals are alkylating agents. However, at appropriate doses, they induce the formation of protective enzymes,[88] a phenomenon referred to as chemoprotection. As noted in the introductory section, some adaptations to stress confer broad protection against subsequent stress, which could possibly be the mechanism of action of chemoprotection.

Aging has been referred to as the most potent human carcinogen.[89] The relationship between aging and cancer is complex; for example, neuroblastomas, many leukemias, and hormone-dependent tumors have radically different age patterns than most sarcomas and many carcinomas. The incidence of most tumors, in a large variety of species, rises dramatically with age. This rise in cancer incidence with age may be attributed to several factors:[90]

1. Long-term carcinogen exposure increases the risk of initiation, e.g., lung cancer incidence reflects duration of smoking rather than chronological age.
2. The prolonged period required for one malignant cell to multiply and develop into a detectable tumor exists.
3. Aging increases the risk because of, e.g.,:
   - A reduction in natural killer or other immune surveillance function
   - An increase in the activation of procarcinogens
   - Epigenetic instability

The pathogenesis of neoplasia has been thought to be a multistage process comprising initiation, promotion, and progression. Initiation is an irreversible alteration of gene(s) that can result from oxidative damage; promotion is a reversible process of cell proliferation that can be modulated by oxidative damage; and progression is an irreversible process of de-differentiation characterized by aneuploidy and clonal variation. The process involves many variables, including tissue type, hormonal influence, proliferative rates, DNA repair capacity, environmental carcinogen exposure, viral infection, immune surveillance, and genotype. It has been suggested that cell division greatly increases the potential for nonrepairable DNA damage and cancer.[91] Particularly at high doses, many carcinogenic agents may exert their effects indirectly by killing cells, which stimulates the growth of new cells with concomitant DNA damage.

## B. CARDIOVASCULAR DISEASE

A role of oxidant-induced damage in cardiovascular disease is supported by epidemiology studies suggesting that dietary antioxidant intake, particularly of vitamin E, correlates with a decreased incidence of heart disease (Chapter 2).[92] Transient or prolonged occlusion of blood vessels can set the stage of the oxidant damage associated with reoxygenation of ischemic tissue, a phenomenon referred to as reperfusion injury. Among the factors involved in this injury is the activity of xanthine oxidase, a source of superoxide and hydrogen peroxide. Studies with isolated heart cells (cardiomyocytes) have shown a delayed effect of hydrogen peroxide at exposure levels that caused apoptosis of only a fraction of the treated cells. This delayed response comprised a substantial increase in cell volume, consistent with the enlargement of the heart observed in some pathologies (hypertrophy).[93]

## C. NEURODEGENERATIVE DISORDERS

As noted earlier, the brain would seem to be highly vulnerable to oxidative stress. Comprising about 2% of body weight, it is responsible for about 20% of the whole-body oxygen consumption at rest. The brain contains relatively high concentrations of easily oxidizable polyunsaturated fatty acids and catecholamines, has high concentrations of $H_2O_2$-producing monoamine oxidase, has areas with relatively low concentrations of several antioxidants, including glutathione, vitamin E, superoxide dismutase, glutathione peroxidase, and, especially, catalase (the brain has about 10% of the catalase activity found in the liver) as well as regions of high iron content.[94,95] Interestingly, this apparent vulnerability to oxidative stress does not cause an obvious accelerated aging of the brain relative to other tissues. As noted in the section on metabolism, this observation argues against the notion that the rate of aging is a simple function of metabolic rate. Nevertheless, aging is often associated with neurodegenerative diseases, including Alzheimer's disease, Parkinson's disease, cerebrovascular diseases (e.g., stroke), and demyelinating diseases (Chapters 7 and 8).

Alzheimer's disease (AD) is associated with oxidative stress as seen in an elevation of several markers of oxidative damage in autopsy tissues. For example, carbonyl groups visualized by a staining procedure were found to be localized in neurofribrillary tangles (NFT), cytoplasms of neurons, and nuclei of both glia and neurons.[96] Other damage markers found in AD include advanced glycation end products, nitrated tyrosine, and lipid peroxidation products.[97] Iron accumulation was observed in the senile plaques and NFT that are the hallmarks of the disease, and, importantly, the iron was found to be accessible to low molecular weight reducing and oxidizing agents such that it could act catalytically to produce hydroxyl radicals and oxidants derived from them.[98,99] AD may be somewhat responsive to antioxidant therapy — dietary supplementation with α-tocopherol appears to exert a slight benefit in delaying manifestations of the disease, including time of entry into nursing homes and ability to perform routine daily tasks.[100]

Parkinson's disease (PD) is a strongly age-related pathology marked by a selective and progressive loss of pigmented catecholaminergic, particularly dopaminergic, neurons of the *substantia nigra pars compacta*. Analyses of autopsied brain tissue have shown that advanced PD is associated with oxidized lipids, proteins, and DNA consistent with a role of oxidative damage in the terminal stages of the disease.[101] A proposed model for the onset and progression of the disease suggests that oxidation of dopamine, followed by the formation of cysteine conjugates and further oxidation products, leads to mitochondrial dysfunction and cell death. The model is supported by studies of the inhibitory effects of these oxidation products on the energy metabolism of isolated brain mitochondria.[102]

Other neurodegenerative diseases significantly associated with oxidative stress include multiple sclerosis,

Creutzfeldt–Jacob disease and meningoencephalitis. All of these diseases are associated with significant increases in the specific and persistent lipid peroxidation marker $F_2$-isoprostane.[103]

## D. AUTOIMMUNE DISEASE

Because tolerance to self appears to require an active thymic role in the production of T suppressor cells as well as in the deletion of self-reactive clones, one might expect thymic involution and age-related immune dysregulation to result in an age-related increase in autoimmune disease. Further, the emergence of altered self antigens through persistent viral infection, posttranslational modifications, somatic mutation, or even postmaturational development of newly expressed genes could also bring about an increased number of autoimmune reactions with age (Chapter 15).

Autoimmune phenomena like autoantibodies, glomerulonephritis, periarteritis, and probably some classes of senile amyloid increase with age. A major role of autoimmune pathology is played in the aging of some rodents, but not all strains and species seem to be affected. Hence, while some workers have hypothesized that autoimmunity is the major aging process and likened senescence to a chronic graft-versus-host reaction, the evidence is not consistent. In humans, most known or suspected autoimmune diseases, including rheumatoid arthritis, have a peak incidence in middle age, and it is difficult to assess the significance of autoimmune phenomena in human aging. Among rodents, in strains that are clearly autoimmune susceptible, such as NZB mice, antioxidant feeding has produced a delay in disease onset and life span extension. If autoimmunity is a major aging process and not a secondary pathology, then antioxidant feeding may be said to delay the accelerated aging that has been reported in NZB mice.

Several other diseases seem to involve free radicals in their etiology, including atherosclerosis, emphysema, arthritis, cirrhosis, and diabetes.[86]

## VIII CONCLUSIONS AND PROSPECTS

The impact of free radical damage on aging remains to be resolved (Table 5.8). For example, failure of conventional antioxidant supplementation to significantly extend the maximum life spans of mammals has argued against the free radical theory of aging. This failure is particularly troubling, because the administration of antioxidants has clear-cut effects in reducing the extent of lipofuscin accumulation, an indication that the antioxidants protect some cellular targets susceptible to oxidation. However, recent studies have shown that antioxidants interfere with cell signaling, which suggests possible adverse effects that could offset the intended

**General Perspectives**

**TABLE 5.8**
**Are Oxidants the Cause of Aging? Some Pro and Con Arguments**

| Pro | Con |
|---|---|
| Caloric restriction appears to reduce oxidative stress | Vitamin C, a superb free radical scavenger, is not synthesized by long-lived primates |
| Life span extension in mutants is often associated with stress resistance | Chronic ionizing radiation at low doses does not shorten life span, may increase it |
| Knockout mice lacking MnSOD or HO-1 have severely restricted survival | Dietary supplementation with natural antioxidants (vitamins C and E) does not extend life spans |
| Enzyme mimetics extend maximum life spans in some aging models | Tissue comparisons, e.g., brain vs. muscle, seem incompatible with simple oxidant and antioxidant models of aging |
| Certain drugs, PBN, (-) deprenyl, possibly acting as antioxidants have been claimed to extend life spans[104,105] | Exercise, often claimed to increase oxidant stress, exerts beneficial effects at least on mean life spans |

benefits. Antioxidant administration can confer anticarcinogenic and other health benefits, which could affect mean, but not necessarily maximum, life spans. Recent evidence on caloric balance indicates that at least some subcellular fractions prepared from lean rodents generate fewer free radicals and are endowed with enhanced antioxidant protection relative to fractions from ad libitum-fed (obese) animals. Work on mitochondria is consistent with radical-mediated damage of mitochondrial DNA and an age-dependent dysfunction that correlates with increased free radical production and decreased antioxidant capacity. The mitochondrial data generally seem consistent with the free radical theory of aging.

Among the most exciting recent developments in aging research is the successful cloning of animals by transferring nuclei of somatic cells into enucleated oocytes. The cloning of mice to six generations, producing mice that show no signs of accelerated aging in a variety of tests, argues that the nuclear genome of at least some somatic cells suffers very little damage, oxidative or otherwise, during the life span of the animal.[106] On the other hand, age-dependent damage to the general population of the somatic cells used for cloning is indicated by the increasing difficulty of producing viable offspring with sequential transplantations. The nuclear transfer technique is likely to be improved and will undoubtedly produce important new insights into aging mechanisms in the near future.

Another promising development is the discovery that certain enzyme-inducing constituents of vegetables and presumably other plants can confer substantial protection against cancer in animal models and, judging by epidemiologic data, in humans. Specific induction of the antioxidant response element (ARE) by phenolic antioxidants, peroxides, and a variety of natural compounds, suggests strategies for life span extension based on manipulations of the stress response, which may be more successful in protecting against oxidative stress than dietary supplementation with simple antioxidants has

been. Novel approaches to antioxidant enhancement based on synthetic agents like manganese salen compounds, nitroxides, and nitrones also offer promise. Of particular interest is the role that these agents may play in cell signaling as opposed to their obvious direct effects in quenching free radicals. Body weight is receiving more attention as a risk factor in aging. Recent refinements in analyzing human mortality and morbidity statistics have led to a steady lowering of the apparent "optimal body mass index." Some recent analyses that eliminated confounding risk factors like smoking and preexisting disease suggest that human mean life spans correlate with changes in caloric balance qualitatively similarly but quantitatively less dramatically than do those of rodents in the classic "caloric restriction" scenario.

## ACKNOWLEDGMENT

I thank T. Prolla for excellent feedback during the preparation of the manuscript. This work was supported by the Department of Energy under Contract DE-AC03–76SF00098.

## REFERENCES

1. Kirkwood, T.B. and Austad, S.N., Why do we age?, *Nature*, 408, 233, 2000.
2. Binger, C.A.L., Faulkner, J.M., and Moore, R.L., Oxygen poisoning in mammals, *J. Exp. Med.*, 45, 849, 1927.
3. Speit, G. et al., Induction of heme oxygenase-1 and adaptive protection against the induction of DNA damage after hyperbaric oxygen treatment, *Carcinogenesis*, 21, 1795, 2000.
4. Caratero, A. et al., Effect of continuous gamma irradiation at a very low dose on the life span of mice, *Gerontology*, 44, 272, 1998.
5. Harman, D., Aging: A Theory Based on Free Radical and Radiation Chemistry, University of California Radiation Laboratory Report UCRL-3078, 1955.

6. Harman, D., Aging: A theory based on free radical and radiation chemistry, *J. Gerontol.,* 11, 298, 1956.

7. Commoner, B., Townsend, J., and Pake, G.E., Free radicals in biological materials, *Nature,* 174, 689, 1954.

8. Truby, F.K. and Goldzieher, J.W., Electron spin resonance investigations of rat liver and rat hepatoma, *Nature,* 182, 1371, 1958.

9. Beckman, K.B. and Ames, B.N., The free radical theory of aging matures, *Physiol. Rev.,* 78, 547, 1998.

10. Finkel, T. and Holbrook, N.J., Oxidants, oxidative stress and the biology of ageing, *Nature,* 408, 239, 2000.

11. Sohal R.J. and Weindruch, R., Oxidative Stress, Caloric Restriction, and Aging, *Science,* 273, 59, 1996.

12. Knight J.A., The biochemistry of aging, *Adv. Clin. Chem.,* 35, 1, 2001.

13. Mehlhorn, R.J. and Cole, G., The free radical theory of aging: A critical review, *Adv. Free Rad. Biol. Med.,* 1, 165, 1985.

14. Vasques-Vicar, J., Kalyanaraman, B., and Kennedy, M.C., Mitochondrial aconitase is a source of hydroxyl radical, *J. Biol. Chem.,* 275, 14064, 2000.

15. Deamer, D.W. et al., The alloxan-dialuric acid cycle and the generation of hydrogen peroxide, *Physiol. Chem. Phys.,* 3, 426, 1971.

16. Lee, S.H., Oe, T., and Blair, I.A., Vitamin C-induced decomposition of lipid hydroperoxides to endogenous genotoxins, *Science,* 292, 2083, 2001.

17. Jiang, Q. et al., $\gamma$-Tocopherol and its major metabolite, in contrast to $\alpha$-tocopherol, inhibit cyclooxygenase activity in macrophages and epithelial cells, *Proc. Natl. Acad. Sci. USA,* 97, 11494, 2000.

18. Rolfe, D.F.S. and Brown, G., Cellular energy utilization and molecular origin of standard metabolic rate in mammals, *Physiol. Rev.,* 77, 731, 1997.

19. Nishikawa T. et al., Normalized mitochondrial superoxide production blocks three pathways of hyperglycemic damage, *Nature,* 404, 787, 2000.

20. Vidal-Puig, A.J. et al., Energy metabolism in uncoupling protein 3 gene knockout mice, *J. Biol. Chem.,* 275, 16258, 2000.

21. Skulachev, V., Uncoupling: New approaches to an old problem of bioenergetics, *Biochim. Biophys. Acta,* 1363, 100, 1998.

22. Earnshaw, W.C., Apoptosis: A cellular poison cupboard, *Nature,* 397, 387, 1999.

23. Esposito, L.A. et al., Mitochondrial disease in mouse results in increased oxidative stress, *Proc. Natl. Acad. Sci. USA,* 96, 4820, 1999.

24. Brown, G.C., Nitric oxide and mitochondrial respiration, *Biochim. Biophys. Acta,* 1411, 351, 1999.

25. Poderoso, J.J. et al., Nitric oxide inhibits electron transfer and increases superoxide radical formation in rat heart mitochondria and submitochondrial particles, *Arch. Biochem. Biophys.,* 328, 85, 1996.

26. Chance, B., Sies, H., and Boveris, A., Hydrogen peroxide metabolism in mammalian organs, *Physiol. Rev.,* 59, 527, 1979.

27. Cadenas, E. and Davies, K.J.A., Mitochondrial free radical production, oxidative stress, and aging, *Free Rad. Biol. Med.,* 29, 222, 2000.

28. Davies, K.J., The broad spectrum of responses to oxidants in proliferating cells: A new paradigm of oxidative stress, *IUBMB Life,* 48, 41, 1999.

29. Halliwell, B., Oxygen and nitrogen are pro-carcinogens. Damage to DNA by reactive oxygen, chlorine and nitrogen species: Measurement, mechanism and the effects of nutrition, *Mutat. Res.,* 443, 37, 1999.

30. Anson, M.R., Hudson, E., and Bohr, V.A., Mitochondrial endogenous oxidative damage has been overestimated, *FASEB J.,* 14, 355, 2000.

31. Beckman, K.B. et al., A simpler, more robust method for the analysis of 8-oxoguanine in DNA, *Free Rad. Biol. Med.,* 29, 357, 2000.

32. Higuchi, Y. and Linn, S., Purification of all forms of HeLa cell mitochondrial DNA and assessment of damage to it caused by hydrogen peroxide treatment of mitochondria or cells, *J. Biol. Chem.,* 270, 7950, 1995.

33. Marnett, L., Oxyradicals and DNA damage, *Carcinogenesis,* 21, 361, 2000.

34. Holliday, R., The inheritance of epigenetic defects, *Science,* 238, 163, 1987.

35. Cannon, S.V., Cummings, A., and Teebor, G.W., 5-Hydroxymethylcytosine DNA glycosylase activity in mammalian tissue, *Biochem. Biophys. Res. Commun.,* 151, 1173, 1988.

36. Stadtman, E.R., Protein modification in aging, *J. Gerontol.,* 43, B112, 1988.

37. Dreyfus, J.C., Kahn, A., and Schapira, F., Posttranslational modification of enzymes, *Curr. Topics Cell R,,* 14, 243, 1978.

38. Rothstein, M., Recent developments in the age-related alteration of enzymes: A review, *Mech. Ageing Dev.,* 6, 241, 1977.

39. Sitte, N., Merker, K., and Grune, T., Proteasome-dependent degradation of oxidized proteins in MRC-5 fibroblasts, *FEBS Lett.,* 440, 399, 1998.

40. Grune, T., Reinheckel, T., and Davies, K.J.A., Degradation of oxidized proteins in mammalian cells, *FASEB J.,* 11, 526, 1997.

41. Sitte, N. et al., Protein oxidation and degradation during cellular senescence of human BJ fibroblasts: Part I — Effects of proliferative scenescence, *FASEB J.,* 14, 2495, 2000.

42. Treff, W.M., Involutionsmuster des Nucleus dentatus cerebelli, in *Altern,* Platt, D., Ed., Schattauer, Stuttgart, 1974, 37.

43. Yin, D., Biochemical basis of lipofuscin, ceroid, and age pigment-like fluorophores, *Free Rad. Biol. Med.,* 21, 871, 1996.

44. Paris, F. et al., Endothelial apoptosis as the primary lesion initiating intestinal radiation damage in mice, *Science,* 293, 293, 2001.

45. Fleming, J.E. et al., Is cell aging caused by respiration-dependent injury to the mitochondrial genome?, *Gerontology,* 28, 44, 1982.

**General Perspectives**

46. Thompson, R.C. and Mahaffey, J.A., Life-Span Radiation Effects Studies in Animals: What can They Tell Us?, Proceedings of The Twenty Second Hanford Life Sciences Symposium, DOE report CONF-830951, pp. 107–116, 1983.

47. Stevens J. et al., The effect of age on the association between body-mass index and mortality, *N. Engl. J. Med.*, 338, 1, 1998.

48. Melov, S. et al., Extension of life-span with superoxide/catalase mimetics, *Science*, 289, 1567, 2000.

49. Lin, K. et al., daf-16: An HNF-3/forkhead family member that can function to double the life-span of *Caenorhabditis elegans*, *Science*, 278, 1319, 1997.

50. Rogina, B. et al., Extended life-span conferred by cotransporter gene mutations in *Drosophila*, *Science*, 290, 2137, 2000.

51. Lin, S.-J., Defossez, P.-A., and Guarente, L., Requirement of NAD and SIR2 for life-span extension by caloric restriction in *Saccharomyces cerevisiae*, *Science*, 289, 2126, 2000.

52. Migliaccio, E. et al., The p66$^{shc}$ adaptor protein controls oxidative stress response and life span in mammals, *Nature*, 402, 309, 1999.

53. McMillan-Crow, L.A. and Cruthirds, V.D., Manganese superoxide dismutase in disease, *Free Rad. Res.*, 34, 325, 2001.

54. Van Remmen, H. et al., Characterization of the antioxidant status of the heterozygous manganese superoxide dismutase knockout mouse, *Arch. Biochem. Biophys.*, 363, 91, 1999.

55. Copin, J.-C., Gasche, Y., and Chan, P.H., Overexpression of copper/zinc superoxide dismutase does not prevent neonatal lethality in mutant mice that lack manganese superoxide dismutase, *Free Rad. Biol. Med.*, 28, 1571, 2000.

56. Ho, Y.-S. et al., Mice deficient in cellular glutathione peroxidase develop normally and show no increased sensitivity to hyperoxia, *J. Biol. Chem.*, 272, 16644, 1997.

57. Esposito, L.A. et al., Mitochondrial oxidative stress in mice lacking the glutathione peroxidase-1 gene, *Free Rad. Biol. Med.*, 28, 754, 2000.

58. Spector, A. et al., The effect of aging on glutathione-1 knockout mice — resistance of the lens to oxidative stress, *Exp. Eye Res.*, 72, 533, 2001.

59. Sun, X., Zhou, Z., and Kang, Y.J., Attenuation of doxorubicin chronic toxicity in metallothionein-overexpressing transgenic mouse heart, *Cancer Res.*, 61, 3382, 2001.

60. Conrad, C.C. et al., Using MT$^{-/-}$ mice to study metallothionein and oxidative stress, *Free Rad. Biol. Med.*, 28, 447, 2000.

61. Agarwal, A. and Nick, H.S., Renal response to tissue injury: Lessons from heme oxygenase-1 gene ablation and expression, *J. Am. Soc. Nephrol.*, 11, 965, 2000.

62. Sohal, R.S., Svensson, I., and Brunk, U.T., Hydrogen peroxide production by liver mitochondria in different species, *Mech. Ageing Dev.*, 53, 209, 1990.

63. Sohal, R.S., Arnold, L.A., and Sohal, B.H., Age-related changes in antioxidant enzymes and prooxidant generation in tissues of the rat with special reference to parameters in two insect species, *Free Rad. Biol. Med.*, 9, 495, 1990.

64. Sohal, R.S. and Brunk, U.T., Mitochondrial production of pro-oxidants and cellular senescence, *Mutat. Res.*, 275, 295, 1992.

65. Hoffman, D. et al., Tobacco-specific N-nitrosamines and Areca-derived N-nitrosamines: Chemistry, biochemistry, carcinogenicity, and relevance to humans, *J. Toxicol. Environ. Health*, 41, 1, 1996.

66. Jinot, J. and Bayard, S., Respiratory health effects of exposure to environmental tobacco smoke, *Reviews Environ. Health*, 11, 89, 1996.

67. Eiserich, J.P. et al., Dietary antioxidants and cigarette smoke-induced biomolecular damage: A complex interaction, *Am. J. Clin. Nutr.*, 62 (suppl), 1490S, 1995.

68. Frei, B. et al., Gas phase oxidants of cigarette smoke induce lipid peroxidation and changes in lipoprotein properties in human blood plasma. Protective effects of ascorbic acid, *Biochem. J.*, 277, 133, 1991.

69. Maranzana, A. and Mehlhorn, R.J., Loss of glutathione, ascorbate recycling and free radical scavenging in human erythrocytes exposed to filtered cigarette smoke, *Arch. Biochem. Biophys.*, 350, 169, 1998.

70. Powell, G.M. and Green, G.M., Cigarette smoke — a proposed metabolic lesion in alveolar macrophages, *Biochem. Pharmacol.*, 21, 1785, 1972.

71. Fenner, M.L. and Braven, J., The mechanism of carcinogenesis by tobacco smoke. Further experimental evidence and a prediction from the thiol-defence hypothesis, *Br. J. Cancer*, 22, 474, 1968.

72. Campisi, J., Replicative senescence: An old lives' tale?, *Cell*, 84, 497, 1996.

73. Cristofalo, V.J. et al., Relationship between donor age and the replicative lifespan of human cells in culture: A reevaluation, *Proc. Natl. Acad. Sci. USA*, 95, 10,614, 1998.

74. Balin, A.K. et al., The effect of oxygen and vitamin E on the lifespan of human diploid cells *in vitro*, *J. Cell Biol.*, 74, 58, 1977.

75. Packer, L. and Fuehr, K., Low oxygen concentration extends the life span of cultured normal human diploid cells, *Nature*, 267, 423, 1977.

76. Balin, A.K., Testing the free radical theory of aging, in *Testing the Theories of Aging*, Adelman, R.C. and Roth, G.S., Eds., CRC Press, Boca Raton, 1982, p. 137.

77. Chen, Q. et al., Oxidative DNA damage and senescence of human diploid fibroblast cells, *Proc. Natl. Acad. Sci. USA*, 92, 4337, 1995.

78. Yuan, H., Kaneko, T., and Marsuo, M., Relevance of oxidative stress to the limited replicative capacity of cultured human diploid cells: The limit of cumulative population doublings increases under low concentrations of oxygen and decreases in response to aminotriazole, *Mech. Ageing Dev.*, 81, 159, 1995.

79. McArdle, A. and Jackson, M.J., Exercise, oxidative stress and aging, *J. Anat.*, 197, 541, 2000.

80. Vina, J. et al., Mechanism of free radical production in exhaustive exercise in humans and rats; role of xanthine oxidase and protection by allopurinol, *IUBMB Life,* 49, 539, 2000.

81. Davies, K.J. et al., Free radicals and tissue damage produced by exercise, *Biochem. Biophys. Res. Comm.,* 107, 1198,1982.

82. Poulsen, H.E., Loft, S., and Vistisen, K., Extreme exercise and DNA modification, *J. Sports Medicine,* 14, 343, 1996.

83. Holloszy, J.O., Exercise increases average longevity of female rats despite increased food intake and no growth retardation, *J. Gerontol.,* 48, B97, 1993.

84. Ji, L.L., Antioxidants and oxidative stress in exercise, *Proc. Soc. Exp. Biol. Med.,* 222, 283, 1999.

85. Tonkonogi, M., Walsh, B., Svensson, M., and Sahlin, K., Mitochondrial function and antioxidant defence in human muscle: Effects of endurance training and oxidative stress, *J. Physiol.,* 528, 379, 2000.

86. Pryor, W.A., The free radical theory of aging revisited: A critique and a suggested disease-specific theory, in *Modern Biological Theories of Aging,* Butler, R.L. et al., Eds., Raven Press, New York, 1986.

87. Burcham, P.C., Internal hazards: Baseline DNA damage by endogenous products of normal metabolism, *Mutation Research,* 443, 11, 1999.

88. Van Iersel, M.L.P.S., Verhagen, H., and van Bladeren, P.J., The role of biotransformation in dietary (anti)carcinogenesis, *Mutation Res.,* 443, 259, 1999.

89. DePinho, R.A., The age of cancer, *Nature,* 408, 248, 2000.

90. Ebbesen, P., Cancer and normal aging, *Mech. Ageing Dev.,* 25, 269, 1984.

91. Ames, B.N. and Gold, L.S., Chemical carcinogens: Too many rodent carcinogens, *Proc. Natl. Acad. Sci. USA,* 87, 7772, 1990.

92. Jha, P. et al., The antioxidant vitamins and cardiovascular disease. A critical review of epidemiologic and clinical trial data, *Ann. Intern. Med.,* 123, 860, 1995.

93. Chen, Q.M. et al., Hydrogen peroxide dose dependent induction of cell death or hypertrophy in cardiomyocytes, *Arch. Biochem. Biophys.,* 373, 242, 2000.

94. Floyd, R.A., Antioxidants, oxidative stress, and degenerative neurological disorders, *Proc. Soc. Exper. Biol. Med.,* 222, 236, 1999.

95. Calabrese, V., Bates, T.E., and Stella, A.M., NO synthase and NO-dependent signal pathways in brain aging and neurodegenerative disorders: The role of oxidant/antioxidant balance, *Neurochem. Res.,* 25, 1315, 2000.

96. Harman, D., Alzheimer's disease: Role of aging in pathogenesis, *Ann. N.Y. Acad. Sci.,* 959, 384, 2002.

97. Smith, M.A. et al., Oxidative stress in Alzheimer's disease, *Biochim. Biophys. Acta,* 1502, 139, 2000.

98. Smith, M.A. et al., Iron accumulation in Alzheimer's disease is a source of redox-generated free radicals, *Proc. Natl. Acad. Sci. USA,* 94, 9866, 1997.

99. Sayre, L.M. et al., In situ oxidative catalysis by neurofibrillary tangles and senile plaques in Alzheimer's disease: A central role for bound transition metals, *J. Neurochem.,* 74, 270, 2000.

100. Sano, M. et al., A controlled trial of selegiline, alpha-tocopherol, or both as a treatment for Alzheimer's disease. The Alzheimer's Disease Cooperative Study, *N. Engl.J. Med.,* 336, 1216, 1997.

101. Bolton, J.L. et al., Role of quinones in toxicology, *Chem. Res. Toxicol.,* 13, 135, 2000.

102. Xin, W. et al., Oxidative metabolites of 5-S-cysteinyl-norepinephrine are irreversible inhibitors of mitochondrial complex I and the a-ketoglutarate dehydrogenase and pyruvate dehydrogenase complexes: Possible implications for neurodegenerative brain disorders, *Chem. Res. Toxicol.,* 13, 749, 2000.

103. Greco, A., Minghetti, L., and Levi, G., Isoprostanes, novel markers of oxidative injury, help in understanding the pathogenesis of neurodegenerative diseases, *Neurochem. Res.,* 25, 1357, 2000.

104. Saito, K., Yoshioka, H., and Cutler, R.G., A spin trap, N-tert-butyl-alpha-phenylnitrone extends the life span of mice, *Biosci. Biotechnol., and Biochem.,* 62, 792, 1998.

105. Kitani, K. et al., Common properties for propargylamines of enhancing superoxide dismutase and catalase activities in the dopaminergic system in the rat: Implications for the life prolonging effect of (-) deprenyl, *J. Neural Transmission. Suppl.,* 60, 139, 2000.

106. Wakayama, T. et al., Cloning of mice to six generations, *Nature,* 407, 318, 2000.

**General Perspectives**

# 6 Theories of Life Span and Aging

*James R. Carey*
University of California, Davis

## CONTENTS

## I.  INTRODUCTION

The gerontologist George Sacher[1] noted that there are different theoretical frameworks within which biological research on aging can be considered, each of which is associated with what he termed a "primitive" question. The best-known is the conventional aging-oriented approach designed to address the question *"Why do we age?"* Sacher notes that this is an ontogenetic issue, and the attack on it is guided by research paradigms concerned with molecular, genetic, and physiological processes. The basic experimental approach to this question involves the comparison of specific functions or structures in old and young animals, such as mice. However, he observed that this far-reaching correspondence between laboratory rodents and humans concealed a paradox: if these two species are so similar in molecular, cellular, and physiological makeup, the data on the ontogeny of aging in rodents brings us little closer to understanding why a rodent grows as old in 2 years as humans do in 70 years. Thus, the second approach to aging research flows from this paradox and is designed to address the longevity-oriented question: *"Why do we live as long as we do?"* This question cannot be answered within the framework of ontogenetic research on aging but rather requires the development of an evolutionary-comparative paradigm. But, then a third approach to aging research is death-oriented and is designed to answer the question *"Why do we die?"* which is a problem separate from aging and longevity; there is no necessary relation between aging

0-8493-0948-4/03/$0.00+$1.50

and dying. None of these questions by themselves adequately frame the field of aging research, because, although there is obvious overlap, they each possess different conceptual centers — evolutionary (life span), physiological (aging), and death (stochastic).

Distinguishing between longevity- and aging-oriented paradigms is important for several reasons. First, longevity and aging are fundamentally different concepts and, like the mouse aging and life span example, they cannot be completely understood without also considering one in the context of the other. Second, unlike the evolutionary theory of senescence, which is based solely on individual natural selection, these theories include processes of sexual selection and kin selection, bringing life history theory more fully to bear on questions concerned with the latter portion of the life cycle. Third, extending the scope of aging-related theory allows consideration of behaviors that are characteristic of younger and older individuals, including divisions of labor and intergenerational transfers. Because mortality factors unrelated to aging (accidents, acute diseases, socioeconomic factors) can be included in a more general conceptual framework, approaches that extend beyond conventional research strategies can be employed to understand how and why people live as long as they do, why they age, and why they die.

This chapter is organized around two of the concepts that constitute "life's finitude" as originally described by Sacher.[2] The next section is on the theory of longevity and includes conceptual and empirical background information. The last section on the theories of aging includes the theory on why all eukaryotes senesce and die, followed by an overview of the main theories of aging at molecular, cellular, systemic, and evolutionary levels. A brief conclusion on the importance of theory in aging science ends the chapter.

## II. THEORY OF LIFE SPAN

*Life span* is an evolved life history characteristic of an organism that refers to the duration of its entire life course. Application of the concept is straightforward at both individual and cohort levels and specifies the period between birth and death for the former (individual) and to the average length of life or life expectancy at birth for the latter (including real and synthetic cohorts). However, life span applied to a population or a species requires a modifier to avoid ambiguity.[3] Maximum observed life span is the highest verified age at death, possibly limited to a particular population or time period. The overall highest verified age for a species is also called its record life span. The theoretical highest attainable age is known as either maximum potential life span, maximum theoretical life span, or species-specific life span. Depending on the context, maximum life span can refer to either the observed or the potential maximum.

### A. Conceptual Aspects of Life Span

The life span concept is relevant only to species in which an individual exits — an entity circumscribed by distinct birth and death processes.[4] Thus, the concept does not apply to bacteria that reproduce by binary fission, to plant species that reproduce by cloning, or to modular organisms with iterated growth such as coral or honeybee colonies. When a single reproductive event occurs at the end of the life course that results in the death of the individual, then life span is linked deterministically to the species' natural history. This occurs with the seed set of annual plants (grasses), in drone (male) honeybees as a consequence of the mechanical damage caused by mating, in many mayfly species when a female's abdomen ruptures to release her eggs after she drops into a lake or stream, and in anadromous salmon that die shortly after spawning. Life span can be considered indeterminate for species (including humans) that are capable of repeated (iteroparous) reproduction. That life span is indeterminate in many species is consistent with everything that is known about the lack of cut-off points in biology — all evidence suggests that species do not have an internal clock for terminating life.

Changes that occur in organisms that enter resting states such as dormancy, hibernation, and aestivation reduce mortality rates and, thus, increase longevity. This also occurs when individuals are subjected to caloric restriction or when their reproductive efforts are reduced. A species' life course may consist of many phases, such as infant, juvenile, and pre- and postreproductive period, and therefore, a change in overall life span will correspond to a commensurate change in the duration of one or more of the stages. When environmental conditions are drastically improved, such as for animals kept in zoos or laboratories or for contemporary humans, mortality rates usually decrease, and thus, longevity increases. Whereas earlier stages such as prereproductive periods are evolved life history traits, the added segment(s) arising at the end of the life course are by-products of selection for robustness or durability at earlier stages and are thus not evolved traits, per se. Rather, these additional segments are due to "ecological release" and are referred to as "post-Darwinian" age classes.

### B. Comparative Demography of Life Span

The literature contains descriptions of only a small number of life span correlates, including the well-known relationship between life span and body mass and relative brain size[5–7] and the observation that animals that possess armor (e.g., beetles, turtles) or capability of flight (e.g., birds, bats) are often long-lived.[8] But, major inconsistencies exist within even this small set of correlates. For example, there are several exceptions regarding the rela-

**TABLE 6.1**
**The Two General Categories of Factors that Favor the Evolution of Extended Life Span and Examples of Species Within Each**

| Category | Examples |
|---|---|
| Environmentally selected | Tortoises, sea turtles, deep-water tube worms, tuatara; birds, beetles, *Heliconius*; butterflies, tree-hole mosquitoes |
| Socially selected | Elephants, killer whales, dolphins, most primates (including humans), naked mole rats, micro-bats (brown bat, vampire), parrots, hornbills, albatross, termite, ant and bee queens, tsetse flies |

*Source:* From Carey, J.R. and Judge D.S., *Pop. Dev. Rev.*, 27, 411, 2001.

tionship of extended longevity and large body size (e.g., bats are generally small, but most species are long-lived), and this positive relationship may be absent or reversed within orders. Likewise, the observation that flight ability and extended longevity are correlated does not provide any insight into why within-group (e.g., birds) differences in life span exist, and it does not account for the variation in longevity in insects, where adults of the majority of species can fly.

An alternative approach for identifying broad correlates of longevity emerged from an examination of several large-scale databases containing the maximum recorded life spans of vertebrate and invertebrate species.[9–11] Many long-lived species across a wide taxonomic spectrum appeared to cluster within one of two general ecology and life history criteria: (1) species that live in either unpredictable environments (e.g., deserts) or where food resources are scarce (e.g., caves, deep water); or (2) species that exhibit extended parental care and live in groups with complex or advanced social behavior. This led to a classification system regarding the life span determinants of species with extended longevity (Table 6.1) that we believe is general and applies to a wide range of invertebrate and vertebrate species:

1. *Environmentally Selected.* This category includes animals whose life histories evolved under conditions in which food is scarce, and where resource availability is uncertain or environmental conditions are predictably adverse part of the time. The extended longevity of animals in this category evolved through natural selection.
2. *Socially Selected.* This category includes species that exhibit extensive parental investment,

extensive parental care, and eusociality. The extended longevity of animals in this category results from natural, sexual, and kin selection.

This classification system places the relationship of life span and two conventional correlates, relative brain size and flight capability, in the context of life history. That is, brain size is related to the size of the social group and the degree of sociality,[12] which, in turn, is linked to extended life span. And, intensive parental care is linked to flight capability in birds and bats, which, in turn, is also linked to extended life span. No system of classification is perfect, and the one presented in Table 6.1 is no exception — the categories are not mutually exclusive, and therefore, some species could be placed in one or both categories. However, this classification serves as a practical and heuristic tool for considering the evolution of animal life spans. In particular, this system provides a general background for closer examination of specific human attributes, including the evolution of life span, and sets the stage for addressing questions of process.

## C. LIFE SPAN PATTERNS: HUMANS AS PRIMATES

Estimates based on regressions of anthropoid primate subfamilies or limited to extant apes indicate a major increase in longevity between *Homo habilis* (52 to 56 years) to *H. erectus* (60 to 63 years) occurring roughly 1.7 to 2 million years ago. Predicted life spans for small-bodied *H. sapiens* is 66 to 72 years.[13] From a catarrhine (Old World monkeys and apes) comparison group, a life span of 91 years is predicted when contemporary human data are excluded from the predictive equation. For early hominids to live as long or longer than predicted was probably extremely rare; the important point is that the basic Old World primate design resulted in an organism with the potential to survive long beyond a contemporary mother's ability to give birth. This suggests that postmenopausal survival is not an artifact of modern life style but may have originated between 1 and 2 million years ago, coincident with the radiation of hominids out of Africa.

The general regression equation expresses the relationship of longevity to body and brain mass when 20 Old World anthropoid primate genera are the comparative group. The predicted longevity is 91 years for a 50 kg primate with a brain mass of 1250 gm (conservative values for humans) when case deletion regressions methods are employed (each prediction is generated from the equation, excluding the species in question) and 72 years when humans are included within the predictive equation (Table 6.2). Using the predictive equation presented in Table 6.2, a typical Old World primate with the body size and brain size of *Homo sapiens* can be expected to live between 72 and 91 years with good nutrition and protection from predation.

**General Perspectives**

**TABLE 6.2**
**Estimates of Longevity for Fossil Hominids, Based on Hominoid Body Size Relationships Range from 42–44 Years for *Australopithecus* to 50 Years for *Homo erectus*. Incorporation of Brain Mass Increased Estimates for *H. habilis* from 43 Years to 52–56 Years and for *H. erectus* from 50 Years to 60–63 Years**

| Hominid Species | Life Span (Years)[a] | Incremental Change |
|---|---|---|
| Australopithicus afarensis | 46.6 | 8.4 |
| Homo habilis | 55.0 | 7.0 |
| H. erectus | 62.0 | 10.9 |
| H. sapiens (prehistorical) | 72.9 | 49.1 |
| H. sapiens (contemporary) | 122.0 | |

[a] When six genera of apes are used as the comparison group, the regression equation is as follows:

$$\log_{10}LS = 1.104 + 0.072(\log_{10}Mass) + 0.193(\log_{10}Brain)$$

This yields a predicted human longevity of 82.3 years

Source: Judge, D.S. and Carey, J.R., Post-reproductive life predicted by primate patterns, *J. Gerontol.*, 55, B201, 2000.

## D. THEORETICAL MODEL OF LONGEVITY EXTENSION IN SOCIAL SPECIES

Improved health and increased longevity in societies set in motion a self-perpetuating system of longevity extension. This positive feedback relationship is based on the demographic tenet that (all else being equal) increased survival from birth to sexual maturity reduces the number of children desired by parents.[14] Because of the reduced drain of childbearing and rearing, parents with fewer children remain healthier longer and raise healthier children with higher survival rates, which, in turn, fosters yet further reductions in fertility. Greater longevity of parents also increases the likelihood that they can contribute as grandparents to the fitness of their children and grandchildren. And, the self-reinforcing cycle continues.

In an essay on the formation of human capital, Abramovitz[15] noted that the decline in mortality rates during the early stages of industrialization in the United States was probably one of the forces behind the expansion of educational effort and growing mobility of people across space and between occupations. Whereas previous conditions of high mortality and crippling morbidity effectively reduced the prospective rewards to investment in education during the preindustrial period,[16] prolonged expectancy for working life span must have made people more ready to accept the risks and costs of seeking their fortunes in distant places and in new occupations. The

positive feedback of gains in longevity on future gains involves a complex interaction among the various stages of the life cycle with long-term societal implications in terms of the investment in human capital,[15] intergenerational relations,[17] and the synergism between technological and physiological improvements — so-called "technophysio evolution."[18] In other words, long-term investment in science and education provides the tools for extending longevity, which, in turn, makes more attractive the opportunity cost of long-term investments in individual education, and thus, helps humans gain progressively greater control over their environment, their health, and their overall quality of life. This concept is reinforced by the finding of Bloom and Canning[19] who noted that, whereas the positive correlation between health and income per capita is very well known in international development, the health-income correlation is partly explained by a causal link running the other way — from health to income. In other words, productivity, education, investment in physical capital, and what they term "demographic dividend" (positive changes in birth and death rates), are all self-reinforcing — these factors contribute to health, and better health (and greater longevity) contributes to their improvements.

## III. THEORIES OF AGING

In contrast to the life span of a species which is an evolved, positively selected life history trait analogous to a biological "warranty period," aging is a by-product of evolution analogous to the progressive deterioration of any system that increases the likelihood of failure. Inasmuch as the concept for designing a machine is fundamentally different from the concept for understanding its deterioration, it follows that a theoretical framework for understanding life span must necessarily be different from the theoretical framework for understanding aging. Therefore, the purpose of this section is to provide a brief overview of the main theories of aging. I begin with a description of the theory concerning why all sexually reproducing organisms undergo senescence,[20] followed by the more conventional theories ordered hierarchically — molecular, cellular, systemic, and evolutionary (Table 6.3).

## A. EVOLUTIONARY ORIGIN OF AGING

Bell[21] established the deep connection between the two invariants of life — birth and death — by demonstrating that protozoan lineages senesce as the result of an accumulated load of mutations. The senescence can be arrested by recombination of micronuclear DNA with that of another protozoa through conjugation. Conjugation (sex) results in new DNA and in the apoptotic-like destruction of old operational DNA in the macronucleus

**TABLE 6.3**
**Classification and Brief Description of Main Theories of Aging**

| Biological Level/Theory | Description |
| --- | --- |
| **Molecular** | |
| Codon restriction | Fidelity and accuracy of mRNA message translation is impaired with aging due to cell inability to decode the triple codons (bases) in mRNA molecules |
| Somatic mutation | Type of stochastic theory of aging that assumes that an accumulation of environmental insults eventually reaches a level incompatible with life, primarily because of genetic damage |
| Error catastrophe | Errors in information transfer due to alterations in RNA polymerase and tRNA synthetase may increase with age resulting in increased production of abnormal proteins |
| Gene regulation | Aging is caused by changes in the expression of genes regulating both development and aging |
| Dysdifferentiation | Gradual accumulation of random molecular damage impairs regulation of gene expression |
| **Cellular** | |
| Wear-and-tear | Intrinsic and extrinsic factors influence life span |
| Free radical accumulation | Oxidative metabolism produces free radicals that are highly reactive and, thus, damages DNA and proteins and, thus, degrades the system structure and function |
| Apoptosis | Process of systematically dismantling key cellular components as the outcome of a programmed intracellular cascade of genetically determined steps |
| **System** | |
| Rate-of-living | An old theory that assumes that there is a certain number of calories or heartbeats allotted to an individual and the faster these are used, the shorter the life |
| Neuroendocrine | Alterations in either the number or the sensitivity of various neuroendocrine receptors gives rise to homeostatic or homeodynamic changes that results in senescence |
| Immunologic | Immune system reduces its defenses against antigens and thus results in an increasing incidence of infections and autoimmune diseases |
| **Evolutionary** | |
| Antagonistic pleiotropy | Alleles that have beneficial effects on fitness at young ages can also have deleterious effects on fitness later in life |
| Mutation accumulation | The force of natural selection declines at older ages to a point where it has little impact on recurrent deleterious mutations with effects confined to later life |
| Disposable soma | Preferential allocation of energy resources for reproduction to the detriment of maintenance and survival of somatic cells |
| Source: | Yates, F.E., Theories of aging, in *Encyclopedia of Gerontology*, Vol. 2, Academic Press, San Diego, 1996. |

(see Figure 6.1). Thus, rejuvenation in the replicative DNA and senescence of operational DNA is promoted by sexual reproduction. When this is extended to multicellular organisms, sex and somatic senescence are inextricably linked.[20] In multicellular sexually reproducing organisms, the function of somatic cells (i.e., all cells constituting the individual besides the germ cells) is survival and function of the replicative DNA — the germ cells.[20] Prior to bacteria, the *somatic* DNA was the *germ line* DNA; prior to multicellular animals, the *somatic cell* was the *germ cell*. Like the macronuclei in the paramecia, the somatic cells senesce and die as a function of their mitotic task of ensuring the survival and development of the germ cells. The advent of sex in reproduction allowed exogenous repair of replicative DNA,[21] while in multicellular organisms, the replication errors of somatic growth and maintenance are segregated from the DNA, passed on to daughter cells, and discarded at the end of each generation. Senescence is built into the life history concept of all sexually reproducing organisms. Thus,

death rate can be altered by modifying senescence, but death can never be eliminated. This evolutionary argument concerning senescence cannot only be regarded as one of the most basic principles of biogerontology, but also as one of the fundamental canons in the emergence of all sexually reproducing organisms.

## B. MOLECULAR THEORIES OF AGING

## 1. Codon Restriction

A *codon* consists of the three contiguous bases in an mRNA molecule that specify the addition of a specific amino acid to the growing amino acid chain as the ribosome moves along the mRNA.[22] The process by which the enzyme RNA polymerase makes an RNA molecule complementary to the sequence of the template strand of the DNA is known as *transcription*. This transcribed message then undergoes a process known as *translation*, in which a ribosome makes a protein by "reading" the

**General Perspectives**

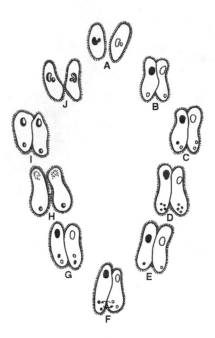

**FIGURE 6.1** Schematic of sexual reproduction in protozoa. (Redrawn from Clark, W.R., *Sex and the Origins of Death*, Oxford University Press, New York, 1996, p. 70.) (A) Two genetically different protozoa, each with a macronucleus (large oval in individual protozoa) and a micronucleus (small oval in protozoa). (B) The two protozoa fuse in the first step of conjugation, and the macronuclei and micronuclei move to opposite ends of the cell. (C) Each micronucleus divides once by meiosis, and (D) the daughter micronuclei each divide again, to produce four haploid micronuclei. (E) Three of the four haploid micronuclei disappear. (F) The remaining micronucleus divides once more, to produce two identical micronuclei, and then (G) the two conjugants exchange one micronucleus. (H) The two haploid micronuclei fuse to produce a single diploid micronucleus. (I) The new micronuclei each direct the production of a new macronucleus; the old macronucleus begins to disintegrate. (J) The two protozoa disengage, and the nuclei assume their starting positions in the cell. The exconjugates are now genetically identical to one another, but they are genetically different from either of the two starting cells. Each will go on to produce genetically identical daughters by simple fission.

sequence of an mRNA codon-by-codon. The *codon restriction theory of aging* is based on the hypothesis that the accuracy of translation is impaired with aging.[23] Experimental evidence in support of this theory includes a study by Ilan and Patel,[24] who reported alterations in mRNA and corresponding synthetases during the development period in a Tenebrionid beetle, from the free-living nematode *Turbatrx aceti*,[25] and from *Drosophila melanogaster*, where it was shown that efficiency of some sythetases of old flies is only 50% that of young flies.[26] The fetal rat liver contains six isoacceptors for tRNA[tyr] compared to the adult, which has only three.[27]

## 2. Gene Regulation

Although every cell contains the same genetic information as every other cell in an organism, a given cell or type of cell expresses only a distinct subset of its genes at any one time. For example, only red blood cells need to produce hemoglobins, and only cells in the retina need to produce light-sensitive proteins. Clearly, cells in the tongue do not need to produce hair. Genes need to be regulated to achieve this diversity of protein function during development. Gene regulation occurs primarily at the level of transcription within a cell, which transcribes only a specific set of genes and not others.

The gene regulation theory, proposed by Kanungo,[28] hypothesizes that senescence results from changes in the expression of genes after reproductive maturity is reached. Recent evidence that gene expression changes with age was reported by Helfand and coworkers[29] using a DNA sequence in *Drosophila* that interacts with "activator proteins" to stimulate gene expression (i.e., called enhancer traps) and which secretes into the cell a stainable marker substance whenever their gene is active.[30,31] The studies enabled the scientists to visualize the gene activity patterns in the antennae of the fly during its adult life. The results revealed that around 10 of 49 genes examined are constitutive, while the remainder showed a changing expression with age. Of these, slightly over half increase their activities from an initially low level, and around a quarter decrease their activity from an initially high level. The activity patterns of the genes were linked to chronological and not to physiological age.[32]

## 3. Dysdifferentiation

A theory of aging that is closely related to the gene regulation theory is what Cutler[33] coined the dysdifferentiation theory — the gradual accumulation of random molecular damages impairs the normal regulation of gene activity, potentially triggering a cascade of injurious consequences.[34] Whereas the gene regulation theory of aging (described above) hypothesizes that mistakes in protein synthesis (and hence, aging) are an outcome of mistakes in the transcription process, the dysdifferentiation theory hypothesizes that the mistakes in protein synthesis are due to molecular damage that causes aberrant expression of genes. Cutler suggested that the cell may synthesize proteins other than those characteristic of its differentiated state due to the lack of stringent gene control. The dysdiffentiation theory postulates that cells from old donors will synthesize more proteins that are uncharacteristic of its differentiated state than will similar cells from young donors. Ono and Cutler[35] showed that there was a twofold age-dependent increase in the amount of alpha- and beta-globin RNA synthesized by mouse brain and liver. However, other studies[36] reported no evidence of age-dependent increases in uncharacteristic globin gene expression in young and old cultures of normal human fibroblast cells. But despite the absence of supporting evidence, the dysdifferentiation theory of aging is of interest partly because it provides testable predictions, and partly because it allows for the stochastic modulation of a programmed process by means of a known genetic mechanism.

### 4. Error-Catastrophe Theory

The basic idea of this theory, first introduced in 1963,[37] is that the ability of a cell to produce its complement of functional proteins depends not only on the correct genetic specification of the various polypeptide sequences, but also on the competence of the protein-synthetic apparatus. This theory differs from DNA-based theories in that it postulates an error in information transfer at a site other than in the DNA. The summation of many small developing errors in the synthetic and enzymatic machinery of the cell mounts to a point beyond which conditions for cell life become impossible.

One of the virtues of this theory is that it is testable — proteins obtained from cells of old donors should exhibit a higher frequency of errors than would proteins taken from cells of younger donors. Rothstein[38] concluded that it is unlikely that transcriptional or translational errors are one of the mechanisms responsible for aging and senescence. However, his conclusions do not rule out errors in the replication of the DNA. Murray and Holliday[39] reported that selected DNA polymerases obtained from older cells have a greater error rate than polymerases taken from younger cells. And Srivastava and coworkers[40] reported that the degree of loss in fidelity is less in calorically restricted mice than in *ad libitum*-fed mice. These observations are thus consistent with the formal theory of error catastrophe.

### 5. Somatic Mutation and DNA Damage

The broad concept of this category of aging theories is that the integrity of the genome is the controlling factor in the aging process, and therefore, that mutations (changes in the polynucleotide sequence that remain uncorrected) or DNA damage (chemical alterations in the double-helical structure that are not fully repaired) underlie the aging process and determine its rate. Both the theory that aging occurs due to either somatic mutation or to DNA damage belongs to the class of stochastic theories of aging that assumes that an accumulation of environmental insults eventually reaches a level incompatible with life, primarily because of genetic damage. Somatic mutation occurs in day-to-day cell replication.[41] It is generally concluded that there is currently little evidence to support the notion that somatic mutation or DNA damage underlies the diseases and dysfunctions of aging.

### C. Cellular Theories of Aging

### 1. Wear and Tear

This theory postulates that ordinary insults and injuries of daily living accumulate, and thus, decrease the organism's efficiency (e.g., loss of teeth leads to starvation). Although there is little question that some wear and tear plays a role in mortality risk and individual life span, this theory has been rejected by most gerontologists as a more general explanation for aging, because (1) animals raised in protected environments still age; (2) many of the minor insults that accumulate are essentially time dependent and thus cannot logically serve as an underlying causal mechanism of the aging process; and (3) the theory is outdated inasmuch as cellular and molecular systems advanced toward more specific mechanisms.

### 2. Free Radical Accumulation

Initially, the mechanistic link between metabolism and aging was unknown. However in the mid-1950s, Harman[42] articulated a "free-radical theory" of aging, speculating that endogenous oxygen radicals were generated in cells and resulted in a pattern of cumulative damage.[43] The standard explanation for this damage is that it is the result of cellular damage caused by free radicals — any number of chemical species that are highly reactive, because they possess an odd number of electrons (Chapter 5). Molecules that have unpaired electrons are thermodynamically unstable because they seek to combine with another molecule to pair off their free electron. The theory postulates that the physiological decrements characteristic of age-related changes can be ascribed to the intracellular damage done by the various free radicals.[44] The net damage to various cell components (e.g., lipids, protein, carbohydrates, nucleic acids) is the result of different types of free radicals present, their production rate, the structural integrity of the cells, and the activity of the antioxidant defense systems present in the organism. Oxidative damage is a candidate for what has been referred to as "public" mechanisms of aging — common mechanisms among diverse organisms that are conserved over the course of evolution.[45] Structural and regulatory genes modulating genesis of free radicals have been identified in a wide range of species ranging from yeast, nematodes, and insects, to mice and humans. The oxidative damage theory is supported in caloric restriction (CR) studies that reveal that CR individuals lower the steady-state levels of oxidative stress and damage and extend life expectancy.

### 3. Mitotic Clock

The majority of cell types grown in laboratory cultures have a finite ability to proliferate. After a number of population doublings, the cell cultures enter the terminally nondividing state referred to as replicative senescence (Hayflick Limit[46]). Many investigations have established a link between aging *in vivo* (live animals) and the proliferative potential of cells in culture. For example, cell cultures derived from one of the longest-lived species of animals, the Galapagos tortoise, doubled up to 130 times, whereas cultures derived from mice with maximum life spans of 3 years were capable of doubling only 10 or fewer times.

One of the most compelling theories for explaining replicative senescence is that incomplete replication of specialized structures at the ends of chromosomes, called telomeres, accounts for the gradual loss of proliferation potential.[47] Telomeres are essential for proper chromosome structure and function, including complete replication of the genome — if organisms could not overcome the "end-replication" problem, they would fail to pass their complete genetic complement from generation to generation. The simplest theory for accounting for the Hayflick Limit is one in which permanent cell-cycle arrest is due to a checkpoint mechanism that interprets a critically short telomere length as damaged DNA and causes cells to exit the cell cycle. This telomere hypothesis of aging provides a molecular mechanism for counting cell divisions in the normal somatic cell. According to the telomere shortening model of cell senescence, to avoid telomere loss and eventual cell cycle arrest, it is necessary for cells to synthesize telomeric DNA by expressing the enzyme telomerase. Therefore, a prediction of this model and one that is borne out in recent studies[48–50] is that telomerase activity will be absent in normal somatic cells but present in germline cells and carcinoma cells. Although telomere length has historically been used as a means to predict the future life of cells, a new model frames the connection between telomere shortening and cellular senescence by introducing the concept of a stochastic and increasing probability of switching to the uncapped/noncycling state.[51]

It is widely believed that senescence evolved to limit the number of available cell divisions, and therefore, that it serves as a brake against the accumulation of the multiple mutations needed for a cell to become malignant (Chapter 4).[52] Results of new studies[53–54] support the notion that DNA repair pathways and antioxidant enzymes are adequate to protect against the accumulation of mutations and development of cancer in small organisms (mice) whose cells do not seem to have a cell division counting mechanism; replicative senescence has evolved in long-lived species to ensure that they would have the greatly increased protection that their longevity necessitated.

### 4. Apoptosis Theory

Apoptosis or programmed cell death, is a process of systematically dismantling key cellular components as the outcome of a programmed intracellular cascade of genetically determined steps (Chapter 4). It describes the orchestrated collapse of a cell and involves membrane dissolution, cell shrinkage, protein fragmentation, chromatin condensation, and DNA degradation followed by rapid engulfment of corpses by neighboring cells. It is an essential part of life at all levels of multicellular organisms; it is conserved as the way cells die from worms to mammals.[55] In Alzheimer's and in stroke, damaged cells die due to apoptosis. Evidence suggests that all cells of multicellular organisms carry within themselves the information necessary to bring about their own destruction, that this process has been evolutionarily conserved. However, the key to understanding this process in the context of aging concerns when and under what conditions the process can be invoked. In particular, the possibility exists that one major function of apoptosis is to serve as a precisely targeted defense mechanism against dysfunctional and potentially immortal (cancer) cells. More generally, apoptosis provides us with a controllable process that is clearly important in regulating cell number. Apoptosis and mitosis are controlled by gene-based signaling systems that can interact at the population and cell levels to bring about the net gain or net loss of cells in a particular tissue.[56]

### D. System Theories of Aging

### 1. Rate of Living

The rate of living hypothesis of aging states that the metabolic rate of a species is inversely proportional to its life expectancy. The original theory makes two predictions,[56] including the following: (1) there is a predetermined amount of metabolic energy available to the organism that can be expressed equally well in terms of oxygen consumption or kilocalories expended per life span, and when this energy is gone, the organism dies; and (2) there is an inverse relationship between metabolic rate and aging. Recent data[57] show that the metabolic potential does not stay at a constant value for different populations of a species. Long-lived strains spend about the same number of calories per day, yet may live significantly longer. Thus, during their lifetime, the long-lived strains expend around 40% more calories than the normal-lived strains.[56]

### 2. Neuroendocrine Control Theory of Aging

The endocrine system evolved to coordinate the activities of cells in different parts of the body by releasing hormones from major endocrine glands and organs into the bloodstream to be transported to other parts of the body, where they affect particular target cells. The major endocrine glands in mammals are the hypothalamus-pituitary-adrenal complex, the thyroid and parathyroid glands, the pancreas, the sex organs, and the adrenal glands.[58] The neurological system is inextricably linked to the endocrine system because of the central role of the hypothalamus in controlling the pituitary gland, which releases hormones such as antidiuretic hormone (ADH), follicle-stimulating hormone (FSH), luteinizing hormone (LH), thyroid-stimulating hormone (TSH), and growth hormone. Normal functioning requires that nervous and endocrine signals be synchronized and responsive to the needs of the many functions they regulate. However, some of the efficiency of the neurological and endocrine systems decreases with age, leading to decreased function and increased frequency of disease (Chapters 10 through 14). Thus, the neuroendocrine control theory of aging hypothesizes that the effectiveness of homeostatic adjustments declines with aging and leads to consequent failure of adaptive mechanisms, aging, and death.[59] Functional loss includes deterioration of reproductive organs and loss of fertility, diminished muscular strength, lesser ability to recover from stress, and impairment of cardiovascular and respiratory activities.

## 3. Immunological Theory of Aging

The cells and molecules responsible for immunity constitute the immune system, and their collective and coordinated response to the introduction of foreign substances is called the immune response.[60] This includes the response of both the innate (mechanisms that exist before infection) and adaptive (mechanisms that develop after infection; also called specific immunity). Innate and adaptive immune responses are components of an integrated system of host defense with important links: (1) the innate immune response influences the nature of the adaptive responses; and (2) adaptive immune responses use many of the effector mechanisms of innate immunity to eliminate microbes. The immunological theory of aging rests on three key findings:[56] (1) that there is a quantitative and qualitative decline in the ability of the immune system to produce antibodies, (2) that there are age-related changes in the ability to induce particular subsets of T cells and to produce different types of cell-mediated responses, and (3) that there is at least correlative evidence linking these alterations to the involution of the thymus (Chapter 15).

## E. Evolutionary Theories of Aging

The evolutionary biology of aging theory postulates that the force of natural selection will always decrease with age with either replacement-level or positive population growth. This is the theoretical basis for the *disposable soma theory* of aging, named for its analogy with disposable goods with a limiting warranty period.[61] The theory postulates that fitness is maximized at a level of investment in somatic maintenance, which is less than would be required for indefinite survival. The disposable soma theory serves as the conceptual foundation for two population genetic hypotheses of aging. The first is termed *negative pleiotropy* and is based on the concept that alleles that

have beneficial effects on one set of components of fitness also have deleterious effects on other components of fitness.[62,63] It argues for selection for forms of genes that confer beneficial effects early in life but ineffective selection to remove such forms of genes that are associated with deleterious effects late in life (postreproduction). The underlying concept is one of trade-offs — a beneficial effect at young ages may have a deleterious effect at older ages. The declining force of natural selection leads to a tendency for selection to fix alleles that have early beneficial effects but later deleterious effects. This biases evolution toward the production of vigorous young organisms and decrepit old organisms. Support for antagonistic pleiotropy is based on the consistent observation of a negative genetic correlation between early reproduction and longevity in both selection experiments[64] and in manipulative studies such as the one by Srgo and Partridge.[65,66] Support for antagonistic pleiotropy is also derived from the observation that longevity quantitative trait loci (QTLs) often appear to have antagonistic effects on life span in different environments and sexes.[67–69] Although antagonistic pleiotropy is an important possible mechanism for the evolution of aging, it probably plays a limited role in explaining the persistence of genetic variation in fitness components.[70] The second population genetic hypothesis concerning the evolution of senescence is the *mutation accumulation* that arises when the force of natural selection has declined to a point where it has little impact on recurrent deleterious mutations with effects confined to late life.[71–72] This theory hypothesizes that mutations producing a deleterious effect at postreproductive ages are not removed from populations by natural selection. Thus, mutations conferring late age-specific deleterious effects will accumulate in populations, causing aging and, ultimately, mortality.

## F. Conclusions

Theories in aging research, as with theories in all of science, must be considered as means and never as ends. They provide a plan for searching and, even if wrong, a specific theory regarding the mechanism(s) of aging can be useful, provided it is based on new observations and suggests an original path for scientific thought.[73] But, theory can be used, not only in a mechanistic (causal) context, but also to broaden the disciplinary scope as was done by Sacher[74] when he rejected the standard view of biological gerontology as synonymous with *"the biology of aging."* Sacher expanded the domain of the field by defining biological gerontology as *"the biology of the finitude of life, in its three aspects of longevity, aging, and death."* He noted that these three aspects constituted an *"irreducible triad,"* and that the ultimate goals of gerontology cannot be attained if attention is confined to one aspect to the exclusion of the others. The theoretical framework outlined by Sacher is important to aging science for at least two reasons. First, it provides conceptual coherence by linking the evolutionary by-product (aging) with the evolutionary "product" (longevity). Second, it helps to vertically integrate the various discoveries at different biological levels by providing hierarchical context from the molecular and cellular to the organ and organismal.

## REFERENCES

1. Sacher, G.A., Longevity and aging in vertebrate evolution, *Bioscience*, 28, 497, 1978.
2. Sacher, G.A., Longevity and aging in vertebrate evolution, *Bioscience*, 28, 497, 1978.
3. Goldwasser, L., The biodemography of life span: Resources, allocation and metabolism, *Trends Ecol. Evol.*, 16, 536, 2001.
4. Carey, J.R., Life span, in *Encyclopedia of Population*, Macmillan Reference USA, New York, 2002 (in press).
5. Austad, S.N., *Why We Age*, John Wiley & Sons, New York, 1997.
6. Comfort, A., The life span of animals, *Sci. Amer.*, 205, 108, 1961.
7. Hakeem, A. et al., Brain and life span in primates, in *Handbook of the Psychology of Aging*, Birren, J., Ed., Academic Press, New York, 1996.
8. Kirkwood, T.B.L., Comparative life spans of species: Why do species have the life spans they do?, *Am. J. Clin. Nutr.*, 55, 1191S, 1992.
9. Carey, J.R. and Judge, D.S., Life span extension in humans is self-reinforcing: A general theory of longevity, *Pop. Dev. Rev.*, 27, 411, 2001.
10. Carey, J.R., Insect biodemography, *Ann. Rev. Entomol.*, 43, 79, 2001.
11. Carey, J.R. and Judge, D.S., *Longevity Records: Life Spans of Mammals, Birds, Amphibians, Reptiles and Fish*, Odense University Press, Odense, Denmark, 2000.
12. Dunbar, R.I.M., Neocortex size as a constraint on group size in primates, *J. Human Evol.*, 20, 469, 1992.
13. Judge, D.S. and Carey, J.R., Post-reproductive life predicted by primate patterns, *J. Gerontol.*, 55, B201, 2000.
14. Ryder, N.B., Fertility, in *The Study of Population*, Hauser, P.M. and Duncan, O.D., Eds., University of Chicago Press, Chicago, IL, 1959.
15. Abramovitz, M., Manpower, capital, and technology, in *Thinking About Growth and Other Essays on Economic Growth and Welfare*, Abramovitz, M., Ed., Cambridge University Press, Cambridge, 1989.
16. Landes, D.S., *The Wealth and Poverty of Nations*, W.W. Norton & Co., New York, 1998.
17. Kaplan, H., A theory of fertility and parental investment in traditional and modern human societies, *Yearbook Phys. Anthrop.*, 39, 91, 1996.
18. Fogel, R.W., Economic growth, population theory, and physiology: The bearing of long-term processes on the making of economic policy, *Am. Econ. Rev.*, 84, 369, 1994.
19. Bloom, D.E. and Canning, D., The health and wealth of nations, *Science*, 287, 1207, 2000.
20. Clark, W.R., *Sex and the Origins of Death*, Oxford University Press, New York, 1996.
21. Bell, G., *Sex and Death in Protozoa*, Cambridge University Press, Cambridge, 1988.
22. Hawley, R.S. and Morik, C.A., *The Human Genome. A User's Guide*, IAP Harcourt Academic Press, San Diego, 1999.
23. Strehler, B., *Time, Cells and Aging*, 2nd ed., Academic Press, New York, 1977.
24. Ilan, J. and Patel, N., Mechanism of gene expression in *Tenebrio molitor, J. Biol. Chem.*, 245, 1275, 1970.
25. Reitz, M.S. and Sanadi, D.R., An aspect of translational control of protein synthesis in aging: Changes in the isoaccepting forms of tRNA in *Turbatrix aceti, Exp. Gerontol.*, 7, 119, 1972.
26. Hosback, M.A. and Kubli, E., Transfer RNA in aging *Drosophila*: Extent of aminoacylation, *Mech. Ageing Dev.*, 10, 131, 1979.
27. Yang, W.K., Isoaccepting transfer RNAs in mammalian differentiated cells and tumor tissues, *Cancer Res.*, 31, 639, 1971.
28. Kanungo, M.S., A model for ageing, *J. Theor. Biol.*, 53, 253, 1975.
29. Helfand, S.L. et al., Temporal patterns of gene expression in the antenna of the adult *Drosophila melanogaster, Genetics*, 140, 549, 1995.
30. O'Kane, K. and Gehring, W., Detection *in situ* of genomic regulatory elements in *Drosophila, Proc. Natl. Acad. Sci. USA*, 84, 9123, 1987.
31. Freeman, M., First, trap your enhancer, *Curr. Biol.*, 1, 378, 1991.
32. Rogina, B. and Helfand, S.L., Timing of expression of a gene in the adult *Drosopila* is regulated by mechanisms independent of temperature and metabolic rate, *Genetics*, 143, 1643, 1996.
33. Cutler, R.G., The dysdifferentiative hypothesis of mammalian aging and longevity, in *The Aging Brain: Cellular and Molecular Mechanisms of Aging in the Nervous System*, Giacobini, E. et al., Eds., Raven Press, New York, 1982.
34. Sharma, R., Theories of Aging, in *Physiological Basis of Aging and Geriatrics*, 2nd ed., Timiras, P.S., Ed., CRC Press, Boca Raton, FL, 1994.
35. Ono, T. and Cutler, R.G., Age-dependent relaxation of gene repression: Increase of endogenous murine leukemia virus-related and globin-related RNA in brain and liver of mice, *Proc. Natl. Acad. Sci. USA*, 75, 4431, 1978.
36. Kator, K. et al., Dysdifferentiative nature of aging: Passage number dependency of globin gene expression in normal human diploid cells grown in tissue culture, *Gerontology*, 31, 355, 1985.
37. Orgel, L.E., The maintenance of the accuracy of protein synthesis and its relevance to aging, *Proc. Natl. Acad. Sci. USA*, 49, 517, 1963.
38. Rothstein, M., Evidence for and against the error catastrophe hypothesis, in *Modern Biological Theories of Aging*, (Aging, Vol. 31), Warner, H.R. et al., Eds., Raven Press, New York, 1987.

39. Murray, V. and Holliday, R., Increased error frequency of DNA polymerases from senescent human fibroblasts, *J. Mol. Biol.*, 146, 55, 1981.

40. Srivastava, V.K.S. et al., Age-related changes in expression and activity of DNA polymerase alpha: Some effects of dietary restriction, *Mutation Res*., 295, 265, 1993.

41. Yates, F.E., Theories of aging, in *Encyclopedia of Gerontology*, Vol. 2, Academic Press, San Diego, CA, 1996.

42. Harman, D., Aging: A theory based on free radical and radiation chemistry, *J. Gerontol.*, 2, 298, 1957.

43. Finkel, T. and Holbrook, N.J., Oxidants, oxidative stress and the biology of ageing, *Nature*, 408, 239, 2000.

44. Sohal, R.S. and Weindruch, R., Oxidative stress, caloric restriction, and aging, *Science*, 273, 59, 1996.

45. Martin, G.M., Austad, S.N., and Johnson, T.E., Genetic analysis of ageing: Role of oxidative damage and environmental stresses, *Nature Genetics*, 13, 25, 1996.

46. Hayflick, L. and Moorhead, P.S., The limited *in vitro* lifetime of human diploid cell strains, *Exp. Cell Res.*, 25, 585, 1961.

47. Bodnar, A.G. et al., Extension of life-span by introduction of telomerase into normal human cells, *Science*, 279, 349, 1998.

48. Harle, C.B., Futcher, A.B., and Greider, C.W., Telomeres shorten during ageing of human fibroblasts, *Nature*, 345, 458, 1990.

49. Counter, C.M. et al., Telomere shortening associated with chromosome instability is arrested in immortal cells which express telomerase activity, *EMBO J.,* 11, 1921, 1992.

50. Blasco, M.A. et al., Telomere shortening and tumor formation by mouse cells lacking telomerase RNA, *Cell*, 91, 25, 1997.

51. Blackburn, E.H., Telomere states and cell fates, *Nature*, 408, 53, 2000.

52. Shay, J.W. and Wright, W.E., When do telomeres matter?, *Science*, 291, 839, 2001.

53. Tang, D.G. et al., Lack of replicative senescence in cultured rat oligodentrocyte precursor cells, *Science*, 291, 868, 2001.

54. Mathon, N.F. et al., Lack of replicative senescence in normal rodent glia, *Science*, 291, 872, 2001.

55. Miller, L.J. and Marx, J., Apoptosis, *Science*, 281, 1301, 1998.

56. Arking, R., *Biology of Aging*, 2nd ed., Sinauer Associates, Inc., Sunderland, MA, 1998.

57. Arking, R., Genetic analyses of aging processes, in Drosophila, *Exp. Aging Res.*, 14, 125, 1988.

58. Audesirk, T. and Audesirk, G., *Biology: Life on Earth*, Prentice Hall, Upper Saddle River, NJ, 1996.

59. Frolkis, V.V., *Aging and Life-Prolonging Processes*, Springer-Verlag, New York, 1982.

60. Abbas, A.K., Lichtman, A.H., and Pober, J. S., *Cellular and Molecular Immunology*, 4th ed., W.B. Saunders Co., Philadelphia, PA, 2000.

61. Kirkwood, T.B.L. and Rose, M.R., Evolution of senescence: Late survival sacrificed for reproduction, *Phil. Trans. Royal Soc. London*, 332, 15, 1991.

62. Williams, G.C., Pleiotropy, natural selection, and the evolution of senescence, *Evolution,* 11, 398, 1957.

63. Rose, M.R., *Evolutionary Biology of Aging*, Oxford University Press, Oxford, 1991.

64. Harshman, L., Life Span Extension of *Drosophila melanogaster*: Genetic Approaches, *Pop. Dev. Review: Suppl.*, (in review), 2003.

65. Sgro, C.M. and Partridge, L.A., Delayed wave of death from reproduction in Drosophila, *Science*, 286, 2521, 1999.

66. Sgro, C.M. and Partridge, L., Evolutionary responses of the life history of wild-caught *Drosophila melanogaster* to two standard methods of laboratory culture, *Am. Nat.*, 156, 353, 2000.

67. Nuzhdin, S.V. et al., Sex-specific quantitative trait loci affecting longevity, *Proc. Natl. Acad. Sci. USA*, 94, 9734, 1997.

68. Vieira, C.E. et al., Genotype-environment interaction for quantitative trait loci affecting life span in *Drosophila melanogaster*, *Genetics*, 154, 213, 2000.

69. Leips, J. and Mackay, T.F.C., Quantitative trait loci for life span in *Drosophila melanogaster*: Interactions with genetic background and larval density, *Genetics*, 155, 1773, 2000.

70. Curtsinger, J.W., Service, P.M., and Prout, T., Antagonistic pleiotropy, reversal of dominance, and genetic polymorphism, *Am. Nat.*, 144, 210, 1994.

71. Medawar, P.B., *The Uniqueness of the Individual*, Dover Publications, Inc., New York, 1981.

72. Charlesworth, B., *Evolution in Age-Structured Populations*, Cambridge University Press, Cambridge, 1994.

73. Cajal, S.R., *Advice for a Young Investigator, 1916*, Reprinted by MIT Press, Cambridge, MA, 1999.

74. Sacher, G.A., Theory in gerontology, *Ann. Rev. Gerontol. Geriatr.*, 1, 3, 1980.

**General Perspectives**

# Part II

## Systemic and Organismic Aging

# 7 The Nervous System: Structural and Biochemical Changes*

*Paola S. Timiras*
University of California, Berkeley

## CONTENTS

## I. PERSISTENCE OF BRAIN PLASTICITY IN OLD AGE

The human nervous system, its changes during development, adulthood, and old age, and its alterations with disease, represent one of the most challenging and intriguing studies of our time. Despite increasing and rapid advances in our understanding, we still have few direct answers to our many questions concerning the activities of its three major divisions, the peripheral, autonomic, and central nervous systems. Studies of the aging nervous system provide valuable information on aging processes and nervous functions. Comparison of the adult and aging brain, with or without neurologic and psychiatric diseases of old age, reveal specific, morphologic, biochemical, and functional differences under normal and diseased states.

We are witnessing, today, a remarkable shift in the way physiologists think about aging of the central nervous system (CNS). The early view formulated at the beginning of the 20th century, was of severe and inexorably progressive deterioration of structure, biochemistry, and function[1] (e.g., with old age, the dire threat of dementia with longer life expectancy). Currently, it is considered that normal brain function can persist into old age with adaptive, com-

---

* Illustrations by Dr. S. Oklund.

pensatory, and learning capabilities at all ages.[2-8] Extension in old age of mental and neurologic competence is a characteristic of successful aging (Chapter 3).

One of the outstanding properties of the CNS is its "plasticity," that is, its capacity to be "shaped or formed or influenced" by external and internal stimuli as well as to learn and to recover from damage. In response to stimuli, neurons can change their signals — transforming operations to adapt to new requirements. If the neuron has undergone injury or loss of synapses, adaptation is reflected in the sprouting of new dendrites, axons, and synapses as in "reactive synaptogenesis."[9] Glial cells, specifically astrocytes, in addition to increasing neuronal stability, may also be involved in regulating the number of neurons and synapses.[10,11] Plasticity, defined as long-term compensatory and adaptive change, is most effective during CNS development. During embryonal (in humans) and early postnatal (in rodents) development, a number of factors, including thyroid and sex hormones, when present at "critical periods" of ontogenesis, determine CNS differentiation and maturation. In the aging brain, plasticity may be less effective but not entirely lost.

Programs for prevention or treatment of CNS disorders are taking advantage of current advances in molecular biology and genetics to locate, and target for intervention, specific molecules responsible for CNS pathology and to identify susceptibility genes and risk factors for complex CNS diseases (Chapter 3). Embryonic neural transplants, cultured neurons, and stem cell implants to replace lost or impaired neurons are opening promising avenues of treatment.[12-14] For example, embryonic cells implanted in the area of basal ganglia survive, grow, and differentiate into neurons.[15] While restoration of normal function has not yet been achieved,[16,17] these, and similar studies, underline the brain's capability to sustain new cell growth and to promote cell differentiation even in old individuals, and raise new, exciting research prospects.[18] However, this approach may not be as effective as previously thought, and current results suggest that caution should be used.

Aging of the CNS will be presented in this and the following three chapters (Chapters 8, 9, and 10), first under normal aging conditions and, second, in a few diseases prevalent in old age. Location, mechanisms, and consequences of changes with aging are summarized in Table 7.1. They are multiple and involve all levels of organization (Figure 7.1, Box 7.1).

## II.   STRUCTURAL CHANGES

### A.   Brain Weight

Brain weight, size, and volume, and metabolism measured by imaging techniques, do not differ significantly in adult and older individuals without neurologic or mental disorders (Figure 7.2).[19-21] The slight (6 to 11%) decrease in

**TABLE 7.1**
**Changes with Aging in the Nervous System**

Locations
   Regional selectivity
   Neuronal loss/gliosis
   Reduced dendrites and dendritic spines
   Synaptic susceptibility
   Vascular lesions
Mechanisms
   Neurotransmitter imbalance
   Membrane alterations
   Metabolic disturbances
   Intra/intercellular degeneration
   Cell adhesion alterations
   Neurotropin changes
Consequences
   Sensory and motor decrements
   Sleep alterations and EEG changes
   Memory impairment
   Increased neurologic and psychiatric pathology
   Impaired homeostasis

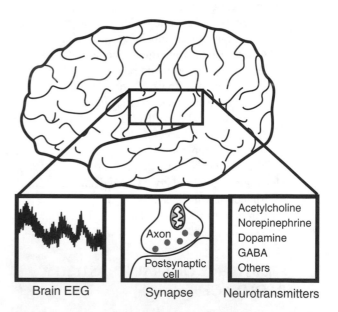

**FIGURE 7.1** Diagram of electrophysiological (brain EEG), functional (synapse), and chemical (neurotransmitters) sites for changes in the brain with aging.

brain weight registered in some healthy elderly individuals contrasts with the severe cortical atrophy (Figure 7.3) reported in many (but not all) patients with Alzheimer dementia.[21-23] The modest shrinkage of a healthy brain in later decades of life does not appear to result in any significant loss of mental ability.

The close association of a larger brain with enhanced functional and behavioral capabilities is well justified in evolutionary and phylogenetic terms.[24] Heavier brain weight (relative to body weight) and expansion of the

## Multiplicity of Changes in CNS Aging

**At the structural level**, aging effects vary depending on the CNS areas and regions, type of cell considered (neurons or glial cells, their number and branching), and site of action, e.g., the synapse, a key site for the reception and transmission of the nervous impulse.

**At the chemical level**, aging influences the metabolism and turnover of several neurotransmitters, substances carrying or transmitting the neural signal at the synapse, and of neurotransmitter receptors. Aging affects plasma and intracellular membranes, where lipid peroxidation and cross-linked proteins accumulate, perhaps due to the continuing action of free radicals (Chapter 5). Alterations of neural membranes impair excitability, fluidity, and molecule transport. Equally important are nuclear alterations in which DNA damage (and failure of DNA repair) may be reflected in disturbed protein synthesis and accumulation of abnormal proteins.

**At the functional level**, aging influences electrical activity and motor, sensory, cognitive, and affective processes. Consequences of aging result in disturbances within the nervous system and in the organism as a whole. Given the key role of the CNS in regulating all body functions, any impairment of its function will impact aging of the entire organism, particularly the efficacy of homeostatic adjustments to the environment.

**At the vascular level**, cerebral blood vessels undergo the same atherosclerotic lesions that involve the entire arterial vasculature (Chapters 16 and 17). Atherosclerosis commences at an early age, progresses with age, and profoundly influences the neural tissue that has a critical dependence on oxygen and glucose for normal metabolism and function.

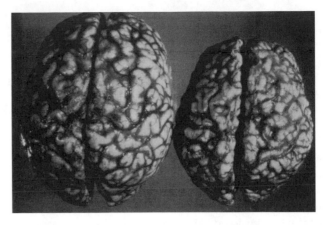

**FIGURE 7.3** Differences in size of normal (left) and Alzheimer's (right) brain. (Courtesy of Dr. L.S. Forno.)

cortex have been implicated in the longer life span of humans compared to other species (Chapter 3). Correlation of size and function began to be appreciated in the late 19th century, when, for example, it was shown that prevalently "visual" animal species had enlarged superior colliculi, and prevalently "auditory" species had enlarged inferior colliculi.[25] However, association of brain size with intelligence does not apply to the relatively minor differences in structure and function among individuals (of different size) within the same species; in humans, given the redundancy of neural cells and their interconnections, comparison of brain size among individuals of adult and old age has little functional significance.

Experimental animals vary with respect to aging and brain size: rats do not show any changes with aging,[26] but dogs and monkeys demonstrate cerebral atrophy.[27]

## B. NEURAL CELL NUMBER

The adult human brain contains approximately $10^{12}$ neurons and 10 to 15 times that number of glial cells. Each neuron has, on average, 10,000 connections (from a few thousand to over 100,000), resulting in an extraordinarily large total number (about $10^{15}$) of connections. It is through these connections that the nervous system exerts its essential role in *communication* with the internal and external environment and between the two. CNS cells include neurons, glial, and endothelial cells (Box 7.2).

Aging affects nervous structures differentially. One striking example of this diversity involves neuronal loss. In the brain of old persons without functional or pathologic deficits, loss of neurons is limited to discrete areas and shows considerable individual variability.[28] In some areas of the cerebral cortex and in the cerebellum, the number of neurons remains essentially unchanged throughout life, except at very old ages, when a loss has been reported in the cerebellum.[29] Other areas where losses may occur are the locus ceruleus (catecholaminergic neurons), the substantia nigra (dopaminergic neurons),

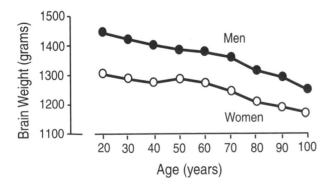

**FIGURE 7.2** Changes in brain weight with aging in human males and females.

**Systemic and Organismic Aging**

the nucleus basalis of Meynert, and the hippocampus (cholinergic neurons). In all brain areas, neuronal loss appears modest compared to the redundancy of existing neurons. However, it may become quite severe in some aging-associated diseases, with a marked loss of cholinergic neurons[30] in the frontal cortex and hippocampus in Alzheimer's disease and of dopaminergic neurons[31] in the substantia nigra in Parkinson's disease.

The number of glial cells increases with aging in most areas, and this increase, or gliosis, may represent a compensatory response not only to the small neuronal loss but also to neuronal impairment.[26] Gliosis is a normal response to neuronal damage at all ages; it persists in old age and is considered to be part of a repertory of compensatory responses that protect neuronal function and plasticity in old age.

Neuronal number in experimental animals decreases slightly with aging, depending on the area considered, the type of cell, and the animal species. For example, in the cerebral cortex of rats, the number of neurons remains unchanged with age.[26] Reduced numbers of neurons have been reported in discrete brain areas of aged monkeys, guinea pigs, and mice.[27] When the number of neurons is examined in these animals throughout the life span, a major reduction in cell number often occurs at young rather than old ages. The glial cells, astrocytes, and oligodendrocytes increase in number, while microglia are unchanged.[26]

## C. DENDRITIC AND SYNAPTIC LOSSES

The normal organization of neuronal networks is maintained in many healthy elderly, but with progressive aging, in old individuals, the number of dendrites and dendritic spines may be reduced (the so-called "denudation" of neurons). For example, the cortex of a young individual shows large pyramidal cells with abundant dendrites rich in dendritic spines. A corresponding zone, in an old individual, shows quite striking loss of dendrites and spines.[32] Dendrites function as receptor membranes of the neurons and represent the sites of excitatory and inhibitory activity. Dendritic spines, tiny and numerous on each dendrite, amplify such activity, and by doing so, control increases in synaptic calcium transport that may serve for induction of information storage. Loss of dendrites and spines results in neuronal isolation and failure of interneuronal communication.

Inasmuch as dendrites undergo a certain degree of renewal continuously, the *denudation of the neurons* may not be a true loss, but rather, a slowing of the renewal process. When the loss of dendrites is viewed in a network of neurons, the consequences of diminished connectivity become apparent.[33,34] In normal aging, with continued environmental stimulation, dendritic loss may be minimal, absent, or even supplemented with a degree

---

> **Box 7.2**
> ### Cells of the Central Nervous System
>
> **Neurons** are regarded as the primary functional cells despite the larger glial cell numbers. Dendrites and axons are cell processes branching from neurons and conducting impulses to or from the cell body. Communication between neurons is through synapses and synaptic spines (the latter, small knobs projecting from dendrites). Synapses are composed of the presynaptic membrane (often enlarged to form terminal buttons or knobs), a cleft, and a postsynaptic membrane. Present in the spines are vesicles or granules containing the synaptic transmitters synthesized in the cell. The axons may be surrounded by a specialized membrane, myelin, or may be unmyelinated. The myelinated fibers transmit the nerve impulse more rapidly than do the nonmyelinated ones.
>
> **Neuroglia**, of ectodermal origin like neurons, comprises astrocytes, which have end-feet that envelop blood vessels and surround neurons, and oligodendrocytes involved in myelin formation. Included in the neuroglia are the **ependymal cells**, which share the same stem cell population and line the cerebral ventricles and the spinal cord central canal. **Microglia**, of mesodermal origin, are part of the immune system and play a role in CNS phagocytosis and inflammatory responses. While the astrocytes have independent functions, they also act as a neuronal-glial unit in such vital actions as neural metabolism and transmission.
>
> **Endothelial cells of cerebral capillaries**, in contrast to those in other tissues, form tight junctions that do not permit the passage of substances through the junctions. Cerebral capillaries are enveloped by the end-feet of astrocytes, a special feature that hinders the exchanges across capillary walls between plasma and interstitial fluid. The resulting, so-called **blood–brain barrier** prevents entry of endogenous metabolites and exogenous toxins and drugs of the blood into the brain, and, reciprocally, the entry of the neurotransmitters into the general circulation.

of dendritic outgrowth. In dementia (Chapter 8), the dendritic loss is severe and progressive (Figure 7.4). With reduced dendrites, synapses are lost, neurotransmission is altered, and communication within and without the nervous system is impaired.

The decreased number of *synapses* in discrete areas of the aging brain follows the corresponding loss of dendrites and dendritic spines. Alterations in synaptic components — membrane, vesicles, and granules — have been variably reported. The "reactive synaptogenesis" or axonal sprouting that follows the loss of a neuron is not entirely lost but decreases with aging.[9] Reactive synaptogenesis represents

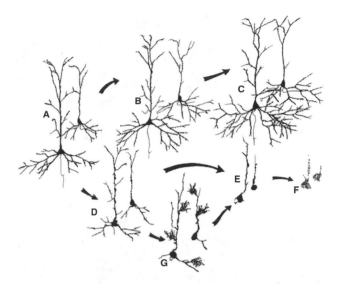

**FIGURE 7.4** Semi-schematized drawing summarizing the changes that may occur in pyramidal neurons of the aging human cerebral cortex. **Sequence A, B, C** follows the changes that may occur in the normal aging cortex under the effects of continued physical and cognitive "challenge" and in the probable presence of continuing small amounts of neuronal loss. Increasing dendritic growth, especially in the peripheral portions of the basilar dentrites, reflects dendritic response to optimal cortical "loading," plus presumed supplementary growth to fill neurophil space left by dendrite systems of those neurons that die. **Sequence A, D, E, F** epitomizes the progressive degenerative changes that characterize senile dementia of Alzheimer type (AD) and includes progressive loss of dendritic spines and dendritic branches culminating in death of the cell. Basilar dendrite loss precedes loss of the apical shaft. **Sequence A, D, G, E, F** represents a unique pattern of deterioration of the dendrite tree found only in the familial type of presenile dementia of Alzheimer. During the period of dendritic degeneration, bursts of spine-rich dendrites in clusters appear along the surface of the dying dendrite shafts. Such changes have been seen in neocortex, archicortex (hippocampus), and cerebellar cortex. The cause and mechanisms of these eruptive attempts at regrowth, even as the neurons are in a degenerative phase, are unknown. (Courtesy of Dr. A.B. Scheibel.)

a compensatory reaction to neuronal loss or damage and is characterized by an increase in the number of synaptic contacts provided by the nearby neurons. Such a compensation, although less efficient, can persist in the old brain.

As "hyperconnectivity" resulting from neurologic causes (e.g., temporal epilepsy) may lead to heightened attention, perceptions, memories, and images, progressive loss of interneuronal communication may cause a downward shifting of neurologic and mental processes. Impaired CNS function in old age may be due to a multifactorial "hypoconnectivity" and increasing "rigidity" (consequent to vascular and metabolic alterations) rather than to regional and moderate neuronal losses. Several functions (e.g., ECG, EEG, pulsatile hormonal secretions) formerly thought to be relatively periodic, show a complex type of variability reminiscent of "chaos" (i.e., unpredictable behavior arising from internal feedback from interconnected loops of non-linear systems).[35] Aging, by reducing interconnectedness and complexity of these systems, might reduce chaos and, thus, restrict plasticity and dynamics of brain processes.[36–38]

A number of other changes, usually quite moderate and with considerable individual variability, may occur in normal aging. These include alterations in the *axon* that may contribute to disruption of neural circuitry, demyelination, axonal swelling, and changes in the number of neurofilaments and neurotubules. Changes occurring with aging in the neural cytoskeleton (as in neurofibrillary tangles), accumulation of amyloid proteins (as in neuritic plaques), impaired cell survival (reduced antiapoptopic synaptic proteins), alterations in the cerebrospinal fluid composition and volume, and other changes, are discussed in relation to their repercussions (e.g., dementia).

## D. NEURAL CELL PATHOLOGY

As for cell and synaptic loss, cell pathology occurs in the normal aging brain, but the number of affected structures is quite small and confined to discrete areas (e.g., the hippocampus). This contrasts with the extensive pathology associated with most neurodegenerative diseases (e.g., Alzheimer's dementia, Parkinson's disease, multi-infarct dementia, etc.).[39]

### MAJOR CELL PATHOLOGIC CHANGES

Major cell alterations are of a degenerative nature (i.e., leading to impaired function and death) and are manifested by the following:

- Intracellular accumulation of abnormal inclusions such as lipofuscin, melanin, Lewy bodies, neurofibrillary and amyloid proteins
- Accumulation of abnormal extracellular amyloid proteins in neuritic plaques and surrounding cerebral blood vessels (perhaps deriving from the systemic circulation through defects of the blood brain barrier permeability), and of ubiquitin, tau proteins, and other abnormal proteins

These changes are usually associated with vascular, atherosclerotic alterations that may induce hemorrhages and infarcts (strokes) consequent to rupture or obstruction of blood vessels (as in multi-infarct dementia).

### 1. Lipofuscin and Melanin

*Lipofuscin* or "age pigment," the by-product of cellular autophagia (self-digestion) and lipid peroxidation due to free radical accumulation (Chapter 5), has a protein and a lipid component. It accumulates with aging in most CNS cells, both neurons and glial cells, where it follows a regional distribution (e.g., in the hippocampus, cerebellum, anterior horn of spinal cord, etc.). It increases linearly with age also in other cells of the body (e.g., cardiac and

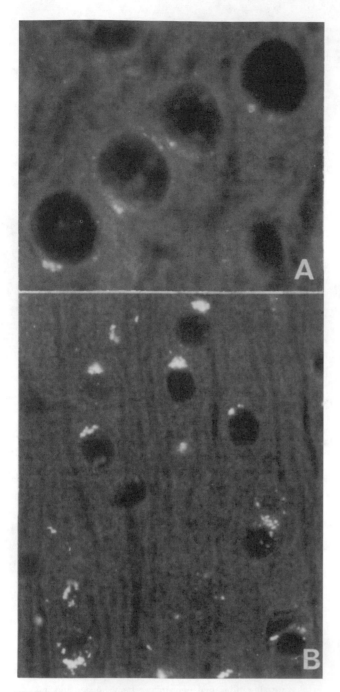

**FIGURE 7.5** Photomicrograph of unstained sections under blue light fluorescence illustrating lipofuscin pigment in neuron somata in cerebral cortex of male Long-Evans rats.[40] (A) Young adults (109 to 113 days) (Original magnification ×32.); (B) aged rats (763 to 972 days). (Original magnification ×14.) (Copyright © Gerentological Society of America. Reproduced with permission of the publisher.)

muscle cells, macrophages, interstitial cells). Lipofuscin can be visualized as autofluorescent material (Figure 7.5) and, with the electron microscope, as dark granules, either scattered in the cytoplasm or clustered around the nucleus (Figures 7.6, 7.7).[40] Its functional significance is unclear. The claims that lipofuscin accumulation interferes with

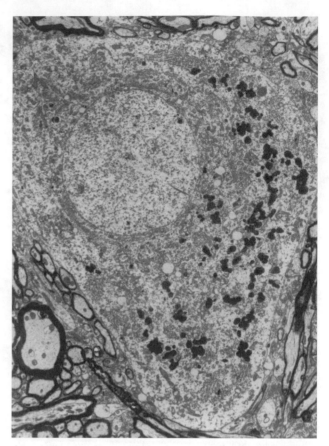

**FIGURE 7.6** A neuron from the cerebral cortex of a 605-day-old male Long-Evans rat.[40] There are numerous dense bodies (lipofuscin) irregularly distributed in the perikaryon. (Original magnification ×1440.) (Copyright © Gerentological Society of America. Reproduced with permission of the publisher.)

intracellular function, or that antioxidants, by reducing brain lipofuscin, may lead to improved neuropsychologic behavior, remain controversial.

*Melanin* pigment is localized primarily in the locus ceruleus and substantia nigra, where it imparts a dark coloring to the catecholaminergic, especially, dopaminergic cells of these regions. Melanin increases until about 60 years of age and then decreases, probably in parallel with the progressive loss of the heavily pigmented cells.

Accumulation of lipofuscin or melanin is not greater in the brain of old people affected by Alzheimer dementia, Down syndrome, progeria, and most degenerative diseases than in old people without these diseases.

## 2. Lewy and Hirano Bodies

The Lewy and Hirano bodies are eosinophilic, cytoplasmic inclusions, derived from the differential expression of particular proteins, responsible for selective neuronal vulnerability.[41] *Lewy bodies* are usually spheroid in shape with a dense central core (Figure 7.8), and the *Hirano bodies* may often appear as spindle-shaped and fusiform. Lewy bodies may be present in aged individuals, 60 years of age and

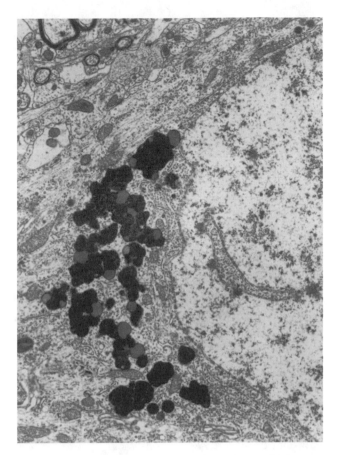

**FIGURE 7.7** A portion of a neuron from the cerebral cortex of a 605-day-old male Long-Evans rat.[40] The lipofuscin granules are clustered at one pole of the nucleus. (Original magnification ×6000.) (Copyright © Gerentological Society of America. Reproduced with permission of the publisher.)

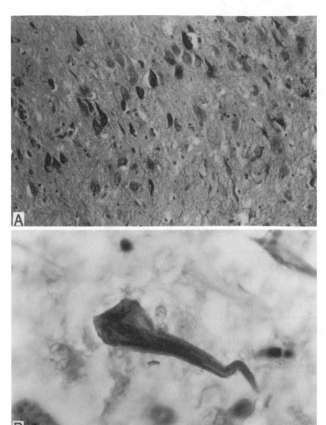

**FIGURE 7.9** Neurofibrillary tangles in hippocampus pyramidal layer: (A) neurofibrillary tangles (low magnification) accumulate in neuronal bodies and stain very darkly; (B) neurofibrillary tangle (high power) shows characteristic flame shape. (Courtesy of Dr. L.S. Forno.)

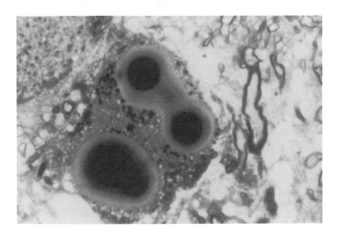

**FIGURE 7.8** Lewy bodies, intraneuronal cytoplasmic inclusions. The dense core is composed of apparently random, tightly packed aggregations of filaments, vesicular profiles, and poorly resolved granular material containing neurofilament antigens. These bodies are found in substantia nigra as well as in other CNS regions. (Courtesy of Dr. L.S. Forno.)

older, without clinical evidence of Parkinson's disease but are more numerous in those individuals affected by this disease. Indeed, the accumulation of such bodies, especially in the dopaminergic cells of the locus ceruleus, and the substantia nigra is considered a hallmark of Parkinson's disease.[42] The accumulation of larger numbers of Lewy bodies in the cerebral cortical neurons of many demented individuals has led to the identification of a "diffuse Lewy body dementia," where the severity of the dementia correlates with the increasing number of Lewy bodies.[43]

Hirano bodies in humans occur as changes related to old age; they are predominant in the hippocampus and more numerous in several subjects with dementing illnesses, including Alzheimer's dementia, Pick's disease, and amyotrophic lateral sclerosis.

## 3. Neurofibrillary Tangles and Neuritic Plaques

*Neurofibrillary tangles* consist of *intracellular* tangled masses of fibrous elements, often in flame-shaped bundles, coursing the entire cell body (Figure 7.9). They are present in the normal aging brain, frequently in the

**Systemic and Organismic Aging**

hippocampus. Their accumulation in the cortex and other brain areas is one of the diagnostic signs of the senile dementia of the Alzheimer type or Alzheimer's disease (AD). In AD, tangles would be in greatest number in the associative areas of the cortex, and their distribution may reflect a primary toxin/infection spread through the olfactory system.[44,45] Their density has also been correlated with the severity of the dementia, of the cell loss, and of the cholinergic deficits.[46,47]

Under the electron microscope, each fiber consists of a pair of filaments wound around each other to form a *paired helical filament or PHF*. Chemically, PHFs are highly insoluble, and their precise protein composition remains unknown. However, they react with monoclonal antibodies specific for neurofilament proteins or for microtubule-associated proteins (MAPs).[48] Because neurofilaments and microtubules are normal constituents of the cytoskeleton, the immunologic reactivity of PHF supports the hypothesis that several cytoskeletal proteins are linked in the formation of these abnormal, helical filaments.[49]

Alterations of *tau protein* (i.e., a microtubular protein that enhances the polymerization of tubulin subunits) and of the enzyme protein kinase that catalyzes its phosphorylation have been implicated in the PHF generation. Tau proteins have attracted special attention because of their crucial role in regulating balance between stability and plasticity of the neuronal cytoskeleton.[50] Tau proteins have several isoforms, and, with aging, more tau would be produced in an attempt to stabilize the aging cytoskeleton; such an attempt would result in the production of predominantly juvenile tau forms and the formation of PHFs.[51-53] However, it is still unknown which proteins make-up the core of twisted filament and whether novel abnormal proteins are present.

The *neuritic plaques* are situated *extracellularly*. Present in normal aging, they are abundant in AD, where they are found in the frontal, temporal, and occipital cortex and the hippocampus. Their distribution is similar to that of the neurofibrillary tangles, and the two structures are often found in close proximity (Figure 7.10). Typically, the plaque consists of a central core of amyloid. Amyloid is a generic term for proteinaceous fibrillar deposits with (1) properties of yellow to green birefringence in crossed polaroid illumination after Congo red dye staining, (2) a large amyloid precursor protein (APP) and its cleavage product β-amyloid peptide, and (3) insolubility and resistance to enzymatic digestion. The amyloid deposits are surrounded by coarse, silver-stained fibers represented, at the electron microscope, as many distrophic axon segments with degenerated mitochondria, synaptic complexes, and dense bodies (Figure 7.10).

Glial and microglial cells react to the plaque and accumulate about the abnormal nerve cell processes. One view is that the development of the plaque proceeds first from

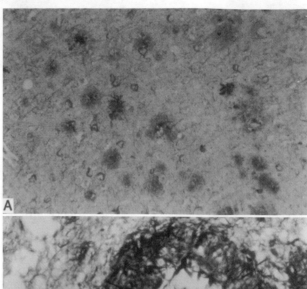
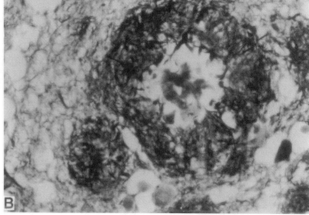

**FIGURE 7.10** Neuritic plaques in frontal cortex, silent area: (A) presence of neuritic plaques, large, and neurofibrillary tangles, small (low power); (B) neuritic plaque (high power). (Courtesy of Dr. L.S. Forno.)

abnormalities of the neural processes, then to deposition of amyloid, and later by stimulation of reactive glial proliferation. Another view is that the primary alteration arises at the level of the brain capillaries, and that amyloid, or a precursor substance of blood origin, leaks from the capillaries and diffuses into the cerebral tissue, where its accumulation leads to the destruction of neurons, glial cell proliferation, and plaque formation. The possible role of amyloid deposition in AD is discussed in the next chapter, with the dementias (Chapter 8).

The presence of abnormal protein aggregates, such as those found in the neurofibrillary tangles and neuritic plaques (and abundant in AD), stimulates the *ubiquitin-dependent protein degradation system*. Thus, ubiquitin accumulates not only in the brain of AD[54] patients but also in Parkinson's disease[55] patients. In normal aging, where fibers, plaques, and Lewy bodies are few, ubiquitin is only minimally activated.[56] Oxidation of proteins is prominent in many cells (Chapter 5), and free-radical-mediated oxidation affects a number of brain enzymes as well. With aging, oxidative damage and glycosylation of proteins result from the decreased rate of protein turnover.

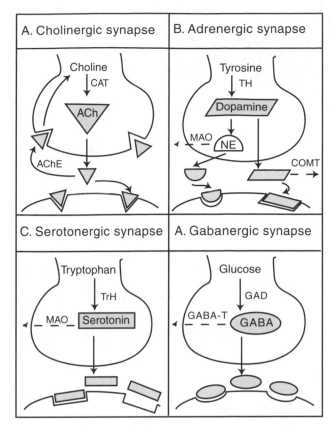

**FIGURE 7.11** Diagrams of established CNS synapses illustrating presynaptic synthesis, storage, metabolism, synaptic release, and postsynaptic binding. (a) Cholinergic synapse. Choline is the precursor of the transmitter acetylcholine (ACh) through the actions of the synthesizing enzyme choline acetyltransferase (CAT). ACh is free in the cytoplasm of the prejunctional axonal ending as well as contained in vesicles. Release of cytoplasmic ACh initiates transmembrane events. Several distinct pre- and postsynaptic ACh receptors have been described. Receptor-bound ACh is inactivated by the enzyme acetylcholinesterase (AChE). (b) Adrenergic synapse. Norepineperine (NE) precursor is tyrosine, and the enzymes involved in NE synthesis are tyrosine hydroxylase (TH), dopa decarboxylase (DOD), and dopamine-beta-hydroxylase (DBH). The primary degrading enzyme is mitochondrial monamine oxidase (MAO). NE is stored in vesicles. Once released, transynaptic degradation is initiated by catechol-*O*-methyl-transferase (COMT). Several pre- and postsynaptic adrenergic receptors have so far been identified. Released NE not degraded by COMT may be taken up by the presynaptic neuron and internalized. Note the simultaneous production of dopamine at its release at the synapse. (c) Serotonergic synapse. The precursor of serotonin is tryptophan, and the synthesizing enzyme is tryptophan hydroxylase (TrH); the degrading enzyme is MAO. Several different types of cell surface receptors are known. (d) Gabanergic synapse. GABA is formed in the GABA-shunt pathway of the Krebs cycle. It is formed by the action of glutamic acid decarboxylase (GAD) from glutamate and is metabolized by deamination. At least two receptors for GABA have been identified in the CNS.

## AMYLOID AND AMYLOIDOSES

Amyloid deposits are not confined to the brain but may occur in several tissues and organs, where their presence is noted in a number of conditions, classified under the general term of amyloidoses, which represent a number of diseases characterized by extracellular accumulation of insoluble fibrillar proteins. These diseases show pronounced age dependence, and their prevalence increases rapidly toward the end of the life span. Inasmuch as amyloid is present in some of the neurodegenerative diseases of old age, considerable attention is being given to understanding the structure of the deposits in the brain and to finding therapies to prevent their formation or to induce their dissolution.

Amyloidoses are classified on the basis of the biochemical composition of the amyloid subunit protein: (1) In primary systemic amyloidosis as in myeloma, amyloid is produced by proteolytic cleavage of the N-terminal variable region of immunoglobulin Ig light chain in phagocytic cells. (2) In secondary amyloidoses occurring in chronic inflammation, accumulation of amyloid A protein occurs. (3) In familial amyloidoses, prealbumin is the major accumulating protein.

Senile cardiac amyloidosis is present in 65% of hearts from persons 90 years of age and older. Focal deposition of this amyloid appears to have few functional consequences, but extensive distribution has been associated with congestive heart failure and fibrillation. Other forms are found in senile lungs, liver, kidneys, and in general, in senescent or injured tissues. Controversy still exists regarding amyloid origin, although it has been suggested that a deficient function of lysosomes may be responsible for its deposition.

## III. BIOCHEMICAL CHANGES

### A. NEUROTRANSMISSION AND CELL COMMUNICATION

Information processing in the nervous system involves neurons "talking" to each other or with target cells. Research in neurotransmission, including neurotransmitter turnover, release, and binding to receptor, is central to our understanding of CNS aging. One of the most studied aspects of aging of the nervous system involves neurotransmitter changes at the synapse. In the healthy elderly, neurotransmitter levels, number and affinity of their receptors, and activity of their metabolic enzymes undergo modest changes circumscribed to specific brain structures and individual neurotransmitter systems. In some diseases, however, a definite relationship exists between loss of one (or several) neurotransmitter(s) with abnormal brain function, for example, dopamine deficit in the nigrostriatal pathway with Parkinson's disease and acetylcholine deficit in the Meynert nucleus with Alzheimer's disease.

Many synapses are classified according to the major neurotransmitter released at their site. Examples of synaptic transmitters include acetylcholine (ACh), norepinephrine (NE), dopamine (DA), serotonin (5-hydroxy-tryptamine), and gamma-aminobutyric acid (GABA), and the synapses

---

**Box 7.3**

**Chemical Classification of Synapses and Neurotransmitters**

Known transmitters include **amines, amino acids, peptides, or gases** (Table 7.2). Of these, some may be viewed as classic, our knowledge of them extending back almost one century; they are **acetylcholine**, the **catecholamines** (norepinephrine, epinephrine, and dopamine), and **serotonin**.

In addition to **GABA**, certain amino acids such as **glycine**, **glutamate**, and **aspartate** have been identified for their important roles. Of these, GABA transmits inhibitory impulses, and aspartate and glutamate usually mediate excitatory synaptic transmission. Their role is crucial to several normal functions of the nervous system, including memory and long-term modification of synaptic transmission, perception of injurious stimuli, activation of membrane receptor, and transmembrane calcium and sodium fluxes.[57]

Among the peptides, a few of the more extensively studied are **enkephalin**, **substance P**, and **hypothalamic neurohormones**. Some peptides such as cholecystokinin and somatostatin, are found in the brain and in the gastrointestinal tract.

The soluble gas **nitric oxide (NO)**, viewed originally as a toxic molecule (in cigarette smoke and smog), is now considered to be a biologic messenger in mammals that acts by nonsynaptic intercellular signaling.[58,59] NO is generated by the enzyme NO synthase (acting on arginine), sensitive to calcium and calmodulin modulation. Unlike other neurotransmitters, NO is not stored or released from vesicles in the neuron, but rather diffuses out from the cell of origin. NO is also found in genitalia (where it is essential for penile erection) and in the gut (where it is involved in peristalsis). **Carbon monoxide (CO),** originally viewed as an exclusively toxic gas,[59] and **metals** such as **zinc**[60,61] are now considered to be putative neurotransmitters in some specific neurons (in the olfactory bulb and hippocampus) often affected by aging.

A membrane-associated, synaptic vesicle protein, **synapsin**, may also influence neurotransmission by increasing the number of synapses, synaptic vesicles, and contacts.[62] There are four synapsins, each with similar amino acid sequences, generated by two alternative gene splicings. All four are associated with the cell cytoskeleton and, like some of the other cytoskeletal proteins, they may undergo phosphorylation during synaptic transmission. Synaptic vesicles may be destabilized in the absence of synaptins, and, during synaptic plasticity, upregulation of release may be defective.

---

are named accordingly. Synthesis, storage, release, postsynaptic interactions, and inactivation of some neurotransmitters are summarized briefly (Figure 7.11, Box 7.3).

Several transmitters, often a classic neurotransmitter and a peptide, may coexist within the same neuron. Such "coexistence," resulting in "co-utilization" of several transmitters, vastly expands the potential diversity of synaptic communications. Indeed, the peptide often "modulates" or "modifies" the action of the classic neu-

rotransmitter. For example, serotonin coexists with substance P and thyrotropin-releasing hormone in neurons of the rat medulla and spinal cord, where the three neurotransmitters collaborate to regulate some motor and behavioral responses.

Synaptic diversification and multiplicity of neurotransmitters result in increasing complexity of neuronal communication and the need for a fine-tuning of all chemical messengers, transmitters, and their modulators

---

**TABLE 7.2**

**Neurotransmitters and Modulators in the Nervous System**

| Amines | Amino Acids | Peptides | Others |
| --- | --- | --- | --- |
| Acetylcholine | Glutamate | Enkephalin | Nitric oxide |
| Catecholamines | Aspartate | Cholecystokinin | Carbon monoxide |
| Norepinephrine | Glycine | Substance P | Zinc |
| Epinephrine | GABA[b] | VIP[c] | Synapsins |
| Dopamine | Taurine | Somatostatin | Cell Adhesion Molecules |
| Serotonin[a] | Histamine | TRH[d] | Neurotropins |
| | | | Others |

[a] Serotonin, 5-hydroxytryptamine, or 5-HT
[b] GABA = gamma-amino butyric acid
[c] VIP = vasoactive intestinal polypeptide
[d] TRH = thyrotropin-stimulating hormone

## TABLE 7.3
### Major Chemical Events at the Synapse

Presynaptic
    Availability of precursors
    Enzymatic synthesis and degradation
    Storage or release
    Transmitter reuptake
    Ionic regulation
Transynaptic
    Free neurotransmitter
    Enzymatic degradation
Postsynaptic
    Receptor binding
    Enzymatic degradation
    Ionic regulation
    Second messenger

---

### Box 7.4
### Analysis of Synaptic Chemical Events: Difficulties of Measurement in Humans

Chemical transmission requires the following series of events:

- Synthesis of the neurotransmitter at the presynaptic nerve terminal
- Storage of some neurotransmitters in vesicles
- Release of neurotransmitter into the synaptic space (between pre- and postsynaptic neurons)
- Presence of specific receptors for the neurotransmitter on the postsynaptic membrane and binding of the neurotransmitter to the receptors, thereby effecting nerve stimulation
- Termination of the action of the released neurotransmitter by its diffusion from the synaptic space, its metabolism, or its reuptake by the presynaptic neuron

In humans, information is often gained by indirect measurement of neurotransmitters and their metabolites in urine, blood, and cerebrospinal fluid and, more recently, directly, in the brain by labeled probes and imaging devices. The use of pharmacological agonists and antagonists of the transmitter under study can provide an alternate approach. Under the best conditions, difficulty in controlling the clinical setting with humans and accurately measuring subtle changes in physiological status suggest that much is to be gained by the use of animal models to study cellular and molecular mechanisms of synaptic function. Even in animals, the information on aging is still scarce and needs to be pursued vigorously.

---

(Table 7.3). Neurotransmitters have excitatory or inhibitory actions, and most important to the function of the nervous system, these actions must remain in balance. With aging, alterations in function are more likely to occur as a consequence of imbalance among neurotransmitters rather than as a consequence of a global alteration of a single neurotransmitter. Such imbalance could involve the classic neurotransmitters for which evidence is already available, or the sustaining of peptides, or both.

In the human brain, knowledge and understanding of age differences in neurotransmitter levels, their synthesizing and degrading enzymes, and receptors are still fragmentary, tentative, and highly incomplete. Levels and activities of neurotransmitters and enzymes decline in many brain regions during normal aging: one classic example is the decreased levels of dopamine due to loss of neurons in the substantia nigra. Receptors may increase in number or affinity for the respective neurotransmitter and its agonists in response to a decrease in concentration of the neurotransmitter, and vice versa (up- and downregulation of receptors). Chemical transmission requires a series of events as described in Box 7.4.

## B. Neurotransmitter Imbalance

Imbalance among neurotransmitters, rather than severe depletion or excess of a particular neurotransmitter, may underlie functional changes with aging. In early studies in rats, neurotransmitters in the cerebral hemispheres, the hypothalamus, and the corpus striatum were compared at progressive ages. Overall, two major changes were noted: serotonin concentrations remained unchanged until very old age when they then increased, and norepinephrine and dopamine concentrations progressively decreased starting at relatively less advanced ages. Thus, in the aging rat brain, the ratio of serotonin to the catecholamines progressively

increases. For the resulting imbalance to become functionally manifest, it is sufficient that a single neurotransmitter be altered rather than several simultaneously.

A final point to emphasize here is that each neurotransmitter has its own regional timetable of aging. For example, in rats, serotonin concentration remains unaltered in the cerebral hemispheres until 3 years of age, while dopamine levels in this same brain region are reduced at 1 year of age.[63] It is this differential change that may be responsible for creating an early neurotransmitter imbalance, well before significant decrements in each transmitter may be detected.

The decline in dopamine content in the substantia nigra, discussed below, and in acetylcholine content in the nucleus basalis of Meynert (Chapter 8) are two examples of neurotransmitter systems that markedly change with aging and that lead to functional disorders.

**Systemic and Organismic Aging**

## C. AGING OF DOPAMINERGIC SYSTEMS AND PARKINSON'S DISEASE

The dopaminergic system undergoes changes with normal aging and with aging-associated diseases. Dopamine is present throughout the brain, but its levels vary significantly from region to region. Currently, four major dopaminergic pathways are recognized, the most prominent originating in the substantia nigra and extending to the corpus striatum and other structures of the basal ganglia; the other dopaminergic pathways involve the limbic system, the hypothalamus, and the cortex (Table 7.4, Box 7.5). Dopamine is also secreted by interneurons in sympathethic ganglia; hence, its administration stimulates several tissues and organs, thereby causing the undesirable side-effects of its therapeutic use.

With normal aging, dopamine content steadily decreases, particularly in the corpus striatum, probably as a consequence of the loss of dopaminergic neurons, perhaps due to free radical damage (Chapter 5).[31,64] This decrease may or may not be associated with motor disturbances, depending on the degree of neuronal loss or the severity of the action of excitotoxins (such as glutamate).[65–68] When neuronal loss is severe, the neurologic alterations become more pronounced and are associated with progressively more severe motor decrements and other neurologic and clinical manifestations, they are categorized as part of Parkinson's disease. This disorder involves rigidity, tremor (at rest), and akinesia (loss of motor function) or bradykinesia (slowed movement).[31,69,70] It is a progressive disease of the elderly, about 85% of those affected are age 70 or older. In this disease, the dopaminergic neurons projecting to the caudate nucleus and putamen degenerate due to reduced stimulation from the damaged substantia nigra (Figure 7.12). This damage is characterized by loss of neurons, gliosis, and the presence of specific inclusions, the so-called Lewy bodies.[42] In addition to Parkinson's disease, several forms of motor disabilities of similar etiopathology but with only some of the motor deficits of Parkinson's disease (for example, lacking the resting tremor), are grouped under the term "parkinsonism."

As already mentioned, Lewy bodies may be found in normal aging, but their incidence is much greater not only in Parkinson's but also in AD patients.[42] There is an overlap in pathology between Parkinson's and Alzheimer's diseases.[41] The prevalence of histological changes of the Alzheimer type is higher in parkinsonians than in an age-matched unaffected population. In Lewy body dementia, mentioned above, the major symptom is cognitive failure, while motor disturbances are relatively minor.[43]

### DRUG-INDUCED PARKINSON'S DISEASE

A contaminant, 1-methyl-4-phenyl-1,2,3,6-tetrahydropyridine (MPTP) of illegally synthesized heroin acts specifically to induce a motor-deficit syndrome in experimental animals (mice, rabbits, monkeys) similar to Parkinson's disease in

---

**TABLE 7.4**
**Dopaminergic Pathways**

Substantia nigra to corpus striatum
    Regulates control of motor function from cortex
Substantia nigra to nucleus accumbens
    Controls locomotion (forward movement)
Limbic system
    Regulates emotion, behavior (sexual and aggressive)
Hypothalamus
    Regulates endocrine releasing and inhibiting hormones
Neocortex
    Regulates locomotor activity

---

**Box 7.5**

**Major Dopaminergic Pathways**

The **nigral-striatal pathway** participates in central control of motor functions by regulating the planning and programming of movement; in broad terms, it is involved in the process by which an abstract thought is converted into voluntary action. It consists of three nuclei — caudate nucleus and putamen (often referred together as corpus striatum) and globus pallidus — and the functionally related substantia nigra. Dopamine released at the caudate nucleus inhibits the stimulatory (cholinergic? glutaminergic?) input from the motor cortex.

In the caudate, a balance is established between the inhibitory and excitatory control of motor activity. A deficit of dopamine and an increase in the cholinergic output result in rigor and tremor, whereas an increase of dopamine or a decrease of the excitatory input results in writhing movements, as in chorea.

Also involved in motor control is the nucleus accumbens in the septal region, which controls locomotion (particularly forward movement).

Within the **limbic system,** other dopaminergic fibers synapse and mediate emotionality, sexual and aggressive behavior, and some neuroendocrine regulation.

In the **hypothalamus,** a rich dopaminergic network regulates several endocrine-releasing and inhibitory hormones. This dopaminergic activity will be discussed with the endocrine system (Chapter 10).

A **neocortical dopaminergic system** is apparently activated by stress. Lesions of the system induce locomotor hyperactivity and inability to suppress behavioral "stereotypic responses." These responses can be blocked by drugs that act as dopamine inhibitors. Uncontrolled stimulation of dopamine receptors would induce a euphoric action or a psychotic action, and might be involved in manic and schizophrenic syndromes.

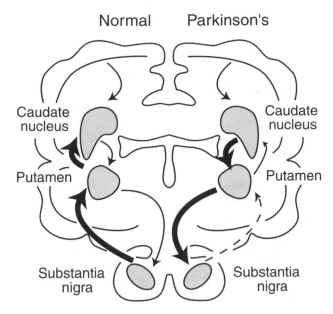

## Normal          Parkinson's

Caudate nucleus

Putamen

Substantia nigra

**FIGURE 7.12** Dopaminergic nigral-striatal pathway in the normal (left dark arrow) and in parkinsonism (right dashed line). In parkinsonism, dopaminergic neurons of the substantia nigra are lost, reducing dopamine (right dashed line) to putamen-caudate (striatum) and subsequent control of cortical stimulatory effects. Increased excitatory transmission to extrapyramidal system (broad open arrow) is associated with tremor and rigor.

humans.[70] In young animals, the syndrome is only incompletely produced, whereas in old animals, the neuropathology and behavior of Parkinson's disease are completely reproduced. Only in primates is the presence of Lewy bodies shown following MPTP administration. Older neurons appear more vulnerable to MPTP, either because of lowered ability to recover from toxic damage or because the damage is cumulative with other damage that has occurred over time. A variety of environmental agents and stresses may be involved in inducing other CNS degenerative diseases, such as senile dementia of the Alzheimer's type, in which age is a key factor (Chapter 8). This MPTP drug model is particularly useful in transplantation studies to assess the efficiency of the transplanted tissue in overcoming the MPTP-induced damage.

While motor abnormalities dominate the clinical picture of Parkinson's disease, current dopamine replacement therapy permits more patients to live long enough to develop more advanced features of the disease, that is, loss of memory, dementia, and postural instability, which are difficult to treat. Indeed, Parkinson's and Alzheimer's diseases may be viewed as two poles of a clinical spectrum of which the common feature is neuronal damage.

## D. TREATMENT OF PARKINSON'S DISEASE

A number of symptomatic treatments are available. The cause of this disease as well as other neurodegenerative ones remains unknown, and no cure or method of prevention has been found. The brain is a most complex structure — even incremental, less than perfect success of therapy

**TABLE 7.5**
**Strategies for Treatment of Parkinson's Disease**

Pharmacological
Neuroprotective
Surgical and electrostimulatory
Cell therapies

should be welcomed. Thus, while most of the neurologic and psychiatric diseases of old age are not yet "curable," they certainly are "treatable" and must be treated. With this in view, current therapeutic approaches are listed in Table 7.5 and are briefly discussed below.

### 1. Pharmacological Strategies

While neurosurgery was the first therapeutic approach, replacement of the lost striatal dopamine adopted in the early 1960s was a welcomed and safer alternative. In the neuronal pathway of catecholamine biosynthesis, the enzyme tyrosine hydroxylase catalyzes the conversion of the amino acid tyrosine to dihydroxyphenylalanine (L-dopa, levodopa), which is then converted to dopamine by the action of dopa-decarboxylase. L-dopa, unlike dopamine, can pass the blood–brain barrier, and when administered as a drug, can then serve as a precursor of neuronal dopamine. Although treatment with L-dopa may not provide a "cure," it ameliorates some of the more disturbing symptoms and has been viewed as a model therapy for aging-associated neurologic disorders.

The original pharmacological treatment had been designed to reduce neural activity with cholinergic inhibitors. The loss of dopaminergic input to the striatum, in the presence of persisting cholinergic excitatory activity, disrupts basal ganglia circuitry. Cholinergic inhibitors are still used as an adjunct to replacement L-dopa therapy.

Despite its beneficial effects, L-dopa treatment has some contraindications, such as motor complications and general and systemic effects, such as gastrointestinal disturbances. In addition, there is a gradual loss of effectiveness after 3 to 5 years of administration, probably due to continuing loss of dopaminergic neurons and reduction of dopa-decarboxylase with consequent deficit in dopamine synthesis. Because neurotoxicity from L-dopa may be due to the generation of oxidative species,[68,71] it is often recommended that antioxidants (e.g., tocopherol) be coadministered to prevent or reduce accumulation of free radicals (eventually capable of destroying the dopaminergic cells) (Chapter 5).

Treatment with L-dopa is often used in combination with other drugs such as dopamine agonists (e.g., bromocriptine),[72,73] or inhibitors of the two enzymes involved in dopamine breakdown, monoamine oxidase (e.g., selegiline)[74] and catechol-0-methyltransferase,[75] thereby

**Systemic and Organismic Aging**

extending the duration of action of L-dopa, or with a variety of anticholinergic drugs.

Current pharmacological therapy will remain essentially ineffectual on a long-term basis until we understand more about the cause of the dopaminergic cells loss. Recent evidence from postmortem brain and animal models of the disease[76] has suggested that early depletion of the antioxidant, glutathione,[77] and iron-mediated oxidative stress[78] may contribute to the loss of dopaminergic neurons of the substantia nigra.[76–78] Basic research into the cause of dopaminergic cell death is needed to prevent the cell loss from occurring.

### 2. Neuroprotective Strategies

Another approach to Parkinson's therapy is the administration of substances that may prevent death of striatal neurons. Such preservation has been accomplished by injection of growth factors into the brain (humans) or by their delivery with a lentiviral vector (in aged rhesus monkeys). One such promising growth factor candidate is of glial origin, the "glial cell line-derived neurotrophic factor" (GDNF), related to other members of the transforming growth factor superfamily.[79,80] GDNF can facilitate the survival of dopaminergic cells *in vitro* and possibly, also *in vivo*.

Other studies have identified a molecule, Nurr1, that plays a critical role during embryonic development in the formation of the group of dopamine-producing cells that are lost in Parkinson's disease. Nurr1 also appears to help keep these cells active throughout life and to assist them in producing appropriate amounts of dopamine. If Nurr1 is essential for dopamine production in adults, it may be possible to find drugs that may increase or decrease its activity and prevent or cure Parkinson's disease.[81]

### 3. Surgical and Electrostimulatory Strategies

Surgical techniques leading to ablation of the pallidum nucleus or the thalamus were employed for many years but were superseded by L-dopa treatment. Currently, focus on surgical interventions has awakened again. Ventromedial pallidotomy reduces overactivity of the globus pallidus that had been implicated in the motor disability associated with Parkinson's disease.[82]

Another approach uses electrical stimulation to specific brain areas with the intent of reestablishing normal electrical circuitry. These techniques include electroshock, high-frequency, deep brain stimulation and transcranial magnetic stimulation.[83–86] However, it is important to bear in mind that such surgeries are drastic measures, as once ablated, a part of the brain cannot be replaced. So, although these therapies may yield symptomatic alleviation of the disease, they may cause other complications later or become ineffectual as dopaminergic neurons continue to be lost.

### 4. Cell Therapies

The promise of cell therapies has captured the imagination of scientists.[87–92] Attempts at replacement of lost dopaminergic cells by various methods (e.g., transplantation/injection/grafting) with fetal nerve tissue, adrenomedullary and carotid-body cells, and stem cells are being actively pursued despite uncertain results and ethical controversy. We already mentioned recent studies in which precursors of dopaminergic cells from fragments of embryonal mesencephalon were transplanted into the brains of Parkinson's patients with mixed results.[15–17] However, the use of transplanted cells, whether pluripotent stem cells, more restricted neuronal and glial precursors, or differentiated neurons with or without the help of growth factors to stimulate cells proliferation and survival, despite some achieved technological progress,[18] is just beginning. It is important to understand that the implanted cells must retain their functional state and that their mere survival does not indicate success or efficiency of the treatment.

### E. Cell Adhesion Molecules

Cell adhesion molecules (CAMs) act on cell surfaces and play an important role in all aspects of CNS function, from development to adult maintenance. During development, they are involved in the formation of the neural tube and neural crest, cell migration, axonal outgrowth, and guidance. In adulthood, they regulate synaptic stabilization and plasticity, myelination, and nerve regeneration after injury and in learning.[93,94] They are members of the immunoglobulin (Ig) gene superfamily (Chapter 15) and are subdivided into three groups: the integrins and cadherins found in the nervous tissue and the selectins in the immune system.[94] CAMs interact with each other and with nonadhesive cell surface and cytoplasmic molecules. When acting together, they may often produce effects opposite to those of the individual factors, e.g., encourage movement or growth along a preferred pathway or immobilize a cell and prevent its movement and growth.[95]

While few studies so far have been made on CAMs in old age, those available indicate that in several brain areas (hippocampus, frontal, and occipital cortices) of neurologically normal elderly humans, CAMs levels closely resemble those of adult individuals; however, there were significantly fewer CAM-positive neurons in the frontal cortex of elderly individuals with AD.[96] In rats, the CAM upregulation that usually occurs with learning and regeneration is reduced in adult as compared to younger animals. This reduction would suggest that neural CAMs promote structural and functional remodeling and that this action may be less efficient in old individuals, especially in those affected with AD.[97,98] Additional studies of CAMs in the brain and in other tissues are necessary to elucidate further the role of these molecules in aging.

## F. Neurotropic Agents

The best known among the neurotropic agents is *nerve growth factor (NGF)*, a basic protein, resembling in structure the hormone insulin (Chapter 14). NGF promotes growth and maintenance of sympathetic and sensory neurons as well as neurons in the brain during developing and mature ages.[99] In the brain, NGF is particularly effective in promoting growth and reducing or preventing damage of cholinergic neurons of the basal forebrain (nucleus basalis of Meynert) drastically affected by neuronal loss in AD. In AD, NGF levels, receptors, and mRNA are reduced. Intraventricular administration of NGF, in young and adult rats, prevents these cholinergic neurons from dying after axonal transection or chemical damage and, in fact, may promote regeneration. However, clinical trials with NGF in AD patients were ineffective.

NGF is abundant in salivary glands, where it is taken up by terminals of neurons and transported in retrograde fashion to their cell bodies.[100] NGF has been purified and sequenced, and a number of similar growth-promoting proteins have been identified. A mouse fibroblast cell line capable of secreting recombinant NGF has been established.[99]

Other neurotrophic agents involved in development and maintenance or repair of neuronal tissue in adult and aged animals include, in addition to GDNF discussed above for its action on dopaminergic neurons,[79,80] gangliosides[101,102] and several compounds of the fibroblast and epidermal growth factors (FGF and EGF) family effective on neurons, glial cells, and specialized sensory cells (e.g., photoreceptors[102]) in health and disease and some components of the cytoskeleton. While clinical use of growth factors is still in the experimental stage, the possibility that neurotropic factors may benefit neuronal survival in CNS degenerative diseases represents a physiologic and hopeful therapeutic intervention.

As suggested at the beginning of this chapter, brain plasticity persists in old age but is less effective than at younger ages. An attempt to generalize the current views, suggests that neurons in old age are capable of displaying compensatory responses (reminiscent of those occurring in ontogensis) when they are provided with an appropriate environment or when the conditions that curtail their growth are removed. With aging, surviving neurons are capable of remodeling their configuration (e.g., reactive dendritic sprouting,[9] axonal growth[98]) in response to functional challenges and with the help of neurotropic agents. However, neuronal loss, isolation of one structure from another, or neurotransmitter deficits are more difficult to repair in the aged brain unless growth-promoting factors again become available and inhibitory factors are removed. As neuronal modeling is gene-regulated during ontogenesis, in a similar way, differential gene expression in aged neurons may direct the morphological and biochemical compensatory events that follow aging or damage and lead to functional recovery.

## IV. METABOLIC AND CIRCULATORY CHANGES

In addition to neurotransmitters, other chemical constituents of the central nervous system are known to change with aging. A number of these are listed in Table 7.6.

From clinical and experimental data, some generalizations may be drawn: changes are specific to discrete regions and structures, follow differential timetables, and differ according to the constituent considered. Because of the lack of consistent studies in humans during normal aging, only a few of these constituents will be discussed here. A good model of regional, age, and constituent specificity is presented by some putative amino acids neurotransmitters. For example, in the rat, of the excitatory ones, glutamic acid declines with aging in the cerebral cortex and brain stem but not in the cerebellum and spinal cord; glycine and GABA, inhibitory amino acids, increase with age in the cerebral cortex, cerebellum, and brain stem, but remain unchanged in the spinal cord.[103] These observations are supportive of an imbalance among neurotransmitters with aging.

Other changes with normal aging include:

- Decreased extra- and intracellular water content of the brain (as of most other organs), while electrolyte distribution remains essentially unchanged
- Regional decrease in protein content and synthesis, perhaps associated with slow turnover (to maintain steady state), increased oxidation of proteins with consequent glycosylation of proteins, or increase in complexity of RNA molecules; either an increase (perhaps related to gliosis) or no change in DNA content; and an accumulation of intraneuronal proteins (such as neurofibrillary tangles)
- Decreased lipid synthesis, primarily decreased synthesis of membrane phospholipids due to

---

**TABLE 7.6**
**Possible Targets of Aging Changes in the Central Nervous Sytem**

| | |
|---|---|
| Total water | Carbohydrates |
| Extra- and intracellular spaces | Circulation |
| Lipids | Energy metabolism |
| DNA, RNA, and protein | Oxygen uptake and glucose |
| Amino acids | utlization |
| | Blood–brain barrier |

**Systemic and Organismic Aging**

increased variation in the structure of lipid substrates rather than reduction of synthesizing enzyme activity or concentration of substrates; changes in membrane lipids would alter membrane fluidity and, in turn, nerve conduction and receptor binding

- Circulatory changes related primarily to atherosclerosis and characterized by a progressive reduction in cerebral blood flow and a corresponding decrease in oxygen uptake, glucose utilization, and, consequently, energy metabolism

## A. METABOLIC CHANGES

Parameters of *cerebral energy metabolism* and *circulation*, such as cerebral blood flow, oxygen consumption, and glucose metabolism (as a measure of cerebral metabolic rate) have been studied in humans and experimental animals utilizing a number of techniques [e.g., radioactive tracers, autoradiography, computed tomography (CT) scans, and other imaging techniques]. Energy metabolism is dependent on normal glucose utilization, normal membrane function, and normal supplies of high-energy phosphate compounds. The brain, unlike most other tissues, normally derives almost all of its energy from the anaerobic oxidation of glucose. Metabolic changes vary considerably with age, from prenatal development, to birth, to old age. Abundant data from experimental animals and some data from humans, show that cerebral oxygen consumption and the activity of the associated enzymes are low in fetal life and at birth, then rise rapidly during the period of cerebral growth and development, and reach a maximal level at about the time maturation is completed. Whole brain blood flow and oxygen consumption remain essentially unchanged between young adulthood and old age in the absence of disease; however, in the presence of minimal atherosclerosis, blood flow is significantly reduced, and, in severe atherosclerosis, blood flow and oxygen consumption are lower than in healthy controls of the same age (Table 7.7). Changes vary not only with the health of the subjects, the animal species, and the brain area considered, but also with the sensory, motor, or mental task performed.[104]

The relative stability of the energy metabolism of the aging brain in the absence of neuropathology must be contrasted with the severe deprivation that occurs in the aged brain affected by degenerative diseases such as AD. For example, in AD, the reduction in rate of cerebral formation of adenosine triphosphate (ATP) from oxidized glucose and oxygen, ranges (compared to controls of the same age but without AD) from 7 to 20% in incipient AD, and from 35 to 50% in stable, advanced dementia.[105] The AD brain may be characterized by hypometabolism, oxidative stress (Chapter 5), and alterations of the glucose-fatty acid cycle.[106] The view that, in AD, accumulation of

free radicals, amyloid, and other modified proteins[107] and alterations in neurotransmission may be secondary to the single, progressive cause of energy deprivation/ischemia is not new.[105,106,108] The lack of success so far obtained when attempting to treat, singly, each of the pathologic hallmarks of AD, appears to support the role of an unified and complex etiology. Neurodegenerative diseases may best be considered to be the consequences of sequential alterations of biochemical processes.[108] Glucose deprivation would elicit the counterregulatory mechanisms to preserve glucose for anabolic needs by switching from glucose to ketone body utilization. Soluble amyloid-$\beta$ peptide, which inhibits glucose utilization and stimulates ketone body utilization, would serve as the agent mediating this adaptive switching. However, while this metabolic switch is successful during early brain development, it is not sufficient to maintain normal energy metabolism in the aged brain. With old age, exhaustion of the metabolic adjustments to the stress of glucose deprivation and ischemia would fail, and this failure would precipitate the manifestations of AD (Chapters 8 and 10).

## B. CIRCULATORY CHANGES

Brain circulatory changes are part of the atherosclerotic lesions common throughout the arterial vascular system. One of the consequences of *atherosclerosis* is the production of infarcts — areas of dying tissue and scar formation following interruption of circulation due to obstruction or rupture of blood vessels. The occurrence of *multiple infarcts* leads to progressive destruction of brain tissue, which may be responsible for a form of senile dementia, properly called "multiple-infarct senile dementia," contrasted with Alzheimer's dementia (Chapter 8). Ischemic injury to the brain can result from several processes, ranging from focal ischemia due to occlusion of an artery supplying a region of the brain, to global ischemia, which occurs during cardiac arrest and resuscitation and reflects a transient loss of blood to the entire brain. Hypoxia and hypoglycemia usually accompany ischemic injury. Neurons are more susceptible than glial cells to ischemia, hypoxia, and hypoglycemia. These three conditions activate glutamate receptors and induce excitotoxicity, that is, an abnormal increase in the levels of the excitatory amino acids such as glutamate, with consequent neuronal cell death. The same conditions also may damage the blood–brain barrier.

The cerebral capillaries are much more permeable at birth than in adulthood, and the *blood–brain barrier* develops during the early years of life. This barrier is a functional concept based on specific mechanisms of exchange across the cerebral capillaries different from all other vascular beds. This peculiar type of exchange permits only a few substances to enter the nervous tissue (hence, the term "barrier") and thus, serves a protective function (but it also limits the effectiveness of several medications that cannot cross the barrier and pass from general circulation in the brain).

The efficiency of the blood–brain barrier increases with age during early development. With aging and disease, the blood–brain barrier may again become permeable at least to selected substances, and the passage of blood-borne substances may represent one of the causes of such dementias as that of the Alzheimer type. Studies of the structure of cerebral capillaries at progressive ages reveal an increase in the capillary wall thickness in old rats and a decrease in mitochondrial content of endothelial cells in old monkeys. Although direct evidence for increased permeability of the blood–brain barrier as the primary pathogenic factor in AD is still lacking, defects of the blood–brain barrier may represent a common pathogenic mechanism linking many different risk factors.

# REFERENCES

1. Cajal, S.R. and May, R.T., *Degeneration and Regeneration of the Nervous System*, Hafner, New York, 1959.
2. Filogamo, G. et al., Eds., Brain plasticity, in *Advances in Experimental Medicine and Biology*, Vol. 429, Plenum Press, New York, 1997.
3. Freund, H., Sabel, B.A., and Witte, O.W., Eds., Brain plasticity, in *Advances in Neurology*, Vol. 73, Lippincott-Raven, Philadelphia, 1997.
4. Eriksson, P.S. et al., Neurogenesis in the adult human hippocampus, *Nature Med.*, 4, 1313, 1998.
5. Gould, E. et al., Proliferation of granule cell precursors in the dentate gyrus of adult monkeys is diminished by stress, *Proc. Natl. Acad. Sci. U.S.A.*, 95, 3168, 1998.
6. Fuchs, E. and Gould, E., *In vivo* neurogenesis in the adult brain: Regulation and functional implications, *Eur.J. Neurosci.*, 12, 2211, 2000.
7. Lowenstein, D.H. and Parent, J.M., Brain, heal thyself, *Science*, 283, 1126, 1999.
8. Kesslak, J.P. et al., Learning upregulates brain-derived neurotropic factor messenger ribonucleic acid: A mechanism to facilitate encoding and circuit maintenance? *Behav. Neurosci.*, 112, 1012, 1998.
9. Cotman, C.W., Axon sprouting and regeneration, in *Basic Neurochemistry*, Siegel, G. et al., Eds., Lippincott-Raven, Philadelphia, 1999.
10. Helmut, L., Glia tell neurons to build synapses, *Science*, 291, 569, 2001.
11. Ullian, E.M. et al., Control of synapse number by glia, *Science*, 291, 657, 2001.
12. Bjorklund, A. and Lindvall, O., Cell replacement therapies for central nervous system disorders, *Nature Neuroscience*, 3, 537, 2000.
13. Gage, F., Mammalian neural stem cells, *Science*, 287, 1433, 2000.
14. Vogel, G., Stem cells: New excitement, persistent questions, *Science*, 290, 1672, 2000.
15. Freed, C.R. et al., Transplantation of embryonic dopamine neurons for severe Parkinson's disease, *N. Engl. J. Med.*, 344, 710, 2001.
16. Weber, W. and Butcher, J., Doubts over cell therapy for Parkinson's disease, *Lancet*, 357, 859, 2001.
17. Vogel, G., Parkinson's research: Fetal cell transplant trial draws fire, *Science*, 291, 2060, 2001.
18. Brundin, P. et al., Improving the survival of grafted dopaminergic neurons: A review over current approaches, *Cell Transplantation*, 9, 179, 2000.
19. Waldemar, G., Functional brain imaging with SPECT in normal aging and dementia. Methodological, pathophysiological, and diagnostic aspects, *Cerebrovasc. Brain Metab. Rev.*, 7, 89, 1995.
20. Loessner, A. et al., Regional cerebral function determined by FDG-PET in healthy volunteers: Normal patterns and changes with age, *J. Nucl. Med.*, 36, 1141, 1995.
21. Budinger, T.F., Brain imaging in normal aging and in Alzheimer's disease, in *Studies of Aging*, Sternberg, H. and Timiras, P.S., Eds., Springer, New York, 1999.
22. Newberg, A.B., Alavi, A., and Payer, F., Single photon emission computed tomography in Alzheimer's disease and related disorders, *Neuroimaging Clin. N. Am.*, 5, 103, 1995.
23. De Leon, M.J. et al., The hippocampus in aging and Alzheimer's disease, *Neuroimaging Clin. N. Am.*, 5, 1, 1995.
24. Sacher, G.A., Maturation and longevity in relation to cranial capacity in hominid evolution, in *Antecedents of Man and After*, 1, The Hague, Mouton, 1975, 419.
25. Brazier, M.A.B., *The Historical Development of Neurophysiology*, American Physiological Society, Washington, DC, 1959.
26. Brizzee, K.R., Sherwood, N., and Timiras, P.S., A comparison of cell populations at various depth levels in cerebral cortex of young adult and aged Long-Evans rats, *J. Gerontol.*, 23, 289, 1986.
27. Brizzee, K.R., Neuron aging and neuron pathology, in *Relations Between Normal Aging and Disease*, Johnson, H.A., Ed., Raven Press, New York, 1985, 191.
28. Long, J.M. et al., What counts in brain aging? Design-based stereological analysis of cell number, *J. Gerontol.*, 54, B407, 1999.
29. Sjobeck, M., Dahlen, S., and Englund, E., Neuronal loss in the brainstem and cerebellum — part of the normal aging process? A morphometric study of the vermis cerebelli and inferior olivary nucleus, *J. Gerontol.*, 54, B163, 1999.
30. Muir, J.L., Acetylcholine, aging, and Alzheimer's disease, *Pharmacol. Biochem. Behav.*, 56, 687, 1997.
31. Sian, J. et al., Parkinson's disease: A major hypokinetic basal ganglia disorder, *J. Neural Transm.*, 106, 443, 1999.
32. Scheibel, M.E. et al., Progressive dendritic changes in aging human cortex, *Exp. Neurol.*, 47, 392, 1975.
33. Cho, H.S. et al., Age-related changes of mRNA expression of amyloid precursor protein in the brain of senescence-accelerated mouse, *Comp. Biochem. Physiol. B Biochem. Mol. Biol.*, 112, 399, 1995.

34. Neill, D., Alzheimer's disease: Maladaptive synapto-plasticity hypothesis, *Neurodegeneration*, 4, 217, 1995.

35. Goldberger, A.L., Is the normal heartbeat chaotic or homeostatic? *News Physiol. Sci.*, 6, 87, 1991.

36. Lipsitz, L.A. and Goldberger, A.L., Loss of "complexity" and aging. Potential applications of fractals and chaos theory to senescence, *J. Am. Med. Assoc.*, 267, 1806, 1992.

37. Freeman, W.J., *Societies of Brains. A Study in the Neuroscience of Love and Hate,* Lawrence Erlbaum Associates, Hillsdale, NJ, 1995.

38. Fries, P. et al., Modulation of oscillatory neuronal synchronization by selective visual attention, *Science*, 291, 1560, 2001.

39. Lantos, P.C. and Papp, M.I., Cellular pathology of multiple system atrophy: A review, *J. Neurol. Neurosurg. Psychiatry*, 57, 129, 1994.

40. Brizzee, K.R. et al., The amount and distribution of pigments in neurons and glia of the cerebral cortex. Autofluorescence and ultrastructural studies, *J. Gerontol.*, 24, 127, 1969.

41. Morrison, B.M., Hof, P.R., and Morrison, J.H., Determinants of neuronal vulnerability in neurodegenerative diseases, *Ann. Neurol.*, 44 (3 Suppl. 1), S32, 1998.

42. Forno, L.S., Pathology of Parkinson's disease: The importance of the substantia nigra and Lewy bodies, in *Parkinson's Disease*, Stern, G., Ed., Chapman Hall, London, 1990.

43. Ince, P.G., Perry, E.K., and Morris, C.M., Dementia with Lewy bodies. A distinct non-Alzheimer dementia syndrome? *Brain Pathology*, 8, 299, 1998.

44. Graves, A.B. et al., Impaired olfaction as a marker for cognitive decline. Interaction with apolipoprotein E4 status, *Neurology*, 53, 1480, 1999.

45. Burns, A., Might olfactory dysfunction be a marker of early Alzheimer's disease? *Lancet*, 355, 84, 2000.

46. Lee, V.M., Disruption of the cytoskeleton in Alzheimer's disease, *Curr. Opin. Neurobiol.*, 5, 663, 1995.

47. Finch, C.E. and Cohen, D.M., Aging, metabolism, and Alzheimer's disease: Review and hypotheses, *Exp. Neurol.*, 143, 82, 1997.

48. Schoenfeld, T.A. and Obar, R.A., Diverse distribution and function of fibrous microtubule-associated proteins in the nervous system, *Int. Rev. Cytol.*, 151, 67, 1994.

49. Brady, S.T., Motor neurons and neurofilaments in sickness and in health, *Cell*, 73, 1, 1993.

50. Mandelkow, E.M. et al., Structure, microtubule interactions, and phosphorylation of tau protein, *Ann. N. Y. Acad. Sci.*, 777, 96, 1996.

51. Strittmatter, W.J. et al., Hypothesis: Microtubule instability and paired helical filament formation in the Alzheimer disease brain are related to apoprotein E4 genotype, *Exp. Neurol.*, 18, 163, 1994.

52. Nothias, F. et al., The expression and distribution of tau proteins and messenger RNA in rat dorsal root ganglion neurons during development and regeneration, *Neuroscience*, 66, 707, 1995.

53. Kosik, K.S., Alzheimer's disease: A cell biological perspective, *Science*, 256, 780, 1992.

54. Cole, G.M. and Timiras, P.S., Ubiquitin-protein conjugates in Alzheimer's lesions, *Neurosci. Let.*, 79, 207, 1987.

55. Andersen, J.K., What causes the build-up of ubiquitin-containing inclusions in Parkinson's disease? *Mech. Ageing Dev.*, 118, 15, 2000.

56. Ohtsuka, H., Takahashi, R., and Goto, S., Age-related accumulation of high-molecular-weight ubiquitin protein conjugates in mouse brains, *J. Gerontol.*, 50, B277, 1995.

57. Thomas, R.J., Excitatory amino acids in health and disease, *J. Am. Geriatr. Soc.*, 43, 1279, 1995.

58. Bredt, D.S. and Snyder, S.H., Nitric oxide: A physiological messenger molecule, *Annu. Rev. Biochem.*, 63, 175, 1994.

59. Haley, J.E., Gases as neurotransmitters, *Essays in Biochemistry*, 33, 79, 1998.

60. Ebadi, M., Murrin, L.C., and Pfeiffer, R.F., Hippocampal zinc thionein and pyridoxal phosphate modulate synaptic functions, *Ann. N. Y. Acad. Sci.*, 585, 189, 1990.

61. Golub, M.S. et al., Developmental zinc deficiency and behavior, *J. Nutr.*, 125 (Suppl. 8), 2263S, 1995.

62. Rosahl, T.W. et al., Essential functions of synapsins I and II in synaptic vesicle regulation, *Nature*, 375, 488, 1995.

63. Timiras, P.S., Hudson, D.B., and Segall, P.E., Lifetime brain serotonin: Regional effects of age and precursor availability, *Neurobiol. Aging*, 5, 235, 1984.

64. Andersen, J.K., Does neuronal loss in Parkinson's disease involve programmed cell death? *BioEssays*, 23, 1, 2001.

65. McEntee, W.J. and Crook, T.H., Glutamate: Its role in learning, memory, and the aging brain, *Psychopharmacology*, 111, 391, 1993.

66. Garcia-Ladona, F.J. et al., Exitatory amino acid AMPA receptor mRNA localization in several regions of normal and neurological disease affected human brain. An *in situ* hybridization histochemistry study, *Brain Res. Mol. Brain Res.*, 21, 75, 1994.

67. Gerlach, M., Riederer, P., and Youdim, M.B., Molecular mechanisms for neurodegeneration: Synergism between reactive oxygen species, calcium and excitotoxic amino acids, *Adv. Neurol.*, 69, 177, 1996.

68. Martin, J.B., Molecular basis of the neurodegenerative disorders, *N. Engl. J. Med.*, 340, 1970, 1999.

69. Nadeau, S.E., Clinical decisions: Parkinson's disease, *J. Am. Geriatr. Soc.*, 45, 233, 1997.

70. Gerlach, M. and Riederer, P., Animal models of Parkinson's disease: An empirical comparison with the phenomenology of the disease in man, *J. Neural Transm.*, 103, 987, 1996.

71. Fahn, S., Welcome news about levodopa, but uncertainty remains, *Ann. Neurol.*, 43, 551, 1998.

72. Fox, S.H. and Brotchie, J.M., New treatments for movement disorders, *Trends Pharmacol. Sci.*, 17, 339, 1996.

73. Gerlach, M. and Riederer, P., New approaches in the treatment of Parkinson's disease, in *Topics in Pharmaceutical Sciences*, Crommelin, D.J.A. et al., Eds., Medpharm Scientific Publishers, Stuttgart, 1994.

74. Gerlach, M., Youdim, M.B., and Riederer, P., Pharmacology of selegiline, *Neurology*, 47, S137, 1996.

75. Waters, C., Catechol-*O*-methyltransferase (COMT) inhibitors in Parkinson's disease, *J. Am. Geriatr. Soc.*, 48, 692, 2000.

76. Andersen, J.K., Use of genetically engineered mice as models for exploring the role of oxidative stress in neurodegenerative diseases, *Frontiers Biosci.*, 3, 8, 1998.

77. Jha, N. and Andersen, J.K., Loss of glutathione (GSH) in Parkinson's Disease: How does GSH act to protect dopaminergic neurons of the substantia nigra? in *Recent Research Developments in Neurochemistry*, Vol. 2, Pandalai, S.G., Ed., Research Signpost, Kerala, 1999.

78. Yanteri, F. and Andersen, J.K., The role of iron in Parkinson's disease and MPTP toxicity, *IUBMB Life*, 48, 1, 1999.

79. Choi-Lundberg, D.L. et al., Dopaminergic neurons protected from degeneration by GDNF gene therapy, *Science*, 275, 838, 1997.

80. Kordower, J.H. et al., Neurodegeneration prevented by lentiviral vector delivery of GDNF in primate models of Parkinson's disease, *Science*, 290, 767, 2000.

81. Zetterstrom, R.H. et al., Dopamine neuron agenesis in Nurr1-deficient mice, *Science*, 276, 248, 1997.

82. Lang, A.E. et al., Posteroventral medial pallidotomy in advanced Parkinson's disease, *N. Engl. J. Med.*, 337, 1036, 1997.

83. Dostrovsky, J.O. et al., Electrical stimulation-induced effects in the human thalamus, *Adv. Neurol.*, 63, 219, 1993.

84. Yudofsky, S.C., Parkinson's disease, depression, and electrical stimulation of the brain, *N. Engl. J. Med.*, 340, 1500, 1999.

85. Berardelli, A. et al., Cortical inhibition in Parkinson's disease: A study with paired magnetic stimulation, *Brain*, 119, 71, 1996.

86. Bejjani, B.P. et al., Transient acute depression induced by high-frequency deep-brain stimulation, *N. Engl. J. Med.*, 340, 1476, 1999.

87. Fischbach, G.D., Cell therapy for Parkinson's disease, *N. Engl. J. Med.*, 344, 763, 2001.

88. Olson, L., Combating Parkinson's disease — step three, *Science*, 290, 721, 2000.

89. McKay, R., Stem cells in the central nervous system, *Science*, 276, 66, 1997.

90. Brustle, O. et al., Embryonic stem cell-derived glial precursors: A source of myelinating transplants, *Science*, 285, 754, 1999.

91. Wakayama, T. et al., Differentiation of embryonic stem cell lines generated from adult somatic cells by nuclear transfer, *Science*, 292, 740, 2001.

92. Luquin, M.R. et al., Recovery of chronic parkinsonian monkeys by autotransplants of carotid body cell aggregates into putamen, *Neuron*, 22, 743, 1999.

93. Coleman, D.R., Neurites, synapses and cadherins reconciled, *Mol. Cell. Neurosci.*, 10, 1, 1997.

94. Serafini, T., An old friend in a new home: Cadherins at the synapse, *Trends Neurosci.*, 20, 322, 1997.

95. Walsh, F.S. and Doherty, P., Neural cell adhesion molecules of the immunoglobulin superfamily: Role in axon growth and guidance, *Annu. Rev. Cell. Dev. Biol.*, 13, 425, 1997.

96. Coleman, D.R. and Filbin, M.T., Cell adhesion molecules, in *Basic Neurochemistry*, Siegel, G. et al., Eds., Lippincott-Raven, Philadelphia, 1999.

97. Yew, D.T. et al., Neurotransmitters, peptides, and neural cell adhesion molecules in the cortices of normal elderly humans and Alzheimer patients: A comparison, *Exp. Gerontol.*, 34, 117, 1999.

98. Ronn, L.C. et al., The neural cell adhesion molecule in synaptic plasticity and ageing, *Int. J. Dev. Neurosci.*, 18, 193, 2000.

99. Thoenen, H., Neurotrophins and activity-dependent plasticity, *Brain Res.*, 128, 183, 2000.

100. Mufson, E.J. et al., Distribution and retrograde transport of trophic factors in the central nervous system: Functional implications for the treatment of neurodegenerative diseases, *Prog. Neurobiol.*, 57, 451, 1999.

101. Hadjiconstantinou, M. and Neff, N.H., GM1 ganglioside: *in vivo* and *in vitro* trophic actions on central neurotransmitter systems, *J. Neurochem.*, 70, 1335, 1998.

102. Dreyfus, H. et al., Gangliosides and neurotrophic growth factors in the retina. Molecular interactions and applications as neuroprotective agents, *Ann. N. Y. Acad. Sci.*, 845, 240, 1998.

103. Timiras, P.S., Hudson, D.B., and Oklund, S., Changes in central nervous system free amino acids with development and aging, *Prog. Brain Res.*, 40, 267, 1973.

104. Clarke, D.D. and Sokoloff, L., Circulation and energy metabolism of the brain, in *Basic Neurochemistry*, Siegel, G. et al., Eds., Lippincott-Raven, Philadelphia, 1999.

105. Hoyer, S., Oxidative energy metabolism in Alzheimer's brain. Studies in early-onset and late-onset cases, *Mol. Chem. Neuropathol.*, 16, 207, 1992.

106. Heininger, K.A., A unifying hypothesis of Alzheimer's disease. IV. Causation and sequence of events, *Rev. Neurosci.*, 11, 213, 2000.

107. Gafni, A., Structural modifications of proteins during aging, *J. Am. Geriatr. Soc.*, 45, 871, 1997.

108. Hardy, J. and Gwinn-Hardy, K., Neurodegenerative disease: A different view of diagnosis, *Mol. Med. Today*, 5, 514, 1999.

**Systemic and Organismic Aging**

# 8 The Nervous System: Functional Changes*

*Paola S. Timiras*
University of California, Berkeley

## CONTENTS

## I.   INTRODUCTION

The functional integrity of the nervous system is well maintained in most elderly persons despite the morphologic and biochemical changes described in the previous chapter. This ability to conserve neurologic and intellectual competence into old age attests to the redundancy present in the system and its capability, under certain conditions, of increasing the number of neuronal and glial cells. As noted in Chapter 7, the recent demonstration of new cell growth in the brain of adult and old animals, including humans, suggest that compensatory mechanisms and some regenerative capacity persist in the aging brain. Indeed, ongoing research affords some encouragement that neurodegenerative diseases may eventually be treatable or prevented altogether.

The difficulty for nerve cells, as for most cells of the body, including neurons formed from multipotent stem cells, is that function depends not only on their number but also on their inherent capacity to connect with other cells: they must wire themselves in some fashion, reach their synaptic target, and be able to trigger the appropriate responses.

---

\* Illustrations by Dr. S. Oklund.

Despite the persistence of central nervous system (CNS) plasticity in old age, we must recognize that in many elderly individuals, especially in the very old, the boundaries between health and disease become less precise as the incidence of neurologic and mental impairment increases. Whether normal aging is simply an early stage of pathology without obvious clinical expression, or whether aging-associated diseases represent cases of *"accelerated aging"* or are *"distinctive disease processes"* remains to be evaluated (Chapter 3). The prevalent view is that the aged cell may represent a favorable environment for pathological changes to occur (Chapters 3 and 4). CNS disorders such as impaired memory, intellect, strength, sensation, balance, and coordination account for nearly half the incidence of disability in individuals beyond the age of 65 and underlie more than 90% of the cases of total dependency in this population.

One characteristic of the aging nervous system that came to light in the previous chapter is the resistance of discrete structures or biochemical pathways to the aging process, and, similarly, the differential effects of aging on various CNS functions. In the text that follows, the effects of aging will be discussed for three major functional areas of the nervous system:

- For motor function, gait, and balance
- For electrical activity, sleep
- For cognitive function, memory

Disruption of these functions can lead to neurologic (e.g., locomotor) and mental (e.g., cognitive, as in dementia) deficits. Dementias are characterized by global loss of cognitive function associated with neurologic deficits and vascular disturbances; they have catastrophic consequences for those who are demented, their families, and the society in which they live. Despite some optimistic trends that suggest a current decrease in the prevalence of dementia, this disease was, at the end of the last century, and still is, the major (90%) cause of the cases of total dependency and the fourth leading cause of death in the 75- to 84-year-old age group.

## II. MOTOR CHANGES: POSTURE, GAIT, AND BALANCE

### A. CONTROL OF POSTURE AND MOVEMENT

Posture (the bearing of one's body that provides a stable background for movement) and movement (the ability to change posture and position) are regulated by a number of structures and functions within and without the nervous system (Table 8.1). Here, nervous regulation is considered primarily with respect to CNS control of movement and, in Chapter 9, with respect to control of sensory (e.g., visual, vestibular) inputs. Outside the CNS, major structures

**TABLE 8.1**
**Structures and Systems Controlling Posture, Balance, and Mobility**

**Central nervous system (CNS)**
Cerebral cortex
Basal ganglia
Cerebellum
Vestibular-ocular and proprioceptive pathways
Limbic system
Spinal cord
**Skeletal muscles**
**Bones and joints**
**Hormones**
**Blood circulation**

involved in the execution of movement and balance are the skeleton, joints, and skeletal muscles (Chapter 21) and, indirectly, for metabolic support, the cardiovascular (Chapter 16) and endocrine (Chapters 10 through 14) systems. The role of physical exercise in improving mobility and balance and in promoting a better quality of life and longer survival in old age is discussed in Chapter 24.

#### PYRAMIDAL AND EXTRA-PYRAMIDAL MOTOR PATHWAYS

**Skilled movements** (e.g., finger movements) are regulated in the brain by nerve fibers that originate in the motor cortex and form the "pyramids" in the medulla, hence, the term pyramidal tracts. **Grosser movements and posture** are regulated by CNS areas (e.g., basal ganglia) other than those connected with the pyramidal tracts, hence, by exclusion, the term extrapyramidal pathways. **Coordination, adjustment, and smoothing of movements** are regulated by the cerebellum, which receives impulses from several sensory receptors. Impulses from all of these brain structures ultimately determine the pattern and rate of discharge of the spinal motor neurons and neurons in motor nuclei of cranial nerves, thereby controlling somatic motor activity. Axons from these neurons form motor nerves that travel from the spinal cord to the various skeletal muscles throughout the body. Once the muscle has been reached, motor nerves synapse at the myoneural junction and transmit the nerve impulse to the muscle fibers. **Contraction or relaxation of muscles** will, in turn, direct bone and joint movements to maintain posture and promote mobility.

With aging, skilled motor movements are slowed, and gross movements, particularly those related to maintenance of posture and gait, are altered.[1-5] These alterations affect the speed of movement, which may be accelerated (as in hyperkinesias, tremors, tics) or slowed (as in hypokinesias, akinesia). Alternatively, they may affect the contraction of specific muscles, resulting in abnormal movements (as in myoclonus, chorea, ballism, athetosis) or in abnormal posture (as in dystonias). Alterations of movement and posture lead to imbalance and, thus, to a higher incidence of falls, one of the most frequent and life-threatening accidents of old age.[6-10] Falling

**TABLE 8.2**
**Key Measures of Functional Assessment of Balance and Gait**

| Balance | Gait |
|---|---|
| Sitting balance | Initiation of gait |
| Rising from a chair | Step height/foot clearance |
| Immediate standing balance | Step length |
| Standing balance with eyes closed (feet close together) | Step symmetry |
| Turning balance (360°) | Step continuity |
| Nudge on sternum | Path deviation |
| Neck turning | Trunk stability |
| One leg standing balance | Walking stance |
| Back extension | Turning while walking |
| Reaching up | Width of base |
| Bending down | Arm swing |
| Sitting down | |
| Recovery from spontaneous loss of balance | |

is a serious public health problem among the elderly because of its frequency, the associated morbidity, and the cost of follow-up health care. Serious injuries resulting primarily from falls represent the sixth leading cause of death among individuals 65 years of age and older.[11–15]

With advancing age, the normal adult gait (manner or style of walking) changes to a *hesitant, broad-based, small-stepped gait* with many of the characteristics of early parkinsonism (Chapter 7), often including *stooped posture, diminished arm swinging, and turns performed en bloc* (i.e., in one single, rigid movement). Some frequently used measures of balance and gait are listed in Table 8.2. Rarely do individuals over the age of 75 to 80 walk without the stigmata of age. Indeed, it is the ability to walk without serious limitations or falling that distinguishes a normal-aged gait from a dysfunctional one. Alterations of locomotion are due to CNS impairment; the peripheral changes seen in normal aging, including minor decreases in nerve conduction velocity, decrease in muscle mass, and increased muscle tone (rigidity), are insufficient to account for the disability.

Despite the frequency of these age-related changes in the regulation of gait and balance, relatively little is known of the mechanisms responsible for them. Nevertheless, much can be learned about the overall physiologic competence of the aging individual by observing gait and balance. For example, disturbances of gait and repeated falls, in the very old, particularly, are signs that herald or reflect serious ill health.[11–15]

## B. Changes in Gait

Normal walking is with head erect (without spinal curvature), arms swinging reciprocally (without grabbing at furniture), regular stepping (without stagger or stumbling movements), and feet clearing the ground at each step. Several of these characteristics are altered with aging: the gait slows, step length shortens, and there is increased irregularity and hesitancy in the walking pattern. Such changes are indicative of impaired stability and may depend on multiple neurologic and extraneurologic causes. Sex differences are found in all ethnic groups, with females always performing less well than males on all gait parameters, especially when affected with impaired vision or cognition and under conditions of poor illumination. Demented patients show significant decrements in all gait parameters.

Measurements of gait characteristics, such as speed and length of stride (stride is the distance between two successive contacts of the heel of the same foot with the ground), do not present any technical difficulty and can provide useful information on the degree of competence or damage of the central and peripheral nervous system. Several pyramidal and extrapyramidal structures, including visual, vestibular, and proprioceptive sensors impart finely graded instructions to the muscles of the neck, trunk, and limbs for maintenance of normal posture, balance, and gait. Alteration in any of these functions will reflect impairment and failure of integration of one or more of the CNS structures involved.[16–18] Further, impairment of gait and balance may also indicate disturbances of the vascular and mental status as well as the conditions of the skeleton and joints and orthopedic disorders (Chapter 21).[19,20]

Similarly, gait changes may be useful in providing other clinical insights. For example, gait asymmetry gives a clue to hemiplegia or arthritis; alterations of shoulder movements in walking may be a clue to parkinsonism; increase in stride width relates to cerebellar disease and arthritis; and trunk flexion suggests unstable balance due to alterations in visual, vestibular, and proprioceptive controls.

## C. Changes in Balance, Falls

*Balance* (stable physical equilibrium), as with gait, can be studied clinically by simply observing individuals as they rise from a chair, stand, walk, or turn: Do they sway, sweep, and stagger when performing these movements? In the elderly, the fear of falling and pain or limitation of joint movement are all reflected in their carriage.[21–24] The main adaptation to a balance disorder is the shortening of step length accompanied by slowing of gait and increasing of time between steps. This pattern is particularly noticeable in people who have fallen repeatedly, and indeed, it is called "post-fall syndrome" or "3 Fs syndrome" (fear of further falling).[24] The immediate consequence of alterations in balance is an increased frequency of falls. There is a good deal of reserve in the postural system, and young adults can fairly well tolerate experiments in which they are tested in moving platforms with sensory input (primarily vision) absent. The elderly are much less tolerant of any loss or decline of sensory input — vision, for example.

*Falls* are one of the most common problems of old age, occurring most often indoors in the immediate environment of the home. Trips and accidents account for 45% of falls, but the incidence of falls decreases with further aging, probably due to the reduced mobility of the very old. Some falls occur without any external cause and may be due to alterations in peripheral (ocular, vestibular, and propioceptive) and central (cerebellar and cortical) coordination[25] or, especially in postmenopausal women, to bone fractures due to osteoporosis (Chapters 11 and 21).

Falls are often fatal in the elderly, but even the nonfatal falls have serious consequences, including physical injury, fear (the 3 Fs syndrome), functional deterioration, and institutionalization.[24] Performance-oriented evaluation of falls shows that their occurrence may be related to central information processing, including decrements in selective attention and choice reaction time (e.g., central processing), sensory input (e.g., vision), and effector components (e.g., muscles).[25]

Women are at greater risk for falls and their consequences than men. This greater vulnerability is due to a number of factors, including more severe osteoporosis and bone fragility, especially after menopause (Chapter 11), less muscle strength, more sedentary, physically less strenuous way of life (Chapters 21 and 24), and a greater degree of comorbidity and disability (Chapter 3). Comparison among ethnic groups indicates few differences in the increasing frequency and severity of motor disabilities, fractures, and falls with old age. Mexican-American women present a profile similar to that of non-Hispanic Caucasian women.[26] However, Japanese women (in Hawaii)[27] and African-American women[28] have lower rates of falls and fractures than Caucasian women of the same age.[27] These differences seem, in the main, to depend on a better neuromuscular performance in Japanese women and on lower incidence of osteoporosis in African-American women.

## D. "GET UP AND MOVE: A CALL TO ACTION FOR OLDER MEN AND WOMEN"[29]

There is considerable overall reserve in the locomotor system, and loss of one of the sources of control of postural information or maintenance may be of little consequence. However, in elderly individuals, in whom impairment tends to occur at several functional levels, this reserve is easily depleted. It is not difficult to understand why a fall is the consequence of the simultaneous loss of a number of factors, neurologic (e.g., extrapyramidal damage) and extraneurologic (drugs, cardiovascular, skeletal, psychological). As our methods for measuring gait and balance become more sophisticated and quantitative, the close examination of these two functions becomes increasingly more important as a means of providing a comprehensive picture of the physiologic or pathologic condition of the old individual. At the same time, improving these two functions, if not vital for survival, will considerably increase the well-being and overall health of the elderly.[30] Benefits of physical exercise versus a sedentary life are discussed in Chapter 24. A regimen of regular physical activity is indicated at all ages; when it is started at a young age and continued throughout life, it may confer significant benefits on health and longevity. Among these benefits, not least are those promoting better neurologic and mental activity in old age, as suggested by the effects of running in "boosting" neural cell number.

### RUNNING "BOOSTS" BRAIN CELLS

A reciprocity of regulation exists between the nervous and the skeletomuscular systems: the nervous system regulates motor activity, and skeletomuscular movements such as those involved in running, may stimulate new nerve cell growth.[31,32] While the benefits of physical exercise on longevity and quality of life in old age are well documented, the mechanisms underlying these effects remain speculative (Chapter 24). Studies in mutant mice affected by ataxia telangiectasia, a rare neurogenerative disease, characterized by progressive loss of brain cells first in the cerebellum and then throughout the brain, with consequent loss of motor control, show that when these mice are placed in running wheels, the miles they log correlate directly with an increase in the number of brain cells; in contrast, in the sedentary mice affected by the disease, most brain cells die.[33] Thus, by placing the animals in a running wheel, many more cells, instead of dying, actually become neurons. With the same mutant mice, the multipotent stem cells in the brain were lacking a critical t (Atm) gene and were incapable of becoming neurons or oligodendrocytes but had no difficulty in becoming astrocytes.[33] These and other data in "senior" and genetically slow-learning mice indicate the importance of running on brain cell growth in adult animals and point to the importance of glial–neuronal interactions (Chapter 7). Studies in humans show that new cell growth also occurs in adult and elderly humans.[34,35]

## III. CHANGES IN SLEEP AND WAKEFULNESS

### A. BIOLOGIC CLOCKS AND SLEEP CYCLES

Alterations occurring in several cyclic functions with aging may be ascribed to changes in so-called *biologic clocks* — the biological timepieces that govern the rhythm of several functions, such as the 24-hour (circadian) hormonal rhythms (Chapter 10) and many behaviors.[36,37] The cessation of menstrual cycles and of ovarian function at menopause (Chapter 11) exemplifies alterations occurring in one such rhythmically recurring event. Cyclic functions are thought to depend on specific signals, located primarily in the brain. These signals are required for the clock to progress through the synthesis of gene-regulated molecules (e.g., RNA, proteins, neurotransmitters).

A prime example of a neural cyclic function is the circadian *sleep/wakefulness cycle,* which undergoes characteristic changes with aging. A precise delineation of sleeping changes intrinsic to normal aging is still lacking. However, the incidence of chronic sleep-related abnormalities (e.g., poor sleep efficiency, frequent and prolonged nocturnal arousals, day sleepiness), becomes increasingly prevalent in old age[38–40] and does not show significant differences among ethnic groups.[41] Changes in sleep patterns have been viewed as part of the normal aging process, often associated with altered responses to external (light)[42] or internal (body temperature)[43] cues. Many of these common disturbances may also be related to the development of pathological processes that disturb sleep.[44,45]

As we all know, sleep is a regular and, usually, necessary phenomenon of our daily lives. Lack of sleep leads to a propensity for sleep. The pressure to sleep is strongest at night, particularly in the early hours of the morning. Yet the mechanisms and purpose of sleep still elude us. A popular assumption is that sleep provides a quiescent period during which the body recovers from the strains of the waking hours and that it represents a descent to a lower level of consciousness, but such a hypothesis remains to be proved. There are, however, certain facts that are well recognized:

1. There is a clear relationship of sleep to electrical activity of the brain, as measured by the electroencephalogram (EEG), that is, the recording of the variations in brain electrical activity. There is also a close association of sleep to whole-body visceral manifestations, such as changes in heart rate, respiration, basal metabolic rate, endocrine function, etc.
2. Sleep patterns change with age, that is, during growth and development and, possibly, in old age as well.
3. Sleep patterns may be modified in neuropsychiatric disorders and by the administration of psychotropic drugs.
4. There are two kinds of sleep: the rapid eye movement (REM) sleep, with an EEG resembling the alert, awake state, and a four-phase, non-REM, slow-wave (SW) sleep with unique EEG patterns (Figure 8.1).

## B. Sleep EEG Changes with Aging

The EEG represents the background electrical activity of the brain as characterized by wave patterns of different frequencies: alpha, 8 to 12 waves/sec; beta, 18 to 30 waves/sec; theta, 4 to 7 waves/sec; delta, less than 4 waves/sec. The alpha waves have the highest amplitude. With advancing age, the alpha rhythm (prominent in a

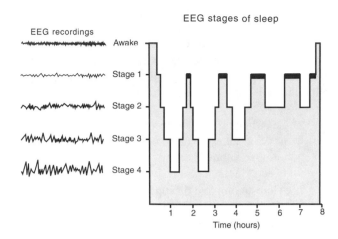

**FIGURE 8.1** *Diagrammatic representation of an 8-h sleep period.* The dark areas represent REM sleep. The EEG recordings (on the left side of the diagram) show the different rhythms that accompany each sleep stage.

person awake with eyes closed and mind at rest), slows throughout the brain as well as focally (temporal region). The beta activity increases in aged persons. With respect to the evoked potential, the data suggest that latency is prolonged after stimulation involving a variety of sensory modalities, perhaps due to decreased conduction velocity or a change in the sensory organs (Chapter 9). These alterations in spontaneous and evoked potential are aggravated in individuals with Alzheimer's disease. As illustrated in Figure 8.1, when people fall asleep, they go from a state of quiet wakefulness through four consecutive stages of SW sleep and several episodes of REM sleep.

In the normal adult, the total amount of sleep per day is approximately 7 h, of which 23% is spent in REM sleep and 57% in stage 2 SW sleep. Of the other stages, stage 1 accounts for 7% and 3 and 4 for about 13% (Figure 8.2). Generally, the episodes of REM sleep occur at about 90-min intervals. They lengthen as the night progresses; this is the stage in which dreaming occurs.

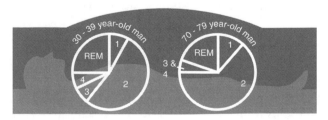

**FIGURE 8.2** *Relative distribution of sleep stages in adult (30 to 39 years) and aged (70 to 79 years) male individuals.* Quiet wakefulness (preceding the onset of sleep) is followed by four (1 through 4) stages of slow-wave sleep with periods of rapid-eye-movement (REM) sleep. In the aged, the period of quiet wakefulness and stages 1 and 2 are lengthened, while stages 3 and 4 of deep sleep have almost completely disappeared.

**Systemic and Organismic Aging**

With aging, the *total* amount of sleeping time changes little. However, nocturnal sleep is interrupted, and sleep time is distributed more widely over the 24-h period by adding short naps. While REM sleep time remains essentially unchanged, there are significant changes in the distribution of the other stages:

1. Lengthening of the period of quiet wakefulness: it takes longer to fall asleep, and the number of awakenings per night increases (sleep fragmentation)
2. Almost complete disappearance of stage 3 and disappearance of stage 4 so that there is little deep sleep time

The marked shortening of SW stages 3 and 4 may explain why older individuals complain subjectively of little sleep despite an overall near-normal daily sleeping time, and why they are much more easily aroused than the young. Additionally, although the statistics vary widely, there seems to be a progressive reduction of sleep time during the night, and about 40% of the elderly suffer from insomnia so defined. *Insomnia,* or "failure to maintain sleep," is a complaint with multiple manifestations: for example, "early sleep insomnia," or the inability to go to sleep before very late at night, or "early sleep offset," awakening very early in the morning. Insomnia has many diverse causes that require different treatments. Hypnotic medications to induce sleep should not be the mainstay of insomnia treatment; often they are ineffective, and they can be habit forming. They may be contraindicated when used with other medications (Chapter 23). The value of hygiene (e.g., regularization of bedtime, other behaviors), good nutrition, and regular physical exercise[46] should not be underestimated in prevention and treatment of insomnia (Chapter 24).

The relationship of the pineal hormone, melatonin, to sleep patterns and its potential value in the treatment of insomnia are discussed in Chapter 13.

## C. Respiratory, Cardiovascular and Motor Changes During Sleep in the Elderly

Periods of *apnea* (cessation of respiration) or *hypoapnea* (slowing of respiration) during the sleeping period increase with aging from an average of five respiratory disturbances per night at 24 years of age, to about 50 per night at 74 years of age. They are brought about by a collapse of the upper airway in the sleeping state and are terminated only by an arousal from sleep that restores the activity of the upper airway muscles. These incidents account for a great deal of the fragmented sleep experienced by the elderly. If matched against young and adult individuals in whom 5 to 8 episodes of apnea per night of sleep are considered the upper limit of normal, a number of elderly judged normal by other criteria would be diagnosed as having

sleep apnea, based on this respiratory pattern. Although the nocturnal hypoxia (low oxygen in blood and tissues) attendant to apnea is transient, changes in oxygen saturation may have adverse effects on brain function. In old age, disordered breathing during sleep, because it leads to hypoxia during the night and to fatigue during the day, may increase the risk of cognitive alterations, even dementia.[47,48] Sleepiness during the day also increases the risk of falls[49] and automobile traffic accidents.[50] Another consequence of impaired function of the upper airway is the increased prevalence of *snoring*. Current statistics indicate that 60% of males and 45% of females in their sixties are habitual snorers.

The significance of respiratory dysfunction associated with sleep in the otherwise normal elderly is not understood.[51] A number of factors appear to be involved, including aging-related alterations in neural and chemical regulation of respiration (Chapter 18). Cardiac arrhythmias (abnormal rhythm of the heartbeat) and pulmonary hypertension are common during apneic periods. Apnea, arrhythmia, and hypertension appear to be most frequent during REM sleep and have been associated with an increased release of norepinephrine from sympathetic stimulation, with mental disorders such as depression, or with the use of some neurotropic medications.[52]

*Sleep-related leg movements* are also common in the elderly. One third experience such "twitches" or leg discomfort every 20 to 40 sec during a large part of the night; they often bring about brief arousal from sleep.[53] The origin of these movements is little known, but they seem to be related to loss of coordination between motor excitation and inhibition. Sleep disturbances are also frequent in individuals suffering from arthritic pains (e.g., knee osteoarthritis),[54] although these interruptions disappear once the arthritic episode is terminated.

## D. The Role of the Reticular Activating and the Limbic Systems

*The reticular activating system*, formed of a network of interconnecting neurons distributed in the core midbrain, controls conscious alertness and, thus, makes sensory, motor, and visceral perception possible. Changes in sleep patterns with aging may be related to alterations in the level of alertness as manifested by the electrical (EEG) alterations discussed above and shifts in neurotransmitters, primarily serotonin (Chapter 7).[55] Serotonin appears to function as an inhibitory neurotransmitter that modulates the effects of light on circadian rhythmicity[56] and regulates several cyclic hormonal secretions.[57,58] It is also the precursor of the pineal hormone melatonin, which has potent sedative and hypnotic (sleep-inducing) activity (Chapter 13). Insomnia, frequent in the elderly, may depend on several factors superimposed on the aging process. Anxiety, depression, and stress, which always affect sleep, are prevalent among the elderly and may account for some of the sleep disturbances.

Changes in the reticular activating system with old age may not only modify sleep patterns but also may alter alertness and behavior. A decrease in sensory input to the higher brain centers may result either from failure of the reticular formation to receive, integrate, and relay signals to the sensory cortex, or from decrements in peripheral sensory perception (Chapter 9), or from both. Any impairment of sensory input would impair motor responses and behavior, decrements that can be detected in EEG recordings and physiologic responses. Such sensory-motor alterations among the aged may explain their decline in *response time,* that is, the speed with which one initiates a motor or behavioral reaction to a sensory stimulus. The greatest slowing of performance may be seen in demented individuals.

### THE ELDERLY AND THE MEDIA

With the lengthening of response time comes a reduced ability to receive and process information when delivered at a relatively fast pace. For example, many elderly individuals have difficulty following and understanding television and radio programs. Their difficulty is not always due to decreased vision or hearing (Chapter 9), but rather, to an overall decline in alertness, attention, and memory. Yet in nursing homes particularly, but also in private homes, television viewing is one of the most frequent pastimes of the elderly.

Research devoted to improving techniques for communicating information to the elderly populations should focus on optimizing delivery as well as providing substantive content. The dual role of mass media to entertain as well as to inform and educate is important to the elderly, who are limited by their isolation, confinement, and sensory deficits, and who are particularly susceptible to mass media influence. It should be the goal of the media services to use this susceptibility to provide understandable and useful programming.

Another brain system affected by old age is the limbic system, which regulates many types of autonomic responses (e.g., blood pressure, respiration) and behavior (e.g., sexual behavior, emotions of rage and fear, motivation). The limbic system consists of a rim of cortical tissue around the hilum of the cerebral hemispheres and of deep structures such as the amygdala and the hippocampus. These structures are involved in memory, mood, and motivation and are frequently affected in old age.

### WARNING

Because depression and insomnia, in some older people, and anxiety, restlessness, and aggressiveness in others always complicate any existing clinical disorders, a variety of drugs are administered to stimulate or tranquilize, depending on the condition. Drugs such as phenothiazines, barbiturates, dibenzazepines, monoamine oxidase inhibitors, etc., are part of the medicinal armamentarium of old people. None of these drugs

exists without side effects, and the metabolism of these drugs — as of most drugs — may be impaired in the elderly (Chapter 23).

## IV. MEMORY AND LEARNING

This and the following section address some of the functions regarded as "the higher functions of the nervous system." While "these functions of the mind" are numerous and range from motivation to judgement, cognition, language, and others, the focus here will be on some aspects of memory and learning (that is, "how we remember and how we learn") and, in the following section, on disorders of cognitive function (e.g., dementias).

*Learning* may be defined as the ability of the individual *to alter behavior on the basis of experience.* Learning depends, in part, on *memory,* which may be defined, in a large context, as a *processing–storage–retrieval function* of the brain and mind, that is, *the ability to recall past events.* While "learning" and "memory" frequently interact, both come in different forms that are thought to depend on different neural mechanisms and sites in the nervous system. It is not surprising, then, that changes in different kinds of learning and memory do not necessarily parallel each other. In addition, adaptation to life events occurs throughout the lifetime and depends on complex interactions between the environment and inherited genes (Chapter 1).

Environmental adjustments modify the nervous system, and as a result, animals can learn and remember. This ability, which can be viewed as an expression of neural plasticity (Chapter 7), is altered with aging.[59-63]

An almost universal complaint among older adults is the experience of not remembering as well as they once did. Impaired memory, from benign forgetfulness to major memory loss, seems to affect a large proportion, but not all, of the elderly. Memory is indispensable to normal cognitive function; hence, in some degenerative diseases of aging such as dementia, all cognitive functions, starting with memory, are lost.

### A. MEMORY ACQUISITION, RETENTION, AND RECALL: CHANGES IN OLD AGE

While disorders of memory may occur at all ages, they occur with increasing frequency in old age. As mentioned earlier, there are various types of memory. Their characteristics and underlying mechanisms are still being elucidated.[59-63] The long-held idea that information storage was widely and equally distributed throughout large brain regions has been displaced by the view indicating that memory is localized in specific areas of the brain. Actually, current theories hold that memory is *localized* in discrete brain areas involved in specific aspects of short-term memory and *widespread,* with many areas articulating to form long-term memory.[59-63] Among the primary areas involved

**Systemic and Organismic Aging**

**TABLE 8.3**
**Classifications of Kinds of Memory**

| Three Stages | Five Stages |
|---|---|
| **Sensory Memory** <br> An image is recorded rapidly, faster than 1 sec | **Nondeclarative Memory** <br> Corresponds to classic conditioning, usually unconscious <br> **The Perceptual Representational Model** <br> Responsible for early processing of sensory and perceptual information |
| **Short-Term Memory or Primary Memory** <br> Information endures several minutes | **Primary Memory** <br> Corresponds to Short-Term Memory |
| **Long-Term Memory or Secondary Memory** <br> May need hours or days to develop but lasts a lifetime | **Episodic Memory** <br> Refers to ability to recollect specific autobiographical events <br> **Semantic Memory** <br> Refers to store of factual knowledge independent from episodic recollection; these two stages together correspond to long-term memory |

are the limbic system (especially the hippocampus), the thalamus, the cerebral cortex (temporal, prefrontal, and frontal lobes) and the cerebellum. The view that memory resides in localized discrete brain areas evokes the concept of cell groups producing specific neurotransmitters, released into the synapse, and involved in the memory process. While neurotransmission is undoubtedly involved in memory and learning processes, the identification of which neurotransmitter or which combination of neurotransmitters is responsible for particular kinds of learning and memory continues to elude us.[64]

Memory is a complex process that involves the ability to sense (visually, audibly, by touch, by smell, etc.) an object or event, to formulate a thought, to retain this information, and to recall it at will. It is not surprising, therefore, that different (and still controversial) types of memory have been identified: two currently accepted classifications are presented in Table 8.3. Additional kinds of frequently used memory classifications are "explicit or declarative memory" (when we consciously recollect previous experiences), "implicit or nondeclarative memory" (when past experiences influence current behavior and performance even though we do not consciously recollect them), "prospective memory" correlated with "executive function" (e.g., the ability to plan, organize, self-monitor, and use strategies to remember to perform future actions), and others.

As for many other functions (Chapter 3), memory and learning show considerable variability among individuals, and this variability increases with advancing old age. Some healthy elderly retain intact memory until very old age (Figure 8.3). However, many aged subjects do not perform as well as younger subjects on many tasks having a significant memory component; in these individuals, the impairment appears to be irrespective of the length of time the information (to be recalled) was retained or of the type of cognitive skills required for its retention.

Memory loss in the elderly appears to be restricted primarily to memory for recent events, leaving immediate and remote memory essentially intact. Age differences, with less proficiency for the elderly, are found for some types of long-term memory (e.g., episodic memory) but not with others (e.g., semantic memory) (Table 8.4).

In the elderly, deficits in memory may be associated with other cognitive deficits and an overall decline in body functions and an increase in comorbidity (Chapter 3).[65] For example, the demented patient shows progressive, severe memory impairment (Figure 8.3). Memory loss for recent events is the first severe problem encountered. This stage is followed by language dysfunction, and, at this point, memory function is difficult to assess.[66] It has been hypothesized that deficits in cognitive functions occurring with dementia progress in the same order as they are acquired in childhood, but in reverse. As functional stages merge and overlap at certain points of human development, the broader lines between the various stages of dementia are subtle rather than abrupt. In severe cases of dementia, the ability to learn and to form new memories is drastically impaired or lost.

## B. Functional and Biochemical Correlates of Memory

Past and current studies have related memory changes in old age to impairment of the mechanisms that promote memory formation, such as reduced speed of elementary cognitive processes and reduced amount of attentional resources (e.g., decrease in alertness and awareness, sensory inputs, and information programming), all of which are required to accomplish complex cognitive tasks.[59–62] Recent reports attempt to explain memory losses in the aged as a failure to remove irrelevant information rather than as a failure of acquiring and retaining new information.[67–69] In other words, information no longer needed would remain active instead of being discarded. The failure to clear irrelevant information could easily disrupt memory.[67–69] Irrespective of whether memory impairment in old age is due to alteration of information acquisition or removal or both, the complexity of memory processes is further complicated by their susceptibility to numerous factors, such as changes in circadian rhythms as they affect

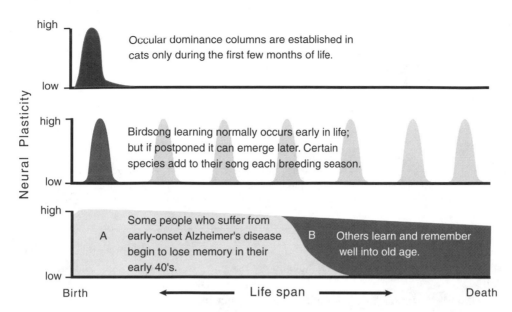

**FIGURE 8.3** *Diversity of temporal forms of plasticity and memory.* While brain plasticity (that is, the capacity to change in response to stimulation and experience) is particularly important during development, acquisition of learning and memory extends throughout the life span. Top: If a kitten's eye is deprived of receiving light during the first 3 months of life, the number of cortical cells is permanently reduced, and vision in that eye is impaired. Middle: Some kinds of plasticity occur at all ages when opportunity arises. Some birds not only remember the song they learned from their father at a very young age, but they can also add new songs to their repertoire during each breeding season. Bottom: In humans, some people continue to learn and remember into old age. However, in some diseases associated with old age (e.g., Alzheimer's Disease), memory and learning are progressively lost. Adapted with permission from *Biological Psychology*, Rosenzweig, M.R., Breedlove, S.M., and Leiman, A.L., Eds., Sinauer Associates, Inc., Sunderland, MA, 2002.

---

**TABLE 8.4**
**Summary of Changes in Human Memory with Old Age**

| Type of Memory | Changes with Old Age |
|---|---|
| *Procedural Memory*<br>Covers learning and retention of motor, counting, spelling, reading, other skills | *Unaffected by aging*<br>However, confusion (false memories) between "real" and "intended" action is more frequent in the elderly |
| *The Perceptual Representational System*<br>Responsible for early processing of sensory and perceptual information | *No proven change*<br>May depend on overall body function, decreasing in parallel with impaired physiologic competence, e.g., declining sensory (visual, acoustic) input |
| *Primary Short-Term Memory*<br>Refers to information held in mind | *Decline with aging is minimal when tasks performed are easy; more severe decline when tasks are complex* |
| *Episodic Memory*<br>Refers to ability to recollect specific autobiographical events that have occurred recently | *Declines starting at 30 years of age and progressively deteriorates to 80 years and older* |
| *Semantic Memory*<br>Refers to storage of factual knowledge independent from episodic recollection | *Declines little with age*<br>However, word-finding failures increase with age, especially retrieving names; spatial memory is reduced |
| *Remote Memory*<br>Refers to memories of the remote past | *Remembering declines with the remoteness of the event, to the same extent in younger and older adults*<br>However, childhood memories often are retained better in older adults |

**TABLE 8.5**
**Neurochemical Correlates of Changes in Memory with Old Age**

- Decreased activity of the brain cholinergic system, especially in hippocampus and midbrain
- Decreased activity of neurotrophic factors, NGF, neuropeptides, hormones, others
- Decreased $N$-Methyl-D-Aspartate (NMDA) NR2 receptors possibly accounting, in part, for impaired memory and learning in adulthood and old age
- Decreased protein synthesis due to:

   blocking of protein kinase activity responsible for amino acid phosphorylation

   in absence of protein phosphorylation, failure of cAMP response element (CRE) to bind to the CRE-binding (CREB) protein that promotes transcription

   impairment of long-term memory by preventing synaptic potentiation of repetitive presynaptic stimuli

- Changes in post-translational modification at the synapse, which may be responsible for impairment of short-term memory in old age
- Decreased cortical activation, as measured by brain-imaging techniques, has been recorded in old individuals in some brain area (frontal, temporal), while, in others (frontal), activation is increased (It is unclear whether increased activation represents recruitment of neurons to compensate for difficulty of task or diffuse, nondifferentiated activity.)

responses that depend on testing time, type of measures chosen to test memory, and aging-associated changes in inhibitory control.[68,69]

The cellular and molecular mechanisms underlying memory processes from early development to old age are being actively studied, and our understanding of them continues to evolve[64] (Table 8.5). Several neurotransmitters have been implicated in modulating and facilitating the acquisition and retention of information. One of the first to be considered was acetylcholine, the major transmitter in the cholinergic neurons of the septohippocampal and entorhinal areas and the nucleus of Meynert (all associated with memory). Indeed, in Alzheimer's dementia (AD), where memory loss is one of the signs of disease, several neurons appear to be missing in these brain areas (Chapter 7). Yet, an increase in the precursor, choline, an increase in the activity of the synthesizing enzyme (choline acetyltransferase), a decrease in the metabolizing enzyme (acetylcholinesterase), with or without supplementation with L-carnitine (an essential cofactor in transporting fatty acids from cytoplasm to mitochondria),[70] or the administration of acetylcholine agonists, have not been shown to normalize impaired memory, whether due to old age, disease, drugs, or experimental lesions. These failures have dampened but not destroyed interest in the role of acetylcholine in memory.[71] Current studies have shown that an inhibitor of acetylcholinesterase, the alkaloid galantamine, acts not only by inhibiting the activity of the enzyme acetylcholinesterase, thereby reducing the breakdown of acetylcholine and prolonging its action at the synapse, but also increases acetylcholine release.[72,73] At this date, preliminary studies in individuals with AD have found that galantamine bestows significant benefits on memory and behavior. Other attempts to bolster cholinergic inputs have involved transplanting embryonal, cultured, and stem cells in critical cholinergic areas, akin to the transplantation of dopaminergic cells in Parkinson's disease (Chapter 7). As with Parkinson's, results are encouraging but still uncertain.

Other neurotransmitters that may play a role in memory processes are serotonin[74] and glutamate.[75] Of these two neurotransmitters, glutamate, the main excitatory transmitter in the brain and spinal cord (Chapter 7), seems to be more involved in memory than serotonin. Glutamate acts by binding to two types of receptors: (1) those that regulate intracellular cyclic AMP (adenosine monophosphate) levels, and (2) those that are ligand-gated ion channels such as the NMDA ($N$-methyl-D-aspartate) receptors (with the subtypes, NR1 and NR2) that permit passage of relatively large amounts of $Ca^{++}$. There is a high concentration of NMDA receptors in the hippocampus, and blockade of these receptors prevents "long-term potentiation," a long-lasting facilitation of transmission in neural pathways following a brief period of high-frequency stimulation. Thus, these receptors may well be involved in memory and learning. Activation of the NMDA receptors may be necessary for converting new memories into long-term memories (Table 8.5). In addition to their role in memory, the possible involvement of excitotoxic amino acids such as glutamate, in a number of neurodegenerative diseases including AD, has been ascribed to their role in free radical accumulation (Chapter 5).

A number of *neuropeptides* and hormones have been implicated in normal memory processes and in AD. Two often proposed candidates are the hypothalamic *vasopressin* (Chapter 10) and the intestinal *cholecystokinin* (Chapter 20); however, there seems to be more substantive evidence that their primary function is in anxiety processes[76–77] and satiety[77] rather than in memory. *Nerve growth factor (NGF)* is another peptide important not only during neural development but also in later life, when it is involved in maintaining CNS plasticity. However, NGF production declines with aging. Replacement therapy with NGF is effective when it is administered directly in the brain,[78] a difficulty that can be circumvented by transplantation of NGF-producing cells or by genetic intervention.[78,79] Secretion of NGF and other neuropeptides has been stimulated in laboratory rats by providing "enriched" conditions[80] and, in the case of the *brain-derived neurotropic factor (BDNF)*, making them run.[81] While some memory improvement has been reported following administration of these neuropeptides as well as that of hormones (e.g., estrogens, see below) in animals, including humans, the efficacy of these *neurotropic factors* in preventing or reducing memory loss in the elderly remains controversial.

Invertebrates (mollusks, flies, worms) offer useful models for the study of memory and learning because of the relative simplicity of their neural networks and genome. Although informative data have been generated by use of these models, it remains to be ascertained how applicable these data are to humans.

A number of studies have tested the effects of stimulating or inhibiting neuronal protein synthesis, known to be implicated in long-term memory.[64,82] Inhibition of protein synthesis occurs in two stages: (1) decrease of the activity of protein kinase enzymes reduces protein phosphorylation; and, (2) when proteins are not phosphorylated, transcription is blocked because of the failure of the cAMP response element (CRE) to bind to the CRE-binding (CREB) protein that promotes transcription (Table 8.5).

While alterations in long-term memory have been explained in terms of changes in genomic (transcriptional) expression, alterations in short-term memory have been ascribed to posttranslation modification (that is, translation of the mRNA signal into a polypeptide) at the synapse.[82]

Another approach to uncovering the mechanisms at work in human memory uses imaging techniques with radioactive tracers (e.g.,14C-labeled 2-deoxyglucose) to measure increases that occur in metabolic activity in response to the presentation of familiar visual clues. The distribution of the metabolic changes is compatible with the view that memory is localized in specific brain areas, but the large number of neurons involved suggests that most "plastic" (i.e., metabolically responsive) cells participate in multiple forms of memory.[64] It is still unclear whether increased activation means that neurons are being recruited to compensate for the difficulty of the task at hand, or whether it simply indicates diffuse, nondifferentiated activity.[83]

Early studies suggested that memory and, especially learning, depended not on stimulation of neurons alone but rather on activation of a neuronal–glial complex, a functional unit including one or several neurons and several neighboring glial cells, primarily, astrocytes.[84] Currently, there is a renewed interest in this area of research, because neuroglial cells play a potentially critical role in brain plasticity and respond to a number of intrinsic (endocrine)[87] and extrinsic (environmental) stimuli.[85–87]

## C.  LEARNING AND LONGEVITY

According to a Japanese proverb: "Aging begins when we stop learning." Indeed, learning has emerged as a factor in prolonging life and reducing disability and disease in old age. Epidemiologic studies have reported that life expectancy varies directly with the amount of schooling one receives.[88,89] The benefit of education persists when "active life expectancy" or life free of disabilities (i.e., health span), is compared to "total life expectancy" or life with disabilities, and this finding is irrespective of sex and race. As Katzman[90] quotes, "Scholars grow wiser with age, but the non-educated become foolish." A higher level of education seems also to be associated with a lower prevalence of Alzheimer's disease.[90,91] In a long-term investigation, started in 1986 and continuing today, 678 Catholic nuns are being studied for the relationship between their writing proficiency at young age and continuing in middle and old age, and their longevity and incidence of dementia.[92,93] Reports to date indicate that those with the higher proficiency lived longer and had a lower incidence of Alzheimer's disease.[92,93]

The beneficial effect of education on longevity, and, indirectly, on normal and abnormal aging, has been ascribed to a number of factors, the most obvious being a better socioeconomic status (i.e., higher family income, greater employment opportunities). Other factors that may be implicated are listed in Table 8.6 and are further discussed in Chapter 24.

One of the most challenging interpretations of these observations is that extended learning may prevent (or at least, protract) the aging-related losses that occur in the nervous system with old age. The active process of learning is thought to build up a "brain reserve" in the form of an increased number and enhanced function of neurons and glial cells, better cerebral blood flow, higher oxygen levels, and glucose metabolism. Such an increased brain reserve may manifest as reduced or delayed neuronal losses or, conversely, reestablishment of neuronal and glial proliferation; increased synaptic density; changes in neurotransmitter levels (e.g., reduced sensitivity to excitotoxins), in receptors (e.g., NGF low and high affinity receptors stimulate and inhibit apoptosis, respectively), and in ion (e.g., $Ca^{++}$) transport and distribution. During normal brain aging as well as

---

**TABLE 8.6**
**Mechanisms of Effects of Increased Education on Successful Aging**

Adequate income
   Better access to medical care
   Better access to recreational activity
Good nutrition
Responsible health behaviors
Moderate alcohol intake
Abstinence from smoking
Possibility of increased brain reserve capacity
   More dendritic branching, more synapses
   Better cerebral blood flow
   Better neural cell efficiency, adaptability, redundancy, survival, and growth

---

**Systemic and Organismic Aging**

in the presence of neurodegenerative diseases, degenerative processes are often associated with adaptive growth and regeneration.[94,95] Continued learning activity may act by inducing and strengthening these adaptive responses at all ages, including old age, thereby giving validity to the well-known adage "use or lose it"[96] (Boxes 8.1 and 8.2).

## V. SENILE DEMENTIAS

### A. DEFINITIONS AND PREVALENCE

*Dementia* (from the Latin de-mens, without mind) refers to a global deterioration of intellectual and cognitive functions characterized by a defect of all five major mental functions — orientation, memory, intellect, judgement, and affect — but with persistence of a clear consciousness. Dementia, caused by a variety of factors, may occur at all ages and may be *reversible* or *irreversible*.

Reversible dementia is generally due to known causes, and once these are removed (e.g., drugs) or cured (e.g., infections), the dementia disappears. A handy mnemonic to remember the main causes of reversible dementia is presented in Table 8.7.

In the elderly, dementia may be secondary to any of the factors listed in Table 8.7. In most cases, it is primary, due to unknown causes, and irreversible (and progressive, that is, worsening with time). According to causes, pathology, and clinical manifestations, dementia has been categorized in several types, the distribution of which is illustrated in Figure 8.4:

- Senile dementia of the Alzheimer type (SDAT or the shorter abbreviation, AD) accounts for 50 to 60% of all senile dementia cases.
- Multiple infarct dementia accounts for 20 to 30% of all cases.
- Reversible dementias, as listed in Table 8.7, account for 10 to 20% of all cases.
- Depression or pseudodementias account for 1 to 5% of all cases.
- The remainder comprise miscellaneous disorders such as Parkinson's disease, Lewy body dementia, Pick's disease, and others.

The relative proportion of the different types of dementias varies somewhat with the population, the age of the patient, and the time period of the study, but overall, AD remains the most frequent.

Differences between normal and pathologic aging of the brain are, for the most part, essentially quantitative.[97] In the normally aging brain, neurofibrillary tangles and neuritic plaques are few (Chapter 7), whereas they are numerous and widely distributed in the brain of the AD patient; in fact, their accumulation is accepted as a definite diagnostic marker of the disease.

---

**Box 8.1**

**Continuing Education Throughout the Life Span**

The beneficial effects of education in prolonging life span and postponing the onset of disability and disease, although intellectually appealing, are still in need of continuing experimental and clinical support. As private research and government health programs attempt to improve our understanding of biologic phenomena, so must all organizations devoted to improving the quality of life at all ages recognize the value of continuing education to human health. By establishing a robust brain reserve at young ages, it is possible to draw from this reserve as we grow old or become disabled or ill. Given our present knowledge that the plasticity of the nervous system persists even in old age, the outlook for at least some degree of regeneration and compensation is brighter than it once was. Combined with other biologic and psychologic evidence, **there is every reason to support education as an important tool in the eternal quest for a better and longer life**.

---

**Box 8.2**

**Memory "Boosters"**

The crucial role of memory in cognition and learning — and, therefore, in most aspects of life — has led to the development of an abundance of memory training techniques, aggressively marketed in radio and television programs and books (e.g., "Mega Memory," "Total Memory Workout," "Page-a-Minute Memory Book," and many more). Nearly all of these techniques can be traced to the Greek **Simonides,** who, more than 2000 years ago, named them mnemonics and suggested them as remedies for specific neurologic problems (for which they still may be useful today) but not for overall memory improvement.

A number of drugs as well are being marketed as **cognitive enhancers** (e.g., Aricept) targeting the neurotransmitter acetylcholine; likewise, **food** as well as **supplements** (e.g., vitamin E and the herbal, ginkgo biloba) are targeting improvement in blood circulation or reduction in free radicals (e.g., Vitamin E) and in inflammation (e.g., nonsteroidal anti-inflammatory drugs, NSAIDs; Chapter 21). These substances have not been demonstrated to provide benefits in healthy or demented individuals and, in fact, may have dangerous side effects.

**TABLE 8.7**
**Underlying and Reversible Causes of Dementia**

D   Drugs
E   Emotional disorders
M   Metabolic or endocrine disorders
E   Eye and ear dysfunctions
N   Nutritional deficiencies
T   Tumor and trauma
I   Infections
A   Arteriosclerotic complications, i.e., myocardial infarction,
     stroke, or heart failure

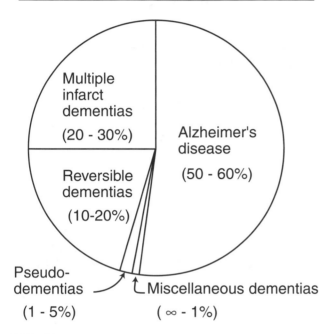

**FIGURE 8.4** Percentage of major forms of dementias in the elderly.

It is important to recognize that dementia is not an inevitable consequence of aging; the greater proportion of the elderly remains lucid and mentally competent until death.[97] Demographic data support this assertion. In 1982, severe dementia affected only 4 to 5% of those aged 65 years and older and 10% of those affected by mild to moderate forms. When the prevalence of dementia was compared at selected intervals (1982, 1988, 1994, 1999), these percentages showed a significant decline.[98] The incidence (that is, the rate of occurrence) of severe dementia increases with age from 0.01% per year at age 65, to 3.5% per year at age 85. Likewise, the *prevalence* (i.e., the number of cases of a disease existing for a given area and at a given time) also increases from less than 1% at ages 65 to 70 to greater than 15% at ages 85 and older. Indeed, *age is the single most important risk factor for dementia* in the older population,[99] even though some studies seem to indicate a slight decline in AD after age 90. Recent studies show a current decline in AD prevalence in the United States (see below).

The relatively young age of the patient, a woman, first described by Alzheimer, indicates that Alzheimer's dementia may also occur at younger ages (40 to 50 years) than those so far discussed. In general, those diagnosed with the *"adult" or "presenile" form of Alzheimer's dementia* present the same symptoms and pathology as those diagnosed with the "senile" form (65 years and older), but their symptoms are generally more severe, and the disease progresses more rapidly.

About two-thirds of the patients in nursing homes have dementia, therefore, two-thirds of the 20 to 30 billion dollars spent in nursing home care in the United States are expended on patients with dementia. Each year, several hundred thousand Americans develop the disease, and at least a hundred thousand die with it. We now have about 1.5 to 2 million patients. AD represents the fourth leading cause of death in the 75 to 84 age group, an estimate considered quite conservative by many. Even in the light of current studies that show a decrease in AD prevalence for the period 1982 to 1999 (see below), the problem of caring for demented patients is a serious one; it represents a significant burden for individual caretakers, health care resources, and society at large.

**POSSIBLE DECLINE IN PREVALENCE OF SEVERE COGNITIVE IMPAIRMENT AMONG OLDER AMERICANS**

Despite some previous pessimistic predictions, a current study allows us to be more optimistic about our ability to reduce the prevalence of severe cognitive impairment, such as AD by various health and societal interventions. Comparison of severe cognitive impairment prevalence among participants of National (U.S.) Long-Term Care Surveys shows a significant decline in cognitive impairment from 1982 to 1999.[98] Contrary to the sobering forecast of a few years ago, of a steady growth to be expected in the number of severely cognitively impaired patients in the United States and Western Europe, the recent data disclose a decline in cognitive impairment prevalence over the last 20 years.

The possible causes for this decline in severe cognitive impairment with aging may be due to the following:

- Continuing improvement in higher education level[99] (see also the section "Learning and Longevity")
- Improved medical care and specific treatments (e.g., better prevention and management of stroke and heart disease, better medications to lower blood cholesterol)[100,101] (Chapter 17)
- Use of known steroid anti-inflammatory drugs (NSAIDs) (commonly utilized in the treatment of arthritis) associated with almost 50% reduction of AD[102]
- Hormone (estrogen) replacement therapy (see below)
- Better nutrition and improved physical activity[103] (Chapter 24)

The decline in severe cognitive impairment prevalence has been associated with a parallel decline in chronic disabilities in the U.S. elderly population.[104–106] Indeed, we can perhaps

**Systemic and Organismic Aging**

be more optimistic about future changes in the prevalence of severe cognitive impairment and other chronic disabilities than had been previously foreseen.

## B. A CAPSULE VIEW OF THE CLINICAL PICTURE

Of the myriad complaints we hear from aged patients and their family members and caretakers, none evokes (or should evoke) as much concern and anxiety as that of a change in mental function. ["I'm lost, lost somewhere in the corridors of my mind...," was the way an Alzheimer's victim, in one of his more lucid moments, explained it to his devoted wife.] Part of the tragedy of this terrible dementia is that it takes hold of fully developed, intelligent, dignified human beings, usually with family and friends, and slowly destroys them. It is a disease that kills the mind years before it takes the body.

Alzheimer's disease begins as simple forgetfulness, something that can be found in a large proportion of the normal healthy elderly. For most, this is the extent of the problem, but in roughly 10% of those over 65, there is progression to a confusional phase of mild to moderate dementia. Many, perhaps most, in this stage, eventually decline further to the point where they can no longer care for themselves and are, frankly, demented. By the time dementia has developed, the life expectancy is about 2.5 additional years. The clinical course of the disease averages about 8 years.

Confusional states in the elderly are associated with a number of conditions that must be differentiated from dementia; indeed, many such conditions can be treated successfully. The most common conditions to be differentiated from primary dementia are listed in Table 8.8. Even when dementia has been diagnosed, AD must be differentiated from several other forms of dementia, the most frequent, in the elderly, is multi-infarct dementia.

## C. MULTI-INFARCT DEMENTIA

This condition results from recurrent infarcts (hence, its name) and sudden and severe vascular lesions of the brain following hemorrhage (stroke) or localized transient ischemia, that is, localized tissue anemia due to obstruction of blood flow;[107] both stroke and ischemia are due to atherosclerosis of cerebral arteries (Chapter 16). The infarcts can sometimes be detected with an x-ray CT scan (computerized tomography) or other imaging and metabolism-measuring devices [magnetic resonance imaging (MRI) and positron emission tomography (PET)], all of which have also been suggested as diagnostic tools for AD.[108] If the infarcts are too small to be visualized, a clinical diagnosis of multi-infarct dementia can be made if the following signs are present (Table 8.9):

---

**TABLE 8.8**

**Types of Cognitive Impairment in the Elderly to be Differentiated from Alzheimer's Disease**

*Delirium*: An acute or subacute alteration of mental status characterized by clouding of consciousness, fluctuation of symptoms, and improvement of mental function after the underlying medical condition has been treated (reversible dementia)

*Depression*: A specific psychiatric entity that can precede or be associated with dementia and that can be differentially diagnosed and treated

*Benign senescent forgetfulness*: Not progressive and not of sufficient severity to interfere with everyday functions

*Paranoid states and psychoses*: With specifically diagnostic psychiatric correlates

*Amnesic syndrome*: Characterized by short-term memory losses without delirium or dementia

---

**TABLE 8.9**

**Characteristics of Multi-Infarct Dementia**

History of abrupt onset or stepwise deterioration
History of transient ischemic attack or stroke
Presence of hypertension or arrhythmia
Presence of any neurologic focal symptoms or signs

---

- A history of abrupt onset or stepwise deterioration related to the transient ischemic attacks and strokes
- The presence of other symptoms of cardiovascular pathology, such as hypertension and arrhythmias
- The presence of focal (localized) neurologic signs or symptoms, circumscribed by the focal nature of the cardiovascular damage

Unfortunately, as in AD, once the diagnosis has been made, therapeutic measures to cure or improve the dementia are not currently available. At this time, preventing stroke and cerebral infarcts through control of hypertension and atherosclerosis and some of the concomitant risk factors such as diabetes mellitus, continues to provide a means to reduce the incidence and progression of this disease (Chapters 14, 16, and 17).

## D. PATHOGENESIS OF ALZHEIMER'S DISEASE

In 1907, Alois Alzheimer, a German neurologist, described a case of a 51-year-old woman with a 5-year history of progressive dementia (i.e., the adult or presenile form) which led to her death. At autopsy, he found many neurofibrillary tangles and neuritic plaques in her cerebral cortex and, particularly, in the hippocampus. Today, despite significant advances in brain imaging techniques and the

**TABLE 8.10**
**Selected Characteristics of Alzheimer's Dementia**

| Anatomo-Histology | Pathology | Metabolism |
| --- | --- | --- |
| Brain atrophy, flattening of gyri, widening of sulci, and cerebral ventricles | Accumulation of cell inclusions: lipofuscin, Hirano and Lewy bodies, altered cytoskeletal Tau proteins, ubiquitin | Decreased oxidative metabolism, slower enzyme activity (Chapter 7) |
| Loss of cholinergic neurons, in nucleus of Meynert, hippocampus, and association cortices | Neurofibrillary tangles, neuritic plaques with amyloid | Free-radical accumulation (Chapter 5) |
| Loss of adrenergic neurons, in locus ceruleus | Perivascular amyloid, distributed throughout the brain, especially in frontal and prefrontal lobes, hippocampus, and association cortices | Impaired iron homeostasis (Chapter 7) Other minerals, zinc, aluminum |
| Denudation of neurons, stripping of dendrites, damage to axons | | Reduced level/metabolism/activity of neurotransmitters |
| Increased microglia | | Increased amyloid ß peptide with accumulation of amyloid proteins |
| | | Increased prion protein |
| | | Altered immune response |

availability of better diagnostic neuropsychological and biochemical tests, the diagnosis of AD remains essentially a clinicopathologic one, requiring a history of dementia in life and the presence of two pathologic lesions, *neurofibrillary tangles and neuritic plaques,* which can only be seen at autopsy. Clinically, it remains a *diagnosis of exclusion,* that is, it is assumed once all other possible causes of mental confusion and dementia have been examined and excluded. Major signs of Alzheimers' disease include morphologic, biochemical, and metabolic alterations as briefly identified in Chapter 7 and listed in Table 8.10.

Despite active research, the origin and nature of the characteristic lesions, tangles, plaques, and perivascular amyloid deposits, remain uncertain. One reason is the small number of animal or *in vitro* AD models; despite several promising transgenic models, none so far have proven entirely satisfactory.[109] While AD clinical and neuropathogical hallmarks may point to a well-defined syndrome, the cellular and molecular defects responsible for the disease are multiple, and their precise nature remains ambiguous. What seems to be generally accepted is that AD is characterized by a "common cascade" of pathologic events that may depend on an interaction between genetic defects and environmental influences (Chapter 3).

Among the most generally accepted mechanisms underlying AD pathology is the *role of free radicals* discussed in Chapter 5. Another theory formulated first by Prusiner[110,111] suggests that the dementia is caused *by unconventional infectious agents such as a virus or virus particles called prions.*[111,112] Spongiform encephalopathies, associated with

dementia in animals (e.g., scrapie in sheep, mad-cow disease in cattle) and in humans (e.g., Creutzfeldt–Jakob disease), resemble some of the morphologic brain lesions in AD, although significant differences exist. The prions would either participate in or induce the formation of the plaque amyloid, they would induce the formation of the paired helical filaments, and they would be responsible for the loss of cholinergic neurons in certain brain areas.[113] Mutation in the prion protein may, in humans, lead to inherited familial "transmissible spongiform encephalopathies," such as the Creutzfeldt– Jacobs disease, with abnormal protein aggregates and dementia.[114] The possible role of prions in AD etiopathology has called attention to the possibility that disorders of immune responses may cause or contribute to AD pathology: the presence of abormal brain proteins, their intra- and extracellular accumulation, and the resulting inflammation and stimulation of brain immune cells, the microglia (Chapter 7), may stimulate immune responses that would further aggravate neural damage.

Still uncertain at present is whether AD lesions originate in the neurons and then spread to the extracellular and perivascular spaces, or whether they are blood-borne and are carried to the brain through a damaged blood–brain barrier. Nerve cell injury is viewed by some researchers as the primary cause, and extracellullar lesions, such as deposition of amyloid fibrils, are viewed as the secondary cause. Intracellular lesions may be consequent to or associated with alterations of cytoskeletal protein phosphorylation by protein kinases and accumulation of brain proteases or of toxic metals such as iron[115] (Chapter 7), zinc,[116] and aluminum.[117,118]

Biochemical analysis of perihelical filaments (PHFs) that accumulate intracellularly to form neurofibrillary tangles, demonstrates that the principal PHF protein subunit is an altered form of the microtubule-associated tau protein.[119,120] Tau protein is known to bind to tubulin and promote the assembly and stability of microtubules. With aging, tau proteins may become hyperphosphorylated (identified as A68 proteins and detected by the specific Alz 50 antibody[121]), perhaps due to increased activity of several kinases. As a consequence, these proteins would no longer be capable of stabilizing the microtubules, and, thus, PHFs and neurofibrillary tangles would form.[122] By far the most popular but still controversial theory of AD pathogenesis today involves the overproduction of amyloid ß protein. The following is a synopsis of the major steps in the so-called "amyloid connection."

## 1.   The "Amyloid Connection"

Amyloid degeneration and amyloidoses have been discussed briefly in Chapter 7. The term "amyloid" given to these deposits over 100 years ago implies erroneously that they are formed of a starch-like substance (Latin *amylum* for starch). Actually, the amyloid molecules are normal or mutated proteins and protein fragments that differ among the various amyloidoses they generate (Chapter 7).

In the brain, amyloid β (Aβ) peptide, the major component of the neuritic plaque amyloid, does not inevitably result from some aberrant reaction, but it is formed in healthy cells as well. What causes its accumulation in AD is not yet known. The Aß peptide, which contains 40 to 42 amino acids, is made as part of a larger protein called the amyloid precursor protein (APP) with 695 amino acids. Complete APP does not harm the cells; it is only when Aβ peptide is clipped out of APP by protein-splitting enzymes, that the smaller molecule may lead to pathology.[122] APP is embedded in the cell membrane vesicles (endosomes), while the Aß molecule sits astride the membrane, where it cannot be reached by the protein-splitting enzymes.[123]

During normal cellular processing, APP is split by the enzyme α–secretase: this α-secretase processing occurs at or near the cell surface and precludes the formation of Aß. However, in some cases, under the action of a β–secretase, APP yields fragments that contain Aß, and such fragments become amyloidogenic (i.e., give rise to amyloid degeneration and accumulation; Chapter 7). This reaction occurs in the lysosomes, the cell organelles that are usually the site of protein breakdown.[123]

In AD, presumably, the balance is shifted away from cleavage through the α–secretase pathway (which would *not* produce amyloid deposition) toward the lysosomal pathway (which would form Aß deposition).[123] A third γ–secretase would further cleave the large APP molecule at the carboxy-terminal segment. The question as to the identity of the factor(s) capable of shifting the balance between the two degradation pathways (nonamyloidogenic and amyloidogenic) remains unanswered. Tentative hypotheses include, most probably: (a) alterations occurring during the early secretory trafficking of APP in the endoplasmic reticulum and Golgi apparatus or in endocytic recycling, (b) presence of an APP mutation, (c) reduced secretase activity with aging, (d) alterations in protein phosphorylation, and (e) abnormalities of lysosomes and others.

Once Aß-peptide has aggregated inside the lysosome, the cell has difficulty getting rid of it. Amyloid ß-peptide accumulation would lead to cell damage and death followed by amyloid accumulation in the extracellular spaces and formation of the neuritic plaque (with the remnants of the neurofibrillary tangles from the dead cell). At this point, microglial cells from the immune system will surround the plaque. Some investigators emphasize that brain damage starts with the formation of amyloid by perivascular macrophages and microglial cells. The damage and loss of neurons would entrain a selective loss of neurotransmitters, with alterations in synaptic signaling. The presence of an inflammatory component in the formation of the neuritc plaque and the accumulation of amyloid provided the rationale for using anti-inflammatory drugs to prevent and treat AD[102] and for attempts to prepare a vaccine against the disease.[124,123]

## 2.   The "Amyloid Connection" and Therapeutic Strategies

Aß peptide or APP cleavage products are neurotoxic:

- They disrupt $Ca^{++}$ homeostasis by increasing neuronal intracellular $Ca^{++}$.
- The increased intracellular calcium may lead to tau protein phosphorylation and formation of PHFs, which, in turn, would form neurofibrillary tangles.
- Accumulation of neurofibrillary tangles would induce neuronal death and formation of neuritic plaques with overproduction and deposition of amyloid.

Based on this amyloid hypothesis, several therapeutic strategies are being proposed. Although therapeutic applications are still under experimentation, a list of these strategies may help to better understand the role of amyloid in AD. Therapeutic strategies intend: (1) to block delivery to the brain, by the bloodstream, of APP molecules responsible for the Aß deposits; (2) to inhibit β- and γ-secretases that cleave APP to produce the Aß peptide; (3) to delay the formation of Aß deposits by interfering with the formation of fibrillar, cytotoxic amyloid filaments; (4) to interfere with activation of macrophages, microglia, and cytokine release that contribute to the inflammatory reaction surrounding the plaques; and (5) to block the Aß molecules on the surface of neurons to prevent their toxic action.

## 3. The Genetic Connection

A few patients with AD have a positive family history with some instances of autosomal dominant inheritance; these cases are designated as *"familial AD."* The majority of AD patients, however, lack familial connections and are designated as affected by *"sporadic AD."* In familial and sporadic AD, the cause of the disease is currently unknown, although, for the familial type, a definite genetic connection is useful for early diagnosis and future prevention and treatment. As for sporadic AD, epidemiological studies would exclude the causal influence of animal contacts, smoking, drinking, dietary habits, prior viral infections, or any correlation with neoplasms. However, they have found an increased frequency of a history of head trauma, an association with Down's syndrome, and a less convincing association with thyroid disease (Chapter 14). The cloning of the gene encoding the Aß-peptide and its mapping to chromosome 21 have strengthened the amyloid connection. Chromosome 21 is altered in Down's syndrome: individuals with this trisomy are severely mentally impaired and develop AD at an early age (Chapter 3). The association of mental disability with AD pathology suggests that overproduction of Aß is the cause of the neural degeneration underlying dementia. [126]

The discovery of an APP mutation (βAPP) on chromosome 14 and the possibility that this mutation may cause AD[127] was followed by the identification on chromosome 1 of two protein mutations, Presenilin 1 (PS1) and Presenilin 2 (PS2). The PS1 would be needed for the development of a major pathway to signal the fate of the cell, for example, whether it will develop in a specific cell type such as muscle fiber or neuron. PS1 and PS2 mutations could lead to disregulation of the γ-secretase activity in a way that would selectively enhance the proteolysis of APP toward the amyloidogenic β-secretase pathway.[128–130] How presenilin mutations increase Aβ peptide accumulation is still unknown, but the effect may be due to the formation, in the endoplasmic reticulum or the Golgi apparatus, of complexes between βAPP and Aβ peptide, with perhaps the addition of other peptides.

The apolipoprotein E allele, E4 (APO E4) on chromosome 19 is not a cause but rather a risk factor for the late-onset form of AD (Chapter 3).[131,132] Fifty percent of AD cases do not carry an APO E4 allele, however, suggesting that other risk factors may exist. One of such factors has been found on chromosome 10[133,134] and may act by modifying Aβ peptide metabolism and increasing its production.

The human brain is an expensive tool with a huge proportion (40% and higher) of human genes involved in constructing and functionally maintaining the CNS.[135] The recent completion of the human genome project will accelerate our knowledge of normal brain function as well as the identification of the genes involved in neurodegenerative diseases such as AD, whether causative or as capable of increasing susceptibility. A list of genes responsible for some forms of familial (inherited) AD or representing factors increasing susceptibility to AD are listed in Table 8.11.

## E. AD MANAGEMENT: MAINTAINING AN OPTIMISTIC OUTLOOK

The etiology of Alzheimer's dementia remains unknown despite several theories briefly surveyed here (e.g., prions, Aβ peptide, specific genes, alterations of microtubular protein tau, abnormal proteins, actions of metals) and in other chapters (5, 13, and 15). Prevention, diagnosis, and effective treatment of AD must wait until we have identified its eti-

**TABLE 8.11**
**Genes Known to Be Linked to Alzheimer's Disease**

| Chromosomal Location | Gene Type | Age of Onset | % Cases Familial | % Cases All |
| --- | --- | --- | --- | --- |
| 1 Presenilin 2 | AD | 40–70 years | 20 | 2–3 |
| 10 | Risk factor | >60 years | — | — |
| 14 Presenilin 1 | AD | 30–60 years | 40–60 | 5–10 |
| 19 Apolipoprotein E4 | Risk factor | >60 years | — | 40–50 |
| 21 APP Mutation (βAPP) Down's Syndrome Trisomy | AD | 45–60 years | 2–3 | <1 |

*Note:* Given the rapid progress in genetics, additional genes may be related to AD.

ology, and from what we know presently, it is possible that there may be multiple etiologies rather than a single one.

Various treatments have been proposed to prevent or slow its progression or to alleviate associated symptoms. Many of these treatments are promising, but to date, none can provide a cure.

The basic medical workup for dementia consists of a complete history and physical examination and a formal mental status exam, in severe cases, with the help of family or friends. Laboratory tests should include, besides the usual blood tests, more advanced neuropsychological testing and brain imaging. CT scan has been recommended in cases where reversible dementia is suspected, and MRI and PET may provide some useful information, particularly with respect to eventual metabolic changes in the brain.

Once reversible causes of dementia have been excluded, the therapeutic modalities currently available for intervening offer little. A wide range of drugs has been, or is being explored, but few drugs appear to be clinically useful. Among those tried (in addition to those already mentioned) are antioxidants (Chapter 5), antiviral and antiprion agents, calcium channel blockers, endorphin blockers, and hormones (e.g., estrogens, see below). A large number of new agents are continually being added to the list of drugs potentially beneficial in AD. The lack of a specific medical therapy for AD, however, should not discourage the physician from helping patients and their families. Some of the basic goals in AD management are outlined in Table 8.12. Many of these approaches will not be easy to implement and will not be successful, but the goal of finding a valid treatment is worth the effort. Dementia is a condition that requires not only the attention of the physician but also that of the other branches of the health care system in order to be effectively managed.

### THE ESTROGEN CONNECTION

Numerous observations on the possible protective actions of estrogens in AD, although still controversial, have been prompted by several epidemiological studies showing that age-adjusted prevalence of AD is significantly higher in women as compared to men — a finding that holds up worldwide and across ethnic groups. Given the fact that AD occurs in women after menopause, when estrogen levels are very low (Chapter 11), this gender difference in AD prevalence has been tentatively related to an estrogen deficiency. In older men, it is possible that the still circulating levels of androgens are converted, by the enzyme aromatase, to estrogens, and thereby maintain sufficient levels of these hormones in the brain.

A possible role for estrogen replacement therapy with respect to AD is supported by the well-established neurologic and behavioral actions of estrogens in healthy animals, including humans (unaffected by AD). Highlights of these actions, include the following: higher brain excitability in female rats,

### TABLE 8.12
### Basic Goals of Alzheimer's Disease Management

- To maintain the patient's safety while allowing as much independence and dignity as possible
- To optimize the patient's function by treating underlying medical conditions and avoiding the use of drugs with side effects on the nervous system
- To prevent stressful situations that may cause or exacerbate catastrophic reactions
- To identify and manage complications that may arise from agitation, depression, and incontinence
- To provide medical and social information to the patient's family in addition to any needed counseling

with a peak at estrus; in epileptic women, higher incidence of convulsive episodes just prior to ovulation, when estrogen levels are highest; regulation of spontaneous or evoked excitability in limbic nuclei (involved in reproductive behavior), and development and maturation of their cyclicity; influence on neural proliferation. survival and death, dendritic density, transcription of microtubular tau proteins, synaptogenesis, and myelinogenesis.

Estrogens appear to improve memory[136–138] and delay the onset of cognitive impairment in AD.[139,140] This latter action may be mediated by several mechanisms presumed to ameliorate dementia, such as: (1) antidepressant effect; (2) improvement of cerebral blood flow; (3) suppression of apolipoprotein E4; (4) direct neuronal and glial stimulation;[87] (5) synergism with nerve growth factor (NGF) for maintenance, survival, and arborization of basal forebrain and hypothalamic neurons; and (6) antioxidant[142] (Chapter 5), antiexcitotopic,[142] and antiamyloid activity.[143] As for other proposed AD treatments, a definite efficacy of estrogens needs further verification.

## REFERENCES

1. Overstall, P.W. and Downtown, J.H., Gait, balance and falls, in *Principles and Practice of Geriatric Medicine*, 3rd ed., Pathy, J.M.S., Ed., Wiley, New York, 1998.
2. Baloh, R.W. et al., Posturography and balance problems in older people, *J. Am. Geriatr. Soc.*, 43, 638, 1995.
3. Alexander, N.B., Gait disorders in older adults, *J. Am. Geriatr. Soc.*, 44, 434, 1996.
4. Slobounov, S.M. et al., Aging and time to instability in posture, *J. Gerontol.*, 53, B71, 1998.
5. Tang, P.F. and Woollacott, M.H., Inefficient postural responses to unexpected slips during walking in older adults, *J. Gerontol.*, 53, M471, 1998.
6. Tinetti, M.E., Doucette, J.T., and Claus, E.B., The contribution of predisposing and situational risk factors to serious fall injuries, *J. Am. Geriatr. Soc.*, 43, 1207, 1995.
7. Tinetti, M.E. and Williams, C.S., The effect of falls and fall injuries on functioning in community-dwelling older persons, *J. Gerontol.*, 53, M112, 1998.

8. Luukinen, H. et al., Rapid increase of fall-related severe head injuries with age among older people: A population-based study, *J. Am. Geriatr. Soc.*, 47, 1451, 1999.

9. VanSwearingen, J.M. et al., The modified Gait Abnormality Rating Scale for recognizing the risk of recurrent falls in community-dwelling elderly adults, *Physical Therapy*, 76, 994, 1996.

10. Covinsky, K.E. et al., History and mobility exam index to identify community-dwelling elderly persons at risk of falling, *J. Gerontol.*, 56, M253, 2001.

11. Tinetti, M.E. et al., Risk factors for serious injury during falls by older persons in the community, *J. Am. Geriatr. Soc.*, 43, 1214, 1995.

12. Bloem, B.R. et al., Idiopathic senile gait disorders are signs of subclinical disease, *J. Am. Geriatr. Soc.*, 48, 1098, 2000.

13. Guralnik, J.M. et al., Lower extremity function and subsequent disability: Consistency across studies, predictive models, and value of gait speed alone compared with the short physical performance battery, *J. Gerontol.*, 55, M221, 2000.

14. Fried, L.P. et al., Preclinical mobility disability predicts incident mobility disability in older women, *J. Gerontol.*, 55, M43, 2000.

15. Begg, R.K. and Sparrow, W.A., Gait characteristics of young and older individuals negotiating a raised surface: Implication for the prevention of falls, *J. Gerontol.*, 55, M147, 2000.

16. Judge, J.O. et al., Dynamic balance in older persons: Effects of reduced visual and proprioceptive input, *J. Gerontol.*, 50, M263, 1995.

17. Sundermier, L. et al., Postural sensitivity to visual flow in aging adults with and without balance problems, *J. Gerontol.*, 51, M45, 1996.

18. Tang, P.F., Moore, S., and Woollacott, M.H., Correlation between two clinical balance measures in older adults: Functional mobility and sensory organization test, *J. Gerontol.*, 53, M140, 1998.

19. Richardson, J.K. and Hurvitz, E.A., Peripheral neuropathy: A true risk factor for falls, *J. Gerontol.*, 50, M211, 1995.

20. McCully, K. et al., The effects of peripheral vascular disease on gait, *J. Gerontol.*, 54, B291, 1999.

21. Thelen, D.G. et al., Age differences in using a rapid step to regain balance during a forward fall, *J. Gerontol.*, 52, M8, 1997.

22. Woolley, S.M., Czaja, S.J., and Drury, C.G., An assessment of falls in elderly men and women, *J. Gerontol.*, 52, M80, 1997.

23. Cho, C.Y. and Kamen, G., Detecting balance deficits in frequent fallers using clinical and quantitative evaluation tools, *J. Am. Geriatr. Soc.*, 46, 426, 1998.

24. Cumming, R.G. et al., Prospective study of the impact of fear of falling on activities of daily living, SF-36 scores and nursing home admission. *J. Gerontol.*, 55, M299, 2000.

25. Shumway-Cook, A. and Woollacott, M., Attentional demands and postural control: The effect of sensory context, *J. Gerontol.*, 55, M10, 2000.

26. Schwartz, A.V. et al., Falls in older Mexican-American women, *J. Am. Geriatr. Soc.*, 47, 1371, 1999.

27. Davis, J.W. et al., Risk factors for falls and serious injuries among older Japanese women in Hawaii, *J. Am. Geriatr. Soc.*, 47, 792, 1999.

28. Rosengren, K.S. et al., Gait, balance, and self-efficacy in older black and white American women, *J. Am. Geriatr. Soc.*, 48, 707, 2000.

29. Blair, S.N. and Garcia, M.E., Get up and move: A call to action for older men and women, *J. Am. Geriatr. Soc.*, 44, 599, 1996.

30. Schlicht, J., Camaione, D.N., and Owen, S.V., Effect of intense strength training on standing balance, walking speed and sit-to-stand performance in older adults, *J. Gerontol.* 56, M281, 2001.

31. Van Praag, H. et al., Running increases cell proliferation and neurogenesis in the adult mouse dentate gyrus, *Nature Neuroscience*, 2, 266, 1999.

32. Van Praag, H. et al., Running enhances neurogenesis, learning, and long-germ potentiation in mice, *Proc. Natl. Acad. Sci.*, 96, 13, 427, 1999.

33. Allen, D.R. et al., Ataxia telangiectasia mutated is essential during adult neurogenesis, *Genes & Development*, 15, 554, 2001.

34. Eriksson, P.S. et al., Neurogenesis in the adult human hippocampus, *Nat. Med.*, 4, 1313, 1998.

35. Fuchs, E. and Gould, E., In vivo neurogenesis in the adult brain: Regulation and functional implications, *Eur. J. Neurosci.*, 12, 2211, 2000.

36. Magri, F. et al., Changes in endocrine circadian rhythms as markers of physiological and pathological brain aging, *Chronobiol. Intl.*, 14, 385, 1997.

37. Halberg, F. et al., Near 10-year and longer periods modulate circadians: Intersecting anti-aging and chronoastrobiological research, *J. Gerontol.*, 56, M304, 2001.

38. Albarede, J.L. et al., Sleep disorders and insomnia in the elderly, in *Facts and Research in Gerontology*, Vol. 7, Springer, New York, 1993.

39. Vitiello, M.V., Sleep disorders and aging: Understanding the causes, *J. Gerontol.*, 52, M189, 1997.

40. Williams, A.J., Sleep, in *Principles and Practice of Geriatric Medicine,* 3rd ed., Pathy, J.M.S., Ed., John Wiley & Sons, New York, 1998.

41. Blazer, D.G., Hays, J.C., and Foley, D.J., Sleep complaints in older adults: A racial comparison, *J. Gerontol.*, 50, M280, 1995.

42. Campbell, S.S., Dawson, D., and Anderson, M.W., Alleviation of sleep maintenance insomnia with timed exposure to bright light, *J. Am. Geriatr. Soc.*, 41, 829, 1993.

43. Campbell, S.S. and Murphy, P.J., Relationships between sleep and body temperature in middle-aged and older subjects, *J. Am. Geriatr. Soc.*, 46, 458, 1998.

44. Van Hilten, J.J. et al., Nocturnal activity and immobility across aging (50–98 years) in healthy persons, *J. Am. Geriatr. Soc.*, 41, 837, 1993.

45. Maggi, S. et al., Sleep complaints in community-dwelling older persons: Prevalence, associated factors, and reported causes, *J. Am. Geriatr. Soc.*, 46, 161, 1998.

**Systemic and Organismic Aging**

46. Alessi, C.A. et al., A randomized trial of a combined physical activity and environmental intervention in nursing home residents: Do sleep and agitation improve? *J. Am. Geriatr. Soc.*, 47, 784, 1999.

47. Dealberto, M.J. et al., Breathing disorders during sleep and cognitive performance in an older community sample: The EVA study, *J. Am. Geriatr. Soc.*, 44, 1287, 1996.

48. Bliwise, D.L., Is sleep apnea a cause of reversible dementia in old age? *J. Am. Geriatr. Soc.*, 44, 1407, 1996.

49. Brassington, G.S., King, A.C., and Bliwise, D.L., Sleep problems as a risk factor for falls in a sample of community dwelling adults aged 64–99 years, *J. Am. Geriatr. Soc.*, 48, 1234, 2000.

50. Teran-Santos, J., Jimenez-Gomez, A., and Cordero-Guevara, J., The association between sleep apnea and the risk of traffic accidents, *N. Engl. J. Med.*, 340, 847, 1999.

51. Saunders, N.A. and Sullivan, C.E., Eds., *Sleep and Breathing*, 2nd ed., Marcel Dekker, New York, 1994.

52. Newman, A.B. et al., Sleep disturbance, psychosocial correlates and cardiovascular disease in 5201 adults: The Cardiovascular Health Study, *J. Am. Geriatr. Soc.*, 45, 1, 1997.

53. Youngstedt, S.D. et al., Periodic leg movements during sleep and sleep disturbances in elders, *J. Gerontol*, 53, M391, 1998.

54. Wilcox, S. et al., Factors related to sleep disturbance in older adults experiencing knee pain or knee pain with radiographic evidence of knee osteoarthritis, *J. Am. Geriatr. Soc.*, 48, 1241, 2000.

55. Frazer, A. and Hensler, G., Serotonin, in *Basic Neurochemistry: Molecular, Cellular and Medical Aspects*, 6th ed., Siegel, G.J. et al., Eds., Lippincott-Raven Publishers, Philadelphia, 1999.

56. Meyer-Bernstein, E.L. and Morin, L.P., Differential serotonergic innervation of the suprachiasmatic nucleus and the intergeniculate leaflet and its role in circadian rhythm modulation, *J. Neurosci.*, 16, 2097, 1996.

57. Mobbs, C.V., Neuroendocrinology of aging, in *Handbook of the Biology of Aging*, 4th ed., Schneider, E.L. and Rowe, J.W., Eds., Academic Press, San Diego, CA, 1995.

58. Wise, P.M., Krajnak, K.M., and Kashon, M.L., Menopause: The aging of multiple pacemakers, *Science*, 273, 67, 1996.

59. Hultsch, D.F. et al., *Memory Change in the Aged*, Cambridge University Press, New York, 1998.

60. Park, D.C. and Schwarz, N., Eds., *Cognitive Aging: A Primer*, Psychology Press, Philadelphia, PA, 1999.

61. Anderson N.D. and Craik, F.I.M., Memory in the aging brain, in *The Oxford Handbook of Memory*, Tulving, E. and Craik, F.I.M., Eds., Oxford Press, New York, 2000.

62. Clark, R.E. and Squire, L.R., Classical conditioning in brain systems: The role of awareness, *Science*, 289, 77, 1998.

63. Rosenzweig, M.R., Breedlove, S.M., and Levinson, A.L., *Biological Psychology: An Introduction to Behavioral, Cognitive, and Clinical Neuroscience*, 3rd ed., Sinauer Associates, Sunderland, MA, 1999.

64. Agranoff, B.W., Cotman, C.W., and Uhler, M.D., Learning and memory, in *Basic Neurochemistry: Molecular, Cellular and Medical Aspects*, 6th ed., Siegel, G.J. et al., Eds., Lippincott-Raven Publishers, Philadelphia, PA, 1999.

65. Mayes, A.R., Selective memory disorders, in *The Oxford Handbook of Memory*, Tulving, E. and Craik, F.I.M., Eds., Oxford Press, New York, 2000.

66. Hodges, J.R., Memory in the dementias, in *The Oxford Handbook of Memory*, Tulving, E. and Craik, F.I.M., Eds., Oxford Press, New York, 2000.

67. Hasher, L., Quig, M.B., and May, C.P., Inhibitory control over no-longer-relevant information: Adult age differences, *Mem. Cognit.*, 25, 286, 1997.

68. Luszcz, M.A. and Bryan, J., Toward understanding age-related memory losses in late adulthood, *Gerontology*, 45, 2, 1999.

69. Hasher, L., Zachs, R.T., and Rahhal, T.A., Timing, instructions, and inhibitory control: Some missing factors in the age and memory debate, *Gerontology*, 45, 355, 1999.

70. Gallagher, M. and Colombo, P.J., Ageing: The cholinergic hypothesis of cognitive decline, *Curr. Opin. Neurobiol.*, 5, 161, 1995.

71. Rani, P.J. and Panneerselvam, C., Protective efficacy of L-carnitine on acetylcholinesterase activity in the aged brain, *J. Gerontol.*, 56, B140, 2001.

72. Woodruff-Pak, D.S., Vogel, R.W. III, and Wenk, G.L., Galantamine: Effect on nicotinic receptor binding, acetylcholinesterase inhibition and learning, *Proc. Natl. Acad. Sci.*, 98, 2089, 2001.

73. Blesa, R., Galantamine: Therapeutic effects beyond cognition, *Dement. Geriatr. Cogn. Disord.*, Suppl. 1, 28, 2000.

74. Michael, D. et al., Repeated pulses of serotonin required for long-term facilitation activate mitogen-activated protein kinase in sensory neurons of Aplysia, *Proc. Natl. Acad. Sci. USA*, 95, 1864, 1998.

75. Shimizu, E. et al., NMDA receptor-dependent synaptic reinforcement as a crucial process for memory consolidation, *Science*, 290, 1170, 2000.

76. Lydiard, R.B., Brawman-Mintzer, O., and Ballenger, J.C., Recent developments in the psychopharmacology of anxiety disorders, *J. Consult. Clin. Psychol.*, 64, 660, 1996.

77. Liddle, R.A., Cholecystokinin cells, *Annu. Rev. Physiol.*, 59, 221, 1997.

78. Conner, J.M. et al., Nontropic actions of neurotrophins: Subcortical nerve growth factor gene delivery reverses age-related degeneration of primate cortical cholinergic innervation, *Proc. Natl. Acad. Sci. USA*, 98, 1941, 2001.

79. Martinez-Serrano, A. et al., Long-term functional recovery from age-induced spatial memory impairments by nerve growth factor gene transfer to the rat basal forebrain, *Proc. Natl. Acad. Sci. USA*, 93, 6355, 1996.

80. Ickes, B.R. et al., Long-term environmental enrichment leads to regional increases in neurotrophin levels in rat brain, *Exp. Neurol.*, 164, 45, 2000.

81. Neeper, S.A. et al., Physical activity increases mRNA for brain-derived neurotropic factor and nerve growth factor in rat brain, *Brain Res.*, 726, 49, 1996.

82. Bailey, C.H., Bartsch, D., and Kandel, E.R., Toward a molecular definition of long-term memory storage, *Proc. Natl. Acad. Sci. USA*, 93, 13445, 1996.

83. Grady, C.L. and Craik, F.I., Changes in memory processing with age, *Curr. Opin. Neurobiol.*, 10, 224, 2000.

84. Vernadakis, A. and Roots, B.I., *Neuron-Glia Interrelations during Phylogeny: Plasticity and Regeneration*, Vernadakis, A. and Roots, B.L., Eds., Humana Press, Totowa, NJ, 1995.

85. Helmut, L., Glia tell neurons to build synapses, *Science*, 291, 569, 2001.

86. Ullian, E.M. et al., Control of synapse number by glia, *Science*, 291, 657, 2001.

87. Higashigawa, K. et al., Effects of estrogens and thyroid hormone on development and aging of astrocytes and oligodendrocytes, in *Neuroglia in the Aging Brain*, De Vellis, J.S., Ed., Humana Press, Totowa, NJ, 2001.

88. Pappas, G. et al., The increasing disparity in mortality between socioeconomic groups in the United States, 1960 and 1986, *N. Engl. J. Med.*, 329, 103, 1993.

89. Guralnik, J.M. et al., Educational status and active life expectancy among older blacks and whites, *N. Engl. J. Med.*, 329, 110, 1993.

90. Katzman, R., Education and the prevalence of dementia and Alzheimer's disease, *Neurology*, 43, 13, 1993.

91. Stern, Y. et al., Influence of education and occupation on the incidence of Alzheimer's Disease, *J. Am. Med. Assoc.*, 271, 1004, 1994.

92. Snowdon, D., *Aging with Grace*, Bantam Books, New York, 2001.

93. Kemper, S. et al., Language decline across the life span: Findings from the nun study, *Psychology and Aging*, 16, 227, 2001.

94. Timiras, P.S., Education, homeostasis and longevity, *Exp. Gerontol.*, 30, 189, 1995.

95. Jacobs, B., Schall, M., and Scheibel, A., A quantitative dendritic analysis of Wernicke's area in humans: II. Gender, hemispheric and environmental factors, *J. Comp. Neurol.*, 327, 97, 1993.

96. Swaab, D.F., Brain aging and Alzheimer's disease, "wear and tear" vs. "use it or lose it," *Neurobiol. Aging*, 12, 317, 1991.

97. Woodruff-Pak, D.S., *The Neuropsychology of Aging*, Blackwell, Malden, 1997.

98. Corder, L.S., Corder, E.H., and Manton, K.G., Changes in the prevalence of severe dementia in the U.S. age 65 and older population from 1982 to 1999, *Gerentologist*, 41 (special issue: Program Abstracts), 158, 2001.

99. Evans, D. et al., The impact of Alzheimer's disease in the United States population, in *The Oldest-Old*, Suzman, R., Willis, D., and Manton, K.G., Eds., Oxford University Press, New York, 1992.

100. Jick, H. et al., Satins and the risk of dementia, *Lancet*, 356, 1627, 2000.

101. Wolozin, B. et al., Decreased prevalence of Alzheimer disease associated with 3-hydroxy-3-methyglutaryl coenzyme A reductase inhibitors, *Arch. Neurol.*, 57, 1439, 2000.

102. McGeer, P.L., Schulzer, M., and McGeer, E.G., Arthritis and anti-inflammatory agents as possible protective factors for Alzheimer's disease: A review of 17 epidemiologic studies, *Neurology*, 47, 425, 1996.

103. Ericksson, P.S. et al., Neurogenesis in the adult human hippocampus, *Nature Med.*, 4, 1313, 1998.

104. Manton, K.G., Corder, L., and Stallard, E., Chronic disability trends in the U.S. elderly populations 1982 to 1994, *Proc. Natl. Acad. Sci. USA*, 94, 2593, 1997.

105. Singer, B. and Manton, K.G., The effects of health changes on projections of health service needs for the elderly population of the United States, *Proc. Natl. Acad. Sci. USA*, 95, 15,618, 1998.

106. Manton, K.G. and XiLiang, Gu., Changes in the prevalence of chronic disability in the United States black and nonblack population above age 65 from 1982 to 1999, *Proc. Natl. Acad. Sci USA*, 98, 6354, 2001.

107. Nyenhuis, D.L. and Gorelick, P.B., Vascular dementia: A contemporary review of epidemiology, diagnosis, prevention, and treatment, *J. Am. Geriatr. Soc.*, 46, 1437, 1998.

108. Budinger, T.F., Brain imaging in normal and Alzheimer's disease, in *Studies of Aging*, Sternberg, H. and Timiras, P.S., Eds., Springer, New York, 1999.

109. Andersen, J.K. and Jurma, O.P., Use of genetically engineered mice as models for understanding human neurodegenerative disease, *J. Am. Geriatr. Soc.*, 44, 717, 1996.

110. Prusiner, S.B., Novel proteinaceous infectious particles cause scrapie, *Science*, 216, 136, 1982.

111. Prusiner, S.B., Shattuck lecture — neurodegenerative diseases and prions, *N. Engl. J. Med.*, 344, 1516 and 1548, 2001.

112. Scott, M.R. et al., Transgenic models of prion disease, *Arch. Virol.*, Suppl. 16, 113, 2000.

113. Hedge, R.S. et al., Transmissible and genetic prion diseases share a common pathway of neurodegeneration, *Nature*, 402, 822, 1999.

114. Supattapone, S. et al., A protease-resistant 61-residue prion peptide causes neurodegneration in transgenic mice, *Mol. Cell Biol.*, 21, 2608, 2001.

115. Connor, J.R. et al., Regional distribution of iron and iron-regulatory proteins in the brain, in aging and Alzheimer's disease, *J. Neurosci. Res.*, 31, 327, 1992.

116. Ashley, I. et al., Rapid induction of Alzheimer A$\beta$ amyloid formation by zinc, *Science*, 265, 1464, 1994.

117. Mesco, E.R., Kachen, C., and Timiras, P.S., Effects of aluminum on tau protein in human neuroblastoma cells, *Mol. Chem. Neuropathol.*, 14, 199, 1991.

118. Sherrard, D.J., Aluminum — much ado about something, *N. Engl. J. Med.*, 324, 558, 1991.

119. Kirkpatrick, L.L. and Brady, C.T., Cytoskeleton of neurons and glia, in *Basic Neurochemistry*, Siegel, G.J. et al., Eds., Lippincott-Raven Publishers, Philadelphia, PA, 1998.

120. Selkoe, D.J. and Lansbury, P.J., Biochemistry of Alzheimer's and prion diseases, in *Basic Neurochemistry*, Siegel, G.J. et al., Eds., Lippincott-Raven Publishers, Philadelphia, PA, 1998.

## Systemic and Organismic Aging

121. Nukina, N. et al., The monoclonal antibody, Alz 50 recognizes tau protein in Alzheimer's disease brain, *Neurosci. Lett.*, 87, 240, 1988.

122. Lee, V.M., Disruption of the cytoskeleton in Alzheimer's disease, *Curr. Opin. Neurobiol.*, 5, 663, 1995.

123. Wild-Bode, C. et al., Intracellular generation and accumulation of amyloid β-peptide terminating at amino acid 42, *J. Biol. Chem.*, 272, 16,085, 1997.

124. Schenk, D. et al., Immunization with Aβ attenuates Alzheimer-disease-like pathology in the PDAPP mouse, *Nature*, 400, 173, 1999.

125. Games, D. et al., Prevention and reduction of AD-type pathology in PDAPP mice immunized with Aβ 1–42, *Ann. Acad. Sci.*, 920, 274, 2000.

126. Lendon, C.L., Ashall, F., and Goate, A.M., Exploring the etiology of Alzheimer Disease using molecular genetics, *J. Am. Med. Assoc.*, 277, 825, 1997.

127. Goate, A. et al., Segregation of a missense mutation in the amyloid precursor protein gene with familial Alzheimer's Disease, *Nature*, 349, 704, 1991.

128. Duff, K. et al., Increased amyloid Aβ42(43) in brains of mice expressing mutant presenilin 1, *Nature*, 383, 710, 1996.

129. Borchelt, D.R. et al., Familial Alzheimer's Disease linked presenilin 1 variants elevate Aβ1–42/1–40 ratio *in vitro* and *in vivo*, *Neuron*, 17, 1005, 1996.

130. Lopera, F. et al., Clinical features of early-onset Alzheimer Disease in a large kindred with an E280A presenilin-1 mutation, *J. Am. Med. Assoc.*, 277, 793, 1997.

131. Slooter, A.J.C., Apolipoprotein E ε4 and the risk of dementia with stroke, *J. Am. Med. Assoc.*, 277, 818, 1997.

132. Evans, D.A. et al., Apolipoprotein E ε4 and incidence of Alzheimer disease in a community population of older persons, *J. Am. Med. Assoc.*, 277, 822, 1997.

133. Ertekin-Taner, N. et al., Linkage of plasma Aβ42 to a quantitative locus on chromosome 10 in late-onset Alzheimer's disease pedigrees, *Science*, 290, 2303, 2000.

134. Myers, A. et al., Susceptibility locus for Alzheimer's disease on chromosome 10, *Science*, 290, 2304, 2000.

135. Helmuth, L., A genome glossary, *Science*, 291, 1197, 2001.

136. Barrett-Connor, E. and Dritz-Silverstein, D.L., Estrogen replacement therapy and cognitive function in older women, *J. Am. Med. Assoc.*, 269, 2637, 1993.

137. Wickelgren, I., Estrogen stakes claim to cognition, *Science*, 276, 675, 1997.

138. Yaffe, K. et al., Serum estrogen levels, cognitive performance, and risk of cognitive decline in older community women, *J. Am. Geriatr. Soc.*, 46, 816, 1998.

139. Tang, M.X. et al., Effect of estrogen during menopause on risk of age onset of Alzheimer's disease, *Lancet*, 348, 429, 1996.

140. Giacobini, E., Aging, Alzheimer's disease and estrogen therapy, *Exp. Gerontol.*, 33, 865, 1998.

141. Liu, J. et al., Stress, hormones, oxidative damage, and neurodegeneration, *Soc. Neurosci.*, 24, 206 (abstract 8314), 1998.

142. Shy, H., Malaiyandi, L., and Timiras, P.S. Protective action of 17β-estradiol and tamoxifen on glutamate toxicity in glial cells, *Int. J. Devl. Neurosci.*, 18, 289, 2000.

143. Chang, D., Kwan, J., and Timiras, P.S., Estrogens influence growth, maturation and amyloid β-peptide production in neuroblastoma cells and in β-APP transfected kidney 293 cell line, in *Brain Plasticity*, Filogamo, G. et al., Eds., Plenum Press, New York, 1997.

# 9 Sensory Systems: Normal Aging, Disorders, and Treatments of Vision and Hearing in Humans*

*Esmail Meisami, Chester M. Brown, and Henry F. Emerle*
University of Illinois, Urbana-Champaign

## CONTENTS

* Illustrations by Dr. S. Oklund.

0-8493-0948-4/03/$0.00+$1.50
© 2003 by CRC Press LLC

## I.  INTRODUCTION

This chapter describes normal aging changes in the structure and function of the human visual and auditory systems. Aging-associated visual and auditory impairments, disorders, and diseases are also presented, and current efforts for their treatment and rehabilitation are reviewed. The basics of the aging changes in the senses of taste, smell, and somatic sensation are not included in the present chapter, mainly due to lack of space, but can be found in the previous edition of this chapter.[1] In this third edition, some discussion of aging of taste and smell is presented in Chapter 20 with aging of the gastrointestinal system.

Sensory impairments in vision and hearing occur so commonly with aging that they often tend to characterize the aged and the aging process. Some of these impairments are due to intrinsic aging processes occurring in the sense organs and their neural and brain components, others are caused by environmental effects, and still others represent manifestations of aging diseases. The study of the aging process in the human sensory systems, in addition to its importance and applicability to geriatrics, also provides some of the most interesting and challenging cases of gerontological investigation. The elements comprising the various senses and their aging portray the entire spectrum of cellular, tissue, organ, and system aging. The peripheral receptor cells of the ear's cochlea and the eye's retina are permanently established at birth, with no turnover and regeneration in later life, in part contributing to the functional decrements in vision and hearing. The aging changes in the eye's lens provide another interesting model system for the study of the aging process, as they begin so early in life and lend themselves to a wide variety of investigations ranging from molecular biology to physiological optics.

### IMPACT OF SENSORY IMPAIRMENTS ON ELDERLY LIFE AND HEALTH

As shown in Figure 9.1, the incidence of sensory impairments increases markedly in people with aging. More than 25% of the population 85 years or older suffers from visual abnormalities; twice as many suffer from hearing impairments. Impaired vision and hearing reduce the capacity for social communication, one of humans' cardinal needs and functions, resulting in social isolation and deprivation. The impact of hearing and visual impairments on elderly health and mortality has been

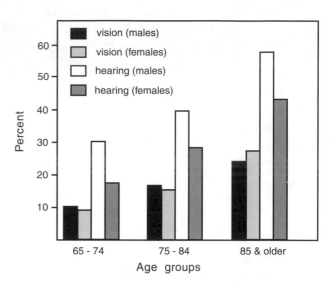

**FIGURE 9.1** Increase in the incidence of visual and hearing impairments in males and females 65 years of age and older in the general U.S. population. Note the higher rate of sensory impairments in males, particularly for hearing. (From National Center for Health Statistics, Publication # (PHS)86–1250, 1986. With permission.)

reviewed recently by Guralnik.[2] Age-related vision and hearing impairments have greater impact on long-term health than previously thought. The results of a 10-year study of 5000 subjects aged 55 to 74 years show that measured (not subjective) visual impairment was predictive of 10-year mortality, but both measured and subjective visual impairments were significant in predicting certain aspects of functional disability. Measured combined visual and auditory impairments lead to the highest risk of functional impairment.[2] The synergistic effect of co-occurring impairments[3] may lead to effective future treatments.[2] The basic aspects of the effects of reduced capacities in somatic, olfactory, and gustatory senses on aging and physical and mental health of the elderly have been reviewed by Meisami.[1] The reduced sensory abilities may lead to depression; in those already suffering from depression, they may hinder the progress of recovery. Because sensory losses in old age are so common and their consequences so widespread, an understanding of these impairments is now essential in geriatrics and elderly care.[1]

## II.  VISION

The eye, with a structure that indicates exquisite adaptation for optical and nervous functions, is the sensory organ for vision (Figure 9.2). The cornea, lens, pupil, and the

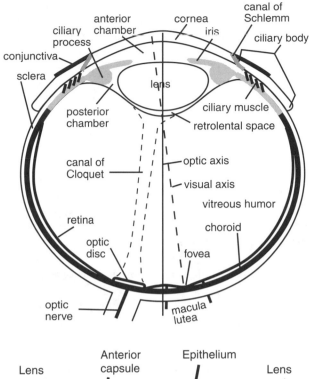

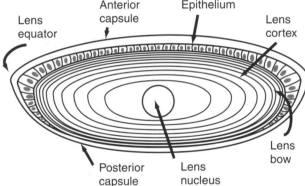

**FIGURE 9.2** *Schematic drawing of the human eye and lens.* [Redrawn from articles by Ordy and Brizee (eye illustration) and Vaughan (lens illustration), in *Sensory Systems and Communication in the Elderly,* Ordy  Brizee, Eds., Raven Press, New York, 1979].

aqueous and vitreous humors participate in the optical functions of the eye, while the retina carries out neural visual functions. Both the optical and neural compartments undergo aging changes, although those of the optical compartment are better known. Some of these changes in the eye's optical apparatus, like those in the lens, start early in life. The changes in the optical compartment are probably the primary causes of decline in the visual capacities of the elderly, while the degenerative changes in the retina are one of the leading causes of old age blindness (Table 9.1).[4–11] In addition to the eye, aging changes in the visual pathways and central visual structures, such as the lateral geniculate bodies and the visual cortical areas, which have been found to be quite extensive in humans, may also be responsible for some of the visual impair-

### TABLE 9.1
### Summary of the Normal Aging Changes in the Human Eye

#### Structural Changes

**Cornea:** Increased thickness; decreased curvature; some loss of transparency; pigment and lipid accumulation (arcus senilis); loss of epithelial cells; reduced epithelial regeneration

**Anterior chamber:** Decreased volume and flow of aqueous humor

**Iris:** Decreased dilator muscle cell number, pigment, and activity; mild increase in density of collagen fibers in stroma

**Lens:** Increased size and anterior-posterior thickness; decreased curvature; increased pigment accumulation (yellowness) and opacity (optical density); decreased epithelial cell number; decreased new fiber formation and antioxidant levels; increased crossover in capsule collagens and lens crystallins; increased hardness in capsule and body and lens nucleus

**Vitreous body:** Increased inclusion bodies; decreased water content; lesser support to globe and retina

**Ciliary body and muscles:** Decreased number of smooth muscles (radial and circular); increased hyaline substance and fiber in ciliary process; decreased ciliary pigment epithelial cells

**Retina:** Decreased thickness in periphery; defects in rod outer segments, and regeneration of discs and rhodopsin; loss of rods and associated nerve cells; some cone loss; reduced cone pigment density; expansion of Muller cells; increased cyst formation; formation of Drusen-filled lesions, and degeneration of macular region in diseased condition

**Pigment epithelium:** Loss of melanin; increased lipofuscin granules

#### Functional Changes

**Corneal and lens functions:** Decreased accommodation power (presbyopia); increased accommodation reflex latency; increased near point of vision; increased lenticular light scattering; decreased refraction; decreased lens elasticity

**Retinal function:** Decreased critical flicker frequency; decreased light sensitivity (increased light thresholds before and after dark adaptation); reduced color vision initially in yellow to blue range and later in the green range

**General optical functions:** Increased papillary constriction (senile miosis); reduced visual acuity; presbyopia

#### Major Pathologies

**Cornea:** "Against the rule" astigmatism
**Lens:** Cataract; hardening and loss of elasticity
**Retina:** Senile macular degeneration; glaucoma; diabetes retinopathy

ments of the elderly, but the knowledge of these aspects is only recently developing.

## A. THE EYE'S OPTICAL COMPONENTS

### 1. Cornea

The *cornea* is the anterior portion of the eye, and its curved surface together with the watery layer of tears is responsible for most of the refraction of the light rays.

During aging, the cornea becomes thicker and less curved, mainly due to an increase in the horizontal diameter of the eye. These changes alter the refractive properties of the cornea, leading to "against the rule" *astigmatism,* a condition characterized by defective corneal curvature and diffusion of light rays.[6] The cornea is also highly sensitive to irritable stimuli, a protective function for the eye. Corneal sensitivity declines by nearly one half between youth and old age.[5]

Other conditions in the cornea associated with aging are the *arcus senilis* (or lipid arc), the *Hudson–Stahli line,* and *spheroidal degeneration.* The arcus senilis increases in frequency and density with aging, particularly after 60. It is a yellowish-white ring around the cornea's outer edge, formed by cholesterol ester deposits derived from plasma lipoproteins *(lipid arcus).* In the dilated pupil, this ring would interfere with passage of light rays; however, because of partial pupillary constriction in the elderly, arcus senilis is not detrimental to visual function.[4–7]

The Hudson–Stahli line is a horizontal brown line formed by iron deposits in corneal basal cells. Its frequency increases from 2% at 10 years to 14% at 30 years and 40% at 60 years but has no detrimental effect on vision. Spheroidal degeneration occurs in cornea's Bowman's layer. It is observed frequently in aged populations exposed to high levels of ultraviolet radiation or ambient light reflected from snow or sand.[5]

The corneal endothelial cells number about one million per cornea at birth; this number declines to 70% by 20 years and to 50 and 30% by 60 and 80 years, respectively. Normally, the pumping action of corneal cells removes water and helps keep the cornea transparent. Because these cells do not divide after birth, their loss due to aging or injury after surgical treatments of the cornea or lens can lead to a decline in corneal transparency. Endothelial cells also secrete the cornea's basement membrane. With aging, warts *(cornea guttata)* appear in this membrane mainly in the cornea's periphery and cause marked increase in corneal permeability. Guttatas are observed with increasing frequency with aging: 20% in youth, 60% in the sixth decade, and nearly 100% in very old age.[5]

### a. Recent Studies on Corneal Biochemical and Structural Aging

Biochemical studies revealed a gradual decline in high-energy metabolism of the aging cornea as shown by linear decreases in phosphomonoesters, phosphocreatine, and ATP accompanied by decreased inorganic orthophosphate.[12] Corneal aging is accompanied by a linear loss of keratocytes paralleling loss in endothelial cell density.[13] A three-dimensional expansion of collagen fibrils along the axial direction occurs in corneal stroma with aging. The expansion is due to decreased molecular tilting angle within fibrils. It reflects an increased number of fibrils due

to expansion of intermolecular Bragg spacing caused by glycation-induced cross-linkages.[14] Previously described morphological aging changes in cornea have been confirmed and better described. Human corneas become less symmetrical with age; increased pupillary dilation and wave-front aberration become more pronounced with aging.[15] Steepening of corneal curvature with aging, as shown by decrease in vortex radius and increase in P-value, reflects a shift to a more spherical surface.[16]

## 2. The Lens

In the process of image formation, the crystalline *lens* of the eye performs two important functions, *refraction* and *accommodation.* For refraction, the lens requires an appropriate crystalline structure and transparency, while for accommodation, it needs to be elastic, amenable to changes in its curvature. The increase in the opacity (optical density) and the hardness (loss of elasticity) of the lens with age are two of the best-known changes in the eye's optical properties that interfere with refraction and accommodation, respectively.[11,17]

Knowledge of the structure and development of the lens is essential for understanding its aging. The biconvex lens is basically a fibrous and relatively acellular structure, consisting of a core surrounded by a capsule (Figure 9.2). Anteriorly, the capsular epithelial cells form the fibers and other lens proteins. The collagen fibers of the lens capsule facilitate changes in lens shape during accommodation. The lens core packed with transparent protein fibers consists of an inner nuclear zone surrounded by a cortex (Figure 9.2).

### a. Structural Aspects of Lens Growth and Aging

The lens is formed during the embryonic period and is fairly spherical in the fetus and newborn. During postnatal development and throughout maturity, the lens continues to grow by addition of new layers of protein fibers laid down by the capsular epithelial cells. As new fibers form, older fibers are pushed into the lens core. This mode of growth results in increased horizontal thickness of the lens together with increased compaction of the fibers in the nuclear zone.[5,17] The lens thickness increases from about 3.5 mm in infancy to 4.5 mm in middle age and to 5.5 mm in old age, growing at a steady, linear rate of 25 μm per year.[5,9,17] Underlying this process of growth are the capsular epithelial cells, which divide and differentiate, losing their nucleus and organelles, and eventually transforming into an inert skeleton of fibrous proteins.[17]

### b. Recent Volumetric and Morphological Studies on Aging Lens

According to Koretz et al.,[18] although total lens volume increases with age, the volume of lens nucleus and the shape of nuclear boundaries do not show any significant changes with aging. The lens center of mass and central clear region move anteriorly with aging.[18] In addition to an increase in lens mass and volume with age, changes occur in point of insertion of the lens zonules.[17] Also, the radius of the lens anterior surface curvature decreases with aging. The increase in sagittal lens thickness with age is caused, in part, by the anterior movement of lens mass and shallowing of the anterior chamber.[5,17]

### c.  Increased Opacity of the Lens

Although many cytoskeletal proteins such as actin, tubulin, and vimentin are found in the lens core, the transparency of the lens is, in principle, due to a particular supramolecular arrangement of the specific lens proteins, $\alpha$-, $\beta$-, and $\gamma$–crystallins, within an ion- and water-free environment.[5,17] During aging, the lens' opacity increases, leading to decreased transparency and increased refraction. Because the crystallin fibers in the lens interior are not regenerated during growth and aging, they undergo many post-translational changes, including glycation, carboamination, and deamidation. These changes increase crossover and interdigitation among crystallins, making them less elastic, more dense, opaque, and yellowish.[5,9,17,19]

Some of these aging changes in the lens proteins occur as a consequence of oxidative damage (Chapter 5) to the protein antioxidants, like glutathione and ascorbate, which diminish in concentrations in the aged lens, while yellow chromophores, particularly metabolites of tryptophan ($\beta$-OH-kynurenine, anthranilic acid, bityrosine), increase in frequency of occurrence and concentration. The net result is a threefold increase in lens optical density (at 460 nm, blue) between 20 to 60 years.[10] This results in decreased transmission and increased light scattering, particularly in the blue and yellow range but much less so in the red range. Percent transmission of light by the eye is about 75% at 10 years and 20% at 80 years. In addition to impairing transparency and refraction of light, these aging changes may also affect color perception. Excess lens opacity as a consequence of extensive accumulation of pigments may result in a pathological condition known as *cataract*, characterized by a cloudy lens.[4–7,17] This condition may cause reduced vision or blindness (see also below). In normal aging, the accumulation of yellow chromophores and the increased refraction of blue light may protect the retina from the damaging effect of blue light, "blue-light-hazard."[5,17]

### d.  Recent Studies on Biochemical and Biophysical Changes in Lens with Aging

Biochemical changes in the human lens with aging include increased insolubilization of nuclear region crystallins, accompanied by formation of high molecular weight aggregates that may underlie the deformity of the lens nucleus. Increased light scattering, spectral absorption, and lens fluorescence are likely causes of the decrease in light transmission with age. Accumulation of glutathione-$\beta$-hydroxykynurenine glycoside causes increased yellowing and fluorescence of the lens, and these may be responsible for accumulation of high molecular weight aggregates.[17] Aging changes were observed in some but not all crystallins, including increased truncation of N-termini, degradation of C-termini, and partial phosphorylation.[20] Other studies show increased $\beta$-crystallins but decreased $\gamma$–crystallins proportions during the postnatal period. A major portion of water-soluble proteins in adult lenses is truncated between $\beta$-B1 and $\beta$-A3/A1 crystallins, and all crystallins are susceptible to deamination with aging.[17,21]

### e.  Phospholipid and Lipid Changes in the Lens Membranes

Phosphatidylcholine decreases with age in epithelial and fiber membranes, but the rate of decrease is higher in the epithelial membranes; both membranes show a steady increase with age in the percentage of sphingomyelin.[22] Epithelial membranes contained about five times more phosphatidylcholine than age-matched fiber fractions.[22] The distribution of 3-$\beta$-OH-cholesterol shows a decrease in the anterior region of the lens relative to the posterior region with aging.[17] Lipid oxidation increases linearly with age.[23] Ganglioside composition changes with age; ganglio-series gangliosides increase with no significant accumulation of sialyl-LewisX gangliosides; however LewisX-containing neolacto-series glycolipids increase with age and cataract progression.[24] Lens aging is accompanied by decreased transport of water and water-soluble low molecular weight metabolites and antioxidants entering the lens nucleus via the epithelium and cortex; this may lead to progressive increase in oxidative damage.[25]

## 3.  Decline in Accommodation and Development of Presbyopia

Lens elasticity is critically important for the operation of the eye's *accommodation reflex*. During accommodation, the lens becomes more spherical in order to focus the image of near objects on the retina. Contraction of the *ciliary muscles* relaxes the lens' *suspensory ligaments* and the lens, allowing it to have a higher curvature (Figure 9.2). This increases the refractive power of the lens and decreases the focal point, resulting in sharp focusing.

With aging, the increased crossover and interdigitation and compaction of collagen fibers in the capsule and crystallins in the lens nucleus result in gradual hardening and reduced elasticity of the capsule and lens' interior.[5,11] These changes make the lens gradually less resilient to accommodate for near vision. Indeed, the *point of near vision,* that is, the minimum distance between the object and the eye for formation of a clear image, increases tenfold during human life, from 9 cm at the age of 10 years to 10 cm at 20 years, 20 cm at 45 years, and 84 cm at 60 years.

**Systemic and Organismic Aging**

The loss of accommodation with aging can also be determined by studying the changes in the eye's *refractive power,* measured in diopters (reciprocal of the principle focal distance of the lens in meters). Thus, the newborn's lens, being more spherical, shows the highest refractive power (about 60 diopters). As shown in Figure 9.3, the accommodation power of the human lens decreases to about 14 diopters at the age of 10, and 5 at 40 years, reaching a minimum of 1 diopter by the sixth decade. At this age, the lens becomes hard and nonresilient, essentially unable to accommodate for near-vision tasks. This condition known as *presbyopia* (Figure 9.3) is of major clinical significance, as practically everyone over 55 years needs corrective convex lenses or eyeglasses for reading and other near-vision tasks.

Environmental effects such as heat and temperature can increase the rate of aging changes in the lens fibers, accelerating presbyopia. People from warmer climates show earlier presbyopia.[6,12] The aging changes in the suspensory ligaments, the ciliary muscles, and their parasympathetic nerve supply and the associated synapses may contribute to the decline in accommodation with aging. Latency of accommodation reflex decreases during aging.

### a.  Recent Studies on Aging of Accommodation Response and Its Causes

The accommodation response shows a decrease in magnitude of fluctuation as well as in amplitude and speed of accommodation with age, indicating a decrease in accommodation dynamics. In older subjects, lens shape contributes little to power, while lens position in the eye significantly influences the power spectrum.[26] The time constant for far-to-near accommodation increases with age at a rate of 7 msec/year, while that for near-to-far accommodation increases at 6 msec/year, supporting a lenticular cause of presbyopia;[26] damping coefficient of the lens increases 20-fold between 15 and 55 years of age.[26]

The static accommodation response with age reveals a slow decline from youth to age of 40, after which the curve slope shows a rapid decline.[27] Also observed is a decrease in tonic accommodation and its amplitude, presumably caused by biomechanical factors.[27] However, subjective depth of focus increases due to increased tolerance to defocus, related to onset of presbyopia.[27] Based on analysis of aging changes in focal length, surface curvature, and resistance to physical deformation, carried out in the isolated human lens, lens hardening appears to be the major cause of presbyopia, and the loss of accommodation cannot entirely explain presbyopia.[28] A magnetic resonance imaging study shows that ciliary muscle contraction remains active during aging, but the accommodated ciliary muscle diameter decreases with increasing lens thickness, indicating that presbyopia may depend on the loss of the ability of the

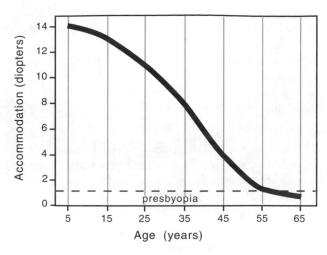

**FIGURE 9.3** Changes in visual accommodation in humans with age. Note that the decline occurs throughout life, resulting in presbyopia in the early fifties. (Adapted from various similar reports, e.g., Duane, A., *Arch. Ophthalmol.*, 5, 1, 1931.)

lens to disaccommodate due to increased lens thickness or inward movement of the ciliary ring, or both.[29]

### 4.  Iris and Senile Miosis

The *iris* is a smooth muscular ring forming the *pupil* of the eye (Figure 9.2). Contraction and dilation of the pupil during the light reflex changes the amount of light entering the eye and is also important in the accommodation reflex. In the elderly, the iris appears paler in the middle, mainly due to loss of pigmentation in the radial dilator muscles. With aging, there is a mild but constant increase in the density of collagen fibers in the iris stroma and noncellular perivascular zone.[5]

A characteristic ocular impairment in the elderly is a persistent reduction in the pupil size (diameter), the so-called senile miosis.[10] Senile miosis is particularly notable in the fully dark-adapted eye; the reduction in diameter occurs gradually with aging, decreasing from a mean of 8 mm in the third decade, to 6 mm in the seventh decade, and to 5 mm in the tenth decade of life[11] (Figure 9.4). Senile miosis results from a relatively higher rate of aging atrophy in the radial dilator muscles, which dilate the pupil, compared to the sphincter constrictor muscles, which constrict it. As a result, the sphincter is constantly dominant in the elderly, causing persistent constriction. Compared to youth, the reduced pupil aperture of the elderly results in a one-third reduction in the amount of light entering the eye.[10]

### 5.  Anterior Chamber and Vitreous Humor

The *anterior chamber* and its fluid, the *aqueous humor,* occupy a space between the cornea and the lens (Figure 9.2). The size and volume of the eye's anterior

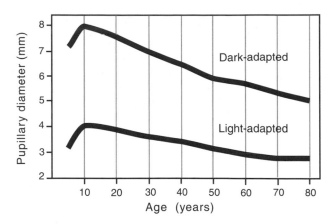

**FIGURE 9.4** Changes with age in pupillary diameter and area, measured in light-adapted and dark-adapted individuals. Note more pronounced effects in dark-adapted eyes. (Redrawn from Weale,[10] based on the original data of Verriest, *Bull Acad. R. Med. Belg.*, 11, 527, 1971.)

chamber decrease with age, mainly due to thickening of the lens. This growth occasionally exerts pressure on the *canal of Schlemm* (Figure 9.2), an outflow channel at the junction between the iris and the cornea, causing decreased flow and increased pressure (intraocular pressure) of the aqueous humor. In normal aging, the increase in intraocular pressure is small and steady. Severe obstruction of the canal of Schlemm caused by degenerative changes in the endothelial cells of the trabecular sheets and meshwork leads to markedly increased intraocular pressure (>22 mm Hg) and the serious eye disease *glaucoma*[4,5,7] (see also below).

The *vitreous humor,* a mass of gel-like substance filling the *eye's posterior chamber,* gives the eye globe its shape and support (Figure 9.2). With age, the vitreous loses its gel-like structure and support, becoming more fluid and pigmented. The increasing inhomogeneity in its gel structure, a process called *syneresis,* can lead to vitreous collapse or its detachment from the retina; often vitreous floaters *(inclusion bodies)* are released in the process, which are responsible for occasional visual flashes. These physical changes in the vitreous may also be due to aging changes in its collagenous fibrous skeleton, which has attachments to the retina, particularly in the vitreous base near the periphery. These attachments change with age, moving posteriorly and decreasing in number.[6,7,9]

## 6. Aging Changes in the Retina

The human *retina* shows considerable age-related structural changes, particularly in its *peripheral zones,* although the *macula* and its *fovea centralis* are not spared. The aged retinal periphery is thinner (10 to 30 μm), containing a lesser number of *rods* and other nerve cell types. The aging loss of rods appears to be a slow process,

beginning in the third and fourth decade, and may be related to accumulated damage due to physiological exposure to light.[5–7] With aging rods, outer segments shorten and disengage from the microvilli of pigment epithelium, resulting in lesser amounts of *membranous discs* and their major constituent *rhodopsin,* the rods' photoreceptor molecule. These events may be related to changes in the turnover of rod discs with aging.[5] Normally, the entire population of rod discs turns over every 2 weeks. This process and packing orderliness of discs slowly declines with aging, perhaps due to changes in the function of pigment epithelial cells that regulate the turnover of photoreceptor cells. The result is reduced efficiency of phototransduction (see below).

### a. Retinal Cells and Optic Nerve Aging

#### i. Early Studies

Retinas of humans and monkeys lose cones at a rate of 3% per decade.[5] The turnover rate of cones was believed to be about a year, making them susceptible to accumulated effects of light damage and post-translational modification of their photoreceptor proteins. Cone pigment density decreases with aging, presumably as a result of cone cell loss.[6] Remaining cones increase in average diameter. Because size and geometry of foveal cones are important in visual acuity, these changes may contribute to the observed losses in visual acuity with age (see below). Nerve cell loss during aging in other neuronal types of retina, i.e., *bipolar, amacrine,* and *horizontal cells,* is in the range of 30%. The loss for the *ganglion cells* is believed to be higher, about 50%. *Mueller cells* (a type of glia cell in the retina) take over the space left by the lost neurons and form cysts, which are common in the aged retina.[4,5,8] The degeneration of macula in advanced age is discussed below under diseases of eyes in the aged.

#### ii. Recent Studies

Panda-Jones et al.[30] found a decrease in photoreceptor density with age at a rate of 0.2 to 0.45% per year, with rods showing a more marked loss than cones. Cone mosaics are more organized in temporal regions compared to peripheral regions and show no change with aging, but aging distorts the regularity of the nasal peripheral cone mosaic.[31] While foveal cone numbers remain stable with age, the parafoveal rods show a decrease of 30% by old age. This condition leads to greater loss of scotopic sensitivity compared to phototopic sensitivity.[32] Short, middle, and long wave-sensitive cones were analyzed for their sensitivity. A decrease in sensitivity, that is, an increase in threshold, was observed for all three types of cones.[33]

### b. Optic Nerve

A study on the optic nerve head region in relation to open-angle glaucoma revealed a decrease in neuroretinal rim area at a rate of 0.28 to 0.39% per year; vertical optic cup diameter and optic cup areas increased with age, as well as mean cup/disc diameter ratio (0.1 from 30 to 70 years).[34] For more discussion of optic nerve changes with aging, see below under glaucoma.

### c.  Aging of Retinal Pigment Epithelium (RPE)

#### i.  Cell Number and Regional Distribution

Earlier studies indicated that retinal pigment cells do not divide after maturity and decrease in number in advanced age.[5] Recent studies show that aging causes an increase in pigment cell area, peripherally, but a decrease, centrally, while total cell number does not change markedly.[35,36] The distribution of cells in the RPE of older retinas becomes more heterogeneous, and RPE cell density from fovea to mid-periphery and to peripheral fundus regions decreases with aging at a rate of 0.3% per year.[37]

#### ii.  Lipofuscin and Lysosomal Activity

RPE cells in the foveal region show increased accumulation of the aging pigment lipofuscin during aging.[5] Increased lipofuscin formation and accumulation in postmitotic cells may be related to increased autophagocytosis and decreased intralysosomal degradation or exocytosis. The RPE layer shows cell loss, pleomorphic changes, decreased content of intact melanin, and metabolic changes.[38] Lysosomal activity in RPE increases, as shown by an increase in cathespin-D and B-glucoronidase; but arylsulfatase-B and A-mannosidase show no change with age.[38] In another study, the latter two and other glycosidases were found to decrease in aging.[39]

## B.  Aging of Visual Functions

## 1.  Eye Movements, Visual Thresholds, Critical Flicker Frequency, Field, and Spatial Vision

### a.  Smooth Pursuit and Saccadic Movements

Moschner and Baloh found a decrease in smooth pursuit movement gains with aging, while saccadic reaction times increased.[40] Saccadic reaction times were slowest in people older than 60 years, compared to young adults.[41] Visual sensitivity to motion decreases with age, and the change is more pronounced in the central visual field compared to other areas.[5]

### b.  Sensitivity and Visual Threshold

As discussed above, the pupillary aperture decreases with age, resulting in lesser light input. The decline in the number of photoreceptor cells (rods) and other aging changes in the retina result in reduced availability and regeneration capacity of photoreceptor pigment (rhodopsin), leading to reduced light utilization in the aged eye. As a result, the *visual threshold,* i.e., the minimum amount of light necessary to see an object, increases with age. This is tested by measuring the change in visual threshold as a function of time spent in darkness *(dark adaptation).* It is known that the threshold for light detection decreases

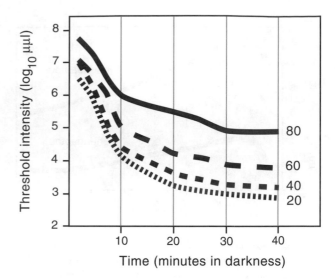

**FIGURE 9.5** Decline in light sensitivity with age. Data show changes in visual threshold after dark adaptation in different age groups (20, 40, 60, 80 years). (Redrawn from Marsh,[19] based on original data of McFarland et al. J. Gerontol., 15, 149, 1960.)

with increased duration of dark adaptation, because rhodopsin regeneration is enhanced in the dark. With advancing age, this regeneration is presumably deficient, resulting in higher light thresholds (i.e., lower sensitivity).

As illustrated in Figure 9.5, the enhanced light sensitivity after dark adaptation is markedly reduced with aging; in fact, the visual thresholds in the completely dark-adapted eyes of the aged group (80 years) is 100 times higher than that of the young group (20 years). However, as evident from the data, the pattern of change in sensitivity during dark adaptation is basically similar in the different age groups. This indicates that retinal function is quantitatively, but not qualitatively, impaired. The decline in threshold may be due to reduced oxygenation of the retina and the rods in the aged.[4,5]

#### i.  Recent Studies on Scotopic Sensitivity

Scotopic sensitivity shows a 0.5 log unit decrease with age; the loss is enhanced in the perimacular region.[42] The area of scotopic spatial summation (Rico's area) was measured in adults in the age range of 20 to 85 and was found to increase with age.[43]

### c.  Critical Flicker Frequency

The rate at which consecutive visual stimuli can be presented and still be perceived as separate is called the *critical flicker frequency* (CFF). Determination of CFF in different age groups provides one way by which changes in visual function with age can be measured. These tests reveal a decline in CFF with aging, from a value of 40 Hz (cycles/s) during the fifth decade to about 30 Hz in the eighth decade.[44] The persistent miosis in the elderly must contribute to this decline, because the decline is less marked with fully dilated pupils. Foveal

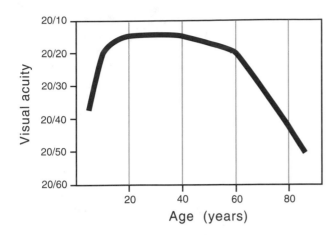

**FIGURE 9.6** Changes in visual acuity measured in Snellen index with age. Note that the aging decline begins after 60 years, resulting in serious deficits in acuity after 80 years. (Redrawn from Pitts, D.G., in *Aging and Visual Function*, Sekuler, R. et al., Eds., Alan R. Liss, New York, 1982.)

flicker frequency sensitivity in healthy aging eyes showed a decrease in foveal temporal contrast sensitivity coupled with losses in amplitude but not temporal resolution. The mean rate of loss was 0.78 decilog per decade after the age of 45 years.[45]

### d.  Aging Changes in Visual Field and Spatial Vision

With aging, a loss in the size of *visual field* is observed, ranging from 3 to 3.5% in middle age to two and four times as much at 60 and >65 years, respectively. Disturbances in visuo-motor performance are particularly observed when changes in "useful or effective visual field" are measured, in contrast to measures of visual field under standard clinical conditions.[46] *Visual acuity* reflects the ability to detect details and contours of objects. It is defined as the minimum separable distance between objects (usually fine lines) and is one of the measures of the visual system's *spatial* discrimination ability. As shown in Figure 9.6, visual acuity declines, commencing at 50 years, particularly worsening during and after the eighth decade, when it becomes detrimental to vision.[5,46]

### e.  Recent Studies on Spatial Vision and Visual Acuity

A recently devised measure of spatial visual ability is contrast sensitivity, where the test format and conditions more closely resemble real world conditions.[46] Aging studies in this category indicate that at low spatial frequencies, such as when grating of wide bars are presented, little aging changes are observed, while at high frequencies (fine bars), a marked decline in contrast sensitivity is observed with aging, beginning at 30 years. This deficit may underlie certain reading disorders in the elderly, such as reading of very small or very large repetitive characters,

where the elderly show nearly 70% deficit.[46] A recent analysis on visual acuity changes with age revealed no significant change in high visual acuity, but all aspects of spatial vision showed a decrease, particularly under conditions of decreased contrast, luminance, and glare. Also found was reduced stereopsis, poor color discrimination, and decreased peripheral vision, particularly when divided attention was required.[47]

The decline in visual acuity and contrast sensitivity with age may be the result of alterations in the following optical and neural factors: (1) altered refraction of the light rays by the cornea and the lens; (2) decline in accommodation; (3) decreased light input due to constricted pupils (senile miosis); (4) decline in the density, number, and function of visual receptor cells, particularly in the fovea; and (5) aging changes in central neural structures of the visual system. It is important to note that elimination of glare and improved illumination can help enhance visual acuity substantially in the aged population.

### f.  Impaired Vision and Everyday Tasks as Shown by Changes in "Useful Field of Vision"

Early studies indicated that impaired visual capacity is a major factor in performance of the elderly in such everyday visual tasks as driving.[46] Incidence of driving accidents was four times higher in elderly with reduced "useful field of view," compared to the age-matched group with maximum useful field of view. Indeed, in the category of accidents occurring at intersections, the incidence rate was 16 times higher in individuals with reduced useful field of vision.[46] Recent studies on the effects of aging on useful field of view by Sekuler et al.[48] confirmed earlier findings of a decrease in this parameter with age; the decrease begins at about the age of 20 years. The reduced ability is more pronounced when conditions require divided attention between central and peripheral tasks.[48] A similar increase in number of errors for localization of peripheral targets indicates a decrease in the useful field of view.[49]

## 2.  Color Perception

Beginning with the fourth decade of life, color perception shows a progressive decline with age. Women are relatively less affected than men. During the fourth and fifth decades, the deficiency in color perception is mostly in the short wavelength range, i.e., yellow to blue. This is explained by the changes in the lens, as it becomes more yellow with age. In the later decades, after the sixties, a deficiency in the green range also becomes manifested, probably due to retinal or more central factors, as evident even after removal of the lens.[19,46] Aging defects in color perception are exaggerated under reduced illumination.

**Systemic and Organismic Aging**

### 3. Changes in Central Visual Pathways

With advancing age, considerable changes in the *optic nerve*[50] and *visual cortex*[51] have been reported. Changes in the visual cortex have included thinning and cell loss. Earlier electrophysiologic investigations indicated marked flattening of the *evoked potential* responses to light flashes in the different parts of the cortex. In a more recent study, it was found that the amplitude of some of the components of visual-evoked potentials are markedly reduced, while the latency of response is increased.[52] The latter changes are significantly more marked in elderly men.

#### a. Recent Findings on Central Visual Impairments in Aging

Justino et al.[53] investigated visual electrophysiologic responses, such as electroretinograms (ERGs) and cortical visual evoked potentials (VEPs), in healthy elderly subjects (>70 years), compared to young adults, using stimuli biased toward the functioning of the intensity-sensitive magnocellular and color-sensitive parvocellular subdivisions of the visual system. The elderly subjects showed a decrease of ERG and VEP amplitudes, as well as an increase in VEP latency. These age-related effects were magnified when stimulus conditions combined magnocellular and parvocellular pathways, indicating that normal aging affects the functioning of both visual pathways. A PET (positron emission tomography) study by Levine et al.,[54] measuring regional blood flow in the brain, indicated that young adults and the elderly utilize different pathways for form perception. Older subjects activate occipital and frontal regions while perceiving forms, while younger people utilize the occipito–temporal pathway preferentially, indicating possible reorganization of visual processing during human aging.

### C. Aging and Eye Diseases

Various degrees of vision loss and blindness, commonly caused by senile cataract, glaucoma, macular degeneration, and diabetic retinopathy, represent the extreme consequences of age-related ocular pathologies (Table 9.1). Diabetic retinopathy is discussed in Chapter 14. In the United States, for patients in the age range of 75 to 85 years, the prevalence of cataracts is 46%, macular degeneration 28%, and glaucoma 7.2%.[6] Screening for these disorders includes testing visual acuity, ophthalmoscopic examination, and checking intraocular pressure.[5,8] As a result of increased occurrence of eye diseases with age, the incidence of blindness shows a 25-fold increase with age, from about 0.1% in the middle age group to 2.5% in the elderly (>75 years).[5]

### 1. Cataract

In some individuals, the accumulated normal aging changes in pigment and protein composition in the lens take pronounced and pathological dimensions, leading to a condition known as cataract.[4,5,7] Among the causes of cataract is the glycation of lens proteins, such as the $Na^+$-$K^+$-ATPase, resulting in abnormal ion–water balance, swelling, and breakdown of crystallin lens proteins. Also, water-insoluble crystallins and yellow chromophores accumulate excessively, particularly in the nuclear type.[5]

As a result of the above pathologies in senile cataract, the lens interior becomes cloudy and opaque, light refraction is greatly reduced, and light scattering is markedly increased. These effects lead to loss of visual acuity, reduced patterned vision, and eventually to functional blindness with only a degree of light perception remaining. The occurrence of lens opacities markedly increases with age from about 4% in the middle age group to 12% at 50 years to nearly 60% in the seventh decade.[5] In the same age groups, the percent of individuals showing visual loss as a result of cataract increases from about 1% to 3% and 30%, respectively. Cataract occurs more frequently in diabetic individuals, especially females. Severe cataract is the third leading cause of blindness in the Western world, after macular degeneration and glaucoma. In general, 50% of people over 65 years of age have cataract.

Cataracts may occur in the lens periphery (*cortical cataract*) or center (*nuclear cataract*). The incidence of the nuclear type is always higher than that of the cortical type (65% compared to 28% in the >75 years group). Cortical cataract is particularly detrimental to visual acuity, while nuclear cataract interferes more with color perception. A mild cataract can be managed with periodic examination and use of eyeglasses. However, when the reduction in visual acuity interferes with the patient's daily activities, cataract surgery may be necessary[4–6] (see below).

#### a. Recent Studies on Genetic and Racial Factors Influencing Cataract Development

Genetic factors are important in development of cataract. Incidence of cortical cataract is much higher among mono- or dizygotic twins compared to the general population.[55] However no correlation between size at birth and development of age-related cataract has been found.[56] A recent Japanese study, in support of genetic causes, shows that galactokinase deficiency, which occurs at a rate of 4% in Japanese with normal vision, is associated with 8% of bilateral cataract cases in Japan and 3% in Korea; no correlation is seen among white or black Americans.[56]

#### b. Recent Studies on Role of Antioxidants, Glycation, Multilamellar Bodies, Nuclear Compaction and Light Scattering

Decreased uptake of protective antioxidative fat-soluble nutrients (carotenoids, retinal and tocopherol) in the aged lens and also differential localization of nutrients to different regions of the lens may account for differential risk of developing cataract in a different lens region.[57]

Similarly, increased intake of vitamin C, an antioxidant component, has a protective effect against age-related cataract, acting possibly to prevent lens tissue oxidative damage seen in cataracts. Sweeney and Truscott report decreased glutathione (GSH) levels to correlate with cataract development.[58] GSH is the principal antioxidant, and older lenses show reduced levels in the central nuclear region. It is proposed that with aging, a barrier develops in the lens, preventing diffusion of GSH to central lens regions, thus allowing oxidation of nuclear proteins and cataract.[58]

According to Zarina et al.,[59] increased glycation end products (GEPs) occur in human senile and diabetic cataractous lenses. More glycation occurs in the nucleus than in the cortex of the lens, and the diabetic group showed still higher amounts compared to the normal senile group.[59] Dark iris color increases the risk of age-related cataract due to an increase in lens optical density. Gilliland et al.[60] show that multilamellar bodies seen in cataractous lenses may be the light-scattering objects responsible for forward light-scattering seen in nuclear cataracts. Another explanation for this abnormality is nuclear compaction in cataractous lenses;[61] cataractous lenses also show higher levels of sodium, magnesium, and potassium ions compared to normal age-matched lenses.[62] Telomerase activity in lens epithelial cells is increased in cataractous lenses; but the decline in cell number in the lens epithelium is not related to cataract development.

### c. Surgical Removal of Cataractous Lens

Although certain drug treatments can alleviate some of the symptoms of cataract in some patients, surgical removal of the cataractous lens is the option of choice in recent years. In most cases, after lens removal, a plastic lens implant is placed instead of the old lens. In the United States, nearly a million cataract operations are performed per year, resulting in improved vision in 97% of the cases. All cataract operations involve removal of opacified lens via a corneal incision followed by fine suturing. In most cases, a plastic lens, called a "lens implant" is inserted in place of the removed lens (see below). The operation, usually performed on an "outpatient" basis, is done under local anesthesia and sedation. The postoperative patient requires means for optical corrections due to loss of lens. Various options available are summarized in Box 9.1.

## 2. Glaucoma

Another treatable eye disease is *glaucoma,* which is characterized by the following triad: the intraocular pressure increases (>21 mm Hg); this leads to progressive excavation of the optic disc, the site where the optic nerve leaves the eye; and cupping of the disc. This leads to ischemic damage to the optic nerve fibers, which may result in blindness. Two types of glaucoma are known, *chronic open-angle glaucoma* (COAG), which is the frequently

---

> **Box 9.1**
>
> **Options Available for Visual Corrections after Removal of Cataractous Lens**
>
> 1. *Eyeglasses:* These are thick and heavy, increasing object size by 25%; they induce optical distortions and interfere with peripheral vision; although they provide good central vision, they cannot be used after surgery, if the other eye is normal.
> 2. *Contact lenses:* Hard or soft extended-wear contact lenses have been used; they are more difficult to use, and eyeglasses are required for reading; however, they correct central and peripheral vision, increase image size by only 6%, and can be used after surgery on one or both eyes.
> 3. *Intraocular implant lens*: This is surgically placed in front or behind the iris at the time of cataract surgery; it requires the use of bifocal eyeglasses and has a higher incidence of surgical and postsurgical complications; however, it increases image size by only 1%, corrects central and peripheral vision, and can be used on one or both eyes; lens implants are made of silicone, acrylic, or hydrogel materials; the lens implants are either placed in front of the iris (intracapsular) or behind the iris (extracapsular).
> 4. *Refractive keratoplasty:* In this method, the cornea is cut and reshaped by making surgical or laser incisions; this method has become effective and popular in the recent years in correcting for far- and near-sightedness and astigmatism, but its applications for correcting for presbyopia or cataract are still in the experimental stage.

---

diagnosed type, and the *closed-angle* type, which is rare.[4–7] The incidence of COAG in the population increases rapidly with age, from a low of 0.2% in the fifth decade to about 1% in the seventh decade, 3% in the eighth decade, and 10% in the ninth decade. Open-angle glaucoma is characterized by an insidious and slow onset, initially asymptomatic and then gradually leading to blindness, if unchecked. Closed-angle glaucoma is rare and is characterized by an acute attack of severe eye pain and marked loss in vision due to a rapid increase in intraocular pressure compressing the entire retina.

### a. Recent Research on Glaucoma

Systemic blood pressure and hypertension correlate with intraocular pressure (IOP) and high-tension glaucoma, but no link appears to exist between blood pressure and normal-tension glaucoma.[62] Prospects of genetic intervention

**Systemic and Organismic Aging**

in primary open-angle glaucoma have been investigated, and five primary open-angle genes have been mapped. Understanding these genes and their products may help in finding better treatments.[63] Collagen degradation appears to play a role in glaucoma.[64] Imbalance in rates of extracellular matrix production and turnover may be important in open-angle glaucoma, and matrix metalloproteinases are likely to be used in its treatment.[64]

### b.  Pharmacological Treatment for Glaucoma

Whereas cataracts are usually treated surgically, glaucoma is often treated medically with eye drops containing various drugs:

- Miotics (substances that constrict the pupil) such as pilocarpine are most commonly used.
- ß-blockers, which block specific sympathetic innervation, such as Timolol, are also used, but caution is warranted, because systemic side effects such as heart failure may occur.
- Sometimes systemic medications are used, such as carbonic anhydrase inhibitors (see below).

Modern pharmacological therapy for glaucoma may be pursued in three stages.[65] First-line treatments are ocular hypotensive agents. ß-blockers such as Timilol and Levobunolol are common choices here. If ß-blocker treatment is ineffective, prostaglandin and carbonic anhydrase inhibitors and α-2-adrenergic agonists (brimonidine, dipiyayl epinephrine) are used. Miotics (e.g., pilocarpine and carbachol) are suggested as second or third lines of treatment; also along the third line of treatments are the cholinesterase inhibitors with fewer side effects. Therapy is often enhanced when drugs are used in combination.[65]

Dorzolamide (dorzolamide hydrochloride), a topical carbonic anhydrase inhibitor, is highly effective in the management of glaucoma and ocular hypertension. It reduces intraocular pressure by decreasing aqueous humor formation. Effects are additive when used with topical ß-adrenergic antagonists. Side effects are bitter taste and transient local burning and a stinging sensation. Prostaglandin analogs, acting as ocular hypotensive drugs, are also among the new therapeutic agents.[66] The two major drugs in this category are Latanoprost, a highly potent and widely used drug, and Unoprostone, which is used less widely, has mild ocular side effects and rare systemic ones. Better hypotensive effects are obtained when Latanoprost is used in conjunction with other routine glaucoma medications.

### c.  Surgery and Laser Therapy

In case of an acute attack of closed-angle glaucoma or if medical treatment is ineffective, surgery may be necessary. Recently, *argon laser trabeculoplasty* (ALT) has been increasingly used, instead of surgery, for the treatment of open-angle glaucoma that could not be controlled by drugs. ALT treatment consists of tiny laser burns evenly spaced around the trabecular meshwork. This laser treatment is sometimes used for controlling glaucoma or as a preventive measure, although pharmacological treatment of the eye can continue.

### 3.    Age-Related (Senile) Macular Degeneration (ARMD)

An important cause of visual impairment in the elderly, often leading to legal blindness within 5 years after onset, is *senile macular degeneration* or *age-related macular degeneration* (ARMD).[4–7,67,68] The disease accounts for nearly half of the registered (legal) cases of blindness in the United States and England. The incidence of ARMD increases with increasing age, from about 4% in the 66 to 74 years age group, to 17% in the 75 to 84 years age group, and 22% in the >84 years age group.

### a.  Aging Changes in the Macula

The *macula lutea* is an area of retina 6 mm in diameter located at the posterior end of the eye's *visual axis* (Figure 9.2). Through its high density of cones and involvement in day and color vision, the macula, and in particular, its central zone, the fovea, provide the structural basis for high visual acuity. Hence, macular degeneration, more than any other eye disease, affects visual acuity and central vision. This disease occurs generally in both eyes and more often (50%) in women. It is believed to be a hereditary disorder, not caused by simple aging of the retinal nerve cells, but largely related to manifestation of inherited pathologies in the nonneural retinal elements such as the pigment epithelium. The patients also show an increased incidence of hyperopia (farsightedness). The disease may result from disturbances in the walls of subretinal capillaries or in the thickness of subretinal membrane or the retinal pigment epithelium.[4–9]

### b.  Recent Biochemical and Pathological Findings in ARMD

Tissue inhibitors of metalloproteinases (TIMP-3) have been implicated in aging and in ARMD. TIMP-3 levels in Bruch's membrane in macula increase with age, and these levels are higher in ARMD subjects compared to normal age-matched individuals.[69] Gelatinase-A (MMP-2) levels are highest in the interphotoreceptor matrix and vitrous and do not increase with age; however, levels are twice higher in the matrix of retinal pigment epithelium associated cases.[70] A summary of the pathological hallmarks and possible sequence of events leading to ARMD is listed in Table 9.2.

### c.  Risk Factors and Treatment of ARMD

Incidence of ARMD is higher in monozygotic twins compared to dizygotic ones, suggesting the importance of genetic factors.[71] Alcohol consumption does not increase

**TABLE 9.2**

**Summary of the Pathology and Sequence of Events Leading to Age-Related Macular Degeneration (ARMD)**

### Pathological Hallmarks

Senile macular degeneration may be accompanied by any of the following pathologic changes depending on the stage and extent of the disease[4,66,67]

1. White excrescences in the subretinal membrane, called *Drusen* (nodules), hyaline deposits ranging in size from punctuate lesions to dome-shaped structures 0.5 mm in diameter, often observed during the early stages but can be found dispersed throughout the fundus
2. Atrophy of retinal pigment epithelium (RPE)
3. *Serous detachment* of RPE
4. Subretinal neovascularization
5. *Disciform scars* which result from RPE detachment

### Proposed Sequence for Development of ARMD

Following is a proposed sequence of events leading to ARMD:

1. Dysfunction of retinal pigment epithelium
2. Accumulation of abnormal intracellular and extracellular material (basal lamina) in RPE
3. These deposits lead to the following changes in Bruch's membrane:
   - Altered composition such as increased lipids and protein cross-linking
   - Decreased membrane permeability to nutrients
4. These events result in decreased diffusion of water-soluble molecules (vitamins and antioxidants) and finally metabolic distress to RPE

*Source:* Zarbin, M.A., *Eur. J. Ophthalmol.*, 8, 199, 1998.

the risk of ARMD,[72] but cigarette smoking poses a significant risk.[73] Unfortunately, treatment of macular degeneration is not nearly as successful as that of cataract and glaucoma.[68] Although a protective role for antioxidants and trace minerals against ARMD development is widely believed, the use of antioxidants for treatment of ARMD is not widely supported.[74]

*d. Laser Therapy in ARMD*

The only realistic goal of treatment for ARMD is to prevent subretinal detachment and hemorrhage, disorders that lead to an acute loss of vision. To accomplish this, the sites of subretinal neovascular formation are located using fluorescin angiography followed by laser photocoagulation. According to Ciulla et al., laser photocoagulation of *choroidal neovascular membranes* (CNVMs) is currently the only well-studied and widely accepted treatment.[75] New treatments may be divided into four major categories: (1) photodynamic therapy, (2) pharmacologic inhibition of CNVM formation with antiangiogenic agents, (3) surgical intervention, and (4) radiation therapy.

## VISUAL DYSFUNCTIONS IN ALZHEMIER'S DISEASE (AD)[76–78]

The AD patients perform significantly worse on tests measuring static spatial contrast sensitivity, visual attention, shape-from-motion, color, visuospatial construction, and visual memory.[76] Correlation analyses showed strong relationships between visual and cognitive scores. The findings show that AD affects several aspects of vision, and the resulting effects are compatible with the hypothesis that visual dysfunction in AD may contribute to performance decrements in other cognitive domains. The pattern of involvement indicates that AD affects multiple visual neural pathways and regions. A better understanding of vision-related deficits can lead to better diagnosis and interventions that may, in turn, help improve functional capacity in patients with dementia and AD. Decreased attention may severely limit cognitive performance in the elderly. Rizzo et al.[77] found deficits in sustained, divided, and selective attention, as well as in visual processing speed, occurring early in AD, which significantly correlated with diminished overall cognitive function. These findings indicate that assessment of visual attentional ability could prove useful for diagnosis of AD and lead to more precise measures of useful perception in AD patients with normal visual fields and acuity.

Also, in mild to moderate cases of AD, a significant effect on perception of structure from motion occurs with relative sparing of motion direction discrimination. This problem is likely to have a cerebral basis and has the potential to affect navigation and the recognition of objects in relative motion, as encountered during walking or automobile driving.[76,77] Certain atypical AD patients show characteristic visual abnormalities. Visual association pathways are usually not as disrupted in the more common form of AD.[78] These visual symptoms were the most identifiable sign of this particular form of AD. In these patients, visual association areas in the occipito-temporo-parietal junction and posterior cingulate cortex as well as primary visual cortical areas demonstrated high concentrations of lesions, while the prefrontal cortical regions had fewer lesions than found typically in Alzheimer's disease patients.[78]

## III. HEARING

Changes in auditory functions with age provide some of the classic and important case studies in gerontology and the physiology of aging. Incidence of hearing disorders rapidly increases with aging, afflicting nearly a third of people over 65 and one half of those over 85 years (Figure 9.1). Hearing impairments are 20% more frequent in elderly men than women. Hearing disorders interfere with the perception of one's own speech and that of others, creating behavioral and social disabilities. These conditions may lead to social withdrawal and isolation, particularly under the extreme condition of deafness.[2] As in the visual system, the age-related problems of the auditory system may stem from structural and functional disorders of the peripheral auditory components, the central neural aspects, or both. Here, too, more is known about the aging of the ear than the central auditory system (Table 9.3).

**Systemic and Organismic Aging**

**TABLE 9.3**

**Summary of the Normal Aging Changes in the Human Ear**

### Structural Changes

**Hair Cell Degeneration**

*Basal cochlea*: frequent, especially in first quadrant; diffuse and patchy; main cause of sensory presbycusis

*Apical cochlea*: infrequent

**Nerve Cell Degeneration**

Observed in spiral ganglia often with basal cochlear hair cell loss but not with apical cases (involved in neural presbycusis); is accompanied by loss of myelinated auditory nerve fibers

### Atrophic Changes

Generally occur in nonneural components (vascular and connective tissue) of cochlea and lead to strial or conductive types of presbycusis

In stria vascularis; frequent in the middle and apical turns of cochlea

In spiral ligaments; accompanied with devascularization

In inner and outer spiral vessels

In Reissner's membrane; due to vacuolization in basilar membrane leading to mechanical damage

### Central Neural Changes

Little neuronal loss in lower auditory centers; heavy loss in conical auditory centers; dendritic degeneration of cortical pyramidal neurons

Increased latency and decreased amplitude of auditory-evoked potentials; effects more marked in elderly males than in females

### Functional Changes

**Pure Tone Hearing**

Loss of hearing in the high-frequency range (presbycusis): loss progressively worsens with age; effects more pronounced in males; noise exposure enhances loss

Decline in maximum sound frequency capable of being heard, from 20 kHz at 10 years to 4 kHz at 80 years

**Speech Perception**

Diminished ability to hear consonants; speech is heard but unintelligible

Diminished ability to perceive reverberated and interrupted speech

**Sound Localization**

Diminished ability to localize sound source, particularly at high frequencies

## A. AGE-RELATED HEARING LOSS (PRESBYCUSIS)

Age-related hearing loss is called presbycusis. Numerous studies have reported loss of hearing with age, particularly for sounds in the high-frequency range.[19,79,80] Presbycusis occurs in both ears but not necessarily at the same time. To assess auditory loss with age, pure tone audiograms of subjects in different age groups are determined. That is, the hearing threshold in decibels (unit of sound intensity, dB) for sounds of increasing frequency is determined, and the results are presented as relative loss of decibels at different frequencies.

### 1. Progressive Hearing Loss with Aging in the High-Frequency Range

In the low-frequency range (0.125 to 1 kHz), young subjects have essentially no hearing deficits, while old subjects show deficits of about 10 to 15 dB. In the high-frequency range, hearing loss for the young group is mild, while in the old group, it becomes progressively worse with increasing sound frequency (Figure 9.7).[19,80] Typical magnitudes of hearing loss are 30 dB at 2 kHz, increasing by about 10 dB for each additional kHz. In octogenarian men living in urban areas, the hearing loss may be as much as 80 dB. Another way of determining age-related hearing loss is by measuring the maximum frequency of sound capable of being heard. This frequency is 20 kHz for children (10 years) decreasing to only 4 kHz for the elderly (80 years); the decline is steady and linear, occurring at the rate of about 2.3 kHz per decade (Figure 9.7).

### 2. Pure Tone Audiograms and Effects of Sex, Environment, and Blood Lipids

Typical pure tone audiograms for normal men and women at different age groups are shown in Figure 9.7. Hearing loss is higher in men than women; this sex difference, which begins after the age of 35 years, possibly reflects the effects of higher exposure of men to work-related and environmental noise. Rosen et al.,[81] have also shown that the elderly people from noise-free rural areas of Sudan show lesser hearing loss compared to those in the urban environment, suggesting that urban environmental noise is one of the determinants of presbycusis (see, however, a recent Italian population study by Megighian et al. discussed below). Another cause of presbycusis may be vascular, such as atherosclerosis and similar disorders related to elevated blood lipids (hyperlipoproteinemia). Incidence of hearing loss and inner ear diseases may be high in patients with elevated cholesterol levels.[82]

### 3. Presbycusis Related to Ear Structure

The major structures of the ear include the following:

- The external ear (pinna. external auditory canal, and tympanic membrane)
- The middle ear (the ossicular chain)
- The inner ear (cochlea and the organ of Corti)

The organ of Corti contains the hair cells, which are the auditory receptors and the mechanoelectrical transducing organs (Figure 9.8a). The cell bodies of the primary auditory neurons are in the cochlea's spiral ganglia, and the axons comprising the auditory nerve enter the medulla to synapse with central auditory neurons.

Auditory pathways in the brain include the medullary and midbrain centers for signal transmission and

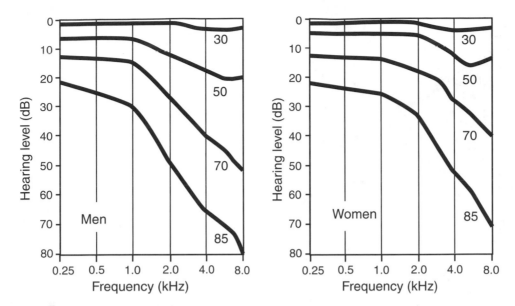

**FIGURE 9.7** Amount of hearing loss in decibels for different pure tones in various age groups of men and women. Note a greater loss in higher frequencies and older ages and a more pronounced hearing loss in men. (Redrawn from Marsh,[19] based on original data of Spoor, A., *Inter. Audiol.*, 6, 48, 1967.)

auditory reflexes, the inferior colliculi, and the auditory cortex in the temporal lobe of the cerebral cortex. The selective nature of presbycusis indicates that it is probably not associated with aging changes in the outer or middle ear (tympanic membrane and ossicles) but more likely due to changes in the inner ear (cochlea) or the central auditory system.[79,80,83]

### 4.  Types of Presbycusis

Presbycusis may occur due to damage to different parts of the auditory systems. Based on the source of damage, four types of presbycusis are recognized: sensory, neural, metabolic (or strial), and cochlear conductive.[80,83] The onset of presbycusis may be anytime from the third to sixth decade of life, depending on type. Individuals suffering from these disturbances show distinct and differing audiograms (Figure 9.8), which are clinically used to diagnose types of impairment. More complicated audiograms are produced when the pathology involves a combination of these disorders (Figure 9.8). The standard type of presbycusis with hearing loss at high Hz is often associated with *neural* or *sensory presbycusis* (Figure 9.8).

#### a.  Sensory Presbycusis

Individuals with sensory presbycusis show a major and sudden loss of hearing in the high-frequency range (>4 kHz), indicating a selective deficit in transduction mechanisms of high-frequency sounds (Figure 9.8b). Speech discrimination is normal. Although the hearing deficit is observed from middle age, the histopathological problems believed to be mainly associated with the cochlear hair cells may start much earlier. Cochleas of humans with sensory

presbycusis typically show loss of outer hair cells and less often of the inner hair cells of the organ of Corti.[80,83] (Figure 9.8a). The loss is diffuse or patchy and mainly limited to the first quadrant of the cochlea's lower basal turn. This part of the cochlea is specialized for detection of high-frequency sounds. The affected sensory hair cells and other supporting cells (Hensen's and Claudius' cells) show accumulation of aging pigment lipofuscin, the amount of which corresponds with the degree of sensory deficits.

#### b.  Neural Presbycusis

In this disorder, hearing of pure tones for all frequencies are affected, but the extent of hearing loss increases with increasing frequency of sound, the deficit being about 40 dB at 1 kHz and nearly 100 dB for high frequencies (>8 kHz) (Figure 9.8c). As a result, speech discrimination is reduced to 60% of the normal level. In the aging auditory system, the first-order sensory neurons are adversely affected. This damage ranges from synaptic structures between the hair cells and the dendrites of the auditory nerve fibers, accumulation of lipofuscin, or signs of degeneration in the cell bodies of the spiral ganglion neurons. Disruption of myelin sheath of the auditory nerve fibers can cause disordered transmission, even if the nerve cells were present.[80,84]

#### c.  Metabolic Presbycusis

In this case, also called the *strial* type, the audiogram is flat (Figure 9.8d), indicating a loss of about 30 to 40 dB at all frequencies, This type is believed to be associated with atrophic changes in the vascular supply, stria vascularis, to the cochlea.[79,80,83,84] The extent of hearing loss is correlated to the degree of degeneration in stria vascularis.[80]

**Systemic and Organismic Aging**

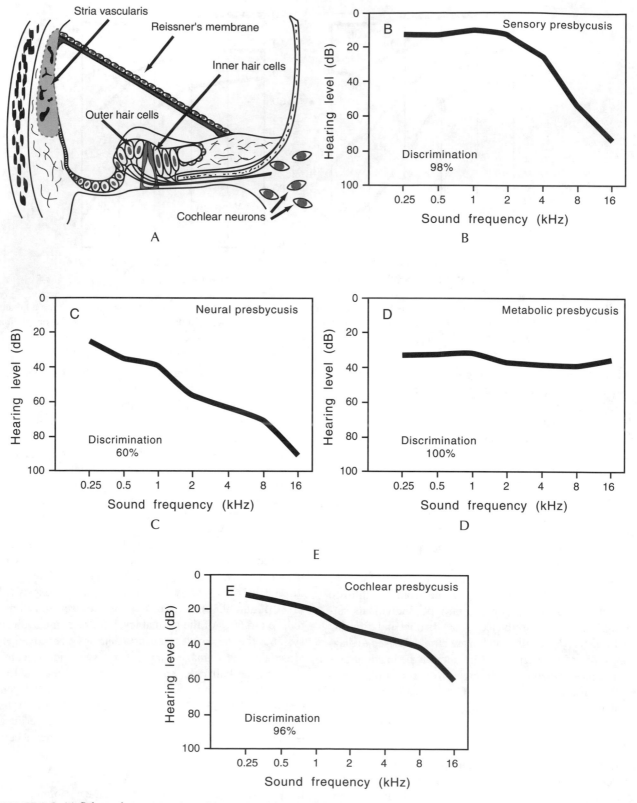

**FIGURE 9.8** (a) Schematic cross section of human cochlea; (b–e) typical pure-tone audiograms of aged individuals suffering from different types of presbycusis (sensory, neural, cochlear conductive, and strial). Each type of presbycusis reflects damage to a different component of the peripheral auditory system: sensory, to hair cells; neural, to primary auditory neurons; cochlear, to basilar membrane; and strial, to cells of the stria vascularis. (Redrawn from Gulya, A.J., *Oxford Textbook of Geriatric Medicine*, Evans, J.G. and Williams, T.F., Eds., Oxford University Press, Oxford, 1992, p. 580.).

### d. Cochlear Conductive Presbycusis

In this type (also called mechanical presbycusis), the audiogram indicates some hearing loss at all frequencies with the loss magnitude increasing linearly with increasing tone frequency (Figure 9.8e). The magnitude of hearing loss is about half of that seen in the neural type, and speech discrimination is only slightly affected in most cases (96% of normal). This type of presbycusis is believed to be due to changes in the mechanical properties of cochlea's basilar membrane, hence, its other name, "mechanical presbycusis." The basilar membrane is markedly thickened, especially at the basal cochlea, and shows calcified, hyaline, or fatty deposits. There is usually no change in the hair cells and sensory neurons.[83]

#### RECENT STUDIES ON PRESBYCUSIS IN KOREAN, ITALIAN, AND AFRICAN-AMERICAN POPULATIONS

A 20-year-long study of pure tone audiometry, by Kim et al., on 6000 Koreans, aged 65 and older,[85] found a high incidence (~ 40%) of presbycusis as well as a significant difference in hearing threshold between men and women aged 65 and older (males > females). Megighian et al.[86] analyzed the presbycusis data of over 13,000 Italian males and females, aged 60 and over, living in either city or rural environments in the Veneto region, and found typical trends of the audiometric curve in presbycusis. Hearing loss was less severe in females than males, especially at higher auditory frequencies. No significant differences emerged between subjects from the urban or the rural environments. According to Bazargan et al.,[87] poor hearing among aged African-Americans was associated with a decreased level of psychological well-being, but this relationship was mediated by the effects of hearing loss on ability to function. Poor vision, on the other hand, is independently associated with a decreased evaluative level of well-being.

### e. Sound Localization

Tests of sound localization indicate a decline in this ability with aging, beginning in the fourth decade.[19] It is known that localization of low-frequency sounds depends on *temporal* discrimination (i.e., time of sound arrival) between the two ears, while in the high frequency range, localization depends on discrimination of sound *intensity* between the two ears. Aging changes occur in both ears, but the rate of aging may be different in the two ears.[79,83] Thus, deficits in localization of sound will be apparent for all frequencies but may be more marked for sounds of higher frequency, as the perception of these are particularly impaired in the old age (see above).

### f. Other Auditory Functional Changes Related to Presbycusis

*Middle Ear Resonant Frequencies* were measured in ~500 adults 48 to 90 years of age, and no significant age-group trends were found; older women, however, had a slight but significantly greater middle ear resonant frequency

than older men.[88] According to Oeken et al.,[89] a major and significant decrease in *Distortion-Product Otoacoustic Emissions* (DPOAE) amplitude occurs with aging, caused by deterioration in pure-tone threshold. It is suggested that delineation of the effects of presbycusis on DPOAE is important if they are to be used in the diagnosis of inner ear disorders.

### B. PERIPHERAL AGING CHANGES IN AUDITORY STRUCTURES

#### 1. Cochlear Ultrastructural Changes

Electron microscopic studies on extracted human temporal bones from individuals 53 to 67 years old, with high-tone hearing loss of unknown cause, revealed changes to the ultrastructure of the stereocilia, pillar cells, stria vascularis, spiral ligament and the cuticular plate, the latter showing the greatest structural degradation.[90] Thus, cuticular plate deformation may play a role in hearing loss, although so far, this has only been proven in guinea pigs.[91] How age relates to otopathic function of the inner ear is still undetermined.

#### 2. Spiral Ganglion and Other Neural Pathways

A quantitative study by Felder et al.,[92] of neuron and fiber number in the auditory pathways of aged individuals with presbycusis, showed clear evidence of cell and fiber loss, with more loss in peripheral structures than central ones; normal controls showed no significant loss. Retrograde degeneration from the periphery to the spiral ganglion may be happening in presbycusis. Because interneuronal connections occur in human spiral ganglions, a trophic supply from other neurons at the spiral ganglion level may slow central axon degeneration in these cases.

### C. TREATMENT OF AGE-RELATED HEARING LOSS

#### 1. Conventional Hearing Aids Versus Cochlear Implants for the Elderly with Presbycusis

Currently no systematic plan for rehabilitation and fitting of hearing aids exists. Further research and efforts are needed to develop better hearing aids that satisfy more precisely the needs of the elderly.[93] Older adults and their immediate family can benefit from information and advice about the serious consequences of presbycusis. For older patients, hearing aids remove many of the sensory, speech perceptual and psychological handicaps of hearing impairment. Few older adults with hearing loss actually use hearing aids. Intervention and screening programs need to be improved to better identify older adults in need of amplification, whose quality of life would be improved with such treatment.[94] Nursing home patients have been found to consistently use

**Systemic and Organismic Aging**

their hearing aids. Amplification alone, however, does not serve all of the communication needs of the nursing home residents. Environmental modification as well as improved assistive listening devices would increase quality of life and the ability to communicate for these patients.[95]

## 2.  Increased Use of Assistive Hearing and Amplification Devices, and Cochlear Implants

An important development of the future is to support the open use of amplification devices, which comes with greater acceptance of the realities of hearing loss.[95] A strong preference of conventional hearing aid use among the elderly show that, even though the sound quality of modern *assistive listening devices* is preferred over conventional aids, subjects usually are unwilling to endure all of the difficulties associated with the use of remote–microphone devices.[95,96]

Cochlear implants[97] and a new generation of "intelligent hearing aids"[98] represent the latest devices. Cochlear implants clearly help improve quality of life for elderly patients with profound presbycusis.[97] Through cochlear implants, elderly patients have demonstrated improved audition as well as evaluative quality of life improvement.[97] Audiographic measures are not sufficient to properly provide treatment for hearing loss in the elderly. A holistic method has been developed to determine the appropriate treatment to improve hearing based on communicative, physical, social, and psychological aspects of the patient, as well as a diagnostic flowchart to best match high-tech devices with appropriate candidates.[96] For more details on cochlear implants, see Box 9.2.

## 3.  A New Generation of "Intelligent" Hearing Aids for Improving Speech Recognition

Speech recognition is difficult for the elderly in the presence of background sounds. Conventional hearing aids offer little help in this regard. A new system being developed by Dr. Albert Feng and collaborators at the University of Illinois, Urbana-Champaign, using binaural microphones, digital converting devices, and complex auditory processing models, offers to enhance speech recognition by removing background noise and amplifying significant speech signals[98] (see Box 9.2 for details).

---

### Box 9.2
### Cochlear Implants and "Intelligent" Digital Hearing Aids
#### Cochlear Implants

Cochlear implants are increasingly used as hearing aids in all age groups. These are prosthetic replacements for the cochlea. Bypassing the inner ear, cochlear implants directly stimulate the auditory nerve, through tiny wires inserted into the cochlea. Sounds are received by an external microphone placed behind the ear and are processed by a pocket speech processor that converts sounds to electrical impulses which are transmitted (via FM signals) by a coil placed behind the ear to an implanted receiver or stimulator. This stimulates the appropriate electrode, which, in turn, stimulates certain auditory nerve fibers. The use of multiple electrodes allows for stimulation of the auditory nerve at multiple terminal locations. This allows for pitch information, important for speech perception. A typical 22-channel cochlear implant transmits up to 22 unique pitch signals to the inner ear.

#### Next Generation of Hearing Aids

Elderly listeners have difficulty listening to speech in the presence of background sounds. Age-related defects in the auditory system such as cochlear hair-cell degeneration (leading to high-frequency hearing loss), problems with sound localization, and temporal gap detection in speech contribute to this aspect of speech recognition. Conventional hearing aids offer little help in this regard and cannot group all the sounds from the source of interest into one coherent auditory percept, separated from background noise sound streams.

A new system being developed by Dr. Albert Feng and collaborators[98] at the University of Illinois, Urbana-Champaign, looks promising as the next generation of hearing aids allowing for the sounds of interest to stand apart from interfering sounds. Pilot tests have shown this system to be effective in many listening circumstances. This system, which is based on use of a two-microphone system, is derived from the binaural signal processing used in animals that use Interaural Time Differences (ITD) to localize sound sources. The two signals are converted to frequencies and analyzed for left–right coincidences. A nonlinear operation, then temporal integration, improves the coincidence information. Estimates of the sound source azimuths are determined by "integration of the coincidence locations across the broadband of frequencies in speech signals." Use of this novel pattern recognition method improves sonic localization by eliminating background noise, which is regarded as any coincidences with delays greater than 2 pi. Computer simulation experiments and anechoic chamber tests reveal that this method can localize four to six unique simultaneous sound sources effectively.[98]

---

**Box 9.3**

**Genes, Chromosomes, and Presbycusis**

**MHC Genes and Presbycusis**

Gene haplotypes in the MHC domain may be associated with pathogenesis of certain types of hearing loss, including presbycusis. An extended MHC haplotype is identified for unrelated people, in a cohort study with strial presbycusis and other types of hearing disorders. In this investigation, 44% of subjects express this MHC haplotype, as opposed to 7% of the general population.[119]

**DFNA25**

Linkage analysis by Greene et al.[120] has found a novel, dominant locus, DFNA25, for delayed-onset progressive high-frequency, nonsyndromic sensorineural hearing loss over many generations of a U.S. family of Swiss descent. The 20 cM region of chromosome 12q21–24 with possible candidate genes [ATP2A (yeast-like F0F1-ATPase alpha subunit), ATP2B1 (ATPase, $Ca^{++}$ transporting, plasma membrane 1 gene), UBE3B (a member of the E3 ubiquitin ligase family), or VR-OAC (Vanilloid receptor-related osmotically activated channel, a candidate vertebrate osmoreceptor)] are implicated.

**DFNA5**

According to Van Lear et al.,[121] over 40 loci for nonsyndromic hearing loss have been mapped to the human genome, but only a small number of these genes have been identified. One mutation found in an extended Dutch family is linked to chromosome 7p15 (DFNA5). This mutation has been defined as an insertion or deletion at intron 7 that removes five G triplets from the 3-prime end of the intron while not affecting intron–exon boundaries. This mutation that is associated with deafness in this family causes a skip of intron 8, with consequent termination of the open reading frame, and is expressed in the cochlea, but the physiological function is still unknown.

**mtDNA4977 Deletion**

Studies of mitochondrial DNA (mtDNA) by Ueda et al. from cochlear sections of temporal bones of control subjects and those with presbycusis found a specific mtDNA4977 deletion in ~80% of those with presbycusis, compared to half as much in controls cochlea.[102] Therefore, some of the advanced sensorineural hearing loss cases of presbycusis should be categorized as mitochondrial oxidative phosphorylation diseases, thus offering novel possibilities for treatment and prevention of sensorineual hearing loss, including presbycusis.[102]

---

## 4.  Genetic Aspects of Presbycusis

Recent studies have focused on genetic aspects of presbycusis. Specifically, several DFNA loci have been determined to exist in relation to inherited deafness and presbycusis, as well as several mitochondrial associated mutations such as mtDNA4977 deletion (see Box 9.3 for details).

### a.  General Inheritance of Presbycusis

In a cohort study by Gates et al.[99] to find the prevalence of age-related hearing loss inheritance, genetically unrelated (spouses) and genetically related (siblings/parent–child) pairs were compared in regards to patterns of hearing level groupings. A familial aggregation was definitely found to exist for sensory and strial presbycusis, as well as for those with normal hearing ability. Women showed a stronger aggregation than men, and the strial heritability estimate was greater than the sensory phenotype.[99]

### b.  Mitochondrial Genetics and Sensorineural Hearing Loss (SNHL Presbycusis)

Acquired mitochondrial defects have been postulated as being key to aging, and especially aging of neuromuscular tissues. Mutations in the mitochondrial cytochrome oxidase II gene were found to be common in the spiral ganglion and membranous labyrinth of archival temporal bones of five patients with presbycusis, implicating mitochondrial mutations in at least a subgroup of presbycusis.[100] Keithley et al.[101] point out, however, that the cytochrome oxidase defect cannot entirely account for presbycusis. In addition, an association between mitochondrial DNA and presbycusis has been found.[102] Patients with SNHL had an increased prevalence for mitochondrial DNA deletion mtDNA4977 (75%) versus 30% in controls. It is proposed that certain SNHL subtypes should be categorized as diseases of mitochondrial oxidative phosphorylation.[102] (see Box 9.3).

**Systemic and Organismic Aging**

## D. SOUND LOCALIZATION AND SPEECH PERCEPTION

### 1. Sound Localization

Tests of sound localization indicate a decline in this ability with aging, beginning in the fourth decade.[24] It is known that localization of low-frequency sounds depends on *temporal* discrimination (i.e., time of sound arrival) between the two ears, while in the high frequency range, localization depends on discrimination of sound *intensity* between the two ears. Aging changes occur in both ears, though the rate of aging may be different in the two ears.[79,83] Thus, deficits in localization of sound will be apparent for all frequencies but may be more marked for sounds of higher frequency, as the perception of these are particularly impaired in old age (see above).

### 2. Early Studies on Hearing Deficits and Speech Perception

#### a. Hearing of Consonants Versus Vowels

The marked hearing deficits in the high-frequency range have important bearing on impaired speech perception in the elderly.[80,103] Thus, *vowels,* generated by low-frequency sounds, are heard better than the *consonants* produced by high-frequency sounds. Similarly, voices of men are heard better that those of women and children, which have a characteristic high pitch. Because consonants make speech intelligible, while vowels make it more audible, a common complaint of the elderly is that they cannot understand spoken words, although they can hear them.[19] Fortunately, lip reading, which is associated more with the expression of consonants, can greatly help this deficiency.

#### b. Speech Speed and Comprehension

As indicated in the data of Figure 9.9, if speech is presented to the aged subjects too fast or with reverberations (i.e., echoing or booming), its comprehension declines markedly. The greatest decline in comprehension is observed if speech is presented with repeated interruptions, such as eight times per second, as is the case with many modern telephone systems.[19] This presents an unfortunate situation for the elderly, who rely so much on the telephone for their communication with the outside world.

#### c. Sound Masking and Speech Comprehension

The ability to mask sounds, important for speech comprehension in a crowd of talking people, is considerably diminished in the elderly. Indeed, tests of hearing loss for pure tones provide usually conservative estimates of hearing deficits, because the tests are usually performed under quiet laboratory conditions, eliminating the need for masking. One example of loss of masking ability with age is the increase, with advancing age, in reporting the incidence of ringing in the ear *(tinnitus),* even though this problem is manifested at all ages. Perhaps the elderly cannot mask these unusual sounds, while the young can.[19]

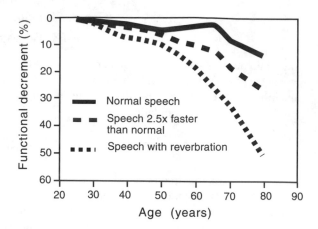

**FIGURE 9.9** *Deficits in speech comprehension at different ages.* Note that normal aging deficits are exaggerated when speech is presented faster or with reverberations. (Redrawn from Marsh[19] based on original data of Bergman et al. *J. Gerontol.,* 31, 533, 1976).

### 3. Recent Studies on Hearing Loss and Speech Recognition

According to Humes et al.,[104] hearing loss is the most significant factor related to individual differences in speech recognition in the elderly, being responsible for 70 to 75% of speech recognition performance variance. In contrast, auditory processing and cognitive function measures account for little, if any, variance in speech recognition.

#### a. Identifying and Recalling Speech in Babble

According to Pichora-Fuller et al.,[105] reallocatable processing resources are used to aid in auditory processing when listening becomes difficult due to noise or age-related auditory deterioration. Thus, fewer resources are available for storage and retrieval aspects of working memory; therefore, "upstream" processing of auditory information is negatively impacted.[105]

#### b. Speech Recognition in Presbycusis

Frisina and Frisina[106] show that peripheral auditory defects, as measured by decreased sensitivity for speech pure tones and substance, contribute to increased thresholds for speech recognition in a quiet environment as well as reduced speech recognition through noise. Cortical portions of the speech and language perception system are not a part of speech-understanding problems in the elderly who are able to use supporting context in speech perception as well as the young. Dysfunction of speech-recognition in noise still exists in the elderly, even when cognitive and peripheral audition are not effected. However, auditory brainstem or cortical temporal-resolution defects account for the loss of speech recognition in noisy environments.

### c.  Sound Lateralization, Speech Gap and Adaptation Effects

Sound lateralization is affected more than speech discrimination ability in presbycusis.[107] Schneider and Hamstra[108] have shown that elderly subjects with early presbycusis have a decreased ability in detecting *gap durations* characteristic of human speech sounds, compared to young adults. Also, these elderly experience differential adaptation effects.

## 4.  Aging Changes in Central Auditory Pathways

In addition to the frequently encountered degenerative aging changes in the neurons of spiral ganglia discussed above,[83,84] varying degrees of aging changes in the central auditory structures have been reported.[109] These include loss of myelin and neuropil (dendrites and synapses) and degeneration and loss of neurons and lipofuscin accumulation in auditory structures, such as the cochlear nuclei and superior olives of medulla, medial geniculate of thalamus, and the inferior colliculi of the midbrain.[80]

### a.  Auditory Cortex Changes

In contrast to the lower centers, the human auditory cortex shows very marked and clear degenerative aging changes, consisting of a general thinning of the cortex and disrupted organization of the vertical columns; this is caused by heavy neuronal loss and degeneration. In his well-known study of human cortex aging, Brody[110] found the highest degree of cell loss to occur in the superior temporal gyrus, the anatomical locus of the auditory association cortex; the loss, amounting to about 50% between the ages of 20 to 90, occurred across the cortical thickness. Golgi studies of pyramidal neurons of the auditory cortex by Scheibel et al.[111] revealed marked degenerative changes in the basal dendrites of layer III and V neurons in the old subjects (eighth to tenth decades).

### b.  Auditory-Evoked Potentials

Allison et al.[52] initially described significant changes in the auditory-evoked potentials in the elderly, including increases in the latency and decreases in the amplitude of the response. The changes are more marked in males than females and occur generally after 60 years. Undoubtedly these degenerative changes contribute to the observed disorders of hearing and speech comprehension in the elderly, although an exact cause-and-effect relationship is still not established. In a more recent study by Oku & Hasegewa,[112] the amplitude of auditory brainstem evoked potentials (Wave I), measured with electroencepholography, was found to decrease in subjects with presbycusis compared to age-matched controls. In addition, the lengthening and shortening of other waves' inter-peak intervals, was thought to reflect aging related electrophysiological changes in the central auditory pathways.

## E.  CENTRAL AND COGNITIVE ASPECTS OF AUDITORY AGING

### 1.  Interaural Asymmetry Of Event Related Potentials (ERPS)

According to Jerger et al.,[113] upon presentation of syntactically and semantically anomalous words in continuous speech, an auditory ERP is evoked, characterized by a positive peak, 600 to 1000 msec in latency. In a dichotic listening task for normal children and young adult subjects, the target word is usually isolated to each side (left and right ear) with equal frequency. Also, amplitude and latency measures of this task were equal for the target-left and target-right conditions. Elderly subjects with presbycusis showed significantly greater latency in the target-left than the target-right tasks; similarly, maximal positivity was significantly higher in the left-attended condition. These findings support past findings of interaural asymmetry in dichotic listening tasks in the elderly.[113]

### 2.  Gap Detection Thresholds

Normal adults and elderly in early stages of presbycusis were asked to determine the presence or absence of a gap between two tones of equal duration. Older adults in early stages of presbycusis were less able to detect a gap between short tonal markers (with durations characteristic of speech sounds) of less than 250 msec. These results point to differential adaptation effects between young and older adults.[108]

### 3.  Sound Lateralization and Speech Discrimination in Sensorineural Presbycusis

Sound lateralization and speech discrimination are linked to central auditory system functionality. Sound lateralization may be affected more than speech discrimination in patients with presbycusis and sensorineural hearing loss of unknown etiology.[114]

### 4.  Influence of Age on Measures of Hearing Disability

A statistically significant interaction was found between age and hearing loss.[115] Elderly with mild to moderate hearing loss reported less handicapping effects of loss of hearing than younger subjects with the same degree of hearing loss.[115]

## ALZHEIMER'S DISEASE (AD), DEPRESSION, AND THE AUDITORY SYSTEM

### Alzheimer's Disease Lesions in Central Auditory Nuclei

Individuals with AD show a specific and consistent degeneration, including senile plaques and neurofibrillary tangles

**Systemic and Organismic Aging**

in such auditory structures as central nucleus of the inferior colliculus and ventral nucleus of the medial geniculate body, but not in cochlear nuclei; but no such changes occur in the age-matched controls (see Sinha et al.[116] for details). The AD–associated degenerative changes suggest that neuronal loss could include all frequency ranges. This is in contrast with typical clinical presbycusis, affecting high-frequency loss, due to peripheral auditory lesions. More recently, Sinha et al.[116] compared temporal bones from patients with and without Alzheimer's disease and found significant difference in the number of hair cells remaining, spiral ganglion cells, and peripheral processes, in the basal cochlear turn only. Lack of degeneration in the entirety of the cochlea, as well as lack of neurofibrillary tangles and neuritic plaques in the peripheral auditory systems of AD subjects, further confirms that the peripheral auditory system is not involved in AD lesions.

### Auditory Memory Changes in AD

A recent study by Pekkonen[117] measured auditory memory in AD patients. Automatic stimulus discrimination in the auditory system is represented by the auditory mismatch negativity (MMN) type of event–related potential (ERP). MMN is utilized as a measure of auditory sensory memory by changing the interstimulus intervals (ISI). Auditory aging is associated with reduced amplitude of MMN to duration deviance at short ISI, while MMN to frequency deviance attenuation at long ISI is caused by age-related memory trace decay. This conclusion is supported by the finding that automatic discrimination for frequency change was not affected in the early stages of AD, while the memory trace decayed faster in AD patients.

### Depression Onset and Sensorineural Hearing Loss

Patients with late onset depression demonstrated more hearing deficits than early onset depressives. Also, age at onset of depression had a significant effect on pure-tone thresholds for the 0.5 to 4 kHz range and on word recognition in a noisy environment in the subject's better ear.[118]

## REFERENCES

1. Meisami, E., Aging of the sensory systems, in *Physiological Basis of Aging and Geriatrics*, 2nd ed., Timiras, P.S., Ed., CRC Press, Boca Raton, FL, 1994, 115.
2. Guralnik, J.M., The impact of vision and hearing impairments on health in old age, *J. Am. Geriat. Soc.*, 47, 1029, 1999.
3. Rantenen, T. et al., Coimpairments: Strength and balance as predictors of severe walking disability, *J. Gerontol. Med. Sci.*, 54, M172, 1999.
4. Graham, P., The eye, in *Principles and Practice of Geriatric Medicine*, 2nd ed., Pathy, M.S.J., Ed., John Wiley & Sons, London, 1991, 985 (see also 3rd ed., 1998).
5. Bron, A.J., The aging eye, in *Oxford Textbook of Geriatric Medicine*, Evans, J.G. and Williams, T.F., Eds., Oxford University Press, Oxford, 1992, 557 (see also 2nd ed., 2000).
6. Leighton, D.A., Special senses — aging of the eye, in *Textbook of Geriatric Medicine and Gerontology*, Brocklehurst, J.C., Ed., Churchill-Livingstone, Edinbrough, 1985, chap. 21 (see also 5th ed., 1998, authored by Tallis, R., Fillit, H., and Brocklehurst J.C.).
7. Stefansson, E., The eye, in *Principles of Geriatric Medicine and Gerontology*, Hazzard, W.R., et al., Eds., McGraw-Hill, New York, 1991, Chap. 42 (see also 4th ed., 1999).
8. Kuwabara, T,, Age related changes of the eye, in *Special Senses in Aging*, Han, S.S. and Coons, D.H., Eds., Institute of Gerontology, University of Michigan, Ann Arbor, 1979, 46.
9. Sekuler, R., Kline, D., and Dismuskes, K., Eds., *Aging and Human Visual Functions*, Alan R. Liss, New York, 1982.
10. Weale, R.A., *Focus on Vision*, Harvard University Press, Cambridge, MA, 1982, Chap. 3.
11. Vaughan, W.J., Schmitz, P., and Fatt, I., The human lens: A model system for the study of aging, in *Sensory Systems and Communication in the Elderly*, Ordy, J.M. and Brizzee, K., Eds. Raven Press, New York, 1979, 51.
12. Lass, J.H. et al., The effects of age on phosphatic metabolites of the human cornea, *Cornea* 14, 89, 1995.
13. Moller-Pedersen, T., A comparative study of human corneal keratocytes and endothelial cell density during aging, *Cornea*, 16, 333, 1997
14. Daxer, A. et al., Collagen fibrils in the human corneal stroma: Structure and aging, *Invest. Ophthalmol. Vis. Sci.*, 39, 644, 1998.
15. Oshika, T. et al., Changes in corneal wavefront aberrations with aging, *Invest. Ophthalmol. Vis. Sci.*, 7, 1351, 1999.
16. Pardhan, S. and Beesley, J., Measurement of corneal curvature in young and older normal subjects, *J. Refract. Surg.*, 15, 469, 1999.
17. Bron, A.J. et al., The aging lens, *Ophthalmologica*, 214, 86, 2000.
18. Koretz, J.F. et al., Aging of the human lens: Changes in lens shape at zero-diopter accommodation, *J. Opt. Soc. Am. A. Opt. Image Sci. Vis.*, 18, 265, 2001.
19. Marsh, G., Perceptual changes with aging, in *Handbook of Geriatric Psychiatry*, Busse, E.W. and Blazer, D.G., Eds., Van Nostrand, New York, 1980, p. 147.
20. Lampi, K.J. et al., Age-related changes in human lens crystallins identified by two-dimensional electrophoresis and mass spectrometry, *Exp. Eye Res.*, 67, 31, 1998.
21. Ma, Z. et al., Age-related changes in human lens crystallins identified by HPLC and mass spectrometry, *Exp. Eye Res.*, 67, 21, 1998.
22. Borchman, D. et al., Regional and age-dependent differences in the phospholipid composition of human lens membranes, *Invest. Ophthalmol. Vis. Sci.*, 35, 3938, 1994.
23. Borchman, D. and Yappert, M.C., Age-related lipid oxidation in human lenses, *Invest. Ophthalmol. Vis. Sci.*, 39, 1053, 1998.
24. Ogiso, M. et al., Age-related changes in ganglioside composition in human lens, *Exp. Eye Res.*, 60, 317, 1995.
25. Moffat, B.A. et al., Age-related changes in the kinetics of water transport in normal human lenses, *Exp. Eye Res.*, 69, 663, 1999.

26. Heron, G., Charman, W.N., and Gray, L.S., Accommodation responses and aging, *Invest. Ophthalmol. Vis. Sci.*, 40, 2872, 1999.

27. Mordi, J.A. and Ciuffreda, K.J., Static aspects of accommodation: Age and presbyopia, *Vision Res.*, 38, 1643, 1998.

28. Glasser, A. and Campbell, M.C., Biometric, optical and physical changes in the isolated human crystalline lens with age in relation to presbyopia, *Vision Res.*, 39, 1991, 1999.

29. Strenk, S.A. et al., Age-related changes in human ciliary muscle and lens: A magnetic resonance imaging study, *Invest. Ophthalmol. Vis. Sci.*, 40, 1162, 1999.

30. Panda-Jonas, S., Jonas, J.B., and Jakobczyk-Zmija, M., Retinal photoreceptor density decreases with age, *Ophthalmology*, 102, 1853, 1995.

31. Kimble, T.D. and Williams, R.W., Structure of the cone photoreceptor mosaic in the retinal periphery of adult humans: Analysis as a function of age, sex, and hemifield, *Anat. Embryol.* (Berl.), 201, 305, 2000.

32. Curcio, C.A., Photoreceptor topography in aging and age-related maculopathy, *Eye*, 15, 376, 2001.

33. Werner, J. S. et al., Senescence of foveal and parafoveal cone sensitivities and their relations to macular pigment density, *J. Opt. Soc. Am. A. Opt. Image Sci. Vis.*, 17, 1918, 2000.

34. Garway-Heath, D.F. et al., Aging changes of the optic nerve head in relation to open angle glaucoma, *Br. J. Ophthalmol.*, 81, 840, 1997.

35. Panda-Jonas, S. et al., Retinal pigment epithelial cell count, distribution, and correlations in normal human eyes, *Am. J. Ophthalmol.*, 121, 181, 1996.

36. Boulton, M. and Dayhaw-Barker, P., The role of the retinal pigment epithelium: Topographical variation and aging changes, *Eye*, 15, 384, 2001.

37. Harman, A.M. et al., Development and aging of cell topography in the human retinal pigment epithelium, *Invest. Ophthalmol. Vis. Sci.*, 38, 2016, 1997.

38. Verdugo, M.E. and Ray, J., Age-related increase in activity of specific lysosomal enzymes in the human retinal pigment epithelium, *Exp. Eye Res.*, 65, 231, 1997.

39. Cingle, K.A. et al., Age-related changes in glycosides in human retinal pigment epithelium, *Curr. Eye Res.*, 15, 433, 1996.

40. Moschner, C. and Baloh, R.W., Age-related changes in visual tracking, *J. Gerontol.*, 49, M235, 1994.

41. Munoz, D.P. et al., Age-related performance of human subjects on saccadic eye movement tasks, *Exp. Brain Res.*, 121, 391, 1998.

42. Jackson, G.R. et al., Aging and scotopic sensitivity, *Vision Res.*, 38, 3655, 1998.

43. Schefrin, B.E. et al., The area of complete scotopic spatial summation enlarges with age, *J. Opt. Soc. Am. A. Opt. Image Sci. Vis.*, 15, 340, 1998.

44. McFarland, R.A. et al., Alterations in critical flicker frequency as a function of age and light-dark ratio, *J. Exp. Psychol.*, 56, 529, 1958.

45. Kim, C.B. and Mayer, M.J., Foveal flicker sensitivity in healthy aging eyes. II. Cross-sectional aging trends from 18 through 77 years of age, *J. Opt. Soc. Am. A.*, 11, 1958, 1994.

46. Sekuler, R. and Sekuler, A.B., Visual perception and cognition, in *Oxford Textbook of Geriatric Medicine*, Evans, J.G. and Williams, T.F., Eds., Oxford University Press, Oxford, 1992, p. 575 (see also 2nd ed., 2000).

47. Haegerstrom-Portnoy, G. et al., Seeing into old age: Vision function beyond acuity, *Optom. Vis. Sci.*, 76, 141, 1999.

48. Sekuler, A.B. et al., Effects of aging on the useful field of view, *Exp. Aging Res.*, 26, 103, 2000.

49. Seiple, W. et al., Age-related functional field losses are not eccentricity dependent, *Vision Res.*, 36, 1859, 1996.

50. Dolman, C.L., McCormick, A.O., and Drance, S.M., Aging of the optic nerve, *Arch. Ophthalmol.*, 98, 2052, 1980.

51. Brody, H. and Vijayashankar, N., Anatomical changes in the nervous system, in *Handbook of the Biology of Aging*, Finch, C.E. and Hayflick, L., Eds., Van Nostrand, New York, 1977, p. 241.

52. Allison, T., Hume, A.L., Wood, C.C., and Goff, W.R., Developmental and aging changes in somatosensory, auditory and visual evoked potentials, *Electroenceph. Clin. Neurophysiol.* 58, 14, 1984.

53. Justino, L. et al., Changes in the retinocortical evoked potentials in subjects 75 years of age and older, *Clin. Neurophysiol.*, 112, 1343, 2001.

54. Levine, B.K. et al., Age-related differences in visual perception: A PET study, *Neurobiol. Aging*, 21, 577, 2000.

55. Hammond, C.J. et al., The heritability of age-related cortical cataract: The twin eye study, *Invest. Ophthalmol. Vis. Sci.*, 42, 601, 2001.

56. Okano, Y. et al., A genetic factor for age-related cataract: Identification and characterization of a novel galactokinase variant, "Osaka," in Asians, *Am. J. Hum. Genet.*, 68, 1036, 2001.

57. Yeum, K.J. et al., Fat-soluble nutrient concentrations in different layers of human cataractous lens, *Curr. Eye Res.*, 19, 502, 1999.

58. Sweeney, M.H. and Truscott, R.J., An impediment to glutathione diffusion in older normal human lenses: A possible precondition for nuclear cataract, *Exp. Eye Res.*, 67, 587, 1998.

59. Zarina, S., Zhao, H.R., and Abraham, E.C., Advanced glycation end products in human senile and diabetic cataractous lenses, *Mol. Cell. Biochem.*, 210, 29, 2000.

60. Gilliland, K.O. et al., Multilamellar bodies as potential scattering particles in human age-related nuclear cataracts, *Mol. Vis.*, 22, 120, 2001.

61. Al-Ghoul K.J. et al., Structural evidence of human nuclear fiber compaction as a function of aging and cataractogenesis, *Exp. Eye Res.*, 72, 199, 2001.

62. Dielemans, I. et al., Primary open-angle glaucoma, intraocular pressure, and systemic blood pressure in the general elderly population. The Rotterdam Study, *Ophthalmology*, 102, 54, 1995.

63. Wirtz, M.K. et al., Prospects for genetic intervention in primary open-angle glaucoma, *Drugs Aging*, 13, 333, 1998.

64. La Rosa, F.A. and Lee, D.A., Collagen degradation in glaucoma: Will it gain a therapeutic value? *Curr. Opin. Ophthalmol.*, 11, 90, 2000.

**Systemic and Organismic Aging**

65. Hoyng, P.F. and Van Beek, L.M., Pharmacological therapy for glaucoma: A review, *Drugs*, 59, 411, 2000.

66. Linden, C., Therapeutic potential of prostaglandin analogues in glaucoma, *Expert Opin. Investig. Drugs*, 10, 679, 2001

67. Lewis, R.A., Macular degeneration in the aged, in *Special Senses in Aging*, Han, S.S. and Coons, D.H., Eds., Institute of Gerontology, University of Michigan, Ann Arbor, 1979, p. 93.

68. Sarks, S.H., Aging and degeneration in the macular region. A clinicopathological study, *Br. J. Ophthalmol.*, 60, 324, 1976.

69. Kamei, M. and Hollyfield, J.G., TIMP-3 in Bruch's membrane: Changes during aging and in age-related macular degeneration, *Invest. Ophthalmol. Vis. Sci.*, 40, 2367, 1999.

70. Plantner, J.J. et al., Increase in interphotoreceptor matrix gelatinase A (MMP-2) associated with age-related macular degeneration, *Exp. Eye. Res.*, 67, 637, 1998.

71. Meyers, S.M. et al., A twin study of age-related macular degeneration, *Am. J. Ophthalmol.*, 120, 757, 1995.

72. Ajani, U.A. et al., A prospective study of alcohol consumption and the risk of age-related macular degeneration, *Ann. Epidemiol.*, 9, 172, 1999.

73. Chan D., Cigarette smoking and age-related macular degeneration, *Optom. Vis. Sci.*, 75, 476, 1998.

74. Fekrat, S. and Bressler, S.B., Are antioxidants or other supplements protective for age-related macular degeneration?, *Curr. Opin. Ophthalmol.*, 7, 65, 1996.

75. Ciulla, T.A. et al., Age-related macular degeneration: A review of experimental treatments, *Surv. Ophthalmol.*, 43, 134, 1998.

76. Rizzo, M. et al., Vision and cognition in Alzheimer's disease, *Neuropsychologia*, 38, 1157, 2000.

77. Rizzo, M. et al., Visual attention impairments in Alzheimer's disease, *Neurology*, 54, 1954, 2000.

78. Hof, P.R. et al., Atypical form of Alzheimer's disease with prominent posterior cortical atrophy: A review of lesion distribution and circuit disconnection in cortical visual pathways, *Vision Res.*, 24, 3609, 1997.

79. Schuknett, H.F., Further observations on the pathology of presbycusis, *Arch. Otolaryngol.*, 80, 369, 1964.

80. Gulya, A.J., Disorders of hearing, in *Oxford Textbook of Geriatric Medicine*, Evans, J.G. and Williams, T.F., Eds., Oxford University Press, Oxford, 1992, p. 580 (see also 2nd ed., 2000).

81. Rosen, S. et al., Presbycusis study of a relatively noise free population in the Sudan, *Ann. Otol.*, 71, 727, 1962.

82. Spencer, J.T., Jr., Hyperlipoproteinemia in the etiology of inner ear disease, *Laryngoscope*, 83, 639, 1973.

83. Schuknett, H.F., *Pathology of the Ear*, Harvard University Press, Cambridge, MA, 1974, p. 388 (see also 2nd ed., 1993).

84. Johnsson, L. and Hawkins, J.E., Jr., Age related degeneration of the inner ear, in *Special Senses in Aging*, Han, S.S. and Coons, D.H., Eds., Institute of Gerontology, University of Michigan, Ann Arbor, 1979, p. 119.

85. Kim, H.N. et al., Incidence of presbycusis of Korean populations in Seoul, Kyunggi and Kangwon provinces, *J. Korean Med. Sci.*, 15, 580, 2000.

86. Megighian, D. et al., Audiometric and epidemiological analysis of elderly in the Veneto region, *Gerontology*, 46, 199, 2000.

87. Bazargan, M. et al., Sensory impairments and subjective well-being among aged African-American persons, *J. Gerontol.*, 56, 268, 2001.

88. Wiley, T.L. et al., Aging and middle ear resonance, *J. Am. Acad. Audiol.*, 10, 173, 1999.

89. Oeken, J. et al., Influence of age and presbycusis on DPOAE, *Acta Otolaryngol.*, 120, 396, 2000.

90. Scholtz, A.W. et al., Selective aspects of human pathology in high-tone hearing of the aging inner ear, *Hear. Res.*, 157, 77, 2001.

91. Anniko, M. et al., Recent advances in inner ear cytochemistry — microanalytical and immunomorphological investigations, *Prog. Neurobiol.*, 30, 209, 1988.

92. Felder, E. et al., Quantitative evaluation of cochlear neurons and computer-aided three-dimensional reconstruction of spiral ganglion cells in humans with a peripheral loss of nerve fibers, *Hear. Res.*, 105, 183, 1997.

93. Popelka, M.M. et al., Low prevalence of hearing aid use among older adults with hearing loss: The epidemiology of hearing loss study, *J. Am. Geriatr. Soc.*, 46, 1075, 1998.

94. Jupiter, T. and Spivey, V., Perception of hearing loss and hearing handicap on hearing aid use by nursing home residents, *Geriatr. Nurs.*, 18, 201, 1997.

95. Jerger, J. et al., Comparison of conventional amplification and an assistive listening device in elderly persons, *Ear Hear.*, 17, 490, 1996.

96. Johnson, C.E. et al., A holistic model for matching high-tech hearing aid features to elderly patients, *Am. J. Audiol.*, 9, 112, 2000.

97. Buchman, C.A. et al., Cochlear implants in the geriatric population: Benefits outweigh risks, *Ear Nose Throat J.*, 78, 489, 1998.

98. Liu, C. et al., A two-microphone dual delay-line approach for extraction of a speech sound in the presence of multiple interferers, *J. Acoust. Soc. Am.*, 110, 3218, 2001.

99. Gates, G.A. et al., Genetic associations in age-related hearing thresholds, *Arch. Otolaryngol. Head Neck Surg.*, 125, 654, 1999.

100. Fischel-Ghodsian, N. et al., Temporal bone analysis of patients with presbycusis reveals high frequency of mitochondrial mutations, *Hear. Res.*, 110, 147, 1997.

101. Keithley, E.M. et al., Mitochondrial cytochrome oxidase immunolabeling in aged human temporal bones, *Hear. Res.*, 157, 93, 2001.

102. Ueda, N. et al., Mitochondrial DNA deletion is a predisposing cause for sensorineural hearing loss, *Laryngoscope*, 108, 580, 1998.

103. Jerger, J. and Hays, D., Diagnostic speech audiometry. *Arch Otolaryngol.*, 103, 216, 1977.

104. Humes, L.E. et al., Factors associated with individual differences in clinical measures of speech recognition among the elderly, *J. Speech Hear. Res.* 37, 465, 1994.

105. Pichora-Fuller, M.K. et al., How young and old adults listen to and remember speech in noise, *J. Acoust. Soc. Am.*, 97, 593, 1995.

106. Frisina, D.R. and Frisina, R.D., Speech recognition in noise and presbycusis: Relations to possible neural mechanisms, *Hear. Res.* 106, 95, 1997.

107. Kubo, T. et al., Sound lateralization and speech discrimination in patients with sensorineural hearing loss, *Acta Otolaryngol. Suppl.*, 538, 63, 1998.

108. Schneider, B.A. and Hamstra, S.J., Gap detection thresholds as a function of tonal duration for younger and older listeners, *J. Acoust. Soc. Am.*, 106, 371, 1999.

109. Feldman, M.L. and Vaughan, D.W., Changes in the auditory pathway with age, in *Special Senses in Aging,* Han, S.S. and Coons, D.H., Eds., Institute of Gerontology, University of Michigan, Ann Arbor, 1979, p. 143.

110. Brody, H., Organization of the cerebral cortex III. A study of aging in the human cerebral cortex, *J. Comp. Neurol.,* 102, 511, 1955.

111. Scheibel, M., Lindsay, R.D., Tomiyasu, V., and Scheibel, A.B., Progressive dendritic changes in aging human cortex, *Exp. Neurol.,* 47, 392, 1975.

112. Oku, T. and Hasegewa, M., The influence of aging on auditory brainstem response and electrocochleography in the elderly, *ORL J. Otorhinolaryngol. Relat. Spec.*, 59, 141, 1997.

113. Jerger, J. et al., Effect of age on interaural asymmetry of event-related potentials in a dichotic listening task, *J. Am. Acad. Audiol.*, 11, 383, 2000.

114. Kubo, T. et al., Sound lateralization and speech discrimination in patients with sensorineural hearing loss, *Acta Otolaryngol. Suppl.*, 538, 63, 1998.

115. Gordon-Salant, S. et al., Age effects on measures of hearing disability, *Ear Hear.*, 15, 262, 1994.

116. Sinha, U.K. et al., Temporal bone findings in Alzheimer's disease, *Laryngoscope*, 106, 1, 1996.

117. Pekkonen, E. et al., Auditory sensory memory and the cholinergic system: Implications for Alzheimer's disease, *Neuroimage*, 14, 376, 2001.

118. Kalayam, B. et al., Age at onset of geriatric depression and sensorineural hearing deficits, *Biol. Psychiatry*, 15, 649, 1995.

119. Bernstein, J.M. et al., Further observations on the role of the MHC genes and certain hearing disorders, *Acta Otolaryngol.*, 116, 666, 1996.

120. Greene, C.C. et al., DFNA25, a novel locus for dominant nonsyndromic hereditary hearing impairment, maps to 12q21-24, *Am. J. Hum. Genet.*, 68, 254, 2001.

121. Van Laer, L. et al., Nonsyndromic hearing impairment is associated with a mutation in DFNA5, *Nature Genet.*, 20, 194, 1998.

**Systemic and Organismic Aging**

# 10 The Adrenals and Pituitary*

*Paola S. Timiras*
University of California, Berkeley

## CONTENTS

## I.   INTRODUCTION

Together with the nervous system, the endocrine system coordinates homeostatic responses to environmental signals for the protection of the individual organism and regulates reproductive functions for the perpetuation of the species. With aging, the main expressions of declining functional competence are as follows:

- Cessation of reproduction in women
- Diminishing capacity to adapt to external demands, especially under stress conditions (Chapters 1 and 3)

Observations of changes in physiologic competence with aging have been related, since the end of the 19th century, to endocrine deficiency because of the broad range of hormonal actions, many of them necessary for reproduction and survival, and because of the possibility of restoration of "vitality" and "well-being" by hormonal replacement therapy. At early and late stages of the life

---

* Illustrations by Dr. S. Oklund.

span, development and aging affect endocrine function and, in turn, are affected by hormonal changes. It is unclear if aging of key endocrine systems is the cause or the consequence of the cellular and molecular changes responsible for senescence and death; nevertheless, it is evident that hormonal manipulations may influence functional decrements, disabilities, diseases of old age, and length of the life span. Previous reviews of the aging of endocrines have been presented in earlier editions of this book[1,2] and in several endocrinology texts.[3–6] This and the following four chapters highlight the salient aging changes in the classical endocrine organs and their hormones, as well as in other neuroendocrine and chemical messengers that mediate paracrine and autocrine actions.

In recent years, a variety of new tools has been added to the classical clinical, physiologic, and biochemical measurements of endocrine function. These new tools, adopted from the fields of genetic engineering and molecular and structural biology, have provided new techniques such as the study of mutations in humans and the use of genetic disruption in transgenic or knock-out animals.[7,8] These methods have particular significance for systems, such as the hypothalamo-pituitary-adrenal system, discussed in this chapter, in which feedback mechanisms are essential to normal physiology. Although new technologies have so far been utilized primarily in the areas of growth, development, and reproduction, correlation of human endocrine pathology in old age with observations of animal gene mutations are providing new insights in the current study of endocrine aging and identify the areas of interest for future research.

## II. ENDOCRINE GLANDS, HORMONES, AND CHEMICAL MESSENGERS

### A. ENDOCRINE AND OTHER CHEMICAL MESSENGERS

The wide distribution and the multiplicity and diversity of secretory products (e.g., chemical mediators, Box 10.1) responsible for communication among cells within the same organism and between the organism and its surrounding environment, testify to the critical role of endocrine regulation of body functions. This important role certainly applies to the hypothalamo-pituitary-adrenocortical axis discussed in this chapter. Among the endocrine glands, the *adrenals* regulate certain aspects of metabolism, behavior, and nervous and immune functions and, thus, play a key role in homeostasis (see below). The *pituitary* (or hypophysis) secretes several hormones, proteins and peptides that stimulate peripheral targets, either other endocrine glands or specific tissues and organs; the pituitary location, in close vicinity of the *hypothalamus* with which it articulates through a vascular net, *the portal system*, and through *direct neuronal connections*, gives the pituitary the unique function of intermediary between

---

**Box 10.1**

**Intercellullar Communication by Chemical Mediators**

**Endocrine glands** are organs that produce and secrete hormones. Reciprocally, hormones are molecules synthesized and secreted by specialized (endocrine) cells and released into the blood. They act on target cells at a distance from their site of origin. Well-recognized glands include the pituitary, adrenals, thyroid and parathyroids, pancreas, testes, and ovaries

Other cells or groups of cells act by **paracrine communication**, that is, they do not form discrete organs but are interspersed with other cells; they produce and secrete hormones in the extracellular fluid and affect neighboring cells. Examples of paracrine hormone-producing cells, are those of the pancreas (with both endocrine and paracrine secretions) and the intestinal mucosa and those producing prostaglandins. In addition, cells may act by **autocrine communication**: they secrete chemical messengers that bind to receptors on the same cell that secreted the messenger. A number of cells may act by **juxtacrine communication**: they express multiple repeats of growth factors and bind simultaneously to receptors on other cells.

Some of the neurotransmitters, previously described (Chapter 7), are also considered chemical messengers. Other important messengers (cytokines, thymic hormones, membrane receptors, and growth or apoptotic factors) regulate immune and hematopoietic functions (Chapters 15 and 18). Chemical messengers are, for the most part, amines, amino acids, steroids, polypeptides, proteins, and, in a few instances, other substances. In several parts of the body, the same chemical messenger can function as a neurotransmitter, a paracrine mediator, and a neurohormone.

---

nervous and endocrine systems. As all endocrine glands, the adrenals and pituitary do not act in isolation but are dependent on *neuroendocrine signals*, usually relayed through or initiated in the hypothalamus,[9–11] and on the functional status of the *target cells*.[12] Thus, aging of endocrine functions depends on changes in the following:

- The gland
- Other endocrines
- Other body systems (e.g., nervous, immunologic, cardiovascular)
- Metabolism and body composition
- Cellular and molecular responses

## TABLE 10.1
## Factors Influencing Evaluation of Endocrine Function in Aging

### Biologic Factors

Physiologic factors
  Metabolic state
  Body composition
  Dietary regimen
  Physical exercise
  Exposure to stress (environmental and psychosocial)
Relations with other endocrines and body systems
Secretory cells and their rates of secretion
Transport of the hormones to target cells
Metabolism of the secreted hormones
  Metabolites may be more or less biologically active than the secreted hormones [e.g., more active, conversion of testosterone (T) to dihydrotestosterone (DHT); less active, conversion of thyroxine (T4) to reverse 3,5,3'-triiodothyronine (T3)]
Number and affinity of hormone receptors
Intracellular postreceptor molecular events
Occurrence of disease and use of medications

### Experimental Design Factors

Sample size
Health status of subjects
The meaning of "old"
Comorbidity
Gender
Subjects under "steady state" or under stress
  Intensity, strength, timing, duration of stress
Outcome of study
What parameter was measured?
For how long? (long- versus short-term experiments)

- The complicating occurrence of disease, medications, and drugs
- The contributing influence of diet and exercise

## B. ASSESSMENT OF ENDOCRINE FUNCTION

A few generalities have emerged from a consideration of biologic and experimental design concomitants that influence evaluation of endocrine function at all ages:

- Hormonal actions affect all body functions
- Hormones regulate responses generated by internal (genes) and external (environmental) signals to promote reproduction (survival of the species) and to maintain homeostasis (survival of the individual)
- The repertory and efficiency of integrative hormonal responses available during adulthood diminish with advancing age and compromise the strategies for adaptation and survival

Evaluation of endocrine function in humans often relies on measurements of blood levels of appropriate hormones under basal (resting or steady state) conditions and under stress (i.e., increased environmental demands on body functions). Such a restricted assessment often leads to incomplete and erroneous conclusions. An adequate endocrine evaluation must assess several levels of endocrine action, including relations with other endocrine and body systems (primarily the nervous and immune systems), hormone receptor interactions at the target cell, and molecular events inside the cell as listed in Table 10.1. While none of these aging-related changes, by itself, may be sufficient to alter physiologic competence, a cascade of minor alterations may desynchronize the appropriate signal at the target cell or molecule and alter hormonal actions. Factors involved in the design of the experimental protocol to assess endocrine function (e.g., sample size, health, gender) (Table 10.1) may also influence the evaluation of changes with old age and may limit the feasibility of long-term experiments, or may invalidate some of the conclusions reached or restrict them to a single animal species.

An ideal "global" approach to the study of endocrine aging cannot be currently undertaken in humans. It can be explored more easily in experimental animals and in cultured tissues or cells; such *in vivo* and *in vitro* models represent an important corollary to human studies. As illustrated diagrammatically in Figure 10.1, changes with aging may occur at all levels of the endocrine system:

- *At the endocrine gland level*, weight loss, with atrophy and vascular changes, and fibrosis occur in most glands, with or without the accompaniment of glandular tumors (adenomas).
- *Basal blood plasma hormone levels* (free, biologically active hormones or hormones bound to plasma proteins) in man and animals are generally not influenced by age, although some hormones decrease significantly.
- *Hormone release* depends on nervous and environmental stimuli as well as positive and negative feedback from circulating hormones.
- Some hormones act exclusively on *one target*, and other hormones act on *many cell types (targets) and by several mechanisms*. Thus, the same hormone may have different actions in different tissues.
- With aging, one of the many actions or *one of the many targets may be selectively affected*, while other actions and targets are preserved.
- *Secretory and clearance rates* often decrease, although it is not clear whether the primary defect involves hormone secretion or hormone clearance; the pertinent question is to what extent

## Systemic and Organismic Aging

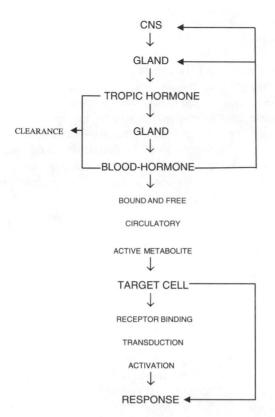

**FIGURE 10.1** Diagrammatic representation of a typical sequence of hormone action and regulation.

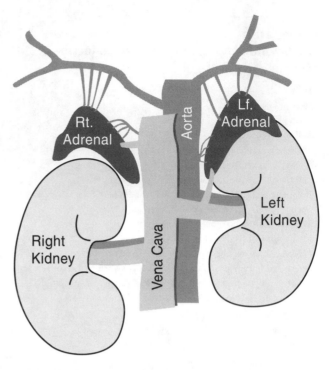

**FIGURE 10.2** Diagram of the kidney and adrenals.

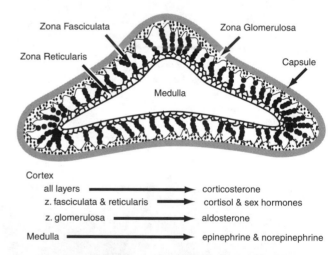

**FIGURE 10.3** Diagram of a section of the adrenals illustrating the various zones and hormones.

the capacity to maintain stable levels of plasma hormones is preserved. The study of the generation of active metabolites may reveal how their rate of production is altered with advancing age.

- *Receptors* located on target cells mediate the specificity of hormones on the particular cells, and their number may increase (upregulation) or decrease (downregulation) depending on the stimulus. Hormones–receptor complexes may be *internalized by endocytosis*, bind to the nucleus, and stimulate or repress transcription of selected RNAs or activity of specific enzymes. Cellular responses are determined by the *genetic programming* of the particular cell. With aging, receptor binding and intracellular responses vary greatly depending on the hormone and the target cell.

## III. THE ADRENAL CORTEX

The adrenals are paired glands that lie above the kidneys (Figure 10.2). They have an inner medulla and an outer cortex (Figure 10.3). The medulla is considered a sympathetic ganglion: it secretes the catecholamines, epinephrine and norepinephrine. The cortex secretes several steroid hormones distinguished into three categories:

- Glucocorticoids, e.g., cortisol, the principal glucocorticoid in humans, and corticosterone, in rats (for major actions, see Box 10.2)
- Sex hormones, e.g., dehydroepiandrosterone (DHEA) (Box 10.2)
- Cortisol and DHEA are secreted by the cells of the zona fasciculata and zona reticularis, and corticosterone is secreted by these and also the zona glomerulosa
- Mineralocorticoids, e.g., the principal hormone, aldosterone, are secreted by the cells of the zona glomerulosa (for major actions, see Box 10.3)

---

### Box 10.2
### Major Actions of Glucocorticoids and Adrenal Sex Hormones

**Glucocorticoids have metabolic actions including:**

- Increased amino acid uptake and gluconeogenesis in liver
- Decreased amino acid uptake and protein synthesis in muscle
- Inhibition of somatic growth due to a general catabolic (decreased protein synthesis) action
- Suppression of pituitary growth hormone secretion
- Anti-insulin effect with consequent exacerbation of diabetes
- Mobilization of serum lipids and cholesterol

**Glucocorticoids have an anti-inflammatory activity.** When administered in supraphysiologic doses, they inhibit inflammatory and allergic reactions. These immunosuppressive actions, utilized clinically for the prevention of transplant rejections and the symptomatic treatment of allergies, have been explained by several mechanisms (e.g., inhibition of leukocyte migration from blood to tissues, decreased number of circulating lymphocytes, and reduced antibody production).

**Glucocorticoids also have effects on the nervous system**, primarily of an excitatory nature, evidenced by the presence of hormone receptors on neurons and glial cells, induction of neurotransmitter enzymes, EEG changes, and toxicity for specific (hippocampal) neurons.

Another important action of glucocorticoids is **their negative feedback on the hypothalamus and pituitary** (see below). Secretion of glucocorticoids and sex hormones from the adrenal cortex is stimulated by the adrenocorticotropic hormone, ACTH, secreted from the anterior pituitary, the secretion of which, in turn, is stimulated by cortiocotropin-releasing hormone, CRH, from the hypothalamus. Glucocorticoids inhibit the secretion of ACTH and of CRH: the higher the levels of the circulating glucocorticoids, the lower the secretion of ACTH and CRH; the lower the glucocorticoid levels, the greater the release of ACTH and CRH.

**Other actions of glucocorticoids include: retention of sodium at renal tubules (but much less efficient than aldosterone), the resorption of bone and altered calcium metabolism, promotion of appetite and increased stomach acid and pepsin secretion, and maintenance of work capacity.**

**Sex hormones** include androgens (with masculinizing and anabolic actions) and estrogens (with feminizing actions). Their secretion decreases markedly in old age.

---

### Box 10.3
### Major Actions of Mineralocorticoids

**The major mineralocorticoid, aldosterone, is indispensable for survival.** It increases sodium reabsorption from the renal tubular fluid, saliva, and gastric juice. In the kidneys, aldosterone acts on the epithelium of the distal tubules and collecting ducts, where it facilitates the exchange of sodium (reabsorbed) for potassium and hydrogen ions (excreted). It may also increase potassium and decrease sodium in muscle and brain cells.

Secondary actions include: maintenance of blood pressure (as a consequence of sodium–water retention and increased blood volume) and moderate potassium diuresis and increased urine acidity (sodium taken up is exchanged for potassium and hydrogen ions which are excreted).

Aldosterone secretion is stimulated only to a minor degree by ACTH, and removal of the pituitary does not produce atrophy of the outer adrenal layer (glomerulosa). Aldosterone secretion is regulated in part by the circulating levels of sodium and potassium: that is, low sodium levels and high potassium levels stimulate secretion. **Most important in the control of aldosterone secretion is the renin-angiotensin system**: the hydrolytic enzyme renin secreted by the juxtaglomerular cells of the kidneys act on the circulating hepatic protein, angiotensinogen, to produce the decapeptide, angiotensin I, which, in turn, is transformed (primarily in the lungs) into the octapeptide, angiotensin II, which, then, stimulates the release of aldosterone.

**Excess of glucocorticoids**, as in Cushing's disease, leads to abnormalities of intermediary metabolism; **excess aldosterone**, as in Conn's disease, leads to hypertension and electrolyte disturbances; **adrenocortical insufficiency**, involving gluco- and mineralo-corticoids, as in Addison's disease, leads to hypotension, shock (from sodium loss), and death.[13]

---

**Systemic and Organismic Aging**

Because of the complex interrelationships of hypothalamus, anterior pituitary, and adrenal cortex, it is necessary in evaluating function to consider the entire axis as an entity. The effects of old age on the hypothalamo–pituitary–adrenal (HPA) axis (with adrenal including cortex and medulla) have been studied extensively, given their importance in the maintenance of homeostasis (see below).

## A. CHANGES WITH AGING IN ADRENOCORTICAL HORMONES UNDER BASAL AND STRESS CONDITIONS

With aging, the adrenal cortex undergoes some *structural changes*. Weight is decreased in humans and in the various animal species that have been examined; nodules (i.e., localized hyperplastic changes perhaps reactive to a reduced blood supply or consequence of multifocal adenomas) occur frequently. The adrenocortical cells, typical secretory cells, rich in mitochondria and endoplasmic reticulum with numerous lipid droplets where the steroid hormones are stored, undergo several changes. Of these, the most frequent are accumulation of lipofuscin granules (Chapters 3 and 7) and the thickening of the connective support tissue as shown by the thick capsule and the fibrous infiltrations around blood vessels. Functional changes with old age are grouped under the three major categories of corticoids and are presented in the following sections.

### 1. Glucocorticoids

Under basal conditions, that is, *under steady-state conditions,* despite the above mentioned anatomical and histologic changes, the following parameters remain *essentially unchanged* in men and women well into old age[12,14]:

- Plasma levels of cortisol and ACTH
- Circadian rhythm of ACTH release
- Cortisol release
- Responses of ACTH and cortisol to administered CRH
- Number of glucocorticoid receptors in target cells or affinity of these receptors for cortisol

A number of early studies suggested that secretion of cortisol is reduced in old age, but that this reduction would be compensated for by a decreased clearance (i.e., reduced metabolism and excretion). Such metabolic compensatory mechanisms would remain operative into old age, with the body adapting to decreasing production rates of the hormone by reducing the rate of removal or vice versa and, thus, maintaining normal circulating levels. However, more recent studies indicate that production and clearance of cortisol are unchanged if the subjects are in good health.[14] Still other studies have reported that in some species (rats,[15] vervet monkeys,[16] tree shrews,[17] baboons[18]),

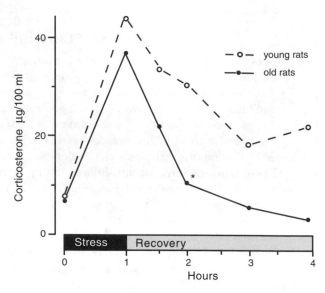

**FIGURE 10.4** Corticosterone titers in young (3 to 5 months) and aged (24 to 28 months) Fischer 344 rats during 1 h of immobilization stress, followed by 4 h of post-stress recovery. *Indicates time when titers are no longer significantly elevated above baseline (determined by two-tailed paired to test). In the case of young subjects, this was after 1 h of the recovery period; for aged subjects, such recovery did not occur within the monitored time period. (Reproduced with permission from Sapolsky, R. et al., *Endocr. Rev.*, 7, 284, 1986.)

glucocorticoid levels are slightly increased in some (but not all) animals (thereby illustrating the individual variability of endocrine changes in the elderly).

Stress stimulates the entire HPA axis, resulting in increased secretion of CRH, ACTH, and glucocorticoids; it also stimulates the sympathetic nervous system and the adrenal medulla to secrete primarily epinephrine and norepinephrine (see below). In some animal species under conditions of stress (physical or psychological), glucocorticoid levels are more elevated in old than young and adult animals: thus, levels of corticosterone (the principal glucocorticoid in the rat) are not only high but persist higher for longer periods in old than in young rats of some (but not all) strains (Figure 10.4). These persistently higher corticosterone levels after stress have been interpreted to indicate a loss of resiliency of the HPA axis that fails to appropriately decrease ACTH release in response to the elevated plasma levels of corticosterone and, in turn, to inhibit corticosterone release when the stress subsides.[15,19] In the rat, glucocorticoids at high levels are toxic to some cells, particularly those of the hippocampus,[20] rich in glucocorticoid receptors. Hippocampal cells, under no-stress conditions, inhibit CRH release; therefore, their corticoid-induced loss results in increased secretion of CRH, and, consequently, increased secretion of ACTH and glucocorticoids, as formulated in the "glucocorticoid cascade hypothesis of aging."[15] Young rats stressed for several weeks or treated with high glucocorticoid doses show hippocampal cell loss and changes in HPA axis function resembling those in old stressed animals.[15,19,20] In humans, competence of the HPA axis does not appear to change with increasing age, but long-term studies are still needed to ascertain the role of this axis in aging.

In contrast to the persistence, in steady state, of normal levels of glucocorticoids or of their increase after stress, levels of the other adrenocortical steroids appear to decline with aging: this is the case of aldosterone, with values almost undetectable past age 65,[21] and DHEA, with values at age 60 and older approximately one third of those at age 30.[22,23]

## 2. Adrenal Sex Steroids and DHEA Replacement Therapy

Dihydroepiandrosterone (DHEA), the principal adrenal androgen, is considered here as a prototype of adrenal sex hormones. DHEA follows a characteristic life cycle:

- Is very high in the fetus
- Is low in childhood
- Rises before puberty
- Is high in the adult
- Undergoes progressive decline to low or negligible rates by age 70

DHEA secretion is regulated by ACTH. Under conditions of stress, the secretion of cortisol and DHEA is increased, but the ratio of DHEA to cortisol falls as the enzymatic pathways to biosynthesis of both hormones use the same intermediates with preferential formation of cortisol.[22,23] The reduced plasma levels together with the lower response of DHEA to ACTH administration have led to the suggestion that DHEA may have some anti-aging effects, perhaps attributable to an antiglucocorticoid action; DHEA levels in men are inversely related to mortality due to myocardial pathology, and severely atherosclerotic individuals compared to normal individuals have lower DHEA levels.[24–26] Hence, there is the claim that DHEA replacement may prevent some of the functional decrements and pathology of old age. It may be recalled that the physiologist Charles-Édouard Brown-Séquard, by early 1889, recognized an association between aging and secretory actions attributed to an organ (the testis); he extolled the anti-aging properties of testicular secretions (androgens).[27] Testicular transplants and administration of androgens have been used repeatedly as possible rejuvenating measures to delay or reverse aging but with little success; indeed, high levels of androgens in aging men may aggravate the incidence and severity of prostate hypertrophy and cancer (Chapter 19).

Effects of DHEA replacement therapy have been examined in animals. Long-term DHEA administration in old mice has reduced the incidence of mammary cancer,

has increased survival, and has delayed onset of immune dysfunction.[28] DHEA administration would also lead to decreased food intake and body weight loss, suggesting that, despite its minor anabolic activity, DHEA may act in a manner similar to food restriction in extending the life span and in retarding tumorogenesis and immunosenescence[29–31] (see also Chapter 24).

## 3. Mineralocorticoids

Secretion, blood levels, and clearance rates of aldosterone are markedly decreased in the elderly.[21] This decrease has been attributed to a declining adrenergic receptor activity.[32] Persistence of normal plasma electrolyte balance despite lower aldosterone levels demonstrates the efficiency of compensatory mechanisms even in old age.

## 4. Adrenal Steroid Receptors

Adrenal steroids exert their cellular and molecular actions by binding to cytoplasmic (cytosolic) and nuclear receptors, the degree of cellular responsiveness being directly proportional to the number of agonist-occupied specific receptor molecules (Figure 10.5). Hormone-mediated responses are controlled partly by the binding of the hormone to specific intracellular receptors and, then, by translocation of the hormone–receptor complex to nuclear receptor sites (Figure 10.6).

Adrenocortical steroid receptors are members of the steroid hormone and nuclear receptor family comprised of the vitamin D receptor, retinoid receptor, and thyroid hormone receptor, as well as a number of so-called "orphan" receptors, because their ligand and function are not well identified.[33–35] All classical steroid receptors (androgen AR, estrogen ER, glucocorticoid GR, mineralocorticoid MR, and progesterone PR) are phosphoproteins that, in the absence of the activating signal, are associated with heat-shock proteins.[36] They all act as transcriptional regulatory proteins and are able to interact with select target genes.[35–37]

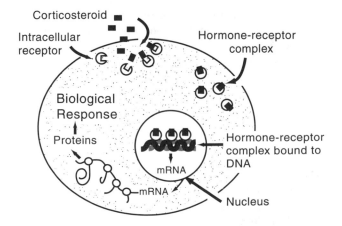

**FIGURE 10.5** Schematic diagram of corticosteroid action in a target cell.

**Systemic and Organismic Aging**

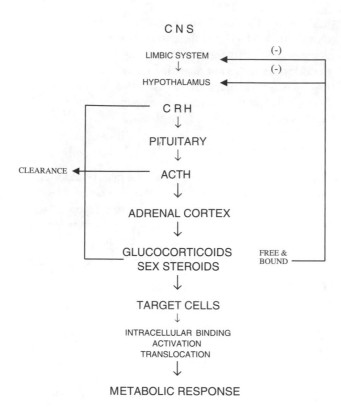

**FIGURE 10.6** Diagrammatic representation of the hypo-thalamo–pituitary–adrenocortical (HPA) axis.

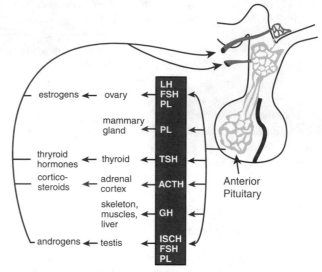

**FIGURE 10.7** Diagrammatic representation of the anterior pituitary and its major hormones, the target endocrines, and their hormones. Note the rich vascularization representing the portal vessels (see text).

Numerous mechansims account for this selectivity, such as interaction with DNA-bound transcription factors, the presence of chaperones, phosphorylation, and subnuclear trafficking pathways that facilitate receptor scanning of the genome.[35–38] Several steroid receptors can be activated in the absence of the hormone; this is the case of ERs that bind competitively to antagonist or agonist nonsteroidal molecules (i.e., Selective Estrogen Receptor Modulators, SERMs), but this does not seem to be the case for glucocorticoids[38] despite some recent studies of the binding of the antagonist RU486.[39] The finding that some of the receptors may be activated by signal transduction pathways in the absence of the specific hormone, although not immediately applicable to adrenocortical receptors, may be worth pursuing in future studies in view of the current progress in our understanding of the role of *coactivators* and *corepressors* in modulating action of estrogen and progesterone receptors.[39,40]

All molecular events in the hormone-cellular response pathway subsequent to receptor binding are subject to alteration with age, although the nature and magnitude of these age-related changes are variable depending on the hormone, the target cell, and the animal species. Overall, the concentration of corticosteroid receptors decreases either in early adulthood or during senescence.[41] For example, in the rat brain, glucocorticoid receptors are detectable on day 17 of gestation, they increase gradually after birth

to adult levels by 15 days of age but then are significantly reduced in aged animals (24-month-olds).[42] The concentration of cytosolic corticosterone receptors in the primary glucocorticoid concentrating region of the brain, the hippocampus, decreases with aging, with no change in receptor affinity or capacity for nuclear translocation.[42,43] Some of the receptor physicochemical properties (e.g., activation, transformation) seem to be more susceptible to aging than receptor number. Such age-related changes have been reported in glucocorticoid receptors in liver, skeletal muscle, and cerebral hemisphere. Aging changes in corticosteroid receptors that alter the responsiveness of target cells and molecules to hormones may contribute to the decline in the effectiveness of adrenocortical responses to stress.

## 5. Regulation of Adrenocortical Secretion

As illustrated in Figure 10.7, circulating levels of adrenocortical hormones depend on a hierarchy of regulation, from the hypothalamus to the pituitary and to the adrenal gland (Box 10.4) and, ultimately, to the target tissues, cells, or molecules. With aging and under conditions of stress, a disruption of this complex regulatory system at one or more levels may result in failure of homeostasis and adaptation.

*Corticotropin-releasing hormone, CRH,* a polypeptide, released from neurons in the median eminence of the hypothalamus, is transported via the portal system to the corticotropes of the anterior pituitary, where it stimulates synthesis and release of the *adrenocorticotropic hormone, ACTH.* ACTH, a protein, stimulates cells of the two inner zones of the adrenal cortex to synthesize and release the glucocorticoids and sex hormones. Thus, after ablation of the pituitary,

---

**Box 10.4**

**Feedback Mechanisms Applicable to Hypothamo–Pituitary–Endocrine Axes and the Portal Pituitary Blood Vessels**

Hypothalamo–pituitary–endocrine axes use feedback signals to regulate their secretory activity to or around a **set point value** necessary for homeostasis. The set point is maintained by negative feedbacks operating in a manner similar to an engineering control system with a set point, a controlling element, a variable element, an integrator, and a feedback signal.

In almost all physiologic systems, if a discrepancy arises between the set point and the variable element, then an error signal is delivered to the controlling element to produce an adjustment in the direction opposite to the original deviation from the set point. This type of control system, in which a variable provides a signal for compensatory reduction in the value of the variable, is referred to as a **negative feedback mechanism**. In the case of the HPA axis, CRH, ACTH, and glucocorticoid secretions are inter-regulated by feedbacks operating at each level: low plasma glucocorticoid levels increase CRH secretion, and CRH stimulates ACTH release, which, in turn, stimulates adrenal cortex glucocorticoid secretion; high levels of plasma glucocorticoid levels inhibit CRH and ACTH secretion and, consequently, decrease adrenal glucocorticoid secretion. In each case, the needed result is the return of glucocorticoid levels to the original, normal, "set point" level.

Signals are relayed from one component of the axis to the other and from the periphery to the axis, by **short- and long-term loops**. The short-term-loop feedback signals are carried through the portal blood vessels from the hypothalamus to the pituitary and vice versa (by retrograde flow). In the long-term loop, feedback signals are relayed from the peripheral endocrine gland and the target tissues to the pituitary and hypothalamus through the general blood circulation.

Portal pituitary vessels represent a direct vascular link between the hypothalamus and the anterior pituitary. On the ventral surface of the hypothalamus, capillary loops from the carotid arteries and the circle of Willis form a vascular plexus that carries blood down the pituitary stalk to the capillaries of the anterior pituitary. This arrangement constitutes a **blood portal system** beginning and ending in capillaries without going through the heart and the general circulation. Hypothalamic hypophysiotropic hormones are carried without dilution in the peripheral blood, directly to the anterior pituitary, where they stimulate synthesis and release of pituitary hormones.

---

these two zones atrophy, and the circulating levels of the corresponding hormones decrease. Conversely, in tumors of the pituitary in which ACTH levels are increased (as may occur in Cushing's disease), the two adrenocortical zones hypertrophy, and the hormonal levels increase.[13]

ACTH is secreted in irregular bursts throughout the 24-hour day, the pulses being more frequent in early morning and least frequent in the evening. The resulting circadian (diurnal) rhythm in cortisol secretion is preserved during aging in humans. However, sustained nighttime cortisol levels (i.e., blunting of the nocturnal drop in cortisol levels compared with daytime values) have also been reported and have been ascribed to the following:

- Reduced renal clearance of the hormone
- Reduced muscle mass and generally reduced basal metabolism
- Alterations in sleep patterns and insomnia (Chapter 7)

## B. Role of the Hypothalamo–Pituitary–Adrenal (HPA) Axis in Response to Stress

A "cornerstone" of stress physiology (implicating stimulation of the HPA axis) is that widely different (e.g., physical, social, emotional) stressors converge to induce a series of responses depending on adrenal stimulation. Responses to stress involve the participation of the hormones of the adrenal cortex together with those of the adrenal medulla as discussed further in the respective sections of this chapter. Most animal species so far studied, including humans, show increased glucocorticoid levels in response to administered CRH or ACTH or to endogenously increased levels of these hormones as a consequence of stress. Conversely, after removal of the adrenal gland in experimental animals or deficiency of the adrenal cortex in some diseases in humans (i.e., Addison's disease), defense mechanisms to stress fail to take place, and this failure will lead to death. Some characteristics of stress are listed in Table 10.2.

The glucocorticoid levels in old animals in response to stress may be lower or higher than in younger animals, but, in many, differences with aging are absent or small. In humans, all studies indicate that an increased plasma cortisol in response to ACTH is preserved in old individuals. In contrast, as already mentioned above, the response of DHEA to ACTH appears significantly reduced with aging. Similarly, stimulation of aldosterone secretion by sodium restriction is less efficient in old subjects compared to young ones.

*In vitro* experiments show that corticosterone secretion from isolated adrenal cortical cells of old (24 months

**Systemic and Organismic Aging**

**TABLE 10.2**
**Some Characteristics of Stress**

Stress induces defense mechanisms for maintenance of homeostasis in response to environmental challenges

Types of stress known to stimulate the hypothalamo-pituitary-adrenal (HPA) axis
  Physical stress:
    Hypoglycemia
    Trauma
    Exposure to extreme temperatures
    Infections
    Heavy exercise
  Psychological stress:
    Acute anxiety
    Anticipation of stressful situations
    Novel situations
    Chronic anxiety

Exposure to stress generates:
  Specific responses (varying with the stimulus and generating different responses with each stimulus)
  Nonspecific responses (always the same, regardless of the stimulus, and mediated through stimulation of neural, endocrine, and immune axes)

---

of age) rats are less responsive than those of young (3 months of age) rats to a synthetic subunit of ACTH and to cAMP, suggesting intracellular changes with aging involving either impaired steroidogenesis or receptor function, perhaps secondary to increased free radical production and membrane alterations (Chapter 5).

Other tests (e.g., metyrapone-induced decreases in plasma cortisol) indicate that the feedback mechanism for ACTH control as well as the stimulatory action of ACTH on adrenocortical secretion are maintained in old men and women. Given the extreme heterogeneity of aging processes, in adrenocortical as in many other functions, it appears that some individuals, but not all, retain a functional competence under basal and stress conditions. The challenge here is to identify those factors that separate successful aging from aging with impaired function and increased pathology (Chapter 3).

## IV. THE ADRENAL MEDULLA

The adrenal medulla is part of the sympathetic division of the autonomic nervous system, and, as such, functions in unison with the other sympathetic structures (for major functions, see Box 10.5). It is interrelated anatomically and functionally with the adrenal cortex by a rich vascular network in which blood flowing from cortex to medulla provides high concentrations of glucocorticoids which induce, in the medulla, some of the enzymes for catecholanine

---

**Box 10.5**
**Structure and Function of Adrenal Medulla**

The **medullary cells** (or chromaffin cells) are considered to be postganglionic neurons (innervated by preganglionic cholinergic fibers) that have lost their axons and become secretory cells. Major secretions include the catecholamines, **norepinephrine (NE)**, also produced and released by neurons in the CNS and by sympathetic neurons, **epinephrine (E)**, formed primarily in the medulla by the methylation of norepinephrine, and some **dopamine (DA)**, also a CNS neurotransmitter (Chapter 7). **Opioid peptides** are also secreted (most of the circulating encephalin comes from the medulla), although their functional significance is uncertain, as they do not cross the blood–brain barrier. NE and E mimic the effects of sympathetic discharge, stimulate the nervous system, and exert metabolic effects that include glycogenolysis in liver and skeletal muscle, mobilization of free fatty acids, and stimulation of metabolic rate. They act by binding to **two classes of G protein-coupled receptors**: α **(1 and 2) receptors**, which induce vasoconstriction in most organs and β **(1 and 2) receptors**, which mediate the metabolic effects and stimulate rate and force of cardiac contraction ($\beta^1$) and dilate blood vessels in muscle and liver ($\beta^2$). Their secretion is under neural control and is also influenced by other hormones, primarily glucocorticoids and thyroid hormones.[45]

---

synthesis (e.g., the enzyme phenylethanolamine-*N*-methyltransferase, PNMT, that catalyzes the formation of epinephrine from norepinephrine). In addition, glucocorticoids have some metabolic interaction with medullary hormones (e.g., mobilization of free fatty acids in emergency situations).

The sympathetic system, including the adrenal medulla, is not essential for life. However, dysfunction of the autonomic nervous system, comprised of sympathetic and parasympathetic divisions, is a well-recognized, although poorly understood, consequence of old age.[44] One of the anatomical characteristics of the autonomic sympathetic division is its organization into a paravertebral sympathetic ganglion chain that, under emergency conditions of stress, can discharge as a unit, as in "rage and fright," when sympathetically innervated structures are affected simultaneously over the entire body. The heart rate is accelerated, the blood pressure rises, red blood cells are poured into the circulation from the spleen (in certain species), the concentration of blood glucose rises, the bronchioles and pupils dilate and, on the whole, the organism is prepared for "fight or flight" (Table 10.3). The contribution of the adrenal medulla to "the emergency function" of the sympatho-adrenal system varies with the

## TABLE 10.3
## "Fright, Flight, or Fight" Responses to Stress

- Increased blood pressure
- Increased heart rate
- Increased force of heart contraction
- Increased heart conduction velocity
- Shift of blood flow distribution away from the skin and splachnic regions and more to heart, skeletal muscle, and brain
- Contraction of spleen capsule (increased hematocrit[a])
- Increased depth and rate of respiration
- Mobilization of liver glycogen to glucose (glycogenolysis)
- Mobilization of free fatty acids from adipose tissue (lipolysis)
- Mydriasis (widening of pupil)
- Accommodation for far vision (relaxation of ciliary muscle)
- Widening of palpebral fissure (eyelids wide open)
- Piloerection (erection of the hair)
- Inhibition of gastrointestinal motility and secretion, contraction of sphincters
- Sweating (cold sweats as skin blood vessels are constricted)

[a] Hematocrit = a measure of volume of red blood cells to volume of whole blood; that is, a measure of volume of cells and plasma in blood

animal species and the type of stress. It also seems to vary with the age of the animal, reaching optimal efficiency in adulthood and showing some selective decline in old age.

### A. VARIABILITY OF CHANGES WITH AGING

Autonomic function appears to be modified with aging, but its eventual impairment may be viewed more as a complication than as a consequence of aging. Under basal conditions, in humans, plasma levels and urinary excretion of catecholamines are highly variable. With aging they may do the following:

- Remain unchanged
- Show a reduction in absolute and averaged circadian amplitude
- Show an increase, the increase being greater after standing and isometric exercise

Plasma and urinary catecholamine elevation, reported after a variety of stimuli, have been interpreted as compensatory reactions to the apparently increasing refractoriness with aging (perhaps due to receptor downregulation) of target tissues to catecholamines.[32,45] However, the apparent increase in norepinephrine plasma levels does not occur with all stimuli, and, additionally, the return to baseline levels is prolonged in the elderly. Also, the two catecholamines often show differential responses: plasma norepinephrine may be elevated in elderly subjects during a mental stress test, plasma epinephrine is not; this suggests a specific hyperactivity of the sympathetic system in general rather than of the adrenal medulla.

### B. TARGET DIFFERENTIAL RESPONSIVENESS

An unresolved issue is whether the high levels of catecholamines in the elderly are due to their higher release from the adrenal medulla or whether the peripheral clearance of the hormones is reduced. One proposed explanation for the overall increase of sympathoadrenal activity with aging would be an increased refractoriness (or decreased sensitivity) of target tissues to catecholamines due to alterations in transport and binding. A decrease in adrenergic receptor number with aging has been reported in some organs and cells (e.g., cerebellum, adipocytes) and a decrease in receptor affinity in others (e.g., lung), but in many cases, changes are not observed. Although current findings do not support the view that receptor numbers are decreased with aging, several concomitant factors (e.g., the use of agonist drugs and of lymphocyte adrenoreceptor density as models, aging-associated decrease in cell membrane fluidity, and influence of other humoral factors) may mask changes in receptor numbers. Clinical and experimental observations have been conducted primarily of $\beta_1$ receptors, and they, as well as the other $\beta_2$ and $\alpha$ adrenergic receptors, show great variability in responsiveness to adrenergic stimulation. Decreased responsiveness may be found in the diminished efficiency of hemodynamic and cardiovascular responses to changes in posture (tilting, standing, and to exercise) and in the case of slower dark adaptation of pupillary size (Chapter 9). Conversely, increased responsiveness is manifested in those organs and tissues regulating blood pressure, which progressively increases with aging, leading to hypertension. Other indices of activity show different responses to aging. In rats, mice, and rabbits, the ability of dopamine to stimulate adenylate cyclase in the corpus striatum decreases progressively with age, as does dopamine binding, but remains unchanged in adipocytes or is increased in hepatocytes.

The lack of uniformity in response patterns has been ascribed to the differential responsiveness of tissues to catecholamines. Sensitivity to catecholamines may also be influenced by other hormones, such as glucocorticoids and thyroid hormones known to affect catecholamine metabolism, while receptors undergo age-related changes. Other factors such as alterations in structure and function of membranes and intracellular molecules may also be operative in changing cell responsiveness to adrenal medullary hormones.

## V. THE PITUITARY GLAND

The pituitary gland (also called hypophysis), through its tropic hormones, regulates the activity of several peripheral endocrine organs (adrenal cortex, thyroid, gonads) and other target tissues (Box 10.6). It has close functional ties with the hypothalamus (and, indirectly, other

**Systemic and Organismic Aging**

## Box 10.6
### Structure and Major Hormones of the Pituitary

The pituitary is situated on the superior surface of the sphenoid bone (in the so-called sella turcica) in close proximity of the hypothalamus. It is comprised of three, more or less separate endocrine glands that produce a relatively large number of hormonally active substances. These include six well-characterized hormones secreted from the **anterior lobe of the gland**: ACTH, discussed above, the two gonadotropins, thyroid stimulating hormone, growth hormone, and prolactin (Figure 10.7). In addition, the anterior lobe secretes lipotropin and pro-opio-melanocortin, precursors of ACTH, opioid peptides, and melanocyte-stimulating hormone (MSH).

The **intermediate lobe**, practically indistinguishable in the human adult and comprised of cells scattered in the anterior lobe, secretes MSH, lipotropin, opioid peptides, and the large precursor, pro-opio-melanocortin.

The **posterior lobe** secretes two peptides, vasopressin (also called antidiuretic hormone, ADH) and oxytocin (Figure 10.8). These hormones are synthesized in the large (magnocellular) neurons of the supraoptic and paraventricular nuclei of the hypothalamus; they are transported down the axons to their endings in the posterior lobe, where they are stored and from which they are secreted into the general circulation after appropriate stimuli (Figure 10.9).

The hormones tropic to peripheral endocrine glands are discussed in the appropriate chapters pertaining to those glands, for example, ACTH with the adrenal cortex in this chapter. Of the other pituitary hormones, growth hormone, prolactin, ADH, and oxytocin, are briefly discussed below.

## Box 10.7
### Major Hypothalamic (Hypophysiotropic) Hormones

The hypothalamus has many important functions that regulate visceral reflexes, complex behavioral and emotional reactions, and synthesis and release of several hormones.[46] It secretes six hypophysiotropic hormones (Figure 10.9) that reach the anterior pituitary by a portal system of blood vessels (i.e., the hormone is carried directly to the pituitary without passing through the general circulation). They are as follows: **corticotropic releasing hormone, CRH; growth hormone releasing hormone, GRH**, and **somatostatin** (also called GIH, inhibitor of growth hormone release); **prolactin releasing hormone, PRH**, and **prolactin inhibitory hormone, PIH** (all of the above are discussed in this chapter); **gonadotropin releasing hormone (GnRH**, Chapters 11 and 12); and **thyrotropin releasing hormone (TRH**, Chapter 13). These hypophysiotropic hormones stimulate or inhibit the synthesis and release of the corresponding pituitary hormones: ACTH, GH, PL, FSH, LH, and TSH, which, in turn, stimulate the synthesis and release of the hormones (corticosteroids, gonadal steroids, thyroid hormones) of their respective endocrine targets, or, in the case of GH and PL, act directly on target tissues.

### TABLE 10.4
### Some Characteristics of Changes in Pituitary Function with Aging

Great variability among individuals of the same species and of different species

Differential alterations with aging — some functions greatly affected and others scarcely altered

Selective rather than global nature of aging-related changes and their importance for survival under stress

The possibility of modifying aging-related changes and influencing the life span by hormonal replacement therapy

The formulation of neuro-immuno-endocrine theories of aging based on the hypothesis that dysregulation of hypothalamo-pituitary-immune signals may cause, or be closely involved in aging processes

CNS centers) that produces a number of peptides with hormone-precursor and behavioral activities[46,47] (Box 10.7). Relatively few studies have explored systematically whether aging-associated changes involve the gland globally, or if they are the discrete and selective ones that are crucial to the aging process. From previous and current studies, a few tentative general considerations can be gleaned, as summarized in Table 10.4.

## A. STRUCTURAL CHANGES

In the absence of overt disease, the pituitary gland robustly withstands major challenges of aging. *In the anterior lobe*, changes with aging are relatively few, and include those cellular changes (e.g., accumulation of lipofuscin) characteristic of aging cells, in general, and of neural and secretory cells (e.g., changes in dendritic and synaptic morphology, in neurotransmitters), in particular (Chapter 7). The numbers of specific cells that secrete its specific hormones are not markedly altered, except for the gonadotropes (i.e., gonadotropin-secreting cells) which show, in experimental animals, changes similar to those observed after castration. Tumor incidence, illustrated primarily by prolactinomas, increases with aging in rats and mice, more so in females than in males.

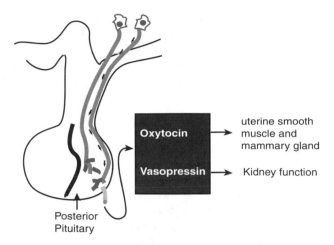

uterine smooth muscle and mammary gland

**Oxytocin**

**Vasopressin** → Kidney function

Posterior Pituitary

**FIGURE 10.8** Diagrammatic representation of the posterior pituitary with its hormones and target organs.

*In the posterior pituitary* (Figure 10.8), histochemical studies in old rats reveal a decrease in neurosecretory material, more marked, according to some, in oxytocin- than ADH-secreting cells but the reverse according to others. Despite these contradictions, with increasing age, decreased amounts of neurohypophyseal hormones are stored in the pituitary, testifying to the depressed functional state of the gland. This conclusion is justified on the basis of quantitative ultrastructural analysis of old rodent pituitary that show increased autophagic activity, increased perivascular space, decline in cell volume, decreased size and number of neurosecretory granules, and reduction in endoplasmic reticulum activity. In humans and other examined animals (e.g., cattle), such changes are not seen or are minimal.

## B. GROWTH HORMONE (GH), GH-RELEASING HORMONE AND SOMATOSTATIN

Aging may be associated, in very general terms, with decreased protein synthesis, lean body mass, and bone formation as well as increased adiposity, suggesting the involvement of GH because of its anabolic (i.e., promoting protein synthesis) and metabolic actions (Box 10.8). Nevertheless, in man and other species, the number of somatotropes (i.e., pituitary cells secreting GH), pituitary content of GH, basal plasma levels of the hormone, and its clearance remain essentially unchanged into old age. Some studies, however, have reported a decrease in GH basal levels in rats and in humans.

---

**Box 10.8**

**Structure and Actions of Growth Hormone (GH)**

GH is a protein encoded by five genes on chromosome 17 that has a high degree of species specificity; it is bound to two proteins in plasma. GH receptor is part of the cytokine receptor superfamily.[48] Major actions of GH include:

1. Before adulthood
   - Growth stimulation
   - Increase in protein anabolism
   - Stimulation of IGFI by tissues
2. In adulthood
   - Stimulation of IGFI by tissues
   - Increased lean body mass and metabolic rate
   - Decreased body fat with increased plasma free fatty acids, thereby providing a ready source of energy for the tissues during hypoglycemia, fasting, and stressful stimuli (see below)
   - Decreased blood cholesterol
   - Increased hepatic glucose output (diabetogenic effect)
   - Anti-insulin effect in muscle
   - Stimulation of pancreatic B cells, thereby making them more sensitive to insulinogenic stimuli with resulting diabetes due to B cell exhaustion

GH was originally thought to produce growth by a direct action on tissues (e.g., bones, muscles), but it is thought today that it acts directly and indirectly through the stimulation of a somatomedin, the **Insulin-like Growth Factor I (IGFI) (in the adult).**

GH secretion is controlled via the hypothalamus, which secretes into the portal blood, both **growth hormone-releasing hormone (GRH)**, that stimulates GH secretion from the anterior pituitary, and **growth hormone-inhibiting hormone (GIH)**, or somatostatin, that inhibits its secretion. GH secretion is under feedback control, as are the other anterior pituitary hormones. Additional factors can stimulate (e.g., hypoglycemia, fasting, stress) or inhibit (e.g., glucose, free fatty acids) its secretion.

---

**Systemic and Organismic Aging**

Obesity is known to suppress circulating GH levels in young individuals, and the frequent obesity of the elderly may contribute to the decreased GH levels, although not all reports agree. Contradictory changes in GH levels have been reported in the elderly after being exposed to a variety of stimuli that are known to cause GH release: in rats, GH elevation after stimulation by a variety of factors is less marked in old than in young animals, whereas in man, levels may be decreased or remain unaltered. However, in response to stress, surgical trauma, exercise, and arginine stimulation, the expected increase of GH secretion may be considerably blunted.[49]

GH secretion in humans undergoes a nocturnal peak during the first 4 hours of sleep, coinciding with stages III and IV of slow-wave sleep; these are the stages most affected in aging (Chapter 8). Studies in older persons have shown a decrease in sleep-related GH secretion with an occasional decrease in the nocturnal peak that has been attributed to low levels of GRH and high levels of somatostatin.[50] However, the relation of GH changes with sleep remains as controversial as that with obesity.

## 1. Growth Hormone Replacement Therapy

Some elderly men have low blood levels of GH as well as low IGFI levels.[51] Relatively long-term administration (6 months) of biosynthetic human GH increases muscle and bone mass and decreases adipose tissue in older men and women. These beneficial effects are small (10 to 14% change) and last only as long as the hormone is administered. Despite the small and temporary benefits of GH administration reported so far, the hormone may decelerate the decline in muscle and bone with aging and, thus, prevent the falls and bone fractures that are major causes of disability and mortality (Chapter 21).[52,53] These observations, together with those of the possibly beneficial effects of other hormone replacement therapies (Chapters 10 through 14), have led to speculation about whether reduced GH levels in old age may contribute to physiologic decline and increased pathology, and that, vice versa, GH administration may delay aging and prolong the life span.

A number of experiments in animals on the possible effects of GH on longevity are controversial.[54] For example, in calorie-restricted rats, mice, and nonhuman primates, caloric restriction reduces GH and IGFI levels in young animals, but in old animals, GH and IGFI levels are higher than in controls of the same age fed *ad libitum* (Chapter 24). Inasmuch as animals on the calorie-restricted diet live longer than their controls, the above data have been interpreted to mean that the delay in aging induced by caloric restriction postpones or delays the decline in GH and IGFI levels characteristic of old age. The same data may also be interpreted to mean that the increased GH and IGFI plasma levels in old rats on

calorie-restricted diets may mediate the effect of caloric restriction in delaying aging (Chapter 24). This latter interpretation would, then, justify the use of GH and IGFI administration in the elderly to delay the onset of functional impairment and to prolong life. This view was further supported by studies in GH receptor knock-out dwarf mice with some phenotypic characteristics (e.g., improved antioxidant defenses, increased insulin sensitivity, reduced metabolic rate and body temperature, and number of cell divisions) that may be related to their prolonged survival.[54] GH administration has also been tested in a variety of conditions in humans, including cardiac failure, healing of burns and other wounds, and Alzheimer's disease, all with variable, negligible, or even life-endangering outcomes.[55]

In humans, it is well known that GH excess has detrimental effects, such as acromegaly, diabetes mellitus, arthritis, and hypertension. In the interventions conducted so far in the elderly, administration of GH to individuals with low or normal GH and IGFI levels has resulted in increased morbidity (e.g., joint swelling and pain, cardiac arrhythmia, insulin resistance). These unwanted side effects and the possibility of increased mortality together with the lack of solid evidence of beneficial effects diminish the usefulness of GH treatment.[55] There remain to be investigated the effects of long-term treatment, appropriate dosages, gender differences, age at onset of treatment, route of administration, and others, before GH administration can be approved as replacement therapy for the elderly. Bone and muscle mass can be improved with good nutrition and physical exercise at all ages, and therefore, these measures should be encouraged in preference to other, less efficacious and more risky interventions (Chapter 24).

## 2. Somatostatin (GIH)

Although the major function of GH is to promote growth of whole-body and several organs during childhood and adolescence, GH, GRH, and somatostatin continue to be secreted throughout life, with GH exerting metabolic effects and GRH and somatostatin regulating GH secretion. Somatostatin levels in plasma are negligible, and its actions are exerted locally in vicinity of the secretory cells; as such, it is representative of that large group of substances produced by diffuse (paracrine) cells in several locations. Somatostatin inhibits TSH as well as GH release. Somatostatin secreted from the hypothalamus and other tissues has multiple biologic actions (in addition to inhibition of GH secretion, the hormone secreted in the pancreas and intestine inhibits secretion of pancreatic and intestinal hormones). Basal plasma levels (originating primarily from the pancreas and intestinal wall neurons) are higher in the elderly (70 to 90 years) than in young adults (22 to 30 years), but daytime variations and responses to test meals are lower in amplitude.[56] Somatostatin levels in brain areas of elderly individuals do not change with advancing age but are decreased in patients with Alzheimer's and Parkinson's diseases in brain regions that also have cholinergic deficits[57] and in the cerebrospinal fluid[58] (Chapter 8). In rats, aging-related changes may occur in the caudal hypothalamus and median eminence and in the corpus striatum. Somatostatin secretion is also influenced by neurotransmitter release, e.g., stimulated by increased norepinephrine and dopamine discharge from brain catecholaminergic neurons. It is not clear whether the decline in these neurotransmitters which occurs during aging (Chapter 8) accounts for the decrease in GH by way of a reduction in GRH release or an increase in somatostatin release, and more research is needed to identify GRH changes with aging. Somatostatin-secreting or -containing tumors have equal incidence in men and women with a peak in the fifth decade, but half of affected subjects have other endocrinopathies.[59]

## C. PROLACTIN (PRL)

Prolactin (PRL) stimulates lactation and has anabolic, diabetogenic, and lipolytic actions.[60] In male animals and humans, but not in females, plasma PRL levels increase with aging, perhaps consequent to the reduction in hypothalamic dopamine (the hypothalamic inhibitor of hypothalamic PRL release) with aging and the high incidence of pituitary PRL-secreting tumors. Low PRL levels in old women may be attributable to low estrogen levels after menopause.[61] Dampening of day and night PRL levels, combined with declining rhythmicities of GH, adrenal, thyroid, pituitary, and pineal secretions and the cessation of ovarian cyclicity, contribute to the progressive failure of chronobiologic regulations with aging (Chapter 14).

## D. VASOPRESSIN (ANTIDIURETIC HORMONE, ADH) AND OXYTOCIN

These two hormones are nonapeptides secreted by neurons of the supraoptic and paraventricular nuclei of the hypothalamus and transported within the axons to the posterior lobe of the pituitary, where they are stored before being released in the circulation (Figure 10.8). Within each hypothalamic nucleus, some neurons produce oxytocin and others vasopressin, or ADH. The nonapeptides are synthesized as part of larger precursor molecules and are stored in association with neurophysins, which are also part of the precursor molecule.

In aging rats, except, perhaps in the very old, vasopressin- and oxytocin-secreting neurons are relatively spared the morphologic changes that occur in neurons in most other hypothalamic nuclei.[62,63] Some of the changes that occur have been related to functional decrements in the secretory activity of neurons in the hypothalamic

paraventricular and supraoptic nuclei. Degenerating neurons often alternate with hypersecreting neurons which compensate for cell loss and impaired function: this diverse aging pattern may account for the disparate observations of low, high, or unchanged hormone levels.[62,63]

The principal action of vasopressin (hence, its alternative name of ADH) is to promote the retention of water by the renal distal tubules and collecting ducts, and changes with aging are discussed with aging of renal function (Chapter 19). The principal actions of oxytocin in stimulating contraction of smooth muscle of the lactating mammary gland and pregnant uterus are thought to be relevant primarily to the reproductive years. However, both hormones have significant CNS actions: they are reported to ameliorate long-term memory and attention, neural connectivity, and social behaviors.[64] These behaviors are susceptible to change with aging and may benefit from administration of these hormones.[62–64]

## VI. STRESS AND ADAPTATION THROUGH HOMEOSTASIS AND ALLOSTASIS

A prerequisite for survival of each organism is an environment favorable to the optimal expression of its function. This is crucial for single-cell organisms and even more so for multicellular, complex organisms. For humans, the environment is represented both by external conditions (e.g., atmospheric temperature, food availability, and social interrelations) and conditions of the internal environment (metabolism, coordination of regulatory signals among body systems, and integration of multiple cellular and molecular functions). Thus, while reproduction is the key function to ensure continuation of the species, individual survival depends on the maintenance of a constantly stable internal environment, in response to internal or external challenges (Table 10.2). *Homeostasis* is the term given to the concept of an "ideal" steady-state, that is, a constant internal environment permitting optimal function and survival (see below "Historical Notes"). In order to survive, an organism must vary all parameters of its internal environment and match them appropriately to environmental demands. Hence, "homeostasis" can be achieved only through dynamic adjustments reflecting repeated fluctuations of various physiologic systems. The term of allostasis refers to these fluctuations and represents the sum of adjustments necessary for maintaining homeostasis (Table 10.5).[65] Cumulative effects of stress may affect health and longevity of the individual organism. The goal of "stability through change" is attained by paying a price in terms of decreased function and increased pathology. Thus, we refer to the cumulative, multifactorial view of physiologic toll as the so-called *allostatic load*, which may be exacted on the body through attempts at adaptation. Given the multitude and variety of

**Systemic and Organismic Aging**

## TABLE 10.5
## Stress, Homeostasis, and Allostasis

An organism must vary all parameters of its internal milieu and match them appropriately to environmental demands through the following:

*Homeostasis*: steady state and optimal set points are achieved; it is obtained by repeated fluctuations of various physiological systems (allostasis) and/or long-term exposure to elevated levels of physiologic activity

*Allostasis*: Emphasis is on optimal operating ranges of physiologic systems; it represents stability obtained through change

*Allostatic load*: The cumulative, multisystem view of physiologic toll that may be exacted on the body through attempts at adaptation

environmental changes, allostasis needs always to be operational and continuously directed to maintain homeostasis. With aging, the allostatic efficiency may decline or deviate from the optimal, thereby endangering homeostasis and generating abnormal (pathologic) responses. Ultimately, then, the severity and persistence of the allostatic load will cause or contribute to the increasing morbidity and mortality of old age (Tables 10.5 and 10.6).

The classical concept of homeostasis and the more recent one of allostasis have been pivotal in formulating some of the tenets of modern and contemporary biology.[65] Initiated in the second half of the 19th century, these concepts have been validated through several important experimental

## TABLE 10.6
## Pathophysiologic Responses During and After Stress

### During Stress

Energy storage ceases because of the following:

↑ sympathetic activity (i.e., increased vigilance and arousal)

↓ parasympathetic activity

↓ insulin secretion

Access to energy storage is facilitated, and energy storage steps are reversed because of the following:

↑ glucocorticoid secretion

↑ epinephrine and norepinephrine secretion

↑ glucagon secretion

### After Stress

If physiologic responses are insufficient and adaptation is incomplete, symptoms of poor health are registered (e.g., loss of energy when freeing energy from storage and returning to storage).

The following are examples of consequences:

  Muscle wasting

  Diabetes (Type 2)

  Ulcers, colitis, diarrhea

  Inhibition of growth (in childhood)

  Osteoporosis (in old age)

    ↓ LHRH, ↓ testosterone

and clinical discoveries from the middle of the 19th century to the present. With the recent decoding of the human genome, they are expected to provide, through physiologic genomics (i.e., gene expression that articulates the return to an optimal state after stress), the important integrative link between a gene and its function in the intact organism as it relates to the environment in which the organism lives.

The impact of the hormonal changes with aging on the ability of the individual to adapt and survive have generated a number of theories based on the hypothesis that specific endocrine signals, together with neural and immune signals (probably genetically linked) may direct aging and death. Neuroendocrine and, more recently, neuro-immuno-endocrine theories propose that aging is not due to intrinsic deteriorative processes in all cells or molecules but rather to a programmed regulation by "pacemaker cells," perhaps situated in the brain and acting through neural, immune, and endocrine signals.[1,2] These signals would orchestrate the passage from one stage of the life span to the other, thereby timing the entire life cycle, including development, growth, maturation, and aging. This, as other extant theories of aging (Chapter 6), while awaiting verification, offer useful guidelines and viable tools to understand aging and promote longevity.

### HISTORICAL NOTES: THE "MILIEU INTERIEUR," HOMEOSTASIS, AND THE GENERAL ADAPTATION SYNDROME

In 1878, the physiologist, Claude Bernard, published, among many previous important texts, a book on the similarity of requirements for life in animals and plants.[66] In this book, he formulated what was to be his greatest scientific contribution and a landmark in the history of modern physiology, that is, the concept of the "*milieu interieur*" (internal environment). Accordingly, the preservation of the constancy of the internal environment is essential to the stability of the living organism, notwithstanding any external change. He further stated that all vital body functions, varied as they are, act, in concert, to preserve constant the conditions of life in the internal organization.

In 1932, another physiologist, Walter B. Cannon, published *The Wisdom of the Body*,[67] in which he analyzed the mechanisms of the internal regulation of body activity and coined the word *homeostasis* to indicate a relatively stable state (steady state) of equilibrium (or a tendency toward such a state) among the various functions of an organism in its response to changes in the environment. He also pointed to the important role of the autonomic, specifically, the sympathetic nervous system in the "fright, flight, or fight" reaction of an individual facing a threat.

In 1948, still another physiologist, Hans Selye, published the first of several books on stress[68] in which he described the important role of the hypothalamo–pituitary–adrenal (HPA) axis in mediating the responses of the organism to different types of stress in such a way as to preserve constancy and stability of body functions. Selye suggested that the responses to stress included: the mobilization of physiologic defenses designed to counterbalance the damaging effects of stress (*specific responses*) and, simultaneously, the stimulation of the adrenal cortex to secrete larger amounts of its hormones (*nonspecific responses*).

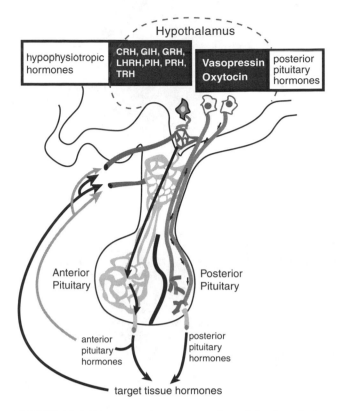

**FIGURE 10.9** Diagrammatic representation of the hypothalamus–pituitary connections with the major hypophysiotropic hormones and posterior pituitary hormones. Note the rich vascularization representing the portal vessels (see text).

The efficiency of these two sets of responses would depend on the success or failure of the organism to adapt to the challenges of stress. In the absence of the adrenal gland (e.g., by surgical removal or pathologic insufficiency), neither of these responses would occur, and the animal would not adapt but would die. According to Selye, these responses to stress could be grouped into three sequential phases as part of a *general adaptation syndrome* characterized by:

- An initial phase in which defense mechanisms are acutely challenged (alarm reaction)
- A period of enhanced adaptive capacity (stage of resistance)
- The loss of the capacity to adapt (stage of exhaustion)

Repeated exposures to stress may lead to *successful adaptation* with more efficient ability to withstand subsequent stress or may induce pathology, the so-called "*diseases of adaptation*" (e.g., cardiovascular diseases, immunosenescence, and shortened life). Furthermore, Selye compared the trajectory of life to that of the general adaptation syndrome:

- Young age would correspond to the so-called "alarm reaction," when many functions are acutely challenged
- Adulthood, with optimal efficiency of responses to stress, would correspond to the period of resistance and successful adaptation
- Old age, with a declining ability to adapt to stress, would correspond to the period of exhaustion

This general response pattern assumes that some form of energy is necessary for the performace of adaptive work and can be exhausted under the influence of stress or lead to disease (Figure 10.10).

## A. RESPONSES TO STRESS MEDIATED THROUGH THE HYPOTHALAMO-PITUITARY-ADRENAL (HPA) AXIS

Exposure to stress elicits physiologic responses that are directed specifically to correct the type of stress considered. These would be simultaneously accompanied by a group of responses (e.g., the fright, fight, or flight responses, Table 10.3) which would be the same, irrespective of the type of stress (Table 10.2), and would depend on stimulation or inhibition by hormonal signals (Table 10.7 and Figure 10.9). While many of these allostatic responses may be regulated by the HPA axis, other organs and systems may also be operative, and, when involved, may be responsible for some of the consequences of stress (Tables 10.8 and 10.9). Thus, homeostatic competence provides a "panoramic view" of overall physiologic performance. This explains the interest among physiologists to identify control mechanisms responsible for this integrated activity.[67–72]

As discussed above in this chapter, under resting (basal) conditions, few changes occur in HPA function in old age. However, under stress, evidence of decreased physiological competence is plentiful. The increased risk of death following stress in the elderly is well acknowledged and needs no documentation here. Older people are less resistant than younger individuals to excessively cold or warm temperatures because of the progressive deterioration of thermoregulatory mechanisms that occurs with age (Chapter 13). Equally diminished is the capacity of old people to adapt to infections, hypoxia, traumatic injury, exercise, and physical work, all representing types of stresses that require complex physiologic adjustments. Emotional stress in the old is also capable of triggering or aggravating a series of physical ailments that, superimposed on an already debilitated state, may be conducive to disease and death (Figure 10.11). Similarly, neuroendocrine and immune systems have many cross-over responses, and both work together to preserve homeostasis[73–79] (Box 10.9).

## B. ALLOSTATIC LOAD AS A COMORBIDITY FACTOR

In considering the *morbidity* of old age, the suggestion was made that prognosis for disability and mortality depended in part on the number of illnesses that simultaneously affected the same individual (Chapter 3). Thus, lists of comorbidity indices were and are being utilized to assist in evaluating the so-called "morbidity load."[80] Likewise, lists of best comprehensive indices of physiologic competence have been formulated, and ten of such parameters are presented in Table 10.8. Deviation of these parameters from normal values during repeated exposures to stress may constitute an allostatic load and represent risk factors endangering health and

**Systemic and Organismic Aging**

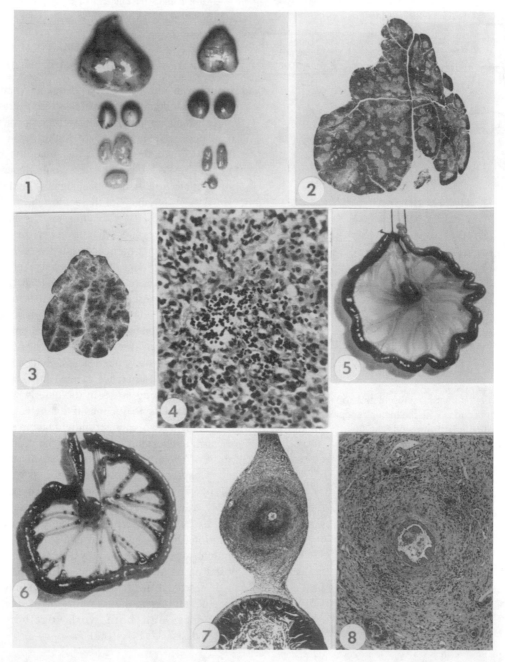

**FIGURE 10.10** Some physiologic and pathologic responses to stress in selected organs. The experimental animal chosen is the rat. (1) (left) Naked view of the adrenal and lymphatic organs of normal rat; (right) same organs during alarm reaction. Three iliac lymph nodes (bottom) and thymus (top) are significantly decreased in size in the stressed animal, whereas adrenals are enlarged and hyperhemic. (2) Low magnification of cross section of the thymus of a normal rat and (3) of a stressed rat. Note the inversion of the thymus pattern (light-dark areas) due to depletion of the cortex of thymocytes and migration of thymocyte debris into the medulla in the stressed animal. (4) Higher magnification of an area of thymus during the alarm reaction, dark patches represent nuclear debris from the degenerate thymocyte. Before direct chemical measurement of adrenocortical hormones in the blood was possible, the involution of the thymus and lymph nodes was taken as an index of adrenocortical activity. (5–8) Hormonally induced vascular lesions approximate those occurring in diseases of adaptation. (5) Normal mesenteric vessels. Macroscopic view of a normal intestinal loop of a rat. Note thin and regular mesenteric vessels. (6) Abnormal mesenteric vessels. Macroscopic view from an animal chronically treated with the adrenocoritcol steroid hormone, desoxycorticosterone acetate. Note numerous beadlike periarteritis nodosa (resembling atherosclerosis) nodules along the mesenteric vessel. (7 and 8) Low and high magnification of periarteritis nodosa nodules in the mesenteric vessels. Note the thick layer of hyalinized fibrin lining vascular lumen and the partial necrosis of the arterial wall.[68]

**TABLE 10.7**
**Functions Stimulated or Inhibited by Physical and Psychological Stress**

| Functions Stimulated by Stress | Functions Inhibited by Stress |
|---|---|
| Cardiovascular | All functions not immediately necessary for defense and survival are decreased: |
| • Increased cardiac rate | • Decreased growth |
| • Elevated blood pressure | • Decreased appetite (anorexia) |
| • Increased blood coagulation | • Decreased reproductive function and sex drive |
| • Redistribution of blood from peripheral (skin) and internal systems (gastrointestinal) to heart, skeletal muscles, brain | • Decreased circulation in tissues not involved in stress response |
| Respiratory | • Decreased response to pain |
| • Increased respiratory ventilation | • Decreased immune function |
| Metabolic | • Decreased thymus size |
| • Increased glycogen mobilization | • Decreased thymic hormones and cytokines |
| • Increased glycemia | |
| • Increased lipolysis | |
| Hormonal | |
| • Increased CRH, ACTH, glucocorticoids | |
| • Increased vasopressin, NGF[a] | |
| • Increased catecholamines (E and NE) | |

[a] NGF = Nerve Growth Factor, taken as an example of a growth-promoting, paracrine factor (Chapter 7)

**TABLE 10.8**
**Selected Parameters of Allostasis**

1, 2. Systolic and diastolic blood pressure (indices of cardiovascular activity)

3. Waist–hip ratio (index of long-term metabolic and lipid deposition)

4, 5. Serum High-Density Lipoprotein (HDL) and total cholesterol levels (indices of atherosclerotic risk)

6. Blood plasma levels of total glycosylated hemoglobin (index of glucose metabolism)

7. Serum dehydroepiandrosterone (DHEA) sulfate levels [index of hypothalamo–pituitary–adrenal (HPA) inhibitor or antagonist]

8. 12-hour urinary cortisol excretion (index of 12-h integrated HPA activity)

9, 10. 12-hour urinary norepinephrine and epinephrine excretion levels (index of 12-h integrated sympathetic activity)

*Source:* Adapted from Seeman, T.E. et al., *Arch. Intern. Med.*, 157, 2259, 1997.

**TABLE 10.9**
**Risk Factors (Allostatic Load) Endangering Health and Shortening Life Span**

**Elevated Physiologic Indices (At Risk)**
• Systolic blood pressure: ≥148 mm Hg
• Diastolic blood pressure: ≥83 mm Hg
• Waist–hip ration: ≥0.94
• Total cholesterol–High-Density Lipoprotein ratio: ≥5.9
• Total glycosylated hemoglobin level: ≥7.1%
• Urinary cortisol level: ≥25.7 mg/g creatinine
• Urinary epinephrine level: ≥5 mg/g creatinine
• Urinary norepinephrine level: ≥48 mg/g creatinine

**Lowered Physiologic Indices (At Risk)**
• HDL cholesterol level: ≤1.45 mmol/L
• DHEA (dehydroepiandrosterone) level: ≤2.5 μmol/L

## C. STRATEGIES FOR "COPING WITH STRESS"

With better understanding of the role of the adrenals in adaptation and survival, and of the nature of the physiologic toll the organism pays for adaptation, it is possible to envision strategies for prevention and delay of the pathologic consequences of stress.[19] As indicated for many other functions, many individuals age successfully and maintain an adequate capability for adaptation well into old age (Chapter 3). A study of these "successful agers" shows that the quality of the environment during the early stages of development may be particularly important; thus, "pick-up the right sort of infancy" may be good advice. For example, in rats, "handling" from birth until weaning induces long-term changes in adrenocortical secretion

shortening the life span (Table 10.9).[65] Such risk factors are operative in increasing the incidence of cardiovascular disease, the decline in memory, and in the decline in physical performance.[65] The concept of an allostatic load generating a risk factor for morbidity and mortality in old age provides a physiologic basis for the overall wear-and-tear theory of aging (Chapter 6) and support for the role of oxidative damage in aging processes (Chapter 5) and of the loss of hippocampal neurons due to glucocorticoid toxicity in the glucocorticoid cascade hypothesis of aging.[15]

**Systemic and Organismic Aging**

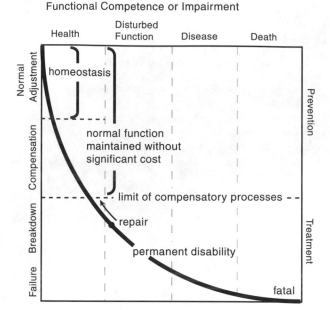

**FIGURE 10.11** Progressive stages of homeostasis from adjustment (health) to failure (death). In the healthy adult, homeostatic processes ensure adequate adjustments in response to stress, and, even for a period beyond this stage, compensatory processes are capable of maintaining overall function without serious disability. When stress is exerted beyond the compensatory capacities of the organism, disability ensues in rapidly increasing increments to severe illness, permanent disability, and death. When this model is viewed in terms of homeostatic responses to stress imposed on the aged, and to aging, a period when the body can be regarded as at the point of "limit of compensatory processes," it is evident that even minor stresses are not tolerable, and the individual moves rapidly into stages of breakdown and failure.

(lower basal levels, faster recovery after stress, and enhaced sensitivity of glucocorticoid feedback) which become manifest in adulthood.[81] The possibility is raised that "stress management" may lead to "salutary states of adaptation" by several approaches: physical (e.g., optimal physical exercise conditioning, responsible diet, Chapter 24), chemical (e.g., administration of neurotransmitter agonists and antagonists, and calcium blockers), and psychological (e.g., improvement of social network, finding outlets for frustration, and manipulation of feelings). This type of intervention is just beginning to be considered on a scientific and rational basis and remains quite controversial. Given the great human heterogeneity, it seems unrealistic to seek a uniform treatment for everyone; rather, a better approach would be a customized set of hygienic habits through a specific knowledge of our genome and our environment. I like to believe that "learned optimism" can be attained through education: *"carpe diem,"* enjoy the day-by-day life, may yield considerable practical results in the near future for delaying pathology and improving life in old age. The recognition

---

**Box 10.9**

**Neuroendocrine-Immune Responses to Stress and Aging**

The role of the neuroendocrine system in regulating metabolic and behavioral responses is discussed throughout this chapter. The function of the immune system is to control, and eventually eliminate, foreign organisms and substances, and this function is altered with aging (Chapter 15).

**Neuropeptides** are present in the immune system and mediate intraimmune communication and communication between immune and neuroendocrine systems.

**Cytokines** are present in neural and endocrine tissues and influence neuroendocrine activity. In response to stress, immune and endocrine systems can muster specific and nonspecific responses, in the case of the immune system, innate and acquired, respectively.

**Hormones**, vasopressin, TSH, ACTH, GH, PRL, **endorphins, and others control many important immune functions** (e.g., cytokine and antibody production, lymphocyte cytotoxicity and proliferation, macrophage function). Reciprocally, **cytokines directly affect neuroendocrine functions**: for example, interleukin-1 (IL-1) activates the HPA by stimulating CRH and ACTH secretion and may also act on TSH, GH, PRL, and LH release.

---

that the plasticity of our neural, endocrine, and immune responses persists well into old age, gives me, as a physiologist, the belief that we can be educated to change and ameliorate "the ways we cope."

## REFERENCES

1. Timiras, P.S., Ed., *Physiological Basis of Aging and Geriatrics,* Macmillan, New York, NY, 1988.
2. Timiras, P.S., Ed., *Physiological Basis of Aging and Geriatrics*, 2nd ed., CRC Press, Boca Raton, FL, 1994.
3. Timiras, P.S., Quay, W.D., and Vernadakis, A., Eds., *Hormones and Aging*, CRC Press, Boca Raton, FL, 1995.
4. Mobbs, C.V. and Hof, P.R., Eds., *Functional Endocrinology of Aging*, Karger, Basel, 1998.
5. Meikle, A.W., Ed., *Hormone Replacement Therapy*, Humana Press, Totowa, NJ, 1999.
6. Morley, J.E. and Van den Berg, L., Eds., *Endocrinology of Aging*, Humana Press, Totowa, NJ, 2000.
7. Shupnik, M.A., Ed., *Gene Engineering in Endocrinology*, Humana Press, Totowa, NJ, 2000.
8. Matzuk, M., Brown, C.W., and Kumar, T.R., Eds., *Transgenics in Endocrinology*, Humana Press, Totowa, NJ, 2001.

9. Meites, J., Ed., *Neuroendocrinology of Aging, Plenum Press, New York, 1983.*

10. Wise, P.M., Neuroendocrine correlates of aging, in *Neuroendocrinology in Physiology and Medicine,* Conn, P.M. and Freeman, M.E., Eds., Humana Press, Totowa, NJ, 1999.

11. Conn, P.M. and Freeman, M.E., Eds., *Neuroendocrinology in Physiology and Medicine,* Humana Press, Totowa, NJ, 2000.

12. Seeman, T.E. and Robbins, R.J., Aging and hypothalamic–pituitary–adrenal response to challenge in humans, *Endocr. Rev.,* 15, 233, 1994.

13. Margioris, A.N. and Chrousos, G.P., Eds., *Adrenal Disorders,* Humana Press, Totowa, NJ, 2001.

14. Barton, R.N. et al., Cortisol production rate and the urinary excretion of 17-hydroxycorticosteroids, free cortisol, and 6β-hydroxycortisol in healthy elderly men and women, *J. Gerontol.,* 48, M213, 1993.

15. Sapolsky, R.M., *Stress, the Aging Brain, and the Mechanisms of Neuron Death,* MIT Press, Cambridge, MA, 1992.

16. Uno, H. et al., Hippocampal damage associated with prolonged and fatal stress in primates, *J. Neurosci.,* 9, 1705, 1989.

17. Uno, H. et al., Degeneration of the hippocampal pyramidal neurons in the socially stressed tree shrew, *Soc. Neurosci.* Abstr., 17, 129, 1991.

18. Sapolsky, R. and Altmann, J., Incidences of hypercortisolism and dexamethasone resistance increase with age among wild baboons, *Biol. Psychiat.,* 30, 1008, 1991.

19. Sapolsky, R.M., *Why Zebras Don't Get Ulcers: An Updated Guide to Stress, Stress-Related Diseases, and Coping,* W.H. Freeman and Co., New York, 1998.

20. Elliott, E. and Sapolsky, R., Corticosterone impairs hippocampal neuronal calcium regulation: Possible mediating mechanisms, *Brain Res.,* 602, 84, 1993.

21. Flood, C. et al., The metabolism and secretion of aldosterone in elderly subjects, *J. Clin. Invest.,* 46, 961, 1967.

22. Migeon, C. et al., Dehydroepiandrosterone and androsterone levels in human plasma, Effect of age and sex, day to day and diurnal variation, *J. Clin. Endocrinol. Metab.,* 17, 1051, 1957.

23. Liu, C.H. et al., Marked attenuation of ultradian and circadian rhythms of dehydroepiandrosterone in postmenopausal women: Evidence for reduced 17,20-desmolase enzymatic activity, *J. Clin. Endocrinol. Metab.,* 71, 900, 1990.

24. Barrett-Connor, E., Khaw, K.T., and Yen, S.S., A prospective study of dehydroepiandrosterone sulfate, mortality, and cardiovascular disease, *N. Engl. J. Med.,* 315, 1519, 1986.

25. Slowinska-Srzednicka, J. et al., Decreased plasma levels of dehydroepiandrosterone sulphate (DHEA-S) in normolipidaemic and hyperlipoproteinaemic young men with coronary artery disease, *J. Intern. Med.,* 230, 551, 1991.

26. Shafagoj, Y. et al., Dehydroepiandrosterone prevents dexamethasone-induced hypertension in rats, *Am. J. Physiol.,* 263, E210, 1993.

27. Timiras, P.S., Neuroendocrinology of aging, retrospective, current, and prospective views, in *Neuroendocrinology of Aging,* Meites, J., Ed., Plenum Press, New York, 1983.

28. Araneo, B.A., Woods, M.L., and Daynes, R.A., Reversal of the immunosenescent phenotype by dehydroepiandrosterone: Hormone treatment provides an adjuvant effect on the immunization of aged mice with recombinant hepatitis B surface antigen, *J. Infect. Dis.,* 167, 830, 1993.

29. Clearly, M.P., The antiobesity effect of dehydroepiandrosterone in rats, *Proc. Soc. Exp. Biol. Med.,* 196, 8, 1991.

30. Svec, F. et al., The effect of DHEA given chronically to Zucker rats, *Proc. Soc. Exp. Biol. Med.,* 209, 92, 1995.

31. Nelson, J.F. et al., Neuroendocrine involvement in aging: Evidence from studies of reproductive aging and caloric restriction, *Neurobiol. Aging,* 16, 837, 1995.

32. Roth, G.S., Mechanisms of altered hormone-neurotransmitter action during aging: From receptors to calcium mobilization, *Annu. Rev. Gerontol. Geriatr.,* 10, 132, 1990.

33. Tsai, M.J. and O'Malley, B.W., Molecular mechanisms of action of steroid/thyroid receptor superfamily members, *Annu. Rev. Biochem.,* 63, 451, 1994.

34. Mangelsdorf, D.J. et al., The nuclear receptor superfamily: The second decade, *Cell,* 83, 835, 1995.

35. DeFranco, D.B. et al., Nucleocytoplasmic shuttling of steroid receptors, in *Vitamins and Hormones,* Vol. 51, Litwack, G., Ed., Academic Press, New York, 1995.

36. Smith, D.F. and Toft, D.O., Steroid receptors and their associated proteins, *Mol. Endocrinol.,* 7, 4, 1993.

37. Chandran, U.R. and DeFranco, D.B., Subnuclear trafficking of glucocorticoid receptors, in *Gene Engineering in Endocrinology,* Shupnik, M.A., Ed., Humana Press, Totowa, NJ, 2000.

38. Nordeen, S.K., Moyer, M.L., and Bona, B.J., The coupling of multiple signal transduction pathways with steroid response mechanisms, *Endocrinology,* 134, 1723, 1994.

39. Nordeen, S.K., Bona, B.J., and Moyer, M.L., Latent agonist activity of the steroid antagonist, RU486, is unmasked in cells treated with activators of protein kinase A, *Mol. Endocrinol.,* 7, 731, 1993.

40. Schreihofer, D.A., Resnick, E.M., and Shupnik, M.A., Steroid receptor actions, in *Gene Engineering in Endocrinology,* Shupnik, M.A., Ed., Humana Press, Totowa, NJ, 2000.

41. Kalimi, M., Glucocorticoid receptors: From development to aging — A review, *Mech. Ageing Dev.,* 24, 129, 1984.

42. Sharma, R. and Timiras, P.S., Changes in glucocorticoid receptors in different regions of brain of immature and mature male rats, *Biochem. Int.,* 13, 609, 1986.

43. Kitraki, E., Alexis, M.N., and Stylianopoulou, F., Glucocorticoid receptors in developing rat brain and liver, *J. Steroid Biochem.,* 20, 263, 1984.

44. Collins, K.J., The autonomic nervous system, in *Principles and Practice of Geriatric Medicine,* 3rd ed., Vol. 2, Pathy, J.M.S., Ed., John Wiley & Sons, New York, 1998.

**Systemic and Organismic Aging**

45. Insel, P.A., Adrenergic receptors — evolving concepts and clinical implications, *N. Engl. J. Med.,* 334, 580, 1996.

46. Halasz, B., The hypothalamus as an endocrine organ: The science of new endocrinology, *Neuroendocrinology in Physiology and Medicine,* Conn, P.M. and Freeman, M.E., Eds, Humana Press, Totowa, NJ, 2000.

47. Fink, G., Neuroendocrine regulation of pituitary function: General principles, *Neuroendocrinology in Physiology and Medicine,* Conn, P.M. and Freeman, M.E., Eds, Humana Press, Totowa, NJ, 2000.

48. Conn, P.M. and Bowers, C.Y., A new receptor for growth hormone-release peptide (GHRP), *Science,* 273, 923, 1996.

49. Sonntag, W.E. et al., Decreased pulsatile release of growth hormone in old male rats, *Endocrinology,* 107, 1875, 1980.

50. Prinz, P.N. et al., Plasma growth hormone during sleep in young and aged men, *J. Gerontol.,* 38, 519, 1983.

51. Papadakis, M.A. et al., Insulin-like growth factor 1 and functional status in healthy older men, *J. Am. Geriatr. Soc.,* 43, 1350, 1995.

52. Rudman, D., Growth hormone, body composition, and aging, *J. Am. Geriatr. Soc.,* 33, 800, 1985.

53. Kaplan, S.L., The newer uses of growth hormone in adults, *Adv. Intern. Med.,* 38, 287, 1993.

54. Bartke, A. et al., Genes that prolong life: Relationships of growth hormone and growth to aging and life span, *J. Gerontol.,* 56A, B340, 2001.

55. Takala, J. et al., Increased mortality associated with growth hormone treatment in critically ill adults, *N. Engl. J. Med.,* 341, 785, 1999.

56. Rolandi, E. et al., Somatostatin in the elderly: Diurnal plasma profile and secretory response to meal stimulation, *Gerontology,* 33, 296, 1987.

57. Davies, P. and Terry, R.D., Cortical somatostatin-like immunoreactivity in cases of Alzheimer's disease and senile dementia of the Alzheimer type, *Neurobiol. Aging,* 2, 9, 1981.

58. Raskind, M.A. et al., Cerebrospinal fluid, vasopressin, oxytocin, somatostatin, and beta-endorphin in Alzheimer's disease, *Arch. Gen. Psychiatry,* 43, 382, 1986.

59. Quay, W.B., Diffuse endocrines and chemical mediators, in *Hormones and Aging,* Timiras, P.S., Quay, W.B., and Vernadakis, A., Eds., CRC Press, Boca Raton, FL, 1995.

60. Tucker, H.A., Neuroendocrine regulation of lactation and milk ejection, *Neuroendocrinology in Physiology and Medicine,* Conn, P.M. and Freeman, M.E., Eds., Humana Press, Totowa, NJ, 2000.

61. Rossmanith, W.G., Szilagyi, A., and Scherbaum, W.A., Episodic thyrotropin (TSH) and prolactin (PRL) secretion during aging in postmenopausal women, *Horm. Metab. Res.,* 24, 185, 1992.

62. Zbuzek, V. and Zbuzek, V.K., Vasopressin and aging, in *Regulation of Neuroendocrine Aging,* Everitt, A.V. and Walton, J.R., Eds., Karger, Basel, 1988.

63. Insel, T.R., Oxytocin — a neuropeptide for affiliation: Evidence from behavioral, receptor autoradiographic, and comparative studies, *Psychoneuroendocrinology,* 17, 3, 1992.

64. Moore, F.L., Evolutionary precedents for behavioral actions of oxytocin and vasopressin, *Ann. N.Y. Acad. Sci.,* 652, 156, 1992.

65. Seeman, T.E. et al., Price of adaptation – allostatic load and its health consequences. MacArthur studies of successful aging, *Arch. Intern. Med.,* 157, 2259, 1997.

66. Bernard, C., *Leçons sur les phénomènes de la vie communs aux animaux et aux végétaux,* 2 Vols., J.B. Baillière et fils, Paris, 1878–1879.

67. Cannon, W.B, *The Wisdom of the Body,* W.W. Norton & Co., New York, 1932.

68. Selye, H., The Physiology and Pathology of Stress; A Treatise Based on the Concepts of the General-Adaptation-Syndrome and the Diseases of Adaptation, Acta Inc., Montreal, Canada, 1950.

69. Lamberts, S.W.J. et al., The endocrinology of aging, *Science,* 278, 419, 1997.

70. Everitt, A.V. and Burgess, J.A., *Hypothalamus Pituitary and Aging,* Charles C. Thomas, Springfield, IL, 1976.

71. Finch, C.E., Regulation of physiological changes during mammalian aging, *Quart. Rev. Biol.,* 51, 49, 1976.

72. Timiras, P.S., Biological perspectives on aging: Does a genetically programmed brain-endocrine master plan code for aging processes? *Am. Sci.,* 66, 605, 1978.

73. Wick, G., Sgonc, R., and Lechner, O., Neuroendocrine-immune disturbances in animal models with spontaneous autoimmune diseases, *Ann. N.Y. Acad. Sci.,* 840, 591, 1998.

74. Savino, W. et al., Neuroendocrine control of the thymus, *Ann. N.Y. Acad. Sci.,* 840, 470, 1998.

75. Hadden, J.W., Thymic endocrinology, *Ann. N.Y. Acad. Sci.,* 840, 352, 1998.

76. Fabris, N., Biomarkers of aging in the neuroendocrine-immune domain, *Ann. N.Y. Acad. Sci.,* 663, 335, 1992.

77. Volpe, R., Ed., *Autoimmune Endocrinopathies,* Humana Press, Totowa, NJ, 1999.

78. Pennisi, E., Neuroimmunology: Tracing molecules that make the brain-body connection, *Science,* 275, 930, 1997.

79. Clevenger, C.V. and Flanagan-Cato, L.M., Neuroendocrine immunology, *Neuroendocrinology in Physiology and Medicine,* Conn, P.M. and Freeman, M.E., Eds., Humana Press, Totowa, NJ, 2000.

80. Charlson, M.E. et al., A new method of classifying prognostic co-morbidity of longitudinal studies, development, and validation, *J. Chronic Dis.,* 40, 373, 1987.

81. Meaney, M., Aitken, D., and Sapolsky, R., Postnatal handling attenuates neuroendocrine, anatomical and cognitive dysfunctions associated with aging in female rats, *Neurobiol. Aging,* 12, 31, 1991.

# 11 The Female Reproductive System

*Phyllis M. Wise*
University of California, Davis

## CONTENTS

## I. INTRODUCTION

Women undergo the menopause at approximately 51 years of age, and the timing of this dramatic physiological change has remained essentially constant since medical records have been kept. The menopause occurs at the time of the exhaustion of the ovarian follicular reserve. Because the ovarian follicles are not only the source of germ cells, but also are the primary source of estrogens, plasma estrogen concentrations drop precipitously during the postmenopausal years and remain low for the remainder of a woman's life, unless she chooses to take hormone replacement therapy. In recent years, we have come to appreciate that estrogens are not only

0-8493-0948-4/03/$0.00+$1.50
© 2003 by CRC Press LLC

reproductive hormones, but that they are pleiotropic hormones that play roles in a wide variety of nonreproductive functions as disparate as bone and mineral metabolism, memory and cognition, cardiovascular function, and the immune system. Thus, the end of reproductive life has far-reaching implications for women, because the hypoestrogenic state brings so many physiological repercussions. With the substantial increase in the average life span of humans from approximately 50 to 80 years that has occurred during the last century (Chapter 2), and the relatively fixed age of the menopause, the number of women who will spend over one-third of their lives in the postmenopausal state has increased dramatically. It is not surprising then that an increasing number of clinical and basic science studies have focused on understanding fully the physiological changes that accompany the menopause, the mechanisms that drive reproductive aging, and the impact of these changes on women's health. It is our hope that a better understanding will be important to gerontologists, because the female reproductive system deteriorates early during the aging process, in the absence of pathological changes that often confound gerontological studies. Therefore, concepts derived from our understanding of the menopause and the aging reproductive system may apply to the process of the biology of aging of other systems.

## II. DEFINITIONS

### A. TERMINOLOGY USED TO CHARACTERIZE THE HUMAN MENOPAUSE

There are several terms that have been used to describe the period of a woman's life that surrounds the end of her reproductive life: premenopause, perimenopause, climacteric, menopause, and postmenopause. The "premenopause" period includes the years between puberty and the menopause. For the majority of women, the production of reproductive hormones during this time is cyclic and predictable, and the interval between their menstrual periods is fairly regular. The "perimenopause," sometimes called the "climacteric," begins before the menopause. It is a more encompassing term

that is defined as the interval of the entire transition from reproductive to the postreproductive period of a woman's life.[1,2] Symptoms we associate with the menopause, such as hot flashes, and irregular menstrual cycles may start to appear. The average perimenopause lasts for 4 years. It continues, by definition, through the 12 months following the last menstrual period. The "menopause" is the permanent cessation of menstruation. Spontaneous menopause occurs in women at approximately 51 years of age and is associated with the depletion of the ovarian follicular reserve. The term "surgical menopause" refers to cessation of menstrual cyclicity that results from the removal of a woman's ovaries when she would not normally undergo reproductive aging. Although the ultimate result of normal menopause and surgical menopause is the same (i.e., cessation of menstrual cycles and infertility), the repercussions on a woman's health may be different because the age of the normal menopause is considerably greater than that of surgical menopause. Hence, changes in reproductive hormones in an older individual may result in a broader spectrum of effects than when they occur in a younger woman who is experiencing fewer age-related changes in other systems. The definitions that are recommended by the World Health Organization are summarized in Table 11.1.

### B. "MENOPAUSE" IN SPECIES OTHER THAN HUMANS

It was originally thought that the menopause was restricted to human females. However, an increasing body of work[3–6] shows that several species of nonhuman primates undergo a process similar to that which women experience across the menopause. However, these changes occur at a more advanced age relative to their total life span; therefore, the postmenopausal period is considerably shorter in these nonhuman primates species compared to women. At the present time, it is unknown whether this is due to the fact that the menopause is truly delayed in nonhuman primates compared to humans or whether the primate species that have been studied have not been maintained under optimal environmental conditions, and therefore, their postmenopausal life span can be extended. The populations of non-

---

**TABLE 11.1**
**Definitions**

| Terminology | Definition |
| --- | --- |
| Menopause | Permanent cessation of menstruation associated with loss of ovarian follicular activity |
| Perimenopause or climacteric | Period immediately prior to and at least 1 year after the menopause; characterized by physiological and clinical features of altered ovarian function |
| Postmenopause | Period of life remaining after the menopause |
| Premenopause | The reproductive period prior to the menopause |

human primates that are available for study are small and limit our ability to use them to better understand the human menopause. However, there are distinct advantages to using these species as models: longitudinal characterization from a population of animals that have been followed through their reproductive life span should be possible; intensive monitoring of hormonal changes through urinary samples is feasible; and records of reproductive history, in terms of numbers of pregnancies and live young, can be obtained. In addition, it is possible to perform longitudinal studies that are invasive and sometimes terminal in these species. Together, these considerations mean that their use will allow us to probe the underlying mechanisms that drive the menopausal transition. Such studies are expensive and labor intensive, however, because nonhuman primate colonies have been maintained in captivity at several research centers, such studies will provide new data in the next several years. It is clear that the hormonal characteristics of the pubertal transition and the adult reproductive years in several nonhuman primate species are similar, but not identical to those found in human females. This has led investigators to speculate that similarities may also exist during the peri- and postmenopausal periods. In fact, several studies establish that similarities exist between nonhuman primates and women during female reproductive aging.

The studies that have been performed to date suggest that the hormonal dynamics during the transition from regular to irregular menstrual cycles are similar in older female rhesus monkeys and women. Variable intermenstrual intervals and delay of ovulation interspersed with breakthrough uterine bleeding were found in perimenopausal rhesus monkeys[4] and women in the fifth decade.[7] However, not all of the hallmarks of the human menopause punctuate the menopausal transition in the rhesus monkey. Importantly, the harbinger of impending reproductive decline, a selective rise in FSH concentrations in the absence of any change in LH levels, does not appear to occur in monkeys approaching the menopause, prior to overt changes in menstrual cycle length. In addition, rhesus monkeys do not exhibit frequent periods of high, unopposed estrogen in association with the transition to the menopausal state, as has been observed in women.[4,8] Studies have been initiated to determine whether the baboon may be a better nonhuman primate model of the human menopause; however, aging colonies of this species are even less common than rhesus monkeys. Considerably more work must be performed before we will know whether any other animal species can be used to model the human menopause.

The origin of the menopause and its potential role in longevity are topics of active discussion. In the wild state, the postreproductive period is theoretically irrelevant to natural selection. Consequently, selection against genes that have potentially deleterious effects later in life will be weak. The life span of a species depends upon a balance between positive selection pressures, which favor long life, and negative factors, which cause a decline in reproductive function in later life. The evolutionary value of an individual depends upon its ability to produce the maximum number of surviving offspring; therefore, selective pressures favor survival of young reproductively active individuals. According to this argument, the menopause may have evolved because of one or more evolutionary pressures that exist during our younger life: (1) women have survived to an age where natural selection is no longer favored as the maintenance of reproductive capacity, or alternatively, (2) the menopause may be a pleiotropic effect of a gene or gene cluster that has selective value earlier in life.

The contrasting argument suggests that the menopause may actually carry an advantage to the female. As early hominids increased population growth, the greater number of children required more parental care and protection.[9] In the hunter–gatherer hominids, this would have required the presence of nonreproductive group members. Selection in older females favored cessation of reproduction, and these nonreproductive women became the "surrogate" mothers. Interestingly, recent work suggests that the advantage for humans to have a longer life span may be a by-product of longevity-enabling genes that maximize the length of time during which women can bear children, and during which they can increase survival probabilities of their children and grandchildren.[10] Counterarguments to this theory would state that the costs of the menopause, i.e., increased mortality and morbidity of women and their offspring in the period of declining reproduction, would far outweigh any benefits that might accrue from grandmother behavior: "the menopause is an artifact of human civilization which emerged when our growing mastery of the environment increased our survivorship."[11]

## C. Age of the Menopause

The World Organization and the U.S. Census Bureau figures show that the average age of the menopause is 51 years of age. However, analytical methods may interfere with accurate determination of the age of the menopause. Serious sampling errors may occur in a population study, and two populations may not be comparable on the basis of age and birth cohort. Random sampling is essential. Hospital patient populations provide a biased sample, because they are already self-selected on the basis of illness. Retrospective methods, in which investigators ask women to recall the date of their last menstrual period, can be biased by memory error, understating age at menopause, recalling age at last birthday in menopause year rather than actual date of last menstrual flow, and the tendency to round up age to the nearest 5 years, resulting

**Systemic and Organismic Aging**

in an artificially high incidence of the menopause at ages 40, 45, and 50.

A woman's reproductive history (number of pregnancies, use of steroidal contraceptives to block follicular development, and ovulation), race, nutritional state, marital status, age at parity, and smoking do not greatly influence the age at which she reaches the menopause. It is thought that these factors do not affect the ongoing rate of ovarian follicular reawakening, development, differentiation, maturation and selection of the dominant follicle, and atresia. Therefore, these factors do not ultimately influence the rate of follicular loss from the endowment of primordial follicles that a woman is born with. Because the exhaustion of the follicular pool is the final determining factor of the menopause, these factors do not have dramatic effects on the age of the menopause.

## III. THE ENDOCRINOLOGY OF THE PERIMENOPAUSE AND POSTMENOPAUSE

There are numerous studies that have quantified hormone concentrations and their pulsatile (ultradian), diurnal, and monthly patterns of secretion during the perimenopausal and postmenopausal period. Clearly, by the time that the postmenopausal period is established, multiple hormones exhibit dramatic changes in concentration. Estrogens decrease dramatically, and androgens decrease to a lesser magnitude (Chapter 12).[12] In similar fashion, LH and FSH concentrations are higher because of the decrease in negative feedback from ovarian steroids and peptides (Figure 11.1). Other pituitary hormones, such as adrenocorticotropin (ACTH), growth hormone (GH), thyroid stimulating hormone (TSH), and

prolactin show age-related changes, but the timing of their changes does not correlate with the timing of the perimenopause or establishment of the postmenopause. Therefore, it is thought that they are related to aging but do not cause and are not a repercussion of the menopause or the endocrinological changes that occur in the reproductive axis. Some structural and physiologic characteristics of the adult female reproductive system are summarized in Box 11.1.

### A. Change in Reproductive Hormone Patterns

### 1. Gonadotropins

Numerous investigations of reproductive aging have revealed a subtle rise in the concentrations of FSH unaccompanied by a rise in LH that becomes evident before any overt changes in menstrual cycle length.[13,14] This sentinel change is considered a predictor of impending reproductive decline and the marker that irregularity in menstrual cycle length will soon appear. This selective rise in FSH is evident during the entire menstrual cycle in older ovulatory women and is unaccompanied by any change in mean circulating LH concentrations (Figure 11.2).[8,13] Hence, it has been called a "monotropic" FSH increase. This early monotropic rise in FSH is not caused by any loss in bioactivity of the FSH molecule, as there is no difference observed in the bioactive:immunoactive FSH ratio between women in their early twenties compared to those in their early forties.[13] As reproductive aging progresses, LH levels also increase, and changes in the pattern of LH secretion have been documented prior to the onset of the perimenopause.[8]

The pulsatile pattern of LH secretion has been monitored by several investigators with variable and contradic-

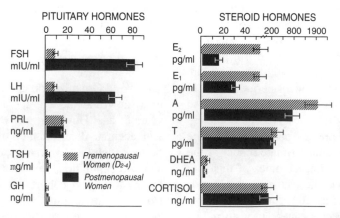

**FIGURE 11.1** Plasma hormone concentrations in premenopausal (days 2 to 4, first week of menstrual cycle) and in postmenopausal women. FSH, follicle-stimulating hormone; LH, luteinizing hormone; PRL, prolactin; TSH, thyroid stimulating hormone; GH, growth hormone; E2 and E1, estrogens; A, androstenedione; T, testosterone; DHEA, dehydroepiandrosterone; and cortisol. Barograms represent mean values, and the bracketed lines represent the standard error of the mean. (From Yen, S.S.C., *Reproductive Endocrinology. Physiology, Pathophysiology and Clinical Management*, Yen, S.S. C., Jaffe, R.B. and Barbieri, R.L., Eds., W.B. Saunders Company, Philadelphia, 1999. With permission.).

---

**Box 11.1**

**Structural and Physiologic Characteristics of the Female Reproductive System**

**The major components of the female reproductive tract** are the two ovaries, the two oviducts (or fallopian tubes), the uterus, the cervix, and the vagina. The ovaries, located in the pelvic cavity, are small (in humans, walnut-sized), oval structures containing the germinal cells, the ova, and the endocrine cells that secrete the two major steroid hormones, estrogen(s) and progesterone. Estrogens are secreted from granulosa and thecal cells lining the follicles and by cells of the corpus luteum; these latter cells are formed at the site of the ruptured follicle at ovulation and produce estrogens and progesterone. Thecal cells also secrete weak androgens. The ovaries lie on either side of the uterus and are covered by the fimbriae, the fringed ends of the oviducts, the tubes that lead to the uterus and in which fertilization occurs. The uterus serves as a gestation sac for the developing embryo and fetus, and the vagina as the receptive organ during intercourse and the birth canal at parturition. The ovary is the primary sex organ or gonad, and the other structures are secondary sex organs, which also include the breast or mammary gland and the external genitalia (vulva). Dependent on the ovarian steroid hormones are the secondary sex characteristics, which include hair distribution, voice pitch, adipose tissue distribution, stature, muscle development, and so forth.

**The major hormones regulating sexual and reproductive function** are operative at four different levels, which are all affected by aging:

- In the brain and, particularly, the hypothalamus, the gonadotropin-releasing hormone, GnRH, is a polypeptide secreted into the portal blood vessels that carry it to the anterior pituitary, where it stimulates the synthesis and release of the two gonadotropins, follicle-stimulating hormone (FSH) and luteinizing hormone (LH). The midbrain, the limbic system (specifically outputs from the amygdala), sustains ovulation and an increase in LH secretion.
- In the anterior pituitary, the two glycoproteins FSH and LH are synthesized and released in response to stimulation by the GnRH and are also stimulated or inhibited by the positive and negative feedback from the plasma ovarian hormones.
- In the ovary, the major hormones are steroids, estrogens (in humans, the most potent estrogen is 17-ß estradiol followed in decreasing order of potency by estrone and estriol), progesterone, and androgens (e.g., testosterone, in very small amounts). Androgens, especially weak androgens, i.e., dehydroepiandrosterone (DHEA), are synthesized by the ovary; this is especially true in the postmenopausal ovary. During pregnancy, the ovary also secretes the polypeptide relaxin.
- In the periphery, ovarian steroid hormones are bound to plasma proteins, metabolized in the liver, and excreted in the urine. At the target cells, estrogens and progesterone follow the same mechanism of action as the other steroid hormones (from the adrenal cortex) and bind to the nuclear receptors and stimulate the RNA and protein synthesis responsible for the many actions of these hormones.

In the female, but not in the male, the reproductive function shows cyclic changes viewed as periodic preparations for fertilization and pregnancy and is associated with underlying cyclic endocrine and behavioral changes. The reproductive cycle of varying length in lower animals is called the estrous cycle; in primates, the reproductive cycle is called the menstrual cycle with ovulation at midcycle. Cyclicity of normal functions is also discussed in Chapters 10 and 12.

---

tory results. On the one hand, Matt et al.[15] showed that the duration of LH pulses increases and the frequency of pulses decreases during the mid-follicular phase of the menstrual cycle of regularly cycling middle-aged women before any discernible alterations in their menstrual cycles (Figure 11.3). On the other hand, Reame and colleagues [16] observed an enhanced LH pulse amplitude in the late luteal phase and a subtle increase in LH pulse frequency in both the early follicular and late luteal phases in older (40 to 50 years) compared to the youngest (19 to 35) age groups, despite no differences in estradiol or progesterone secretion. Finally, other laboratories have failed to detect

any difference in the LH pulse frequency or amplitude, which would eliminate a functional change in the GnRH pulse generator as an explanation for the monotropic FSH rise in older reproductive-aged women. To our knowledge, the work of Matt et al.[15] is the only study performed in regularly cycling premenopausal women. None of the other studies in humans have controlled for changing cycle length during the perimenopausal period when they monitored LH pulses, and this may underlie the discrepancy among studies in humans.

Once the perimenopause begins and cyclic irregularity has begun, decreasing responsiveness to the positive feed-

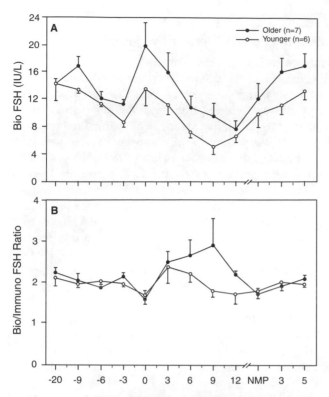

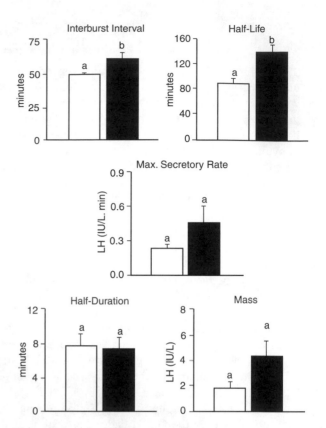

**FIGURE 11.2** Values for FSH levels normalized to the day of the LH surge in older (mean age 42.8 years) and younger (mean age 22.6 years) women. (a) Bioactive FSH concentrations across the menstrual cycle relative to the LH surge (day 0). (b) Bioactive/immunoreactive FSH ratio across the menstrual cycle according to the day of the cycle relative to the LH surge. No differences between older and younger women were observed. Mean values are represented by dots, and bracketed lines indicate standard error of the mean. (From Klein, N.A. et al., *J. Clin. Endocrinol. Metab.*, 81, 1038, 1996. With permission.)

back effects of estradiol becomes apparent: the hypothalamus/pituitary axis becomes less capable of mounting an LH surge, despite normal follicular phase estradiol secretion.[14,17] Van Look and colleagues[17] showed that estradiol was able to induce LH surges of attenuated amplitude in only a small portion of the women studied. The results of these studies suggest that during the early stages of the perimenopausal transition, prior to any decrease in circulating estradiol, decreased responsiveness to estrogen may lead to attenuated preovulatory gonadotropin surges that may lead to an inadequate secretion of progesterone during the luteal phase.

## 2. Ovarian Steroids and Peptides

Women who exhibit regular menstrual cycles, continue to have normal circulating concentrations of estradiol and progesterone after the age of 40 years and the onset of the monotropic FSH rise.[13,16] In fact, as older ovulatory women near the end of this phase of regular ovulation, preovulatory estradiol levels tend to be elevated and

**FIGURE 11.3** Comparison of LH intersecretory burst interval, estimated half-life, burst maximal secretory rate, burst half duration, and burst mass between young and middle-aged women. Deconvolution analysis procedure was utilized from concentration–time series obtained from 8 h of sampling during the mid to late follicular phase of the menstrual cycle in younger (open bars) and middle-aged (solid bars) women. Barograms represent mean values, and bracketed lines indicate standard error of the mean: (a) no statistical significance and (b) statistically significant difference between young and middle-aged women. Middle-age women showed a significantly prolonged LH intersecretory burst interval and estimated half-life compared to younger women. Significant differences were not found for burst maximum secretory rate, half-duration, or mass. (From Matt, D.W. et al., *Am. J. Obstet. Gynecol.*, 178, 504, 1998. With permission.)

to rise earlier in the menstrual cycle than in younger women.[8] Santoro and colleagues measured daily urinary steroid metabolites, estrone conjugates, and pregnanediol glucuronide, in premenopausal women 43 years old and older and compared these hormones to women between 19 to 38 years of age. They found that the older women had elevated follicular and luteal phase estrone excretion (Figure 11.4). Because FSH levels were also elevated in these women, it is possible that the change in this hormone led to an earlier selection and development of the dominant follicle that is destined to ovulate, which, in turn, may lead to relative increases in early follicular phase estradiol secretion.[13] In fact, elevated estradiol during the early follicular phase rise (i.e., day

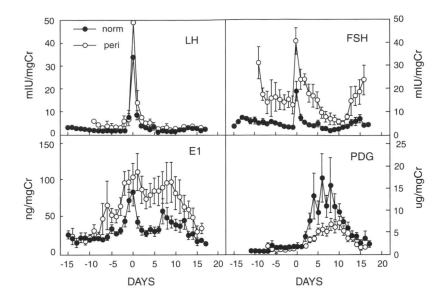

**FIGURE 11.4** Daily urinary gonadotropin and sex steroid excretion patterns in perimenopausal women aged 43 to 52 years (¡) compared to those in mid-reproductive women aged 19 to 38 years (λ). Urinary FSH and estrone metabolites (E1) are significantly higher in perimenopausal women compared to young controls. Data are standardized to day 0, the putative day of ovulation, and expressed as mean ± standard error (bracketed lines). Urinary hormonal excretion was measured relative to Cr (creatinine excretion), LH (luteinizing hormone), FSH (follicle-stimulating hormone), E1 (estrogens), and PDG (pregnanediol glucuronide). (From Santoro, N. et al., *J. Clin. Endocrinol. Metab.*, 81, 1495, 1996. With permission.)

3 of the menstrual cycle) predicts a poor response to controlled ovarian hyperstimulation, a treatment for infertility; and this could contribute to decreasing fertility and fecundity in women of this age.[18]

The endocrinology of the perimenopause has been studied in several longitudinal studies that each involves relatively few women who have been willing to provide frequent blood or urine samples over a prolonged interval of time.[8,14] The results show that, as the perimenopausal period becomes established, hormone patterns become highly erratic with unpredictable relationships between gonadotropins, steroids, the incidence of ovulation, and menstrual bleeding (Figure 11.5). Often, there are periods of hypoestrogenemia, similar to that observed in postmenopausal women, which are punctuated with ovulatory and anovulatory episodes of estrogen secretion. Furthermore, as women traverse the perimenopausal period, the luteal phase becomes shortened, and luteal phase progesterone levels become inadequate to maintain a normal corpus luteum.[8] Plasma concentrations of testosterone, dehydroepiandrosterone, and its sulfated form decline with advancing age.[12] Because this decrease occurs in both women and men, changes in these steroids are thought to be only peripherally related to the menopause.

The ovary synthesizes several peptides, including inhibin, activin, and follistatin. Inhibin is synthesized by granulosa cells of the ovarian follicle and selectively inhibits FSH through direct actions on the pituitary gland. It is a heterodimeric protein consisting of an α- and β-polypeptide chain connected by disulfide bonds. Two forms of inhibin (A and B) exist: the α-subunits are identical, and the β-subunits are different in inhibin A and inhibin B. Activin, which selectively stimulates FSH secretion, is a combination of the β-subunits of inhibin among themselves via disulfide bond linkages. It is now clear that there are three forms of activins resulting from the differential dimerization of β-subunits. Follistatin is another group of proteins that are made by the ovary, which inhibits FSH by binding to activins with high affinity and, thus, neutralizes their biological activity. Of these factors, only inhibin has been studied within the context of aging and the menopause. Initial studies that used nonspecific assays, which did not differentiate between the α- and β-subunits of inhibin, yielded contradictory data. More modern, new two-site specific assays for dimeric inhibin A and B that have recently been developed, demonstrate that these two hormones exhibit two very distinct and different patterns of secretion across normal menstrual cycles. With the use of these specific assays for dimeric inhibin A and B, it has become clear that total immunoreactive inhibin decreases during the perimenopausal and postmenopausal period and becomes undetectable after menopause.[19] In addition, inhibin B decreases during the early follicular phase secretion in older women concomitant with the monotropic FSH rise (Figure 11.6).[20] These data have led to the conclusion that the decrease in the number of primordial and early antral follicles remaining in the ovaries of older women leads to decreased inhibin B concentration, which, in turn, leads to the initial selective increase in FSH concentrations. As discussed later in

**Systemic and Organismic Aging**

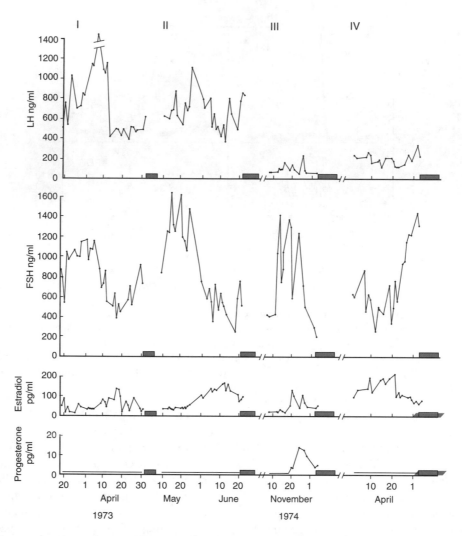

**FIGURE 11.5** Daily concentrations of serum LH (luteinizing hormone), FSH (follicle-stimulating hormone), estradiol, and progesterone during four cycles in one 49-year-old subject during the menopausal transition. Hormone levels are arrayed by calendar date, and the hatched area indicates menstruation (mean ± S.E.). During the menopausal transition, hormone patterns show discordant regulation relative to the normal menstrual hormone relationship: FSH concentrations are elevated and not inhibited by high estrogen levels, whereas LH levels are in the normal or low range (From Sherman, B.M., West, J.H., and Korenman, S.G., *J. Clin. Endocrinol. Metab.*, 42, 629, 1976. With permission.)

this chapter, changes in the pattern of FSH secretion could result from changes in the pattern of GnRH secretion. However, whether this underlies the change in FSH in women is virtually impossible to test, because, to date, assays are not sufficiently sensitive to detect GnRH in peripheral plasma. Whether changes in GnRH secretion are important during aging will have to rely on studies performed in nonhuman primates and other animal models of the menopause.

## B. REPRODUCTIVE CYCLICITY AND DECREASING FERTILITY AND FECUNDITY

As discussed above, the endocrinology of the perimenopausal period is marked by instability. The reproductive hormone profiles of the menstrual cycles immediately preceding the menopause exhibit a great deal of variability and lack predictability. However, women appear to maintain ovulatory cycles for several years after endocrinologic signs of reproductive aging first appear, usually for several years after the onset of the marked decline in fertility. As women age, the average length of the menstrual cycle becomes more and more variable. Data suggest that the initial change may be a lengthening of the cycle; however, rapidly, cycles become shortened. The most thorough studies were performed by Treloar and colleagues.[21,22] They analyzed the menstrual records of 25,825 woman years in over 2700 women and demonstrated that there is a progressive, age-related shortening of the menstrual cycle from a median of 28 days at 20 years of age to 26 days at 40 years of age. Prior to menopause when women are at a median age 45.5

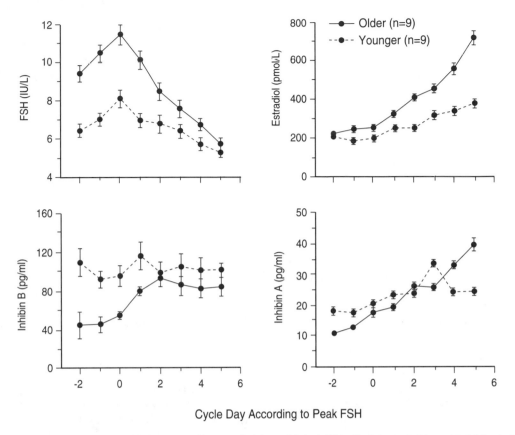

**FIGURE 11.6** Plasma concentrations of FSH (top left), estradiol (top right), inhibin B (bottom left), and inhibin A (bottom right) normalized to the day of maximal FSH secretion in older (solid lines) and younger (dashed lines) women. In older women compared to younger women, FSH (follicle-stimulating hormone) concentrations are higher. whereas inhibin B concentrations are lower. Values are mean values, and bracketed lines represent the standard error of the mean. (From Klein, N.A. et al., *J. Clin. Endocrinol. Metab.*, 81, 2742, 1996. With permission.)

years, cycle irregularity becomes more and more common and exhibits an overall increase in menstrual cycle length. Endocrine studies of steroid and gonadotropin secretion across the menstrual cycle demonstrate that the cycle shortens due to a progressive shortening of the length of the follicular phase without a significant change in length of the luteal phase.[13,14] The decrease in cycle length continues until the onset of oligomenorrhea when both average cycle length and the standard deviation increase.[21] Studies using ultrasound of normal older women, 40 to 45 years of age who continue to ovulate confirm that the shortened follicular phase occurs concomitantly with development and ovulation of an apparently normal follicle.[13,23] Metcalf demonstrated that ovulatory frequency remains high (95%) in menstruating women aged 40 to 55 years until the onset of oligomenorrhea, when the percentage of ovulatory cycles drops significantly.[24] During this time, fertility decreases, because the ovulated ova may be less fertilizable. Longitudinal studies of urinary excretion of progesterone metabolites indicate that, during this period of oligomenorrhea, ovulations occur sporadically. Sometimes luteal phase progesterone is normal; however, sometimes

progesterone levels are low,[14] perhaps resulting in less than optimal uterine function and lowered success in implantation. Some cycles show evidence of follicular phase estrogen secretion that is not followed by subsequent progesterone secretion. Whether this results from decreased responsiveness of the hypothalamus and pituitary to the positive feedback effects of estradiol or inadequate follicular development is unclear.

The fecundity of a couple can be predicted by the age of the female more than the male.[25,26] Women reach a peak in fecundity at approximately 24 years of age, and this declines gradually thereafter with a rapid decline beginning after age 35 years. Studies of donor insemination recipients (which control for age and fertility of the male partner and coital frequency) also reveal that cumulative pregnancy rates significantly decrease as women age:[27,28] women who were artificially inseminated with donor sperm exhibited significantly reduced cumulative conception rates after the age of 30 years (Figure 11.7).[28] In addition, assisted reproductive technologies, such as *in vitro* fertilization (IVF), are much less successful in women who are older than 40 years. This decline in IVF success goes beyond the ovulatory response, because only

**Systemic and Organismic Aging**

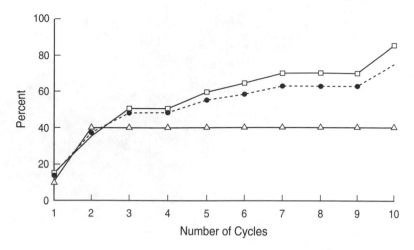

**FIGURE 11.7** Cumulative conception rates in women with no detectable infertility factors with laparoscopic confirmation. $\lambda$ = all ages; ¨ = 19 to 34 years; $\Delta$ = 35 to 45 years. (From Stovall, D.W. et al., *Obstet. Gynecol.*, 77, 33, 1991. With permission.)

women with adequate follicular development who progressed to oocyte recovery were considered: implantation rates decline as a function of age, the risk of spontaneous abortion increases, and birth defects increase. These changes may be related to the age of the oocyte (discussed below). Therefore, infertility in older women is much more complex than simple ovulatory inefficiency.

## C. CHANGES IN STRUCTURE AND OVARIAN FUNCTION

Clinical evidence suggests that the predominant effect of age on fertility is due to abnormalities present in the older oocyte. Oocytes from normal younger and older reproductive-aged women, examined at the second metaphase of meiosis, exhibited distinct differences in the placement of microtubules and chromosomes.[29,30] Immunofluorescent staining and confocal microscopic methods have revealed that the meiotic spindle has chromosomes displaced from the metaphase plate in the majority of oocytes from older women (40 to 45 years) compared to younger women (20 to 25 years), in whom the majority of oocytes exhibited chromosomes that were completely aligned within the metaphase plate.[29] In addition, karyotypes of unfertilized oocytes reveal a higher rate of single chromatid abnormalities in oocytes from older infertile women.[31] It is possible that abnormalities in the meiotic spindle account for the higher rate of aneuploidy observed in preimplantation embryos[32] and in live offspring from older reproductive-aged women. The most compelling evidence for an effect of the aging oocyte on female fertility comes from clinical studies of donor oocyte in IVF programs. Pregnancy and delivery rates are much more strongly correlated with age of the donor than age of the recipient.[25]

In women, a large body of evidence points to the conclusion that declining ovarian follicular reserve is the major factor that underlies the physiologic basis of the perimenopause and menopause. Thus, ultimately, the permanent cessation of menstrual cyclicity can be largely attributed to changes within the ovaries. From the original endowment of over 2 million primordial follicles, which is set during fetal development, the number gradually declines until relatively few, poorly responsive follicles remain at the time of the menopause. In women, a rather constant percentage of follicles continually matures from primordial to secondary follicles, from fetal life through childhood into the reproductive years and ending at the menopause. These very early stages of follicular development appear to be independent of outside hormonal influences, such as gonadotropin levels or patterns of secretion or intraovarian concentrations of steroids or ovarian peptides. However, later stages of follicular development and differentiation are hormone dependent, and if the proper amount, pattern of secretion, or sequence of hormonal events does not occur, the follicle becomes atretic. Throughout a woman's lifetime (until menopause), groups of primordial follicles will spontaneously develop into primary follicles; however, the great majority of primary follicles undergo atresia. Only during the reproductive years are there sufficient levels of LH and FSH to save an occasional immature follicle and nurture its development into a dominant follicle that eventually ovulates. Current knowledge regarding follicle depletion is based on very few but thorough studies.[33–35] Together, these studies report morphometric estimates of primordial follicle number from females from 6 to 55 years of age. Mathematical modeling has been used on this combined database to estimate the number of follicles at birth and to hypothesize that follicular depletion is biexponential, with an accelerated loss occurring at age 38 years (Figure 11.8). It is the dominant, mature follicle, present only during the reproductive years, that produces the majority of steroid hormones that are noticeably deficient after the menopause.

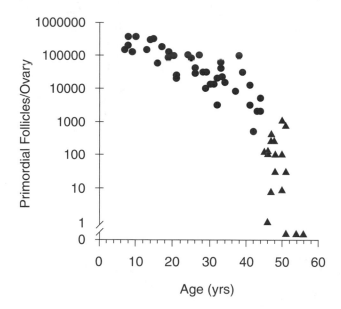

**FIGURE 11.8** Age-related decrease in the total number of primordial follicles within both human ovaries from birth to the menopause. As a result of recruitment, the number of follicles decrease in a log linear fashion until approximately 37 years of age (●), then the rate of decrease in the follicular pool accelerates (▲), such that there are virtually no follicles left in the ovary at the time of the menopause. (From Richardson, S.J., Senikas, V., and Nelson, J.F., *J. Clin. Endocrinol. Metab.*, 65, 1231, 1987. With permission.)

## IV.  SYMPTOMS OF THE MENOPAUSE AND REPERCUSSIONS OF LACK OF OVARIAN STEROIDS

### A.  HOT FLASHES

Hot flashes are the most common symptom of the perimenopausal and early postmenopausal years. Recent data suggest that hot flashes occur in 70 to 80% of women in the United States, with a median age of onset of approximately 51 years of age.[36] The prevalence of hot flashes in young oophorectomized (surgically menopausal) women is even higher. Studies of the risk factors have found few strong correlations, although thinner women who smoke seem to report a higher incidence than heavier women who do not. Cultural factors affect the reporting of hot flashes: Asian women report hot flashes at rates of only 10 to 25%. Whether this is due to a genuinely lower frequency that results from genetics or diet, or whether non-Western cultures are acculturated not to report them is unknown.

The hot flash can vary in intensity from the occasional, transient sensation of warmth to periodic episodes of heat that lead to disturbed sleep, fatigue, and irritability. At its most intense, a hot flash is characterized by a sudden sensation of intense warmth or intense heat that spreads over the body, particularly on the face, neck, and

chest, drenching perspiration, and tachycardia. The symptoms are similar to a heat-dissipation response, when it includes sweating and peripheral vasodilation, and these symptoms are frequently followed by chills. Kronenberg[37] reported that of 506 women reporting symptoms, 87% reported daily hot flashes, and of these, one third reported more than one occurrence per day. The duration of a flash is generally 1 to 5 min, however, some can last several minutes.

There are several physiological characteristics of the hot flash. Sweating and increased skin conductance, an electrical measure of sweating, are considered the measures that define a hot flash. This index increases at a time designated as "0" time and remains elevated for several minutes (Figure 11.9). Until recently, core body temperature was thought not to change during hot flashes. However, Freedman[38] used radiotelemetry capsules to more sensitively measure core body temperature and found that significant increases could be detected 15 min prior to sweating. Respiratory exchange ratio ($CO_2$ production/$O_2$ consumption), an estimate of metabolic rate, increases just prior to the onset of sweating. As well, skin temperature increased at about the same time due to peripheral vasodilation. Freedman[38] also found that central noradrenergic activation, measured by plasma 3-methoxy-4-hydroxyphenylglycol, increased significantly. A central origin of flushing is suggested by the concordance of hot flashes with pulses in LH secretion (Figure 11.10).[39] It was thought that this concordance could result if GnRH were responsible for both hot flashes and LH pulses, however, this was proven not to be the case because administration of a GnRH agonist that blocked LH pulses did not affect the occurrence of hot flashes.[40] Furthermore, administration of clonidine, an α2-adrenergic receptor agonist, reduces noradrenergic activation and hot flashes;[41] whereas, yohimbine, an α2-adrenergic receptor antagonist, increases central noradrenergic activation and the incidence of hot flashes.[42] Together, these data support the hypothesis that the central noradrenergic system is involved in initiating hot flashes and are consistent with the concept that increases in brain norepinephrine lead to vasodilation and hot flashes. It has been suggested that the thermoneutral zone, within which sweating, peripheral vasodilation, and shivering do not occur, is virtually nonexistent in symptomatic women but is normal (about 0.4°C) in asymptomatic women. The results suggest that small temperature elevations preceding hot flashes acting within a reduced thermoneutral zone constitute the triggering mechanism in symptomatic women. Central sympathetic activation is also elevated in symptomatic women, which has been shown to reduce the thermoneutral zone in animal studies. Clonidine reduces central sympathetic activation, widens the thermoneutral zone, and ameliorates hot flashes (Figure 11.11).

**Systemic and Organismic Aging**

Although hot flashes clearly accompany the decline in ovarian estrogen at menopause, and this clearly plays an important role, decreased levels of estrogen alone are not responsible for hot flashes because levels do not differ between symptomatic and asymptomatic women.[43] Further, clonidine reduces the frequency of hot flashes without influencing estrogen levels. Nevertheless, estrogen replacement therapy (ERT) virtually eliminates hot flashes through unknown mechanisms.

## B. Effects on Organs of the Reproductive Tract and Other Target Organs

Other target organs, such as the vagina, cervix, uterus, and oviducts, undergo atrophy after the menopause. The vagina undergoes significant atrophy of the epithelial cell layer, blood flow, and secretions, leading to overall thinning and greater vulnerability to infection. Dryness can also result in painful sexual intercourse and decreased libido. The uterus loses weight and volume, decreases in the thickness of the endometrial lining, and undergoes atrophy of the epithelium and glands; however, this is not thought to bring about a vulnerability to diseases. The ovaries also undergo atrophy and become more fibrotic as the number of follicles diminishes at an increasing rate after the age of 35. The stromal tissue becomes hyperplasic, probably due to chronic stimulation by elevated gonadotropins. Postmenopausally, the surface of the ovary becomes convoluted with deep clefts and cysts that often form at the surface of the epithelium. Degenerating corpora lutea can be seen as scar tissue composed of collagen fibers.

The skin, bladder, and hair also appear to contain estrogen receptors and undergo changes after the menopause. Sebaceous glands of the skin atrophy, and together with a thinning of the epidermis, leads to increased sensitivity to temperature, humidity, and trauma. There is a reduction in collagen synthesis with decreased estrogen, leading to a decrease in collagen content of approximately 2% per year. Muscle tone in the bladder decreases, which can lead to urinary incontinence and increased urinary frequency and nocturia. Body hair frequently undergoes redistribution with changes in hormone balance, resulting in increased growth and stiffer and darker hair on the face, while hair in the armpits and pubic region becoming thinner.

## V. EFFECTS OF THE MENOPAUSE ON NONREPRODUCTIVE TARGETS

### A. Skeletal System

Osteoporosis is a major cause of morbidity and mortality in the elderly (Chapter 21).[44] It is defined as a state of reduced skeletal mass accompanied by microarchitec-

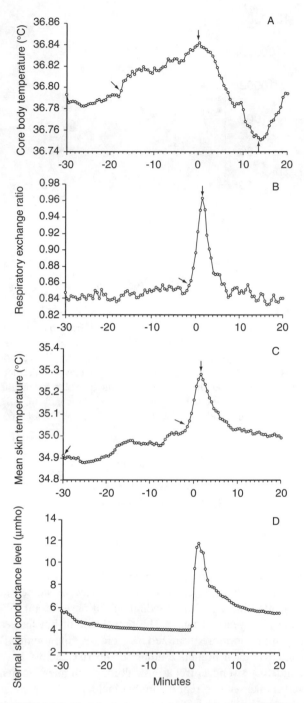

**FIGURE 11.9** Core body temperature (a), respiratory exchange ratio (b), skin temperature (c), and sternal skin conductance (d) during a hot flash. Time 0 is the beginning of the sternal skin conductance response. Intervals between the arrows are significantly different. (From Freedman, R.R., *Fertil. Steril.*, 70, 332, 1998. With permission.)[38]).

tural deterioration of the skeleton with a consequent increase in skeletal fragility, leading to an increased risk of fractures. Increased bone fragility and increased risk of fractures affects predominantly the hip, spine, and wrist. However, fractures can occur in almost any bone,

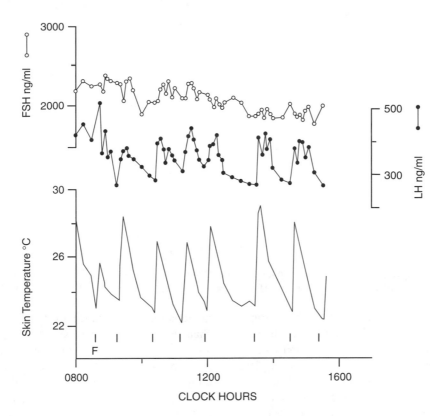

**FIGURE 11.10** Changes in skin finger temperature, serum LH (●), and FSH (○) levels during an 8-h study in postmenopausal women. Note the close temporal relationship between each subjective hot flash (indicated by the vertical bars), cutaneous temperature elevation, and pulsatile LH (luteinizing hormone) release. FSH (follicle-stimulating hormone) levels did not show a consistent pattern of change in relationship to hot flashes. (From Tataryn, I.V. et al., *J. Clin. Endocrinol. Metab.*, 49, 152, 1979. With permission.)

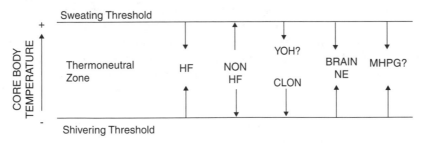

**FIGURE 11.11** Schematic showing factors that influence the "thermoneutral zone," leading to hot flashes. Elevated brain norepinephrine (NE) in animals reduces this zone. Yohimbine (YOH) elevates brain norepinephrine and should narrow the zone. Conversely, clonidine (CLON) should widen it. HF = women that exhibit hot flashes; Non-HF = asymptomatic women. The effects of MHPG (3 methoxy-4-hydroxyphenylglycol), the primary brain NE metabolite, are uncertain. (From Freedman, R.R., *Menopause: Biology and Pathobiology,* Lobo, R.A., Kelsey, J., and Marcus, R., Eds., Academic Press, San Diego, 2000. With permission.)

because the factors that affect these bones also influence the entire skeletal system. Sex steroids play an essential role in the maintenance of bone health throughout life, and adverse effects of hormone deficiency are clear in older men and young surgically menopausal and older postmenopausal women. Several risk factors lead to decreased bone mass in the elderly, including decreased physical activity, inadequate calcium and vitamin D intake, excessive alcohol and caffeine consumption, cigarette smoking, and the use of a variety of pharmacological agents, particularly glucocorticoids. Lifelong

healthy bone remodeling requires the interplay of multiple factors including cytokines, growth factors, mechanical stimuli, and hormones, particularly estrogens, parathyroid hormone, and vitamin D. We will only review the role of estrogen because this is the only one listed that changes dramatically at the menopause.

### 1. Lifetime Changes in Bone Mass

Bone mass increases until menarche, when the rate of increase in bone mass decreases rapidly (Chapter 21).

**Systemic and Organismic Aging**

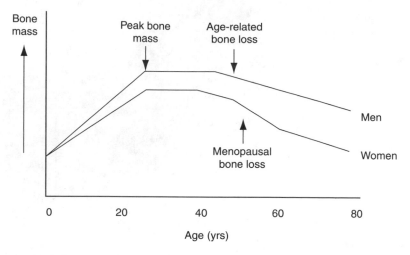

**FIGURE 11.12** Lifetime changes in bone mass in men and women. Bone mass decreases in both sexes after approximately 25 years of age. Loss accelerates in women after the menopause. (From Compston, J.E., *Physiol. Rev.*, 81, 419, 2001. With permission.)

Peak mass appears to be maintained through the fourth decade, when bone loss sets in (Figure 11.12). The exact timing, the causes of the onset of bone loss, and whether bone loss occurs at different rates in areas of the skeletal system at different times have not been well studied. However, the loss of bone in the spine, proximal femur, and forearm is well documented in cross-sectional and prospective studies.[45] This loss accelerates at the menopause for approximately 5 to 10 years. Loss during the first 5 to 10 years after the menopause is thought to be responsible for hip fractures in over 250,000 women each year.[46] Genetic factors are important determinants of peak bone mass, however, the gene(s) that control this are not known at the present time. Nutrition, calcium intake, physical activity, and hormone status also influence peak bone mass attained in young adults.

Sex steroids influence bone growth and the attainment of peak bone mass. They are clearly involved in the sexual differences between bone mass that become apparent at puberty. Estrogen is critical for normal closure of the epiphyseal plates in both sexes, and hypogonadism or abnormal estrogen receptor action adversely affects the attainment of peak bone mass and closure of epiphyseal plates. Delayed puberty is associated with reduced bone mineral density. Amenorrhea associated with anorexia nervosa, hyperprolactinemia, and other disorders results in decreased bone density.

## 2. Age-Related Bone Loss and the Relationship with Estrogen

Estrogen deficiency of the postmenopausal period is the major factor in the development of osteoporosis. Bone remodeling goes on constantly throughout life and is a preventative program that maintains the youth, vibrancy, and strength of the skeleton by regulation of bone resorption and bone formation (Chapter 21). With age and the menopause, the balance of formation and resorption changes: there is an increased rate of skeletal remodeling with an increase in resorption that is greater than the rise in bone formation. There is also a loss of the close coupling between formation and resorption, which is the key to adequate replacement of resorbed bone. With the loss of bone tissue and of the microarchitectural integrity of trabecular bone, the weakened bone cannot withstand even normal stress and may fracture even after trivial trauma.

Estrogen replacement at the time of the menopause or during the postmenopausal period prevents bone loss and results in increased bone mineral density during the first year of treatment. Although the precise mechanisms are still not understood, estrogens regulate the bone remodeling process. Estrogens reduce the rate of activation of bone remodeling and may correct the imbalance between resorption and formation. They may also influence responsiveness to parathyroid hormone. The consequences are an increase of between 3 to 10% during the first 1 to 2 years of replacement therapy that is especially obvious at sites of high turnover, such as the spine, and a reduction in the loss of bone mass. This leads to a reduction in fracture risk in the hip, spine, and wrist; however, continued treatment must be maintained for the beneficial effects to remain. Long-term data using bone density determinations as the primary outcome show that postmenopausal women are protected from loss of bone mass. A prospective study has shown prevention of vertebral factures with long-term use of estrogen, and a variety of epidemiologic studies indicate that estrogens reduce the risk of hip fractures.[46]

The effects of ERT on bone are dose dependent. However, recent data suggest that even low levels of estrogen are helpful. Thus, in postmenopausal women, even the small amounts of estrogen produced endogenously determine bone mineral density and fracture risk. In a large population-based study, it was shown that women aged 65 and older with serum estradiol levels between 10 to 25 pg/mL had significantly higher bone mineral density in the hip, spine, and other areas than those with estradiol below 5 pg/mL. Furthermore, women with undetectable serum estradiol levels had significantly increased risk of hip and vertebral fractures compared with those with levels above 5 pg/mL. These interesting and unexpected data challenge the perception that endogenous estrogen production in postmenopausal women does not have the physiological skeletal effects and emphasize the potential protective actions of relatively low concentrations of the hormone.

## B.  CENTRAL NERVOUS SYSTEM

### 1.  Psychological Effects

Psychological effects of the menopause have been reported in several studies. Anxiety and depressive moods are often common symptoms and are thought to be associated with decreased estrogen. Hormone replacement therapy may decrease vulnerability to postmenopausal depression.[47] Whether or not these symptoms are related more directly to the feelings associated with the impact of the menopause on quality of life, to discomfort and changing sexual activity, to the unpredictability of vasomotor symptoms, or to the actual

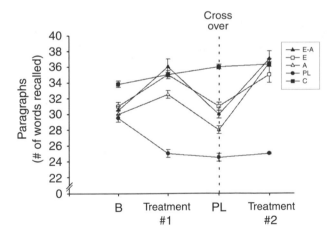

**FIGURE 11.13** Number of words recalled in written paragraphs by premenopausal women who had total abdominal hysterectomy and ovariectomy for benign disease with and without hormone therapy after surgery. Women, prior to surgery, were tested using the Paragraph Recall Test and designated B on the x-axis. Following surgery, subjects received estrogen plus androgen (E-A), estrogen alone (E), androgen alone (A), or placebo (PL) for the first 3 months. During the fourth month, all subjects received placebo, following which, they were randomly crossed over to a different treatment for an additional 3-month period of time. A fifth group consisted of women who underwent hysterectomy but were not ovariectomized (c); (Mean ± S.E., bracketed lines). (From Sherwin, B.B., *Psychoneuroendocrinology*, 13, 345, 1988. With permission.)

decrease in estrogen is not clear; however, it is clear that some women are predisposed to suffer from these effects of the menopause. Estrogen has been shown to influence mood and affect in a number of studies.[47-49] Therefore, there is reason to believe that hormone replacement therapy may improve mood disorders associated with the postmenopausal transition.

### 2.  Effects on Cognition and Memory

Clinical studies demonstrate that estrogen influences memory and cognition[50,51] and can protect against neurodegenerative diseases such as Alzheimer's disease (Chapter 8).[52-55] These findings, however, are not without controversy. As the results of more clinical studies become available, we are beginning to appreciate that the protective actions of estrogen do not apply in all situations.[56-58] It is important to differentiate between studies designed to test whether estrogens can affect the onset of declining cognitive function in healthy women (initiation phase) from studies designed to test whether estrogens can slow a pathological neurodegenerative process when it has already been initiated (propagation phase). Together, the results of many studies begin to point to the possibility that estrogens can protect the brain during the initiation phase but are less efficacious in preventing or altering the rate of the propagation phase of neurodegenerative diseases.

Many studies have examined whether ERT improves cognitive function in young and older healthy women. The majority of these studies show that estrogen can enhance cognitive function (Figure 11.13). It appears that by maintaining normal cognitive function, estrogen may further act to decrease the risk and delay the onset of Alzheimer's disease. Because memory is a broad term describing several distinct neural functions, many of which originate in different and overlapping regions of the brain, it is not surprising that estrogen can influence specific subtypes of memory (Chapter 8). For example, ERT seems to enhance immediate and delayed recall but not spatial or visual memory.[50,59] Therefore, it is possible that when studies group differing components of memory, results may exhibit a "masking effect" with regards to estrogen action on cognitive function.

Studies performed using experimental animal models support the conclusion that estradiol exerts trophic actions during adulthood. The approach to assess memory function in rats has utilized the radial arm maze, active avoidance, and the Morris water maze. Luine[60] found that estradiol treatment resulted in enhanced ability of male rats to perform spatial memory tasks, tested through radial arm maze tasks. Singh et al.[61] reported that estradiol treatment of ovariectomized rats caused a significant elevation in the level of active avoidance performance and accelerated the rate of learning relative to ovariectomized rats (Figure 11.14). Both of these investigators demonstrated that the improvement in learning and memory was correlated with enhanced levels of choline acetyltransferase levels, the rate-limiting enzyme in acetylcholine synthesis, high affinity choline uptake, and acetylcholinesterase.[61,62] Acetylcholine is clearly involved in learning and memory.[63] Therefore, these results suggest that estradiol-dependent enhancement of acetylcholine dynamics may mediate improvements in learning and memory (for review, see Gibbs and Aggarwal[64]).

Several lines of evidence suggest that estrogen replacement therapy may reduce the likelihood for cerebrovascular stroke by modifying risk factors that underlie stroke and coronary heart disease.[65] For example, estrogen may protect by exerting beneficial effects on diabetes and on the serum lipid profile. Interestingly, coronary heart disease doubles the risk for stroke, and ERT greatly reduces the risk for coronary heart disease (by 30 to 40%). It follows that estrogen may decrease the risk for stroke in parallel with its protective actions in coronary heart disease. Investigators who use animal models of cerebrovascular stroke have presented evidence that estradiol decreases the extent of injury (reviewed in Wise et al.[66]) (Figure 11.15). In contrast to protection, estrogen may, under some circumstances, impose an increased risk for stroke by influencing coagulation and fibrinolysis. Concerns of the thrombotic potential of estrogen arose from early observations that oral contraceptives appeared to increase the risk of venous thrombosis, stroke, and pulmonary embolism.[67] These observations led to speculation that estrogen replacement may also impose a risk for ischemic stroke in postmenopausal women.

## C. Cardiovascular System

Cardiovascular disease is the leading killer of women in most Western countries (Chapters 2, 16, and 17).[68] Furthermore, once affected by ischemic heart disease, women do worse than men following myocardial infarction,[69,70] coronary intervention,[71] and coronary artery bypass surgery.[72] The incidence of cardiovascular disease is 1 in 9 in women aged 45 to 64, but rises to 1 in 3 in women over 65.[73] At menopause, the rate of myocardial infarction increases threefold, irrespective of age at menopause,[74] and the incidence of cerebrovascular disease rises rapidly,[75] suggesting that endogenous estrogen plays a protective role. Both of these incidences increase in association with a series of events that increase coronary risk factors.

Atherosclerosis and the lesions of the vascular system underlie much of the etiology of cardiovascular disease (Chapter 16). It is a complex disease that involves the proliferation of smooth muscle cells in response to injury in an attempt to repair the damage. Plaque formation

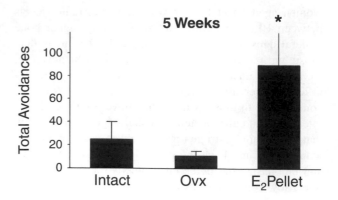

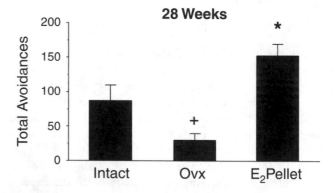

**FIGURE 11.14** Active avoidance performance after 5 and 28 weeks of ovariectomy in gonadally intact (Intact), ovariectomized (OVX), and estradiol-replaced (E2 pellet) rats. Intact and estradiol-replaced rats performed better than ovariectomized rats on the two-way active avoidance paradigm. Barograms represent mean values, and bracketed lines indicate standard errors. *P = 0.05, E2 vs. Intact; +P = 0.05, OVX vs. Intact. (From Singh, M. et al., *Brain Res.*, 644, 305, 1994. With permission.)

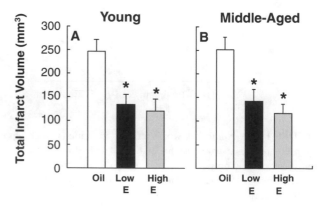

**FIGURE 11.15** Neuroprotective effects of estradiol against occlusion of the middle cerebral artery in young and middle-aged rats. Both ages (A = young, B = middle-aged) remain equally responsive to low and high physiological doses of estradiol replacement. Barograms represent mean values, and the bracketed lines represent the standard error of the mean; * indicates statistically significant difference between estrogen-treated and untreated control. (From Dubal, D.B. and Wise, P.M., *Endocrinology*, 142, 43, 2001. With permission.)

results due to the accumulation of lipids that fill the smooth muscle cells and begin to occlude the arteries. These plaques can rupture, causing a thrombus that can increase stenosis or lead to total occlusion. Estrogens appear to decrease the progress of atherosclerosis and maintain arteriolar tone and integrity, thus decreasing the risk of cardiovascular disease and cerebrovascular stroke.

Several epidemiological studies suggest that hormone replacement therapy reduces the risk of coronary artery disease;[76,77] although it does not appear to protect against the disease process once it is initiated.[78] These observational studies suggest that hormone replacement therapy reduces the risk of coronary artery disease by approximately 50%. The cardiovascular benefit of hormone replacement therapy was initially attributed to an indirect effect that involves improvement of the lipid profile, however, more recent studies found that hormone replacement therapy may decrease cardiovascular morbidity and mortality through direct mechanisms that involve inhibition of intimal hyperplasia and other components of vessel pathology and pathophysiology.[79] Certainly, routes of administration, doses, and preparations used may influence the efficacy of the effects of hormone replacement therapy on the cardiovascular system. Unfortunately, at the present time, the optimum preparations are not known.

## 1. Serum Lipids and Lipoproteins and the Menopause

Effects of the menopause and hormone replacement therapy on lipids and lipoproteins have received considerable attention in the study of atherosclerotic risk (Chapter 17). During menopause, levels of total and low-density lipoprotein (LDL) cholesterol and triglycerides increase, while the levels of high-density lipoproteins (HDL) decrease (Table 11.2).[80] Estrogen deficiency leads to decreased catabolism of LDL and decreased production of HDL cholesterol. Hormone replacement therapy reverses lipid changes, and that type of estrogen and progesterone and the route of delivery are important in influencing these changes. It appears that oral estrogens decrease levels of total and LDL cholesterol and increase HDL and triglycerides.[81] Conjugated oral estrogens and transdermal estradiol appear to have slightly different effects.[79] Although the mechanisms behind lipid changes are not completely understood, differences depend largely upon the route of delivery. High concentrations of estrogen in the portal system alter hepatic metabolism of lipids.

Hormone replacement may also affect cardiovascular disease by altering levels of apolipoproteins and blood pressure (Chapter 17). High levels of apolipoprotein A-I, the major protein component of HDL, are associated with a more favorable cardiac risk profile; whereas lipoprotein(a) is linked to atherosclerosis and is a predictor of coronary artery disease.[82] Levels of apolipoprotein A-I decrease with the menopause. Apolipoproteins A-I and lipoprotein(a) respond to hormone replacement therapy (Table 11.2). Estrogen reduces systolic and diastolic blood pressure so that after the menopause, blood pressure rises. Hemostatic changes lead to increases in plasma fibrinogen, factor VII, and antithrombin III, which combine to increase cardiovascular risk.

## 2. Vascular Effects and Endothelial Cell Factors Associated with the Menopause

In addition to the metabolic effects, estrogens have direct effects on the arterial wall (Chapter 16).[83] Alteration in endothelial function is a factor that contributes to the decline in cardiac function after the menopause. Investigations revealed that premenopausal women have enhanced arterial vasodilation. Furthermore, postmeno-

---

**TABLE 11.2**
**Adjusted Cardiovascular Risk Factors in Women**

| | Current Nonusers | | Current Users | | |
| | Past | Never | Estrogen | Estrogen + Progestin | p (Users Versus Nonusers) |
|---|---|---|---|---|---|
| Triglycerides | 123 | 120 | 141 | 131 | <0.001 |
| LDL-C | 141 | 141 | 125 | 127 | <0.001 |
| HDL-C | 58 | 58 | 67 | 66 | <0.001 |
| HDL$_2$-C | 17 | 16 | 21 | 21 | <0.001 |
| Apo A-1 | 141 | 140 | 159 | 156 | <0.001 |
| Apo B | 95 | 95 | 91 | 92 | <0.001 |
| Lipoprotein | 114 | 116 | 101 | 101 | <0.001 |
| Fibrinogen | 3.1 | 3.2 | 3 | 3 | <0.001 |
| Factor VII | 126 | 125 | 136 | 127 | <0.001 |
| Factor VIII | 133 | 136 | 134 | 132 | NS |

*Source:* Adapted from Wild, R.A., *Obstet. Gynecol.*, 87, 27S, 1996.

**Systemic and Organismic Aging**

pausal women given oral estradiol had improved endothelium-dependent, flow-mediated vasodilation.

Estrogens are antioxidants and appear to protect the endothelial cell wall from injury (Chapter 5). They may do this by protecting LDL cholesterol against oxidative modifications. In addition, estrogens may inhibit endothelial cell expression of adhesion molecules, leading to inhibition of endothelial cell expression of adhesion molecules, which, in turn, leads to decreased platelet aggregation and adhesion that is often observed in the early stages of atherosclerosis (Chapter 16). This may be important in vascular thrombosis formation and tone. In addition estrogens reduce arterial cholesterol ester influx and hydrolysis, reduce lipoprotein-induced arterial smooth muscle proliferation, and inhibit foam cell formation. Estrogens also influence coronary vasoactivity, which has been implicated in myocardial ischemia. Estrogens also influence the production and effect of nitric oxide (NO). Endothelial cells produce NO, which induces relaxation of arterial myocytes and results in vasodilation. One action of estrogen appears to be stimulation of the release of NO.

We now understand that the dose of estrogen administered and the route of estrogen delivery are key components in determining its effects in the cardiovascular system because of its effects on clotting potential. At higher doses, oral estrogen, which enters the body via the enterohepatic system, stimulates the production of thrombogenic factors,[84,85] predominantly through its actions on the liver. Alternatively, lower doses of estrogen, delivered orally or transdermally, do not significantly affect hemostasis.[84,86–88] Importantly, transdermal delivery of estrogen bypasses enterohepatic circulation and thus prevents estrogen-mediated stimulation of thrombogenic factors in the liver. Collectively, these findings highlight the importance of low physiological doses in estrogen replacement of postmenopausal women and the possible differential effects that depend upon route of delivery.

## VI.  REPRODUCTIVE DECLINE IN FEMALE RODENTS

Many studies to elucidate the mechanisms that govern reproductive aging are impossible to perform in humans or in primates, because they are invasive, involve large numbers of animals, and require experimental manipulations that are not feasible in women or nonhuman primates. Therefore, investigators have turned to using laboratory rodents to assess the role of the hypothalamus, pituitary, and ovary in the decline of reproductive function and the impact of hypoestrogenemia on physiological function. It is important to bear in mind that these species do not undergo the menopause, in the strict

sense, because they never experience menses during their reproductive life span. It appears that despite this fundamental difference, there are significant parallelisms between middle-aged female rats and pre- and perimenopausal women. First, women and laboratory rodents examined thus far exhibit a monotropic rise in FSH concentrations prior to any interruption in regular reproductive cycles.[20,89] In humans, the change is prominent during the periovulatory phase of the menstrual cycle. In a similar manner, middle-aged rats exhibit elevated FSH levels during estrous afternoon. Second, during middle age, menstrual cycle length and estrous cycle length become highly variable in women and rats, respectively.[14,90] Cycles of increased and decreased length have been reported in women between the ages of 37 to 45 years old as they enter the perimenopausal transition. Likewise, middle-aged rats (9 to 15 months of age) exhibit highly variable estrous cycles with prolonged periods of estrus or diestrus between each preovulatory LH surge. Third, women and laboratory rats experience a period of elevated estrogen levels around the time of the transition to the perimenopausal state.[8,91] Finally, estradiol's ability to induce LH surges is compromised in perimenopausal women and middle-aged rats.[17,92] For all of the above reasons, rodents may serve as excellent models to examine the factors that initiate the process of reproductive aging during middle age. Information gained from these species may be helpful in predicting where to concentrate studies in nonhuman primates and humans. Hopefully, studies in rodents will allow us to uncover and explore concepts that can be generalized to human reproductive aging.

### A.  CHANGES IN THE HYPOTHALAMIC/PITUITARY AXIS

One of the earliest changes in middle-aged rats is the pattern of the secretion of the preovulatory LH and FSH surge.[91,93–95] In middle-aged (7 to 9 months old) rats that had not exhibited any change in estrous cycle length and still maintained normal regular 4-day estrous cycles, the LH surge was delayed, and peak concentrations were dampened (Figure 11.16). These changes were likely due to changes in the neurotransmitter-induced activation of GnRH neurons at the time of the surge. Strict maintenance of the temporal pattern and synchrony of neurotransmitter input to GnRH neurons is critical to stimulate the necessary pattern of GnRH synthesis and secretion that leads to the LH surge.[96]

Several studies suggest that the circadian organization and synchrony of the neurotransmitters' rhythm in various aspects of neural activity are altered in middle-aged rats as they begin the transition to reproductive senescence. A central role of a circadian pacemaker in regulating precise timing of the events leading to the LH surge was beautifully revealed by the work of Everett

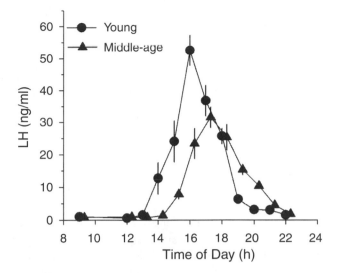

**FIGURE 11.16** Plasma luteinizing hormone (LH) levels in young (●) and middle-aged (▲) rats on proestrus. The preovulatory LH surge is attenuated in middle-aged (7 to 9 months old) compared to young (3 to 4 months old) rats. These changes were evident prior to any irregularity of estrous cyclicity or plasma estradiol concentrations (Mean ± S.E.). (From Wise, P.M., *Proc. Soc. Exp. Biol. Med.,* 169, 348, 1982. With permission.)

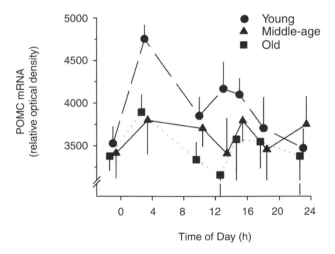

**FIGURE 11.17** Proopiomelanocortin (POMC) mRNA concentration in the arcuate nuclei of young (●), middle-aged (▲), and old (■) rats. In young rats, POMC mRNA levels exhibit a diurnal rhythm. This rhythm disappears by the time rats become middle-aged (Mean ± S.E). (From Weiland, N.G., Scarbrough, K., and Wise, P.M., *Endocrinology*, 131, 2959, 1992. With permission.)

and Sawyer.[97] They showed that blockade of a neural circadian pacemaker abolishes the surge. Thus, the daily rhythmicity in the activity of neural events serves as a foundation for the orderly timing of the GnRH surge and, hence, the preovulatory release of LH. In a series of studies, Wise and colleagues measured the diurnal

rhythm of monoamine turnover rates,[98,99] neurotransmitter receptor densities,[100] and neuropeptide mRNAs[101–103] (Figure 11.17) and found that virtually all of the neural rhythms were dampened or altered by the time female rats reached middle age and had begun the transition to irregular estrous cycles. Often, the attenuation in rhythmicity was progressive, and changes were more exaggerated as animals aged. The change in rhythmicity of any single neurotransmitter must be considered to be subtle because often, the overall average did not change, and investigators who measured these endpoints at only one time of day would be unlikely to detect a significant age-related change. Yet, together, disruption of the synchrony and coordination of multiple neural signals that govern the precise timing of GnRH secretion may ultimately lead to profound changes in the ability of rats to maintain regular reproductive cyclicity. Whether age-related changes in the integrity of the biological clock underlie changes in the diurnal rhythm of multiple outputs of the clock is not clear (Chapters 10 and 12). The rhythm in local cerebral glucose utilization, an index of overall cellular activity, changes with age, in that the rise in glucose utilization occurs earlier and is blunted in middle-aged rats.[104]

## 1. Aging Influences the Balance of Stimulatory and Inhibitory Inputs to GnRH Neurons

The ultimate pattern of GnRH secretion is determined by the balance of stimulatory and inhibitory inputs. Most work has focused on stimulatory factors because clearly, these are critical to the GnRH surge (and because it has always been assumed that these were the most important). Recent studies examined the possible role of decreased glutamate input to GnRH neurons in aging rats[105,106] because glutamate is a neurotransmitter that exerts important direct stimulatory effects on GnRH neurons.[107] However, we are beginning to appreciate more deeply that the amplitude and timing of the preovulatory LH surge depends upon a decrease in inhibitory tone.[108,109] Opioid peptides and GABA are critical inhibitory neurotransmitters that normally restrain GnRH secretion during the estrous cycle. A decrease in their activity normally occurs on proestrous afternoon, permitting stimulatory factors to maximally influence GnRH neurons. Researchers found that unless the inhibitory inputs to GnRH neurons are suppressed, the effects of norepinephrine and other stimulatory factors do not result in LH surges of normal amplitude or timing (for review see Kalra and Kalra[110]). In addition, pharmacological blockade of the inhibitory tone early on proestrus results in a premature LH surge. Until now, few studies using aging rodents have focused their attention on the possibility that aging involves an increase in inhibitory influences on GnRH neurons.

**Systemic and Organismic Aging**

## 2. Steroid Feedback Regulation of the Hypothalamus

One of the most consistent changes that occurs with age is a decreased ability of steroids to stimulate various neural outputs of the hypothalamus. In young ovariectomized rats, administration of estradiol elicits a daily LH surge that is timed relative to the light:dark cycle. Wise and colleagues[111] have shown that in middle-aged rats, equivalent doses of estradiol induce a significantly attenuated and delayed LH surge, and that fewer animals respond to estradiol. This is strikingly similar to the decreased steroid responsiveness observed in older women.[17] This age-related difference is not due to a decrease in the amount of GnRH or the density of GnRH receptors or estradiol receptors. Instead, it appears that the lack of responsiveness may be due to decreased ability of estradiol to induce a diurnal pattern in the neurotransmitters that are required for GnRH secretion (reviewed in Wise et al.[112]).

A diminished ability of estradiol to induce LH secretion may not be the only change in responsiveness to estradiol. Whether or not these changes are due to neurotoxic effects of repeated cyclic increases in estradiol is unknown. It is clear that chronic estrogen implants abolish the ability of young rats to produce a surge in LH in response to administered steroids. High pharmacological levels of estradiol administration alter dopamine neurotransmission in association with neurodegeneration with astrocytic and microglial hyperactivity.[113]

## B. Extension and Reactivation of Regular Estrous Cycles: Evidence for a Role for the Central Nervous System in Reproductive Aging

Some of the earliest evidence suggesting that the hypothalamus plays a role in reproductive aging came from two classical experimental approaches. First, transplantation of ovaries of old animals to the kidney capsule of young ovariectomized female hosts demonstrated that a permanently aged ovary could not explain the lack of cyclicity and ovulation. They revealed that follicular development and ovulation occurred under the influence of neuroendocrine signals of the young host.[114,115] Second, pharmacological methods showed that administration of drugs that restored the level of activity of monoaminergic neurotransmitters could restore cyclicity, albeit temporarily. Further, progesterone treatment, thought to act centrally, or electrochemical stimulation of the preoptic area of old rats resulted in restoration of estrous cyclicity. These results implicate changing hypothalamic function as a crucial element in reproductive decline. More recent studies that focused more on the middle-age transition period, suggest that hypothalamic changes may contribute

to the onset of irregular cycles that ultimately lead to acyclicity (for review see Wise et al.[112]).

## VII. STEROIDAL AND NONSTEROIDAL HORMONE REPLACEMENT THERAPY

### A. Steroid Replacement Therapy

Throughout this chapter, we discussed how estrogen and progesterone replacement therapy may act in postmenopausal women to maintain bone, cardiovascular, and brain function. In addition, the beneficial effects of estrogen replacement on hot flashes and improving sleep patterns may lead to the reported actions of estrogen in improving mood, depression, and insomnia. As our understanding of the breadth of estrogen and progesterone action in the body grows, the broad and profound impact of hormone replacement therapy on postmenopausal women becomes increasingly clear. We now know that steroids act in a broad spectrum of tissues to promote health and overall well-being.[46,83,116–118] However, we must be mindful that estrogens can be contraindicated in some women.[119] Concern has been expressed regarding the potential dangers of uninterrupted estrogen therapy in women with a family history of estrogen-responsive cancers or in women with a history of thrombotic disease. The results of the most recent studies show that hormone replacement therapy is not effective when a disease process has already been initiated.

### B. Nonsteroidal Therapies

Because steroid hormone replacement therapy is contraindicated in women who may be genetically predisposed to estrogen-dependent cancers and women with a previous history of cardiovascular disease or stroke, alternative nonsteroidal therapies are important to consider. Most recently, drugs with tissue-specific estrogen activity have been developed and used with considerable success. The development of selective estrogen receptor modulators (SERMs) during the past decade has been remarkable. These compounds, which may act as estrogen agonists in selective tissues and antagonists in others, provide us the opportunity of using these drugs to treat women who cannot tolerate ERT. Two compounds that have been tested extensively are tamoxifen and raloxifene. Tamoxifen is an antiestrogen or pure antagonist in breast tissue and can be used to treat breast cancer; but it acts as an estrogen agonist in bone.[120,121] The newest SERM, raloxifene, functions similarly to tamoxifen in bone, breast, and cardiovascular system, but exhibits minimal agonist activity in the uterus.[122] It is thought that these compounds interact with estrogen receptors and force them into different biological activities. They induce conformational changes intermediate between those that are considered pure agonist or pure antagonists. Thus, SERMs bind with receptors to create a shape that is slightly different than that of estrogen, which implies that they will have different biological consequences in different tissues.[123–125]

SERMs may manifest different activities in different target cells due to interactions of the compound–receptor interactions with other proteins. These "adaptor" proteins can facilitate or suppress the receptor interaction with DNA. Adaptor proteins may be expressed differentially in various tissues, for example, bone but not breast or vice versa. By taking advantage of the unique cellular distribution of various adaptor proteins, it may be possible to develop SERMs with actions that are restricted to specific cell types. It is clear that the potential for tissue-specific actions and variation in responses increases further.

These studies clearly show that the classification of estrogen agonists and antagonists is complex and tissue dependent. Further, these tantalizing studies suggest that future replacement therapy with nonsteroidal compounds may be critical for women who cannot take estrogen replacement during their postmenopausal years.

## ADDENDUM

Recent studies suggest that the use of a combination of conjugated equine estrogens and medroxyprogesterone acetate may be associated with beneficial effects on hip fracture but may increase risk of invasive breast cancer and deep vein thrombosis. Effects of the risks of stroke and coronary artery disease were less clear; however, risk had a tendency to increase. The effects of this preparation on cognition and memory were not tested. It is critical that we perform studies to determine whether the detrimental effects of hormone replacement are attributable to this particular preparation of estrogens and progestins and/or the dose that was used.[130]

## REFERENCES

1. Lobo, R.A., The perimenopause, *Clin. Obstet. Gynecol.*, 41, 895, 1998.
2. Prior, J.C., Perimenopause: The complex endocrinology of the menopausal transition, *Endocrine Reviews*, 19, 397, 1998.
3. Hodgen, G.D. et al., Menopause in rhesus monkeys: Model for study of disorders in the human climacteric, *Am. J. Obstet. Gynecol.*, 127, 581, 1977.
4. Gilardi, K.V.K. et al., Characterization of the onset of menopause in the rhesus macaque, *Biol. Reprod.*, 57, 335, 1997.
5. Gould, K.G., Flint, M., and Graham, C.E., Chimpanzee reproductive senescence: A possible model for evolution of the menopause, *Maturitas*, 3, 157, 1981.
6. Graham, C.E., Kling, O.R., and Steiner, R.A., Reproductive senescence in female non-human primates, in *Aging in Non-Human Primates*, Bowden, D.M., Ed., New York: Van Nostrand Reinhold, New York, 1979.
7. Shideler, S.E. et al., Ovarian-pituitary hormone interactions during the perimenopause, *Maturitas*, 11, 331, 1989.
8. Santoro, N. et al., Characterization of reproductive hormonal dynamics in the perimenopause, *J. Clin. Endocrinol. Metab.*, 81, 1495, 1996.
9. Mayer, P.J., Evolutionary advantage of the menopause, *Hum. Ecol.*, 10, 477, 1982.
10. Perls, T.T. and Fretts, R.C., The evolution of menopause and human life span, *Ann. Hum. Biol.*, 28, 237, 2001.
11. Gosden, R.G., *The Biology of the Menopause: The Causes and Consequences of Ovarian Ageing*, Academic Press, London, 1985.
12. Zumoff, B. et al., Twenty-four hour mean plasma testosterone concentration declines with age in normal premenopausal women, *J. Clin. Endocrinol. Metab.*, 80, 1429, 1995.
13. Klein, N.A. et al., Reproductive aging: Accelerated ovarian follicular development associated with a monotropic follicle-stimulating hormone rise in normal older women, *J. Clin. Endocrinol. Metab.*, 81, 1038, 1996.
14. Sherman, B.M. and Korenman, S.G., Hormonal characteristics of the human menstrual cycle throughout reproductive life, *J. Clin. Invest.*, 55, 699, 1975.
15. Matt, D.W. et al., Characteristics of luteinizing hormone secretion in younger versus older premenopausal women, *Am. J. Obstet. Gynecol.*, 178, 504, 1998.
16. Reame, N.E. et al., Age effects on follicle-stimulating hormone and pulsatile luteinizing hormone secretion across the menstrual cycle of premenopausal women, *J. Clin. Endocrinol. Metab.*, 81, 1512, 1996.
17. Van Look, P.F.A. et al., Hypothalamic-pituitary-ovarian function in perimenopausal women, *Clinical Endocrinology*, 7, 13, 1977.
18. Licciardi, F., Liu, H., and Rosewaks, Z., Day 3 estradiol serum concentrations as prognosticators of ovarian stimulation response and pregnancy outcome in patients undergoing *in vitro* fertilization, *Fertil. Steril.*, 64, 991, 1995.
19. MacNaughton, J. et al., Age related changes in follicle stimulating hormone, luteinizing hormone, oestradiol and immunoreactive inhibin in women of reproductive age, *Clin. Endocrinol.*, 36, 339, 1992.
20. Klein, N.A. et al., Decreased inhibin B secretion is associated with the monotropic FSH rise in older, ovulatory women: A study of serum and follicular fluid levels of dimeric inhibin A and B in spontaneous menstrual cycles, *J. Clin. Endocrinol. Metab.*, 81, 2742, 1996.
21. Treloar, A.E. et al., Variation of the human menstrual cycle through reproductive life, *Int. J. Fertil.*, 12, 77, 1967.
22. Treloar, A.E., Menstrual activity and the pre-menopause, *Maturitas*, 3, 249, 1981.
23. Klein, N.A. et al., Ovarian follicular development and the follicular fluid hormones and growth factors in normal women of advanced reproductive age, *J. Clin. Endocrinol. Metab.*, 81, 1946, 1996.

**Systemic and Organismic Aging**

24. Metcalf, M., Incidence of ovulatory cycles in women approaching the menopause, *J. Biosoc. Sci.*, 11, 39, 1979.

25. Klein, N.A. and Soules, M.R., Endocrine changes of the perimenopause, *Clin. Obstet. Gynecol.*, 41, 912, 1998.

26. Menken, J., Trussell, J., and Larsen, U., The fecundity of a couple can be predicted by the age of the female more than the male, *Age and Infertility Science*, 233, 1389, 1986.

27. Schwartz, D. and Mayauz, M., Female fecundity as a function of age: Results of artificial insemination in 2193 nulliparous women with azoospermic husbands, *N. Engl. J. Med.*, 306, 404, 1982.

28. Stovall, D.W. et al., The effect of age on female fecundity, *Obstet. Gynecol.*, 77, 33, 1991.

29. Battaglia, D. et al., Influence of maternal age on meiotic spindle assembly in oocytes from naturally cycling women, *Hum. Reprod.*, 11, 2217, 1996.

30. Battaglia, D.E., Klein, N.A., and Soules, M.R., Changes in centrosomal domains during meiotic maturation in the human oocyte, *Mol. Hum. Reprod.*, 2, 845, 1997.

31. Angell, R., Aneuploidy in older women, *Hum. Reprod.*, 9, 119, 1994.

32. Benadiva, C., Kligman, I., and Munne, S., Aneuploidy 16 in human embryos increases significantly with maternal age, *Fertil. Steril.*, 66, 248, 1996.

33. Block, E., Quantitative morphological investigations of the follicular system in women: Variations at different ages, *Acta Anat.*, 14, 108, 1952.

34. Gougeon, A., Ecochard, R., and Thalabard, J.C., Age-related changes of the population of human ovarian follicles: Increase in the disappearance rate of non-growing and early-growing follicles in aging women, *Biol. Reprod.*, 50, 653, 1994.

35. Richardson, S.J., Senikas, V., and Nelson, J.F., Follicular depletion during the menopausal transition: Evidence for accelerated loss and ultimate exhaustion, *J. Clin. Endocrinol. Metab.*, 65, 1231, 1987.

36. Freedman, R.R., Menopausal hot flashes, in *Menopause: Biology and Pathobiology*, Lobo, R.A., Kelsey, J., and Marcus, R., Eds., Academic Press, San Diego, CA, 2000.

37. Kronenberg, F., Hot flashes: Epidemiology and physiology, *Ann. N.Y. Acad. Sci.*, 592, 52, 1990.

38. Freedman, R.R., Biochemical, metabolic, and vascular mechanisms in menopausal hot flashes, *Fertil. Steril.*, 70, 332, 1998.

39. Tataryn, I.V. et al., LH, FSH and skin temperature during the menopausal hot flash, *J. Clin. Endocrinol. Metab.*, 49, 152, 1979.

40. DeFazio, J. et al., Induction of hot flashes in premenopausal women treated with a long-acting GnRH agonist, *J. Clin. Endocrinol. Metab.*, 56, 445, 1983.

41. Laufer, L.R. et al., Effect of clonidine on hot flashes in postmenopausal women, *Obstet. Gynecol.*, 60, 583, 1982.

42. Freedman, R.R., Woodward, S., and Sabharwal, S.C., α2-Adrenergic mechanism in menopausal hot flushes, *Obstet. Gynecol.*, 76, 573, 1990.

43. Hutton, J.D. et al., Relation between plasma estrone and estradiol and climacteric symptoms, *Lancet*, 1, 671, 1978.

44. Melton, L.J.I., How many women have osteoporosis now? *J. Bone Miner. Res.*, 10, 175, 1995.

45. Compston, J.E., Sex steroids and bone, *Physiol. Rev.*, 81, 419, 2001.

46. Lindsay, R., The menopause and osteoporosis, *Obstet. Gynecol.*, 87, 16S, 1996.

47. Halbreich, U., Role of estrogen in postmenopausal depression, *Neurology*, 48, S16, 1997.

48. Mortola, J.F., Estrogens and mood, *J. Soc. Obstet. Gynecol. Canada*, 19, 1, 1997.

49. Palinkas, L.A. and Barrett-Connor, E., Estrogen use and depressive symptoms in postmenopausal women, *Obstet. Gynecol.*, 80, 30, 1992.

50. Sherwin, B.B., Estrogen effects on cognition in menopausal women, *Neurology*, 48, S21, 1997.

51. Schmidt, R. et al., Estrogen replacement therapy in older women: A neuropsychological and brain MRI study, *J. Amer. Geriat. Soc.*, 44, 1307, 1996.

52. Henderson, V.W. and Paganini-Hill, A., Estrogen and Alzheimer's Disease, *J. Soc. Obstet. Gynecol. Canada*, 19, 21, 1997.

53. Kawas, C. et al., A prospective study of estrogen replacement therapy and the risk of developing Alzheimer's disease: The Baltimore Longitudinal Study of Aging, *Neurology*, 48, 1517, 1997.

54. Asthana, S. et al., Cognitive and neuroendocrine response to transdermal estrogen in postmenopausal women with Alzheimer's Disease: Results of a placebo-controlled, double-blind, pilot study, *Psychoneuroendocrinology*, 24, 657, 1999.

55. Waring, S.C. et al., Postmenopausal estrogen replacement therapy and risk of AD: A population-based study, *Neurology*, 52, 965, 1999.

56. Henderson, V.W. et al., Estrogen for Alzheimer's disease in women: Randomized, double-blind, placebo-controlled trial, *Neurology*, 54, 295, 2000.

57. Marder, K. and Sano, M., Estrogen to treat Alzheimer's disease: Too little, too late? So what's a woman to do? *Neurology*, 54, 2035, 2000.

58. Wang, P.N. et al., Effects of estrogen on cognition, mood, and cerebral blood flow in AD: A controlled study, *Neurology*, 54, 2061, 2000.

59. Barrett-Connor, E. and Kritz-Silverstein, D., Estrogen replacement therapy and cognitive function in older women, *J. Am. Med. Assoc.*, 269, 2637, 1993.

60. Luine, V.N., Steroid hormone influences on spatial memory, *Ann. N.Y. Acad. Sci.*, 743, 201, 1994.

61. Singh, M. et al., Ovarian steroid deprivation results in a reversible learning impairment and compromised cholinergic function in female Sprague-Dawley rats, *Brain Res.*, 644, 305, 1994.

62. Luine, V.N., Estradiol increases choline acetyltransferase activity in specific basal forebrain nuclei and projection areas of female rats, *Exptl. Neurol.*, 89, 484, 1985.

63. Wenk, G. et al., Neurotransmitters and memory: Role of cholinergic, serotonergic and noradrenergic systems, *Behav. Neurosci.*, 101, 325, 1987.

64. Gibbs, R.B. and Aggarwal, P., Estrogen and basal forebrain cholinergic neurons: Implications for brain aging and Alzheimer's disease related cognitive decline, *Horm. Behav.*, 34, 98, 1998.

65. Paganini-Hill, A., Hormone replacement therapy and stroke: Risk, protection or no effect? *Maturitas,* 38, 243, 2001.

66. Wise, P.M. et al., Estrogens: Trophic and protective factors in the adult brain, *Front Neuroendocrinol.*, 22, 33, 2001.

67. Gillum, L.A., Mamidipudi, S.K., and Johnston, S.C., Ischemic stroke risk with oral contraceptives, *J. Am. Med. Assoc.*, 284, 72, 2000.

68. Wenger, N.K., Women's heart research: It's about time, *Journal of Women's Health*, 4, 459, 1995.

69. Vaccarino, V. et al., Sex differences in mortality after myocardial infarction. Is there evidence for an increased risk for women? *Circulation*, 91, 1861, 1995.

70. Bueno, H. et al., Influence of sex on the short-term outcome of elderly patients with a first acute myocardial infarction, *Circulation*, 92, 1133, 1995.

71. Greenberg, M.A. and Mueller, H.S., Why the excess mortality in women after percutaneous transluminal coronary angioplasty? *Circulation*, 87, 1030, 1993.

72. Khan, S.S. et al., Increased mortality of women in coronary artery bypass surgery, *Ann. Int. Med.*, 112, 561, 1990.

73. Rosano, G.M.C. et al., Cardioprotective effects of ovarian hormones, *Eur. Heart J.*, 17, 15, 1996.

74. Matthews, K.A., Kuller, L.H., and Sutton-Tyrrell, K., Changes in cardiovascular risk factors during the peri- and postmenopausal years, in *Proceedings of the International Symposium on the Biology of the Menopause*, Bellino, F.L., Ed., Springer, Norwell, MA, 2000.

75. Kannel, W.B. and Thom, T.J., The incidence, prevalence and mortality of cardiovascular disease, in *The Heart*, Schlant, R.C. and Alexander, R.W., Eds., McGraw-Hill, New York, 1994.

76. Bush, T.L. et al., Cardiovascular mortality and noncontraceptive use of estrogen in women: Results from the lipid research clinics program follow-up study, *Circulation*, 75, 1102, 1987.

77. Stampfer, M.J. et al., Postmenopausal estrogen replacement therapy and cardiovascular disease. Ten-year follow-up from the Nurses' Health Study, *N. Engl. J. Med.*, 325, 756, 1991.

78. Hulley, S. et al., Randomized trial of estrogen plus progestin for secondary prevention of coronary heart disease in postmenopausal women, *J. Am. Med. Assoc.*, 280, 605, 1998.

79. Sites, C.K., Hormone replacement therapy: Cardiovascular benefits for aging women, *Coron. Artery Dis.*, 9, 789, 2001.

80. Poehman, E.T., Toth, E.J., and Ades, P.A., Menopause-associated worsening of plasma lipids and blood pressure: A longitudinal study, *Eur. J. Clin. Invest.*, 27, 322, 1997.

81. Walsh, B.W. et al., Effects of postmenopausal estrogen replacement on the concentrations and metabolism of plasma lipoproteins, *N. Engl. J. Med.*, 325, 1196, 1991.

82. Armstrong, V.W. et al., The association between serum Lp(a) concentrations and angiographically assessed coronary atherosclerosis: Dependence on serum LDL levels, *Atherosclerosis*, 62, 249, 1986.

83. Wild, R.A., Estrogen: Effects on the cardiovascular tree, *Obstet. Gynecol.*, 87, 27S, 1996.

84. Scarabin, P.-Y. et al., Effects of oral and transdermal estrogen/progesterone regimens on blood coagulation and fibrinolysis in postmenopausal women. A randomized clinial trial, *Arterioscler. Thromb. Vasc. Biol.,* 17, 3071, 1997.

85. Luyer, M.D.P. et al., Prospective randomized study of effects of unopposed estrogen replacement therapy on markers of coagulation and inflammation in postmenopausal women, *J. Clin. Endocrinol. Metab.*, 86, 3629, 2001.

86. Schwartz, S.M. et al., Stroke and use of low-dose oral contraceptives in young women. A pooled analysis of two US studies, *Stroke*, 29, 2284, 1998.

87. Giltay, E.J. et al., Oral, but not transdermal, administration of estrogens lowers tissue-type plasminogen activator levels in humans without affecting endothelial synthesis, *Arterioscler. Thromb. Vasc. Biol.,* 20, 1396, 2001.

88. Petitti, D.B. et al., Stroke in users of low-dose oral contraceptives, *N. Engl. J. Med.*, 335, 8, 1996.

89. DePaolo, L.V., Age-associated increases in serum follicle-stimulating hormone levels on estrus are accompanied by a reduction in the ovarian secretion of inhibin, *Exp. Aging Res.*, 13, 3, 1987.

90. Fitzgerald, C.T. et al., Age-related changes in the female reproductive cycle, *Brit. J. Obstet. Gynaecol.*, 101, 229, 1994.

91. Lu, J. K.H., Changes in ovarian function and gonadotropin and prolactin secretion in aging female rats, in *Neuroendocrinology of Aging*, Meites, J., Ed., Plenum Press, New York, 1983.

92. Wise, P.M., Estradiol-induced daily luteinizing hormone and prolactin surges in young and middle-aged rats: Correlations with age-related changes in pituitary responsiveness and catecholamine turnover rates in microdissected brain areas, *Endocrinology*, 115, 801, 1984.

93. Wise, P.M., Alterations in proestrous LH, FSH, and prolactin surges in middle-aged rats, *Proc. Soc. Exp. Biol. Med.*, 169, 348, 1982.

94. van der Schoot, P., Changing pro-oestrous surges of luteinizing hormone in ageing 5-day cyclic rats, *J. Endocrinol.*, 69, 287, 1976.

95. Nass, T.E. et al., Alterations in ovarian steroid and gonadotrophin secretion preceding the cessation of regular oestrous cycles in ageing female rats, *J. Endocrinol.*, 100, 43, 1984.

96. Lloyd, J.M., Hoffman, G.E., and Wise, P.M., Decline in immediate early gene expression in gonadotropin-releasing hormone neurons during proestrus in regularly cycling, middle-aged rats, *Endocrinology*, 134, 1800, 1994.

97. Everett, J.W., Sawyer, C.H., and Markee, J.E., A neurogenic timing factor in control of the ovulatory discharge of luteinizing hormone in the cyclic rat, *Endocrinology*, 44, 234, 1949.

**Systemic and Organismic Aging**

98. Wise, P.M., Norepinephrine and dopamine activity in microdissected brain areas of the middle-aged and young rat on proestrus, *Biol. Reprod.*, 27, 562, 1982.

99. Cohen, I.R. and Wise, P.M., Age-related changes in the diurnal rhythm of serotonin turnover in microdissected brain areas of estradiol-treated ovariectomized rats, *Endocrinology*, 122, 2626, 1988.

100. Weiland, N.G. and Wise, P.M., Aging progressively decreases the densities and alters the diurnal rhythm of alpha-1-adrenergic receptors in selected hypothalamic regions, *Endocrinology*, 126, 2392, 1990.

101. Cai, A. and Wise, P.M., Age-related changes in the diurnal rhythm of CRH gene expression in the paraventricular nuclei, *Am. J. Physiol.*, 270, E238, 1996.

102. Krajnak, K. et al., Aging alters the rhythmic expression of vasoactive intestinal polypeptide mRNA, but not arginine vasopressin mRNA in the suprachiasmatic nuclei of female rats, *J. Neurosci.*, 18, 4767, 1998.

103. Weiland, N.G., Scarbrough, K., and Wise, P.M., Aging abolishes the estradiol-induced suppression and diurnal rhythm of proopiomelanocortin gene expression in the arcuate nucleus, *Endocrinology*, 131, 2959, 1992.

104. Wise, P.M. et al., Aging alters the circadian rhythm of glucose utilization in the suprachiasmatic nucleus, *Proc. Natl. Acad. Sci. USA*, 85, 5305, 1988.

105. Zuo, Z. et al., Decreased gonadotropin-releasing hormone neurosecretory response to glutamate agonists in middle-aged female rats on proestrus afternoon: A possible role in reproductive aging? *Endocrinology*, 137, 2334, 1996.

106. Gore, A.C. et al., Neuroendocrine aging in the female rat: The changing relationship of hypothalamic GnRH neurons and NMDA receptors, *Endocrinology*, 141, 4757, 2000.

107. Brann, D.W., Glutamate: A major excitatory transmitter in neuroendocrine regulation. *Neuroendocrinology*, 61, 213, 1995.

108. Akabori, A. and Barraclough, C.A., Gonadotropin responses to naloxone may depend upon spontaneous activity in noradrenergic neurons at the time of treatment, *Brain Res.*, 362, 55, 1986.

109. Akabori, A. and Barraclough, C.A., Effects of morphine on luteinizing hormone secretion and catecholamine turnover in the hypothalamus of estrogen-treated rats, *Brain Res.*, 362, 221, 1986.

110. Kalra, S.P. and Kalra, P.S., Opioid-adrenergic-steroid connection in regulation of luteinizing hormone secretion in the rat, *Neuroendocrinology*, 38, 418, 1984.

111. Wise, P.M., Rance, N., and Barraclough, C.A., Effects of estradiol and progesterone on catecholamine turnover rates in discrete hypothalamic regions in ovariectomized rats, *Endocrinology*, 108, 2186, 1981.

112. Wise, P.M. et al., Aging of the female reproductive system: A window into brain aging, *Rec. Prog. Hormone Res.*, 52, 279, 1997.

113. Brawer, J.R. and Sonnenschein, C., Cytopathological effects of estradiol on the arcuate nucleus of the female rat: A possible mechanism for pituitary tumorigenesis, *Am. J. Anat.*, 144, 57, 1975.

114. Aschheim, P., Relation of neuroendocrine system to reproductive decline in female rats, in *Neuroendocrinology of Aging*, Meites, J., Ed., Plenum Press, New York, 1983.

115. Peng, M.-T. and Huang, H.-H., Aging of hypothalamic-pituitary-ovarian function in the rat, *Fertil. Steril.*, 23, 535, 1972.

116. Hammond, C.B., Menopause and hormone replacement therapy: An overview, *Obstet. Gynecol.*, 87, 2S, 1996.

117. Sherwin, B.B., Hormones, mood, and cognitive functioning in postmenopausal women, *Obstet. Gynecol.*, 87, 20S, 1996.

118. Sullivan, J.M. and Fowlkes, L.P., The clinical aspects of estrogen and the cardiovascular system, *Obstet. Gynecol.*, 87, 36S, 1996.

119. Speroff, L., Postmenopausal hormone therapy and breast cancer, *Obstet. Gynecol.*, 87, 44S, 1996.

120. Love, R. et al., Effects of tamoxifen on cardiovascular risk factors in post-menopausal women, *Ann. Int. Med.*, 115, 860, 1991.

121. Love, R. et al., Effects of tamoxifen on bone mineral density in postmenopausal women with breast cancer, *N. Engl. J. Med.*, 326, 852, 1992.

122. Yang, N.N. et al., Estrogen and raloxifene stimulate transforming growth factor-β3 gene expression in rat bone: Potential mechanism for estrogen- or raloxifene-mediated bone maintenance, *Endocrinology*, 137, 2075, 1996.

123. Allan, G.F. et al., Hormone and antihormone induce distinct conformational changes which are central to steroid receptor activation, *J. Biol. Chem.*, 267, 19513, 1992.

124. Beekman, J. M. et al., Transcriptional activation by the estrogen receptor requires a conformational change in the ligand binding domain, *Mol. Endocrinol.*, 7, 1266, 1993.

125. McDonnell, D.P. and Norris, J.D., Analysis of the molecular pharmacology of estrogen receptor agonists and antagonists provides insights into the mechanism of action of estrogen in bone, *Osteoporosis Int.*, 7, S29, 1997.

126. Yen, S.S.C., The human menstrual cycle: Neuroendocrine regulation, in *Reproductive Endocrinology. Physiology, Pathophysiology and Clinical Management*, Yen, S.S.C., Jaffe, R.B., and Barbieri, R.L., Eds., W.B. Saunders Company, Philadelphia, PA, 1999.

127. Sherman, B.M., West, J.H., and Korenman, S.G., The menopausal transition: Analysis of LH, FSH, estradiol, and progesterone concentrations during menstrual cycles of older women, *J. Clin. Endocrinol. Metab.*, 42, 629, 1976.

128. Sherwin, B.B., Estrogen and/or androgen replacement therapy and cognitive functioning in surgically menopausal women, *Psychoneuroendocrinology*, 13, 345, 1988.

129. Dubal, D.B. and Wise, P.M., Neuroprotective effects of estradiol in middle-aged female rats, *Endocrinology*, 142, 43, 2001.

130. Writing Group for the Women's Health Initiative 2002, Risks and benefits of estrogen plus progestin in healthy postmenopausal women, *JAMA*, 288, 321, 2002.

# 12 The Ensemble Male Hypothalamo–Pituitary–Gonadal Axis

*Johannes D. Veldhuis*
Mayo Clinic and Graduate School of Medicine

*Michael L. Johnson and Daniel Keenan*
University of Virginia School of Medicine

*Ali Iranmanesh*
Veterans Affairs Medical Center

## CONTENTS

## I.   INTRODUCTION

The causes of progressive hypoandrogenemia (low blood levels of androgens) in the aging male arise at multiple levels in the gonadal axis. Pathophysiological mechanisms embody a threefold composite of the following:

- Diminished hypothalamic gonadotropin-releasing hormone (GnRH) release

- Impaired feedforward drive by pituitary luteinizing hormone (LH) of testicular androgen biosynthesis
- Altered sex steroid-dependent feedback control of GnRH/LH secretion[1–4]

The foregoing tripartite perspective is reviewed here.

## II. OVERVIEW OF THE GNRH–LH–TESTOSTERONE FEEDBACK AXIS

Neuroendocrine axes or "ensembles" comprise specialized arrays of interacting neural structures and corresponding glands. Neuroglandular communication occurs via blood-borne hormonal signals, which maintain stability and homeostasis of the system.[5–8] Autoregulation is mediated by negative feedback and feedforward (i.e., positive feedback) linkages within the axis. Feedbacks represent negative (inhibitory) responses (Chapter 10); for example, testosterone *feeds back* on the hypothalamus and pituitary by inhibiting secretion of GnRH and LH; GnRH feeds forward on LH by stimulating its release and, in turn, LH feeds forward on testicular Leydig cells by driving testosterone secretion. In a general model, hypothalamic neuronal products regulate the synthesis and secretion of glycoprotein hormones by the anterior pituitary gland; secreted pituitary hormones, in turn, drive biosynthetic responses in remote target glands; and peripherally acting peptides and steroids feed back on the brain and pituitary gland to repress neurohormone output (Chapter 10). These time-delayed and nonlinear signaling interactions confer the overall dynamics of a particular axis, as illustrated further here for classical "closed-loop" adaptive control of the male reproductive axis.[9]

The androgenic component of the male gonadal system comprises three pivotal regulatory nodes: (1) 800 to 1200 mediobasal hypothalamic GnRH-secreting neurons; (2) clusters of anterior-pituitary (gonadotrope) cells, which synthesize and release LH molecules; and (3) gonadal interstitial (Leydig) cells, which produce testosterone and small amounts of estradiol (Figure 12.1).[8–11] These neuroglandular elements communicate by way of[12–14] time-lagged feedforward and feedback signals.[5,6,9]

Interactive control is conferred via corresponding interglandular interfaces.[15] According to this ensemble perspective, no one component of the axis acts in isolation.[7,8] Some structural and physiologic characteristics of the male reproductive system are presented in Box 12.1.

### A. REPRODUCTIVE AGING

Reproductive aging impacts the general physical and psychosocial health of women and men.[1–4,13–14,16–31] Relative physical frailty, sarcopenia, osteopenia, diminished exercise capacity, impaired quality of life, dyslipidemia, reduced cognitive abilities, and sexual dysfunction are recognized concomitants of aging and sex-hormone deficiency.[17,18,29,32–36] In the male, a declining availability of androgens is the primary manifestation of gonadal-axis aging. In contrast, spermatogenesis appears to be relatively preserved in older men.[21,37] Accordingly, the following review of reproductive aging focuses on the mechanisms that subserve impoverished testosterone production.

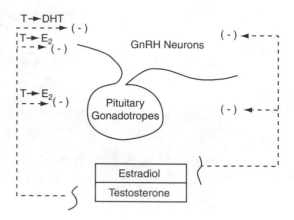

**FIGURE 12.1** Simplified construct of key regulatory interactions that drive the dynamics of the human male hypothalamo–pituitary–testicular axis. The ensemble comprises pulsatile gonadotropin-releasing hormone (GnRH) secretion by mediobasal hypothalamic neurons, episodic luteinizing hormone (LH) release by anterior-pituitary gonadotrope cells, and intermittent testosterone (T) and estradiol ($E_2$) production by gonadal Leydig cells. Interrupted lines denote negative feedback (-) connections. Autofeedback of GnRH on electrically excitable neurons also occurs in some species. Testosterone inhibits hypothalamic GnRH pulse frequency and burst amplitude and impedes GnRH's feedforward on pituitary LH secretion. Certain actions of T are mediated via its principal metabolites $E_2$ or $5\alpha$-dihydrotestosterone (DHT). (Adapted from Keenan, D.M. and Veldhuis, J.D., *Am. J. Physiol. Regul. Integr. Comp. Physiol.*, 280, R1755, 2001.)

### B. RELATIVE TESTOSTERONE DEFICIENCY IN AGING MEN

Healthy older men exhibit reduced spermatic-vein and systemic concentrations of testosterone, as affirmed by a recent meta-analysis.[38] The longitudinal New Mexico Aging Process Study in healthy men estimated that the serum total testosterone concentration declines by 110 ng/dL per decade after age 61 years.[39] Independent cross-sectional analyses in the European SENIEUR and Massachusetts Male Aging cohorts identified a greater age-related fall in bioavailable (nonSHBG-bound) testosterone concentrations.[19,40,41] The latter measure distinguishes healthy cohorts of older and young men more vividly than serum total testosterone concentrations (Figure 12.2). Hypoandrogenemia in elderly men can be exacerbated by intercurrent acute illness, underlying chronic disease, physical and psychological stress, hospitalization, multiple medications use, and nutritional inadequacy.[42]

### C. DISRUPTION OF LH AND TESTOSTERONE SECRETION IN AGING

Older men manifest consistent blunting of the amplitude of LH and testosterone secretory pulses and disruption of their orderly release patterns (Figure 12.3).[13,23,43]

### Box 12.1
### Structural/Physiologic Characteristics of the Male Reproductive System

The reproductive system in human males includes the primary organ, the two testes, which produce germ cells or sperm, as well as the principal male hormones, and a number of secondary sex organs (epididymis, vas deferens) which serve for transport of the sperm and others (seminal vescicles, prostate) for the production of the seminal fluid.

As for the ovary, the hormones involved in testicular function are operative at four different levels:

- In the hypothalamus, gonadotropin-releasing hormone, GnRH, a peptide, is secreted by neurons and carried, through the portal vessels (Chapter 10), to the anterior pituitary, where it stimulates the secretion and release of the two gonadotropins, FSH and LH (the latter, also called, in the male, interstitial cell-stimulating hormone ICSH).
- The gonadotropins, FSH and LH, are proteins similar to those secreted from the anterior pituitary of the female; they are regulated by GnRH and, through a negative feedback, by circulating levels of testosterone for LH and inhibin for FSH.
- In the testis, the major hormones are steroids: testosterone, secreted by the interstitial Leydig cells, and extremely small amounts of estradiol. Inhibin, a protein, is secreted by the Sertoli cells, and its secretion is stimulated by FSH. High circulating levels of testosterone and inhibin inhibit (negative feedback) release of LH and FSH, respectively, from the anterior pituitary.
- At the periphery, testosterone, as the ovarian steroids, is transported in the plasma, partly bound (99%) to the steroid-hormone-binding-globulin (SHBG), and partly, in the free, active form (1%); it is metabolized in the liver and excreted in the urine. In some of the target tissues, such as the prostate, testosterone is converted to dihydrotestosterone (DHT) by the enzyme 5α-reductase; DHT binds to the same receptors as testosterone but more strongly, thereby being more biologically efficient in amplifying the actions of testosterone at the target tissues.

It should be recalled that androgens (i.e., male sex steroids) are also secreted by the adrenal cortex (Chapter 10), and, in fact, these androgens represent the major portion of the adrenal sex hormones; however, they have 20% less androgenic activity than that of testosterone. Major actions of androgens include, in addition to development and maintenance of secondary sex organs and sex characteristics (e.g., type of voice, hair distribution, behavior), an important protein-anabolic, growth-promoting effect.

Irregular hormone secretion patterns emerge in the elderly male even in the presence of normal young-adult serum concentrations of LH and total testosterone.[1,13,24,43] There is concomitant deterioration of the expected synchrony among the secretion of LH, testosterone, FSH, and prolactin and oscillations of sleep-stage and nocturnal penile tumescence (NPT).[44] The multiplicity of altered reproductive neurohormone outflow in the older male points to more general erosion of central nervous system (CNS) control.[23] Indeed, aging also is marked by disruption of the orderly secretion of GH, ACTH, cortisol, and insulin (Chapters 10 and 14).[45] Whether a single unifying basis underlies such inferred loss of neurointegrative function is not known.[46]

Whereas the (incremental) LH pulse amplitude is blunted in the aging male, the acute stimulatory action of exogenous GnRH on LH release is preserved. Moreover, pulsatile intravenous infusion of GnRH for 14 days normalizes pulsatile, entropic, and 24-hour rhythmic LH secretion in older men but does not restore young-adult serum concentrations of total or bioavailable testosterone (Figure 12.4).[1] The latter age-related defect denotes impaired gonadal steroidogenic responsiveness to endogenous LH stimulation. Injection of hCG (human Chorionic Gonadotropin, a placental hormone strongly resembling LH) as an LH surrogate also fails to evoke maximal Leydig-cell androgen biosynthesis in the elderly male.[20,47] Thus, available clinical studies delineate impaired hypothalamic GnRH stimulation on LH secretion and reduced Leydig-cell responsiveness to LH/hCG in aging.

### D. Feedback Alterations in the Aging Male

Negative-feedback control of the GnRH-LH secretory unit by androgens and estrogens is disrupted in the aging male. This notion derives from observations of:

- abnormal LH secretory responses to the imposition of androgen negative feedback by the intramuscular, intravenous, or transscrotal administration of testosterone and intravenous or transdermal delivery of 5α-dihydrotestosterone
- abnormal LH secretory responses to the withdrawal of sex steroid-hormone negative feedback by exposure to antagonists of the androgen or estrogen receptor or to an inhibitor of adrenal and gonadal steroidogenesis[24,48-52]

**Systemic and Organismic Aging**

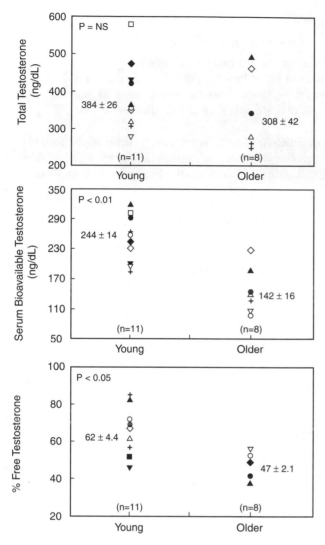

**FIGURE 12.2** Distinctions among measurements of serum total (top), bioavailable (middle), and percentage free/bioavailable (nonsex steroid hormone-binding globulin [SHBG]-bound) testosterone concentrations in healthy, community-dwelling, ambulatory unmedicated volunteers. Assays were applied to 24-h pooled sera in 11 young (ages 18 to 25 years) and eight older (ages 60 to 71) men. (Adapted from Mulligan, T. et al., *Europ. J. Endo.*, 141, 257, 1999.)

The precise nature and extent of altered sex-steroid negative feedback remain controversial. Indeed, in light of current knowledge of young-adult androgen secretion rates, many earlier feedback studies employed pharmacological amounts of sex steroids, thus limiting valid mechanistic interpretations.[6,9]

Plasma LH bioactivity has been measured by *in vitro* Leydig cell assays in the young and older male. Discrepant results in aging men probably reflect procedural differences, volunteer heterogeneity, and limited sample collection protocols.[24,42,53–55] The absence of elevated plasma LH bioactivity in many hypoandrogenemic older men would suggest impaired hypothalamo-pituitary

secretory drive and excessive negative feedback on the GnRH/LH unit (below).[6,30,41,42]

Experimental observations are consistent with a multilevel pathophysiology of androgen deficiency in the aging male rodent, similar to that of older men. For example, the senescent rat exhibits:

- Impoverished in situ hypothalamic GnRH synaptology, i.e., reduced number of synapses, reduced *in vitro* hypothalamic GnRH release
- Blunted *in vivo* LH pulse amplitude
- Preserved pituitary responsiveness to GnRH stimulation
- Impaired *in vitro* and *in vivo* Leydig-cell steroidogenesis

In addition, partial desensitization of effector-target tissue interfaces could contribute to hypogonadism in the aging male, e.g., downregulation of GnRH's drive of gonadotropes and of LH's stimulation of Leydig-cell androgen biosynthesis.[56] Available clinical studies have not yet explored the foregoing issues in detail.[45,57–59]

## III. PULSATILE LH RELEASE IN HEALTHY OLDER MEN

Compared with young adults, older men secrete LH:

- At a reduced pulse amplitude
- At higher pulse frequency
- With greater pattern irregularity[1,4,6,13,23,43,60]

A fundamental but unresolved issue in aging men is the precise basis for the inferred disruption of GnRH neuroregulation and the extent to which attendant Leydig-cell failure confounds such alterations.[6,28,39]

Hypoandrogenemia (decreased blood testosterone) in some older men evolves without measurable hypergonadotropism (increased secretion of gonadotropins) as assessed by radioimmunoassay, immunoradiometric, or immunofluorometric assay and *in vitro* bioassay of serum LH concentrations.[1,4,6,20,23,24,28,40,43,53–55,61] In a prospective study of healthy aging men, the mean decline in serum bioavailable testosterone concentrations over 15 years was 16%, and the percentage rise in serum LH concentrations was 10%.[40] Acute androgen deprivation amplifies pulsatile LH secretion by 100 to 200% in young men within 2 to 5 days[48,49,62] but by only 50 to 100% in older men.[49] In particular, experimental testosterone depletion does not stimulate incremental LH pulse amplitude fully in the elderly male (Figure 12.5).

Steroidogenic blockade in young men induces low-amplitude, high-frequency, and disorderly patterns of LH release, which reflects an "open-loop" feedback state

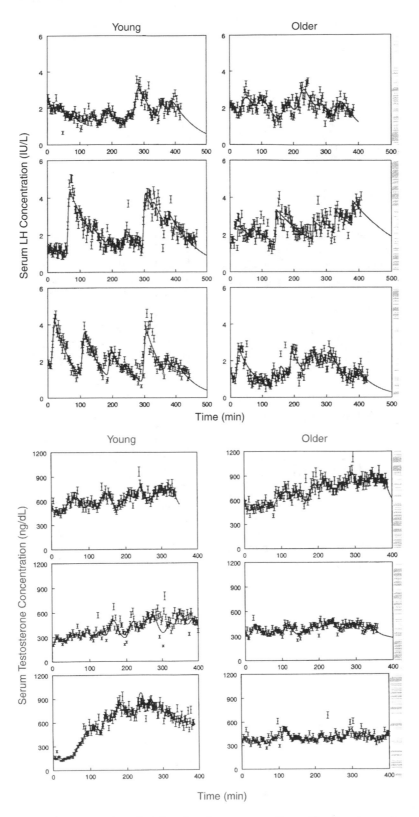

**FIGURE 12.3** Illustrative paired serum LH (Upper Panel) and total testosterone (Lower Panel) concentration profiles in healthy young and older men. Volunteers underwent repetitive (2.5 min) blood sampling overnight, beginning at sleep onset (time zero). Sera were submitted to chemiluminescence-based assay. Continuous curves are predicted by multiparameter deconvolution analysis (text). Each data point is defined further by the within-sample standard deviation (vertical bars). (Adapted from Mulligan, T. et al., *J. Clin. Endocrinol. Metab.*, 80, 3025, 1995.)

**Systemic and Organismic Aging**

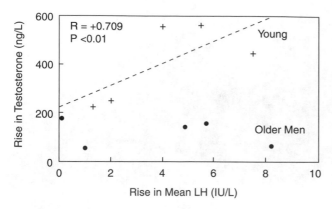

**FIGURE 12.4** Failure of pulsatile intravenous GnRH infusions for 14 days to elevate serum total testosterone concentrations equivalently ("Rise in Testosterone," y-axis) in older (solid circles) and young (plus signs and regression line) men, despite comparable increases in mean serum LH concentrations ("Rise in Mean LH," x-axis). (Adapted with permission from Mulligan, T. et al., *Europ. J. Endo.*, 141, 257, 1999.)

(Figure 12.6). The resultant neuroendocrine features closely mimic gonadotropin output observed at baseline in aging individuals (above).[49] Thus, the extent to which reduced testosterone bioavailability in older men accentuates their abnormal LH secretory phenotype will be important to establish.

## IV. EVIDENCE OF PARTIAL GNRH DEFICIENCY IN OLDER MEN

Hypothalamic GnRH deficiency has been postulated but not proven in aging men, as reviewed in Urban et al.[5] and Table 12.1.[6] The clinical hypothesis of an hypothalamic GnRH deficiency derives from a variety of indirect considerations:

1. The incongruity of normal or minimally elevated mean serum LH concentrations in the face of reduced testosterone bioavailability in many healthy ambulatory, community-living, and unmedicated older men[13,14,20,24,26,29,30,41,42,54,63]
2. An analogous disparity in the aged male mouse, rat, and horse[45,57,64]
3. A blunted rise in maximal LH production in old rodents following orchidectomy or during restraint stress[9,64]
4. The impairment of LH pulse-amplitude responses to pharmacological androgen depletion in older men (above)
5. The preservation of acute gonadotrope responsiveness to intravenous GnRH stimulation (Figure 12.7)[1,55,61]
6. The diminished high-amplitude LH pulses *in vivo* in older men and rodents

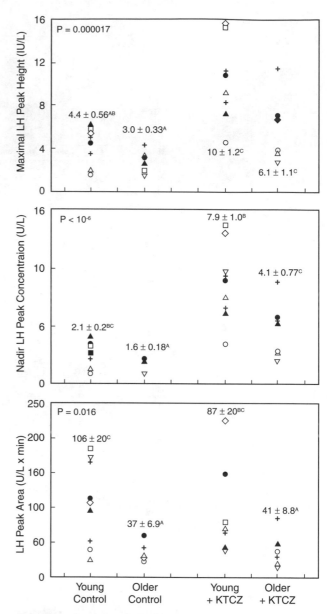

**FIGURE 12.5** Impaired amplification of pulsatile LH release in older men administered the steroidogenic inhibitor ketoconazole (KTCZ) to deplete systemic testosterone. Discrete peak-detection analysis (cluster) was used to quantitate serum LH concentration peak heights (top), interpulse nadir values (middle), and incremental LH peak areas (bottom). Numerical values are the mean ± SEM (N = 9 young and N = 7 older men). Unshared alphabetic superscripts identify significantly different group means. Adapted from Veldhuis, J.D., Zwart, A.D., and Iranmanesh, A., *Am. J. Physiol.*, 272, R464, 1997. With permission.)

7. The reduced *in vitro* GnRH secretion by hypothalamic explants isolated from aged male rats or mice[59]
8. The disruption of synapses from GnRH-secreting neurons in the hypothalamus of aging rats[58]

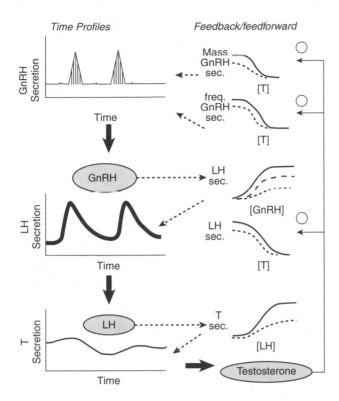

FIGURE 12.6 Dose-dependent interface functions encapsulating GnRH's drive of LH secretion (LH sec, top, and middle), LH's stimulation of testosterone (T) biosynthesis (bottom), and testosterone's feedback (-) on GnRH burst frequency and amplitude as well as LH pulse amplitude (top). Interrupted lines in the inset panels illustrate arbitrary shifts in dose-response properties, which could emerge in selected pathophysiologies. Adapted with permission from Keenan, D.M. and Veldhuis, J.D., *Am. J. Physiol.*, 275, E157, 1998.)

9. The restoration, by hypothalamic neuronal transplantation, of sexual activity in the impotent old male rat[43]

Hypothalamic hypogonadotropism with normal or heightened acute GnRH action as observed in older men would seem paradoxical. However, this anomaly is also evident:[6,61]

- In the ewe with surgical hypothalamo-pituitary disconnection
- In men with reversible fasting-induced hypogonadism
- In women with anorexia nervosa, exertional amenorrhea, and hyperprolactinemia[9,26,65]

Limited histopathological studies indicate that the pituitary content of immunoreactive gonadotropins is increased in aging men, thus potentially explicating the normal or heightened actions of exogenous GnRH. Alternatively, reduced sex-steroid negative feedback

### TABLE 12.1
### Indirect Evidence for Disrupted Hypothalamic Regulation of the Gonadotropic Axis in Aging Men

1. Loss of high-amplitude LH pulses despite normal or increased pituitary LH stores[1,13,43]
2. Accelerated frequency of LH secretory events[1,6,8,13,43]
3. Restitution of high-amplitude LH pulses by pulsatile, intravenous infusions of GnRH[1]
4. Limited LH secretory response to opiate-receptor blockade
5. Impaired LH secretory rise following antiestrogen exposure[24,]
6. Anomalous LH secretory response to antiandrogen administration[51,52]
7. Altered feedback restaint of LH secretion by testosterone or 5α-dihydrotestosterone[27,]
8. Increased disorderliness of spontaneous LH secretion patterns in older men[1,2,23]
9. Disrupted synchrony among LH, FSH, prolactin, testosterone, nocturnal penile tumescence (NPT), and sleep[2,3,23]
10. Normal or enhanced GnRH-stimulated LH, FSH, and α-subunit secretion[61]
11. Normal or low LH bioactivity in the face of reduced testosterone bioavailability[24,54]
12. Blunted 24-h rhythmicity of LH production[1,22]

due to partial Leydig-cell failure could enhance the effects of GnRH on pituitary gonadotropin secretion in aging men. [3,105]

## V.  ALTERED SEX-STEROID NEGATIVE FEEDBACK IN THE OLDER MALE

Clinical studies have reported normal, blunted, and accentuated inhibition of LH secretion by exogenous androgens in the aging male. Most investigations used pharmacological amounts of testosterone or 5α-dihydrotestosterone delivered nonphysiologically (e.g., transdermally, intramuscularly, or by continuous intravenous infusion).[27,50,66] Administration of a selective androgen-receptor antagonist in older men failed to normalize pulsatile LH secretion, thereby identifying androgen-independent alterations in aging.[5,6,67,68] On the other hand, the concentration of sex-steroid receptors falls in the brain and pituitary gland of the aged rat and in genital fibroblasts of older humans.[63,69] Reduced receptor number would predict relative resistance to androgen negative feedback, as inferred in two other recent clinical reports.[66] Likewise, cross-correlation analyses of paired serum LH and testosterone concentration time series revealed muted, rather than augmented, negative feedback by testosterone on LH release in elderly men.[14] In view of such widely conflicting inferences, further investigations will be needed to establish the nature of androgen negative-feedback signaling to GnRH and LH in the aging male.

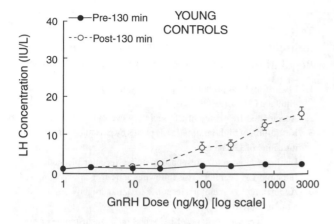

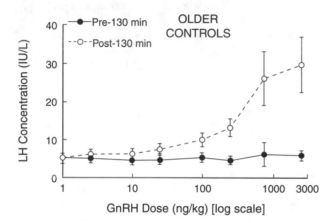

**FIGURE 12.7** Intravenous GnRH dose-gonadotropin secretory response curves observed in young and older men. Serum LH (y-axis) concentrations were analyzed by immunoradiometric assay. Data are the mean ± SEM obtained by sampling blood every 10 min before GnRH injection (pre-130 min) and after GnRH injection (post-130 min) in eight young (top panels) and seven older (bottom panel) men. Volunteers were studied at 8 h on separate days. GnRH doses were infused in randomly assigned order. Weight-adjusted GnRH doses are given on a logarithmic scale (x-axis). (Unpublished compilation of coauthors and Dr. Thomas Mulligan, Geriatrics and Extended Care, McGuire Veterans Affairs Medical Center, Richmond, Virginia.)

## VI. IMPAIRED LEYDIG-CELL TESTOSTERONE PRODUCTION IN THE AGING MALE

The number of interstitial-Leydig cells decreases in aged men,[9,37] and steroidogenic responsiveness to hCG (human chorionic gonadotropin) declines.[9,30,47] Nonetheless, earlier hCG stimulation studies may be criticized on several experimental grounds. First, variable concomitant (endogenous) LH exposure among study cohorts could confound interpretations. Second, the half-life of hCG is multifold longer than that of native LH; *viz.*, 18 to 30 h for hCG versus 0.75 to 1.5 h for LH.[70,71] A kinetically sustained hCG stimulus

would not mimic normal LH pulsatile drive of Leydig cells.[72] Third, unlike LH, hCG binds to the LH/hCG receptor nearly irreversibly and readily downregulates gonadal steroidogenesis.[9,73] And, fourth, a supraphysiological hCG stimulus would explore only maximal testis responsiveness. Accordingly, androgenic responses to hCG injection may hold little, if any, relevance to the clinical mechanism of testosterone deficiency in aging men, who maintain normal or only slightly elevated serum LH concentrations.

An alternative clinical research strategy to appraise Leydig-cell steroidogenic responsiveness is to infuse midphysiological pulses of recombinant human (rh) LH during short-term suppression of endogenous LH output via pretreatment with a potent GnRH-receptor agonist or antagonist. The latter paradigm of GnRH-receptor antagonist administration is illustrated in Figure 12.8, as applied initially in young men. The former experimental model of GnRH agonist pretreatment unveiled a 50% reduction in pulsatile rh LH-stimulated testosterone secretion in older, compared with young men when endogenous GnRH secretion was inhibited by leuprolide (a peptide, analog of GnRH that first stimulates and then inhibits pituitary secretion of LH).[74] Analogously, continuous subcutaneous delivery of LH in the aged male rat failed to normalize subsequent acute *in vitro* Leydig cell steroidogenic responsiveness.[57] Whether more prolonged and near-physiological pulsatile LH stimulation will restore *in vivo* gonadal androgen production in older men remains unknown.

## VII. MULTIPLICITY OF PRESUMPTIVE HYPOTHALAMIC REGULATORY ALTERATIONS IN THE OLDER MALE

A contemporary hypothesis of attenuated or disrupted hypothalamic GnRH secretion in older men[6,13,20,26,43,67] cannot be established (or dismissed) without directly quantitating the episodic secretion of GnRH and mediobasal hypothalamic (arcuate nucleus) multiunit electrophysiological activity in aged (male) animals. However, recent statistical analyses document deterioration of the orderly release of LH and testosterone, as well as of GH, ACTH, cortisol, and insulin in older individuals.[1,4,23,62] Regularity analysis via the approximate entropy statistic provides a barometer of altered within-axis feedback control. This concept differs from conventional pulsatility assessments (Figure 12.9 [left panal]). The approximate entropy metric quantitates consistently more irregular LH secretion patterns in older than young men (Figure 12.9 [right panel]). This distinction defines disruption of physiological linkages within the GnRH–LH–testosterone ensemble in older men.[23,62,75]

The complementary statistical tools of cross-correlation and cross-approximate entropy analyses have delineated asynchronous release of LH and testosterone,[14,23]

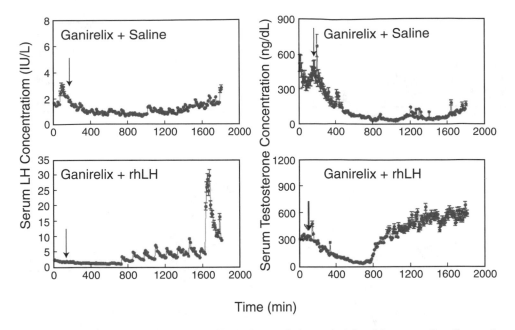

**FIGURE 12.8** Ability of recombinant human (rh) LH (50 IU, Serono Laboratories) [top] but not saline [bottom] to restore serum LH [left] and testosterone [right] concentrations in young men pretreated with a selective GnRH-receptor antagonist peptide (ganirelix) to block endogenous GnRH-dependent LH secretion. Serum concentrations of LH and testosterone were assayed in blood sampled every 10 min beginning 2 h before ganirelix injection. Ten hours after ganirelix treatment, seven successive (6-min squarewave) pulses of rh LH were infused i.v. every 2 h for 14 h. A supramaximal GnRH stimulus (500 µg) was administered at 1930 min (x-axis) to release pituitary LH stores. (Unpublished compilation with Dr. Christopher Fox, University of Virginia, Charlottesville, Virginia).

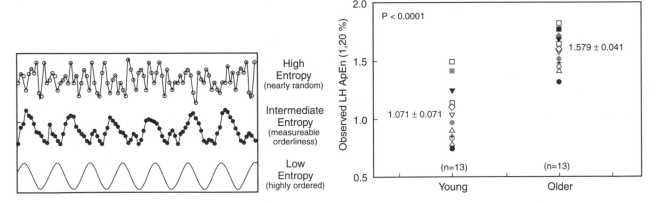

**FIGURE 12.9** Left Panel: Concept of approximate entropy (ApEn) analysis to quantitate the orderliness of neurohormone secretory patterns, as a barometer of network-level control. Profiles illustrate irregular (high ApEn, top), intermediate (middle), and reproducible (low ApEn, bottom) time-series patterns. Right Panel: Observed LH ApEn values in thirteen young and older men. Elevated ApEn in elderly individuals denotes a more disorderly LH release process, which, in turn, predicts impaired network (feedforward and feedback) coordination.[23,62,75] Serum LH concentrations were determined by high-precision immunoradiometric assay of blood sampled every 10 min for 24 h. ApEn (1, 20%) denotes a normalized parameter set of m = 1 (window length) and r = 20% (de facto threshold for the detection of pattern recurrence, defined as a percentage of each series standard deviation). Numerical values are the group mean ± SEM. (Unpublished compilation by the authors.)

LH and prolactin,[3] and LH and FSH,[2] and uncoupled oscillations of LH and nocturnal penile tumescence (NPT)[3] and LH and sleep stage in older men (Figure 12.10).[4] Thus, clinical data point to extensive disruption of neurohormone outflow in aging, which would putatively mirror deterioration of CNS-directed integrative control in the aging male.[1,2,4,6,23,62,67,75,76]

## VIII. ENSEMBLE CONCEPT OF THE INTERLINKED GNRH–LH–TESTOSTERONE FEEDFORWARD AND FEEDBACK AXIS

The dynamic nature of the male gonadal axis (Figures 12.1 and 12.5) can be encapsulated by deterministic

**Systemic and Organismic Aging**

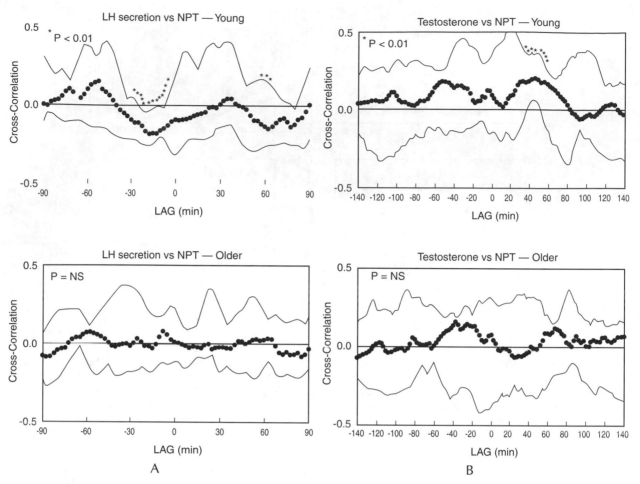

**FIGURE 12.10** Cross-correlation plots of paired (deconvolution-estimated) sample LH secretory rates and nocturnal penile tumescence (NPT) oscillations [Panel A] and matching NPT and serum total testosterone concentrations [Panel B]. Data are the median and range of cross-correlation coefficients (y-axis) in young (top) and older (bottom) men. Lag times (x-axis) represent the observed delay (min) between changing LH secretion and NPT or NPT and testosterone concentrations. A positive lag time indicates that the LH secretion rate rises before NPT declines [Panel A] or that the testosterone concentration increases before NPT increases [Panel B], and vice versa. In young men, LH secretion and NPT oscillations are coupled inversely over a 55 to 62.5 min NPT time lag and over a 7.5 to 32.5 min LH time lag (top, Panel A). Testosterone changes are linked positively to NPT variations 32.5 to 55 min later (top, Panel B). Both correlation features are abolished in aged individuals (bottom, panels A and B, where P = NS denotes P > 0.05). (Adapted with permission from Veldhuis, J.D. et al., *J. Clin. Endocrinol. Metab.*, 85, 1477, 2000.)

linkages and stochastic (apparently random) elements (Figure 12.11).[7,8,44,67,68,77] An analogous perspective applies to the somatotropic and corticotropic axes.[78,79] Stochastic features emerge at several levels in the reproductive system:

1. The random timing of successive GnRH pulse generation
2. The potentially inconstant feedforward and feedback linkages among GnRH, LH, and testosterone secretion
3. The variably time-delayed hormone action on target cells
4. The experimental uncertainties inherent in measuring hormone concentrations

A comprehensive formulation of GnRH–LH–testosterone dynamics thus includes stochastic variability superimposed on deterministic feedback and feedforward connections.[7,8,44,67,68,77,78] This biomathematical construct predicts a reduction in LH secretory burst mass and an acceleration in GnRH/LH pulsing frequency but normal LH elimination kinetics in healthy older men (Figure 12.12). Further development of such biomathematical forms should allow an exploration of endogenous hormone effector-target cell response interfaces.[7,8,44,67,68,77]

## A. MONOHORMONAL SECRETORY ASSESSMENT

Recently validated clinical research tools permit evaluation of monohormonal secretory contrasts in young and older men. They include:

1. Procedural/experimental uncertainties

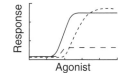

2. Cellular/secretory nonuniformity; blood admixture

3. Feedback/feedforward linkages

**FIGURE 12.11** *Stochastic (apparently random) variability inherent in monitoring in vivo neurohormone secretion. Each source of variation can modify the observed serum hormone concentration. (Original representation.)*

1. High-frequency and extended blood-sampling paradigms to monitor pulsatile LH and testosterone secretion[5,80–82]
2. Precise, sensitive, specific, and reliable automated assays of serum LH and testosterone concentrations[43,82–84]
3. Objective peak-detection methods and deconvolution-based techniques to quantitate neurohormone pulsatility and secretion[85–88]
4. Approximate-entropy (regularity) statistic to appraise the orderliness of feedback-dependent hormone patterns[23,75]
5. Selective inhibitors of gonadal and adrenal steroidogenesis to induce reversible hypoandrogenemia[48]
6. Pulsatile intravenous infusion of GnRH to impose a "hypothalamic GnRH clamp"[60]

7. Measurements of free alpha subunit as surrogate markers of LH pulses[61]

## 1. Frequent Blood Sampling

The functional GnRH–LH unit operates via intermittent rather than continuous drive.[9,11,56] Because GnRH triggers episodic LH output,[5] quantitation of pulsatile LH release can provide a window to GnRH secretion.[82] Validation experiments affirm that sampling blood at 5- or 10-min (but not 20- to 30-min) intervals for 12 to 24 h will capture the majority of detectable LH secretory episodes in healthy humans (Figure 12.13).[5,80,81] Suitable sampling protocols document low-amplitude LH pulses in older men.[1,6,13,28,30,43] Highly intensive (2.5-min) overnight monitoring of the gonadal axis has corroborated this finding.[1,13,23] Based on the ensemble concept illustrated above (Figure 12.6), impoverishment of high-amplitude LH secretory bursts could denote:

- Diminished hypothalamic GnRH drive
- Heightened negative-feedback restraint by androgens
- Reduced gonadotrope-cell responsiveness to physiological amounts of GnRH

Available clinical data largely exclude this last consideration.

Intensive blood-sampling schedules have unveiled not only lower-amplitude but also higher-frequency LH pulse generation in older men.[1,6,13,43,60] Accelerated LH pulsatility could represent a secondary rise in hypothalamic GnRH pulse-generator activity due to reduced negative feedback imposed by lower concentrations of bioavailable testosterone in the aging male (Figure 12.14). This hypothesis is based on clinical studies in young men that document that excessive androgen

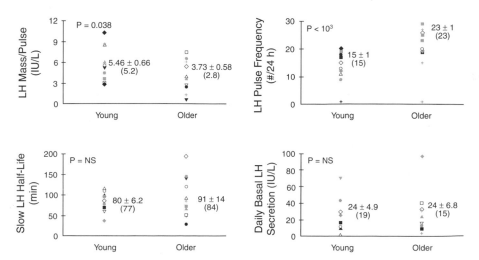

**FIGURE 12.12** Age contrasts in estimated LH secretory burst mass (top left) and LH pulse frequency (top right) in 13 young and older men, as predicted by a stochastic differential-equation (SDE-based) model of combined neurohormone secretion and elimination (text). The slow (delayed) phase LH half-life (bottom left) and the total daily LH secretion rate (bottom right) did not differ by age.[44]

action suppresses[83,89] and reduced androgen availability stimulates LH pulse frequency.[48,49] Alternatively, aging may be marked by a primary defect in GnRH pulse-generator control, which coexists with failure of Leydig-cell steroidogenesis.[1,6,9,47]

## 2. LH Assays

Quantitation of serum LH concentrations requires reliable, specific, valid, sensitive, and precise assay technology. Immunoradiometric, immunofluorometric, and chemiluminescence-based assays meet these requirements and often correlate well with *in vitro* LH bioassays.[5,6,9,24,51,53,71] To ensure high precision, one should use a fully automated (robotics) assay system and model-free data reduction procedures.[5,44,75,86]

## 3. Enumerating Serum LH Concentration Peaks and Quantitating Underlying LH Secretory Bursts

Discrete peak-detection techniques are used to quantitate pulsatile hormone release objectively.[5,85] One such validated method is model-free cluster analysis (Figure 12.15).[85] This particular computer-assisted methodology has been validated for by a combination of biomathematical LH-pulse simulations, LH infusions in leuprolide (an LH analog)-downregulated men, exogenous GnRH-stimulated LH pulses in patients with Kallmann's syndrome (isolated GnRH deficiency), repetitive sampling of GnRH in hypothalamo-pituitary portal blood of sheep, and recording of mediobasal hypothalamic electrophysiological correlates of GnRH release in the rhesus monkey.[1,5,10] The concept of the discrete peak-detection (cluster) approach is illustrated in Figure 12.15.

LH peaks in the blood are generated jointly by underlying secretion and elimination, namely:

- The amount (mass) of LH secreted within each burst
- The endogenous LH half-life
- Any concurrent basal (or nonpulsatile) hormone secretion
- Any prior secretory output that continues to decay during the observation interval[86,87]

Deconvolution analysis provides a family of techniques to quantitate such (unobserved) secretory and kinetic contributions to hormone pulses.[86,87] Two complementary classes of deconvolution procedures exist; *viz.*, model-specific and waveform-independent methods (Figure 12.16).[5,6,81,86,87] Several classes of deconvolution analysis corroborate an anomalous low-amplitude and high-frequency pattern of pulsatile LH secretion in older men.[6,13,43]

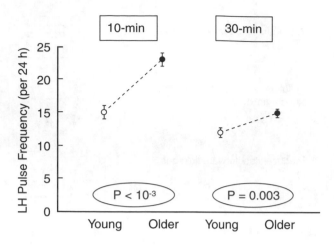

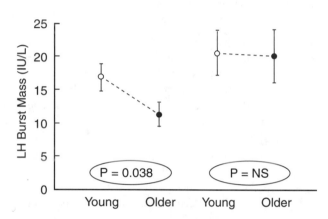

**FIGURE 12.13** Censoring effects of a 30 min versus 10 min frequency of blood sampling on SDE-based estimates of LH secretory burst frequency (top) and mass (bottom) in 13 young and older men (see Figure 12.12). Data are the mean ± SEM (unpublished compilation).

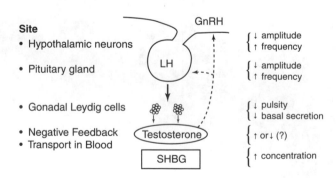

**FIGURE 12.14** Schematized summary of inferred neuroendocrine defects in the aging male GnRH–LH–testosterone–SHBG (sex hormone-binding globulin) feedback axis. (Original representation.)

## 4. Approximate Entropy of Hormone-Release Patterns

In addition to pulsatile features, LH time series exhibit measurable regularity properties. Regularity denotes the

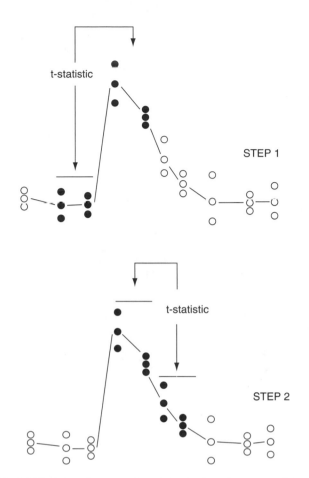

**FIGURE 12.15** Concept of statistically based two-step "cluster" analysis of serum hormone concentration peaks. This model-free algorithm scans the data series with a pooled-variance t-test to mark significant upstrokes (Step 1) and downstrokes (Step 2) which jointly demarcate a discrete neurohormone pulse.[85]

order-dependent reproducibility of moment-to-moment (nonpulsatile) subpatterns in the data (Figure 12.9). Orderliness of monohormonal release is quantitated via regularity statistics, such as approximate entropy (ApEn).[62,75,76] The ApEn metric is applicable to hormone profiles containing as few as 30 to 100 consecutive measurements. On theoretical grounds, ApEn provides a sensitive barometer of feedback and feedforward alterations within an interactive network.[2,23,62,75] For example, ApEn calculations have unveiled more irregular patterns of GH, ACTH, prolactin, and aldosterone secretion by neuroendocrine tumors;[26,75] more disorderly release of GH in the female than male rat and human; and more variable LH, testosterone, ACTH, cortisol, GH, and insulin output in aging men and women.[1,2,6,23]

Cross-ApEn provides an analogous two-variable regularity statistic, which quantitates the degree of synchronous release of paired hormones. Cross-ApEn analysis of matching LH and testosterone profiles has identified marked deterioration of bihormonal synchrony in the aging male.[1,23] To establish the basis for loss of

coordinate within-axis control requires additional pertinent "localizing" experiments, which test specific feedforward and feedback connections. In this regard, elevated cross-ApEn of each of LH-prolactin, LH-FSH, LH-NPT, LH-sleep, and NPT-sleep linkages in older men suggests globally altered reproductive neurohormone outflow in aging (Figure 12.17).[4]

## 5. Enforced GnRH "Clamp" Studies

Infusion of synthetic GnRH provides a clinical probe of the hypothalamic basis of hypogonadotropism, such as in pathological hyperprolactinemia, anorexia nervosa, Kallmann's syndrome (isolated GnRH deficiency), and fasting-induced male hypogonadism.[1,20,60,61] For example, 90-min pulsatile intravenous infusions of GnRH evoked comparable pulsatile, diurnal, and entropic measures of daily LH secretion in older and young men.[1] However, GnRH failed to stimulate 24-h testosterone production equivalently in older men, pointing to defective Leydig-cell steroidogenesis or reduced biological activity of LH in aging.[1] The latter hypothesis was rendered unlikely by comparable estimates of plasma LH concentrations by *in vitro* Leydig-cell bioassay.[1]

## 6. Ketoconazole-Induced Steroidogenic Blockade to Enforce Androgen Withdrawal

Ketoconazole, an antifungal agent, has been shown to block Leydig cells and adrenal steroidogenesis; it is being used to induce an acute and reversible decline in androgens, as a type of temporary pharmacologic castration in healthy men. Inhibition of adrenal and gonadal steroidogenesis by short-term administration of ketoconazole will invoke reversible castration-like unleashing of pulsatile LH secretion.[48,49] Testosterone production falls by approximately 85%, and this response triggers a reciprocal rise in LH production, thus allowing appraisal of endogenous GnRH release and actions.

## 7. Free Alpha-Subunit Measures

At very high gonadotropin pulse frequencies, short-lived free alpha subunit peaks are sometimes more vivid then LH pulses.[61] Alternatively, more frequent blood sampling and deconvoluton analysis will resolve rapid pulse frequencies.[48,82,86]

## B. Paradigmatic Tests of Altered System Feedback Control

The precise mechanisms that mediate the reduction in LH secretory burst mass, elevation in LH pulse frequency, and more irregular LH release patterns in aging men are not known. However, the foregoing alterations are specific, because FSH secretion is significantly elevated

**Systemic and Organismic Aging**

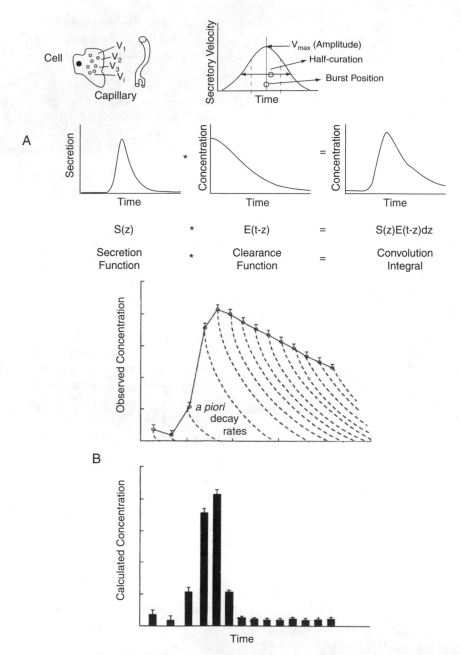

**FIGURE 12.16** Illustrative deconvolution procedures to dissect underlying hormone secretory activity based on a "Secretory Burst Model" (waveform-specific multiparameter deconvolution analysis, Panel A) or a "Waveform-Independent" methodology (nonparametric technique, Panel B). (Adapted from Veldhuis, J.D. and Johnson, M.L., *Methods in Neurosciences*, 28, 25, 1995..)

(Figure 12.18), and prolactin secretion is reduced in older men (Figure 12.19). Emergent hypotheses of aging pathophysiology can be considered and tested provisionally by computer-assisted models of the interactive GnRH–LH–testosterone axis.[7,8,67,68] One biomathematical construct incorporates a diurnal trend in LH-driven testosterone secretion and relevant feedback and feedforward (dose-response) interfaces to define connectivity among GnRH, LH, and testosterone signaling (Figure 12.1). The impact of postulated alterations in the aging (male) axis can be appraised in this formulation by pulsatility, approx-

imate entropy (ApEn), and 24 h rhythmic analyses of simulated pulse trains. Although informative, model-based approaches confront several unresolved issues, as highlighted in Table 12.2.

We have applied an ensemble-like construct of GnRH, LH, and testosterone linkages to explore selected clinical postulates in the aging male, *viz.*: (a) augmented negative feedback by testosterone on GnRH/LH output (left shift of the corresponding dose-response function), (b) blunted feedforward of LH on testosterone secretion (reduced maximum of the LH–testosterone interface function), (c)

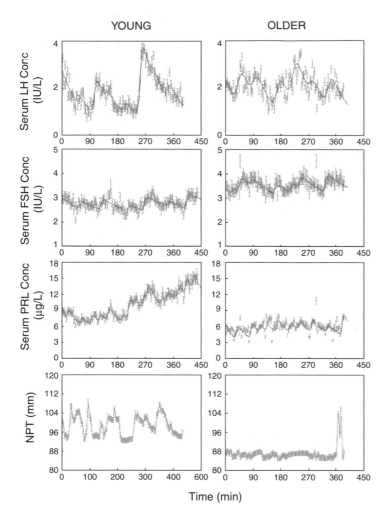

**FIGURE 12.17** Relationships among LH, FSH, prolactin, and nocturnal penile tumescence (NPT) oscillations in one young (left) and older (right) man monitored every 2.5 min overnight.[3]

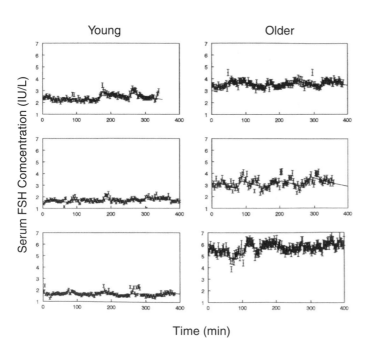

**FIGURE 12.18** Serum FSH concentrations in three young and older men sampled every 2.5 min overnight illustrate the rise in basal and pulsatile FSH output in the healthy aging male. (Reprinted with permission from Veldhuis, J.D. et al., *J. Clin. Endocrinol. Metab.,* 84, 3498, 1999.)

**Systemic and Organismic Aging**

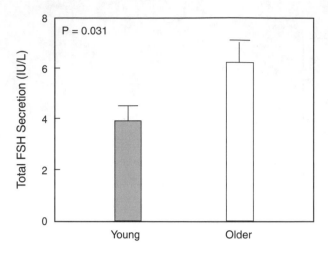

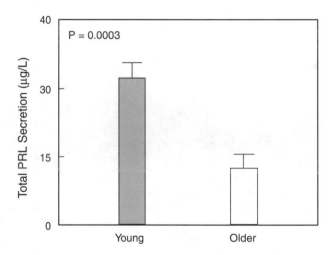

**FIGURE 12.19** Contrasts in overnight FSH (upper) and prolactin (lower) secretion in young and older men. In the aging male, FSH production rises, whereas that of prolactin (PRL) falls. Data are the mean ± SEM (N = 8 young and 7 older men). Compiled jointly with permission from Iranmanesh, A., Mulligan, T., and Veldhuis, J.D., *J. Clin. Endocrinol. Metab.*, 84, 1083, 1999 *J. Clin. Endocrinol. Metab.*, 84, 3506, 1999.)

**TABLE 12.2**

**Selected Neuroendocrine Modeling Issues in the Aging Male Reproductive Axis**

1. How do feedforward and feedback dose-response *interfaces* differ in aging?
2. How does *GnRH pulse-generator timing* behavior change with age?
3. Which alterations in the aging male reproductive axis reflect pathophysiology versus secondary adaptations?
4. What mechanisms account for greater *intersubject heterogeneity* in aging of GnRH-LH testosterone secretion?

a more rapid hypothalamic GnRH pulse-firing rate; (d) attenuated negative feedback of testosterone on GnRH/LH release, and (e) a combination of heightened feedback by testosterone on GnRH/LH output and impaired LH feedforward on testosterone secretion.[5,8,9,67,68] Only a simulated model involving restricted LH feedforward on testosterone secretion (alone or combined with impeded testosterone feedback repression of GnRH/LH) mimicked the known elevations of LH and testosterone ApEn (denoting more irregular monohormonal release patterns) and of LH-testosterone cross ApEn (identifying asynchronous LH and testosterone secretion) in aging men.[1,23] Inferentially impaired LH feedforward on testosterone secretion could result from diminished Leydig-cell steroidogenic capabilities; reduced testosterone feedback on GnRH/LH could, in turn, mirror impoverished testosterone bioavailability in the older male (above). More disorderly LH release in a low (testosterone) feedback state would accord with clinical findings in the renin–angiotensin and GH–IGF–I axes, wherein feedback withdrawal also drives secretory irregularity.[62]

## IX. CONCLUSIONS

Neuroendocrine axes maintain physiological homeostasis by adaptive signaling among interconnected glands. According to this general thesis, the dynamics of the aging male gonadotropic axis can be examined in relation to its ensemble interactive properties. Thus, hypothalamic neurons secrete the decapeptide GnRH in discrete bursts under feedback repression by testosterone. Intermittent pulses of GnRH drive time-delayed episodes of pituitary LH secretion. Fluctuating LH concentrations in systemic blood stimulate gonadal testosterone output via a dose-response interface. The present chapter underscores the interlinked nature of this network and highlights specific pathophysiological hypotheses of multilevel axis disruption in aging men.

## ACKNOWLEDGMENTS

We thank Elizabeth Lovestrand for skillful preparation of the manuscript; Paula P. Azimi for the deconvolution analysis, data management, and graphics; Brenda Grisso and Ginger Bauler for performance of various immunoassays; the nursing staff at the University of Virginia General Clinical Research Center for conduct of the research protocols; and Drs. Steven M. Pincus and Martin Straume for their mathematical contributions. This work was supported in part by NIH Grant MO1 RR00847 to the General Clinical Research Center of the University of Virginia Health Sciences Center, the Center for Biomathematical Technology, and NIH RO1 AG14799 and 1RO1AG 19695–01. This focused chapter necessarily omits many primary references because of editorial constraints. The authors therefore acknowledge numerous colleagues, who have made earlier foundational observations.

# REFERENCES

1. Mulligan, T. et al., Two-week pulsatile gonadotropin releasing hormone infusion unmasks dual (hypothalamic and Leydig-cell) defects in the healthy aging male gonadotropic axis, *Europ. J. Endo.*, 141, 257, 1999.

2. Pincus, S.M. et al., Effects of age on the irregularity of LH and FSH serum concentrations in women and men, *Am. J. Physiol.,* 273, E989, 1997.

3. Veldhuis, J.D. et al., Disruption of the young-adult synchrony between luteinizing hormone release and oscillations in follicle-stimulating hormone, prolactin, and nocturnal penile tumescence (NPT) in healthy older men, *J. Clin. Endocrinol. Metab.*, 84, 3498, 1999.

4. Veldhuis, J.D. et al., Older men manifest multifold synchrony disruption of reproductive neurohormone outflow, *J. Clin. Endocrinol. Metab.*, 85, 1477, 2000.

5. Urban, R.J. et al., Contemporary aspects of discrete peak detection algorithms. I. The paradigm of the luteinizing hormone pulse signal in men, *Endocr. Rev.*, 9, 3, 1988.

6. Veldhuis, J.D., Recent insights into neuroendocrine mechanisms of aging of the human male hypothalamo-pituitary-gonadal axis, *J. Androl.*, 20, 1, 1999.

7. Keenan, D.M. and Veldhuis, J.D., A biomathematical model of time-delayed feedback in the human male hypothalamic-pituitary-Leydig cell axis, *Am. J. Physiol.*, 275, E157, 1998.

8. Keenan, D.M. and Veldhuis, J.D., Explicating hypergonadotropism in postmenopausal women: A statistical model, *Am. J. Physiol. Regul. Integr. Comp. Physiol.*, 278, R1247, 2000.

9. Veldhuis, J.D., Male hypothalamic-pituitary-gonadal axis, in *Reproductive Endocrinology*, 4th ed., Yen, S.S.C., Jaffe, R.B., and Barbieri, R.L., Eds., W.B. Saunders Co., Philadelphia, PA, 1999.

10. Clarke, I.J. and Cummins, J.T., The temporal relationship between gonadotropin-releasing hormone (GnRH) and luteinizing hormone (LH) secretion in ovariectomized ewes, *Endocrinology*, 111, 1737, 1982.

11. Belchetz, P.E. et al., Hypophysial responses to continuous and intermittent delivery of hypothalamic gonadotropin-releasing hormone, *Science*, 202, 631, 1978.

12. Foresta, C. et al., Specific linkages among luteinizing hormone, follicle stimulating hormone, and testosterone release in the peripheral blood and human spermatic vein: Evidence for both positive (feedforward) and negative (feedback) within-axis regulation, *J. Clin. Endocrinol. Metab.*, 82, 3040, 1997.

13. Mulligan, T. et al., Amplified nocturnal luteinizing hormone (LH) secretory burst frequency with selective attenuation of pulsatile (but not basal) testosterone secretion in healthy aged men: Possible Leydig cell desensitization to endogenous LH signaling — a clinical research center study, *J. Clin. Endocrinol. Metab.*, 80, 3025, 1995.

14. Mulligan, T. et al., Aging alters feedforward and feedback linkages between LH and testosterone in healthy men, *Am. J. Physiol.*, 42, R1407, 1997.

15. Davies, T.F. and Platzer, M., The perifused Leydig cell: System characterization and rapid gonadotropin-induced desensitization, *Endocrinology*, 108, 1757, 1981.

16. Neaves, W.B. et al., Leydig cell numbers, daily sperm production, and serum gonadotropin levels in aging men, *J. Clin. Endocrinol. Metab.*, 59, 756, 1984.

17. Barrett-Connor, E., Goodman-Gruen, D., and Patay, B., Endogenous sex hormones and cognitive function in older men, *J. Clin. Endocrinol. Metab.*, 84, 3681, 1999.

18. Davidson, J.M. et al., Hormonal changes and sexual function in aging men, *J. Clin. Endocrinol. Metab.*, 57, 71, 1983.

19. de Lignieres, B., Transdermal dihydrotestosterone treatment of "andropause," *Ann. Med.*, 25, 235, 1993.

20. Winters, S.J. and Troen, P., Episodic luteinizing hormone (LH) secretion and the response of LH and follicle-stimulating hormone to LH-releasing hormone in aged men: Evidence for coexistent primary testicular insufficiency and an impairment in gonadotropin secretion, *J. Clin. Endocrinol. Metab.*, 55, 560, 1982.

21. Morley, J.E. and Kaiser, F.E., Hypogonadism in the elderly man, *Adv. Endocrinol. Metab.*, 4, 241, 1993

22. Tenover, J.S. et al., Age-related alterations in the circadian rhythms of pulsatile luteinizing hormone and testosterone secretion in healthy men, *J. Gerontol.*, 43, M163, 1988.

23. Pincus, S.M. et al., Older males secrete luteinizing hormone and testosterone more irregularly, and jointly more asynchronously, than younger males, *Proc. Natl. Acad. Sci. (USA)*, 93, 14100, 1996.

24. Urban, R.J. et al., Attenuated release of biologically active luteinizing hormone in healthy aging men, *J. Clin. Investigation*, 81, 1020, 1988.

25. Iranmanesh, A., Mulligan, T., and Veldhuis, J.D., Mechanisms subserving the physiological nocturnal relative hypoprolactinemia of healthy older men: Dual decline in prolactin secretory burst mass and basal release with preservation of pulse duration, frequency, and interpulse interval, *J. Clin. Endocrinol. Metab.*, 84, 1083, 1999.

26. Bergendahl, M. et al., Fasting suppresses pulsatile luteinizing hormone (LH) secretion and enhances the orderliness of LH release in young but not older men, *J. Clin. Endocrinol. Metab.*, 83, 1967, 1998.

27. Winters, S.J. and Atkinson, L., Serum LH concentrations in hypogonadal men during transdermal testosterone replacement through scrotal skin: Further evidence that aging enhances testosterone negative feedback, *Clin. Endocrinol.*, 47, 317, 1997.

28. Veldhuis, J.D. et al., Joint basal and pulsatile hypersecretory mechanisms drive the monotropic follicle-stimulating hormone (FSH) elevation in healthy older men: Concurrent preservation of the orderliness of the FSH release process, *J. Clin. Endocrinol. Metab.*, 84, 3506, 1999.

29. Tenover, J.S., Testosterone in the aging male, *J. Androl.*, 18, 103, 1997.

**Systemic and Organismic Aging**

30. Vermeulen, A., The male climacterium, *Annals of Medicine*, 25, 531, 1993.

31. Wise, P.M. et al., Neuroendocrine concomitants of reproductive aging, *Exp. Gerontol.*, 29, 275, 1994.

32. Katznelson, L. et al., Increase in bone density and lean body mass during testosterone administration in men with acquired hypogonadism, *J. Clin. Endocrinol. Metab.*, 81, 4358, 1996.

33. Kahn, E. and Fisher, C., REM sleep and sexuality in the ages, *J. Geriatr. Psychiatry*, 2, 181, 1969.

34. Morley, J.E. et al., Effects of testosterone on replacement therapy in old hypogonadal males: A preliminary study, *J. Am. Geriatr. Soc.*, 41, 149, 1993.

35. Urban, R.J. et al., Testosterone administration to elderly men increases skeletal muscle strength and protein synthesis, *Am. J. Physiol.*, 269, E280, 1995.

36. Cunningham, G.R. et al., Testosterone replacement therapy and sleep-related erections in hypogonadal men, *J. Clin. Endocrinol. Metab.*, 70, 792, 1990.

37. Tillinger, K.G. et al., The steroid production of the testicules and its relation to number and morphology of Leydig cells, *Acta Endocrinologica*, 19, 340, 1955.

38. Gray, A. et al., An examination of research design effects on the association of testosterone and male aging: Results of a meta-analysis, *J. Clin. Epidemiol.*, 44, 671, 1991.

39. Morley, J.E. et al., Longitudinal changes in testosterone, luteinizing hormone, and follicle-stimulating hormone in healthy older men, *Metab. Clin. Exp.*, 46, 410, 1997.

40. Madersbacher, S. et al., Serum glycoprotein hormones and their free α-subunit in a healthy elderly population selected according to the SENIEUR protocol. Analyses with ultrasensitive time resolved fluoroimmunoassays, *Mech. Aging Devel.*, 71, 223, 1993.

41. Gray, A. et al., Age, disease, and changing sex hormone levels in middle-aged men: Results of the Massachusetts male aging study, *J. Clin. Endocrinol. Metab.*, 73, 1016, 1991.

42. Kaiser, F.E. and Morley, J.E., Gonadotropins, testosterone, and the aging male, *Neurobiology of Aging*, 15, 559, 1994.

43. Veldhuis, J.D. et al., Attenuation of luteinizing hormone secretory burst amplitude is a proximate basis for the hypoandrogenism of healthy aging in men, *J. Clin. Endocrinol. Metab.*, 75, 52, 1992.

44. Keenan, D.M., Veldhuis, J.D., and Yang, R., Joint recovery of pulsatile and basal hormone secretion by stochastic nonlinear random-effects analysis, *Am. J. Physiol.*, 44, R1939, 1998.

45. Bonavera, J.J. et al., In the male brown-Norway (BN) male rat reproductive aging is associated with decreased LH-pulse amplitude and area, *J. Androl.*, 18, 359, 1997.

46. Roth, G.S. and Heiss, G.I., Changes in the mechanisms of hormone and neurotransmitter action during aging: Current status of the role of receptor and post-receptor alterations, *Mech. Ageing Dev.*, 20, 175, 1982.

47. Harman, S.M. and Tsitouras, P.D., Reproductive hormones in aging men. I. Measurement of sex steroids, basal luteinizing hormone, and Leydig cell response to human chorionic gonadotropin, *J. Clin. Endocrinol. Metab.*, 51, 35, 1980.

48. Veldhuis, J.D., Zwart, A.D., and Iranmanesh, A., Neuroendocrine mechanisms by which selective Leydig-cell castration unleashes increased pulsatile LH release in the human: An experimental paradigm of short-term ketoconazole-induced hypoandrogenemia and deconvolution-estimated LH secretory enhancement, *Am. J. Physiol.*, 272, R464, 1997.

49. Schnorr, J.A., Bray, M.J., and Veldhuis, J.D., Aromatization mediates testosterone's short-term feedback restraint of 24-hour endogenously driven and acute exogenous GnRH-stimulated LH and FSH secretion in young men, *J. Clin. Endocrinol. Metab.*, 86, 2600, 2001.

50. Vermeulen, A. and Deslypere, J.P., Long-term transdermal dihydrotestosterone therapy: Effects on pituitary gonadal axis and plasma lipoproteins, *Maturitas*, 7, 281, 1985.

51. Veldhuis, J.D., Urban, R.J., and Dufau, M.L., Evidence that androgen negative-feedback regulates hypothalamic GnRH impulse strength and the burst-like secretion of biologically active luteinizing hormone in men, *J. Clin. Endocrinol. Metab.*, 74, 1227, 1992.

52. Veldhuis, J.D., Urban, R.J., and Dufau, M.L., Differential responses of biologically active LH secretion in older versus young men to interruption of androgen negative feedback, *J. Clin. Endocrinol. Metab.*, 79, 1763, 1994.

53. Marrama, P. et al., Decrease in luteinizing hormone biological activity/immunoreactivity ratio in elderly men, *Maturitas*, 5, 223, 1984.

54. Mitchell, R. et al., Age-related changes in the pituitary-testicular axis in normal men; lower serum testosterone results from decreased bioactive LH drive, *Clin. Endocrinol.*, 42, 501, 1995.

55. Kaufman, J.M. et al., Influence of age on the responsiveness of the gonadotrophs to luteinizing hormone-releasing hormone in males, *J. Clin. Endocrinol. Metab.*, 72, 1255, 1991.

56. Marshall, J.C. and Kelch, R.P., Gonadotropin-releasing hormone: Role of pulsatile secretion in the regulation of reproduction, *N. Engl. J. Med.*, 315, 1459, 1986.

57. Grzywacz, F.W. et al., Does age-associated reduced Leydig cell testosterone production in brown Norway rats result from under-stimulation by luteinizing hormone?, *J. Androl.*, 19, 625, 1998.

58. Witkin, J.W., Aging changes in synaptology of luteinizing hormone-releasing hormone neurons in male rat preoptic area, *Neurosci.*, 22, 1003, 1987.

59. Sortino, M.A. et al., Different responses of gonadotropin-releasing hormone (GnRH) release to glutamate receptor agonists during aging, *Brain Research Bulletin*, 41, 359, 1996.

60. Aloi, J.A. et al., Pulsatile intravenous gonadotropin-releasing hormone administration averts fasting-induced hypogonadotropism and hypoandrogenemia in healthy, normal-weight men, *J. Clin. Endocrinol. Metab.*, 82, 1543, 1997.

61. Zwart, A.D. et al., Contrasts in the gonadotropin-releasing dose-response relationships for luteinizing hormone, follicle-stimulating hormone, and alpha-subunit release in young versus older men: Appraisal with high-specificity immunoradiometric assay and deconvolution analysis, *Eur. J. Endocrinol.*, 135, 399, 1996.

62. Veldhuis, J.D. et al., Secretory process regularity monitors neuroendocrine feedback and feedforward signaling strength in humans, *Am. J. Physiol.*, 280, R721, 2001.

63. Vermenlen, A., Deslypere, J.P., and Kaufman, J.J., Influence of antiopioids on luteinizing hormone pulsatility in older men, *J. Clin. Endocrinol. Metab,* 68, 68, 1987.

64. Sartin, J.L. and Lamperti, A.A., FSH and LH response to testosterone and gonadotropin-releasing hormone in aging male rats, *Exp. Aging Res.*, 10, 183, 1984.

65. Veldhuis, J.D. et al., Altered neuroendocrine regulation of gonadotropin secretion in women distance runners, *J. Clin. Endocrinol. Metab.*, 61, 557, 1985.

66. Muta, K. et al., Age-related changes in the feedback regulation of gonadotropin secretion by sex steroids in men, *Acta Endocrinol. (Copenh.)*, 96, 154, 1981.

67. Keenan, D.M. and Veldhuis, J.D., Hypothesis testing of the aging male gonadal axis via a biomathematical construct, *Am. J. Physiol. Regul. Integr. Comp. Physiol.*, 280, R1755, 2001.

68. Keenan, D.M., Sun, W., and Veldhuis, J.D., A stochastic bioma thematical model of the male reproductive hormone system, *SIAM J. Appl. Math*, 61, 934, 2000.

69. Ono, K. et al., Age-related changes in glucocorticoid and androgen receptors of cultured human pubic skin fibroblasts, *Gerontology*, 34, 128, 1988.

70. Yen, S.S.C. et al., Disappearance rates of endogenous luteinizing hormone and chorionic gonadotropin in man, *J. Clin. Endocrinol. Metab.*, 28, 1763, 1968.

71. Veldhuis, J.D. et al., Metabolic clearance of biologically active luteinizing hormone in man, *J. Clin. Investig.*, 77, 1122, 1986.

72. Winters, S.J. and Troen, P.E., Testosterone and estradiol are co-secreted episodically by the human testis, *J. Clin. Investig.*, 78, 870, 1986.

73. Glass, A.R. and Vigersky, R.A., Resensitization of testosterone production in men after human chorionic gonadotropin-induced desensitization, *J. Clin. Endocrinol. Metab.*, 51, 1395, 1980.

74. Mulligan, T., Iranmanesh, A., and Veldhuis, J.D., Pulsatile IV infusion of recombinant human LH in leuprolide-suppressed men unmasks impoverished Leydig-cell secretory responsiveness to midphysiological LH drive in the aging male, *J. Clin. Endocrinol. Metab.*, 86, 5547, 2001.

75. Veldhuis, J.D. and Pincus, S.M., Orderliness of hormone release patterns: A complementary measure to conventional pulsatile and circadian analyses, *Eur. J. Endocrinol.*, 138, 358, 1998.

76. Pincus, S.M., Approximate entropy as a measure of system complexity, *Proc. Natl. Acad. Sci. (USA)*, 88, 2297, 1991.

77. Keenan, D.M. and Veldhuis, J.D., Stochastic model of admixed basal and pulsatile hormone secretion as modulated by a deterministic oscillator, *Am.J. Physiol. Regul. Integr. Comp. Physiol.*, 273, R1182, 1997.

78. Keenan, D.M., Licinio, J., and Veldhuius, J.D., A feedback-controlled ensemble model of the stress-responsive hypothalamo-pituitary-adrenal axis, *Proc. Natl. Acad. Sci. (USA)*, 98, 4028, 2001.

79. Farhi, L.S. et al., A construct of interactive feedback control of the GH axis in the male, *Am. J. Physiol.*, 281, R38, 2001.

80. Veldhuis, J.D. et al., Performance of LH pulse detection algorithms at rapid rates of venous sampling in humans, *Am. J. Physiol.*, 247, 554E, 1984.

81. Veldhuis, J.D., Lassiter, A.B., and Johnson, M.L., Operating behavior of dual or multiple endocrine pulse generators, *Am. J. Physiol.*, 259, E351, 1990.

82. Mulligan, T. et al., Validation of deconvolution analysis of LH secretion and half-life, *Am. J. Physiol.*, 267, R202, 1994.

83. Wang, C. et al., Graded testosterone infusions distinguish gonadotropin negative-feedback responsiveness in asians and white men — a clinical research center study, *J. Clin. Endocrinol. Metab.*, 83, 870, 1998.

84. Wu, F.C.W. et al., Patterns of pulsatile luteinizing hormone secretion from childhood to adulthood in the human male: A study using deconvolution analysis and an ultrasensitive immunofluorometric assay, *J. Clin. Endocrinol. Metab.*, 81, 1798, 1996.

85. Veldhuis, J.D. and Johnson, M.L., Cluster analysis: A simple, versatile and robust algorithm for endocrine pulse detection, *Am. J. Physiol.*, 250, E486, 1986.

86. Veldhuis, J.D. and Johnson, M.L., Specific methodological approaches to selected contemporary issues in deconvolution analysis of pulsatile neuroendocrine data, *Methods in Neurosciences*, 28, 25, 1995.

87. Veldhuis, J.D., Carlson, M.L., and Johnson, M.L., The pituitary gland secretes in bursts: Appraising the nature of glandular secretory impulses by simultaneous multiple-parameter deconvolution of plasma hormone concentrations, *Proc. Natl. Acad. Sci. USA,* 84, 7686, 1987.

88. Veldhuis, J.D. et al., Assessing temporal coupling between two, or among three or more, neuroendocrine pulse trains: Cross-correlation analysis, simulation methods, and conditional probability testing, *Methods in Neurosciences*, 20, 336, 1994.

89. Veldhuis, J.D. et al., Role of endogenous opiates in the expression of negative feedback actions of estrogen and androgen on pulsatile properties of luteinizing hormone secretion in man, *J. Clin. Investig.*, 74, 47, 1984.

**Systemic and Organismic Aging**

# 13 The Thyroid, Parathyroid, and Pineal Glands*

*Paola S. Timiras*
University of California, Berkeley

## CONTENTS

## I.   INTRODUCTION

In the present chapter, aging of the thyroid gland and its possible consequences on the whole organism and on specific functions such as basal metabolism and thermoregulation are considered first. As in the case of other neuroendocrine systems (Chapters 10 to 12), the thyroid will be examined within the framework of the hypothalamo–pituitary–thyroid axis. Second, changes with aging of the parathyroids and of the thyroid C-cells secreting calcitonin are included by virtue of their intimate anatomic association with the thyroid gland. Third, included in this chapter is a discussion of the aging of the pineal gland and its role as a photoneuroendocrine transducer, a synchronizer of the sleep–wake and night–day cycles, and as a possible contributor to those homeostatic networks that regulate the life span.

---

* Illustrations by Dr. S. Oklund.

0-8493-0948-4/03/$0.00+$1.50
© 2003 by CRC Press LLC

## II.  THE THYROID GLAND

### A.  THYROID FUNCTION AND REJUVENATION?

The hormones of the thyroid gland, primarily thyroxine (T4) and triiodothyronine (T3), have traditionally engaged the interest of investigators seeking hormonal determinants of aging. Early studies in humans suggested that certain signs of aging resembled those of thyroid insufficiency or hypothyroidism. Hypothyroid individuals develop a number of signs that might be interpreted as "precocious senility," including a reduced metabolic rate, hyperlipidemia and accelerated atherosclerosis, early aging of skin and hair, slow reflexes, and slow mental performance.[1] Because such patients improve markedly after hormonal replacement therapy, it was argued that similar symptoms in normally aging individuals may represent effects secondary to thyroid involution with old age. The well-known action of thyroid hormones in controlling whole-body growth and brain development[1] made it reasonable to suspect that these hormones might also control the rate or site of aging.

As early as the last decade of the 19th century, researchers optimistically attempted rejuvenation or prolongation of life through hormone administration (analogous to the current hormone replacement therapy). In later studies, while thyroid hormone administration to some animals (mice) seemed to shorten rather than prolong life,[2] in others (fowl), it caused an apparent dramatic rejuvenation.[3] Delayed and impaired growth and maturation associated with a significant prolongation of the life span were also reported in rats made hypothyroid at an early (first postnatal week) age[4,5] and mimicked the effects reported after food restriction (Chapter 24). However, the aging process in euthyroid (with normal thyroid state) elderly humans was never significantly slowed or altered by the administration of these hormones, and the life span was not affected by hypo- or hyperthyroidism.[6,7]

The capacity to maintain an euthyroid (normal) state continues into old age and is well preserved in centenerians despite a number of changes in various aspects of thyroid hormone production, secretion, and action.[8,9] However, as for all other endocrines, normal thyroid function in the elderly may easily be endangered by repeated challenging demands and stress; given the physiologic importance of thyroid hormones, the ensuing dysthyroid (abnormal) state might lead to decreased overall functional competence, disease, and aging.

During development, the thyroid gland is necessary for whole-body and organ growth and development and maturation of the central nervous system. In adulthood, it has an essentially metabolic function, regulating tissue oxygen consumption, thereby maintaining metabolic rate and body temperature. Although the thyroid gland is not essential for life, in its absence, there is poor resistance to cold, mental and physical slowing, and, in children, mental retardation and dwarfism. Reciprocally, in hyperthyroidism, metabolic and behavioral alterations threaten well-being and survival.[1] Some morphologic, structural, and physiologic characteristics of the thyroid gland and the hypothalamo–pituitary–thyroid axis are presented in Boxes 13.1 and 13.2, Table 13.1, and Figures 13.1 to 13.3.

### B.  STRUCTURAL CHANGES WITH AGING

From maturity to old age, the size of the thyroid gland decreases, although in some cases, the weight of the gland may remain unchanged or even increase. The increase is due most often to the presence of nodules (small lumps or masses of hyperplastic and hypertrophic cells) or to mild endemic goiter (hypothyroidism due to iodine deficiency).[10–12] The general shape of the gland remains unchanged, although microscopic changes are frequent and suggest reduced function (Table 13.2).

When the cells are inactive, as in the absence or deficiency of TSH or in some old individuals, follicles are large, more colloid is accumulated in the follicular cavity due to reduction of endocytosis, and the lining cells are flat and display signs of reduced secretory activity (Table 13.2). However, in acute illness, thyroid hormonal output may be greatly increased (see below).[13–15] When thyroid cells are active, as following TSH stimulation, follicles are small and cells are cuboidal or columnar with signs of active endocytosis, i.e., transport of the colloid into the cell.

#### NODULES

Nodules, primarily involutional in older individuals, are characterized by localized tissue proliferation with overlapping areas of cell involution (reduction in size and function) or stimulation.[15] The prevalence of micronodules and clinically palpable nodules may both increase and decrease with aging. The variability of this finding reflects geographic differences in the prevalence of endemic goiter (overall enlargment of the thyroid) due essentially to iodine deficiency.[11,15,16] Endemic goiter is more prevalent in young women than men; it may be caused by a relative iodine deficiency associated with poor dietary habits and nutritional deficiencies that may occur at adolescence or the increased nutritional demands created by pregnancy. Overall, the prevalence of thyroid nodules increases with age. Nodules are present in 90% of women over the age of 70 and 60% of men over the age of 80, whereas the prevalence of goiter tends to decrease (Figure 13.4). Multinodular goiter is also associated with increased antithyroid antibodies. The much lower prevalence of goiter and nodules in men than women remains essentially unchanged into old age.

### C.  CHANGES WITH AGING IN THYROID HORMONE SERUM LEVELS

Changes with aging in the serum levels of thyroid hormones are ambiguous due to fluctuations in response to a variety of stimuli and disease. Total and free T4 serum levels may be unchanged, slightly decreased, or increased depending upon age, sex, and general health. There is also a greater

**Box 13.1**

**Some Structural Characteristics of the Thyroid Gland and the Hypothalamo–Pituitary–Thyroid Axis**

The thyroid gland is a bilateral organ that bridges the lower larynx and upper trachea with a narrow isthmus. A third pyramidal lobe, remnant of the thyroglossal duct, is not unusual. As one of the most vascularized endocrines, it receives blood from the superior thyroid arteries, branches off the external carotid artery, and is drained by corresponding veins into the internal jugular vein (Figure 13.1). In normal individuals, vascularity, size, and microscopic structures vary with TSH levels, nutrition, temperature, sex, and age.

The functional units of the thyroid gland are multiple, variable-sized follicles (or cavities), filled with a colloid rich in a glycoprotein, thyroglobulin, which contains the thyroid hormones, T3 and T4, in precursor forms. Each follicle is lined by a single layer of epithelial cells resting on a basement membrane (Figure 13.2). The cell surface lining the follicle is rich in microvilli (visible at high magnification) that project into the follicular lumen where the colloid is secreted; hormones are secreted into the blood at the opposite basal cell pole adjoining the rich capillary net.

The function of the thyroid gland in the elderly, as in individuals of all ages, can be best viewed in the context of:

- Regulation by the hypothalamo–pituitary–thyroid axis
- Thyroid hormone metabolism
- Interaction of thyroid hormones with receptors at target cell

These connected levels of integration (Figure 13.3) can be briefly outlined as follows:

1. At the level of the hypothalamus, thyrotropin-releasing hormone (TRH), a tripeptide, is synthesized in neurons of the hypothalamus and secreted from their nerve endings into the primary capillary plexus of the pituitary portal system (Chapter 10) in the median eminence. TRH is delivered by portal vessels to the thyrotropes (cells secreting TSH) of the pituitary anterior lobe. TRH stimulates these thyrotropes to synthesize and secrete thyroid stimulating hormone (TSH). TRH synthesis and secretion is regulated by events in the nervous system.
2. At the level of the pituitary anterior lobe, TSH, a glycoprotein secreted by the thyrotropes, stimulates, as its name implies, the cells of the thyroid gland to synthesize and secrete the thyroid hormones, T4, and T3. Synthesis and secretion of TSH is regulated in a dual fashion: it is stimulated by hypothalamic TRH and inhibited by the thyroid hormones, in a so-called negative feedback loop (Chapter 10), i.e., the higher the serum levels of thyroid hormones, the lower TSH release and vice versa. TSH, in addition, maintains the integrity of the structure and growth and vascularity of the gland; in the absence or deficit of TSH, the gland atrophies. Thyroid hormones, their receptors, and function at target cells are summarized in Box 13.2.

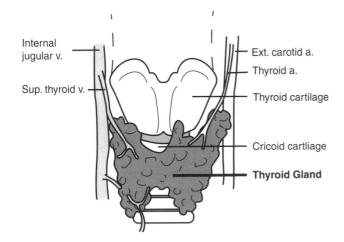

**FIGURE 13.1** Diagrammatic representation of the thyroid gland. The thyroid and cricoid cartilages and the major blood vessels are included. Note that four parathyroid glands are imbedded in the superior and inferior poles of the thyroid gland. C cells secreting calcitonin are dispersed throughout the thyroid gland.

individual variability among the elderly than young subjects. Despite this greater variability among elderly than young subjects, it may be stated that values in the elderly are in the lower normal range for T4 and T3 and in the higher normal range for TSH; borderline abnormal values are more usual in women than men (Table 13.2).[17–19]

The observation that free and bound T4 levels remain unchanged in old age is reminiscent of the case of the adrenocortical hormones, where a slower metabolism (in the liver and tissues) and excretion (by the kidney) of the hormones compensate for the reduced secretion and maintain the levels essentially constant (Chapter 10). A reported significantly reduced free and bound serum T3 in some aged individuals may reflect:

1. Impaired T4 to T3 conversion in tissues
2. Increased T3 degradation (metabolism or excretion)
3. Decreased T3 secretion from the thyroid gland due to either failure of stimulation by TSH or intrinsic, primary alteration of the gland

**Systemic and Organismic Aging**

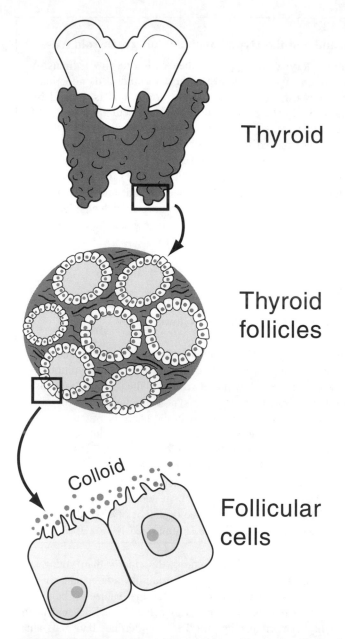

**FIGURE 13.2** Diagrammatic representation of the thyroid (on the top), the thyroid follicles (in the middle), and the follicular cells (at the bottom): the microvilli from the follicular cells project into the follicular colloid.

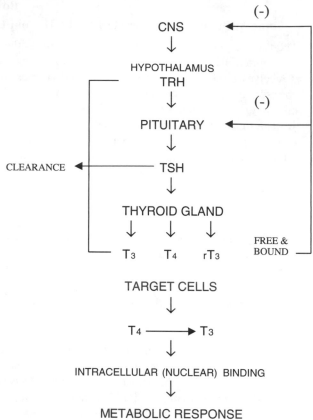

**FIGURE 13.3** Diagram of the interrelations of the hypothalamo–pituitary–thyroid axis.

---

**TABLE 13.1**

**Some Critical Aspects of Thyroid Hormone Regulation**

1. The major source of circulating T3 is not from thyroid gland secretion, but from peripheral deiodination of T4
2. The negative feedback at the pituitary anterior lobe is mainly through T4 taken from the circulation and converted in the thyrotrope to T3 by thyrotrope T4 deiodinase
3. The peripheral deiodination of T4 depends on the physiological state of the organism; it allows an autonomy of response of the tissues to the hormones
4. Deiodination can convert T4 (a less biologically active hormone) to T3 (a more active hormone); this conversion depends on activities of the various deiodinating enzymes

---

Based on the current data, it is impossible to distinguish among the foregoing possibilities. However, T4 to T3 conversion falls with age, at least in experimental animals (rat).[20] With respect to laboratory animals, not only are there differences from humans but also intraspecies variation. In the Sprague-Dawley rat, one of the most used animals for aging research, serum T4 rather than T3 is markedly reduced with aging.[21]

A certain number of elderly show elevated TSH with a T4 that falls within the "normal" range (as established by values obtained in the younger adult population).

Under normal feedback conditions, TSH levels are elevated in response to low T4 values. In a good percentage of elderly, high TSH in the presence of normal T4 levels may indicate alterations in thyroid–pituitary–hypothalamus feedback.[22–24] It may also suggest that a significant portion of the elderly are hypothyroid and may benefit from hormone replacement therapy. Inasmuch as hormonal levels are measured in immunologic assays, high

---

**Box 13.2**

**The Thyroid Hormones**

At the level of the thyroid gland, under stimulation by anterior pituitary lobe TSH, the thyroid follicular cells synthesize and secrete, through their apical borders into the follicular lumen, the high molecular weight glycoprotein, thyroglobulin. Iodide, trapped into the cells from blood capillaries by the basal cell membrane, then oxidized by the enzyme thyroid peroxidase (TPO) to the state of molecular iodine at the luminal cell surfaces, iodinates some of the tyrosyl residues of thyroglobulin. The magnitude of the iodide trap and the activity of TPO are augmented by TSH. Again, under the influence of TSH and TPO, some of the iodotyrosyl residues are coupled to iodothyronyll residues (incipient T4 and T3). Iodinated thyroglobulin is hydrolyzed in the follicular lumen, T4 and T3 liberated from peptide linkage, and secreted in the blood. Monoiodotyrosine and diiodotyrosine are deiodinated intracellularly, and the resulting iodide is recycled to iodinate thyroglobulin. Small amounts of thyroxine, T4, triiodothyronine, T3, and to a much lesser extent, reverse T3, rT3, are secreted from the gland. Binding of TSH to its receptors on the basal membranes of the follicular cells stimulates the enzyme, adenyl cyclase, and causes an increase in intracellular cAMP that mediates most of the stimulatory actions of TSH on these cells. Some thyroid actions of TSH are, in addition, due to stimulation of cell membrane phospholipids.

The major secreted hormonal product of the thyroid gland is T4; T3 is secreted only in small amounts, and its concentration in the blood derives mainly from the peripheral deiodination of T4. One third of circulating T4 is converted to T3 in peripheral tissues. Both hormones are present in serum bound to proteins or in the free state. T3 is less tightly bound to plasma proteins than is T4 and is, therefore, more readily available for cellular uptake. The free hormone is biologically active and interacts with specific receptors localized in the membrane, mitochondria, cytoplasm, and nucleus of responsive cells. T3 binds to nuclear receptors more tightly than T4, hence, T3 is more rapidly and biologically active than T4. T3 and T4 are deiodinated and deaminated in the tissues. In the liver, they are conjugated, passed into the bile, and excreted into the intestine. Conjugated and free hormones are also excreted by the kidney. Some critical aspects of thyroid hormone regulation are summarized in Table 13.1.

---

**TABLE 13.2**

**Some Morphologic and Secretory Changes in the Thyroid Gland with Aging**

**Morphologic**

Distension of follicles

Variant color of colloid (after staining)

Flattening of the follicular epithelium suggestive of reduced secretory activity

Fewer mitoses

Increased fibrosis of interstitial connective tissue and parenchyma

Vascular changes of atherosclerotic nature, suggestive of decreased transport of hormones between cells, blood, and follicles

**Secretory**

Lower circulating T3 levels but generally within the normal (lower) range

Simultaneously decreased secretion and metabolic clearance of T4 with resulting essentially normal levels

Decreased peripheral conversion of T4 to T3

Failure of upregulation of T3 nuclear receptors

Elevated TSH levels in 10% of the elderly, associated with an increase in antithyroid antibodies, present even in the absence of manifestations of hypothyroidism

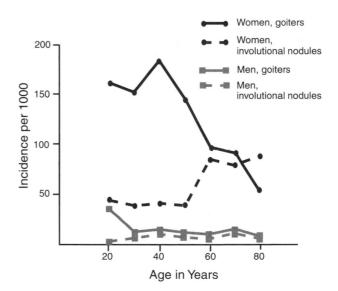

**FIGURE 13.4** Incidence of goiter and involutional nodules in the thyroid gland as a function of age in women and men.

### D. Changes with Aging in the Hypothalamo–Pituitary–Thyroid Axis

As described earlier, circulating TSH, the principal regulator of thyroid gland function, is regulated by the hypothalamic-releasing hormone (TRH) and by direct inhibitory feedback of high circulating thyroid hormone levels (Figure 13.3). In this way, the pituitary–thyroid axis is

TSH vis-à-vis normal T4 levels may be due to altered TSH molecules that retain immunoreactivity but are less biologically active, thus requiring higher TSH levels to maintain normal T4 (see below).

**Systemic and Organismic Aging**

capable of bringing about the appropriate adjustments to internal and external environmental changes. Other regulators of thyroid function include the following:

1. Direct autonomic inputs (thyroid follicular cells are innervated by the sympathetic nervous system and are sensitive to sympathetic signals)
2. Possible influence of the neurotransmitter, serotonin, perhaps involved in the regulation of circadian TSH periodicity
3. Possible involvement of the cytokines from the immune system (Chapter 15) that may be responsible for some of the thyroid alterations in nonthyroidal disease, with mechanisms that are still unclear
4. The inhibitory role of neuropetide Y, a member of the pancreatic polypeptide family (Chapter 14) capable of inhibiting TSH secretion[20]

Another little-known local regulation may be at the thyroid by some still unidentified iodinated compounds.

The major regulatory pathway remains the TRH–TSH axis. When TRH–TSH levels are low, the ability of the thyroid gland to secrete thyroid hormones is reduced, and vice-versa, when levels are high, the activity of the gland is increased. TRH–TSH action on the thyroid gland involves:

- Increase in iodide trapping and binding
- Stimulation of T3 and T4 synthesis
- Promotion of thyroglobulin secretion into colloid at the thyroid level
- Promotion of colloid endocytosis into thyroid cells
- Increase in blood flow at the thyroid level

Prolonged stimulation of the thyroid gland by TSH results in cell hypertrophy with overall enlargement of the gland or goiter.

With aging, TSH circulating levels are usually elevated when T4 or T3 circulating levels are lowered. However, in rats, qualitative changes also occur. TSH, a glycoprotein, is present in several forms with different molecular weights and immunoreactivity. In old rats, there is a progressive decrease in the quantitatively major TSH and also biologically most active form, and an increase in high and low molecular weight species.[21] The biological significance of this age-related increased TSH polymorphism remains speculative. Polymorphism also occurs in other glycoproteic hormones of the pituitary, such as the gonadotropins, in which the proportion of polymorphism increases with aging. In rats, not only are TSH levels and TSH polymorphism increased with aging, but the typical circadian cyclicity of the hormone is abolished as well.[22] The functional significance of TSH rhythmicity is still obscure; however, the loss of specific pulsatile signals may suggest a progressive failure with aging of fine tuning of thyroidal function (for neuroendocrine control of hormone pulsatility see Chapter 12). Such a failure is supported by the observation that in some species, the low T3 and T4 serum levels are associated with normal or reduced (rather than elevated) TSH levels. Inasmuch as the major amounts of pituitary T3 (effective in the negative feedback inhibition of TSH) may be derived from intrapituitary deiodination of T4, it is possible that the low thyroid hormone levels are due to decreased systemic deiodination (without affecting the levels of pituitary T3) or that the thyroid hormones-pituitary feedback is impaired. For TSH as well as for hypothalamic TRH, current results are in conflict, particularly in relation to sex differences, a decrease in TRH having been reported exclusively in either males or females, depending on the investigator. Increased TSH responsiveness to TRH has been reported in apparently healthy elderly subjects in whom T4 levels are unchanged and basal TSH levels are slightly increased.

## E.  THYROID HORMONE RECEPTORS

Thyroid hormone receptors are within the same family of nuclear receptors as those for glucocorticoids (Chapter 10). An analogy between the two classes of hormones may be useful: glucocorticoid levels do not change significantly in elderly individuals, and yet, the ability to withstand stress is decreased, and the probability of the ensuing pathology (e.g., diseases of adaptation) is increased. Similarly, the relative adequacy of the pituitary–thyroid axis into old age seems to belie occurrence of symptoms of altered thyroid state of the elderly. One possible explanation is that alterations with aging occur essentially at the peripheral level, primarily within the cell, and involve the hypothalamo–pituitary–thyroid axis to a much lesser extent. The study, then, of intracellular hormone receptors and their eventual changes with aging and their impact on target cell function may prove useful in the elucidation of intracellular alterations.

Thyroid hormone receptors, primarily for T3, have been identified in nuclei, mitochondria, plasma membranes, and cytosol of target cells (Figure 13.5). The biologic actions of the hormone are by interaction of T3 with the nuclear receptors that bind to regulatory regions of genes (the thyroid hormone-response elements), thereby modifying gene expression and stimulating protein synthesis.[22] There are two classes of nuclear T3 receptors, each subdivided into two subgroups, the α (α1 and α2) and β (β1 and β2) receptors located on chromosomes 17 and 3, respectively.[25] They are expressed in almost all tissues, except for β2 receptors limited to the brain. Both α1 and β receptors, when occupied, activate a T3 response element *in vitro*, but they differ in potency; the α2 receptor does not bind to the T3 receptor and may inhibit binding of T3 receptors to DNA.[26,27]

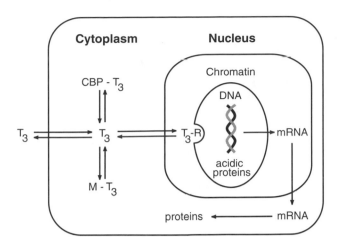

**FIGURE 13.5** Model for T3 interaction with target cell: CBP = cytosol binding proteins, M = mitochondria, R = nuclear receptor. As T3 enters the cell, it may be bound to a cytosol-binding protein, in reversible equilibrium with a small pool of free T3 that can interact reversibly with T3 receptors in the nucleus and perhaps also with receptors in the mitochondria.

In general, the number of receptors appears to be inversely related to hormone levels, so that receptors increase in number when the hormone level is low, as in hypothyroidism (upregulation), and decrease in number when the hormone level is high, as in hyperthyroidism (downregulation). *In vitro* studies of T3 nuclear binding in rat liver and brain show that receptor number remains unchanged with aging, although low circulating T4 levels *in vivo* might have forecast an upward regulation. *In vivo*, the major source of intracellular T3 is local conversion from T4; thus, nuclear receptor binding is dependent on the availability of T4 and T3 to the cell, and hormone availability is decreased with aging.[28]

After *in vivo* administration of radiolabeled T4, cytoplasmic free T4 and T3 values are lower, while bound values are higher in brain (cerebral hemispheres) of old as compared to young rats. Furthermore, the amount of T3 derived intracellularly from conversion of T4 is significantly reduced in the old animals, and nuclear binding of T3 and, to a lesser extent, T4 is also reduced.[28] These data indicate a reduction with aging of intracellular production of T3 (from T4). At the same time, plasma T4 levels decrease with aging in Long-Evans rats. These two factors — reduced secretion of T4 from the thyroid gland and reduced intracellular conversion of T4 to T3 — would be responsible for a reduced intracellular T3 binding and consequent reduced T3 actions (Figure 13.6).

## F. CHANGES WITH AGING IN METABOLIC AND THERMOREGULATORY ACTIONS

Of the major actions of thyroid hormones listed in Table 13.3, those that direct growth and development cease when adulthood is reached. In some instances,

however, thyroid hormones appear to promote growth of adult tissues, even those that reach maturity early in life, such as the nervous tissue. For example, a facilitatory action of thyroid hormones has been proposed — although currently disputed — in the recovery of spinal cord and peripheral nerve injuries in humans and other animals. Catch-up whole-body growth and rehabilitation of some behaviors, impaired by thyroid deficiency at an early age, have been described in adult rats after return to the euthyroid state. Other actions of thyroid hormones that may be relevant to the aging process and that may affect the life span are summarized in Table 13.4. Metabolic rate, calorigenesis, and cholesterol metabolism will be considered here briefly.

## 1. Basal Metabolic Rate

Basal metabolic rate (BMR) is the rate at which oxygen is consumed and carbon dioxide is produced by the organism under conditions of physical and mental rest and in moderate ambient temperature. BMR, significantly influenced by thyroid hormones, decreases consistently with aging, but the magnitude of the fall varies with the study and the criterion of measurement (body weight versus lean body mass). BMR is customarily calculated in terms of oxygen consumption in kilocalories per unit of time and of body surface. With this method, the value for a healthy (euthyroid) male weighing 80 kilos and aged 20 to 30 years is 39.5 kilocalories per square meter of body surface/hour.

With aging, from adolescence to old age, metabolic rate decreases progressively, in both men and women — in the latter, values are always slightly lower than in men — from 46 kilocalories in men and 43 in women at 14 to 16 years of age to 35 and 33, respectively, at 70 to 80 years. The cause of this progressive decline in BMR is not definitely known. It may be associated, at least in part, with the age-related declining levels of thyroid hormones, primarily T3, although, as described above, this decline appears generally slight and does not occur in all aged individuals.

The decreased BMR with aging has also been ascribed to a progressive increase in adipose body mass (less metabolically active) relative to lean body mass (more metabolically active). The metabolically active cytoplasmic mass decreases with aging, as shown by an age-related decline in total body potassium (the major intracellular electrolyte) and urinary excretion of creatine (a product of muscle metabolism). However, these metabolic alterations could also result from the changing thyroid hormone levels.

## 2. Are Thyroid Changes with Aging Due to Altered Metabolism and Tissue Demand Rather than Inadequate Secretion?

As previously discussed in relation to the adrenal hormones, and as will be seen in relation to the pancreas and

**Systemic and Organismic Aging**

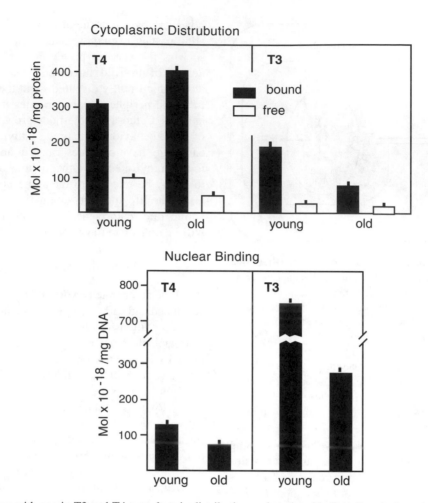

**FIGURE 13.6** Changes with age in T3 and T4 cytoplasmic distribution and nuclear binding. Protein bound cytosolic T4 and T3 is higher in the brain of old (24-month-old) than young (2-month-old) male Long-Evans rats. Lower T3 levels in older animals suggest depressed T4 conversion to T3 with aging. The reduced free hormone availability is manifested in lower nuclear binding, particularly for T3. Similar results were also reported in the liver. (Drawn from Margarity, M., Valcana, T., and Timiras, P.S., *Mech. Ageing Dev.*, 29, 181, 1985.)

---

**TABLE 13.3**
**Major Actions of Thyroid Hormones**

Thyroid hormones regulate the following:
Calorigenesis
Metabolism
Brain maturation
Behavior
Growth and development

---

insulin, the peripheral metabolism of thyroid hormones and tissue demand for these hormones appears to be altered with aging. Thus, with old age, T4 and corticosteroid turnover rates decline, and insulin resistance develops. Several authors have concluded that it is the peripheral demand for the metabolic hormones that declines with increasing age, while the circulating hormone levels are maintained constant. The most common explanation offered for this decline in demand would be the loss of

---

**TABLE 13.4**
**Some Metabolic and Cardiovascular Actions of Thyroid Hormones**

- Stimulation of O2 consumption (calorigenesis) in almost all metabolically active tissues (except brain, testes, spleen, others)
- Stimulation of Na+K+ ATPase actvity: increased Na+ transport may be responsible, in part, for increased energy consumption
- Stimulation by high doses of thyroid hormones of protein catabolism, body temperature, mentation (irritability, restleness, perhaps mediated by neurotransmitter responses)
- Stimulation of intermediary metabolism, particularly, cholesterol metabolism
- Stimulation of cardiac output due to the increased sensitivity to catecholamine effects on cardiac rate and contraction strength
- Decrease of peripheral circulatory resistance due to increased cutaneous vasodilation
- Lowering of circulating cholesterol levels

metabolically active tissue mass with aging. However, this explanation has several flaws: endocrine changes begin early in life, before significant changes in body composition; changes in body composition do not occur in all individuals; and, when those changes occur, they may be secondary to the endocrine changes.

If a smaller lean body mass cannot account for the reduction in peripheral thyroid hormone metabolism, perhaps an explanation may be found elsewhere, for example, in some metabolic alteration in the target tissues, such as an overall decline in protein synthesis with aging. Thus, endocrine changes that occur with aging may ultimately be associated with a decline in overall protein synthesis which begins at the end of adolescence, when growth ceases, and persists throughout the duration of the life span. Thyroid hormones with their widespread actions on cell metabolism represent an ideal model to illustrate age-related changes in metabolic-endocrine interrelations.

## 3. Hypothyroidism as a "Protective" Response in the Elderly

Thyroid hormones, necessary for growth and development when energy requirements are high, may become detrimental when the only energy needed is for homeostasis. A selective reduction of the general anabolic actions of thyroid hormones may occur with cessation of growth and subsequent aging without a concomitant slowing of catabolic effects and the building up of free radical accumulation (Chapter 5). For example, injections of high doses of T4 in young animals are tolerated quite well; the animals respond with increased appetite and more rapid growth. The same doses injected to the adult animal result in muscle wasting and weight loss. These and other findings have led to the suggestion that "there is an homeostatic wisdom in the arrangement whereby conversion of T4 to T3 is inhibited when catabolism is already overactive."[29] From this perspective, the age-related decline in thyroid hormone metabolism would reflect not so much a reduced demand for tissue utilization as an increased need for protection against the catabolic actions of these hormones.[30] Therefore, the reduced intracellular availability of T3 for nuclear binding and consequent effects on protein synthesis may represent a compensatory beneficial response to the decreased metabolic needs of the aging individual.

### THYROID STATUS AND LONGEVITY

As noted previously, the lower circulating T3 in the elderly and T4 in the aged rat, and the reduced conversion of T4 to T3 in target tissues may represent a beneficial compensation to the catabolic actions of the hormones; a certain "homeostatic wisdom," reflecting the changing metabolic needs of the organism. This proposal is supported by studies in which rats made hypothyroid neonatally outlived the corresponding controls by about 4 months, for males, and two months, for

females.[4,5] In these experiments, the mortality of hypothyroid rats was similar to that of controls until 24 months old but was markedly lower thereafter. Maximum life duration was 35 months for male hypothyroid and 31 for male controls; it was 38 months for female hypothyroid and 36 months for female controls. The sex difference may be related to the higher T3 and T4 levels in males than in females and their greater reduction through neonatal intervention. The life-extending effects of hypothyroidism resemble those found after pituitary ablation.[20] They also resemble effects produced by food restriction, for the body weight of hypothyroid animals is significantly reduced, and the inhibition of growth may act as an antiaging factor (Chapter 24). When the thyroid hormone levels are increased by administration of exogenous T4 over many months (12 and 22 months), the average life span is significantly shortened in the treated animals.[4,5] If T4 treatment is initiated at an already senescent age (26 months), the life span is not affected. Thus, the life-shortening effects of excess thyroid hormones are not due to the direct action of the hormones to initiate or promote old age diseases, the direct cause of death, but rather to accelerate the aging process, which may result, in turn, from a more rapid timetable of development. Thyroid hormones seem to act as pacemakers capable of controlling not only certain key events during development (the metamorphosis of tadpoles into frogs) but also intervene in initiating the aging processes.[20]

## 4. Calorigenesis

Thyroid hormones stimulate oxygen consumption in almost all tissues, with the exception of adult brain, testes, uterus, lymph nodes, spleen, and anterior pituitary. Increased oxygen consumption leads to increased cellular metabolic rate, and this leads to increased heat production or calorigenesis.

The magnitude of the calorigenic effect of thyroid hormones depends on several factors, such as interaction with catecholamines and initial metabolic rate — the higher the levels of catecholamines and the lower the metabolic rate at the time of T3 administration, the greater the calorigenic effect. The calorigenic effect of thyroid hormones contributes to the maintenance of body temperature together with a number of other metabolic and neural adjustments. With aging, thermoregulation is progressively impaired, and this decline may be due in part to alterations in thyroid function.

## 5. Thermoregulatory Changes

Thermoregulation, in homeothermic (warm-blooded) animals, including humans, involves a series of adjustments destined to maintain body temperature. These adjustments maintain a balance between heat production (stimulated by muscular exercise, assimilation of food, and hormones regulating basal metabolism) and heat loss (induced by conduction and radiation heat loss through the skin and mucosae, sweat, respiration, urination, and defecation). The balance depends on a group of reflex and hormonal

**Systemic and Organismic Aging**

**TABLE 13.5**
**Thermoregulatory Insufficiency in the Elderly**

In the elderly, thermoregulatory insufficiency results from the
following:
- Decreased heat production
- Decreased body mass
- Reduced muscle activity
- Less efficient shivering
- Reduced sweating response
- Less efficient vasomotor responses
- Decline in temperature perception

responses that are integrated in the hypothalamus and operate to maintain body temperature within a narrow range, despite wide fluctuations in environmental temperatures.

The well-documented increased susceptibility of older people to hypothermia and heat stroke reflects the less efficient temperature regulation that is commonly associated with aging (Table 13.5). In the elderly, thermoregulatory inefficiency to cold or heat is usual and results from deficits at several levels of thermoregulation: in peripheral temperature sensation, at hypothalamic autonomic and neuroendocrine centers, and in higher cerebro-cortical centers, which control temperature perception and coordinate the multiple inputs that determine the effectiveness of adaptative adjustments. The higher mortality of the elderly during "hot" or "cold waves" is well documented and may be attributed to physiologic decrements and to economic (inadequate diet, poor clothing, housing) and emotional and mental (depression, dementia) impairments.

In rodents immersed in cold water, body temperature falls lower and takes longer to return to normal in old rats than in younger animals. Dietary restriction (i.e., tryptophan deficiency), initiated at weaning, continued to 1 year of age and followed by a normal diet, improved thermoregulatory responses; the fall in temperature after cold exposure was less severe and the return to normal more rapid.[31] This persistence of efficient thermoregulation in old animals has been interpreted as a benefit of the delayed aging induced by dietary restriction (Chapter 24).

Responses to heat are also impaired with aging. For example, the onset of sweating is slower in the elderly. As a consequence of the impaired thermoregulatory competence, the elderly also have a reduced fever response. Fever is the most universal hallmark of disease and depends on a "resetting" of the hypothalamic thermostat in response to a variety of agents, such as bacteria responsible for the production of interleukin-1 which acts on lymphocytes (Chapter 15). Interleukin-1 enters the brain, where it stimulates the local production of prostaglandins, activating the fever response. Fever responses are dampened in old humans and old animals.

## 6. Cholesterol Metabolism

Serum cholesterol levels rise with aging in the human and in the rat, despite a fall in hepatic cholesterol synthesis, indicating a net reduction in overall turnover (Chapter 17). A similar situation is obtained in hypothyroidism, prompting several attempts to treat hyperlipidemia with thyroid hormones and their analogs. Several studies suggest that a causal relationship exists between decreased thyroid activity and elevated serum cholesterol. In young rats, thyroidectomy results in reduced cholesterol turnover and elevated serum cholesterol levels, while in old rats, there is no significant postoperative change. Similar results have been observed in a wide variety of mammalian species. The implication is that a reduction in thyroid hormone secretion may be associated with a decrease in the sensitivity of cholesterol metabolism to thyroid hormones. In humans, a hyperbolic fall in serum cholesterol occurs with increasing serum T3 concentrations, and a similar inverse correlation is seen between declining serum T3 and the age-related increase in serum cholesterol levels. Taken together, these results support the inference that age changes in cholesterol metabolism may be secondary to age changes in the thyroid axis.

The specific defect in cholesterol metabolism that develops in the hypothyroid state would result from a reduction in turnover of the serum low-density lipoproteins (LDL), elevated in both hypothyroidism and aging, while metabolism of high-density lipoproteins (HDL) remains unchanged. The significance of altered lipoprotein metabolism in the development of atherosclerosis (Chapter 17) suggests that early attempts by investigators to relate declining thyroid function to factors predisposing to atherosclerotic lesions may not be entirely without foundation.

## G. ABNORMAL THYROID STATES IN THE ELDERLY

The precariousness of the euthyroid state in the elderly is translated into a greater frequency of thyroid disorders with old age. This age-related increase is often masked by the atypical manifestations of the disease. Thus, although modalities for treatment of thyroid disease are readily available and straightforward, the subtleties of diagnosis in the elderly often provide a challenge for the clinician. A diagnosis of thyroid disease in the elderly may be delayed or missed, because the possibility of thyroid disease is overlooked. In the elderly, signs and symptoms of thyroid disease frequently are minimal or atypical, and it is commonly assumed that they are caused by the normal aging process or other diseases.

At all ages, thyroid disease is 3 to 14 times more frequent in women than men. Abnormalities of thyroid function occur more often in older patients who are institutionalized or ill than in the healthier elderly who live in the community. Overall, in the elderly, the prevalence of hypothyroidism falls between 0.5 to 4.4% and that of hyperthyroidism between 0.5 and 3%. As already mentioned, in addition to individual variability, the thyroid state is greatly influenced by extrathyroidal factors, such as general health, disease, drugs, and nutrition, as discussed in the following section.

## 1. Effect of Nonthyroidal Disease on Thyroid Function

In case of illnesses of various etiologies, the most common finding is a lowering of free T3 serum concentration, whereas free T4 levels may be low, normal, or high, and total rT3 levels are high.[7,32] Often, TSH levels, which would be expected to be elevated in response to T3 low levels, are decreased. Any nonthyroidal illness, that is, any systemic illness (e.g., liver disease, HIV infection), acute psychiatric condition, or postoperative state may modify the metabolism of thyroid hormones without implying primary thyroid dysfunction. In addition to medical conditions, a number of common medications (e.g., salicylates) may affect thyroid function. The presence of antibodies to the TSH receptor may also be diagnostically useful and reveal different forms of autoimmune thyroid disease.

## 2. TSH Antibodies and Thyroid Autoimmune Diseases

The frequency of antithyroid antibodies, primarily directed against the cell-surface receptors for TSH on cells of the thyroid gland, increases with aging.[33–36] This increased incidence may be due to the decline in the self-recognition ability of the immune system (Chapter 15) as well as to the increased TSH polymorphism.[21]

The incidence of two autoimmune diseases of the thyroid — Graves' disease or toxic diffuse goiter, and Hashimoto's disease or chronic lymphocytic thyroiditis — increases with increasing age.[34–36] Graves' toxic goiter is associated with hyperfunction or hyperthyroidism (with low TSH, high T3 and T4) (Table 13.6). Hashimoto's thyroiditis is associated with hypofunction or hypothyroidism (with high TSH, low T3 and T4). Both are prevalent in women, in whom the incidence reaches a peak around 40 years of age. TSH levels are not detectable in Graves' disease because of continuous stimulation of the thyroid receptors by TSH antibodies and consequent high T4 and T3 secretion. TSH levels are either normal or elevated in Hashimoto's disease as a consequence of low T3 and T4 levels and TSH blockage by TSH antibodies.

Although the events that trigger the disease are not known, the immune system is strongly implicated (Chapter 15). Graves' and Hashimoto's diseases may represent the extremes of a continuum of signs and symptoms resulting from a deranged immune system with hyperthyroidism at one end and hypothyroidism at the other. Antibodies against TSH receptors or thyroid cell components are present in both disorders, although in different proportions; they may lead to cell death, which, in the case of Hashimoto's disease, occurs by apoptosis (Chapter 4).[35] TSH-receptor antibodies compete with TSH at the receptor site on the thyroid cell. Some of these antibodies stimulate T3 and T4 secretion as in Graves', others block access of TSH at the receptor sites and reduce T3 and T4 secretion, as in Hashimoto's.

TSH-receptor antibodies are not the only antithyroid antibodies present in autoimmune disorders of the thyroid. The most common antibodies, important clinically, are those directed against thyroglobulin (the thyroid cell protein, precursor of T3 and T4), those against thyroidal microsomal and nuclear components, and those against T3 and T4. All these antibodies lead to thyroid destruction; they are more frequent in Hashimoto's disease. Another feature of Graves's and Hashimoto's disorders is invasion of the thyroid tissue by lymphatic tissue. This infiltration is more aggressive in Hashimoto's, with extensive destruction of the thyroid gland.

## 3. Hyperthyroidism

Thyroid hormone excess or hyperthyroidism is caused by either Graves' disease (i.e., diffuse toxic goiter) or nodular toxic goiter. Because the signs and symptoms in older

**TABLE 13.6**
**Autoimmune Diseases of the Thyroid Gland**

| Characteristics | Graves' Disease | Hashimoto's Thyroiditis |
| --- | --- | --- |
| Thyroid status | Hyperthyroid | Hypothyroid |
| TSH | Generally undetectable | Normal to elevated |
| T4, T3 (serum) | Above normal | Below normal |
| Antibodies (ABs) | Stimulatory ABs compete with TSH at receptor sites Loss of TSH control over thyroid function | Some ABs block TSH actions |
| Autoantibodies against thyroglobulin, T3, and T4 destroy thyroid microsomal and nuclear components | Generally present | Generally present |
| Lymphocytic Invasion | Limited | Marked |
| Female:Male ratio | As high as 10:1 | As high as 10:1 |

**TABLE 13.7**
**Common Signs and Symptoms of Hyperthyroidism in the Elderly**

- Cardiovascular abnormalities
  - Congestive heart failure symptoms
  - Atrial fibrillation
  - Angina (coronary heat disease)
  - Pulmonary edema
- Tremor
- Nervousness
- Weakness
- Weight loss and anorexia (poor appetite)
- Palpable goiter (may not be present)
- Eye findings (may not be present)
- Thyroid nodules (nonspecific)

patients are often attributed to another illness, the term "masked hyperthyroidism" has been introduced (Table 13.7). Cardiovascular abnormalities are quite common in older patients with hyperthyroidism. In one study, 79% of subjects had an abnormal cardiovascular examination, 67% had symptoms of congestive heart failure, 39% were in atrial fibrillation, 20% had symptoms of angina, and 8% presented with pulmonary edema. Tachycardia, or fast heartbeat, even in the presence of atrial fibrillation, is less impressive in older than in younger hyperthyroid patients. Typical signs and symptoms such as tremor, nervousness, and muscular weakness may be ignored, because they are considered common in the elderly. Weight loss, another common symptom in the elderly, may lead to an evaluation for malignant gastrointestinal lesion. A goiter may or may not be palpable, and the eye signs of Graves' disease may be absent. Because thyroid nodules are common in the elderly, it is difficult to arouse suspicion of significant thyroid disease solely on the basis of their presence.

Apathetic hyperthyroidism is another term used for a thyroid disease in the elderly, the characteristics of which could potentially be confused with a hypothyroid state. This condition is characterized by the following:

1. Blunted affect, i.e., withdrawal behavior with apathy and depression
2. Absence of hyperkinetic motor activity
3. Slowed mentation
4. Weakness of shoulder muscles
5. Diarrhea
6. Edema of the lower extremity
7. Droopy eyelids
8. Cardiovascular abnormalities

Laboratory diagnosis of hyperthyroidism includes assay of serum T4 and T3, and a radionuclide thyroid scan can help differentiate Graves' disease from toxic multinodular goiter.

Treatment of hyperthyroidism includes medical, radiological, and surgical approaches, singly or combined.[12,37] In the medical treatment, the goal is to reduce thyroid hormone production. The most commonly used antithyroid medications are oral propylthiouracil and methimazole: both are thioureylenes that inhibit the formation of thyroid hormones by interfering with the incorporation of iodine into tyrosyl residues of thyroglobulin, perhaps by inhibiting thyroid peroxidase. The major disadvantage of these inhibitors of thyroid hormone synthesis is the recurrence of hyperthyroidism after treatment is stopped. Administration of radioiodide is usually the preferred treatment in the elderly. The isotope is actively collected by the thyroid gland, particularly in the regions of active proliferation, and the fast-growing and secreting tissue is destroyed. Radioiodide can be administered in conjunction with medical treatment because the effect of the former is gradual and not usually complete until 3 months after treatment. Close follow-up is necessary in view of the possibility of post-radioiodide hypothyroidism. Surgery of the thyroid gland may be necessary for large multinodular glands resistant to radioiodide.[12]

### 4. Hypothyroidism

Thyroid hormone deficiency, or hypothyroidism, is caused by autoimmune thyroiditis or is a consequence of treatment for hyperthyroidism. Just as "masked hyperthyroidism" exists in the elderly, so does "masked hypothyroidism" (Table 13.8). Typical symptoms of hypothyroidism such as fatigue, weakness, dry skin, hair loss, constipation, mental confusion, depression, and cold intolerance may be attributed to old age instead of thyroid disease. In addition, the insidious onset and slow progression of hypothyroidism makes it more difficult to diagnose. An elevated TSH is one of the most reliable symptoms of hypothyroidism because T4 may or may not be decreased in mild cases.

**TABLE 13.8**
**Frequently Missed Common Signs and Symptoms of Hypothyroidism in Elderly Patients**

- Cardiovascular abnormalities
  - Dyspnea (shortness of breath)
  - Chest pain
  - Enlarged heart
  - Bradycardia (slow heartbeat)
- Anorexia (poor appetite) and constipation
- Muscular weakness
- Mild anemia
- Depression
- Cold intolerance
- Joint pain

---

**Box 13.3**
**Major Actions of Parathyroid Hormone and Calcitonin**

**Parathyroid hormone**, PTH, a polypeptide secreted in response to hypocalcemia, raises the concentration of plasma calcium by:

1. Increasing renal calcium resorption
2. Mobilizing calcium from bones by stimulating osteoclastic activity (i.e., destruction of bone cells, osteocytes)
3. Stimulating the absorption of calcium from the small intestine in the presence of adequate amounts of vitamin D
4. Lowering levels of plasma inorganic phosphate by inhibiting renal resorption of phosphate

In general, serum calcium remains unchanged well into old age, despite a few reports of a slight decline. This maintenance of serum calcium is all the more remarkable in view of the decline in calcium dietary intake and intestinal absorption associated with old age (Chapter 21). One of the major regulators of calcium balance is PTH secretion: high concentrations of calcium inhibit PTH secretion, and low concentrations stimulate secretion. The effects of PTH on mineral metabolism are mediated by the binding of the hormone to the PTH receptor in target tissues.[41] The PTH receptor belongs to the family of receptors for light, odorants, catecholamines, and other peptides.[42]

Another protein, parathyroid hormone-related protein (PTHRP), is synthesized in cartilage and in other tissues (mammary gland, brain, renal glomeruli); it is particularly abundant in fetal life, when it may function as a growth factor.[43] Although PTHRP is not a true hormone, but a distant homologue of PTH, its local release activates the type 1 PTH receptor for which it has the same affinity as does PTH with a resulting inhibition of PTH action.[43,44]

**Calcitonin**, also a polypeptide, is secreted not only in response to increased calcium levels but also to gastrointestinal hormones such as glucagon. CT receptors are found in bones and the kidneys. The major actions of this hormone involve regulation of plasma calcium levels by:

1. Lowering plasma calcium and phosphate levels by inhibiting bone resorption
2. Increasing calcium excretion in urine

CT also has some minor action on water and electrolytes and decreases gastric acid secretion.

---

Common signs and symptoms in elderly hypothyroidism are listed in Table 13.8. Of these, cardiovascular abnormalities and anorexia are common, muscular weakness and mild anemia are found in roughly one half of patients, depression and cold intolerance in 60% of patients, and joint pain may also be present.

Treatment of hypothyroidism is thyroid replacement in the form of oral L-thyroxine or a combination of T4 and T3.[38] Follow-up determination that elevated TSH levels have returned to normal is the best indication that adequate replacement has been achieved. Overreplacement is dangerous in older patients, especially in those with cardiac disease. Safe replacement should be started with low initial doses (0.025 to 0.05 mg of L-T4 per day). Maintenance dose in the elderly is generally lower than in younger patients, and once established, usually remains constant (around 0.1 mg per day).

In view of the difficulties in clinically diagnosing thyroid disease in the elderly, the role of laboratory screening becomes important. Although arguments have been made that such screening, including a serum TSH, in the healthy ambulatory population of elderly is not worthwhile, certain guidelines can be established. Thyroid disease states, although potentially difficult to diagnose clinically in the elderly, can be detected if the possibility of such disease is entertained. The relative ease and success of treatment makes such detection well worthwhile.

## III. THE PARATHYROID GLANDS AND THE THYROID C CELLS

The four parathyroid glands, in humans, are imbedded in the thyroid gland. They secrete the parathyroid hormone (PTH). The so-called C cells, dispersed throughout the thyroid gland, the parathyroids, and thymus secrete another hormone, calcitonin (CT). Both hormones play a significant role in the maintenance of calcium homeostasis (Chapter 21).[39,40] With aging, they do not appear to be consistently altered in their function or to undergo increased pathology.[40] Major actions of PTH and CT are summarized in Box 13.3.

### A. CHANGES WITH AGING

#### 1. The Parathyroid Glands

Structural changes in the aging human parathyroids are few. Studies from laboratory animals reveal only minor changes, such as the presence of degenerating cells containing colloid and of mitochondria showing bizarre patterns.[45] Studies of immunoassayable PTH levels show an increase or decline, depending on gender and race: they decline after 60 years of age in white men but not white women.[46] Increased PTH levels may reflect impaired hormonal renal clearance[47] or accumulation of immunoassayable but biologically inactive fragments.[46] Both PTH and CT are derived by the action of proteases on larger pre-prohormones, which, in both cases, are less active biologically than immunologically similar to the hydrolytically formed hormones. With aging, the processing of the precursor hormones may be altered, resulting in the secretion of less biologically active pre-prohormone molecules. Another interpretation relates PTH changes with aging to alterations in vitamin D. PTH and vitamin D coordinate the regulation of calcium homeostasis by the intestine, bone, and kidney. Vitamin D receptors have been identified in parathyroid cells, and vitamin D metabolites reduce the synthesis of pre-pro-PTH mRNA.[48–50]

While serum calcium levels are maintained throughout the life span, the mechanism by which they are regulated changes markedly with aging. At young and adult ages, calcium levels are maintained by the calcium ingested minus the calcium excreted (in feces and urine) without chronic loss of bone mineral. At older ages, serum calcium levels are maintained by resorption of calcium from bone rather than by intestinal absorption of dietary calcium or its renal retention. With aging, possible mechanisms responsible for this shift include a decreased capacity of PTH to stimulate renal production of the biologically active form of vitamin D, which stimulates intestinal absorption of calcium, and a decreased capacity of active vitamin D to simulate intestinal absorption of calcium.[39]

## 2. Thyroid C Cells

Structural changes of C cells are rare, except for the finding, in old rats, of an higher ratio of C cells to thyroid follicular cells.[45] Little is known of the changes with aging in CT and whether and how they affect the aging of bone.[51] A decrease in CT reported in humans is greater in men than women, but not in rats, in which there seems to be an increase. Because CT decreases bone resorption, its potential usefulness in the therapy of aging-related bone demineralization and osteoporosis has been explored but has not yielded significant preventive or therapeutic benefits.[51] Further discussion of the potential role of these and other hormones in aging of bone is presented in Chapter 21.

## IV. THE PINEAL GLAND

The pineal gland (also called epiphysis) was believed by the 17th century philosopher, mathematician, and physician, Rene Descartes, to be "the seat of the human soul." This early role was followed by a wide variety of other equally imaginative functions ascribed to the gland. It is now known to secrete melatonin, an indolamine synthesized from the neurotransmitter serotonin (Chapter 7) in the cells (pinealocytes) of the pineal gland and, to a lesser extent, other body tissues.[52–56] Melatonin secretion is regulated by circadian rhythm (i.e., 24 hour, daily cycles) and helps entrain and synchronize internal body functions with the day–night cycle and with annually changing seasons (with consequent shifts in respective length of light/dark cycles). In recent years, numerous books aimed at the lay public have promoted melatonin as a "wonder drug," successful in treating, alleviating, or preventing a wide variety of human ailments, from lengthening of the life span, to strengthening of the immune system, to reducing the risk of cancer. As with other hormones, the rationale for taking melatonin is to restore its decreased levels in old age; but, as is the case for "replacement therapy" with some other hormones (Chapters 10 and 11), melatonin benefits have not been conclusively validated, and its potential side-effect has not been entirely excluded. In the United States, melatonin is classified as a food supplement (Chapter 24) and, as such, can be purchased without need of a physician's prescription. Therefore, rigorous scientific studies in humans and other animals are needed to evaluate its therapeutic effects and eventual potentially adverse effects due to its indiscriminate use. Major characteristics of structure and function of the pineal gland are summarized in Box 13.4.

### A. CHANGES WITH AGING

With aging, the human pineal weight may increase due to concretion accumulation (see Box 13.4) and the presence of cavities that vary considerably among individuals and mammalian species (such enlargement does not appear to affect function negatively). Biochemical changes vary individually, although both night and day melatonin levels decrease in most elderly, probably due to alterations (e.g., axon swelling) of sympathetic innervation.[53,58]

The functional importance of melatonin resides in its regulation or fine tuning of circadian rhythmicity. For example, exogenous melatonin may stimulate or inhibit gonadal function, depending on the animal species and the time of administration. This variability has been interpreted to mean that it is not the change in melatonin per se that causes the gonadal change but, rather, the consequent alteration in the timing signal that coordinates body function with the light–dark cycle in the environment. This interpretation may be correct, especially in seasonally breeding animals responsive to changes in day length. With respect to the gonadal action of melatonin in humans, pineal tumors may be associated with sexual precocity; but that occurs only when the tumors are large enough or localized in such a way to produce hypothalamic damage, which then, would be responsible for the delayed onset of puberty.

Decline of melatonin levels at older ages may impair the orchestration of such rhythms that may be restored by

**Box 13.4**

**Structure and Function of the Pineal Gland**

Arising from the roof of the 3<sup>rd</sup> ventricle, the pineal gland is an unpaired structure located almost at the center of the brain, where it is connected by a stalk to other midbrain structures. The gland contains neuroglial cells and secretory cells (pinealocytes) and has abundant blood supply provided by highly permeable fenestrated capillaries. At early ages in young animals and infants, the gland is large, with the cells arranged in alveoli. It begins to involute before puberty and to accumulate calcium concretions, called "pineal sand." The pineal synthesizes and secretes the indolamine melatonin which derives from the neurotransmitter serotonin (Chapter 7) and its precursor, the amino acid L-tryptophan.[57] Synthesis and secretion are high during the night period and low during the light period of the day. Melatonin is metabolized in the liver and excreted in the urine. Its actions are mediated by its binding to low- and high-affinity receptors. The high-affinity ones belong to the family of G-protein-coupled receptors; they are abundant not only in the pineal but also:

1. In the hypothalamic suprachiasmatic nuclei, which contain the dominant pacemakers for many circadian rhythms in the body (e.g., rhythm in ACTH secretion, Chapter 10),
2. In the retina, where they may relay photoperiod information
3. In other cortical and subcortical areas of the brain, where the receptors may be involved, in some species (e.g., humans), in mediating the sleep-inducing effects of melatonin

In some fishes, amphibians, and reptiles, the pineal gland has an extracranial component, a "third eye," that can act as a photoreceptor. However, in mammals, the pineal gland does not respond directly to light stimuli, but rather, indirectly via a multisynaptic pathway that originates in the retina.

its administration (e.g., effects of melatonin in improving sleep and alleviating jet-lag symptoms). The importance of this kind of biologic regulation for longevity has been questioned repeatedly, but its testing has been largely limited to insects, whose life span is shortened by disruption of light–dark cycles.

In humans, reduced activity of the aging pineal gland may play a role in the progressive impairment of immunomodulatory, metabolic, oncostatic, and autonomic functions.[59] There is now evidence that melatonin has an hypnotic (sleep-producing) effect:

1. Peak melatonin concentrations coincide with sleep
2. Melatonin administration in doses that mimic night levels can promote and sustain sleep
3. Exogenous melatonin may also influence circadian rhythms and reduce fatigue and re-establish the timing of sleep when night–light cycles are disrupted[60–62]

According to several studies, melatonin would be a potent scavenger of the highly toxic hydroxyl radical and other oxygen-centered radicals (Chapter 5), thereby possessing a protective action against oxidative damage.[63–65] While improvement of pineal function and, particularly, normalization of melatonin levels, may ameliorate the deficits of some aging-dependent functions, its actions in preventing or reducing a variety of aging-related diseases, as well as its contribution to longevity, await further confirmation.

## REFERENCES

1. Braverman, L.E. and Utiger, R.D., Eds., *Werner and Ingbar's the Thyroid*, 8th ed., Lippincott Williams & Wilkins, Philadelphia, PA, 2000.
2. Robertson, T.B., The influence of thyroid alone and of thyroid administered together with nucleic acids upon the growth and longevity of the white mouse, *Aust. J. Exp. Biol. Med. Sci.*, 5, 69, 1928.
3. Crewe, F.A.E., Rejuvenation of the aged fowl through thyroid medication, *Proc. Roy. Soc.*, 45, 252, 1924–1925.
4. Ooka, H., Fujita, S., and Yoshimoto, E., Pituitary-thyroid activity and longevity in neonatally thyroxine-treated rats, *Mech. Ageing Dev.*, 22, 113, 1983.
5. Ooka, H. and Shinkai, T., Effects of chronic hyperthyroidism on the lifespan of the rat, *Mech. Ageing Dev.*, 33, 275, 1986.
6. Timiras, P.S., Hormones of the thyroid and parathyroid glands, in *Hormones and Aging*, Timiras, P.S., Quay, W.D., and Vernadakis, A., Eds., CRC Press, Boca Raton, FL, 1995.
7. Lazarus, J.H., Thyroid disorders, in *Principles and Practice of Geriatric Medicine*, 3rd ed, Vol. 2, Pathy, M.S.J., Ed., John Wiley & Sons, New York, 1998.
8. Mariotti, S. et al., Complex alteration of thyroid function in healthy centenarians, *J. Clin. Endocrinol. Metab.*, 77, 1130, 1993.
9. Anderson-Ranberg, K. et al., Thyroid function, morphology and prevalence of thyroid disease in a population-based study of Danish centenarians, *J. Am. Geriatr. Soc.*, 47, 1238, 1999.
10. Püllen, R. and Hintze, G., Size and structure of the thyroid in old age, *J. Am. Geriatr. Soc.*, 45, 1539, 1997.
11. Hurley, D.L. and Gharib, H., Thyroid nodular disease: Is it toxic or nontoxic, malignant or benign? *Geriatrics*, 50, 24, 1995.

**Systemic and Organismic Aging**

12. Kinder, B., Surgical diseases of the thyroid and parathyroid glands, in *Principles and Practice of Geriatric Surgery*, Rosenthal, R.A., Zenilman, M.E., and Katlic, M.R., Eds., Springer, New York, 2001.

13. Trivalle, C. et al., Differences in the signs and symptoms of hyperthyroidism in older and younger patients, *J. Am. Geriatr. Soc.*, 44, 50, 1996.

14. Mariotti, S. et al., The ageing thyroid, *Endocrine Rev.*, 16, 686, 1995.

15. Hintze, G. et al., Prevalence of thyroid dysfunction in elderly subjects from the general population in an iodine deficiency area, *Aging*, 3, 325, 1991.

16. Hermus, A.R. and Huysmans, D.A., Treatment of benign nodular thyroid disease, *N. Engl. J. Med.*, 388, 1438, 1998.

17. Hershman, J.M. et al., Serum thyrotropin and thyroid hormone levels in elderly and middle-aged euthyroid persons, *J. Am. Geriatr. Soc.*, 41, 823, 1993.

18. Szabolcs, I. et al., Factors affecting the serum free thyroxine levels in hospitalized chronic geriatric patients, *J. Am. Geriatr. Soc.*, 41, 742, 1993.

19. Lindeman, R.D. et al., Subclinical hypothyroidism in a biethic, urban community, *J. Am. Geriatr. Soc.*, 47, 703, 1999.

20. Everitt, A.V., The thyroid gland, metabolic rate and aging, in *Hypothalamus, Pituitary and Aging*, Everitt, A.V. and Burgess, J.A., Eds., Charles C. Thomas, Springfield, IL, 1976.

21. Choy, V.J., Klemme, W.R., and Timiras, P.S., Variant forms of immunoreactive thyrotropin in aged rats, *Mech. Ageing Dev.*, 19, 273, 1982.

22. DeVito, W.J., Neuroendocrine regulation of thyroid function, in *Neuroendocrinolgy in Physiology and Medicine*, Conn, P.M. and Freeman, M.E., Eds., Humana Press, Totowa, NJ, 2000.

23. Veldhuis, J.D., The neuroendocrine control of ultradian rhythms, in *Neuroendocrinolgy in Physiology and Medicine*, Conn, P.M. and Freeman, M.E., Eds., Humana Press, Totowa, NJ, 2000.

24. Murialdo, G. et al., Circadian secretion of melatonin and thyrotropin in hospitalized aged patients, *Aging*, 5, 39, 1993.

25. Brent, G.A., The molecular basis of thyroid hormone action, *N. Engl. J. Med.*, 331, 847, 1994.

26. Katz, D. and Lazar, M.A., Dominant negative activity of an endogenous thyroid hormone receptor variant ($\alpha 2$) is due to competition for binding sites on target genes, *J. Biol. Chem.*, 268, 20904, 1993.

27. Schueler, P.A. et al., Binding of 3,5,3'-triiodothyornine (T3) and its analogs to the *in vitro* translational products of c-erbA protooncogenes: Differences in the affinity of the $\alpha$- and $\beta$-forms for the acetic acid analog and failure of the human testis and kidney $\alpha$-2 products to bind T3, *Mol. Endocrinol.*, 4, 227, 1990.

28. Margarity, M., Valcana, T., and Timiras, P.S., Thyroxine deiodination, cytoplasmic distribution, and nuclear binding of thyroxine and triiodothyronine in liver and brain of young rats, *Mech. Ageing Dev.*, 29, 181, 1985.

29. Chopra, I.J., *Triiodothyronines in Health and Disease*, Monographs in Endocrinonology, Vol. 18, Springer-Verlag, New York, 1981.

30. Utiger, R.D., The thyroid: Physiology, thyrotoxicosis, hypothyroidism, and the painful thyroid, in *Endocrinology and Metabolism*, 3rd ed., Felig, P., Baxter, J.D., and Frohmar, L.A., Eds., McGraw-Hill, New York, 1995.

31. Segall, P.E. and Timiras, P.S., Age-related changes in thermoregulatory capacity of thryptophan-deficient rats, *Fed. Proc.* 34, 83, 1975.

32. Utiger, R.D., Altered thyroid function in nonthyroidal illness and surgery: To treat or not to treat? *N. Engl. J. Med.*, 333, 1562, 1995.

33. Furmaniak, J. and Rees Smith, B., The structure of thyroid autoantigens, *Autoimmunity*, 7, 63, 1990.

34. Szabolcs, I., Bernard, W., and Horster, F.A., Thyroid autoantibodies in hospitalized chronic geriatric patients: Prevalence, effects of age, nonthyroidal clinical state, and thyroid function, *J. Am. Geriatr. Soc.*, 43, 670, 1995.

35. Williams, N., Autoimmunity: Thyroid disease — a case of cell suicide? *Science*, 275, 926, 1997.

36. Armengol, M.P., Thyroid autoimmune disease: Demonstration of thyroid antigen-specific B cells and recombination-activating gene expression in chemokine-containing active intrathyroidal germinal centers, *Am. J. Pathol.*, 159, 861, 2001.

37. Franklyn, J.A., The management of hyperthyroidism, *N. Engl. J. Med.*, 330, 1731, 1994.

38. Toft, A.D., Thyroid hormone replacement — one hormone or two? *N. Engl. J. Med.*, 340, 469, 1999.

39. Armbrecht, H.J., Age-related changes in calcium homeostasis and bone loss, in *Principles and Practice of Geriatric Medicine*, 3rd ed, Vol. 2, Pathy, M.S.J., Ed., John Wiley & Sons, New York, 1998.

40. Marx, S.J., Hyperparathyroid and hypoparathyroid disorders, *N. Engl. J. Med.*, 343, 1863, 2000.

41. Jüppner, H. et al., Parathyroid hormone and parathyoid hormone-related peptide in the regulation of calcium homeostasis and bone development, in *Endocrinology*, 4th ed., DeGroot, L.J., Ed., W.B. Saunders, Philadelphia, 2001.

42. Marchese, A. et al., Novel GPCRs and their endogenous ligands: Expanding the boundaries of physiology and pharmacology, *Trends Parmacol. Sci.*, 20, 370, 1999.

43. Strewler, G.J., The physiology of parathyroid hormone-related protein, *N. Engl. J. Med.*, 342, 177, 2000.

44. Jüppner, H., Receptors for parathyroid hormone and parathyroid hormone-related peptide: Exploration of their biological importance, *Bone*, 25, 87, 1999.

45. Blumenthal, H.T. and Perlstein, I.B., The biopathology of aging of the endocrine system: The parathyroid glands, *J. Am. Geriatr. Soc.*, 41, 1116, 1993.

46. Endres, D.B. et al., Age-related changes in serum immunoreactive parathyroid hormone and its biological action in healthy men and women, *J. Clin. Endocrinol. Metab.*, 65, 724, 1987.

47. Liang, C.T. et al., Regulation of renal sodium/calcium exchange by PTH: Alteration with age, *Environ. Health Perspect.*, 84, 137, 1990.

48. Fujita, T., Calcium, parathyroids and aging, *Contrib. Nephrol.*, 90, 206, 1991.
49. Peacock, M., Interpretation of bone mass determinations as they relate to fracture: Implications for asymptomatic primary hyperparathyroidism, *J. Bone Miner. Res.*, 6, S77, 1991.
50. Silver, J., Russell, J., and Sherwood, L.M., Regulation by vitamin D metabolites of messenger ribonucleic acid for preproparathyroid hormone in isolated bovine parathyroid cells, *Proc. Natl. Acad. Sci. USA,* 82, 4270, 1985.
51. Clissold, S.P., Fitton, A., and Chrisp, P., Intranasal salmon calcitonin. A review of its pharmacological properties and potential utility in metabolic bone disorders associated with aging, *Drugs Aging*, 1, 405, 1991.
52. Yu, H.S. and Reiter, R.J., Eds., Melatonin: Biosynthesis, physiological effects, and clinical applications, CRC Press, Boca Raton, FL, 1993.
53. Quay, W.B. and Kachi, T., Amine secreting endocrines in *Hormones and Aging*, Timiras, P.S., Quay, W.D., and Vernadakis, A., Eds., CRC Press, Boca Raton, FL, 1995.
54. Urbanski, H.F., Influence of light and the pineal gland on biological rhythms in *Neuroendocrinology in Physiology and Medicine*, Conn, P.M. and Freeman, M.E., Eds, Humana Press, Totowa, NJ, 2000.
55. Wayne, N.L., Neuroendocrine regulation of biological rhythms in *Neuroendocrinology in Physiology and Medicine*, Conn, P.M. and Freeman, M.E., Eds., Humana Press, Totowa, NJ, 2000.
56. Malpaux, B., The neuroendocrine control of circadian rhythms in *Neuroendocrinology in Physiology and Medicine*, Conn, P.M. and Freeman, M.E., Eds., Humana Press, Totowa, NJ, 2000.
57. 57, Gastel, J.A. et al., Melatonin production: Proteasomal proteolysis in serotonin N-acetyltransferase regulation, *Science*, 279, 1358, 1998.
58. Reuss, S. et al, The aged pineal gland: Reduction in pinealocyte number and adrenergic innervation in male rats, *Exp. Gerontol.*, 25, 183, 1990.
59. Brzezinski, A., Melatonin in humans, *N. Engl. J. Med.,* 336, 186, 1997.
60. Dawson, D. and Encel, N., Melatonin and sleep in humans, *J. Pineal Res.*, 15, 1,1993.
61. Garfinkel, D. et al., Improvement of sleep quality in elderly people by controlled-release melatonin, *Lancet*, 346, 541, 1995.
62. Haimov, I. et al., Melatonin replacement therapy of elderly insomniacs, *Sleep*, 18, 598, 1995.
63. Reiter, R.J. et al., Melatonin as a free radical scavenger: Implications for aging and age-related diseases, *Ann. N.Y. Acad. Sci.,* 719, 1, 1994.
64. Reither,R.J., The role of the neurohormone melatonin as a buffer against macromolecular oxidative damage, *Neurochem. Int.*, 27, 453, 1995.
65. Srinivasan, V., Melatonin, oxidative stress and ageing, *Current Science*, 76, 46, 1999.

# 14 The Endocrine Pancreas, Diffuse Endocrine Glands and Chemical Mediators*

*Paola S. Timiras*
University of California, Berkeley

## CONTENTS

## I.   INTRODUCTION

The regulatory action of hormones on metabolism is well established and has been briefly considered in the preceding four chapters (Chapters 10 through 13). Similarly well established is the fact that this regulatory action undergoes significant changes with aging, but that the onset, extent, and pathologic consequences of these changes vary with the endocrine gland considered. There is also considerable individual heterogeneity in the ability of the elderly to maintain metabolic balance in response to internal and external challenges disruptive of homeostasis. One illustrative example of these aging-associated changes is the altered regulation of carbohydrate metabolism by the pancreatic hormones, insulin and glucagon. Carbohydrates, such as sugar and starch, are derived from aldehyde and ketone compounds and represent, with proteins and fats (lipids), the major components of human nutrition (Chapter 24). Interest in insulin and its regulation of glucose metabolism in aging arises from the relatively high percentage of the elderly, 65 years and older, who have lost the ability to maintain glucose homeostasis. The mechanisms of this age-related impairment in carbohydrate economy with aging have been extensively investigated, especially with regard to diabetes mellitus, a chronic disorder characterized by impaired metabolism of glucose and the late development of complications involving the cardiovascular and nervous systems. There are different types of diabetes mellitus but, regardless of the type, the disease is associated with a common hormonal defect, that is, insulin deficiency that can be total, partial, or relative and may or may not be associated with insulin resistance.[1–3]

In this chapter, the causes and consequences of insulin deficiency on carbohydrate metabolism (e.g., diabetes mellitus) and life span will be examined first. In this context, a number of other hormones, termed "counterregulatory," that oppose the metabolic action of insulin will be considered as well. Second, we will

---

* Illustrations by Dr. S. Oklund.

0-8493-0948-4/03/$0.00+$1.50
© 2003 by CRC Press LLC

---

**Box 14.1**
**Structure and Function of the Pancreas**

The pancreas lies inferior to the stomach, in a bend of the duodenum. It is both an endocrine and an exocrine gland. Four islet cell types have been identified, each producing a specific polypeptide hormone:

- A cells secrete glucagon
- B cells secrete insulin
- D cells secrete somatostatin
- F (also called PP) cells secrete pancreatic polypeptide

Of these pancreatic hormones, insulin is secreted only by the B cells, whereas the other three are secreted also by the gastrointestinal mucosa, and somatostatin is also found in the brain.

Insulin and glucagon are important in the regulation of carbohydrate, protein, and lipid metabolism:

- Insulin, an anabolic hormone, increases the storage of glucose from ingested nutrients, glucose, fatty acids, and amino acids in cells and tissues.
- Glucagon, a catabolic hormone, mobilizes glucose, fatty acids, and amino acids from stores into the blood.

In the pancreas, Somatostatin may regulate, locally, the secretion of the other pancreatic hormones; in brain (hypothalamus) and spinal cord, it may act as a neurohormone and neurotransmitter (Chapters 7 and 10). The function and origin of the pancreatic polypeptide are still uncertain, although the hormone may influence gastrointestinal function and promote intra-islet homeostasis. A diagram of an islet of Langerhans is presented in Figure 14.1, and a list of the pancreatic hormones can be found in Table 14.1. Each cell type secretes its hormone into the blood capillaries surrounding the islet, and each hormone has direct access in the islet extracellular space, to the other cell types, and can influence the secretory activity of the others (paracrine effects). Somatostatin is the chief inhibitory paracrine mediator of the islets and locally inhibits insulin and glucagon release. Some islet mediators, for example, serotonin (discussed as a neurotransmitter in Chapter 7) contributes, in B cells, to the regulation of insulin synthesis and release (thereby providing autocrine-type mediation).

---

discuss some changes with aging in another group of hormones with metabolic regulatory activity. These are molecules with classical hormonal activity but also with autocrine, paracrine, neurotransmitter, and other actions. They are grouped under the term "diffuse endocrine and chemical mediators."

## II.  AGING OF THE ENDOCRINE PANCREAS

Several hormones participate in the regulation of carbohydrate metabolism. Four of them are secreted by the cells of the islets of Langerhans in the pancreas: two, insulin and glucagon, have major actions on glucose metabolism and two, somatostatin and pancreatic polypeptide, exert modulating actions on insulin and glucagon secretion. Other hormones affecting carbohydrate metabolism include: epinephrine, thyroid hormones, glucocorticoids, and growth hormones. Some of their actions have been discussed in the respective chapters (10 and 13). Major structure and physiologic characteristics of the endocrine pancreas are summarized in Box 14.1; the exocrine functions are concerned with digestion and are discussed in Chapter 20.

## A.  Morphological and Functional Changes with Aging in Pancreatic Islets

### 1.  Morphological Changes

With aging, few morphologic changes have been reported in the endocrine pancreas in humans.[4,5] Among these are the following:

- A certain degree of atrophy
- An increased incidence of tumors
- The presence of amyloid material and lipofuscin granules (signs of abnormal cellular metabolism)

In old rats, contrary to the atrophy in old humans, islets are larger than in young rats and contain more B cells and more insulin per B cells. This increase in B cells has been interpreted as a possible compensatory mechanism for the decreased responsiveness of tissues to insulin.[6] However, in some cases, despite the larger B cell mass in old rats, maximum glucose-stimulated insulin secretion is lower with aging in both males and females when expressed in terms of islet weight. At either end of the age range, although female rats have smaller islets than male rats, the former secretes more

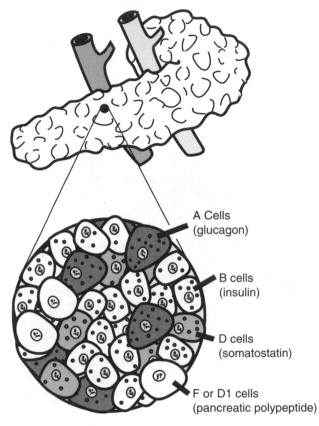

**FIGURE 14.1** Diagrammatic representation of the pancreas with pancreatic cells.

A Cells (glucagon)

B cells (insulin)

D cells (somatostatin)

F or D1 cells (pancreatic polypeptide)

## TABLE 14.1
### Major Pancreatic Hormones

| Pancreatic Site | Hormone | Alternate Source |
|---|---|---|
| B cells | Pre-proinsulin Proinsulin Insulin (+ connecting C-peptide) | — |
| A cells | Proglucagon Glucagon (+ glicentin) | GI mucosa |
| D cells | Somatostatin | GI mucosa CNS |
| F, D, or PP cells | Pancreatic Polypeptide | GI mucosa |

GI = gastrointestinal system.

CNS = central nervous system.

insulin per unit amount of islet tissue, a gender difference still unexplained. Somatostatin levels increase with aging and treatment of islets with antisomatostatin antibodies, by removing the inhibitory action of somatostatin,

## TABLE 14.2
### Major Actions of Insulin

**Insulin:**

- Lowers blood glucose by facilitating transport of glucose into muscle and adipose cells
- Simultaneously stimulates intracellular metabolic use of glucose
- Increases glycogen synthesis in liver and muscle cells
- Slows gluconeogenesis (in liver)
- Facilitates intracellular transport of amino acids and lipids and promotes protein and triglyceride synthesis
- Promotes overall body growth (general effect)

impairments of the glucose-stimulated insulin response may be partially reversed.[7]

The totality of the reported morphologic changes in the pancreas is, in most cases, minor and cannot account for the significantly frequent metabolic consequences in a number of elderly individuals in whom glucose homeostasis is impaired. Rather, several factors — changes in pancreatic hormones and in the responsiveness of extra-pancreatic targets to insulin — have been implicated. Major actions of insulin and glucagon are listed in Tables 14.2 and 14.3; the well-recognized regulators of insulin and glucagon secretion shown in Table 14.4, and a number of pancreatic and extra-pancreatic causes for altered glucose metabolism with aging are shown in Table 14.5.

### INSULIN SECRETION AND ACTION

Insulin is synthesized in the B cells as part of a larger (86 amino acids) single-chain polypeptide, proinsulin, derived from its pre-pro progenitor molecule. Cleavage of two pairs of dibasic amino acids of proinsulin and removal of a so-called connecting strand (C-peptide) in secretory granules converts proinsulin to the double-chain polypeptide insulin (51 amino acids), the two chains being held together by two disulfide linkages. Upon appropriate stimulation of the B cell, insulin and the C-peptide of the secretory granules are secreted, along with some remaining unprocessed proinsulin, into the portal circulation that drains venous blood from the gastrointestinal tract to the liver (Chapter 20).

The plasma concentration of glucose is the key regulator of insulin secretion. Glucose is first transported, by a glucose transporter protein (GLUT 2), in the B cell, where it is phosphorylated and metabolized. Insulin secretion is stimulated by blood glucose levels even slightly above the fasting level of 75 to 100 mg/dL, and it is reduced when blood glucose is low. Other stimulatory factors include several amino acids, intestinal hormones, glucagon, acetylcholine (parasympathetic stimulation), and others. Secretory inhibitory factors include somatostatin, norepinephrine (sympathetic stimulation), and others (Table 14.4).

Circulating insulin is degraded within minutes in the liver and kidneys. Antibodies to components of islet B cells detected in a high proportion of patients with Type 1 diabetes,

**Systemic and Organismic Aging**

**TABLE 14.3**
**Action of Insulin and Glucagon on Metabolically Important Target Tissues[a]**

| | Insulin Actions | | |
|---|---|---|---|
| **Blood** | **Muscle** | **Adipose Tissue** | **Liver** |
| Glucose | Uptake of:<br>　Glucose<br>　K+<br>　Amino acids<br>　Ketones | Uptake of:<br>　Glucose<br>　K+ | |
| | Synthesis of:<br>　Protein | Synthesis of:<br>　Fatty acids | Synthesis of:<br>　Glycogen<br>　Lipid<br>　Protein |
| | Gluconeogenesis<br>Protein breakdown | Activity of:<br>　Lipoprotein lipase | Gluconeogenesis<br>Glucose output<br>Cyclic AMP<br>Ketogenesis |
| | **Glucagon Actions** | | |
| Glucose | Intropic action on the heart | Synthesis of free fatty acids | Glycogenolysis<br>Gluconeogenesis<br>Ketogenesis<br>Cyclic AMP |

[a] *Ketogenesis* is the formation of ketone bodies, metabolites derived from fatty acid oxidation. *Gluconeogenesis* is the synthesis of glucose in the liver from noncarbohydrate precursors, such as amino acids, glycerols, or lactate. *Lipoprotein lipase* is an enzyme that catalyzes the hydrolysis of glycerides. *Glycogenolysis* is the breakdown of glycogen. *Cyclic AMP* is adenosine 3',5' or 2',3'-monophosphate, an important cellular regulator.

**TABLE 14.4**
**Major Chemical Signals Regulating Insulin and Glucagon Secretion**

| Source of Signals | Insulin Release by B Cells | | Glucagon Release by A Cells | |
|---|---|---|---|---|
| | **Stimulation** | **Inhibition** | **Stimulation** | **Inhibition** |
| Nutrition | Glucose<br>Amino acids<br>Fatty acids | | Protein and amino acids | Glucose (hyperglycemia)<br>Free fatty acids<br>Ketones |
| GI tract | GI peptide hormones<br>　(gastrin, secretin, CCK[a],<br>　etc.) | | CCK[a] and gastrin | Secretin |
| Pancreas | Glucagon | Somatostatin | | Somatostatin<br>Insulin |
| Autonomic signals | Acetylcholine<br>Cholinergic mediators | Catecholamines<br>Adrenergic mediators | Catecholamines<br>Adrenergic mediators | Acetylcholine<br>Cholinergic mediators |
| Local tissue autocrine and paracrine signals | | Serotonin | Hypoglycemia | Catecholamines mostly |
| | | Prostaglandin E | Strenuous exercise | ∝-adrenergic mediators |

[a] CCK, cholecystokinin.

## TABLE 14.5
### Some Factors Responsible for Glucose Intolerance[a] with Aging

**Insulin alterations:**
- Unchanged or elevated plasma levels of insulin
- Alteration in insulin receptors and their internalization in target tissues
- Decreased number of glucose transporter units in target cells
- Alterations in activities of cellular enzymes involved in postreceptor cellular responses
- Increased secretory ratio of proinsulin (less biologically active) to insulin (more biologically active)

**Carbohydrate metabolism alterations:**
- Decrease of body's muscle mass and increase in adiposity
- Diminished physical activity
- Increased fasting plasma free fatty acids that inhibit cellular glucose oxidation
- Increased liver gluconeogenesis

[a] Glucose intolerance is measured by impaired ability to lower blood glucose after a standard dose of glucose. Fasting blood glucose levels remain unchanged or may increase with aging.

---

attack these cells, leading to their extensive destruction and to insulin deficiency. The resulting type, Type 1 diabetes, is initiated by genetic mechanisms.

Insulin binds with specific membrane receptors forming an insulin-receptor complex that is taken into the cell by endocytosis.[8,9] Insulin receptors are found in almost all cells of the body. The insulin-receptor is a heterodimer made up of two α and two β chains with disulfide bridges. Recent studies show that the β subunit, a protein kinase, catalyzes the phosphorylation of two insulin receptor substrates, named IRS-1 and IRS-2, located just inside the cell membrane.[10–12] Thus phosphorylated, these two proteins can serve as "docking sites" for a number of other proteins, one of which is the "glucose transporter," the protein carrier of glucose into the cell. The second pathway activated by the IRS complex is called RAS complex, and its role in diabetes type 2 has been postulated but little studied to date.

Because intracellular free glucose concentration is low (due to its rapid, efficient phosphorylation), some glucose enters cells from the blood down its concentration gradient, even in the absence of insulin. With insulin, however, the rate of glucose entry is increased in insulin-sensitive tissues (muscle and fat) due to facilitated diffusion mediated by the transporters.

The insulin-receptor complex enters the lysosomes where it is cleaved, the hormone internalized, and the receptor recycled.[13] Increased circulating levels of insulin reduce the number of receptors — downregulation of receptors — and decreased insulin levels increase the number of receptors — upregulation. The number of receptors per cell is increased in starvation and decreased in obesity and acromegaly; receptor affinity is decreased by excess glucocorticoids.

## 2. Changes in Metabolic Effects of Insulin with Aging

As the principal mediator of storage and metabolism of nutrient fuels (carbohydrate, protein, and fat), insulin promotes the entrance of glucose and amino acids in insulin-sensitive cells, such as muscle and adipose tissue (Table 14.3). After a meal, the increased blood glucose levels stimulate the release of insulin from the B cells. Glucose is transported into cells by facilitated diffusion along an inward gradient created by low intracellular free glucose and by the availability of specific carriers, the transporters. In the presence of insulin, the rate of movement of glucose into the cell is greatly stimulated in a selective (according to tissue) fashion. As a consequence of activated glucose transport from blood into cell, blood glucose falls to preprandial levels. If insulin is administered in inappropriately high doses, or at a time when glucose levels are low, then concentration of glucose in the blood falls below the normal limit (hypoglycemia) with severe consequences for glucose homeostasis and danger to survival (e.g., coma and neurologic and mental symptoms).

In the liver, insulin does not directly affect the movement of glucose across the cell membrane but facilitates glycogen deposition and decreases glucose output (Table 14.3). Consequently, there is a net increase in glucose uptake. Insulin directly and indirectly induces or represses the activity of many enzymes. For example, insulin suppresses the synthesis of key gluconeogenic enzymes and induces the synthesis and increases the activity of key glycogenetic enzymes and of enzymes involved in lipogenesis.

When insulin levels are low, as occurs several hours after a meal or during fasting, cellular availability of ingested nutrient fuels is depressed, blood glucose is low (hypoglycemia), and mobilization of previously stored fuels is accelerated. Major adjustments to maintain glucose levels within normal range, include the following:

- Decrease in glucose uptake by insulin-sensitive tissues (muscle and fat)
- Preservation of glucose uptake in nonsensitive tissues, primarily brain, critically dependent on glucose for energy metabolism
- Decreased liver glycogen synthesis
- Mobilization of glucagon, a counterregulatory hormone released in response to low insulin levels, to stimulate endogenous glucose release into the blood from glycogenolysis (i.e., glycogen breakdown) and gluconeogenesis (synthesis of glucose from noncarbohydrate precursors such as amino acids and lactate)

**Systemic and Organismic Aging**

During and soon after a meal, when blood glucose is high (hyperglycemia), glucose balance is maintained by several adjustments:

- Insulin secretion is increased
- Endogenous production of glucose is suppressed
- Cellular uptake of glucose (generated by nutrition) is increased, primarily in muscle
- Utilization (in muscle and adipose cells) and storage (in liver as glycogen) of glucose, fat, and amino acids arriving in the blood from the gastrointestinal tract are promoted

A number of functional tests suggest impaired competence of glucose metabolism with aging.[14,15] Many clinical studies indicate a slight (about 1 mg/dL/decade) aging-related increase in fasting blood glucose levels in healthy individuals — not significantly affected by gender. This aging-related effect may be less marked or does not occur in nonobese, physically active elderly.[16,17] However, with age, there is often a striking change in response to a glucose challenge measured by the typical "oral glucose tolerance test." (Figure 14.2)[18,19] In young adults, fasting plasma glucose levels range from 76 to 110 mg/dL; after oral administration of a standard glucose dosage, glucose levels rise to a peak of about 120 mg/dL after 30 min but return to preglucose levels within 1 hour.

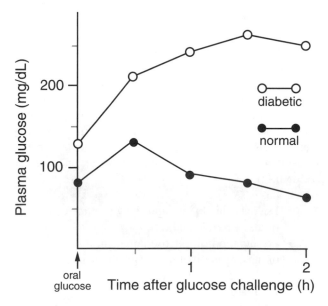

**FIGURE 14.2** Normal and abnormal glucose tolerance tests. After measuring fasting blood glucose levels, the subject is given a glucose drink. In a normal subject, blood glucose level increases about 30 min after ingestion of the glucose. After about 1 h, the blood glucose level is normal. In the diabetic, the fasting glucose level is higher and increases more after oral glucose than in normal controls. Furthermore, the blood glucose level remains significantly high 2 h after ingestion.

In older individuals, blood glucose levels may be higher and take longer to return to normal compared to those in younger individuals. In persons with diabetes mellitus, fasting glucose is greater than 115 mg/dL and the 1, 1.5, and 2 hours postglucose values are greater than 185, 165, and 140 mg/dL, respectively. The inability of a person to lower blood glucose after a standard glucose challenge, designated "glucose intolerance," has been used to diagnose diabetes mellitus. Considerable controversy in the literature has arisen from the apparent necessity, on the basis of this criterion, to label as diabetic more than half of the 65 years and older population. Glucose intolerance as measured above, may be due to causes other than low insulin in the presence of severe hyperglycemia. Hence, other mechanisms have been suggested to explain the decreased sensitivity of glucose metabolism to insulin with aging, taking into account insulin and carbohydrate metabolism alterations (Table 14.5).

## 3. Changes in Glucagon and Counterregulatory Hormones with Aging

Glucagon, secreted by the A cells of the islets of Langherans, like insulin, derives from a larger polypeptide precursor and is degraded in the liver. Factors affecting its release from A cells and its actions are, in general, antagonistic to those of insulin (Tables 14.3 and 14.4).

Glucagon:

- Promotes breakdown of glycogen and lipids
- Promotes conversion of nonglucose molecules to glucose and ketone bodies
- Raises blood glucose levels
- While insulin serves as a hormone of fuel storage, glucagon serves as a hormone of fuel mobilization through glycogenolytic, gluconeogenic, lipolytic, and ketogenic actions in the liver

Glucagon acts by binding to protein kinase receptors identified in various tissues. The increase in plasma glucagon reported in a number of pathologic conditions (e.g., trauma, infections, burns, myocardial infarctions) has suggested a protective (?) role for glucagon as the hormone of insult or injury (e.g., increasing the force and energy of cardiac contractions or inotropic effects); however, this role and its possible relation to aging, remain unclear.

The few studies investigating changes of glucagon actions in the elderly do not indicate any gross abnormality.[20] However, in experimental animals and patients with Type 2 diabetes, an apparently paradoxical rise in glucagon has been reported after glucose administration. This response may contribute to the complex arrays of alterations in glucose metabolism considered as part of the counterregulatory mechanisms.

Counterregulatory hormones (e.g., gastrointestinal peptides prostaglandins, growth hormone [GH], epinephrine, cortisol) serve relatively short-term and adaptive functions (for example, in response to stress, Chapter 10) and oppose the hypoglycemic action of insulin. They generally have a hyperglycemic action, and, in some older persons, may induce diabetes mellitus. Severe hypoglycemia may induce stress. In elderly diabetics, the response of counterregulatory hormones to hypoglycemia, often consequent to insulin administration, is impaired. Because the sensory and sympathomedullary responses to stress (Chapter 10) may also be compromised, the subject may be unaware of the life-threatening hypoglycemia and fail to undertake the necessary metabolic adjustments (e.g., ingestion of glucose) for maintaining homeostasis.

## B.  Insulin Resistance in the Elderly

Insulin resistance refers to failure of insulin to stimulate glucose uptake by peripheral tissue. If an altered responsiveness to insulin exists in the healthy elderly, it is minor and insufficient to account significantly, alone, for the observed impairment in glucose tolerance. Insulin secretion, its metabolism, hepatic extraction of insulin, insulin half-life, insulin clearance, etc., are not significantly changed in the elderly. In fact, in some cases, the higher blood glucose levels in the elderly are associated with higher insulin levels than in younger subjects. What, then, are the mechanisms of the loss of glucose tolerance with aging? The answer has been sought at:

1. The pancreas level, where insulin secretion may be depressed
2. The peripheral level, where resistance of target tissues to insulin may be increased due to a defect in insulin receptors
3. A disruption of the delicate balance between insulin production and tissue responsiveness to insulin involving primarily the liver

The fact that glucose metabolism is compromised in the face of normal insulin secretion and metabolism, even in advanced age, is a cogent argument that peripheral tissues become resistant to the actions of insulin with aging. For this reason, it has been proposed that decreased glucose uptake (by approximately one third) in the elderly, compared to young persons, was due to a decreased number of insulin receptors. More recently, several investigators, suggested that insulin resistance is not due to an "insulin receptor problem."[1-3,21-24] In experiments with knockout mice, it appears that tissue resistance to insulin may result from a defect in the signaling pathway linked to the IRS-2 genes (see above) and that stimulation of this pathway by appropriate compounds may effectively abolish the insulin resistance that is the hallmark of diabetes type 2.[10-12]

Other explanations of the defect or defects in peripheral glucose uptake have focused on postreceptor reactions, including the involvement of a number of cytoplasmic processes such as receptor-mediated phosphorylation and dephosphorylation events or the generation of intracellular mediators following receptor-ligand interactions.[25] The observation that minimal receptor occupancy is required for proper insulin action further supports the suggestion that postreceptor defects may be responsible for apparent "resistance to insulin."

A corollary to a decreased insulin secretion hypothesis with aging is a proportional increase in the secretion of unprocessed proinsulin, which has considerably less biologic activity than insulin.[26,27] Plasma proinsulin levels in the basal state do not differ significantly in the elderly compared to younger controls, but after glucose loading, the amount of proinsulin relative to that of insulin is greater in some older individuals. Considerable attention has been directed to the enzymes that process prohormones.[26,27] Enzymes for cleaving proinsulin to insulin and C-peptides might become less efficient with aging, hence, the high circulating levels of the prohormone. Alternatively, the increased resistance of peripheral tissues to insulin with aging may create a greater demand for insulin with consequent insufficient time before secretion for cleavage of the prohormone. Irrespective of its causes, the elevated proinsulin to insulin ratio with aging appears to be too small to account for the observed alterations in glucose metabolism.

In conclusion, although the mechanisms underlying glucose intolerance with aging are still a subject of debate, the prevalent view is that they reflect a major reduction in peripheral tissue responsiveness to glucose and insulin. The possible existence of defects in the cascade of postreceptor reactions is being actively investigated.

## C.  Additional Mechanisms Responsible for Glucose Intolerance in the Elderly

Current research, in the main, supports the concept that, with aging, defects in glucose homeostasis (also referred to as glucose intolerance) may be due to a multiplicity of factors (Table 14.5), including:

1. Loss of hepatic sensitivity to insulin and reduced glycogenesis
2. Increased glucagon levels
3. Changes in diet and exercise regimen
4. Loss of lean muscle mass and impaired insulin-mediated glucose uptake in skeletal muscle
5. Increase in adipose tissue (e.g., obesity), and impaired insulin-mediated glucose uptake in adipose tissue

**Systemic and Organismic Aging**

**TABLE 14.6**
**Comparison of Some Defects of Glucose Metabolism in Aging, Obesity, and Diabetes Type 2**

| Metabolic Parameter | Aging | Obesity | Diabetes Type 2 |
|---|---|---|---|
| Glucose | | | |
|   Basal level | Variable | Variable | Variable or ↑ |
|   Tolerance | ↓ | Variable | ↓↓↓ |
| Insulin | | | |
|   Basal level | Variable | ↑ | Variable |
|   With glucose loading: | | | |
|     Initial response | Delayed | ↑ | Delayed |
|     Later response | ↑ | ↑↑↑ | Variable |
| Probable site of defect: | | | |
|   Number of insulin receptors | Variable | ↓ | Variable or ↓ |
|   In postreceptor events | Yes | ? to yes | Yes |
| Lipids | | | |
|   Triglyceride | Variable | ↑ | ↑ |

↑ = increase; ↓ = decrease; multiple arrows indicate much greater response. Variable = no consistent change.

The latter factor, that is, increase in adipose tissue, may contribute to the decreased ability of insulin to facilitate cell glucose uptake (Table 14.6). With aging, while muscle cells undergo sarcopenia, adipocytes increase in number and in size (Chapter 24). Cell enlargement reduces the concentration of receptors on the cell surface.[28] This, coupled with a possible reduction in the absolute number of receptors (suggested above), could lead to reduced insulin binding and decreased cell response to the hormone.[29] A comparison of the metabolic characteristics in aging, obesity, and diabetes type 2 reveals the complexity and the subtlety of the changes and underlies the specificity of the changes.[29] In addition, this comparison may explain why obesity contributes or predisposes to diabetes type 2. The relationship of obesity to diabetes is reflected in the high incidence of the two conditions in specific ethnic groups, such as the Pima Indians of the state of Arizona.[30]

Caloric (food) restriction in rats reduces adipose tissue mass and adipocyte size and maintains responsiveness to insulin and glucagon into old age. However, caloric restriction also retards the aging process, in general, and persistence of normal insulin response may merely reflect this antiaging affect, also involved in the maintenance of small-size adipocytes (Chapter 24).

Despite the close relationship between aging and obesity and insulin resistance, it must be realized that adipose tissue is responsible for the removal of less than 5% of administered glucose in humans and rats. Liver and muscle are the primary sites of glucose uptake. It is in these tissues that the age-related defect in transport and intracellular metabolism must develop, and several factors must intervene simultaneously to explain the increased insulin resistance of the elderly.

It must be kept in mind that, besides alterations of the endocrine pancreas and insulin receptor abnormalities, hyperglycemia and obesity may be caused by a number of endocrine disorders (e.g., Cushing's syndrome, acromegaly). This multifactorial etiology may be profitably studied in animal models such as the obese ob/ob mouse, which shows hyperinsulinemia and insulin resistance. The etiology of the syndrome has been related to a number of neuroendocrine defects involving the hypothalamic satiety center, the neurotransmitter serotonin, temperature regulation, and alterations in glucocorticoid and thyroid hormones. Despite these metabolic and endocrine disorders, the life span of the ob/ob mouse is little affected, and as the animal ages, there is a remission of symptoms. Spontaneous remission with aging of diabetic or prediabetic symptoms is also a common finding in humans.

## D. Diabetes Mellitus Type 2

### 1. Prevalence and Classification of Diabetes Mellitus Types

Diabetes mellitus is a widespread health problem in humans. A recent survey (1999) shows that it affects almost 7% of the population in the United States, with a health care cost of $100 billion. In the last ten years, its prevalence has been rapidly escalating due to changes in demographic age distributions (Chapter 2), with an increasing proportion of obese and elderly in the population. It has been predicted that the prevalence of type 2 diabetes will double by the year

2010. In the United States and Western Europe, the disease affects about 16 to 20% of those aged 65 and over, and it is likely that the size of the diabetic population will increase.

In countries with more limited food availability, the number of diabetics is much lower (as low as 2%). These low rates have been attributed to a variety of factors, among which the diet is most important, but exercise, lifestyle, and heredity may also contribute to the low risk and reduced severity of the disease. In all cases, this disabling disease has a tremendous impact not only on the immediate health of the affected individuals, but also on their long-term viability.[20–22] Diabetes mellitus has been classified into several types and subtypes according to a variety of etiologies —endogenous or genetic and exogenous or environmental. An abbreviated classification,[31] restricted to the most frequent types, considers the following four categories:

- Type 1 is due to little or no secretory capacity of insulin. It is primarily found in children (juvenile diabetes) but is also known to occur occasionally at other ages. It is considered an autoimmune disease (Chapter 15) due to its association with specific immune response (HLA) genes and the presence of antibodies that destroy the islets cells.
- Type 2 is found in over 90% of the diabetic patient population. Insulin secretory capacity is partially preserved, but plasma insulin levels, sometimes quantitatively high, are inappropriately low in relation to the magnitude of insulin resistance and hyperglycemia. The previous classification as "non-insulin dependent diabetes mellitus (NIDDM)" should be abandoned. Type 2 diabetes generally appears after the age of 40 years, hence, its alternative name of "late-onset diabetes."
- Gestational diabetes occurs, as the name indicates, in women during pregnancy.
- Maturity-onset diabetes of the young (MODY) (types 1, 2, and 3) is characterized by genetic defects of B cell function (e.g., mutations of several genetic loci in chromosomes 12, 13, and 17).

Of these classes, late-onset diabetes mellitus (type 2), the most frequent form of the disease as well as the one related to aging, is the only one that will be briefly considered here.

## 2. Pathogenesis of Diabetes Mellitus Type 2

As indicated above, diabetes mellitus type 2 is the most common form of diabetes. The onset of the disease occurs years before symptoms are appreciated, therefore, it is important to screen high-risk individuals systematically (e.g., every 3 years). Personal promoting factors are well established and include:

- Increasing age (usually 40 years and older)
- Reduced physical activity
- Obesity, especially in those individuals with central or upper body obesity and with genetic susceptibility (100% concordance rates in identical twins), the expression of which is modified by environmental factors (see below and Chapter 3)
- Ethnicity differences: in the United States, diabetes mellitus type 2 occurs at an earlier age (in some cases during adolescence), and it is more frequent in Native Americans (e.g., the Pima Indians[30]), Mexican descendents, and blacks. Worldwide, there is a propensity for people of Asian descent, Polynesians, and Australian Aborigines, to develop the disease when they migrate to westernized surroundings (underlying the importance of changes in diet and degree of physical activity).

Diabetes mellitus types 1 and 2 share similar clinical and laboratory characteristics, as summarized in Table 14.7. Some characteristics specific to type 2 are compared with those of aging and obesity in Table 14.6. In general, type 2 is characterized by:

- Levels of insulin that are usually normal or increased but low, relative to the severe hyperglycemia, and that further decline with the increasing severity of the hyperglycemia
- Impaired insulin action that may be due to increased insulin resistance and extra-insulin aging-associated changes (see above)
- Given the failure of insulin to control hyperglycemia, the consequent "glucotoxicity" may

---

**TABLE 14.7**
**Characteristics of Diabetes Mellitus**

| | |
|---|---|
| Decreased glucose uptake | Hyperglycemia |
| | Decreased glycogenesis |
| | Increased hepatoglucogenesis |
| | Glycosuria |
| | Polyuria |
| | Polydipsia |
| | Polyphagia |
| Increased protein catabolism | Increased plasma amino acid |
| | Increased gluconeogenesis |
| | Weight loss, growth inhibition |
| | Negative nitrogen balance |
| Increased lipolysis | Increased free fatty acids |
| | Ketosis |
| | Acidosis |
| Vascular changes | Microangiopathies |

**Systemic and Organismic Aging**

impair B cell response to glucose, perhaps through the glucose metabolite, glucosamine, that would impair the insulin-mediated translocation of GLUT to the cell membrane

As already discussed, insulin action is a complex process involving multiple steps, and most of these are affected in diabetes. At present, it appears unlikely that a single pathogenic mechanism may be responsible for type 2 diabetes.[2,3,32,33]

## 3.   Consequences of Diabetes Mellitus Type 2

A variety of "diseases of old age" (e.g., coronary heart disease, glomerulonephrosis, retinopathy, limb gangrene, stroke, and cataract) are consequences or complications of diabetes and major causes of ill health and mortality.[2,3,32,33] While some diabetics live long lives with little indisposition, in general, the rate of disability of diabetics is two to three times greater than that of nondiabetics. For example, in diabetics, blindness is about ten times[34] and gangrene (with consequent limb amputation) about 20 times more common,[35,36] and 14% of diabetics (usually the elderly) are bedridden for an average 6 weeks per year.[37]

Types 1 and 2 of diabetes mellitus are major risk factors for the pathologic consequences of atherosclerosis;[38,39] for unclear reasons, the risk increase is greater in women than men (Chapter 16). The major reason for this increased risk is that diabetes induces pathologic changes in the arterial wall, a specific type of microangiopathy,[40] which aggravates the vascular aging-related atherosclerosis (Chapter 16). Atherosclerosis involving the arteries of the heart, lower extremities, and brain is the major cause of death from diabetes.[38,39] The atherosclerotic process is undistinguishable from that affecting nondiabetics, but it begins earlier and may be more severe. Diabetes also increases the severity of hypertension and promotes higher LDL levels (Chapter 16), thereby, worsening the atherosclerotic lesions.[41] The causes of the microangiopathy, hypertension, and high LDL remain poorly understood: clinical studies indirectly support the hypothesis that hyperinsulinemia (high blood levels of insulin) either due to endogenous causes, such as in type 2 diabetes, or to exogenous administration of insulin for diabetes therapy, may contribute to the microangiopathy, perhaps by stimulating vascular smooth muscle proliferation (Chapter 16). Another suggestion focuses on the high levels of glucose, transitorily present even under excellent conditions of therapeutic management. Under these conditions, glycosylation of cellular polypeptides and proteins might modify their enzymatic and structural properties and cause widespread cellular alterations that could contribute to long-term complications.[42,43]

**TABLE 14.8**
**Diabetes and Accelerated Aging**

| Diabetes | Aging |
|---|---|
| Microangiopathy | — |
| Cataracts | Cataracts |
| Neuropathy | Neuropathy |
| Accelerated atherosclerosis | Atherosclerosis |
| Early decreased fibroblast proliferation | Decreased fibroblast proliferation |
| Autoimmune involvement | Autoimmune involvement |
| Skin changes | Skin changes |

### RELATIONSHIP BETWEEN DIABETES AND AGING

While, historically, research has focused on the increased incidence of diabetes with aging, recent years have witnessed a reversal of this focus, the question asked being whether there might not be an acceleration of aging in diabetes (Table 14.8). Patients with diabetes display an increased incidence of several features commonly associated with aging: cataracts (Chapter 9), microangiopathy (Chapter 16), neuropathy (Chapter 8), dystrophic skin changes (Chapter 22), and accelerated atherosclerosis (Chapter 16). Accelerated atherosclerosis is a major feature of the various genetic syndromes reported to resemble premature aging, and all of these syndromes include abnormal glucose tolerance (Chapter 3).[44] Further, in normal aging, in patients with progeria and in diabetics, the proliferative capacity of cultured fibroblasts is reduced, perhaps due, in part, to a reduced response to insulin and to growth factors. Insulin resistance has been reported in cells from patients with Werner's syndrome and progeria (Chapter 3). In addition, in both juvenile-onset and maturity-onset diabetics, the rate of collagen aging (the aging of collagen having been represented as the fundamental aging process) is accelerated. And, finally, the putative autoimmune etiology of juvenile onset diabetes, observations of immune dysfunction in aging and in diabetes, and the reports of increased pancreatic amyloidosis in senile humans and animals have excited the interest of proponents of an immunogenesis of aging.[31] Collectively, these studies point to the possibility of an intriguing relationship between diabetes and aging.

The great deal of genetic variability in human populations with respect to the life span also applies to predisposition to diabetes. Studies of twins have demonstrated a high degree of concordance in late-onset diabetics which would make the incidence of a putative recessive diabetic gene greater than 40%.[45] However, genetic factors in diabetes are at present poorly understood, probably owing to the considerable heterogeneity and complexity of this disorder. Juvenile-onset diabetes has been recently linked to several specific major histocompatibility complex (HLA) phenotypes that may predispose selected individuals to viral infection or autoimmune reactions (Chapter 15), but the basis of genetic factors in late-onset diabetes remains unresolved. The apparent high heritability of late-onset diabetes may indicate a pathology complicated by the effects of "normal aging," or alternatively, diabetes may represent an acceleration of basic aging processes in a large, genetically predisposed percentage of the population.

Whatever genetic and pathological components ultimately prove to underlie selected specific manifestations of aging or the evident heterogeneity in diabetes and atherosclerosis, it is perhaps significant in this regard to note that the currently recommended treatment for late-onset diabetes mellitus is carefully restricted diet and regular exercise, and both are commonly considered to be the normal healthy individual's best defenses against atherosclerosis and senility, while the best assurance of a long life span remains the subjects' thoughtful choice of long-lived parents.

## 4. Management and Pharmacologic Interventions

Treatment of diabetes mellitus, in general, and, particularly of type 2 diabetes, involves:

- Changes in the lifestyle, which can be viewed as the cornerstone of treatment, especially in the early stages of the disease
- Pharmacologic interventions that include oral glucose-lowering agents and insulin therapy; irrespective of the medication used, the goal of the therapy should be to lower blood glucose as close to normal levels as possible

In diabetes mellitus type 1, the primary focus in treatment is to replace the lacking insulin; lifestyle changes are also beneficial, because they facilitate insulin therapy and improve general health. As stated above, in diabetes type 2, changes in lifestyle — diet[47–49] and physical exercise[50–53] — are the cornerstone of treatment, especially in the early stages of the disease. Pharmacologic interventions represent a secondary therapeutic strategy and include not only insulin but also a variety of agents capable of stimulating insulin secretion or reducing tissue resistance to insulin.[2,3,46,54] In type 1 and type 2 diabetes, success in treatment requires careful follow-up and rapid interventive responses to the continuously progressing dysfunction of glucose homeostasis. For most diabetics, glucose control deteriorates over time, thus necessitating more intensive pharmacologic interventions.[55] The synopsis of treatment presented in Box 14.2 is not intended to give practical information on the complex management of diabetes, but rather to emphasize some of the major characteristics of the disease by pointing out the sites and mechanisms of therapeutic interventions.

## III. DIFFUSE ENDOCRINE AND CHEMICAL MEDIATORS

A diverse group of mediators, less important than insulin and glucagon, contributes to regulation of nutrient metabolism. Some of these change in level and significance in older age.[56] The so-called counter-regulatory hormones have been discussed. A disparate group of similar media-

tors, some of them opposing the hypoglycemic actions of insulin, are discussed briefly here. Among these are hormones of the diffuse endocrine system of the gastrointestinal tract, generally peptide in nature, which include: gastrin, involved in stimulation of gastric acid and pepsin secretion; secretin, involved in stimulation of gallbladder contraction; and cholecystokinin (CCK), which stimulates delivery of pancreatic digestive enzymes and bicarbonate to the lumen of the intestines. Others, such as gastric inhibitory polypeptide (GIP) potentiate insulin release mediated by glucose or amino acid. The concentration of some of these hormones changes with aging, in fasting, and during glucose tolerance tests.[56]

Other local intercellular mediators include prostaglandins, a large class of bioactive lipids with multiple actions (e.g., inhibition of platelet aggregation, increase in vascular permeability, and promotion of smooth muscle contraction).[57] Some prostaglandins elicit a selective decrease in the counter-regulatory response of glucagon to hypoglycemia. Calcium ion is an important regulator of the cellular actions of insulin. Physiologic insulin concentrations augment intracellular calcium concentrations, and experimental lowering of intracellular calcium compromises some of the metabolic actions of insulin.[58] Insulin regulates the enzyme $Ca^{++}/Mg^{++}$ ATPase, which is also an important regulator of intracellular calcium ion. The activity of this enzyme is decreased in several tissues of diabetic or obese rats and in human diabetics.[59] Stimulation of this ATPase lowers the increased insulin resistance characteristic of type 2 diabetes. These and other data suggest that lowered insulin sensitivity of target cells is related to a decrease in intracellular calcium ion. A large number of other factors that influence the actions of calcium and possible relations to insulin action have not yet been evaluated in aging.

It may be useful to recall here that the role of nutrients is not only to satisfy the metabolic needs of the cells under steady state conditions, but also to provide greater energy under conditions of increasing demand, such as during the period of growth or under the challenge of stress. Thus, carbohydrates, absorbed primarily as glucose, are used, immediately, for energy through aerobic pathways and, secondarily, for lipoprotein synthesis, conversion to fat, and for storage as glycogen. The actions of most hormones, including insulin and glucagon, the major hormonal regulators of carbohydrate metabolism, as well as those of the counter-regulatory hormones, glucocorticoids, thyroid hormones, and growth hormones, may occur in a time-frame ranging from seconds to hours; for example, insulin may initiate stimulation of glucose transport in cells in a few seconds, but it may require minutes and hours before the full extent of its actions are completed. Under conditions of acute stress, when an immediate and robust response is indispensable for survival, the

**Systemic and Organismic Aging**

---

**Box 14.2**

**Treatment of Diabetes Type 2**

**Nonpharmacologic Measures — Lifestyle Changes**

1. Diet

   Weight loss, especially in obese persons induces:
   - Decline in blood glucose levels
   - Decline in LDL profile (Chapter 16)
   - Improvement of blood pressure
2. Weight loss:
   - Is difficult to maintain for a long period of time
   - May be achieved by different types of diet restriction, provided the caloric deficit is about 500 kcal/day
   - May benefit from changed composition of dietary fat from saturated to unsaturated (e.g., Mediterranean diet, Chapter 24)
3. Exercise
   - Moderate exercise (Chapter 24):
     - Improves insulin action
     - Facilitates weight loss
     - Lowers cardiovascular risk

**Pharmacologic Measures**

Oral Glucose-Lowering Agents

Sulfonylurea drugs act by:
- increasing insulin secretion by binding to their receptor associated with ATP-dependent K channels on the surface of B cells

Benzoic acid derivatives act as sulfonylureas:
- To stimulate insulin secretion by binding to a different portion of the sulfonylurea receptor

Biguanides act by reducing:
- Liver glucose production
- Body weight

Thiazolidinediones reduce insulin resistance:
- Probably through activation of a nuclear receptor (the peroxizome proliferator–activated receptor γ) that regulates the transcription of several insulin-responsive genes

α-Glucosidase inhibitors:
- Reversibly inhibit α-glucosidases (intestinal enzymes that break down complex carbohydrates)

Insulin Therapy
- Once or twice daily, injection of intermediate (Lente) or long-acting (Ultralente) insulin

---

rapid activation of neurotransmitters and local hormones with paracrine action may serve to provide the metabolic energy needed to support the necessary adjustments for adaptation (Chapter 10).

## REFERENCES

1. Quay, W.B., Pancreatic regulation of nutrient metabolism, in *Hormones and Aging*, Timiras, P.S., Quay, W.B., and Vernadakis, A., Eds., CRC Press, Boca Raton, FL, 1995.
2. Sinclair, A.J., Diabetes mellitus, in *Principles and Practice of Geriatric Medicine*, Pathy, M.S.J., Ed., John Wiley & Sons, New York, 1998.
3. Sherwin, R.S., Diabetes mellitus, in *Cecil Textbook of Medicine,* 21st ed., Goldman, L. and Bennett, C., Eds., W.B. Saunders, Philadelphia, PA, 2000.
4. Andres, R., Aging and diabetes, *Med. Clin. North America*, 55, 835, 1981.
5. Sugawara, K. et al., Marked islet amyloid polypeptide-positive amyloid deposition: A possible cause of severely insulin-deficient diabetes mellitus with atrophied exocrine pancreas, *Pancreas*, 8, 312, 1993.
6. Raven, E.P. et al., Effect of age and diet on insulin secretion and insulin action in the rat, *Diabetes,* 32, 175, 1983.
7. Chaudhuri, M., Sartin, J.L., and Adelman, R.C., A role for somatostatin in the impaired insulin secretory response to glucose by islets from aging rats, *J. Steroid Biochem.*, 38, 431, 1983.

8. Golfine, I.D., The insulin receptor: Molecular biology and transmembrane signaling, *Endocrine Rev.*, 8, 235, 1987.

9. Rosen, O.M., Structure and function of insulin receptors, *Diabetes,* 38, 1508, 1989.

10. Alper, J., New insights in type 2 diabetes, *Science*, 289, 37, 2000.

11. Withers, D.J. and White, M., Perspective: The insulin signaling system: A common link in the pathogenesis of type 2 diabetes, *Endocrinology*, 141, 1917, 2000.

12. Kido, Y. et al., Tissue-specific insulin resistance in mice with mutations in the insulin receptor, IRS-1, and IRS-2, *J. Clin. Investig.*, 105, 199, 2000.

13. Marshall, S. and Olefsky, J.M., Separate intracellular pathways for insulin receptor recycling and insulin degradation in rat adipocytes, *Cell Physiol.*, 117, 195, 1983.

14. Davidson, M.B., The effect of aging on carbohydrate metabolism: A review of the English literature and a practical approach to the diagnosis of diabetes mellitus in the elderly, *Metabolism*, 28, 688, 1979.

15. Taylor, R. and Agius, L., The biochemistry of diabetes, *Biochem. J.*, 250, 625, 1988.

16. Lindberg, O, Tilvis, R.S., and Strandberg, T.E., Does fasting plasma insulin increase by age in the general elderly population? *Aging*, 9, 277, 1997.

17. Garcia, G.V. et al., Glucose metabolism in older adults: A study including subjects more than 80 years of age, *J. Am. Geriatr. Soc.*, 45, 813, 1997.

18. Harris, M.I. et al., Prevalence of diabetes, impaired fasting glucose, and impaired glucose tolerance in U.S. adults. The Third National Health and Nutrition Examination Survey, 1988–1994, *Diabetes Care*, 21, 1236, 1998.

19. Halter, J.B., Diabetes mellitus in older adults: Underdiagnosis and undertreatment, *Am. J. Geriatr. Soc.*, 48, 340, 2000.

20. Simonson, D.C. and DeFronzo, R.A., Glucagon physiology and aging: Evidence for enhanced hepatic sensitivity, *Diabetologia,* 25, 1, 1983.

21. Jackson, R.A., Mechanisms of age-related glucose intolerance, *Diabetes Care,* 13 (Suppl. 2), 9, 1990.

22. Raven, G.M., Role of insulin resistance in the pathophysiology of noninsulin dependent diabetes mellitus, *Diabetes/Metabolism Rev.,* 9 (Suppl. 1), 5S, 1995.

23. LeRoith, D., Insulin-like growth factors, *N. Engl. J. Med.*, 336, 633, 1997.

24. Polonsky, K.S., Sturis, J., and Bell, G.I., Non-insulin dependent diabetes mellitus — a genetically programmed failure of the beta cell to compensate for insulin resistance, *N. Engl. J. Med.*, 334, 777, 1996.

25. Fink, R.I. et al., The role of the glucose transport system in the postreceptor defect in insulin action associated with human aging, *J. Clin. Endocrinol. Metab.*, 58, 721, 1984.

26. Gold, G., Reaven, G.M., and Reaven, E.P., Effect of age on pro-insulin and insulin secretory patterns in isolated rat islets, *Diabetes*, 30, 77, 1981.

27. O'Rahilly, S. et al., Brief report: Impaired processing of hormones associated with abnormalities of glucose homeostasis and adrenal function, *N. Engl. J. Med.,* 333, 1386, 1995.

28. Olefsky, J.M. and Reaven, G.M., Effects of age and obesity on insulin binding adipocytes, *Endocrinology*, 96, 1486, 1975.

29. Stout, R.W., Ageing and glucose tolerance, in *Diabetes in Old Age*, Finucane, P. and Sinclair, A.J., Eds, John Wiley & Sons, New York, 1995.

30. Knowler, W.C. et al., Diabetes incidence and prevalence in Pima Indians: A 19-fold greater incidence than in Rochester, Minnesota, *Am. J. Epidemiol.*, 108, 497, 1978.

31. American Diabetes Association: Clinical Practice Recommendations 1998. An up-to-date summary of the current classifications of diabetes and standards of care for the management of diabetic care of patients, including the goals of treatment. *Diabetes Care,* 21 (Suppl. 1), 1998.

32. Morley, J.E. and Kaiser, F.E., Unique aspects of diabetes mellitus in the elderly, *Clin. Geriatr. Med.*, 6, 693, 1990.

33. Singh, I. and Marshall, M.C., Diabetes mellitus in the elderly, *Endocrinol. Metab. Clin. North Am.*, 24, 255, 1995.

34. Dornan,T.L. et al., A community survey of diabetes in the elderly, *Diabetic Med.*, 9, 860, 1992.

35. Pecoraro, R.E., Reiber, G.E., and Burgess, E.M., Pathways to diabetic limb amputation. Basis for prevention, *Diabetes Care*, 13, 513, 1990.

36. Larsson, J. et al., Decreasing incidence of major amputation in diabetic patients: A consequence of multidisciplinary foot care team approach? *Diabetic Med.,* 12, 770, 1995.

37. Damsgaard, E.M., Froland, A., and Green, A., Use of hospital services by elderly diabetics and fasting hyperglycemic patients aged 60–74 years, *Diabetic Med.,* 4, 317, 1987.

38. 38. Kannell, W.B. and McGee, D.L., Diabetes and cardiovascular disease: The Framingham Study, *J. Am. Med. Assoc.*, 241, 2035, 1979.

39. Haffner, S.M. et al., Mortality from coronary heart disease in subjects with type 2 diabetes and in nondiabetic subjects with or without prior myocardial infarction, *N. Engl. J. Med.*, 339, 229, 1998.

40. Siperstein, M.D., Unger, R.H., and Madison, L.L., Studies of muscle capillary basement membranes in normal subjects, diabetic and prediabetic patients, *J. Clin. Invest.*, 47, 1973, 1968.

41. Pyorala, K., Diabetes and atherosclerosis: An epidemiological view, *Diabetes and Metabolism Review*, 3, 463, 1987.

42. Cerami, A., Vlassara,, H. and Brownlee, M., Glucose and aging, *Sci. Amer.*, 256, 91, 1987.

43. Cerami, A., Vlassara, H., and Brownlee, M., Role of advanced glycosylation products in complications of diabetes, *Diabetes Care*, 11 (Suppl. 1), 73, 1988.

44. Goldstein, S., Human genetic disorders that feature premature onset and accelerated progression of biological aging, in *The Genetics of Aging*, Schneider, E.L., Ed., Plenum Press, New York, 1978.

45. Tattersall, R.B. and Pyke, D.A., Diabetes in identical twins, *Lancet*, 2, 1120, 1972.

## Systemic and Organismic Aging

46. Physician's Guide to Non-Insulin-Dependent (Type II) Diabetes: Diagnosis and Treatment, 3rd ed., American Diabetes Association, Alexandria, Virginia, 1998.

47. Rendell, M., Dietary treatment of diabetes mellitus, Editorial, *N. Engl. J. Med.*, 342, 1440, 2000.

48. Chandalia, M. et al., Beneficial effects of high dietary fiber intake in patients with type 2 diabetes mellitus, *N. Engl. J. Med.*, 342, 1392, 2000.

49. Shorr, R.I. et al., Glycemic control of older adults with type 2 diabetes: Findings from the Third National Health and Nutrition Examination Survey, 1988–1994, *J. Am. Geriatr. Soc.*, 48, 264, 2000.

50. Ryan, A.S. et al, Resistance training increases insulin action in postmenopausal women, *J. Gerontol.*, 51, M199, 1996.

51. Dela, F. et al., Training induced enhancement of insulin action in human skeletal muscle: The influence of aging, *J. Gerontol.*, 51, B247, 1996.

52. Pratley, R.E. et al., Aerobic exercise training-induced reductions in abdominal fat and glucose-stimulated insulin response in middle-aged and older men, *J. Am. Geriatr. Soc.*, 48, 1055, 2000.

53. Caruso, L.B. et al., What can we do to improve physical function in older persons with type 2 diabetes? *J. Gerontol.*, 55, M372, 2000.

54. Turner, R.C. et al., Intensive blood glucose control with sulphonylureas or insulin compared with conventional treatment and risk of complications in patients with type 2 diabetes, *Lancet*, 352, 837, 1998.

55. Sinclair, A.J., Turnbull, C.J., and Croxson, S.C.M., Document of care for older people with diabetes, *Postgrad. Med. J.*, 72, 334, 1996.

56. McConnell, J.G. et al., The effect of age and sex on the response of enteropancreatic polypeptides to oral glucose, *Age Ageing*, 12, 54, 1983.

57. Giugliano, D. et al., Normalization by sodium salicylate of the impaired counterregulatory glucagon response to hypoglycemia in insulin-dependent diabetes. A possible role for endogenous prostaglandins, *Diabetes*, 34, 521, 1985.

58. Draznin, B., Cytosolic calcium and insulin resistance, *Am. J. Kidney Dis.*, 21 (Suppl. 3), 32, 1993.

59. Levy, J. et al., Effects of food restriction and insulin treatment on ($Ca^{2++}$ $Mg^{2+}$)-ATPase response to insulin in kidney basolateral membranes of non-insulin-dependent diabetic rats, *Metab. Clin. Exp.*, 39, 25, 1990.

# 15 The Immune System

*Lia Ginaldi*
University of L'Aquila

*Hal Sternberg*
BioTime Inc.

## CONTENTS

## I. INTRODUCTION

Understanding the mechanisms of the immune function and the influence of aging thereon is not a simple matter. Immunological reactions are highly complex (Table 15.1) and require the participation of numerous humoral factors, cell types (Table 15.2), tissues, and organs (Table 15.3). The immune system has the enormous task of protecting us from disease (Table 15.4) and that which is foreign, while avoiding harm to that which is self. Before one can appreciate the dynamic effects of aging on the immune function that are described in this chapter, it would be constructive to be familiar with some key terms and concepts listed in Table 15.5. It would also be helpful to be aware of the numerous cell types (Table 15.2) involved in immune reactions that are under sensitive control of a variety of humoral factors and molecular interactions (Table 15.6).

**TABLE 15.1**
**Complexity of the Immune System and Immunologic Senescence**

Components are multiple and include the following:
  Organs and tissues
  Cells
  Secretory factors
Differential reactions depend on the following:
  Immunogen (composition, route, dose, half-life)
  Histocompatibility genes
  Prior humoral, psychologic, and nutritional state
  Antigenic history
  Age
Immunologic senescence does not equally affect all components and
  activities of the immune system.
Differences in experimental approaches and subjective interpretation
  of data often lead to contradictory conclusions.

## II. HOW DOES THE IMMUNE SYSTEM FUNCTION?

The immune system exerts defense functions (Table 15.4) that can be differentiated into specific or acquired, and innate responses, which are closely interconnected. Innate immunity provides broad nonspecific host defenses that do not require antigenic specificity or immunologic memory. The specific immune responses are mediated by T and B lymphocytes, clonally distributed, capable of specific reactions with single epitopes of a given antigen, and displaying adaptive immunity.

### A. T LYMPHOCYTES

T helper (Th) cells function as key regulatory elements in the immune response, whereas cytotoxic T lymphocytes (CTL) function mainly as effector cells to eliminate the cells of the host infected by viruses. Progenitor T cells, derived from the bone marrow, must first enter the thymus and rearrange the genes encoding components of the T-cell receptor (TCR) for antigen. As a result, each T cell expresses a unique receptor made up of a combination of DNA regions, selected from a multitude of different copies available. Thereafter, newly generated T cells must undergo positive selection for antigen reactivity and negative selection for reactivity to autoantigens. T cells surviving these sequential processes are then released from the thymus as naïve cells, and may persist for many years until stimulated by a specific antigen. These mature, selected, antigen-specific T cells are then required to recognize antigenic peptides together with self major histocompatibility complex (MHC) products, circumscribed by the immunological rule of "self-restriction." Because only very small numbers of T cells expressing each of the multitude of possible clonotypic antigen-specific receptors

can exist in each individual, a successful immune response is absolutely dependent upon the rapid production of larger numbers of T cells of the same antigenic specificity. This process, called *clonal expansion*, is dependent upon specific antigenic stimulation, nonantigen specific costimulation, and the production and utilization of cytokine growth factors such as interleukin-2.[1] At the termination of the response, the excess cells must be removed from the system. This occurs by programmed cell death (apoptosis). A small fraction of the antigen-specific cells must, however, remain intact to function as *memory cells* in the event that antigen reexposure occurs at a later date. Some memory cells are also maintained in the body in a constant state of activation and slow turnover.

### B. B LYMPHOCYTES

The main characteristic of B lymphocytes is their capability to differentiate into plasma cells and secrete immunoglobulin (Ig) proteins called antibodies. Igs are receptors for antigen on the B-cell surface, and mediators of humoral responses when secreted in a soluble form. B cells derive from a bone marrow precursor that becomes a virgin mature B lymphocyte through a DNA rearrangement of the genes encoding for the Ig variable region, with a process similar to that involving the TCR specificity generation. The variable region can be further subdivided into regions where variability is greatest, called hypervariable or complementary determining regions (CDRs); these regions are truly responsible for antigen binding and antibody diversity.[2] However, *the most potent process in the generation of antibody diversity is somatic mutation*, which occurs in germinal centers following specific antigen stimulation. There are five classes of Ig detectable in serum: IgG, IgA, IgM, IgE, and IgD. Each Ig class or isotype differs by the sequence of its constant region (Table 15.7), and this difference translates into distinct functional capability (Table 15.8).

### C. INNATE IMMUNITY AND SPECIFIC ADAPTIVE IMMUNITY

The capability to cope with pathogens, cancer cells, and other threatening agents resides not only in specific immune mechanisms, but also in *innate immune reactions*, mainly mediated by *polymorphonuclear leukocytes, macrophages, and natural killer (NK) cells*. The natural defense mechanisms include chemotaxis, phagocytosis, natural cytotoxicity, cell–cell and cell–matrix interactions, and production of soluble mediators.[3] Innate immunity is nonspecific, and repeated exposure to the same pathogen will result in the same response. In contrast, adaptive immunity is characterized by specificity and immunologic memory. This means that upon reexposure to a given pathogen, memory cells will be recruited rapidly and generate a rapid and directed immune response.

**TABLE 15.2**
**Cell Types of the Immune System**

**Stem Cells:**
**Pluripotent Cells in Bone Marrow that Differentiate into all Other Leukocytes (White Blood Cells) Involved in Immune Function**

| Lymphocytes | Neutrophils | Antigen Processing Cells (APC) |
|---|---|---|
| **A. T Cells** (originating in thymus)<br><br>1. T-helper (receptor CD4), differ in part based on cytokines<br><br>a. Th0-form by naïve T cells activation; features in common with Th1 and Th2; mature into Th1 and Th2 upon stimulation with antigen<br><br>b. Th1 participates in cell-mediated immunity, provides help for CD8 mediated effector function and IgG2a antibody; often involved in response against intracellular antigens, e.g., Listeria and Mycobacterium tuberculosis<br><br>c. Th2 efficient in aiding B cells to produce IgA, IgE, and IgG1 antibodies (may be involved in allergic responses)<br><br>2. T-Killer-cytotoxic T lymphocytes (CTL), CD8+ subset, with Fc receptors for Igs, destroy malignant and viral infected cells<br><br>**B. B Cells**<br><br>Mature in bone marrow, migrate to other tissues, multiply, and further differentiate (upon appropriate interaction with antigen presenting cells) to form *plasma cells*; each B cell is genetically programmed to secrete one specific antibody<br><br>**C. Naïve (T or B) cells**<br><br>Long-lived quiescent cells never yet activated by antigen interaction; important in preventing new opportunistic infection<br><br>**D. Memory (T or B) cells**<br><br>Upon initial exposure of naïve cells to antigens within lymphoid organs, these cells undergo clonal expansion, generating *memory cells* with various cytokine secretion, responses, and effects<br><br>**E. Thymocytes**<br><br>Immature pre-T cells located in the thymus; stimulated by thymosins (from thymic epithelial cells) to further differentiate into mature T lymphocytes<br><br>**F. Natural killer cells**<br><br>Lyse target cancerous cells or infected by virus; considered a component of the innate immune system and comprise 5 to 15% of leukocytes; release cytotoxic substances upon interaction with abnormal cells | Also called polymorphonuclear leukocytes (PMN); most common leukocytes (50 to 70%); may be considered: *the first line of defense against infection*; possess a dense lobed nucleus and short life; first to migrate and enter infected site; cytoplasm contains granules with lysosomal (degradative) enzymes that lyse engulfed bacteria along with release of oxidizing agents; involved in immune reactions against bacteria, fungi, parasites, viral infections, and tumor cells | Accessory cells that engulf antigen, degrade it into fragments, and present a portion on the cell surface in association with membrane receptors; examples include: **macrophages:** large cells (derived from monocytes) with high quantities of degradative enzymes; **dendritic cells (DC):** involved in primary immune responses within lymphoid organs; *Type DC1:* derived from monocytes and, upon interaction with T helper cells induce the Th1 cell phenotype; *Type DC2:* derived from lymphocytes, and upon interaction with T helper cells induce the Th2 cell phenotype; **Langerhans cells:** specific dendritic cells within the skin |

**Systemic and Organismic Aging**

**TABLE 15.3**
## Major Structures of the Immune System

Lymph Nodes

Gland-like structures, arranged in groups, interspersed throughout the lymphatic circulation. They consist of a fibrillar network in which lymphocytes are organized, mature, and interact. They serve as sites where antigens are trapped and destroyed. Major lymphoid organs include adenoids, appendix, Peyer's patches, spleen, and tonsils.

Bone Marrow

Meshwork of connective tissue and stem cells contained within bone cavities. Stem cells are multipotent and differentiate into leukocytes and reticulocytes.

Lymphatic Vessels

Thin-walled vessels that direct the flow of lymph in a particular direction using valves. The lymph is pumped, upon muscle contraction and osmotic pressure, through the lymphatic system, into the lymphatic duct, and then into the large subclavian veins.

Reticuloendothelial System

Phagocytic cells contained in reticular tissues located in lymph nodes and liver.

Spleen

A relatively large lymphoid organ situated in the upper quadrant of the abdominal cavity. Like the thymus, it has a cortex and medulla. The cortex contains densely packed lymphocytes and germinal centers. The medulla encapsulates the cortex (unlike other lymphoid tissues) and has a variety of leukocytes. Some regions are particularly rich in B lymphocytes and appear important for B-lymphocyte storage and activation.

Thymus

A lymphoid organ under the sternum. It consists of a network of epithelial cells that secrete various polypeptide factors important for the maturation of thymocytes (which are also contained within the thymus) into T cells and immune function in general.

**TABLE 15.4**
## Major Functions of the Immune System

To prevent disease from infection
    Bacterial
    Viral
    Fungal
    Parasitic
To prevent cancer
To prevent immunological destruction of self (autoimmunity)

**TABLE 15.5**
## Key Terms and Concepts

| | |
|---|---|
| Humoral Immunity | Aspect of immune function related to the secretion of antibodies by B cells or plasma cells |
| Cell-Mediated Immunity | Immune reaction elicited by cytotoxic lymphocytes |
| Epitope | Specific region (also called antigenic determinant) on a molecule(s) recognized and bound by an antibody |
| Major Histocompatibility Complex (MHC) | Represents a large cluster of genes, which code for many of the cell surface glycoproteins that play a role in defining the interactions between lymphocytes and macrophages during an immunological reaction |
| Human Leukocyte Antigens (HLA) | Represented by gene clusters on chromosome 6. The gene cluster contains three loci called Class I, II, and III. |
| An individual's HLA genotype has been linked to predisposition to certain diseases | Among the functions of genes encoded are:<br>*Defining Self*: cell surface recognition glycoproteins present on almost every body cell and involved in rejection of allografts (coded as Class I loci, i.e., specifically designated HLA-A, B, and C)<br>*Cell Interaction*: cell surface molecules on B lymphocytes and macrophages important in cellular interactions related to immune function (Class II at loci HLA-D or DR, D-related)<br>*Complement Mediated Lysis*: blood proteins known as complement (i.e., Class III loci), whose major function is lysis of infected cells or bacteria. They are sequentially activated in a cascade-like reaction resulting in cell lysis and other activities required for normal immune function. |

**TABLE 15.6**
**Chemical Mediators or Modulators**

| Cytokines | Surface Receptors | Mediators Involved in Apoptosis | Thymic Secretory Factors |
|---|---|---|---|
| Influence proliferation, differentiation, and survival of lymphoid cells; has numerous actions on other body cells, comprises the following:<br>**Interleukin (IL)** — family, 16 different proteins from IL1 and up; numerous effects on lymphocytes and other cells with IL receptors<br>**Tumor Necrosis Factor (TNF)**<br>TNF-α: cytotoxic against malignant and inflammatory cells; produced primarily by macrophages<br>TNF-β: cytotoxic against malignant cells; enhances phagocytosis; produced primarily by T cells<br>**Interferon (IFN)**<br>IFN-α, β: produced by many cells; antiviral actions<br>IFN-γ: synthesized by activated NK and T cells; involved in activation of macrophages and inflammation<br>**Colony Stimulating Factor (CSF)** — glycoprotein regulating white blood cell production, activity, and survival<br>**Granulocyte-Macrophage Colony Stimulating Factor (GM-CSF)** — regulates hematopoiesis, affects phagocyte function and angiogenesis | **Cell Adhesion Molecules (CAM)**<br>Surface receptors mediating cell–cell and cell–matrix interactions<br>Influence maturation, circulation, and homing<br>Involved in inflammatory response and signal transduction<br>Mediate interaction of killer cells with target cells<br>**CD Receptors**<br>(Cluster of Differentiation)<br>Designate dozens of Human Surface Molecules on leukocytes (CD4, CD8, CD28, etc.)<br>Performs different functions e.g., adhesion, migration, stimulation, activation, proliferation, and cell death<br>**Cytotoxic T Lymphocyte Antigen (CTLA)**<br>Stimulation inhibits T cells with opposite effects than CD28<br>**Selectins** — family of membrane components on lymphocytes for adhesion, migration, and homing to capillary endothelial cells | **Fas (CD95)** — membrane receptors involved in triggering apoptosis upon interaction with Fas-ligand (membrane receptors on Natural Killer and activated T cells)<br>**Bcl-2 (B-cell Leukemia)** — protein family derived from chromosome 18; involved in apoptosis and prosurvival (i.e., preventing cell death from various stresses)<br>**Caspases** — protein families (intracellular cysteine proteases) act on apoptosis and inflammation<br>**ICE** — interleukin 1β converting enzyme triggers apoptosis by overproducing a family of caspases | **Factor Thymique Serique (FTS)** — small nonapeptide involved in pre-T cell maturation<br>**Thymostimulin and Thymopoietin** — small proteins involved in T-cell maturation and immune regulation<br>**Thymosins** — small polypeptides secreted by thymic epithelial cells; regulate maturation and proliferation of pre-T lymphocytes.<br>Thymosin α1 is the most studied with regard to therapeutic potential toward infectious disease (hepatitis, HIV, and cancer) |

The innate immune system is intermixed and collaborates with clonally distributed T and B cells and represents a first line of defense against different pathogens. An example of this strict collaboration is the antigen recognition process that requires the presentation of the epitope to T lymphocytes on the surface of antigen presenting cells (APCs), such as macrophages and dendritic cells, in association with the MHC class II (in the case of T helper cells) or class I (for T-cytotoxic lymphocytes) molecules.[4] The adaptive immune system is dependent upon the functional integrity of the innate immune system, without which antigen presenting cells cannot be primed to present antigen in a stimulatory form to T cells. Humoral immunity is, in turn, dependent upon intact T cell responsiveness for B cell generation, differentiation, and antibody production.

*Cytokines* (Table 15.6) are responsible for differentiation, proliferation, and survival of lymphoid cells. Other soluble mediators are proinflammatory agents and play an important role not only in specific immune responses but also in inflammation. Cytokines include interleukins (IL), colony-stimulating factors (CSF), chemokines, and others, such as tumor necrosis factors (TNF). These molecules constitute a complex network and act by binding to specific membrane receptors. The immune orchestra depends on a subtle and well-tuned network of these humoral mediators.[5]

**Systemic and Organismic Aging**

**TABLE 15.7**
**Some Characteristics of Immunoglobulin (Ig) Classes**

| IgG | IgM | IgA | IgD | IgE |
|---|---|---|---|---|
| Most abundant antibody (80%) | Largest Ig (multivalent pentamer comprising 5 to 10% of total Ig) | May exist as a monomer or dimer and has two subclasses (IgA1 and IgA2) | Present in very low concentration in serum | Often produced in response to exposure to parasite antigens |
| About half in the blood | Secreted upon initial exposure to antigen | Can tolerate adverse conditions | Serves as a highly specific membrane receptor on B cells | Receptors on mast cells (called Fc receptors); binds the IgE class and |
| Consists of four subgroups: IgG1-IgG4 | Like IgG can trigger complement- mediated lysis | Commonly found in the respiratory tract, tears, intestine, and stomach | Interaction of antigen with IgD receptors can trigger the mechanism of antigen processing and presentation | Binding of antigen to IgE on mast cells can trigger the release of the vasodilator histamine causing redness and inflammation (allergic reactions) |
| Found in mammary glands | | | | |
| Can cross placental barrier providing prenatal Ig protection | | Produced by plasma cells localized in these tissues | | |
| IgG4, primary antibody to enter tissues and become involved in Ig reactions | | Predominant Ig in milk, thereby helping to provide postnatal immune protection | | |
| IgG1 class is the major Ig produced by B cells | | | | |
| IgG class can also trigger complement-mediated lysis | | | | |

**TABLE 15.8**
**Some Key Immunoglobulin (Ig) Characteristics**

Polypeptides (with small amounts of carbohydrate on heavy chain)
Bivalent: two identical sides
Four-chain monomers consisting of two identical *heavy chains* (each about 50,000 MW) and two identical *light chains* (each about 25,000 MW)
IgM is a pentamer.
IgA is a dimer.
Each *heavy chain* is linked to a *light chain* by *disulfide* linkages, and the two *heavy chains* are bound to each other by *disulfide linkages*
Constant region (relatively conserved amino acid sequence) that defines functional capabilities and distinguishes the five different *heavy chain* classes, IgG, IgA, IgM, IgE, IgD, and two types of *light chains* κ and λ
Antigen-binding portion (Fab) has a variable amino acid sequence near the amino terminus, while the constant region (Fc) is near the carboxy ends
Complementary determining regions (CDR) are portions of *heavy and light chains* with hypervariable amino acid sequence responsible for determining antigen-binding specificity

*Cell adhesion molecules (CAMs)* are surface receptors mediating cell–cell and cell–matrix interactions (Table 15.6). Cell adhesion is fundamental in lymphocyte functions including maturation, circulation and homing, generation of inflammatory responses, and interaction of killer cells with their targets. There are different families of CAMs, and many of them also have regulatory functions and signal transduction properties. The modification of the number of CAMs on the cell surface, in addition to alterations of their affinity and avidity, provide the molecular basis for the interaction between different cells.

## III. IMMUNE FUNCTION AND LONGEVITY

The optimal functioning of the immune system has crucial importance for survival and aging (Chapter 6).[6] There are several links between longevity and genes involved in determining immune responsiveness, mainly *histocompatibility complex*. For example, an increased frequency of HLA-B16 and DR7 and a decreased frequency of HLA-B15, B8, and DR4 have been demonstrated in elderly subjects.[7]

The causes of death in the very old often involve infectious agents (pneumonia, influenza, gastroenteritis, bronchitis).[8] This suggests that failing immunity primarily contributes to the increased incidence of those diseases that the immune system is designed to protect. In a longitudinal study of the very old, nonsurvival was predicted by the clustered parameters of poor T-cell proliferative responses, high CD8 (cytotoxic/suppressor) cell fraction, and low CD4 (helper/inducer) and CD19 (B) cells.[9] In a Dutch cross-sectional study, it was demonstrated that CD4 lymphopenia in the oldest, resulted in an increased mortality risk over the first two years following diagnosis.[10]

A variety of theories exist to explain the immune modifications occurring during aging. The immune system is supposed to collapse with age, and several changes have been considered as paradigms of a defective responsiveness, such as:

- Increased susceptibility to infectious diseases and cancer
- Increased levels of autoantibodies
- Higher incidence of autoimmune manifestations
- Decreased antibody production to nonself antigens
- Defective natural killer activity
- Thymus involution
- Decreased T lymphocyte proliferation[11]

However, aging is not simply the cause or the result of immune deterioration, but it depends also on *a network of interconnected cellular defense mechanisms*, including DNA repair, antioxidants, production of heat, shock, and stress proteins, and apoptosis.[12] All of these antiaging tools, acquired during evolution, are variably regulated in different species and different individuals of the same species. Accordingly, it has been hypothesized that individuals who have *survived in good health to the maximum life span are equipped with optimal cell defense mechanisms.*

In the past, the great majority of studies simply compared immune parameters from young and old subjects, including people over 65 years; however, the human life span is potentially longer, i.e., around 100 years. Furthermore, the effects attributable purely to aging can be difficult to dissect from the effects that are secondary to exogenous factors,[13] such as underlying diseases, frequent in aging, or the use of medication, which might influence the immune system. For this reason, the use of strict biochemical and clinical inclusion and exclusion criteria for immunogerontological studies in man, known as the Senieur protocol has been proposed.[14] Thus, studies with healthy centenarians reveal a great capacity of the immune system to maintain its defensive function in even the very old.[15] Centenarians are, therefore, the best example of successful aging because they have escaped major age-related diseases and have reached the extreme limit of human life.

A complex reshaping of the immune system occurs with age. The remodeling theory of aging[16] suggests that immunosenescence is the result of the continuous adaptation of the body to the deteriorative changes occurring over time. According to this hypothesis, body resources are continuously optimized, and immunosenescence must be considered a dynamic process. Some immune parameters decline and deteriorate in the elderly, including centenarians, while many others remain unchanged or even increase.[17,18]

Furthermore, accumulation of memory T cells, decrease and exhaustion of naïve T cells,[19] and marked

---

**TABLE 15.9**
**Hallmarks of Immunosenescence**

Atrophy of the thymus:
  Decreased size
  Decreased cellularity (fewer thymocytes and epithelial cells)
  Morphologic disorganization
Decline in the production of new cells from the bone marrow
Decline in the number of cells exported by the thymus gland
Decline in responsiveness to vaccines
Reduction in formation and reactivity of germinal center nodules in lymph nodes where B cells proliferate
Decreased immune surveillance by T lymphocytes and NK cells

---

reduction of T-cell repertoire, mostly regarding CD8+T cells, are apparently some *hallmarks of immunosenescence in humans* (also see Table 15.9) and may be considered potential candidates to predict morbidity and mortality.[20] Immunosenescence, as a manifestation of clonal exhaustion, may also occur in young individuals as a result of chronic antigenic stimulation, and as such, is not only a problem of the elderly. Therefore, research on immunosenescence, already important in the context of increasing numbers of the elderly in society, is also critical in the context of chronic infection, organ transplantation, and possibly cancer, in younger individuals as well. The evidence of interdisciplinarity research on immunosenescence underlines the necessity for an integrative approach to age-associated clinical disease.

## IV. IMMUNOSENESCENCE: REMODELING OF THE IMMUNE SYSTEM

The classical concept of "immunosenescence" considers a generalized age-related unidirectional decline in immune responses, leading to increased susceptibility to infections as well as inflammatory and degenerative diseases, enhanced autoreactivity, and frequent occurrence of tumors. However, recent research, highlighting the complexity and multifaceted effects of aging on the immune system, has led to a reformulation of these original concepts. It is now well established that *immunosenescence does not mean immunodeterioration.*[21,22] Although the capacity to cope with major immunological stresses, such as chronic and acute infections, declines, basal levels of immune function are maintained. Changes in the expression of functionally important cellular receptors can contribute to the remodeling of immune function characteristic of the elderly.[23,24]

Several changes in antigen expression that characterize the elderly are shared by various pathological conditions, *raising the question of the relationship between the aging process and the pathogenesis of such diseases,*

**Systemic and Organismic Aging**

*particularly frequent in the elderly.* For example, the CD3 downregulation is a feature of several T-lympho-proliferative disorders,[25] the CD20 overexpression characterizes various infectious diseases, and the high density of CD5 on B cells is a feature of B chronic lymphocytic leukemias.[26]

Changes in lymphocyte function are not necessarily always toward a lowering of response. For example, aging of immune cells is associated with increased expression of several CAMs (cell adhesion molecules), which could likely result in an augmented capability to adhere. Another example is the CD50 overexpression on T cells during aging that may represent an effort of the immune system to partially supply the decreased number of monocytes bearing its ligand LFA-1, which could result in compromised recognition and presentation processes.[27]

*Immunosenescence therefore represents a new and peculiar reequilibrium of the immune system,* in which several changes in the representation and phenotype of lymphocyte subsets are implicated. On the whole, data on immunosenescence indicate that changes occurring over time might be considered the result of global reshaping, where the immune system continuously looks for possible stable points for optimal functioning.[28]

## V. THE CYTOKINE NETWORK

The cytokine network undergoes profound changes with age. Some of these shifts are noted in Table 15.10, however, additional shifts in cytokines and other immune factors are listed in Table 15.11.

Immunologic aging includes a shift toward a type 2 dominant state.[29,30] The cellular and humoral components of the immune response are regulated by cytokines produced by two general subsets of helper cells known as Th1 and Th2.[5] Th1 cells tend to provide help for CD8 mediated cellular effector function and the IgG2a class of antibody. Th2 cells are more efficient in providing help for B cells and the IgA, IgE, and IgG1 classes of antibodies. Type 1 cytokines include IL-2, IFN-γ, IL-12, and IL-15. Type 2 cytokines include IL-4, IL-5, IL-6, IL-10, and IL-13.

Newborn mice and infant humans exhibit impaired cellular-mediated immunity but strong humoral immunity, and are thus possibly in a state in which type 2 responses dominate. Soon thereafter, a type 1 state becomes dominant and persists in healthy mice and humans until mid- to later life, at which time a dominant type 2 cytokine profile may again emerge.[31] The reduced production of cytokines such as IL-2 and perhaps IL-3 by aged naïve T cells will lead to less expansion during effector generation and altered properties of effectors obtained, including a decreased susceptibility to cell death.[32]

It is possible that, with aging, the increase in tumor incidence, the increased rate of infections, and the reappear-

---

### TABLE 15.10
### Some Aging-Related Shifts in Cytokines

Increased proinflammatory cytokines IL-1, IL-6, TNF-α

Increased cytokine production imbalance

Decreased IL-2 production

Increased production of IL-8, which can recruit macrophages and may lead to pulmonary inflammation

Increase in dysfunctional IL-8

Decreased secretion of IFN-γ (interferon)

Altered cytokine responsiveness of NK cells, which have decreased functional abilities

Increased levels of IL-10 and IL-12 upregulated by Antigen Processing Cells

---

### TABLE 15.11
### Additional Aging-Related Shifts in Immune Functions

- Altered membrane fluidity
- Increased apoptosis perhaps due to decline in CD28 expression and IL-2 production
- CD20 overexpression on lymphocytes
- Increased CAMs expression on lymphocytes
- Old cells may have greater levels of messenger RNA for three mitotic inhibitors
- Decreased number of HLA class I and II antigenic sites on lymphocytes
- Increase in activated T-cell-expressing DR molecules
- Decreased proportion of T, B, and NK cells expressing CD62L and increased density per cell of this adhesion receptor expression
- Upregulation of L-selectin per T cell
- Shift in lymphocyte population to contain more CD3-NK cells and CD3+CD56+ T cells
- CD3 downregulation and CD50 upregulation on T cells affecting activation and proliferation
- Increased T-cell death by Fas/Fas-ligand-mediated response in presence of IL-2
- Heightened density of CD5 on B cells
- Decreased number of monocytes with LFA-1
- Decreased ability of dendritic cells to stimulate T-cell secretion of IFN-γ and IL2

---

ance of latent viral infections are the result of decreased cellular immune surveillance due to this cytokine imbalance.

The capability of mononuclear cells to produce proinflammatory cytokines such as IL-1, IL-6, and TNF-α increases with age.[33] The abnormal IL-1 and TNF-α secretion observed in old subjects may be relevant to a number of abnormalities associated with aging. For example, IL-1 increases the secretion of serum amyloid A protein and other acute-phase proteins by the liver and stimulates bone resorption; IL-6 induces fever, activates T and B lymphocytes, and modulates hepatic acute-phase protein synthesis. Moreover, high levels of IL-6 have been referred to as the most powerful predictors of morbidity and mortality in the elderly. IL-6 is one of the pathogenetic elements in inflammatory and age-related diseases, such as rheumatoid arthritis, osteoporosis, atherosclerosis, and late-onset B-cell neoplasia.[8]

## VI. THE PROINFLAMMATORY STATUS IN SUCCESSFUL AGING

The proinflammatory status is seemingly a characteristic of successful and unsuccessful aging. Centenarians have lived in good shape and without disability until very old age,[12] despite the fact that in most of them, the biochemical parameters related to inflammation can reach high values. The new theory of "inflamm-aging" is an interplay between genetic and environmental components. The capability to mount a strong inflammatory process can contribute to fitness and survival, and people characterized by such a capacity have been positively selected. Current data are compatible with the *conceptualization of aging as the result of chronic stress* impinging upon the macrophage as one of the major target cells in this process. *Human immunosenescence can, therefore, be envisaged as a situation in which the most evolutionary recent, sophisticated defense mechanisms deteriorate with age, while the most evolutionary ancestral and gross mechanisms are preserved or negligibly affected and, in some cases, almost upregulated.* The prolonged attrition exerted on the immune system by antigens leads to the production of memory T cells, indicating that the body reacted successfully.[16] However, these physiological responses at the same time lead to a progressive accumulation of clones of memory cells, which fill the entire "immunological space." Together with thymic involution and the consequent age-related decrease of thymic output of new T cells, this situation leaves the body practically devoid of virgin T cells, and thus, likely more prone to a variety of infectious, such as bacterial (*Escherichia coli, Streptococcus pneumoniae, Mycobacterium tuberculosis, Pseudomonas aeruginosa*) and viral (Herpes virus, influenza virus), as well as noninfectious[34] (atherosclerosis, diabetes, osteoporosis and osteoarthritis, dementia, autoimmunity) diseases, where immunity and inflammation play major roles.[4]

## VII. HUMORAL IMMUNITY

There has been the observation, in clinical practice, of an increased frequency with age of pathological processes involving B cells and antibody production. Examples of such pathological conditions include the following:

1. B chronic lymphocytic leukemia
2. Presence of autoantibodies or monoclonal gammopathies
3. Common occurrence of amyloidogenesis

**TABLE 15.12**
**Some Aging-Related Effects on B Cells**

- Decreased number of circulating and peripheral blood B cells
- Alteration in B-cell repertoire (diversity)
- Decreased generation of primary and secondary memory B cells
- General decline in lymphoproliferative capacity

Moreover, a consequence of the altered antibody response may be a marked propensity for:

1. Infectious diseases, particularly pneumonia
2. Recurrent infections
3. Poor responses to vaccines[35]

This results in an increased morbidity and mortality in elderly subjects. Although the presumption has been that much of this change is due to decreased helper T-cell function rather than an intrinsic primary B-cell deficit, in recent years, *aging-associated alterations in B-cell repertoire expression* and in the generation of primary[36] and memory B cells[37] have been documented. Thus, the decreased humoral responsiveness of aged individuals is apparently due not only to alterations in the environment and ancillary stimulatory mechanisms, but also to alterations in B cells per se (Table 15.12).

### A. B Cells

Alterations in B-cell development include (a) a skewing of V-gene utilization (b), a decrease in the generation of various developmental B-cell subsets, and (c) a decrease in the number of pre-B cells.[38] *A reduction in newly emerging cells from the bone marrow is consistent with the alterations cited above.*[39,40]

Bone marrow stromal cell contact and IL-7 produced by stromal cells are essential for B lymphopoiesis. The pre-B-cell receptor (a functional heavy chain with the surrogate light chain) has a ligand on stromal cells that may play a role in inducing the secretion of IL-7. The stromal cells from aged mice are unable to correctly process the signals from the developing B-lineage precursors and are unable to efficiently secrete the IL-7.[41]

There is a significant decrease with age, including centenarians, of peripheral B cells and those B cells coexpressing the CD5 molecule.[23] CD5+ B cells are specifically involved in the development of autoimmune and lymphoproliferative disorders.[42,43] Interestingly, although B cells coexpressing CD5 are decreased in the elderly, they exhibit a higher expression of CD5 molecule at a cell level, and this finding is probably implicated in the dysfunction or hyperfunction of these cells, probably contributing to the increased susceptibility of old subjects to oncological diseases and autoimmunity.[23] The pan B-cell

**Systemic and Organismic Aging**

antigen CD20, increased during B-cell stimulation,[44] is also overexpressed in the elderly, probably reflecting a condition of chronic B-lymphocyte activation that leads to the above-mentioned immune manifestations.

## B.  IMMUNOGLOBULINS

*An unpredicted paradox, i.e., an increase of immunoglobulin serum level and a concomitant decrease in peripheral blood B lymphocytes is observed in the elderly.*[6,18] Both IgG and IgA serum levels significantly increase with age, whereas IgM does not (Table 15.13). A number of different possibilities can be envisaged to explain age-related changes in the level of autoantibodies and B cells, including: an increased number of B and plasma cells in organs other than peripheral blood, an increased life span of B and plasma cells in germinal centers, and an increased production of Igs per cell.[27] Furthermore, the profound changes in cytokine network can be partly responsible for the dysfunction in Ig production.

The frequency of subjects with detectable serum levels of organ-specific or non-organ-specific *autoantibodies* has been presumed to increase with age. However, organ-specific autoantibodies (antithyroperoxidase and antithyroglobulin) are practically absent in the plasma of healthy centenarians; in contrast, nonselected elderly people show an age-related increase in these autoantibodies. Non-organ-specific autoantibodies (anti-dsDNA, antihistones, rheumatoid factor, anticardiolipin) seem to follow a different trend by increasing in healthy aged donors and centenarians.[6]

A widely observed change in the *in vivo* immune response of aged humans or experimental animals is the *diminished ability to generate high-affinity protective antibody responses to immunization against infectious agents or experimental antigens.*[45] It is proposed that a signaling disequilibrium from the aged T cells, which provide less efficient help in quantitative terms, supports the growth of low-affinity B cells. Although age-related declines in IL-2 production and T-cell expansion may contribute to provide poor help for humoral responses, aging also impairs other aspects of the interaction between T cells and B cells, in particular, the contact mediated help. The ligand for CD40 (CD40L or CD154), expressed in Th cells, is a crucial antigen in T-B cooperation. This receptor stim-

ulation develops a cascade of events including B cell CD23 expression, which is important in regulation of Ig synthesis. CD40L expression and its activation pathway are clearly impaired by aging.[2]

The formation of germinal centers during T dependent antibody responses provides an apparently critical environment for the selection of high-affinity antibodies through processes including somatic hypermutation of the Ig variable regions and subsequent selection of B cells expressing high-affinity antibodies. The effective response and the faster neutralization of the pathogen after a secondary challenge is a consequence of this phenomenon.[46] Reduction of germinal center reactivity is a landmark of immunosenescence and contributes to immunological dysfunction in the elderly.[47]

## VIII. CELLULAR IMMUNITY

Although cellular and humoral immune responses are modified with advancing age, the loss of effective immune activity is largely due to *alterations within the T-cell compartment (Table 15.14), which occur, in part, as a result of thymic involution.*[48] Substantial changes in the functional and phenotypic profiles of T cells have been reported with advancing age.[49]

## A.  T-LYMPHOCYTE SUBPOPULATIONS AND THE MEMORY/NAÏVE UNBALANCE

One of the most consistent changes noted in T cells with advancing age is the decrease in the proportion of naïve T cells with a concomitant increase in T cells with an activated/memory phenotype.[28,50] Virgin, unprimed, naïve T cells are CD45RA+CD62L+CD95- T cells. They require a co-stimulatory signal, such as CD28, to optimally proliferate after CD3 stimulation. CD95 antigen is expressed on peripheral blood T cells after TCR/CD3

---

**TABLE 15.13**
**Aging-Related Shifts in Antibodies**

- General decrease in humoral responsiveness: decline in high-affinity protective antibody production
- Increased autoantibodies: organ-specific and non-organ-specific antibodies directed to self
- Increased serum levels of IgG (i.e., IgG1 and IgG3) and IgA; IgM levels remain unchanged

---

**TABLE 15.14**
**Some Aging-Related Effects on T Cells**

- General decline in cell-mediated immunological function
- T-cell population is hyporesponsive
- Decrease in responsiveness in T-cell repertoire (i.e., diversity of CD8+ T-cells)
- Decline in new T-cell production
- Increase in proportion of memory and activated T cells, while naïve T cells decrease
- Diminished functional capacity of naïve T cells (decreased proliferation, survival, and IL-2 production)
- Senescent T cells accumulate due to defects in apoptosis
- Increased proportion of thymocytes with immature phenotype
- Shift in lymphocyte population from T cells to NK/T cells (cells expressing T-cell receptor and NK cell receptors)

stimulation.[12,17] Memory/activated T cells are CD45RA+CD95+CD28- lymphocytes. CD8+CD28- T cells, which are increased in the elderly, have phenotypic and functional features of terminally differentiated armed effector T lymphocytes. The differences in activation requirements and response characteristics of naïve versus memory T lymphocytes may in large part explain the changes in immune responsiveness that occur with advancing age.[51]

The age-related unbalance of virgin and memory cells is found among CD4 and CD8 cell subsets, but it is deeper within the CD8+ T cells. Class I-restricted CD8+ T cells play a major role in infectious diseases caused by pathogens living inside cells, and they constitute an important effector arm for immune surveillance against tumors.[50] Infectious diseases, such as influenza and pneumonia, and cancer are major health problems in older people and represent leading causes of death in this population.[10,52] A very low number of CD95- T cells correlates with a shorter life expectancy. Concomitantly, the progressive expansion of CD28- T cells can be interpreted as a compensatory mechanism. During aging, when the ability to replenish the naïve pool via thymopoiesis is reduced, the immune system tries to compensate for the progressive loss of naïve T cells by increasing thymic-independent pathways, such as the peripheral expansion of mature CD28- T cells, especially within the CD8+ subset.[19]

The age-related increase of T lymphocytes lacking CD7 is consistent with the increase of HLA-DR+ and CD45RO+ T cells. CD3+CD7- T lymphocytes are in fact mainly activated and memory cells and may represent T lymphocytes at a fully differentiated maturational stage.[53] Both the absence of CD7 and the down-expression of CD3 probably contribute to their lower proliferative response compared to CD7+ T cells.[23,28]

## B. T-Cell Dysfunctions

T-cell clonality is a common characteristic of healthy aged persons with clonal expansions found in the CD4+ as well as in the CD8+ memory population.[54] T-cell expansions may derive from latent or repeated infections. They do not represent malignant clonal T-cell expansions. Once expanded, it is possible that these T-cell clones persist, because mechanisms that regulate clonal homeostasis, such as apoptosis, are defective in the elderly.

The T-cell population from aged individuals is hyporesponsive.[55] Unlike the antigen-stimulated memory cells generated in young subjects, which proliferate vigorously and often produce high levels of IL-2 and other cytokines, the memory cells found in aged subjects are hyporesponsive.[48,51] Naïve cells from aged mice respond to antigenic stimulation with decreased cell survival, decreased proliferation, and markedly reduced IL-2 production. Overall

decreases in responses of older organisms to a new antigenic challenge may therefore reflect both a diminished residual population of available antigen-specific cells and a diminished functional capacity of the remaining naïve T cells.[10,32]

Most of the above-mentioned deficits are a consequence of the so-called "replicative senescence," leading to a sort of clonal exhaustion, partially sustained by telomere shortening.[56]

## IX. IMMUNOSENESCENCE AND THYMIC FUNCTION

During the physiological thymus involution that accompanies aging, the organ diminishes in size and cellularity, and its structure becomes disorganized.[10] Accompanying the morphologic changes of the thymus are changes in its biochemistry and physiologic function.

The *number of T cells exported decreases*. There may be a developmental block resulting in *an increased frequency of thymocytes with an immature phenotype (Table 15.14)*. The effect of age-related involution on the kinetics of thymocyte differentiation could depend on an intrinsic defect within the thymocyte population or a deficiency in the ability of stromal cells to support differentiation or defects in lymphokine-driven thymocyte proliferation.[57] *Administration of factors such as FTS, thymostimulin, thymopoietin, and thymosin-1 reduces various aspects of age-related immune dysfunction.*[58]

## A. The Thymus and the Peripheral Microenvironment

A complete naïve T-cell repertoire is produced within the first year of life, and naïve T cells survive for the life of the host and maintain that repertoire in the absence of a thymus.[59] As individuals age, the composition of their T-cell compartment changes, accumulating antigen-experienced or memory cells, while losing naïve cells. This phenomenon[60] is believed to occur gradually over time as a consequence of (a) antigen-driven clonal expansion and maturation, as well as (b) a decline in the output of newly differentiated naïve T cells from the thymus.[58] These processes would result in an immune compartment that has shifted toward memory-like responses, able to respond to previously encountered antigenic determinants but much less competent to respond to novel antigens. Probably, the aged peripheral microenvironment causes an accelerated maturation of newly produced T cells to the memory state. The capacity of the host environment to present Ag, as well as cytokine production, appears to play a role in this process.[61,62]

Immunosenescence is associated with *a loss in the diversity of the T-cell repertoire*.[59] The T-cell repertoire is created within the thymus.[60] With aging, antigen-driven

**Systemic and Organismic Aging**

clonal T-cell expansion, as well as the decreased availability of naïve T cells, is likely to compromise the broad diversity of the T-cell repertoire seen early in life.[54]

## B. Thymus in Centenarians

Recent studies showed that, unexpectedly, age-related changes in the T-cell compartment in healthy centenarians are much less dramatic than might be predicted. *Centenarians who have retained immune responsiveness at levels usually observed in young individuals also show better retention of thymic structure and function.*[12,19]

The presence of a great number of naïve T lymphocytes within the CD4 and the CD8 T-cell subsets in the peripheral blood of centenarians poses the problem of their origin when the thymus has undergone profound involution.[63] Thymic remnants or other lymphoid organs could be able to produce T cells continuously, until the extreme limit of human life.[64] Despite the replacement of the perivascular space with fat, the remaining cortical and medullary tissue in the aging thymus is histologically normal. The phenotypes of all of the expected thymic T-cell intermediates are present, there is evidence of cell proliferation, and there is evidence of ongoing TCR gene rearrangement. Episomal DNA fragments representing the excisional products of the TCR gene rearrangement process are produced during thymopoiesis. These fragments are called TCR rearrangement excision circles (TREC). They are stable, do not duplicate during mitosis, and are therefore diluted with each cellular division, and have been used to successfully quantify thymus function. TREC frequency in peripheral T cells decreases with age.[65] Furthermore, the continued presence of TREC-containing cells within the peripheral blood of elderly subjects could reflect these cells' longevity but might also reflect the sustained output of recent thymic emigrants (RTE) from the thymus.

## X. T-CELL REPLICATIVE SENESCENCE

Replicative senescence describes the characteristic of all normal somatic cells to undergo a finite number of cell divisions before reaching an irreversible state of growth arrest (Chapter 4). The so-called Hayflick limit[66] might be particularly devastating for lymphocytes because the capability of rapid clonal expansion is essential to their function.[4]

Because resting T cells are long lived, they are susceptible to the processes of postmitotic senescence relevant to nonreplicative cells. In contrast, activated T cells proliferate in response to antigens, and these clonally expanding T cells suffer replicative senescence that might lead to clonal exhaustion. Concomitantly, they alter their functional phenotype and surface markers.

## A. Telomere Shortening and Clonal Exhaustion

Telomeres are the repetitive DNA sequences at the end of eukaryotic chromosomes and are critical for genomic stability, the protection of chromosome ends from exonucleolytic degradation, and the prevention of aberrant end-to-end fusion (Chapter 4). An additional aspect of telomere biology is the so-called "end-replication problem." The ends of linear chromosomes cannot be fully replicated during each round of cell division. Thus, somatic cells lose telomeric DNA during replication. A critical telomere length may signal cell-cycle arrest.[56] Therefore, telomeres function as a mitotic clock, and telomere shortening with age may contibute to immunosenescence. Lymphocytes from centenarians and individuals with Down's syndrome and progeria (models of accelerated aging) have telomere lengths in the same range as senescent T-cell cultures. CD4+ memory cells that have experienced several rounds of cell division, show consistently shorter telomeres than naïve cells, and this difference is the same when the cells are isolated from young or old donors.

The characteristics of replicative senescence observed during aging may also be present in other diseases involving chronic antigenic stimulation, such as HIV infection and other infectious diseases, including rheumatoid arthritis[16] and Crohn's disease.[19] These examples illustrate situations in which T-cell replicative senescence may play a part in immune system dysfunction independently from the age of the host. The effects of this type of clonal exhaustion in the elderly may simply be more noticeable than in the young because of thymic involution, which would prevent effective generation of naïve T cells, and because T cells present in the old may already have undergone many rounds of division.

In addition to the number of cell replications, other mechanisms could modulate telomere length. For example, telomerase is a reverse transcriptase that elongates telomeres. Telomere loss is proposed to be a consequence of the downregulation of telomerase activity with age. Because optimal telomerase induction requires optimal stimulation via CD3 and costimulatory receptors, age-associated defects in costimulation may contribute to suboptimal telomerase induction. The introduction of a constitutively expressed telomerase catalytic subunit into cells with a limited life span is sufficient to stabilize their telomeres and extend their life span indefinitely without inducing changes associated with neoplastic transformation.[39]

## B. HIV Infection: A Model for Immunosenescence

Advanced age, characterized by lack of adaptive immune response to new intracellular pathogens, shares similarities with persistent and chronic stimulatory conditions of the immune system by infectious agents such as HIV.[67]

AIDS has characteristics in common with an accelerated aging of the immune system. For example, in HIV-infected patients, the TREC number is very low, similar to that characteristic of centenarians, and the proliferative capability of T lymphocytes is greatly decreased. Immunosenescence and AIDS also share the same cytokine pattern, i.e., the predominance of the Th2 profile, and the same receptor modulation on the lymphocyte membrane, for example, CD3 downregulation on naïve and memory cells and increased CAM expression on leukocytes,[26] concomitantly with increased markers of immune activation. Such a chronic stimulation results in clonal exhaustion, which leads to immunodeficiency marked by telomere shortening.

The loss of naïve T cells, particularly within the CD8+ T-cell compartment, represents a hallmark of immunosenescence as well as AIDS and could provide a useful biomarker in both conditions. In particular, the progressive decrease of naïve CD8+ T cells, the expansion of CD8+CD28- T cells, and the restriction in the CD8+ T-cell repertoire, suggest a typical perturbation of the CD8+ T-cell subset that occurs in HIV disease and advanced aging.[19,62]

On the other hand, age is an important predictor of progression in HIV infections. The more rapid progression appears due to an inability of older persons to replace functional T cells that are being destroyed. The additive effects of HIV infection and age on the immune system contribute to a decreased length of survival as well as more rapid disease progression.[68,69]

## XI. INNATE IMMUNITY

Innate immunity remains relatively intact in the elderly, whereas acquired immunity is primarily affected. However, there are changes in the ability of antigen processing cells (APCs) to communicate with T cells.[3,70] The upregulation of nonspecific proinflammatory responses and downregulation of specific immunity may reflect a compensatory event. Because of the limited capacity of T-cell modulation due to the dramatic involution of the thymus, the potential to modulate innate immunity at the APC-T cell interface could be a key for reversing impaired immune competence, even in the face of multiple external factors and chronic illness.[4]

### A. The Interface Between Innate and Adaptive Immunity: Antigen Presentation

The complex process of immune activation is dependent on the close participation of T cells and APCs. APCs are responsible for uptake, processing, and presentation of antigen to T cells. Impaired ability of APCs to stimulate T cells in elderly has been shown. Expression of co-stimulatory molecules that assist in the efficiency of cell to cell communication may be altered in old subjects and thus alter cytokine production by APCs, which regulates downstream T-cell effector functions.[15,16] However, recent studies have shown enhanced antigen presentation by APCs from healthy elderly, associated with increased levels of IL10 and IL12.[17] It is hypothesized that this upregulation in IL12 production by APCs may be compensatory to an inherent age-related decline in T-cell function to maintain immunocompetence.[71]

### B. Natural Killer Cells

Natural killer cells (NK), originally identified as a population of large granular lymphocytes, are cytotoxic cells that play a critical role in the innate immune response against infections and tumors. NK cells lyse target cells (tumor cells or virus-infected cells) upon initial encounter without the need of prior antigenic sensitization (distinct from cytotoxic T cells) and without MHC restriction. Target cell lysis takes place by a secretory mechanism via exocytosis of cytoplasmic granules containing perforins that damage membrane and granzyme that damages DNA, or by a nonsecretory mechanism via Fas-Fas Ligand interaction.[72]

Age-associated alterations in the number and function of NK cells have been reported (Table 15.15). There is a general consensus that a progressive increase in the percentage of NK cells occurs in the elderly, associated with an impairment of their cytotoxic capacity when considered on a "per cell" basis.[29,72] Furthermore, there is a major shift in lymphocyte population from conventional T cells to NK/T cells with senescence.[23] The increased percentage of NK cells observed in the elderly is mainly due to the increase in the mature CD56+ dim subset.[73] An age-related increase in cells with high NK activity (i.e., CD16+CD57-) has also been described. In contrast, changes in cells with intermediate (CD16+CD57+) or low (CD16-CD57+) NK activity are minor.

*Decreased NK cell function in the very old is associated with increased incidence of infectious diseases.*[74] Old subjects with low numbers of NK cells have three times mortality risk in the first 2 years of follow-up than those with high NK cell levels. Furthermore, other

---

**TABLE 15.15**
**Aging-Related Changes in Natural Killer (NK) Cells**

- General decline in cell function
- Good correlation between mortality risk and NK cell number
- Increase in proportion of cells with high NK activity (i.e., CD16+, CD57-)
- Progressive increase in percentage of NK cells
- Impairment of cytotoxic capacity per NK cell
- Increase in NK cells having surface molecule CD56 dim subset

---

**Systemic and Organismic Aging**

evidence supporting the significance of NK cells in healthy aging comes from the studies in centenarians. In fact, they have a very well-preserved NK cell cytotoxicity.[12] It can be speculated that well-preserved NK activity can help in becoming a centenarian. The shift in lymphocyte population from conventional T cells to CD3- NK cells and CD3+CD56+ T cells, possibly of extrathymic origin, is probably a consequence of the age-related thymic involution.[75]

A main function of NK cells is the capacity to synthesize and release a broad range of cytokines, such as IFN-γ or TNF-α, that participate in the initiation of the Thl-dependent adaptive immune response. The production of IFN-γ secreted by activated NK cells is significantly lower (Table 15.9) in elderly compared to young individuals.[72]

Antibody-dependent cellular cytotoxicity (ADCC) mediated by NK cells when triggered via CD16 is comparable between young and aging subjects.[76] On the contrary, NK cells from elderly donors are defective in their response to cytokines with a subsequent decreased capacity to develop lymphokine activated killer (LAK) cells able to kill NK-resistant cell lines. Decreased NK cytotoxic capacity is associated with defective PKC-coupled (phosphorylase kinase C) transmission signals. Deterioration at the molecular level (perforin and granzymes) in the lytic mechanism may, at least in part, be responsible for the decline in cell-mediated cytotoxicity. This is an important mechanism of tumor control *in vivo*, mediated by both cytotoxic T lymphocytes and NK cells, and both types of cells predominantly use the perforin-dependent pathway. It is possible that the high incidence of tumors in old age could partly be a result of a compromised early spontaneous cytotoxicity mediated by perforin.[77]

## C. Phagocytic Cells: Granulocytes and Macrophages

Macrophages and polymorphonuclear (PMN) cells or neutrophils are important components of the first line of defense because they are the first inflammatory cells recruited to tissue sites in response to inflammation or infection. *The function of macrophages and granulocytes in the elderly is impaired*[78] (Table 15.16). There is evidence that *impaired* in vivo *activation in the early stages of inflammation occurs with aging and could contribute to greater susceptibility to infection.*[79]

Studies on polymorphonuclear neutrophil (PMN) function in older persons have yielded conflicting results. Normal or impaired phagocytosis, chemotaxis, degranulation, and nitroblue tetrazolium (NBT) reduction, and a relatively preserved intracellular killing activity in PMN from the elderly compared to younger individuals have been documented.[80] The flow cytometric analysis of cell surface marker expression in PMN displayed a higher

---

**TABLE 15.16**
**Influence of Aging on Macrophages and Granulocytes**

- General functional impairment of macrophages and granulocytes
- GM-CSF is unable to activate granulocytes from elderly subjects (e.g., superoxide production and cytotoxic abilities)
- Polymorphonuclear neutrophils appear to possess higher levels of surface markers CD15 and CD11b and lesser vesicles containing CD69, which leads to the impairment observed to destroy a bacteria
- In elderly subjects, the monocyte phenotype shifts (i.e., expansion of CD14dim and CD16bright subpopulations, which have features in common with mature tissue macrophages)
- Macrophages of aged mice may produce less IFN-γ, less nitric oxide synthetase, and less hydrogen peroxide

---

level of CD15 (adhesion with endothelial cells and platelets) and CD11b (receptors to C3bi).

Aging is associated with an impairment in the insertion of CD69-containing vesicles into the plasma membrane of human PMN. This impairment in the externalization of CD69-containing vesicles is likely to be related to the impairments in PMN phagocytosis, bacteriacidal activity, and release of reactive oxygen species seen with increasing age.[81]

Cytokines, such as IL-2, and bacterial products (including lipopolysaccharide), rescue PMN cells from undergoing apoptosis or programmed cell death and continue to produce superoxide anions needed to kill pathogens that have been engulfed; this cascade of PMN cell functions has been found to be suppressed in older individuals.[30,71,78]

Granulocyte-machrophage colony-stimulating factor (GM-CSF) is a regulator of granulopoiesis and of granulocyte and mononuclear phagocyte function. GM-CSF is not able to prime granulocytes from elderly subjects for the activation of several parameters, including superoxide production, intracellular calcium flux, antibody-dependent cellular cytotoxicity, and intracellular killing mechanisms.

Macrophages perform several functions, from phagocytosis and killing to cytokine production. Some macrophage functions seem to decline with age, while others remain apparently fully intact. For example, the capacity of macrophages to respond to infection with a virulent intracellular bacterial infection does not seem to be influenced by the increasing age of the host.[82] On the contrary, there is evidence of the following:

1. An increased susceptibility to parasitic infection in aged mice, probably related to impaired IFN-γ production
2. Impaired tumor lysis, or a decrease in nitric oxide synthetase levels from direct activation

of macrophages in aged mice by IFN-γ or lipopolysaccharide[83]

3. Diminished (50%) hydrogen peroxide production from macrophages of aged BALB/c compared to those from young mice in response to stimulation with bacterial products

Also, the monocyte phenotype is consistently modified in the elderly. There is a significant expansion of CD14dim CD16 bright subpopulation of circulating monocytes in elderly subjects that may indicate a state of *in vivo* monocyte activation.

The secretion of IL-8, which recruits macrophages, appears to be dysfunctional in the elderly. The overproduction of IL-8 in elderly men may be detrimental. For example, the recruitment of massive numbers of immune cells to the lungs could result in increased pulmonary inflammation, which may well increase morbidity and mortality among elderly patients. In addition to cytokine dysregulation, hormonal imbalances in the elderly may also affect functioning of the constitutive immune response.

## XII. LYMPHOCYTE PROLIFERATION AND APOPTOSIS

Cell proliferation and cell death are two physiologically active phenomena closely linked and regulated (Chapter 4). A failure of these mechanisms determines profound dysregulations of cell homeostasis with major consequences in immune functioning and the onset of autoimmune diseases and cancer, which apparently increase in old subjects.[84] An important function that has been suggested to deteriorate with age and to play a major role in the aging process is the capability of cells from aged subjects to respond to mitogenic stimuli and, consequently, to undergo cell proliferation.[24,50] However, the cellular activation processes are complex, the proliferative responses can follow different interconnected signal transduction pathways, and only some of them appear to be modified during aging.[85]

Cell growth, immunosenescence, and longevity are strictly interconnected and deeply related to programmed cell death or apoptosis. The cellular equilibrium between cell survival and proliferation on one hand, and programmed cell death on the other, seem to be unbalanced with advancing age, although in each type of immune cell, it could be differentially modulated, resulting in a variety of clinicopathological consequences.[86]

The *impairment of lymphoproliferative capacity* commonly associated with senescence has been attributed to several mechanisms, including reduction in the number of functional cell precursors or their responsiveness to noncognate stimuli and early activation signals,

alteration of intracellular signalling,[87] increased tendency of activated aged lymphocytes to undergo cell cycle arrest, expansion of the memory cell compartment or other lymphocyte subsets that may be functionally restricted, altered pattern of cytokine production, and endocrinologic changes affecting the hormonal milieu within the organism.[88] Various receptors are present on the T-cell surface. Their altered expression could also play an important role in the impairment of proliferative response upon antigenic stimulation.[31]

As previously described, repetitive stimulations may result in ever-decreasing telomerase induction, failure to maintain telomere lengths, and proliferative cessation.[89] A current hypothesis suggests that telomere attrition may trigger growth arrest by activating DNA damage limitation programs. Old resting or stimulated cells have more messages for three mitotic inhibitors (p16, p21, p27) than young cells.[90]

Apoptotic deletion of activated mature lymphocytes is an essential physiological process implicated in both the regulation of the immune response and the control of the overall number of immunocompetent cells. Closely interrelated signaling mechanisms convey activation or death messages achieving the necessary equilibrium between cell proliferation and cell deletion.[91] During the course of aging, numerous alterations of these signaling pathways may shift the balance toward cell death. Diminished synthesis of growth and survival factors, transmembrane signaling defects, default in the expression of particular genes implicated in the control of cell proliferation, and the inability to cope with oxidative stress are all age-associated alterations liable to initiate the apoptotic process in senescent organisms.[56]

CD28 costimulation may protect against apoptosis in several ways, including increased IL2 production. The increased apoptosis with aging may be at least partly explained by the decreased CD28 expression and decreased IL-2 production. However, even in the presence of exogenous IL-2, old cells show increased susceptibility to activation-induced cell death, which is mediated by Fas/Fas-ligand interactions.[85]

The increased susceptibility to apoptosis of T lymphocytes from elderly humans is associated with increased expression of functional Fas receptors.[92] On the other hand, T-cell senescence may also be associated with defective apoptosis.[31,93] In some cases, cells from centenarians were more resistant to apoptosis than were cells from young donors or aged subjects. Resistance to apoptosis could contribute to cellular longevity and possibly to organismic longevity. The data concerning the resistance of lymphocytes from centenarians to undergo apoptosis are consistent with the observation that the intracellular levels of bcl-2 in centenarians are similar to those present in cells from young donors. This protein plays a critical role in protecting cells from several stresses, including oxidation, as well as from cell death.[93]

Although the precise role of apoptosis in the aged immune system remains to be identified, investigators are pursuing at least three lines of inquiry: (1) senescent T cells accumulate due to acquired defects in apoptosis;[92] (2) older T cells may undergo apoptosis at an early stage of activation more rapidly than younger T cells, thus explaining the proliferation defect noted in T cells of older subjects; and (3) defective costimulation through CD28 may result in reduced T cell responses with aging.[90]

In senescent T cells, due to their reduced capacity to execute DNA repair, the frequency and quantity of cells undergoing apoptosis can be increased.[94] For example, the IL-1β-converting enzyme (ICE) family of caspases, the downstream apoptosis-related proteolysis enzymes, are induced after DNA damage, and these proteases are responsible for the breakdown of DNA-repair proteins. However, only cells with functional apoptosis capability would be deleted by this process, whereas certain mutant cells not able to undergo apoptosis can escape from this deletion process and further proliferate into tumor or autoimmune phenotype.

## XIII. IMMUNOSENESCENCE: CLINICAL CONSEQUENCES AND THERAPEUTIC APPROACHES

It appears that the very old who have retained good health are those with relatively well-functioning immune systems.[10,70] It is the status of the immune system that predicts survival of the individual, rather than the state of health of the individual, which determines the integrity of the immune system. Manipulations designed to prevent or delay immunosenescence might, therefore, allow susceptible individuals to achieve their potential life span while remaining in good health.[95]

Infectious diseases, such as bacterial infections (lung, gastrointestinal, urinary tract, skin, and soft-tissue)[96] and viral infections (reactivation of herpes zoster and significantly increased morbidity and mortality due to influenza virus),[97] are a great clinical problem in the elderly (Tables 15.4 and 15.17). In addition, changes in immunity create difficulty in detecting active and inactive tuberculosis. Moreover, atypical presentation of diseases in geriatric patients may result in an infection being recognized later than usual and cause a delay in starting treatment. The points raised above make prevention of disease even more necessary in the elderly. Response to vaccination, which requires intact cell-mediated immunity to drive the humoral response, is clearly diminished.[97] There are worldwide efforts to improve vaccination of the elderly, for example, doubling the vaccine dose or using new adjuvants.[98]

The deleterious effects of malnutrition or the lack of even a small quantity of nutrients on the cellular and humoral immune response have been largely demonstrated.[97] The

**TABLE 15.17**
**Major Diseases Associated with Aging in Immune Function**

1. Increased tumor incidence and cancer
2. Increased incidence of infectious diseases caused by the following:
   a. *Escherichia coli*
   b. *Streptococcus pneumonia*
   c. *Mycobacterium tuberculosis*
   d. *Pseudomonas aeruginosa*
   e. Herpes virus
   f. Gastroenteritis, bronchitis, and influenza
3. Reappearance of latent viral infection
4. Autoimmune diseases and inflammatory reactions as follows:
   a. Arthritis
   b. Diabetes
   c. Osteoporosis
   d. Dementia

experimental supplementation of diet with one or several nutrients, for example, zinc, selenium, vitamin E, etc., has demonstrated an improvement in many immunological parameters in the elderly. Manipulations of dietary lipid intake (N-3 fatty acid fish oil enrichment) can have a significant effect on immune function, perhaps via altered eicosanoid production or influenced membrane fluidity and access to receptors. Caloric restriction (which can extend maximum life span) has not been definitively demonstrated to be effective. Another type of manipulation that seems to be successful in improving the immune function of aged rodents is hormone replacement or supplementation, which has also yielded unconvincing results in man. Melatonin supplementation has been utilized by the general public in an uncontrolled fashion with dubious and unproven benefits. The antioxidant effects of some products (vitamins and melatonin) which help to restore cell redox balance might be responsible for or contribute to the benefits observed in some experimental models and trials.[70,95] The mitochondrial-oxidation theory considers aging as linked to injury of mitochondrial genome by superoxide and other reactive oxygen species formed in the electron transport chain. Thus, the supplementation with thiolic antioxidants prevents mitochondrial degeneration in the aged. Also, the administration of interleukins, such as IL-2, which seems to be defective in the elderly, has been found to have some benefits. IL-12 cytokine immunotherapy, in association with influenza vaccination, could enhance CTL responses and reduce influenza morbidity and mortality among high-risk elderly persons.[57,98]

Future research on strategies to modulate the process of T-cell replicative senescence, for example, by using genetic techniques to enforce telomerase expression or by manipulating CD28 expression, may lead to novel approaches to improve the immune function in the elderly.[8,15]

## REFERENCES

1. Cerottini, J.C. and MacDonald, H.R., The cellular basis of T-cell memory, *Annu. Rev. Immunol.*, 7, 77, 1989.
2. Grabstein, K.H. et al., The regulation of T cell-dependent antibody formation *in vitro* by CD40 ligand and IL-2, *J. Immunol.*, 150, 3141, 1993.
3. Cooper, E.L. et al., When did communication in the immune system begin?, *Int. J. Immunopathol. Pharmacol.*, 7, 203, 1994.
4. Pawelec, G., Immunosenescence: Impact in the young as well as the old?, *Mech. Ageing Dev.*, 108, 1, 1999.
5. Shearer, G.M., Th1/Th2 changes in aging, *Mech. Ageing Dev.*, 94, 1, 1997.
6. Cossarizza, A. et al., Cytometric analysis of immunosenescence, *Cytometry*, 27, 297, 1997.
7. Papasteriades, C. et al., HLA phenotypes in healthy aged subjects, *Gerontology*, 43, 176, 1997.
8. Wick, G. et al., Diseases of aging, *Vaccine*, 18, 1567, 2000.
9. Bender, B.S. et al., Absolute peripheral blood lymphocyte count and subsequent mortality of elderly men, *J. Am. Geriatr. Soc.*, 24, 649, 1986.
10. Remarque, E. and Pawelec, G., T-cell immunosenescence and its clinical relevance in man, *Rev. Clin. Gerontol.*, 8, 5, 1998.
11. Pawelec, G. and Solana, R., Immunosenescence, *Immunol. Today*, 18, 514, 1997.
12. Franceschi, C. et al., The immunology of exceptional individuals: The lesson of centenarians, *Immunol. Today*, 16, 12, 1995.
13. Khanna, K.V. and Markham, R.B., A perspective on cellular immunity in the elderly, *Clin. Infect. Dis.*, 28, 710, 1999.
14. Ligthart, G.J. et al., Necessity of the assessment of health status in human immunogerontological studies: Evaluation of the SENIEUR protocol, *Mech. Ageing Dev.*, 55, 89, 1990.
15. Pamer, E.G., Antigen presentation in the immune response to infectious diseases, *Clin. Infect. Dis.*, 28, 714, 1999.
16. Franceschi, C., Bonafè, M., and Valensin, S., Human immunosenescence: The prevailing of innate immunity, the failing of clonotypic immunity, and the filling of immunological space, *Vaccine*, 18, 1717, 2000.
17. Sansoni, P. et al., Lymphocyte subsets and natural killer cell activity in healthy old people and centenarians, *Blood*, 80, 2767, 1993.
18. Paganelli, R. et al., Changes in circulating B cells and immunoglobulin classes and subclasses in a healthy aged population, *Clin. Exp. Immunol.*, 90, 351, 1992.
19. Fagnoni, F.F. et al., Shortage of circulating naïve CD8+ T cells provides new insights on immunodeficiency in aging, *Blood*, 95, 2860, 2000.
20. Franceschi, C. et al., Biomarkers of immunosenescence: The challenge of heterogeneity and the role of antigenic load, *Exp. Gerontol.*, 34, 911, 1999.
21. Ginaldi, L. et al., Immunological changes in the elderly, *Aging Clin. Exp. Res.*, 11, 281, 1999.
22. Ginaldi, L. et al., The immune system in the elderly. I. Specific humoral immunity, *Immunol. Res.*, 20, 101, 1999.
23. Ginaldi, L. et al., Changes in the expression of surface receptors on lymphocyte subsets in the elderly: A quantitative flow cytometric analysis, *Am. J. Hematol.*, 67, 63, 2001.
24. Wakikawa, A., Utsuyama, M., and Hirokawa, K., Altered expression of various receptors on T cells in young and old mice after mitogenic stimulation: A flow cytometric analysis, *Mech. Ageing Dev.*, 94, 113, 1997.
25. Ginaldi, L. et al., Differential expression of CD3 and CD7 in T-cell malignancies. A quantitative study by flow cytometry, *Br. J. Haematol.*, 93, 921, 1996.
26. Ginaldi, L. et al., Altered lymphocyte antigen expressions in HIV infection: A study by quantitative flow cytometry, *Am. J. Clin. Pathol.*, 108, 585, 1997.
27. De Martinis, M. et al., Adhesion molecules on peripheral blood lymphocyte subpopulations in the elderly, *Life Sci.*, 68, 139, 2000.
28. Ginaldi, L. et al., Immunophenotypical changes of T lymphocytes in the elderly, *Gerontology*, 46, 242, 2000.
29. Ginaldi, L. et al., The immune system in the elderly. III. Innate immunity, *Immunol. Res.*, 20, 117, 1999.
30. Castle, S. et al., Evidence of enhanced type 2 immune response and impaired upregulation of a type 1 response in frail elderly nursing home resident, *Mech. Ageing Dev.*, 94, 7, 1997.
31. Mountz, J.D. et al., Cell death and longevity: Implications of Fas-mediated apoptosis in T-cell senescence, *Immunol. Rev.*, 160, 19, 1997.
32. Linton, P.J. et al., From naïve to effector — alterations with aging, *Immunol. Rev.*, 160, 9, 1997.
33. Franceschi, C. et al., Inflamm-aging. An evolutionary perspective on immunosenescence, *Ann. N.Y. Acad. Sci.*, 908, 244, 2000.
34. Wick, G. et al., Is atherosclerosis an immunologically mediated disease?, *Immunol. Today*, 6, 27, 1995.
35. Arreaza, E.E. et al., Lower antibody response to tetanus toxoid associated with higher auto-anti-idiotypic antibody in old compared with young humans, *Clin. Exp. Immunol.*, 92, 169, 1993.
36. Stephan, R.P., Sanders, V.M., and Witte, P.L., Stage specific alterations in murine B-lymphopoiesis with age, *Int. Immunol.*, 8, 509, 1996.
37. Yang, X., Sterda, J., and Cerny, J., Relative contribution of T and B cells to hypermutation and selection of antibody in germinal centers of aged mice, *J. Exp. Med.*, 183, 959, 1996.
38. Klinman, N.R. and Kline, G.H., The B-cell biology of aging, *Immunol. Rev.*, 160, 103, 1997.
39. Solana, R. and Pawelec, G., Molecular and cellular basis of immunosenescence, *Mech. Ageing Dev.*, 102, 115, 1998.
40. Szabo, P. et al., Maturation of B cell precursors is impaired in thymic-deprived nude and old mice, *J. Immunol.*, 161, 2248, 1998.
41. Stephan, R.P., Reilly, C.R., and Witte, P.L., Impaired ability of bone marrow stromal cells to support B lymphopoiesis with age, *Blood*, 91, 75, 1998.

42. Chen, X. and Kearney, J.F., Generation and function of natural self-reactive B cells, *Semin. Immunol.,* 8, 19, 1996.

43. Ginaldi, L. et al., Levels of expression of CD19 and CD20 in chronic B-lineage leukaemias, *J. Clin. Pathol.,* 51, 354, 1998.

44. Ginaldi, L. et al., Changes in antigen expressions on B lymphocytes during HIV infection, *Pathobiology,* 66, 17, 1998.

45. Song, H., Price, P.W., and Cerny, J., Age-related changes in antibody repertoire: Contribution from T-cells, *Immunol. Rev.,* 160, 55, 1997.

46. Herrera, E., Martinez-A.C., and Blasco, M.A., Impaired germinal center reaction in mice with short telomeres, *EMBO J.,* 19, 472, 2000.

47. Zheng, B. et al., Immunosenescence and germinal center reaction, *Immunol. Rev.,* 160, 63, 1997.

48. Linton, P.J. et al., Antigen-independent changes in naïve CD4 T cells with aging, *J. Exp. Med.,* 184, 1891, 1996.

49. Ginaldi, L. et al., The immune system in the elderly. II. Specific cellular immunity, *Immunol. Res.,* 20, 109, 1999.

50. Pawelec, G. et al., The T cell in the ageing individual, *Mech. Ageing Dev.,* 93, 35, 1997.

51. Flurkey, K., Stadecker, M., and Miller, R.A., Memory T lymphocyte hyporesponsiveness to non-cognate stimuli: A key factor in age-related immunodeficiency, *Eur. J. Immunol.,* 22, 931, 1992.

52. Liu, J. et al., The monitoring biomarker for immune function of lymphocytes in the elderly, *Mech. Ageing Dev.,* 94, 177, 1997.

53. Kukel, S. et al., Progressive increase of CD7- T cells in human blood lymphocytes with ageing, *Clin. Exp. Immunol.,* 98, 163, 1994.

54. Schwab, R. et al., Expanded CD4+ and CD8+ T cell clones in elderly humans, *J. Immunol.,* 158, 4493, 1997.

55. Fernandez-Gutierrez, B. et al., Early lymphocyte activation in elderly humans: Impaired T and T-dependent B cell responses, *Exp. Gerontol.,* 34, 217, 1999.

56. Pawelec, G. et al., T cell immunosenescence *in vitro* and *in vivo, Exp. Gerontol.,* 34, 419, 1999.

57. Thoman, M.L., Early steps in T cell development are affected by aging, *Cell Immunol.,* 178, 117, 1997.

58. Frasca, D. et al., Regulation of cytokine production in aging: Use of recombinant cytokines to upregulate mitogen-stimulated spleen cells, *Mech. Ageing Dev.,* 93, 157, 1997.

59. Thoman, M.L., Effects of the aged microenvironment on CD4+ T cell maturation, *Mech. Ageing Dev.,* 96, 75, 1997.

60. Mackall, C.L. and Gress, R.E., Thymic aging and T cell regeneration, *Immunol. Rev.,* 160, 91, 1997.

61. Timm, J.A. and Thoman, M.L., Maturation of CD4+ lymphocytes in the aged microenvironment results in a memory-enriched population, *J. Immunol.,* 162, 711, 1999.

62. Effros, R.B., Costimulatory mechanisms in the elderly, *Vaccine,* 18, 1661, 2000.

63. Pawelec, G. et al., Extrathymic T cell differentiation *in vitro* from human CD43+ stem cells, *J. Leukoc. Biol.,* 64, 733, 1998.

64. Douek, D.C. and Koup, R.A., Evidence for thymic function in the elderly, *Vaccine,* 18, 1638, 2000.

65. McFarland, R. et al., Identification of a human recent thymic emigrant phenotype, *Proc. Natl. Acad. Sci.,* 97, 4215, 2000.

66. Effros, R.B., Replicative senescence in the immune system: Impact of the Hayflick limit on T cell function in the elderly, *Am. J. Hum. Gen.,* 62, 1003, 1998.

67. Bender, B.S., HIV and aging as a model for immunosenescence, *J. Gerontol.,* 52, 261, 1997.

68. Adler, W.H. et al., HIV infection and aging: Mechanisms to explain the accelerated rate of progression in the older patient, *Mech. Ageing Dev.,* 96, 137, 1997.

69. Bestilny, L.J. et al., Accelerated replicative senescence of the peripheral immune system induced by HIV infection, *AIDS,* 14, 771, 2000.

70. Castle, S.C., Clinical relevance of age related immune dysfunction, *Clin. Inf. Diseases,* 31, 578, 2000.

71. Mbawuike, I.N. et al., Cytokines and impaired CD8+ CTL activity among elderly persons and the enhancing effect of IL-12, *Mech. Ageing Dev.,* 94, 25, 1997.

72. Solana, R. and Mariani, E., NK and NK/T cells in human senescence, *Vaccine,* 18, 1613, 2000.

73. Krishnaraj, R., Senescence and cytokines modulate the NK cell expression, *Mech. Ageing Dev.,* 96, 89, 1997.

74. Wikby, A. et al., Age-related changes in immune parameters in a very old population of Swedish people: A longitudinal study, *Exp. Gerontol.,* 29, 531, 1994.

75. Solana, R., Alonso, M.C., and Pena, J., Natural killer cells in healthy aging, *Exp. Gerontol.,* 34, 435, 1999.

76. Borrego, F. et al., NK phenotypic markers and IL2 response in NK cells from elderly people, *Exp. Gerontol.,* 34, 253, 1999.

77. Rukavina, D. et al., Age-related decline of perforin expression in human cytotoxic T lymphocytes and natural killer cells, *Blood,* 92, 2410, 1998.

78. Bruunsgaard, H. et al., Impaired production of proinflammatory cytokines in response to lipopolysaccharide (LPS) stimulation in elderly humans, *Clin. Exp. Immunol.,* 118, 235, 1999.

79. Di Lorenzo, G. et al., Granulocyte and natural killer activity in the elderly, *Mech. Ageing Dev.,* 108, 25, 1999.

80. Esparza, B. et al., Neutrophil function in elderly persons assessed by flow cytometry, *Immunol. Invest.,* 25, 185, 1996.

81. Noble, J.M., Ford, G.A., and Thomas, T.H., Effect of aging on CD11b and CD69 surface expression by vesicular insertion in human polymorphonuclear leucocytes, *Clin. Sci.,* 97, 323, 1999.

82. Rhoades, E.R. and Orme, I.M., Similar responses by macrophages from young and old mice infected with mycobacterium tuberculosis, *Mech. Ageing Dev.,* 106, 145, 1998.

83. Sadeghi, H.M. et al., Phenotypic and functional characteristics of circulating monocytes of elderly persons, *Exp. Gerontol.,* 34, 959, 1999.

84. Ginaldi, L. et al., Cell proliferation and apoptosis in the immune system in the elderly, *Immunol. Res.*, 21, 31, 2000.

85. Effros, R.B., Insights on immunological aging derived from the T lymphocyte cellular senescence model, *Exp. Gerontol.*, 31, 21, 1996.

86. Souvannavong, V. et al., Influence of aging on B-cell apoptosis and expression of the activation marker alkaline phosphatase, *Aging Immunol. Infect. Dis.*, 6, 197, 1996.

87. Weng, N.P. et al., Tales of tails: Regulation of telomere length and telomerase activity during lymphocyte development, differentiation, activation and aging, *Immunol. Rev.*, 160, 43, 1997.

88. Utsuyama, M. et al., Impairment of signal transduction in T cells from old mice, *Mech. Ageing Dev.*, 93, 131, 1997.

89. Weng, N.P. et al., Human naïve and memory T lymphocytes differ in telomeric length and replicative potential, *Proc. Natl. Acad. Sci. USA*, 92, 11,091, 1995.

90. Phelouzat, M.A. et al., Excessive apoptosis of mature T lymphocytes is a characteristic feature of human immune senescence, *Mech. Ageing Dev.*, 88, 25, 1996.

91. Sansoni, P. et al., T lymphocyte proliferative capability to defined stimuli and costimulatory CD28 pathway is not impaired in healthy centenarians, *Mech. Ageing Dev.*, 96, 127, 1997.

92. Herndon, F.J., Hsu, H.C., and Mountz, J.D., Increased apoptosis of CD45RO⁻ T cells with aging, *Mech. Ageing Dev.*, 94, 123, 1997.

93. Ucker, D.S. et al., Physiological T-cell death: Susceptibility is modulated by activation, aging, and transformation, but the mechanism is constant, *Immunol. Rev.*, 142, 273, 1994.

94. McConnell, K.R., Dynan, W.S., and Hardin, J.A., The DNA-dependent protein kinase (p460) is cleaved during Fas-mediated apoptosis in Jurkat cells, *J. Immunol.*, 158, 2083, 1997.

95. High, K.P., Micronutrient supplementation and immune function in the elderly, *Clin. Infect. Dis.*, 28, 717, 1999.

96. Fein, A.M., Pneumonia in the elderly: Overview of diagnostic and therapeutic approaches, *Clin. Infect. Dis.*, 28, 726, 1999.

97. Fulop, T. et al., Relationship between the response to influenza vaccination and the nutritional status in institutionalized elderly subjects, *J. Gerontol.*, 5, 59, 1999.

98. Betts, R.F. and Treanor, J.J., Approaches to improved influenza vaccination, *Vaccine*, 18, 1690, 2000.

# 16 Cardiovascular Alterations with Aging: Atherosclerosis and Coronary Heart Disease*

*Paola S. Timiras*
University of California, Berkeley

## I.  INTRODUCTION

While cardiovascular disease continues to be the major cause of death in the United States and other industrialized, socioeconomically advanced countries, there has been, during the last 50 years, a marked decrease in total death rates in heart disease (by 56%) and stroke (by 70%). Indeed, it is estimated that 73% of the decline in total death rates over this time period was due to this reduction in cardiovascular disease mortality (Chapter 2). Despite this current, extraordinarily rapid decline in mortality, cardiovascular disease associated with advancing age remains the most important single worldwide cause of death, in old age and in both sexes. This is, in part, because people now live longer and are more susceptible to the occurrence of degenerative diseases. It is also due to some unknown aspects of modern life that are increasing the incidence of atherosclerosis. As outlined in Chapter 2, several concomitant factors may have contributed to the decline in cardiovascular morbidity and mortality such as better control of hypertension, changes in lifestyle (diet, physical exercise), scientific breakthroughs in understanding the disease, improvement in medical care, and decline in cigarette smoking. This chapter will review some of the age-associated changes in structure and func-

---

* Illustrations by Dr. S. Oklund.

tion of the arteries; it will focus on atherosclerosis, the major and universal manifestation of cardiovascular pathology,[1-6] and it will emphasize coronary heart disease as an example of its clinical consequences.[7,8] Aging of the heart, not directly related to atherosclerosis of the coronary arteries, is discussed in Chapter 21.

The question of how long we might live were it possible to prevent atherosclerosis remains unanswered; it would seem that complete prevention of atherosclerosis would add several years to life.[1,5,9,10] Scientists throughout the world representing many disciplines are now attacking the problem in order to understand the nature of the disease and to find means of preventing or curing it. The 1990s have witnessed the beginning of a "new era" of atherosclerosis research in which it is possible not only to prevent but also to arrest or to induce regression of the lesions and their clinical manifestations.[1,11]

### SOME STRUCTURAL AND FUNCTIONAL CHARACTERISTICS OF THE CARDIOVASCULAR SYSTEM

Blood vessels are part of a more complex system, which includes the heart, representing the mechanical pump that propels the blood into the blood vessels, the nervous system that regulates contraction and relaxation of the heart and vessels, and the lymphatic vessels that subserve the immune system. A synoptic list of the major components and functions of the cardiovascular system (also called circulatory system) includes:

- *The blood and blood vessels* that carry gases, metabolic products, and hormones, from all tissues to the heart and back
- *The lymph and lymphatic vessels* that carry lymphocytes, clotting factors, and proteins to lymphatic organs and drain into veins (see Chapter 15)
- *The heart* that pumps the blood through the blood vessels to all organs, tissues, and cells (see also Chapter 21)
- *The brain and the central and peripheral nervous centers* that coordinate cardiovascular activity

## II.  DEFINITIONS

### A.  *Arteriosclerosis and Atherosclerosis*

Arteriosclerosis is a generic term for any vascular damage that leads to progressive thickening and loss of resiliency of the arterial wall. One type of arteriosclerosis is atherosclerosis, which refers to specific alterations occurring in the vascular (arterial) wall, such as atheromas or plaques characterized by a combination of fatty accumulation in the intima and an increase in connective tissue in the subintimal layers. It is this form of arteriosclerosis that is the most widespread and, at the same time, the most threatening, inasmuch as it affects those arteries such as the aorta, the coronary, and the cerebral arteries that are crucial in providing the necessary blood supply for the heart, brain, and other vital organs. Atherosclerosis, then, is the vascular disorder that plays a major role in coronary heart disease

(due to atherosclerosis of the coronary arteries) as well as in stroke (due to atherosclerosis of cerebral arteries).[2-4,7,8]

Arterial diseases may also arise from congenital structural defects, from inflammatory diseases (e.g., syphilitic aortitis), from hypersensitivity or autoimmune diseases which principally affect the smaller vessels and may lead to their occlusion (thromboangiitis obliterans), and from specific capillary lesions as in diabetic microangiopathy (Chapter 14).

### B.  TYPES AND STRUCTURE OF ARTERIES

In its journey from the heart to the tissues, the blood passes through channels of six principal types:

- Elastic arteries
- Muscular arteries
- Arterioles
- Capillaries
- Venules
- Veins

In this system, the arteries show a progressive diminution in diameter as they recede from the heart, from about 25 mm in the aorta to 0.3 mm in some arterioles. The reverse is true for the veins; the diameter is small in the venules and progressively increases as the veins approach the heart. All arteries are comprised of three distinct layers, intima, media and adventitia, but the proportion and structure of each varies with the size and function of the particular artery. The morphology of the arteries, summarized in Box 16.1 is illustrated in Figure 16.1 and represented diagrammatically in Figure 16.2.

### C.  PROGRESSIVENESS AND UNIVERSALITY OF ATHEROSCLEROSIS

Atherosclerosis, as a prototype of cardiovascular changes with age,[12] is characterized by the following:

- Onset at young age
- Progression through adulthood
- Culmination in old age with overt disease manifestations
- Widespread distribution throughout the arterial tree
- Consequences leading to severe disability or death

Today, atherosclerosis must be viewed as a disease that, sooner or later, affects everyone. Working silently over the years from early childhood, it gradually destroys the arteries, ultimately preventing the exchanges of gases and nutrients necessary to keep organs, tissues, and cells alive and functioning normally.

Although a scourge of modern civilization and often discussed in relation to the pressures of an urban techno-

---

**Box 16.1**

**Some Morphologic Characteristics of the Arterial Wall**

A large artery, like the aorta, is comprised of the following layers, going from the lumen to the most external layers:

1. The **intima**, or innermost layer, consists of a layer of endothelial cells separated from the elastin layer underneath by a narrow layer of connective tissue that anchors the cells to the arterial wall.
2. A large layer of elastic fibers forms the **elastica interna** layer.
3. Below this layer, are concentric layers of **smooth muscle cells** intermixed with elastic fibers. Elastic lamellae, smooth muscle cells, and occasional fibroblasts are imbedded in a ground substance rich in proteoglycans (starch-protein complexes) that serve as binding or "cement" material in the interstitial spaces. The outer layer of the media is penetrated by branches of the **vasa vasorum**.
4. Between the smooth muscle layer and the adventitia, there is again another layer of elastic fibers, the **elastica externa**. Layers 2, 3, and 4 form the **media**.
5. The outer layer, or **adventitia**, is formed of irregularly arranged collagen bundles, scattered fibroblasts, a few elastic fibers, and blood vessels which, because of their location, are called vasa vasorum (vessels of the vessels); they provide blood to the adventitia and the outer media layers.
6. In addition to the endothelial, elastic, smooth muscle, and collagen cells, a few cells of the immune system, **monocytes**, are occasionally present in the arterial wall. During the early events of lesion development, monocytes enter the intima in regions of endothelial damage. Cytokines, immune growth factors, trigger the differentiation of monocytes into macrophages that scavenge oxidized low-density lipoproteins (LDL) (Chapter 17). With the progression of the lesion, macrophages accumulate in the arterial wall, degenerate, and die and are transformed into the so-called foam cells.

The structure of the aorta and large arteries serves well with their function as blood reservoir and for stretching or recoiling with the pumping of the heart. The wall of the arterioles contains less elastic fibers but more smooth muscle cells than that of the aorta. The arterioles represent the major site of resistance to blood flow, and small changes in their caliber cause large changes in total peripheral resistance. Muscle cells are innervated by noradrenergic nerve fibers, constrictor in function, and in some cases, by cholinergic nerve fibers that dilate the vessels (Table 16.1).

The structure of **capillaries** shows a diameter just large enough to permit the red blood cells to squeeze through in single file. In the same manner as the intima of the arteries, the capillary wall is formed of a layer of endothelial cells resting on a basement membrane. The major function of the capillaries is to promote exchange of nutrients and metabolites between the blood and the interstitial tissues. Such exchanges are facilitated by the presence of specialized junctions, gaps, or fenestrations.

---

**TABLE 16.1**

**Summary of Factors Regulating Arteriolar Diameter**

| Vasodilator | Vasoconstrictor |
|---|---|
| $\downarrow$ Oxygen tension | $\uparrow$ Norepinephrine |
| $\downarrow$ pH | $\uparrow$ Epinephrine |
| $\uparrow$ $CO_2$ | $\uparrow$ Angiotensin II |
| $\uparrow$ Temperature | $\uparrow$ Vasopressin |
| $\uparrow$ Lactic acid | |
| Histamine release | |
| Potassium ions | |
| Adenosine and nucleotide | |
| Kinins | |

*Note:* With aging, capacity for arteriolar dilation or constriction is reduced due to reduced elasticity, collagen cross linkage, calcification, changes in adrenergic receptor sensitivity, and atherosclerosis.

cratic society, atherosclerosis has been with us from ancient times; the disease has been detected in Egyptian mummies and described in early Greek writings. There are several ways in which atherosclerosis impairs the normal function of the arteries:

- It may corrode the arterial walls to such a degree that they suddenly yield to the pressure of the blood inside and explode in a massive hemorrhage (e.g., rupture of an aneurysm)
- It may set off, in reaction to its destructive processes, a secondary proliferation of its tissues, thereby gradually blocking the arterial lumen (e.g., formation of a thrombus)
- It may induce clotting of the blood within the diseased artery and, in this way, obstruct blood flow (also formation of a thrombus)
- It is a *progressive disease* that develops slowly over years or decades; however, the final

**Systemic and Organismic Aging**

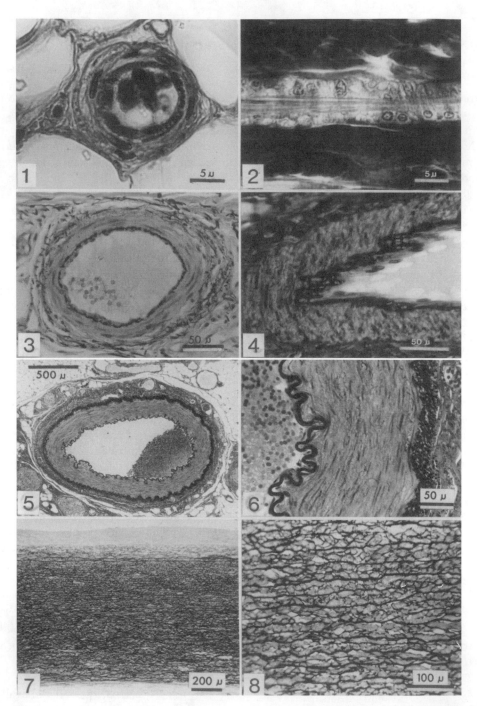

**FIGURE 16.1** Morphology of normal arterial vessels. (1) Arteriole cross section (human mesentery). Stain: iron hematoxylin-aniline blue. Note: Red blood cells in lumen, endothelial nucleus, internal elastic membrane, and smooth muscle cells with elongate nuclei. (2) Arteriole — longitudinal section (cat ileum submucosa). Stain: Mallory-Azan. Note: Smooth muscle cells coiling around endothelial tube that contains the nucleus of an endothelial cell. (3) Small artery — cross section (human external ear, subcutaneous tissue). Stain: Verhoeff and Van Gieson. Note: Distinct internal elastic membrane and smooth muscle of media; elastic fibers are beginning to accumulate from an external layer. (4) Small artery — tangential section showing fenestrated internal elastic membrane (human external ear, subcutaneous tissue). Stain: as in (1). Note: All elastic membranes in the arterial tree bear fenestrae (window-like openings). (5) Medium artery — cross section, low power (human mesenteric artery). Stain: as in (1). Note: All layers of the wall, intima (with internal elastica), media, and adventitia (with externa elastica) are distinct. (6) Medium artery — cross section, high power. Stain: as in (1). Note: Internal elastic membrane, well-developed muscular media, adventitia with external elastic tissue disposed as coarse fibers in helices, hence, cut tangentially. (7) Large artery — cross section, low power (human aorta). Stain: as in (3). Note: Thick intima and high content of elastic tissue (appearing as black lines). (8) Large artery — cross section, high power (human aorta). Stain as in (3). Note: Multiple thick membranes forming concentric tubes interconnected by finer cross membranes. The interstices are filled mainly with collagenous connective tissue and sparse smooth muscle.

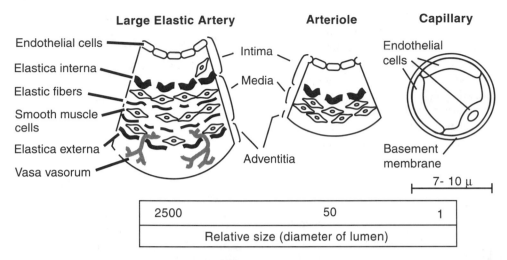

**FIGURE 16.2** Diagrammatic representation of the arterial wall illustrating a large (elastic) artery, an arteriole, and a capillary. (Drawing by Dr. S. Oklund.)

"accidents" for which it is responsible (hemorrhage, thrombosis) may be initiated within only a few seconds

One of the characteristics of atherosclerosis is its *universality*: it is present in almost all animal species, where it has been investigated, and throughout all populations within a species. So insistent and progressive is its onslaught with advancing age that it is generally considered to be an inevitable manifestation of aging — a "wearing out" of the arteries, as in the common saying "a man is as old as his arteries" (undoubtedly applicable also to women[13]). Indeed, death by some of its consequences (e.g., heart attack or stroke) is now so common that we have come to regard it as a natural end of life. At the same time, however, much of the burden of atherosclerosis is the accepted conviction that it kills us prematurely, in the sense that alterations in the arteries are capable of irreparably damaging such vital organs as the heart and brain, at a time when the functional competence of these structures is otherwise sound.[5,11] (From the analogy of the heart or brain as a motor and the arteries as the pipes that convey the fuel to the motor: if the motor is deprived of fuel because of a breakdown in the pipe system, it will stop working, even though the motor is without defect).[14]

Etiopathogenesis of atherosclerosis is similar in all organisms, although there are some individual differences, probably related to the preponderance of one risk factor over the others.[15] Thus, the first effects (early lesions) on the arterial lining may differ in time of onset, rapidity of progression, and severity, whether the risk factor is hyperlipidemia, diabetes, hypertension, smoking, increased lipid deposition (Chapter 17), concomitant endocrine alterations (Chapter 14), or accumulation of free radicals (Chapter 5). Consequences of risk factors

may also vary according the genetic makeup of the affected individual.[16–18] According to studies in twins, the greater role of genetic factors in increasing susceptibility to death from coronary heart disease at younger than older ages implies that the genetic influence decreases as the individual ages.[17,18]

## D. Protocols of Atherosclerosis Studies

Currently, the study of atherosclerosis is being approached from two main directions: studies in humans or in experimental animals and tissue culture.

The study of atherosclerotic lesions in man may involve:

- The timetable of their appearance
- Their location
- Their eventual regression
- Their clinical consequences
- Their adaptability to respond to internal metabolic needs as well as to changes in the external environment of the body
- The analysis of the various morphologic and biochemical components of the lesions
- The identification of genetic factors, directly or indirectly implicated in the disease[19,20]

In humans, progress in the study of atherosclerosis has been greatly facilitated by the use of imaging techniques for the arterial wall *in vivo*, such as *angiography* (x-ray visualization of blood vessels after the injection of radio-opaque, contrast material), associated with computer analysis.[21,22] The arteries most frequently examined by angiography to quantify the lesions and to follow their progression (or regression) under natural conditions or in response to various treatments, are the carotid and coronary arteries as well as the femoral artery. Drawbacks to

**Systemic and Organismic Aging**

angiography include cost and potential for complications (allergic reaction to contrast agents, renal failure consequent to contrast administration, bleeding, thrombus formation, pain during the procedure). All of these potential problems may be minimized by observing proper techniques and medication.

The study of atherosclerotic lesions in animals aims to discover the mechanisms that induce or, conversely, may be capable of preventing or curing such lesions, namely:

- To attempt to reproduce the human disease in animals, by manipulation of the diet, administration of "stress" hormones (e.g., glucocorticoids), exposure to trauma (e.g., mechanical, bacterial or viral injury of arterial wall), alterations of clotting system and platelets, and administration of oxidants (to induce accumulation of free radicals)
- To use genetic manipulations (transgenic, knock-out animals with vascular pathology resembling atherosclerosis)[23,24]

However, it is recognized that observations in animals are not always referable to man and, atherosclerotic lesions, in particular, differ depending on the species.

## III. PATHOLOGY

### A. COURSE OF ATHEROSCLEROSIS

Although the consequences of atherosclerotic lesions become manifest clinically in the fourth decade of life and thereafter, atherosclerosis is not exclusive to advanced age, but rather, it represents the culmination of progressive changes in the arterial wall beginning at a very early age.[25,26] For example, microscopically identifiable vascular changes may start *prenatally* under conditions of impaired fetal development: this is the case of low-birth-weight (small-for-date) newborns who are at higher risk for cardiovascular disease later in life than individuals of the same age but born with normal body weight.[25] Microscopic alterations consist, initially, of intimal thickening, later followed by cell proliferation, and accumulation of proteoglycans (i.e., high molecular weight complexes of proteins and starch). Some of these alterations, but not all, contain lipids and, in this case, can be readily observed as fatty streaks.

Despite species and individual variability, the approximate time sequence involved in the development of atherosclerotic lesions with respect to specific pathologic changes has been generally established (at least in North America) to proceed in the following order (Figure 16.3):

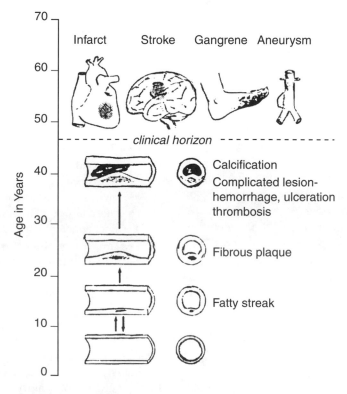

**FIGURE 16.3** Natural history of atherosclerosis shown in this diagrammatic concept of the pathogenesis of human atherosclerotic lesions and their clinical manifestations. (From McGill et al., *Atherosclerosis and Its Origin*, Sandler, M. and Bourne, G.H., Eds., Academic Press, New York, 1963, p. 39. With permission.)

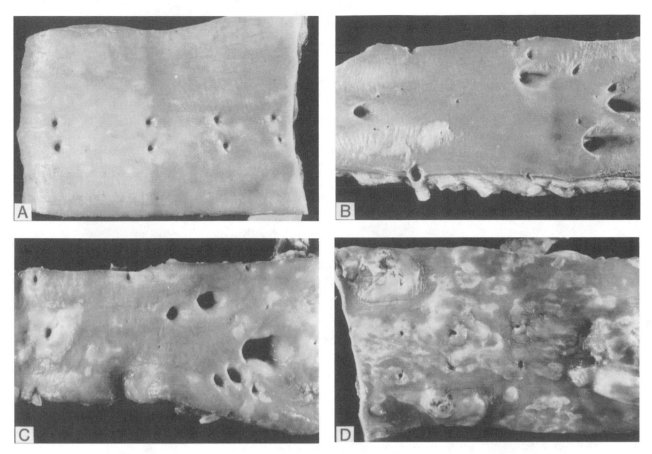

**FIGURE 16.4** Progression of morphologic changes in human aorta from early to advanced atherosclerosis. Aortas were split open and the intima exposed for photography. (a) Aorta (thoracic) of a 32-year-old male showing early fatty plaques (represented by lighter coloration) localized mainly around the orifices of the intercostal arteries. As discussed in the text, because of the contribution of hemodynamic factors in the genesis of the atherosclerotic lesion, orifices of collateral branches are frequently the site at which lesions first appear. (b) Aorta of a 24-year-old female showing early fatty plaques, also around the orifices of collateral branches. (c) Aorta of a 55-year-old male showing advanced plaques characterized not only by a greater amount of fatty material but also by fibrotic thickening of the wall. (d) Aorta of a 65-year-old male showing large, complicated, and calcified plaques. (Courtesy of Dr. O. N. Rambo.)

- The *fatty streaks* appear in the arteries during the first decade of life (as early as the first years or even months or days of life) and continue into the second or third decade
- The *fibrous or "pearly" plaques* appear from the second decade on; clinical consequences (e.g., cardiac infarct, stroke, gangrene, and aneurysm) occur from the fourth decade on[12]

Atherosclerosis seems to develop in several "waves" throughout the life span, inasmuch as early and late stages of the lesions can be found side by side in the same vessel or in different vessels of the same person (Figures 16.4 and 16.5). However, this view assumes that the early stage is represented by the fatty streaks and advanced stages of the same lesion by the transformation of the streaks into plaques; this assumption is challenged by the observation that the two types of lesions are often found in different locations, and that arterial areas with a fair amount of fatty streaks do not subsequently develop a commensurate number of plaques. While the constituents of the lesions are similar and include, primarily, lipid and free radical accumulation, smooth muscle and connective cell proliferation, and migration and proliferation of immune cells reminiscent of inflammation, not all lesions progress to the same degree in all individuals of the same age, and it is possible for the two types of lesions, early and late, to coexist in neighboring regions along the same artery.

## B. Some Characteristics of Blood Flow and Arterial Function

The arterial system provides not only for circulation of the blood as a whole but also, when necessary, for the special needs or functions of a particular organ. Certain organs — brain, heart, kidney — receive a larger proportion of blood than others, and, within the same organ, blood flow varies considerably depending on the degree of activity, as dramatically evidenced in the 30-fold

**Systemic and Organismic Aging**

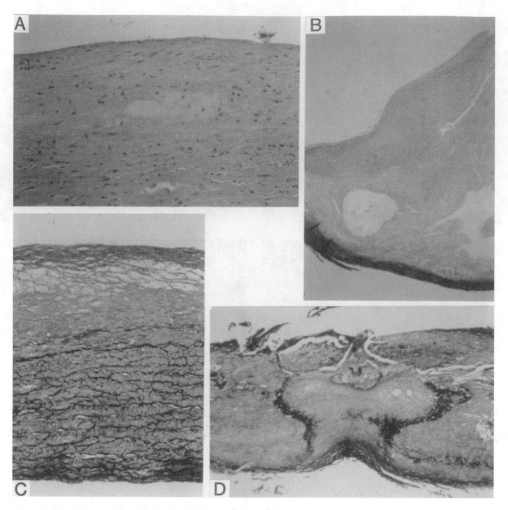

**FIGURE 16.5** Progression of microscopic changes in human arteries from early to advanced atherosclerosis. (a) Earliest fatty plaque in fibrous intima (aorta from 24-year-old male: hematoxylin-eosin stain; original magnification ×152). (b) Large, atheromatous cystic plaque (carotid artery from a 65-year-old male; Verhoeff and Van Gieson stain; original magnification ×24). (c) Atheromatous plaque showing fatty infiltration and alterations of elastic tissue (carotidendarterectomy in 68-year-old male; stain as in (b); original magnification ×95). (d) Large ulcerated calcified plaque with metaplastic bone (carotid artery with occlusion in 79-year-old male; stain as in (b); original magnification ×24) (Courtesy of Dr. O. N. Rambo.)

increase in blood flow to the exercising muscle. The velocity of blood flow declines gradually with the size of the artery, from approximately 8 cm/sec in the medium-sized arteries to 0.3 cm/sec in the arterioles; pressure, on the other hand, remains high in the large- and medium-sized arteries but falls rapidly in the small arteries to low levels of 30 to 40 mm Hg; the magnitude of the drop varies depending on degree of arteriolar constriction.

The mechanisms that regulate arteriolar diameter and, hence, blood flow through the arteries involve local (e.g., release of chemical substances into intercellular space) and systemic, primarily nervous, control (Table 16.1). Nervous control is exerted by groups of neurons in the brain medulla (the vasomotor center), and by peripheral stretch receptors, the baroreceptors,[27] located on the arch of the aorta, at the bifurcation of the common carotid

artery, and in the walls of the cardiac atria and of the large veins. The arterial system as a whole is never static but continuously undergoes structural changes and adaptations that permit the organism to respond to changing requirements for blood supply.

## ROLE OF ENDOTHELIAL CELLS IN AUTOREGULATION OF VASCULAR MOTILITY

Cardiovascular adjustments to continuously changing demands of the organism are made possible, in part, by the capacity of blood vessels to constrict or relax (i.e., dilate) and, thereby, decrease or increase blood flow to a given tissue. Such regulation is under central and peripheral nervous control as well as under control from locally acting agents (Table 16.2). Some of the signals generated locally derive from the endothelial cells of the vascular intima. These cells are of great interest[28,29] because:

## TABLE 16.2
## Major Substances Secreted by the Vascular Endothelium

| Substance | Synthesis, Metabolism, and Receptors | Biological Actions |
|---|---|---|
| • Prostacyclin (PGI$_2$) | • Synthesized from arachidonic acid via cyclooxygenase pathway, associated with ↑ intracellular Ca$^{++}$ | • Contributes to vasodilation<br>• Inhibits platelet aggregation |
| • Endothelium-derived relaxing factor (EDRF), identified also as nitric oxide (NO), present in three forms | • Synthesized from L-arginine by NO synthase (three isoforms: in brain, in macrophages, in endothelial cells) | • Relaxes smooth muscle and induces vasodilation<br>• Has antithrombitic activity<br>• Others: influences brain fuction, stimulates cytotoxicity by macrophages, relaxes gastrointestinal smooth muscle |
| • Endothelins (ET-1, ET-2, ET-3), also in the intestine: vasoactive intestinal constricting peptide | • Derived from a larger pre-proendothelin peptide by the endothelin-converting enzyme: ↑ intracellular Ca$^{++}$; ↑ Ca$^{++}$ channel activation<br>• Two receptors have been cloned: ET$_A$ receptor, specific for ET-1, and ET$_B$, responding to all three ETs | • Induces vasoconstriction<br>• Others: cardiac (coronary constriction) neuroendocrine (e.g., ↑ ANP, renin, aldosterone, catecholamines), ↑ renal vascular resistance, bronchospasm, others |
| • Vascular endothelial growth factor (VEGF) | • Family of four peptides with endothelial cells as specific targets with receptors on endothelial and tumor cells | • Induces mitogenesis and promotes angiogenesis<br>• Promotes embryonic development, wound healing, reproductive functions |
| • Cytokines (see Table 15.6) (also secreted by lymphocytes, neutrophils, and monocytes and expressed on the surfaces of various cells) | • Growth factors with specific receptors lacking tyrosine kinase domains, e.g., interleukin IL-1 to IL-8; platelet-derived growth factor PDGF; fibroblast growth factors FGFa, FGFb, others | • Participate in formation and repair of the vascular wall<br>• Promote cell adhesion, diapedesis, and chemotaxis<br>• Stimulate thrombotic activity |

- They provide the vascular lining
- They have a life span of 10 years and longer
- They undergo aging-associated changes
- They are involved in the pathogenesis of atherosclerosis

Major products of endothelial cells include vasodilators (prostacyclin, endothelium-derived relaxing factor) and vasoconstrictors (endothelins) (Table 16.2). With aging, overall changes are characterized by decreased relaxation and increased constriction, the resulting imbalance of vascular tone leading to increased vasoconstriction with disruption of blood flow. Simultaneously, vasomotor responses to extraendothelial stimuli are altered: for example, the vasodilator actions of acetylcholine and histamine may be diminished, while the vasoconstrictor action of adrenergic stimulation is increased. These changes in vasomotility may contribute significantly to cardiovascular pathology.

Changes with aging in the function of endothelial cells also involve alterations in blood coagulation that would favor conversion from anticoagulatory conditions to a procoagulant, prothrombogenic state. Finally, with aging, endothelial cell proliferation and responsiveness to local growth factors [e.g., epidermal growth factor (EGF), fibroblast growth factor (FGF)] are diminished, and equally diminished, would be cell migration, perhaps due to increased cell adhesion; such changes would underlie the greater susceptibility of the endothelial lining to injury and the longer repair time that occurs in older individuals. A summary of these actions is presented in Table 16.3.

## TABLE 16.3
## Summary of Changes in Vascular Endothelium with Aging

- Endothelial cells undergo significant changes indicative of abnormal function.
- The imbalance of vascular tone is manifested by increased vasoconstriction.
- Vascular integrity (cell proliferation and migration, wall remodeling) and injury repair through local growth factors are impaired.
- Maintenance of blood fluidity is disrupted with increased cell adherence, blood coagulation, and thrombogenic properties.
- These alterations by themselves may induce pathology or may predispose with other factors to atherosclerosis.

It may be noted that vascular endothelial cells secrete factors (e.g., VEGF) that promote angiogenesis (i.e., formation of new blood vessels)[30] and that tumor growth depends on an appropriately sustaining blood circulation of the rapidly dividing carcinogenic cells.[31] Under these conditions, formation of new blood vessels (angiogenesis) as well as the secretion of the vasodilator EDRF are stimulated. These observations have led to numerous attempts to reduce tumor growth by blocking blood supply to the fast-proliferating cells.

Even at birth, arteries vary in structure and in distribution depending upon the hemodynamic conditions under which they operate and continue to change with

**Systemic and Organismic Aging**

maturation and advancing age. In the process of adapting to extrinsic and intrinsic stimuli, structural changes occur at particular sites, and the gross pattern of vascular distribution to an organ or body part may undergo considerable change (as in the development of collateral circulation).

Current evidence from human and experimental observations, argues against a rigid classification of the various components of the arterial wall into definite "species" of cells. Instead, they favor a more versatile view of the cellular configuration of the arteries, in which cells are not irrevocably specialized but, rather, can assume more than one function when the need arises. For example, a muscle cell will not only contract upon stimulation, but under certain conditions, it can also phagocytize lipid (i.e., becoming a foam cell) and even produce collagen or elastic fibers — a potential that might have a bearing on the formation and stability of atherosclerotic lesions.[32] Smooth muscle cell mutations may be triggered by injury to the vascular wall (e.g., mechanical trauma, high blood lipids or oxidized LDL, altered carbohydrate metabolism or immune function, accumulation of free radicals, pollutant toxicity, bacteria, and viruses). Mutations, associated with the local release of growth factors, would be responsible for the abnormal proliferation of smooth muscle cells, with their subsequent migration to the intimal and elastic layers of the vascular wall. Invasion of the intima by smooth muscle cells would represent the first event in the pathogenesis of atherosclerosis; it forms the basis of *the myoclonal theory of atherosclerosis*.[32] This theory is consistent with the role of smooth muscle cells in the formation and stability of the plaque, their potential transformation from a contractile to a different phenotype in response to injury, and their expression of genes that facilitate plaque formation.[32–34]

## C. Blood Supply to the Arterial Wall and Metabolic Exchanges

Blood supply of a nonfunctioning organ or part of an organ can be diminished and, conversely, blood supply of an actively functioning part can be increased by three basic mechanisms:

1. Arteriovenous anastomoses, i.e., short channels that connect arterioles to venules, bypassing the capillaries
2. Specialized muscular arrangements in the walls of arteries, as in the sphincters of hepatic and splenic arteries
3. Arrangement of the capillaries in the capillary bed in such a manner that there is a preferential capillary channel from arteriole to venule (a controversial assumption that presupposes contractile structures in the capillaries)

One of the best examples of blood circulation adjusted to hemodynamic requirements is the establishment of a *collateral circulation* (i.e., circulation that is carried through secondary channels) when the main arterial supply to a specific organ or tissue is reduced or cutoff. Adequate blood supply to the heart is assured at all times by the two (right and left) *coronay arteries* (the first branches of the aorta) and by the presence of a rich system of anastomoses, not only between the two main coronary arteries but also between the coronary arteries and arteries from the pericardium, lungs, thorax, and diaphragm. Thus, in the case of narrowing (stenosis) or occlusion (thrombosis) of one of the coronaries, if the process of occlusion is sufficiently slow, there can be time for competent collateral circulation to be established. This is the rationale for the administration of substances that promote angiogenesis (e.g., vascular-endothelial growth factor, VEGF) and, thus, stimulate collateral circulation in the myocardium.[30]

The extent of *vascularization of the arterial* wall varies from species to species: for example, vascularity is less well developed in man, rabbit, and chicken (species highly susceptible to atherosclerosis) than in the horse, goat, and cow (species less susceptible to the disease).[35] Lack of adequate vascularization may represent a major cause of arterial disease (Table 16.4). The intima and inner media obtain their nutrition by diffusion from the lumen; thus, any process that causes a thickening of the intima or that damages the vasa vasorum might be expected to cause an ischemic-type injury (i.e., due to inadequate blood flow) to the arterial tissue.[36]

Conversely, when the atherosclerotic lesion has reached the stage of plaque, regardless of the causes of plaque formation (e.g., lipid infiltration, thrombosis), the lesion becomes vascularized. Thus, although an inadequacy of blood supply to the arterial walls has been implicated in the initial stages of atherosclerosis, the increased vascularization associated with further development of the lesion has been viewed as an aggravating factor in advanced stages of the disease. Increased vascularization may also be responsible, at least in part, for the sudden accidents such as hemorrhage, with or without thrombosis, that are characteristic of these advanced stages.

---

**TABLE 16.4**
**Localized Factors Contributing to Atherosclerotic Lesions**

- Marginal vascularization of arterial wall
- Relative ischemia
- Limited metabolic exchange
- Blood turbulence and mechanical stress

Efficiency of exchanges of metabolic products between vascular wall and lumen and vice versa is determined by the special structure of the arterial wall. Diffusion, which normally regulates nutritional exchanges between blood and tissues, varies with the layer of the arterial wall. In the intima, for example, because of its proximity to the bloodstream, nutrients diffuse from the blood, and products of arterial tissue metabolism are discharged into the lumen in the reverse direction. The other layers are too thick to be nourished by diffusion; in the adventia and outer media, metabolic needs are met by the vasa vasorum, leaving the inner portion of the media metabolically undersupplied and, therefore, at risk. Hence, any increase in thickness of this layer (due to increased proliferation of smooth muscle cells, macrophages, fibroblasts) or compromise of the circulation of the vasa vasorum, will lead to alterations in metabolic exchanges and accumulation of metabolic byproducts. Once the lesion has been established, metabolic injury will aggravate it and impair eventual recovery processes.[37]

## D. Localization of Lesions

It must be kept in mind that atherosclerotic lesions are focal; they have preferred sites of occurrence and others where they are seldom found. Thus, other factors to be considered in evaluating the etiology and progression of the atherosclerotic lesion is its location along the arterial wall and the influence of *blood turbulence*. Preferred sites are *around the orifices* of arteries branching from a major artery (Figure 16.4), such as the intercostal arteries from the descending aorta; another is at *the bifurcation of a large artery* into two smaller ones, such as the abdominal aorta bifurcating into the two iliac arteries. Blood flows at such *orifices* and *bifurcations*, increases in velocity, exceeds critical velocity, and becomes erratic and turbulent. This turbulence creating a mechanical stress favors onset and progression of the lesion at these sites (Figure 16.4 and Table 16.5).[36]

---

**TABLE 16.5**
**General Characteristics of Atherosclerotic Lesions**

Early onset — progressive
Focal lesions
Early lesions
Advanced lesions
  Damage
  Repair
  Regression
Progression of localized-type lesions influenced by:
  *Local factors* — vessel structure and metabolism, blood turbulence
  *Systemic factors* — diabetes, hypertension, stress, genetic
   predisposition

---

## IV. TIMETABLE OF CHANGES IN ATHEROSCLEROTIC LESIONS

### A. Progression of Lesions

As mentioned above, atherosclerotic lesions begin at an early age and progress continuously throughout life. The localization, rate of progression, or type of lesion varies widely depending on several factors (Table 16.5) that are related to the following:

- The structure and function of the vessel considered
- The specific hemodynamic physiologic requirements
- The number of associated pathologic conditions, either local (e.g., hemorrhages, thrombi) or systemic (e.g., hypertension, diabetes)[37,38]

In all cases, atherosclerosis occurs fundamentally as a localized lesion of the wall. Although its diffusion — in the vessel, in the area of the vessel, and throughout the whole body — may be considerable, the steadily progressing pathologic process is always confined to *one focal point*, the form and size of which depends on local and generalized conditions. The atherosclerotic plaque or atheroma represents the characteristic site at which the histogenesis of the disease can best be analyzed.[16] Although it has not yet been possible to detect the precise beginning of an atherosclerotic lesion, there are a sufficient number of characterizing features in those lesions to assign them the descriptive terms "early" and "advanced" (Figures 16.4 and 16.5).

### B. Early Lesions

For large elastic arteries like the aorta, lesions usually begin as scattered foci in which the innermost layers of the vascular wall show signs of damage accompanied by the growth of repair tissue (Table 16.5). In the smaller arteries, the progression of events is less easily identifiable, but damage and repair processes still represent the most important characteristics of the early lesion in these structures.

Often, the membranes of intimal endothelial cells become "sticky." A simultaneously developing stickiness of circulating monocytes (from the immune system, Chapter 15) facilitates the mutual attraction of these two cell types and the invasion of the arterial wall by the monocytes. In the wall, the monocytes are transformed into lipid-engorging macrophages and become *foam cells* (i.e., cells engorged with lipid droplets that are dissolved in the course of histologic staining and appear as numerous vacuoles resembling foam).

As the intima thickens as a result of an increase in tissue fluid in the intimal ground substance, there is a

**Systemic and Organismic Aging**

disruption and disintegration of the innermost elastic lamellae, followed first by moderate influx and then swelling and flooding of the area with amorphous materials, primarily proteins and sulfated proteoglycans. The proteins probably derived from the blood, as a consequence of the increased permeability of the damaged endothelium (as in all inflammatory edemas), are often coupled with lipids, which become visible when uncoupled. Proteoglycans may derive from the blood or may be formed *in situ*, but, in any case, they are similar to the materials that accumulate in most young repair tissues of the body, where they help to build collagen fibers, the principal component of scars (Chapter 22). The early lesions, at this stage, are essentially proliferative, due to the release of growth factors by endothelial and smooth muscle cells and macrophages; T lymphocytes are present. Inflammatory processes in the arterial wall occur early and persist with the plaque progression. Signs of inflammation would consist of accumulation, in the fatty streak, first, and in the plaque, later, of monocyte-derived macrophages (with or without lipid) and a varying number of T lymphocytes producing proteolytic enzymes.[35,36] Inflammatory reactions associated with early and late atherosclerotic lesions appear to be directly related to the level of C-reactive protein, a marker for systemic inflammation.[41–43] Inflammatory processes would increase the risk of a first thrombolytic event and of myocardial infarction and stroke. These observations have led to the formulation of *an inflammatory theory* of atherosclerosis; they have suggested the use of anti-inflammatory agents such as aspirin for the prevention of cardiovascular diseases.[44] Whether these still-disputed beneficial effects are mediated by aspirin anti-inflammatory action or by other mechanisms (e.g., thrombolytic action) remains to be established.

This phase of the atherosclerotic lesion involves the *repair and protective processes* that characterize any inflammation (Table 16.5). One of the consequences of these processes is the further aggravation of intimal hyperplasia or thickening. Sooner or later, lipid appears in many of these lesions, mostly in their basal portions and not only within the cells (local smooth muscle cells and invading and proliferating macrophages) but also in the matrix and on the disintegrating elastic lamellae.

The lipid material, first in the form of small droplets, gradually fills the cells, imparting a "foamy" appearance in histologic section: its ubiquitous presence is the basis for *the lipid accumulation theory* of atherosclerosis. With the increase in the number of foam cells and extracellular lipid, the fat accumulation becomes visible to the naked eye as tiny yellow spots or streaks in the inner lining of the arteries, the so-called "fatty streaks" (Figures 16.4 and 16.5). Taken as evidence of lesion, these fatty streaks are most commonly found in the aorta of children and younger individuals, although similar foci have been described in octogenarians and centenarians who earlier showed a so-called "juvenile" atherogenic index.

### CHANGES IN VASCULAR GROUND SUBSTANCE WITH AGING

Given the importance of wall thickness in the exchanges between blood and vascular wall, the ground substance (extracellular matrix) and its major constituents, the proteoglycans, have been studied extensively. These show progressive quantitative and qualitative changes as the atherosclerotic lesion progresses. Alone, as part of the larger group of glucosaminoglycans, and in combination with other components of the ground substance, such as hyaluronic acid and embedded substances such as collagen and elastin, the proteoglycans regulate some important viscoelastic and water-binding properties of tissues, including those of the arterial wall, where they serve as lubricants and support elements (Table 16.6). Impairment of these properties with aging would result in alterations of the ground substance with reduced ability to provide mechanical support and hydration. Such alterations result in (a) a weakening of the arterial wall, (b) a decrease in its ability to support compressive load of normal blood pressure, and (c) chemical alterations of transport and binding of water-soluble substances.

### C.  ADVANCED COMPLICATED LESIONS

These lesions, found mainly in adults and elderly persons in whom autopsy examinations are conducted more frequently, have been studied more exhaustively than the early lesions typical of the first decade of life. Detailed morphologic descriptions of advanced human atherosclerosis can be found in most textbooks of pathology and in

**TABLE 16.6**
**Probable Role of Ground Substance in Early Atherosclerotic Lesions**

| Major Components | Properties |
|---|---|
| Glucosaminoglycans (proteoglycans) | Viscoelastic |
| Hyaluronic acid | (impaired with aging, reduced mechanical support) |
| Collagen | Water-binding |
| Elastin | (with aging, reduced hydration, altered transport) |

**TABLE 16.7**

**Percentage of Total Lipids in Human Aortic Intima at Different Ages and in Different Types of Lesions**

| | TYPE OF LESION | | | | |
|---|---|---|---|---|---|
| | Normal Intima | | | | Calcified Fibrous |
| | Age 15 | Age 65 | Fatty Streak[a] | Fibrous Plaque[a] | Plaque[a] |
| **Total lipid** | | | | | |
| mg/100 mg dry tissue | 4.4 | 10.9 | 28.2 | 47.3 | 50.0 |
| **% of total** | | | | | |
| Cholesterol ester | 12.5 | 47.0 | 59.7 | 54.1 | 56.3 |
| Free cholesterol | 20.8 | 12.2 | 12.7 | 18.4 | 22.4 |
| Triglycerides | 24.8 | 16.6 | 10.0 | 11.1 | 6.5 |
| Phospholipids | 41.9 | 24.2 | 17.6 | 16.6 | 14.8 |

[a] Irrespective of age, see text.

*Source:* Adapted from Smith, E.B., *J. Atherosci. Res.*, 5, 224, 1965. With permission.

specific texts dealing with atherosclerosis. Only a brief summary will be presented here.

With the passing of years, more and more lipids — especially cholesterol esters — accumulate in the fatty streaks of the established lesion (Table 16.7); the foam cells increase in number to the extent that those in the center of the arterial wall die — probably due to lack of oxygen and the enormous amounts of fat in their cytoplasm displace or alter the organelles concerned with normal cellular function (Figure 16.5). The lipids released from the disintegrating foam cells, with that already present extracellularly, assemble in large pools, partly as cholesterol crystals and partly as an amorphous mixture of triglycerides, phospholipids, and sterols: these lipid pools have the consistency of a soft paste or gruel (hence, the term "atheroma" from the Greek "ather," indicating a gruel-like substance). An aorta that is riddled with atheromatous plaques may contain as many as several times its normal lipid content (Table 16.7).

The mass of extracellular lipid acts as an irritant to the arterial wall and provokes a proliferative reaction in the surrounding vascular tissue, similar to the inflammatory reaction that occurs in response to any foreign body encountered by the organism. The major lipid component is the *low-density lipoprotein (LDL)*, and particularly, its *oxidized form* (Chapters 5 and 17). While newly formed LDL is relatively benign, oxidized LDL acquires a new configuration, binds to specific receptors, and becomes less susceptible to removal by the high-density lipoprotein (HDL). Monocytes attracted from the circulating blood to the arterial wall and transformed into macrophages become trapped in the wall due to oxidized LDL, which inhibits their motility.[39,40]

The resulting atheroma develops like a sac, encapsulating the gruel, but remaining much thicker on the lumen side of the arterial wall, where it forms a thick barrier between the bloodstream and the gruel (Figure 16.5). At this stage, the atherosclerotic lesion has progressed considerably from its earlier manifestation as a fatty streak; not only is it larger and thicker, but it also seems to rise above the inner surface of the artery like a cushion, resembling an encapsulated abscess. Because of its appearance, the lesion has been given the name of "atheromatous abscess," but it is also called a "raised plaque" or a "fibrous plaque," and its characteristic pearly white color resulting from the high content of collagen fibers in the capsule has also given it the name of "pearly plaque" (Figures 16.4 and 16.5).

One of the main characteristics of the advanced atheroma is its progressiveness. As the plaque grows with fat, it consumes more of the arterial wall underneath it, transforming the cells into foam cells and disintegrating one elastic lamella after another. In this process, the entire media is destroyed, and the atheroma invades the adventitia, which then reacts by setting up a series of inflammatory-like responses, such as hyperemia (due to vascular invasion) and lymphocytic infiltration.[40] Simultaneously, the capsule of the advanced atheroma, perhaps in a compensatory effort, thickens considerably, building a new arterial wall. However, as it does not contain muscle or elastic fibers but contains almost exclusively scar (connective) tissue, this wall becomes functionally less efficient. As the pearly plaque becomes established, calcium deposits precipitate on the gruel, the capsule, or both, in the form of fine granules, thin strips, or huge masses. In the coronaries, for example, with the accumulation of large amounts of calcium over the years, the arteries become exceedingly hard and brittle, hence, the term "hardening" or "sclerosis" of the arteries.

Until this stage, the changes in the atherosclerotic lesion are counterbalanced by repair processes, and no loss of tissue has occurred; indeed, in the sense that the

**Systemic and Organismic Aging**

lesion continues to form scar tissue, it can be viewed as "productive." In this respect, the function of the vessel, although impaired, is not drastically altered inasmuch as there is still a lumen and an intact, though thickened, wall with a relatively smooth lining permitting blood supply to the tissues. Some lesions remain in this stage for an indefinite period of time, whereas others eventually undergo changes that cause the breakdown of the vessel, inviting the perils that have made atherosclerosis a deadly disease. The sequence of events summarized here underlines the important role of lipid accumulation and inflammatory responses in the formation of the atheroma and forms the basis of the "lipid accumulation" and "inflammation" theories of atherosclerosis (see above).

## D. Complications of Advanced Lesions

When the capsule of the atheroma breaks away:

- The plaque is transformed into an ulcer
- Blood clots over the uneven surface (forming thrombi)
- Part of the exposed gruel is carried away by the bloodstream (forming emboli)
- Hemorrhage may occur into the gruel or under the lips of the ulcer
- The ulceration breaks through the remnant of the wall, causing rupture of the artery and massive hemorrhage into the space outside (as in an aneurysm)
- The end result may be thrombosis and embolism, or aneurysm and hemorrhage

Although little is yet known of the precise factors that promote ulceration and hemorrhage in plaques, they are, perhaps, favored by certain local changes in the lesions (e.g., extensive cell necrosis in the capsule) as well as hemodynamic events (e.g., sudden rise in blood pressure). Other complications of advanced atherosclerosis include narrowing or widening of the arterial lumen, *thrombosis*,[45,46] and dilation, with possible rupture, of the arterial wall, *aneurysm*.[47] The view that atherosclerosis narrows and tends to shut down the arterial lumen is applicable mainly to some arteries such as the coronaries, which are embedded into an unyielding environment. Arterial occlusion by stenosis (narrowing) of the lumen is usually a slow process, and collateral circulation, as previously described, frequently has time to establish itself so that a sufficient blood supply reaches the area normally serviced by the stenotic vessel.

Large arteries, such as the aorta, are widened by the atherosclerotic process as a result of the progressive weakening of the wall by the formation of scar tissue, causing it to give way to the mounting pressure within. In these cases, the arteries not only widen but also tend to lengthen, bending and twisting in the process. *Aneurysms,* balloon-like bulges that press upon neighboring structures and often burst (with subsequent hemorrhage), frequently occur in a given spot in the wall that is much weaker than the rest.[47] *Rupture* of the arterial wall that has been weakened by atherosclerosis can also be triggered by hypertension. When the rupture occurs in the relatively small cerebral arteries, the result is a "stroke." When it occurs in the aorta, especially in the descending portion, the usual result is massive bleeding into the abdominal cavity (a consequence surgeons try to avoid by "reconstruction" of the arterial wall).

## E. Thrombosis, Embolism, Platelets and "Clot Busters," and Growth Factors

According to the "Virchow's Triad" formulated by the German pathologist R. Virchow, more than one century ago, three factors — vascular injury, altered blood flow, and changes in blood coagulability — are responsible for vascular thrombosis. Thrombosis represents the process by which a plug of clot, or thrombus, is formed in a blood vessel (or in one of the heart cavities) by coagulation of the blood.[45,46] The *thrombus:*

- Contains few platelets, abundant fibrin, many trapped red blood cells
- Is produced by activation of the plasma coagulation system
- Forms in areas of slow blood flow (stasis)

It is distinguished from the *embolus,* which is also a clot of coagulated blood and often originates from a thrombus detached from the arterial wall where it was formed. The embolus is carried in the blood current, and an "embolism" represents the plugging of an artery by a clot (embolus) that has been brought to its place by the blood current. Thrombi develop more frequently in atherosclerotic than in normal arteries, appearing particularly on the ulcerated plaques,[45,46] on the arterial wall, or wherever there is a crack or fissure in the plaque. "Mural" thrombi are small and flat, develop over the surface of the wall of the large arteries, and are relatively harmless as long as they do not seriously impede the flow of blood through the vessel. When, however, the thrombus is large or develops in a small artery (such as a coronary artery or a cerebral vessel), it can fill its entire lumen and block all flow of blood, with disastrous results for the tissues that are to be supplied by the plugged vessel. Such occlusions generally occur suddenly, within minutes or hours, leaving little or no time for a collateral circulation to become established. The occurrence of thrombosis (forming the basis for *the thrombogenic theory* of atherosclerosis) has been related also to blood chemistry (e.g., high blood lipids) and to hemodynamic changes (e.g., hypertension).

*Platelets* are small, circulating, granulated cell fragments of bone marrow megakaryocytes; their number is about 300,000, their life duration about 10 days, and their production is regulated by colony-stimulating factors (Chapter 18). Platelets contain two types of granules, one type is made up of nonprotein substances involved in platelet activation, the other, protein substances including blood clotting factors and platelet-derived growth factor (PDGF), also produced by macrophages and vascular endothelial cells. The major function of platelets is to participate in the repair of breaks in the vascular wall to prevent blood loss, a complex process called *hemostasis*.[45–48] The first step in this process is the formation of a platelet plug, over the break, by adhesion of platelets to the wall; this is followed by platelet activation and aggregation to form a clot. When the clot is in place, tissue repair begins, triggered by PDGF. Platelets work in concert with the endothelial cells (see above) also damaged by the injury and initiate a *coagulation cascade* involving several factors (e.g., platelet-activating factor [PAF], a cytokine) and leading to the formation of the proteins, fibrinogen and fibrin, to stabilize the platelet plug. With healing of the injury, the clot is removed by enzymatic action (fibrinolysis) and by phagocytosis from immune cells.[49–51]

Given the important role of platelets and endothelial factors in clot and, hence, thrombus formation, *several fibrinolytic drugs* are used to dissolve the clots especially life-threatening in the coronary arteries (see below). Among these, some of the most frequently used are the enzymes *urokinase* (extracted from the kidneys) and *streptokinase* (extracted from bacteria) and the genetically engineered *tissue plasminogen activator (t-PA)*, a natural-occurring promoter of the fibrinolytic enzyme, plasmin.[52] A number of inhibitors of platelet aggregation include aspirin, a weak inhibitor, probably more efficient in decreasing the incidence of myocardial infarctions than in treating acute coronary thrombosis, and inhibitors of the platelet glycoprotein IIb/IIIa complex, that promotes platelet aggregation and is a member of the family of integrins (involved in the adhesion of cells to the extracellular matrix).[52] Current evidence has linked an inherited platelet trait involving one allele on glycoprotein IIIa to coronary artery disease.[53,54]

To be noted here is the increasing list of growth factors participating in atherosclerotic processes and in repair of lesions. Among them, fibroblast growth factor (FGF), platelet-derived growth factor (PDGF), and platelet-activating factor (PAF) have important roles in plaque formation. Factors stimulating or inhibiting proliferation of vascular cells are being investigated actively to determine if this aspect of plaque formation is amenable to pharmacologic interventions.

## V. THEORIES OF ATHEROSCLEROSIS

It is evident that the pathogenesis of the atherosclerotic lesion is an extremely complex, multifactorial process not yet fully understood. At the present state of our knowledge, it may be advantageous to clarify certain specific steps of the process rather than to attempt to formulate a single, all-inclusive theory of atherosclerotic pathology. Such an aim is consistent with the major, currently proposed theories, each dealing with different processes in the arterial wall. Such theories have been referenced throughout this and previous (Chapter 5) or subsequent chapters (Chapter 17) and are summarized in Table 16.8.

## VI. CORONARY HEART DISEASE

One of the major life-threatening consequences of atherosclerosis involves the coronary arteries that supply blood to the heart wall. Atherosclerosis of these vessels leads to

**TABLE 16.8**
**Theories of Atherosclerosis**

| Lipid Accumulation | Myoclonal | Thrombogenic | Inflammation | Free Radicals |
|---|---|---|---|---|
| Alterations of lipoproteins with accumulation of oxidized LDL in arterial wall (Chapter 17) | Chronic smooth muscle proliferation in response to damaging agents | Lesions may be initiated through alterations of endothelial cells with consequent hemorrhage and thrombus formation | Early inflammatory processes in arterial wall followed by macrophage, T lymphocyte migration | Increased accumulation of free radicals and induction of oxidative stress (Chapter 5) |
| Lipid infiltration involves muscle cells, monocytes, macrophages, and T lymphocytes | Smooth muscle cell mutations responsible for abnormal proliferation of smooth muscle cells with migration to intima and elastic layers | Importance of platelets, growth factors, and blood clotting | Associated with increased C-reactive protein and responsive to anti-inflammatory agents | |

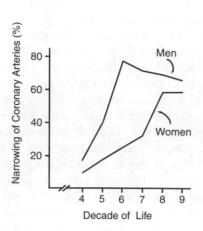
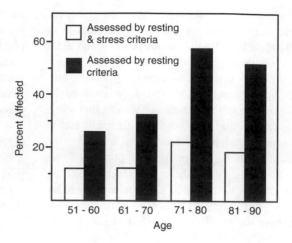

**FIGURE 16.6** Percentage narrowing of coronary artery lumen with aging (on the left). Note sex differences with narrowing coronary artery occurring earlier and being more severe in men than women. Incidence of coronary artery disease with aging (on the right). Note the increase in incidence with age, particularly under conditions of stress.

coronary heart disease (CHD), also called ischemic heart disease, which implies reduced blood flow to the heart and consequent angina pectoris and myocardial infarction. Other possible consequences include arrhythmias due to defects of impulse conduction and electrocardiographic (ECG) changes. All of these conditions lead to heart failure, severe disability, and death.[8,55,56]

CHD continues to be the major cause of disability and death in the United States. Its prevalence increases with age and shows a significant sex difference, women having a lower incidence than men; this sex difference disappears after 70 years of age. Stress is another significant contributing factor to the high CHD incidence in the elderly (Figure 16.6). However, as indicated at the beginning of this chapter and more extensively discussed in Chapter 2, after a rise in the first half of the last century, mortality from this cause has been decreasing since the late 1960s. This decline includes all sectors of the adult population. In the 35 to 74 year age group, the rate of mortality due to CHD has fallen considerably and is all the more remarkable, because this decrease occurred after a sustained period of increase in this disease beginning about 1940. Apparently, we have been doing the "right things" (e.g., improvement of lifestyle and medical advances in control of hypertension and atherosclerosis) during the past decades to decrease mortality from CHD and the overall adult cardiovascular diseases.[55–57]

### CORONARY CIRCULATION

The right and left coronary arteries arising from the aorta as it emerges from the left ventricle, are the major vessels supplying blood to the myocardium (the muscle layer of the heart). Venous drainage is via a superficial system ending primarily in the coronary sinus, and via a deep system that drains the remainder of the heart and empties directly into

## TABLE 16.9
## Local Regulation of Coronary Blood Flow

| Vasodilation | Vasoconstriction |
| --- | --- |
| Low Oxygen | Angiotensin II |
|   High $CO_2$ | Sympathetic stimulation (direct) |
|   High $H^+$ | |
|   High $K^+$ | |
|   High lactic acid | |
|   High prostaglandins | |
|   High adenine nucleotides and | |
|     adenosine | |
| Vagal stimulation | |

Coronary flow at rest (250 mL/min) is 5% of the cardiac output. The heart extracts 70 to 80% of the oxygen from this blood. Therefore, oxygen consumption can be significantly increased on demand only by increasing blood flow. Coronary flow is influenced by several local factors that regulate flow through vasodilation or vasoconstriction (Table 16.9).

Major vasodilators include reduced oxygen and increased carbon dioxide concentration, increased products of metabolism (hydrogen ions, potassium, lactic acid), local agents (prostaglandins, adenine nucleotides, and adenosine) and neural stimulation (stimulation of parasympathetic innervation). Major vasoconstrictors include angiotensin II (Chapter 10) and stimulation of the sympathetic innervation. Cardiac circulation is preferentially preserved, as it is in the brain, when that of other organs is compromised.

### A. CHD CONSEQUENCES

When progressive narrowing of the coronary lumen reduces flow through the coronary artery to the point that the myocardium becomes ischemic (i.e., reduced blood flow, insufficient to sustain function), *angina pectoris* develops. If the myocardial ischemia is severe and pro-

**TABLE 16.10**

**Symptoms of Angina Pectoris and Acute Myocardial Infarction in the Elderly**

**Angina Pectoris**

Pain, less marked than in adult; may present as headache or epigastric distress

**Myocardial Infarction**

Variable presentation with chest pain, including breathlessness, confusion, fainting, GI symptoms, sweating, hypotension, etc.

**TABLE 16.11**

**Major Risk Factors in Coronary Heart Disease**

Age
Genetic predisposition
Hypertension
Diabetes mellitus
Hypercholesterolemia
Cigarette smoking
*Also*:
Obesity
Poor physical fitness and lack of exercise
Personality type (?)

longed, irreversible changes occur in the cardiac muscle, and the result is *myocardial infarction*. The heart stops functioning, and death ensues within a few minutes. If ischemia is less severe, death may not occur, but it may generate permanent functional impairment.

## B. Signs and Symptoms

The major sign of angina pectoris is squeezing or pressure-like pain, retrosternally, radiating to the left shoulder, arm, hand, neck, and jaw. The pain often appears with exertion, emotion, or after a large meal. Anginal pain in the elderly is often less marked, possibly due to reduced activity or altered pain perception; it may be effort-induced or may occur at rest or in bed and may present as headache or epigastric (around the stomach) pain relieved by antacids (Table 16.10).

Major signs and symptoms of myocardial infarction are variable in the elderly, although chest pain remains the commanding feature in patients admitted to coronary care units (Table 16.10). Other symptoms include breathlessness, confusion, behavior change, fainting, palpitations, vomiting, sweating, abdominal pain, and hypotension. The pain may last over 30 min and is not relieved by multiple doses of nitroglycerine. Diagnosis is confirmed by ECG changes. With the establishment of the infarct, irreversible changes occur in the myocardium: muscle cells first become leaky, and the rise in serum enzymes and isoenzymes is a biochemical diagnostic sign of the infarct. The

first enzyme to be elevated is serum glutamic oxaloacetic transaminase (SGOT) followed by creatinine phosphokinase (CPK) and lactic dehydrogenase (LDH).

## C. Risk Factors for CHD

CHD epidemiology, so important in identifying the causes and subsequent prevention and treatment, can best be studied by focusing on the consequences. Indeed, the decreased mortality from CHD (Chapter 2) may be ascribed primarily to identification of risk factors and to their prevention. Studies in various populations identified the following risk factors that predispose to the development of CHD: age, genetic predisposition, hypertension, diabetes mellitus, hypercholesterolemia, and cigarette smoking. Other risk factors include obesity, poor physical fitness, lack of exercise, and personality type (Table 16.11). These lists dictate specific preventive measures such as treatment of hypertension and diabetes, elimination of cigarette smoking, amelioration of dietary habits toward an optimal body weight, and encouragement of measures to improve physical fitness (Chapter 24).

### Elevated Homocysteinemia: A Cardiovascular Risk?

Not all patients with CHD have high blood LDL cholesterol levels (Chapter 17). In fact, the normal cholesterol levels preserved in some patients clash with the current emphasis on high blood cholesterol as a major risk factor for CHD. Another substance that, when elevated, may represent a risk factor for CHD is homocysteine, a nonprotein amino acid and intermediate in methionine metabolism; it is a donor of methyl groups to choline and creatinine and stabilizer of cell membrane fluidity. In early studies, homocysteine blood levels were markedly higher in men who later had CHD than in age-matched controls who remained free of cardiac infarction.[58] Later studies reported that the relationship of homocysteine to cardiovascular pathology persisted even when homocysteine levels were only moderately elevated.[59] Accumulation of mutations in enzymes involved in the elevation of homocysteine blood levels correlates with increased CHD risk and occurrence of thrombogenesis. Hyperhomocysteinemia may damage the blood vessels by:

- Impairing the production of EDRF from endothelial cells[60]
- Stimulating smooth muscle proliferation[61]
- Acting as a thrombogenic agent[62]
- Being inversely related to folic acid levels[63]

The inverse relationship of homocysteine to folic acid and the low folic acid intake in a large percentage (40%) of the population have prompted the recommendation for the widespread screening for homocysteine blood levels and for the use of dietary folic acid supplements.[62] However, the actual clinical benefits of normalization of elevated levels of homocysteine are still unproven. In addition, inasmuch as $B_{12}$, $B_6$, and folic acid are cofactors in the metabolism of homocysteine, folate supplements may mask the danger of anemia due to low levels of $B_{12}$ (Chapter 18).

**Systemic and Organismic Aging**

**TABLE 16.12**
**Major Types of Coronary Heart Disease Treatment**

Medical Treatment
  Diet
  Exercise
  No smoking
  Pharmacologic agents
Surgical Treatment
  Aortocoronary bypass graft
Percutaneous coronary angioplasty with streptokinase/tissue
  plasminogen activator (TPA) anticoagulant therapy

## D. NEW APPROACHES TO MANAGEMENT

The reduction of mortality due to CHD reported in the l980s had arisen from the awareness of contributing factors and their alleviation or amelioration (Table 16.12). New studies demonstrate that appropriate dietary and drug interventions can induce regression and reversal, or at least arrest the progression of atherosclerotic lesions. Reduction of LDL (the atherogenic lipoprotein) blood levels, by administration of drugs or manipulation of the diet, resulted in a significant regression of atherosclerotic lesions in cases of hyperlipoproteinemia or hypercholesterolemia (familial or not), in men and women, and in younger as well as older (65+ years) individuals (Chapter 17). Individuals who have had a myocardial infarction, coronary by-pass surgery, or angioplasty should be treated promptly and aggressively to lower levels of LDL and stabilize plaques.

Current studies show that the administration of antioxidants, by reducing the level of oxidized LDL, may also be beneficial (Chapter 5). So far, promising results with antioxidant therapy have been reported in nonhuman primates and swine. Facilitation of collateral circulation (time permitting) and stimulation of angiogenesis are other interventive procedures under study.[64] The effectiveness of these treatments supports the view that the atherosclerotic process is not, as previously thought, an inexorably progressive condition. Rather, current advances in molecular biology offer new approaches to therapy and prevention by targeting molecules that are as diverse as adhesion molecules and transcriptions factors.[65] While the biological rationale and progress of these therapies is still being chartered, they undoubtedly will contribute substantially to "remodeling" and repairing the arterial wall.

Current medical treatment includes:

1. Treatment of the underlying disease, if any (hypertension, diabetes mellitus, hyperlipidemias)
2. Behavioral therapy: low-cholesterol, low-fat diets, cessation of smoking, reduced stress, and increased physical exercise, especially for cardiac rehabilitation

3. Administration of pharmacologic agents with the intention of:
   a. Reducing LDL and cholesterol blood levels
   b. Decreasing free radical levels (antioxidants)
   c. Increasing cardiac blood flow, reducing cardiac work
   d. Promoting collateral circulation and angiogenesis[64]
   e. Preventing clotting

Surgery has also been successful with aorto-coronary bypass grafts and transluminal coronary angioplasty (mechanical dilation of the area of constriction) with anticoagulant therapy,[66,67] and by the intracoronary injection of the enzyme streptokinase or even better of tissue plasminogen activator (TPA), a recombinant protease (see above).[52–54]

## REFERENCES

1. Braunwald, E., Cardiovascular medicine at the turn of the millenium: Triumphs, concerns, and opportunities, *N. Engl. J. Med.*, 337, 1360, 1997.
2. Powell, C. and MacKnight, C., Epidemiology of heart disease, in *Principles and Practice of Geriatric Medicine*, Pathy, M.S. J., Ed., 3rd ed., John Wiley and Sons, New York, 1998.
3. Smith, E.B., The pathogenesis of atherosclerosis, in *Principles and Practice of Geriatric Medicine*, Pathy, M.S.J., 3rd ed., John Wiley & Sons, New York, 1998.
4. Tresch, D.D. and Aronow, W.S., Eds., *Cardiovascular Disease in the Elderly Patient*, Marcel Dekker, Inc., New York, 1999.
5. Carr, J.J. and Burke, G.L., Subclinical cardiovascular disease and atherosclerosis are not inevitable consequences of aging, *J. Am. Geriatr. Soc.*, 48, 342, 2000.
6. Burke G.L. et al., Factors associated with healthy aging: The cardiovascular health study. CHS Collaborative Research Group, *J. Am. Geriatr. Soc.*, 49, 254, 2001.
7. Stott, J.D., and Williams, B.O., Ischaemic heart disease, in *Principles and Practice of Geriatric Medicine*, Pathy, M.S.J., Ed., 3rd ed., John Wiley and Sons, New York, 1998.
8. Prakash, A., Ed., *Preventing Coronary Heart Disease*, Adis Intern., Hong Kong, 2000.
9. Gresham, G.A., *Reversing Atherosclerosis*, Charles C. Thomas, Springfield, IL, 1980.
10. Morrison, L.M. and Schjeide, O., *Atherosclerosis, Prevention, Treatment and Regression*, Charles C. Thomas, Springfield, IL, 1984.
11. Lakatta, E.G., Cardiovascular aging research: The new horizons, *J. Am. Geriatr. Soc.*, 47, 613, 1999.
12. McGill, H.C., Geer, J.C., and Strong, J.P., Natural history of human atherosclerotic lesions, in *Atherosclerosis and Its Origin*, Sandler, M. and Bourne, G.H., Eds., Academic Press, New York, 1963.

13. Harris, D.J. and Douglas, P.S., Enrollment of women in cardiovascular clinical trials funded by the National Heart, Lung and Blood Institute, *N. Engl. J. Med.*, 343, 475, 2000.

14. Constantinides, P., *Experimental Atherosclerosis*, Elsevier, Amsterdam, 1965.

15. Aranow, W.S. and Frishman, W.H., Risk factors for atherosclerosis in the elderly, in *Principles and Practice of Geriatric Surgery*, Rosenthal, R.A., Zenilman, M.E., and Katlic, M.R., Eds., Springer, New York, 2001.

16. Karathanasis, S.K., Lipoprotein metabolism high-density lipoprotein, in *Monographs in Human Genetics. Vol. 14. Molecular Genetics of Coronary Artery Disease: Candidate Genes and Processes in Atherosclerosis*, Lusis, A.J., Rotter, J.I., and Sparkes, R.S., Eds., Karger, Basel, 1992.

17. Heller, D.A. et al., Genetic and environmental influences on serum lipid levels in twins, *N. Engl. J. Med.*, 328, 1150, 1993.

18. Marenberg, M.E. et al., Genetic susceptibility to death from coronary heart disease in a study of twins, *N. Engl. J. Med.*, 330, 1041, 1994.

19. Dietz, H.C. and Pyeritz, R.E., Molecular biology — To the heart of the matter, *N. Engl. J. Med.*, 330, 930, 1994.

20. Keating, M.T. and Sanguinetti, M.C., Molecular genetic insights into cardiovascular disease, *Science*, 272, 681, 1996.

21. Zierler, R.E., Kohler, T.R., and Strandness, D.E., Duplex scanning of normal or minimally diseased carotid arteries: Correlation with arteriography and clinical outcome, *J. Vasc. Surg.*, 12, 447, 1990.

22. Achenbach, S. et al., Value of electron-beam computed tomography for the noninvasive detection of high-grade coronary stenoses and occlusions, *N. Engl. J. Med.*, 339, 1964, 1998.

23. Warden, C.H. et al., Atherosclerosis in transgenic mice overexpressing apolipoprotein A-II, *Science*, 261, 469, 1993.

24. Breslow, J.L., Mouse models of atherosclerosis, *Science*, 272, 685, 1996.

25. Rich-Edwards, J.W. et al., Birth weight and risk of cardiovascular disease in a cohort of women followed since 1976, *Brit. Med. J.*, 315, 396, 1997.

26. Berenson, G.S. et al., Association between multiple cardiovascular risk factors and atherosclerosis in children and young adults, *N. Engl. J. Med.*, 338, 1650, 1998.

27. Robertson D. et al., The diagnoses and treatment of baroreceptor failure, *N. Engl. J. Med.*, 339, 1449, 1993.

28. Lüscher, T.F., The endothelium and cardiovascular disease — A complex relation, *N. Engl. J. Med.*, 330, 1081, 1994.

29. Timiras, P.S., Changing regulation of vascular endothelium with aging, *Internal Medicine*, 5, 129, 1997.

30. Ferrara, N., VEGF: An update on biological and therapeutic aspects, *Curr. Opin. Biotechnol.*, 11, 617, 2000.

31. St. Croix, B. et al., Genes expressed in human tumor endothelium, *Science*, 289, 1197, 2000.

32. Ross, R. and Glomset, J.A., The pathogenesis of atherosclerosis, *N. Engl. J. Med.*, 295, 369, 1976.

33. Bennett, M.R., Evan, G.I., and Schwartz, S.M., Apoptosis of human vascular smooth muscle cells derived from normal vessels and coronary atherosclerotic plaques, *J. Clin. Invest.*, 95, 2266, 1995.

34. Libby, P., Molecular bases of acute coronary syndromes, *Circulation*, 91, 2844, 1995.

35. Ross, R., Atherosclerosis — an inflammatory disease, *N. Engl. J. Med.*, 340, 115, 1999.

36. Maseri, A., Inflammation, atherosclerosis, and ischemic events — exploring the hidden side of the moon, *N. Engl. J. Med.*, 336, 1014, 1997.

37. Schlichter, J. and Harris, R., The vascularization of the aorta. A comparative study of the aortic vascularization of several species in health and disease, *Am. J. Med. Sci.*, 218, 610, 1949.

38. Patel, D.J. and Vaishnaw, R.N., *Basic Hemodynamics and its Role in Disease Processes*, University Park Press, Baltimore, MD, 1980.

39. Moore, S., Ed., *Injury Mechanisms in Atherosclerosis, The Biochemistry of Disease*, Vol. 9, Marcel Dekker, New York, 1981.

40. Vikhert, A.M. and Zhdanov, V.S., *The Effects of Various Diseases on the Development of Atherosclerosis*, Pergamon Press, New York, 1981.

41. Reynold, G.D. and Vance, R.P., C-reactive protein immunohistochemical localization in normal and atherosclerotic human aortas, *Arch. Pathol. Lab. Med.*, 111, 265, 1987.

42. Liuzzo, G. et al., The prognostic value of C-reactive protein and serum amyloid A in severe unstable angina, *N. Engl. J. Med.*, 331, 417, 1994.

43. Ridker, P.M. et al., C-reactive protein and other markers of inflammation in the prediction of cardiovascular disease in women, *N. Engl. J. Med.*, 342, 836, 2000.

44. Ridker, P.M. et al., Inflammation, aspirin, and the risk of cardiovascular disease in apparently healthy men, *N. Engl. J. Med.*, 336, 973, 1997.

45. Davies, M.J., Mechanisms of thrombosis in atherosclerosis, in *Hemostasis and Thrombosis: Basic Principles and Clinical Practice*, 3rd ed., Colman, R.W. et al., Eds., Lippencott, Philadelphia, 1993.

46. Bang, N.U. et al., *Thrombosis and Atherosclerosis*, Year Book Medical Publishers, Chicago, IL, 1982.

47. Reilly, J.M. and Sicard, G.A., Natural history and treatment of aneurysms, in *Principles and Practice of Geriatric Surgery*, Rosenthal, R.A., Zenilman, M.E., and Katlic, M.R., Eds., Springer, New York, 2001.

48. Hamstein, A., Hemostatic function and coronary heart disease, *N. Engl. J. Med.*, 332, 677, 1995.

49. Antiplatelet Trialists' Collaboraton, Collaborative overview of randomized trials of antiplatelet therapy I. Prevention of death, mycardial infarction, and stroke, by prolonged antiplatelet therapy in various categories of patients, *Brit. Med. J.*, 308, 81, 1994.

50. Harker, L.A., Platelets and vascular thrombosis, *N. Engl. J. Med.*, 330, 1006, 1994.

51. Handin, R.I., Platelets and coronary artery disease, *N. Engl. J. Med.*, 334, 1126, 1996.

**Systemic and Organismic Aging**

52. Ridker, P.M. et al., Endogenous tissues type plasminogen activator and risk of myocardial infarction, *Lancet,* 341, 1165, 1993.

53. Weiss, E.J. et al., A polymorphism of a platelet glycoprotein receptor as an inherited risk factor for coronary thrombosis, *N. Engl. J. Med.,* 334, 1090, 1996.

54. Schneiderman, J. et al., Increased type 1 plasminogen activator inhibitor gene expression in atherosclerotic human arteries, *Proc. Natl. Acad. Sci. USA*, 89, 6998, 1992.

55. Stamler, J., Coronary heart disease: Doing the "right things," *N. Engl. J. Med.*, 312, 1053, 1985.

56. Bild, D.E. et al., Age related trends in cardiovascular morbidity and physical functioning in the elderly: The cardiovascular health study, *J. Am. Geriatr. Soc.*, 41, 1047, 1993.

57. Ornish, D., Can lifestyle changes reverse coronary heart disease? The Lifestyle Heart Trail, *Lancet*, 336, 129, 1990.

58. Stampfer, M.J. and Malinow, M.R., Can lowering homocysteine levels reduce cardiovascular risk? *N. Engl. J. Med.,* 332, 328, 1995.

59. Stampfer, M.J. et al., A prospective study of plasma homocysteine and risk of myocardial infarction in U.S. physicians, *J. Am. Med. Assoc.*, 268, 877, 1992.

60. Stamler, J.S. et al., Adverse vascular effects of homocysteine are modulated by endothelium derived relaxing factor and related oxides of nitrogen, *J. Clin. Invest.,* 91, 308, 1993.

61. Tsai, J.C. et al., Promotion of vascular smooth muscle cell growth by homocysteine: A link to atherosclerosis, *Proc. Natl. Acad. Sci. USA,* 91, 6369, 1994.

62. Selhub, J. et al., Vitamin status and intake as primary determinants of homocysteinemia in an elderly population, *J. Am. Med. Assoc.*, 270, 2693, 1993.

63. Den Heijer, M. et al., Hyperhomocysteinemia as a risk factor for deep vein thrombosis, *N. Engl. J. Med.*, 334, 759, 1996.

64. Lee, S.H. et al., Early expression of angiogenesis factors in acute myocardial ischemia and infarction, *N. Engl. J. Med.,* 342, 626, 2000.

65. Gibbons, G.H. and Dzau, V.J., Molecular therapies for vascular disease, *Science*, 272, 689, 1996.

66. Ricou, F., Percutaneous transluminal coronary angioplasty, in *Principles and Practice of Geriatric Medicine*, 3rd ed., Pathy, M.S.J., John Wiley & Sons, New York, 1998.

67. Camacho, M.T. et al., Cardiac surgery in the elderly, in *Principles and Practice of Geriatric Surgery*, Rosenthal, R.A., Zenilman, M.E., and Katalic, M.R., Eds., Springer, New York, 2001.

# 17 Lipids, Lipoproteins and Atherosclerosis

*John K. Bielicki and Trudy M. Forte*
Lawrence Berkeley National Laboratory

## CONTENTS

## I.  INTRODUCTION

Atherosclerotic cardiovascular disease is the major cause of morbidity and mortality in industrialized nations. The disease is complex, and many factors singly or in combination contribute to its etiology, including genetic predisposition, hypercholesteremia (elevated cholesterol), hypertension, diabetes mellitus, smoking, obesity, and stress. Atherosclerosis, the condition whereby lipids accumulate in the artery wall, starts in the first decade of life and progresses throughout life; its severity is linked to the aforementioned risk factors. As the list of risk factors indicates, cholesterol, and by extension, plasma lipoproteins, have a major role in the process of atherogenesis. The structure, synthesis, and metabolism of lipoproteins are reviewed in this chapter along with their role as positive or negative risk factors in premature atherosclerosis.

## II.  LIPOPROTEIN NOMENCLATURE, STRUCTURE, AND COMPOSITION

Lipids in the plasma including cholesterol, phospholipids, and triglycerides are transported in the form of lipid complexes stabilized by specific proteins called apolipoproteins (apo). These complexes form lipoproteins that transport lipids in the plasma and interstitial fluid. The generalized features of a lipoprotein are illustrated in Figure 17.1. The lipoprotein particle is essentially an oil droplet (hence, its globular shape) stabilized by a surface coat of hydrophilic molecules, including proteins, phospholipids, and unesterified cholesterol. The hydrophobic core of the particle consists of the highly water-insoluble lipids, cholesteryl ester and triglyceride. As the protein (apolipoprotein) content of the particles increases relative to the lipid content, the particles become smaller and denser. The differences in densities of the particles are the fundamental basis for the nomenclature used for defining lipoproteins, e.g., very low density (VLDL), intermediate-density (IDL), low-density (LDL), and high-density (HDL) lipoproteins. Chylomicrons (CM) secreted by intestinal absorptive cells are the largest and most buoyant of the lipoprotein particles. The major classes of lipoproteins and their compositions are summarized in Table 17.1.

### A.  Major Classes of Lipoproteins

As the compositions in Table 17.1 indicate, lipoproteins can be grouped into three major categories based on their composition:

*CM and VLDL*. These are relatively poor in protein, phospholipid, and cholesterol but high in triglyceride (55 to 95% of total particle weight). In more

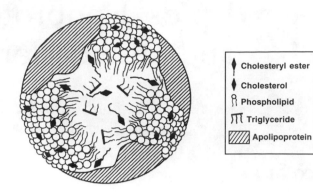

**FIGURE 17.1** Generalized organization of mature plasma lipoproteins. Particles are overall globular in morphology with polar (water-soluble) components on the surface to stabilize the particle in the aqueous plasma environment. The polar constituents are primarily protein (apolipoprotein) and phospholipid. Some cholesterol is also on the surface. The "core" of the particle consists of nonpolar (highly insoluble in the aqueous environment) components, cholesteryl esters, and triglyceride.

general terms, these lipoproteins are referred to as triglyceride-rich lipoproteins.

*IDL and LDL*. These particles are characterized by high levels of cholesterol that are mainly in the form of the highly insoluble cholesteryl ester. Because up to 50% of the LDL mass is cholesterol, it is not surprising that LDL has a significant role in the development of atherosclerotic disease.

*HDL*. The hallmark of this lipoprotein is its high protein content (~50%) and relatively high phospholipid content (~30%). HDL is generally divided into two subclasses, $HDL_2$ and $HDL_3$; of the two, $HDL_2$ is the larger and more buoyant particle.

### B.  Major Apolipoproteins

In addition to their core lipids that help define the lipoprotein particles, the particles are also defined by the

---

**TABLE 17.1**
**Composition (Weight Percent) of the Major Classes of Lipoproteins**

| Lipoprotein | Protein | Phospholipid | Cholesterol | Triglyceride |
|---|---|---|---|---|
| CM | 1–2 | 3–6 | 2–7 | 80–95 |
| VLDL | 5–10 | 15–20 | 10–15 | 55–65 |
| IDL | 19 | 19 | 38 | 23 |
| LDL | 20–25 | 22 | 45 | 10 |
| HDL | 45–50 | 30 | 20 | 5 |

*Note:* CM, Chylomicron; VLDL, very low density lipoprotein; IDL, intermediate-density lipoprotein; LDL, low-density lipoprotein; HDL, high-density lipoprotein

**TABLE 17.2**
**Major Apolipoproteins**

| Apolipoprotein | Molecular Weight (daltons) | Lipoprotein Class in which Found |
|---|---|---|
| ApoA-I | 28,000 | HDL |
| ApoA-II | 17,000 | HDL |
| ApoB-100 | 540,000 | VLDL, IDL, LDL |
| ApoB-48 | 260,000 | CM |
| Apo(a) | 300,000–800,000 | LDL |
| ApoC-I | 8000 | CM, VLDL, HDL |
| ApoC-II | 10,000 | CM, VLDL, HDL |
| ApoC-III | 12,000 | CM, VLDL, HDL |
| ApoE | 34,000 | CM, VLDL, IDL |

apolipoproteins on their surfaces. The apolipoproteins act as markers that determine the metabolic fate of the particles (see Section VI). The major apolipoproteins and the lipoprotein class in which they are found are summarized in Table 17.2. Some apolipoproteins, e.g., apoAs and apoBs, are found uniquely associated with specific classes of lipoproteins, while others like apoCs can associate with almost all classes of particles.

## III. PLASMA CONCENTRATION OF LIPOPROTEINS

Generally, one thinks of lipoproteins in terms of pathologic conditions such as premature atherosclerosis, but in fact, they are macromolecules necessary for the maintenance of cell and tissue function and integrity. Triglycerides transported by lipoproteins are a major source of energy for cells. Cholesterol transported by lipoproteins is utilized by cells for cell division, cell growth, and membrane repair; cholesterol is also essential for the production of steroid hormones, adrenocortical hormones and sex hormones. Another important function of lipoproteins is the transport of fat-soluble vitamins. An overabundance of lipoproteins, however, particularly those carrying cholesterol, can be deleterious by predisposing to premature cardiovascular disease.[1]

The most commonly used clinical indicator for measuring potential risk of premature cardiovascular disease is the level of plasma lipids. Fasting levels of triglyceride, cholesterol, and HDL cholesterol can often be used to identify possible abnormalities. The expected normal adult plasma lipid levels are shown in Table 17.3. Females characteristically have lower triglyceride concentrations than males and have higher HDL cholesterol (55 mg/dL versus 43 mg/dL); it is well known that there is an inverse relationship between HDL cholesterol levels and risk for heart disease; thus, the female has the more protective profile. The ratio of total cholesterol to HDL cholesterol is an important value, because values of 4.0 and above are associated with increased risk for coronary heart disease.

**TABLE 17.3**
**Normal Plasma Lipid Levels (mg/dL)**

| | Triglyceride | Total Cholesterol | HDL Cholesterol |
|---|---|---|---|
| Adult female | 80 | 190 | 55 |
| Adult male | 120 | 200 | 43 |
| Neonate | 38 | 70 | 35 |

Table 17.3 provides, for comparison, the plasma lipid levels from cord blood of normal, full-term newborns. The newborn infant has triglyceride and total cholesterol levels one half to one third those of the adult. The HDL cholesterol concentrations are relatively high (35 mg/dL) in the newborn, where the ratio of total cholesterol to HDL cholesterol is 2 compared with the adult values of 3.5 for females and 4.6 for males. HDL is considered beneficial; that is, it is protective against atherosclerosis, whereas LDL is a positive risk factor (see Table 17.4). The lipid levels in infants are perhaps the most "ideal," as LDL levels are low and HDL relatively high. Except for genetic

**TABLE 17.4**
**Positive and Negative Risk Factors in Atherosclerosis**

| Positive | Negative |
|---|---|
| Age: Males > 45 years Females > 55 years | Elevated HDL cholesterol |
| Family History of early CHD | Low LDL cholesterol |
| Elevated LDL cholesterol (> 130 mg/dl)) | Good genes |
| Low HDL cholesterol (< 35 mg/dl) | Female gender (estrogen) |
| Diabetes mellitus | Exercise |
| Hypertension | |
| Obesity | |
| Smoking | |

*Note:* CHD = coronary heart disease.

**Systemic and Organismic Aging**

abnormalities (such as homozygous familial hypercholes-terolemia), the vascular walls of neonates are free of fatty streaks. Fat accumulation appears, however, in the first years of life, indicating that dietary input and environmental factors probably influence the initiation and progression of atherosclerosis. At birth, no distinction can be seen between male and female infants, because sex hormone concentrations are low and apparently have little metabolic influence at this stage of development.

## IV. RISK FACTORS IN ATHEROSCLEROSIS

Three major indicators of increased risk to atherosclerosis are triglyceride concentrations greater than 200 mg/dL, LDL cholesterol concentrations greater than 130 mg/dL, and HDL cholesterol concentrations less than 35 mg/dL.

Aging is associated with a progressive increase in plasma triglyceride and cholesterol concentrations; thus, the process of aging is an important risk factor in atherosclerosis. As Table 17.4 indicates, besides age, there are other positive and negative factors that affect atherogenesis. In the adult, gender differences have a definite effect on plasma lipid levels, as shown by the higher levels of triglyceride and total cholesterol and lower levels of HDL cholesterol in males as compared to females (Table 17.3). In females before menopause, the most functionally significant difference is the higher HDL cholesterol concentration, 55 mg/dL, as compared to males, 43 mg/dL. This increase in HDL cholesterol is in a specific subclass of HDL, the $HDL_2$.

### A. HIGH-DENSITY LIPOPROTEIN

Epidemiologic studies suggest that elevated $HDL_2$ concentrations play a protective role in cardiovascular disease.[2] In women, the hormone estrogen is important, because it elevates $HDL_2$ and, thus, tends to protect premenopausal women from an early onset of atherosclerosis. Indeed, the incidence of cardiovascular disease increases after menopause.

$HDL_2$ levels can also be increased by physical exercise. It is known that male marathon runners have HDL patterns similar to those of females. An obvious conclusion is that exercise is beneficial in maintaining healthy HDL levels. Cigarette smoking, on the other hand, decreases HDL levels, while cessation of smoking reverses the effect. Other factors that contribute to atherosclerosis include genetic disorders that increase VLDL and LDL and decrease HDL levels, diabetes, obesity, and hypertension.[3] Regulation of plasma lipid levels is clearly complex and involves genetic and environmental components.

### B. SMALL DENSE LDL

Elevated LDL cholesterol is directly linked with an increased risk of cardiovascular disease. However, within the LDL class, one can distinguish specific subclasses, LDL subclass pattern A and LDL subclass pattern B, which correlate differently with regard to their contribution to cardiovascular disease.[4] Pattern A LDL are large (>25.5 nm diameter) buoyant particles, while pattern B LDL are less buoyant and smaller in size (small dense LDL, <25.5 nm diameter). The latter LDL pattern (but not pattern A) is associated with an increased risk of atherosclerosis; pattern B is also associated with a constellation of risk factors that predispose to atherosclerosis, including low HDL, particularly $HDL_2$, elevated apoB concentrations, and elevated triglyceride.[5] LDL subclasses are genetically influenced; pattern B appears to be associated with an autosomal dominant allele(s) that has a rather high population frequency of 25 to 30%. Interestingly, expression of the pattern B phenotype is age dependent. This phenotype is not expressed in males until approximately 20 years of age and in females until menopause.

## V. SYNTHESIS OF LIPOPROTEINS

There arc two major sites of synthesis of lipoproteins: the small intestine and the liver (Figure 17.2).

### A. INTESTINE

Lipoprotein secretion by the intestine is regulated to a great extent by what we eat. Pancreatic lipase in the intestinal lumen hydrolyzes dietary triglycerides to fatty acids and monoglycerides. These moieties, together with dietary cholesterol, form micelles by interaction with bile. The lipid micelles are taken up by absorptive cells in the small intestine, and trigylcerides are resynthesized and assembled into triglyceride-rich lipoproteins, the chylomicrons (CM). The composition and size of the chylomicrons depends on dietary lipids, where more saturated fats yield smaller particles and unsaturated fats yield larger CM. Formation of CM requires the stabilization of the particle surface with apoB-48, a truncated form of apoB-100 necessary for the release of CM from the cell. ApoC-II and C-III are also added to the surface; the secreted CM acquires apoE upon release into the circulation. ApoE and apoC-II are crucial for the rapid removal of these large particles from the circulation.

The intestine is also the site of synthesis for HDL containing apoA-I. Newly secreted HDL, however, are chemically and structurally different from circulating HDL. The newly secreted particles possess mainly phospholipid and unesterified cholesterol and are organized as a bilayer with a disk-like structure stabilized by apoA-I on its rim. Such discoidal HDL are "immature" forms of HDL, whereas the spherical ones are mature HDL. Discoidal HDL are also termed "nascent HDL," and normally, they are rapidly converted to mature HDL by the plasma enzyme, lecithin:cholesterol acyltransferase (LCAT) discussed below.

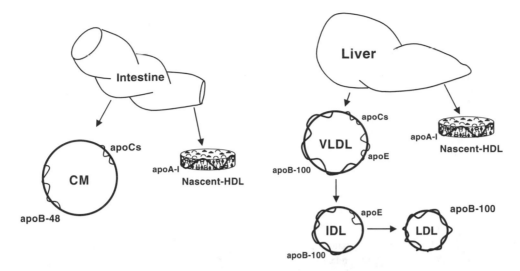

**FIGURE 17.2** Sites of synthesis of the major plasma lipoproteins. The small intestine secretes mainly triglyceride-rich particles in the form of large and small chylomicrons (CM). The intestinal cells also secrete nascent HDL, which are similar in their chemical characteristics to those secreted by the liver. Liver secretes VLDL which, following lipolysis, generates smaller, denser particles including intermediate-density lipoproteins (IDL) and low-density lipoproteins (LDL). High-density lipoproteins (HDL) are also secreted by the liver; however, as indicated on the schematic, the nascent or precursor particle is discoidal in shape rather than spherical. These nascent particles are subsequently transformed in the plasma into the mature, spherical forms by the enzyme LCAT (Figure 17.4).

## B. LIVER

The liver is the major organ regulating cholesterol homeostasis. Parenchymal liver cells synthesize and secrete VLDLs, which are large particles rich in triglycerides; the major protein on their surface is apoB-100. Additional proteins found on the surface of the VLDL include apoE and the apoCs. In the circulation, VLDLs rapidly lose core lipids, mainly triglyceride, through lipolysis and give rise to an intermediate-sized particle, the IDL (also referred to as the VLDL remnant). IDL is relatively rich in cholesteryl ester and possesses apoB-100 and apoE on its surface. Further lipolysis yields the cholesteryl ester rich LDL particle that possesses only apoB-100; obviously, the cascade of events demonstrates that VLDL are precursors of LDL (Figure 17.2).

The liver also synthesizes and secretes nascent or precursor HDL possessing apoA-I. Like those from the intestine, these nascent HDL are discoidal particles that require LCAT for maturation.

## VI. APOLIPOPROTEINS AS DETERMINATORS OF LIPOPROTEIN METABOLISM

Although the lipid moieties of lipoproteins are involved in processes of growth and survival as well as the development of atherosclerosis, it is the protein associated with the lipids that directs the metabolic fate of the lipoproteins. The origin and function of major apolipoproteins are summarized in Table 17.5.

## A. APOLIPOPROTEIN A-I

As previously mentioned, apoA-I is the major protein of HDL constituting approximately 75% of the total protein on mature HDL; the protein is synthesized in the liver and intestine. Epidemiological studies have abundantly shown an inverse relationship between plasma concentrations of apoA-I HDL and risk to atherosclerosis.[6] Transgenic mouse models have also been developed that experimentally demonstrate that apoA-I plays an important role in reducing atherosclerosis.[7] These studies were carried out with the atherosclerosis susceptible C57BL/6 mouse strain that develops aortic atherosclerotic lesions when the mice are maintained on a high-fat, high-cholesterol, atherogenic diet for 3 to 4 months. Transgenic mice expressing elevated levels of human apoA-I had little or no lesions compared to the nontransgenic mice on the same diet. Such studies are good evidence that apoA-I may have a direct role in the prevention of atherosclerosis.

### 1. Apolipoprotein A-I in Reverse Cholesterol Transport

One of the most widely recognized functions of apoA-I is its role in reverse cholesterol transport (RCT), schematically illustrated in Figure 17.3. Reverse cholesterol transport is the process whereby lipid-free apoA-I and specific subclasses of HDL mediate the removal of excess cholesterol from peripheral cells, including those of the artery wall, and transport this cholesterol to the liver for catabolism or to the adrenals for reutilization for steroid hormone synthesis.[8] The first step of RCT involves the

**Systemic and Organismic Aging**

**TABLE 17.5**

**Function and Origin of Major Apolipoproteins**

| Apolipoprotein | Function | Origin |
|---|---|---|
| ApoA-I | Activator of LCAT;[a] cholesterol efflux via ABCA1[b] transporter | Intestine, liver |
| ApoA-II | Modulates LCAT activity | Liver |
| ApoB-100 | Recognition of LDL receptor; triglyceride transport from liver cell | Liver |
| ApoB-48 | Triglyceride transport from intestinal cell | Intestine |
| Apo(a) | Inhibits fibrinolysis | Liver |
| ApoE | Recognition of LDL receptor | Liver, macrophage |
| ApoC-I | Activator of LCAT | Liver |
| ApoC-II | Activator of lipoprotein lipase | Liver |
| ApoC-III | Modulates apoE uptake; lipoprotein lipase inhibitor | Liver |

[a]LCAT = Lecithin:Cholesterol Acyltransferase.
[b]ABCA1 = ATP Binding Cassette protein 1.

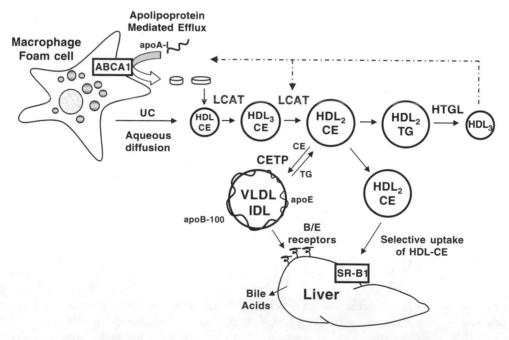

**FIGURE 17.3** Schematic of reverse cholesterol transport (RCT). The initial stages of RCT involve cholesterol efflux from extrahepatic cells such as macrophage foam cells in the artery wall. The process of cholesterol efflux can be broken down into an aqueous diffusion mechanism mediated by mature HDL. This reaction is largely dependent on the capacity of LCAT to esterify cholesterol on the surface of HDL, thus creating a concentration gradient that favors net movement of cholesterol out of the cell membrane and onto the surface of phospholipid-rich HDL. The second mechanism of cholesterol efflux involves the membrane transporter ABCA1 that facilitates the efflux of phospholipid and cholesterol to lipid-free apoA-I. The efflux of lipid to apoA-I results in the formation of nascent discoidal HDL which are substrates for LCAT, thus giving rise to mature spherical particles. As cholesteryl esters (CE) accumulate in HDL, the particles become larger and more bouyant. Cholesteryl esters of large HDL$_2$ can be transferred to apoB-containing VLDL in exchange for triglyceride (TG); this exchange is mediated by cholesteryl ester transfer protein (CETP). Triglycerides accumulating in HDL are hydrolyzed by hepatic triglyceride lipase (HTGL), thus reducing the size of HDL and regenerating HDL$_3$. ApoA-I is also released from the surface thus recycling apoA-I for apolipoprotein-mediated efflux. Alternatively, HDL cholesteryl esters can be directly transferred to liver and adrenal gland by a process called "selective uptake" of CE, which does not require whole particle internalization by cells. Selective uptake is mediated by the cell surface scavenger receptor, SR-B1.

removal of unesterified cholesterol and phospholipid from cells. Cholesterol assimilated into nascent HDL or preformed HDL is then esterified to cholesteryl ester by the enzyme lecithin:cholesterol acyltransferase (LCAT); apoA-I is a cofactor that activates the enzyme.

Reverse cholesterol transport can be initiated by at least two mechanisms.

### a. Aqueous Diffusion

The first mechanism involves the diffusion of membrane cholesterol from the cell surface to preformed $HDL_3$ by a nonspecific, energy-independent process called aqueous diffusion. This mechanism, shown in Figure 17.3, is facilitated by the LCAT reaction on HDL (discussed below) that requires apoA-I as a cofactor.

### b. Apolipoprotein-Mediated Efflux

The second mechanism involves a specific metabolic process whereby lipid-free apoA-I promotes phospholipid and cholesterol efflux from cells to form nascent HDL particles. This process is termed "apolipoprotein-mediated efflux" to distinguish it from the process of aqueous diffusion that requires preformed HDL. Unlike the aqueous diffusion mechanism, apolipoprotein-mediated efflux requires metabolic energy and the ATP binding cassette (ABCA1) transporter. The ABCA1 transporter was recently discovered following the analysis of cholesterol efflux from fibroblasts from Tangier patients. These patients have severe HDL and apoA-I deficiency. The patient's fibroblasts were shown to lack specific apolipoprotein-mediated efflux, but the capacity to release cellular cholesterol to preformed HDL was normal.[9] This observation prompted an intense search for the gene responsible for apolipoprotein A-I-mediated efflux.

### 2. Tangier's Disease and the ABCA1 Transporter

Tangier's is an autosomal recessive disease characterized by orange tonsils and an almost complete absence of apoA-I and HDL in the plasma of affected subjects. The orange tonsils are due to an accumulation of cholesteryl ester and associated carotenoids. Additional clinical features include hepatosplenomegaly, peripheral neuropathy, and premature cardiovascular disease.

The defective gene product responsible for Tangier's disease was shown to be the ABCA1 transporter, which in affected patients, exhibits numerous deletions, insertions, and substitutions.[10] The normal transporter consists of a single polypeptide chain containing two domains, each possessing six helical segments that form "pores" in membranes. Adjacent to each of the "pore" forming domains are two nucleotide binding domains responsible for the hydrolysis of ATP. The energy released from the hydrolysis of ATP facilitates the transport of phospholipid and cholesterol across the plasma membrane. Mutations in this ABCA1 transporter are associated with an inability

of lipid-free apoA-I to remove excess cholesterol from cells with the resulting accumulation of cholesterol in the cells. In Tangier patients, lipid-free apoA-I cannot be adequately lipidated, and the protein is then removed by the kidneys, leading to a severe deficiency of apoA-I and HDL in these patients. Current studies with the ABCA1 transporter indicate that familial hypoalphalipoproteinemia (FHA), a condition associated with low HDL and elevated triglyceride concentrations (both risk factors in atherosclerosis), is also associated with mutations in the ABCA1 transporter.[11] Age is a modifier of the phenotype, where younger FHA individuals (<30 years) have higher concentrations of HDL cholesterol and lower concentrations of triglyceride than FHA individuals >30 years.

### 3. ApoA-I and Lecithin: Cholesterol Acyltransferase

In addition to its role in removal of excess cholesterol from cells, a major function of apoA-I is the activation of the enzyme, lecithin:cholesterol acyltransferase (LCAT). This enzyme, which is synthesized and secreted by the liver, is essential for the normal maturation of nascent HDL into the mature plasma form. The action of LCAT on nascent discoidal HDL is shown in Figure 17.4. LCAT, which associates with the surface of the discoidal HDL particle, is activated by apoA-I. The enzyme removes the acyl chain from the sn-2 position of phospholipid and transfers it to the 3'-hydroxy group on unesterified cholesterol, thus forming the hydrophobic cholesteryl ester molecule. Lysophospholipid is also generated and is rapidly removed by association with albumin. Cholesteryl ester, because of its hydrophobicity, moves into the core of the HDL particle, where it coalesces into a lipid droplet, thus transforming the disk to a sphere. LCAT is a necessary enzyme for normal cholesterol homeostasis, because a deficiency of the enzyme, familial LCAT deficiency (FLD), is known to result in the abnormal accumulation of unesterified cholesterol in cell membranes. The latter alters vital function of cells and is also associated with an increase in premature atherosclerosis in some, but not all, patients.[12] FLD subjects are characterized by exceedingly low levels of HDL and apoA-I and by the presence of nascent HDL in their plasma.

### B. Apolipoprotein A-II

ApoA-II is the second most abundant apolipoprotein on HDL and is produced primarily by the liver. Its functional significance is not completely understood, although it is thought to inhibit LCAT and thereby modulate HDL metabolism by influencing the conversion of free cholesterol to cholesteryl ester. Transgenic mice overexpressing mouse apoA-II become more atherogenic than nontransgenic litter mates, leading investigators to suspect that

### Systemic and Organismic Aging

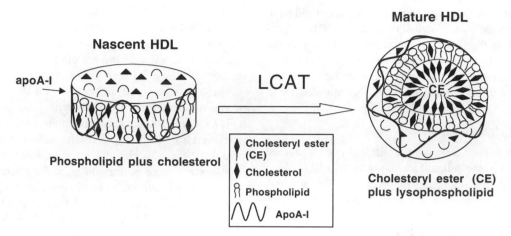

**FIGURE 17.4** The lecithin:cholesterol acyltranstransferase reaction and transformation of nascent discoidal HDL to mature plasma HDL. Nascent HDL particles are discoidal in shape because of the lack of a cholesteryl ester core. Cholesteryl esters are generated as a result of the LCAT reaction. This enzyme removes a fatty acyl chain from the sn2 position of phosphatidylcholine and transfers the fatty acid to the 3′ hydroxyl group of cholesterol. The cholesteryl esters thus formed are highly insoluble and form a lipid droplet between the phospholipid bilayer of discoidal HDL. As the cholesteryl ester droplet expands, HDL assumes its spherical shape. Cholesteryl esters are thus considered a storage and transport form of cholesterol. Lysophospholipid, the other product of the reaction, binds to albumin.

apoA-II may be pro-atherogenic.[13] Recent studies with transgenic mice overexpressing apoA-II suggest that elevated apoA-II increases leptin, thus contributing to obesity and insulin resistance, both known to be risk factors in atherosclerosis.[14]

## C. APOLIPOPROTEIN B-100

ApoB-100 is an important protein synthesized by the liver and associated with VLDL, IDL, and LDL.[15] It is the sole protein on LDL that possesses one molecule of apoB-100 per particle. This apolipoprotein has two important functions:

1.  It is necessary for the assembly and secretion of triglyceride-rich particles by the liver. In a rare genetic disease, aβlipoproteinemia, apoB-100 is not synthesized by the liver, and, as a consequence, this organ becomes fatty because of the intracellular accumulation of triglycerides. The lack of apoB-100 has serious metabolic implications because lipolysis of liver-derived VLDL normally produces LDL. The latter are important transporters of cholesterol required for normal growth and development as well as for steroid hormone production.
2.  It is a ligand for the LDL receptor. This receptor internalizes LDL, thus delivering cholesterol to cells for various cellular functions. An overabundance of LDL and apoB-100 can lead to saturation of the receptors, principally in the liver, with a consequent accumulation of excess cholesterol in the plasma and initiation of the

atherosclerotic process. The concentration of apoB-100 in plasma is a good indicator of atherosclerotic risk; elevated apoB levels correlate with elevated levels of circulating cholesterol.

## D. APOLIPOPROTEIN B-48

ApoB-48 is a truncated form of apoB-100 synthesized in the human intestine but not in the liver.[16] It is required for the assembly and secretion of chylomicrons that transport dietary lipids into the bloodstream. ApoB-48 is a product of the apoB-100 gene, but editing of the mRNA in intestinal cells led to a stop codon that signals premature termination of apoB translation, with the end result that the molecular weight of the apoB is only 48% that of the apoB-100 protein. Because apoB-48 is required for the transport of chylomicrons from the intestinal cell, in aβlipoproteinemia, where apoB is not secreted, the intestinal cells become lipid laden. Such individuals have steatorrhea (presence of excess lipids in stools) and diarrhea along with malnutrition.

ApoB-48 is not recognized by the LDL receptor. Following lipolysis, CM remnants are removed by the liver through the action of apoE (also found on CM), which recognizes the LDL receptor.

## E. APOLIPOPROTEIN (a)

In the 1960s Berg discovered a novel protein associated with LDL in some individuals.[17] This new lipoprotein antigen was named apolipoprotein (a), and the lipoprotein is known as lipoprotein (a) or Lp (a). Apo(a) is a glycoprotein of variable size with molecular weights ranging

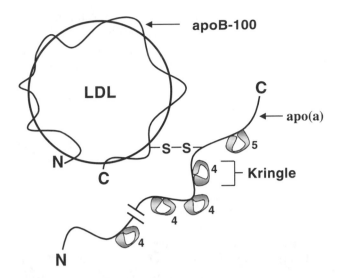

**FIGURE 17.5** Stucture of Lp(a). Apolipoprotein(a) is a glycoprotein of variable size secreted by the liver, which forms a covalently linked disulfide bridge with apoB on the surface of LDL particles. The variability in size of apo(a), 300,000 to 800,000 Daltons, is due to the variation in the number of "kringel" repeats, which are protein subunits of plasminogen. Smaller apo(a) molecules are associated with increased Lp(a) concentrations in plasma and, hence, increased risk for atherosclerosis.

from 300,000 to 800,000 Daltons and is synthesized in the liver. Apo(a) is covalently bound to apoB-100 through a disulfide bridge located in the carboxyl terminal region of apoB-100, as illustrated in Figure 17.5.[18] Lipoprotein (a) is a unique subset of LDL and its lipid composition is like that of LDL. The apo(a) protein has high homology with plasminogen, which hydrolyzes fibrin and aids in the dissolution of clots. Plasminogen possesses five pretzel-shaped protein units called "kringels." Apo(a) contains kringel 5 of plasminogen and variable numbers of the kringel 4 unit as indicated in Figure 17.5. The number of kringel 4 units in the apo(a) structure is responsible for the variation in molecular weight of the protein; it is now accepted that there is an inverse relationship between apo(a) size and Lp(a) concentration. Lp(a) is an independent risk factor for atherosclerosis, and its plasma concentrations are genetically controlled. Individuals with Lp(a) levels greater than 30 mg/dL are at increased risk for heart disease. Lp(a), however, is not correlated with other known risk factors such as concentrations of LDL cholesterol, HDL cholesterol, apoA-I, and apoB. The physiological role of Lp(a) in coronary artery disease is not completely understood but appears to be involved in atherogenesis.[19] The mechanisms whereby Lp(a) contributes to atherosclerosis may involve the following: (1) inhibition of fibrinolysis because the molecule can interfere with plasminogen function and (2) the binding of Lp(a) particles to the extracellular matrix in the subendothelial space of the artery wall, where they are subsequently oxidized. The oxidized Lp(a) particles can be taken up by the scavenger receptor on macrophages, thus contributing to foam cell formation in fatty streaks.

## F. Apolipoprotein E

The liver is the major site of synthesis of apoE, although macrophages, adrenal gland, and brain also synthesize this protein. ApoE is another apolipoprotein recognized by the LDL receptor (also known as the apoB-E receptor). As a ligand for the LDL receptor, apoE has an important role in targeting CM and VLDL cholesteryl ester-rich remnants to the liver for catabolism. A family has been described with apoE deficiency; a genetic defect in which apoE is not synthesized. Homozygous family members have premature coronary artery disease and elevated plasma cholesterol levels, thus suggesting that apoE plays a significant role in cholesterol metabolism.

Some apoE is synthesized by macrophages. During secretion of apoE, excess cell cholesterol is transported from these cells along with the protein, thus reducing the accumulation of intracellular cholesterol and the potential development of foam cells. ApoE, therefore, has an important function in reverse cholesterol transport.

There are several genetic variants of apoE. The normal apoE molecule possesses a single cysteine at amino acid residue 112 and an arginine at residue 158 identifying it as the E3 isoform. The cysteine residue is in the lipid binding region of the molecule, while the arginine is in the LDL receptor binding domain. A point mutation at residue 158 can occur wherein the arginine is replaced by cysteine (E2 isoform). This mutation is associated with an inability of the apoE-containing lipoproteins to recognize the LDL receptor. This results in reduced clearance of cholesteryl ester-rich remnant lipoproteins and the elevation of plasma cholesterol concentrations and is a risk factor for atherosclerosis. Another mutation in apoE exists wherein the cysteine at residue 112 is replaced by an arginine. This isoform, known as apoE4, is associated with risk for familial late-onset Alzheimer's disease (AD). As the number of apoE4 alleles increases, there is an increase in risk for AD and also a decrease in average age of onset.[20]

## G. Apolipoprotein Cs

The apoC proteins are synthesized mainly in the liver and function principally as cofactors in enzyme reactions that have a critical role in lipid metabolism and cholesterol homeostasis (Table 17.5). ApoC-I functions as an activator of LCAT and, in this respect, is functionally similar to apoA-I. It is likely that this protein may have other, as yet unidentified, function(s).

ApoC-II is a cofactor in lipoprotein lipase activation; the protein is essential for the hydrolysis of the triglyceride core in CM and VLDL with the resultant formation of remnant particles (Figure 17.6). A deficiency of plasma

**Systemic and Organismic Aging**

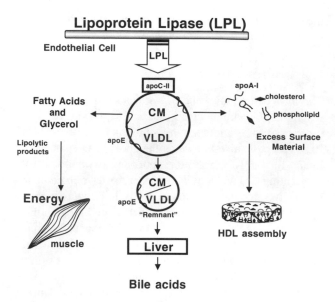

**FIGURE 17.6** Schematic diagram of the physiological function of lipoprotein lipase (LPL). This enzyme is key in clearing TG-rich lipoproteins from plasma. LPL is found on the lumenal surfaces of the capillary endothelium and requires apolipoprotein (apo) C-II for activation. Triglyceride-rich particles including chylomicrons (CM) and very-low density lipoproteins (VLDL) carry apoC-II and activate the enzyme; in so doing, the triglyceride (TG) core is hydrolyzed, and the constitutive parts of TG, free fatty acids and glycerol, are liberated and used to produce energy. Removal of core TG creates remnant particles that are enriched in cholesteryl ester. Most remnants are removed by the liver. Excess surface components (apolipoproteins, phospholipid, and cholesterol) generated by shrinking of the core are used to form HDL.

apoC-II leads to chylomicronemia and a severe elevation of plasma triglyceride concentrations.

ApoC-III has a functional role in two areas of lipoprotein metabolism: (1) it acts as an inhibitor of lipoprotein lipase activity and, hence, modulates triglyceride hydrolysis and (2) it modulates the cellular uptake of apoE-containing lipoprotein particles. An increase in apoC-III can result in the elevation of plasma lipoprotein remnants, thus elevating plasma cholesterol.

## VII. LIPOLYTIC ENZYMES

### A. LIPOPROTEIN LIPASE (LPL)

This enzyme is important for the metabolism of triglyceride-rich lipoproteins. LPL is synthesized by adipocytes, heart, and kidney and migrates from the sites of synthesis to the capillary endothelium, where it is bound to the cell surface. LPL is responsible for catabolism of the large, triglyceride-containing lipoproteins, principally CM and VLDL. It catalyzes the hydrolysis of triglyceride to free fatty acids and glycerol and requires the presence of apoC-II, the obligatory enzyme activator. The action of

LPL in the degradation of CM and VLDL is illustrated in Figure 17.6. VLDL and CM interact with LPL at the endothelial cell surface where hydrolysis occurs. The triglyceride core of these large particles is removed, thus generating fatty acids, excess surface material, and lipoprotein remnants. The fatty acids are utilized by cells for energy; excess fatty acids are taken up by adipocytes and stored as triglycerides. Chylomicron and VLDL remnants generated during lipolysis are rich in cholesteryl ester and contain apoE on their surfaces. Remnants are rapidly cleared from the plasma through the LDL receptors of the liver; this receptor also recognizes apoE (see section VII.A). The removal of apoE-containing remnants is vital for normal cholesterol metabolism.

Excess surface components are produced when the triglyceride core in CM and VLDL is removed. These components, especially phospholipids and free cholesterol, are incorporated into HDL as shown in Figure 17.6.

### B. HEPATIC TRIGLCYERIDE LIPASE (HTGL)

This enzyme, as suggested by its name, is secreted by the liver and is associated with the surface of hepatocytes. The major substrate for hepatic triglyceride lipase (HTGL) is the VLDL remnant that carries a small amount of triglyceride. Lipolysis of triglyceride in remnants results in the formation of LDL. HTGL also has the ability to hydrolyze triglycerides that accumulate in HDL through the action of cholesteryl ester transfer protein (CETP). CETP transfers excess cholesteryl ester from cholesteryl ester-rich $HDL_2$ to apoB-containing lipoproteins in exchange for triglyceride. This exchange is important in reverse cholesterol transport as shown in Figure 17.3. HTGL hydrolyzes the triglyceride in the triglyceride-enriched $HDL_2$, and in so doing, regenerates smaller $HDL_3$ particles. The latter HDL have a crucial role in reverse cholesterol transport, as indicated in Figure 17.3, by acting as acceptors of cholesterol from cells.

## VIII. RECEPTORS IMPORTANT IN CHOLESTEROL METABOLISM

### A. THE LDL RECEPTOR

LDL are the major transporters of cholesterol in the circulation; cholesterol in the form of cholesteryl ester is delivered to cells by internalization of LDL by a specific receptor mechanism, the LDL receptor. This receptor is also referred to as the apoB-E receptor because the receptor recognizes apoE. Although almost all cells possess LDL receptors, the liver possesses the largest number of receptors. Functionally, this receptor is extremely important in the regulation of intracellular cholesterol synthesis and flux, and in addition, it regulates further synthesis of the receptor. Based on the work of Brown and Goldstein,

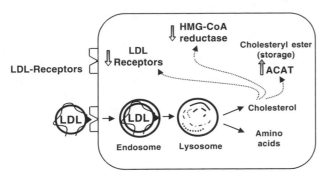

**FIGURE 17.7** Schematic outline of the function of the LDL receptor (also called the apoB/E receptor). This receptor plays an important role in the catabolism of cholesterol-containing lipoproteins. Lipoproteins carrying apolipoproteins (apo) B or E are recognized by the receptor and are bound. The bound particles are internalized in a membrane-bound sac, the endosome. Hydrolytic enzymes invade the vesicle which becomes a lysosome, and the proteins and lipids are broken down to amino acids and unesterified cholesterol. In the cytosol, cholesterol is esterified to cholesteryl esters by the cellular enzyme, acyl coenzymeA:cholesterol acyltransferase (ACAT). Cholesterol coming into the cell downregulates, or decreases, the activity of the cell's own cholesterol-making machinery. The rate-limiting step in the *de novo* synthesis of cholesterol is 3-hydroxy-3-methylglutaryl coenzyme A (HMG Co A) reductase; therefore, this enzyme is the one regulated by receptor-mediated cholesterol accumulation. In addition to a decrease in HMG CoA reductase, accumulation of cellular cholesterol also decreases the number of LDL receptors on the cell surface. Overall, the receptor-mediated process is a finely tuned system in the regulation of cholesterol metabolism.

the LDL receptor function is outlined in Figure 17.7.[21] The liver possesses high-affinity receptors that recognize apoB-100-containing LDL (or apoE-containing remnants). The LDL (or remnant) binds to the receptor and is internalized into a structure referred to as the endosome. The endosome delivers its contents to the lysosome, wherein hydrolysis and degradation of the internalized lipoprotein takes place. The LDL is degraded into its molecular constituents, amino acids, and cholesterol.

In the cell, catabolism of LDL results in the activation of several processes. Accumulation of unesterified cholesterol upregulates the enzyme, acyl-coenzyme A:cholesterol acyltransferase (ACAT), which in turn re-esterifies cholesterol. The accumulation of cellular cholesterol then downregulates the enzyme, 3-hydroxy-3-methylglutaryl coenzyme A (HMG CoA) reductase, which is the rate limiting step in cellular cholesterol synthesis. In other words, the cell slows its machinery for synthesizing cholesterol when adequate amounts are being delivered to it by the lipoproteins. In addition to downregulating the HMG CoA-reductase, degraded LDL regulates synthesis of the LDL receptor and, thus, the cell reduces its uptake of cholesterol. The liver is finely tuned for the maintenance of cholesterol homeostasis.

Receptor-mediated uptake and degradation of apoB- and apoE-containing lipoproteins is important in maintaining normal plasma cholesterol levels. Elevated levels of cholesterol-containing particles such as VLDL remnants and LDL that result from the overproduction of cholesterol by the liver could result in saturation of the LDL receptor and accumulation of excess cholesterol in the plasma and the increased risk to cardiovascular disease. The risk to cardiovascular disease is also enormously increased when the LDL receptor is defective or deficient. This happens in certain genetic defects, where either the receptor protein is not synthesized or the synthesized protein is defective. Patients with such receptor defects are hypercholesterolemic, and their condition is called familial hypercholesterolemia. In the homozygous state, patients can have staggeringly high plasma cholesterol levels, as high as 1000 mg/dL as compared to the normal 150 to 200 mg/dL. The patients have precocious atherosclerosis and, unless managed extremely carefully, will not survive the second decade of life. The disease state is a clear case of underutilization of LDL. Heterozygotes for the disease have decreased numbers of functional receptors, and therefore, have elevated plasma cholesterol and premature coronary artery disease.

## B. THE MACROPHAGE SCAVENGER RECEPTOR AND THE PROCESS OF LESION FORMATION

This receptor plays a key role in the process of atherogenesis in the artery wall and is found on the surface of macrophages in the subendothelial space. The scavenger receptor recognizes oxidatively modified LDLs. Modification of LDL is an important contributor to the atherogenic process; modification can happen through several mechanisms. In the case of hypercholesteremia, such as that resulting from defective LDL receptors, the residence time of LDL in plasma is increased, thus increasing the likelihood of protein alterations by oxidation events. Glycation of apoB protein in diabetes also leads to modification of the LDL particles; it is well known that patients with diabetes mellitus have an increased risk to atherosclerosis. LDL entering the subendothelial space can be retained and bound to the extracellular matrix of the artery wall, where they then undergo oxidation. The oxidatively modified LDLs are recognized and internalized by scavenger receptors on macrophages. As indicated in Figure 17.8, the current theory on the development of the atherosclerotic lesion implicates oxidized LDL as a key player in the process.[22] During the early events of lesion development, monocytes enter the arterial intima in regions of endothelial damage. Cytokines, specialized cell signals, trigger the differentiation of monocytes into macrophages and also stimulate smooth muscle cell proliferation. Macrophages scavenge the oxidatively modified LDL in the artery wall via the scavenger receptor. Uptake

**Systemic and Organismic Aging**

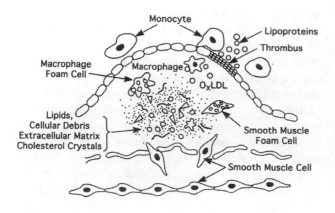

**FIGURE 17.8** Schematic of the role of macrophages and lipoproteins in foam cell and atheroma formation. Early steps in atherosclerotic lesion formation involve the appearance of foam cells in the subendothelial space. Current research suggests that oxidatively modified LDL (OxLDL) plays an important role in the production of foam cells. OxLDL are taken up by the scavenger receptor of macrophages in the artery wall; however, the internalized OxLDL cholesterol is not reutilized, and it accumulates within the cell in the form of lipid droplets. Macrophages with large quantities of accumulated lipid droplets are foam cells. The macrophage foam cells in the artery wall ultimately die and release their residue of cholesterol, which accumulates in the lesion. The macrophages also release factors that stimulate smooth muscle cells to proliferate and accumulate lipids, thus further augmenting the atherosclerotic process.

of the modified LDL is probably a protective function; however, if the insult continues, more macrophages amass at the injury site, and modified LDL accumulate intracellularly in great abundance. The macrophage, however, cannot degrade the extra burden of cholesterol that then accumulates within the cell in the form of lipid droplets, thus transforming the macrophage into a foam cell. These cells eventually die, and the lipids, cell debris, and cholesterol crystals are released into the extracellular space and form the nucleus for more complicated lesions.

As noted in a previous section, HDL and apoA-I are important in reversing the early steps of cholesterol accumulation in macrophages. When reverse cholesterol transport works efficiently, foam cell development is minimized.

## C. THE SR-B1 RECEPTOR

Cholesteryl esters accumulating in $HDL_2$ can be delivered to the liver (and also adrenal cortical cells) by a process of "selective uptake." Selective uptake involves the selective transfer of HDL cholesteryl esters to hepatocytes (or adrenal cells) in the absence of whole particle uptake (i.e., the protein component of HDL). This process is distinct from that of receptor-mediated endocytosis and involves a membrane "docking" protein for HDL particles. This

"docking" protein is the SR-B1 receptor that belongs to a family of Class B scavenger receptors. Cholesteryl esters entering the liver via the SR-B1 receptor are hydrolyzed to form unesterified cholesterol that is converted into bile acids for excretion from the body. Cholesteryl esters delivered to the adrenal gland via the "selective-uptake" mechanism are stored or converted into steroid hormones.

## IX. HYPERLIPOPROTEINEMIA

Lipid metabolism is normally tightly regulated with triglyceride and cholesterol fluxing through the enterohepatic circulation as illustrated in Figure 17.9. Clinical concern arises when concentrations of lipoproteins such as VLDL, IDL, and LDL are abnormally elevated (a condition termed hyperlipoproteinemia), because such elevations can accelerate the development of atherosclerosis. Not only may atheromas form, but deposition of cholesterol may also occur in tendons and skin, producing raised nodules called xanthomas; these are overt manifestations of severe hypercholesterolemia.

Hyperlipoproteinemias are designated primary or secondary (Table 17.6). Primary hyperlipoproteinemias are due to a single gene defect or to a combination of genetic factors; in addition, genetic factors may be exacerbated by environmental or dietary factors. The incidence of the different primary hyperlipoproteinemia ranges from 1 in 250 in the population to 1 in 1 million. Secondary hyperlipoproteinemias are complications of more generalized metabolic disturbances such as diabetes mellitus, hypothyroidism, excessive intake of alcohol, or chronic kidney failure. Knowledge of the plasma concentrations of cholesterol and triglycerides usually reveals the class of lipoproteins that is high, and this is useful in making a diagnosis and in designing proper drug and diet therapy.

Treatment protocols for hyperlipoproteinemia are summarized in Table 17.7. Basic to the treatment of all hyperlipidemias is a diet that maintains normal body weight and that minimizes the plasma cholesterol concentration. If a patient is overweight, weight loss should be attempted and then maintained by a diet low in cholesterol and saturated animal fats and relatively high in mono- and polyunsaturated vegetable oils.

Patients with secondary hyperlipidemia require treatment of the underlying disorder (diabetes, hypothyroidism, excess alcohol consumption, etc.) and should reduce all other risk factors, such as smoking and hypertension, while maintaining physical fitness.

Primary hyperlipoproteinemia requires more aggressive treatment. In addition to a proper diet, drugs that lower plasma lipoprotein concentrations are used.[23] These drugs function by diminishing the production of lipoproteins or by increasing the efficiency of their removal.

The most widely used drugs in treating hypercholesteremia are the statins. This class of drugs including lov-

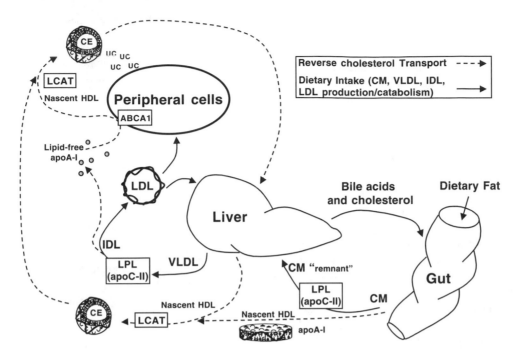

**FIGURE 17.9** Schematic diagram summarizing whole body lipoprotein metabolism and cholesterol transport. Intestinal epithelial cells absorb dietary fat and package the lipid in the form of chylomicrons (CM), which are secreted into the blood. Lipoprotein lipase activated by apoC-II hydrolyzes the core triglycerides within the CM, generating remnants that are taken up by the liver. The liver repackages the lipids in the form of VLDL particles secreted into the bloodstream. Lipoprotein lipase mediates the hydrolysis of VLDL triglycerides to generate IDL and LDL particles as part of a lipolytic cascade. Low-density lipoproteins (LDL) transport cholesterol to peripheral cells and the liver. The cholesterol content in peripheral cells is tightly regulated at least, in part, as a result of cellular cholesterol efflux mediated by lipid-free apoA-I and plasma HDL particles. Nascent HDL generated by the intestine and liver are converted to mature HDL by the action of LCAT. Nascent HDL generated upon the interaction of apoA-I with the ABCA1 transported are also substrates for LCAT giving rise to mature HDL. Mature HDL, with the help of LCAT, can promote the efflux of cholesterol from peripheral cells via the aqueous diffusion mechanism. HDL transports cholesterol in the form of cholesteryl esters back to the liver for production of bile acids that are secreted into the intestine for excretion.

### TABLE 17.6
### Diseases Related to Lipoprotein Abnormalities

| Disorder | Clinical Findings |
|---|---|
| **Primary Hyperlipoproteinemias** | |
| Single gene | Atheromas |
| | Pancreatitis |
| | Xanthomas |
| Multiple genes | Atheromas |
| **Secondary Hyperlipoproteinemias** | |
| Diabetes mellitus | Atheromas |
| | Pancreatitis |
| | Xanthomas |
| Hypothyroidism | Atheromas |
| Estrogen excess (oral contraceptive) | Pancreatitis |
| | Xanthomas |

### TABLE 17.7
### Treatment of Hyperlipoproteinemias

Diet Therapy: low cholesterol, low animal fat, relatively high polyunsaturated fats

| Drug Therapy: | |
|---|---|
| **Class** | **Metabolic effect** |
| HMG CoA-reductase inhibitors (lovastatin, pravastatin, simvastatin, fluvastatin, atorvastatin) | Inhibits cholesterol biosynthesis; enhances LDL clearance; lowers plasma cholesterol |
| Bile acid sequestrants (cholestyramine, colestipol) | Binds and removes bile acids in intestine; increases cholesterol conversion to bile; increases LDL clearance |
| Fibric acid derivatives (clofibrate, gemfibrozil) | Reduces synthesis and increases catabolism of VLDL (increases LPL activity) |
| Nicotinic acid | Reduces synthesis of VLDL; lowers LDL; elevates HDL by reducing its clearance |

**Systemic and Organismic Aging**

astatin, parvastatin, simvastatin, atorvastatin, and fluvastatin, inhibits HMG CoA-reductase and with it cholesterol synthesis. To compensate for a reduction in cholesterol synthesis in the liver, the number of hepatic receptors for LDL increases, and this brings about a reduction in plasma LDL by increasing clearance of circulating LDL.

Nicotinic acid (niacin) has long been used to reduce the production of VLDL and, in so doing, it also lowers LDL. It also increases HDL concentrations by decreasing HDL clearance from the plasma. Nicotinic acid can produce cutaneous flushing and pruritus (itching) involving the face and upper body, but this appears to subside with continued use. It may, however, interfere with compliance, that is, the patient's willingness to continue with the drug.

Fibric acid derivatives such as clofibrate and gemfibrozil are effective in decreasing the synthesis of VLDL. This drug also increases VLDL catabolism by increasing the activity of LPL.

Normally, cholesterol returned to the liver is converted to bile acids that are delivered to the small intestine lumen where some of the cholesterol in the form of bile acid is ultimately excreted and some is reabsorbed. Normally, formation of bile acid has a negative-feedback effect on further production of bile acids. Removing the bile acids so that they no longer exert negative feedback speeds up the conversion of cholesterol to bile acids and reduces body pools of cholesterol, including cholesterol sequestered in xanthomas. Resins, such as cholestyramine and colestipol, have such an effect. They readily bind bile acids in the intestinal lumen and increase the flux of cholesterol from the liver to bile.

Both nicotinic acid and HMG CoA-reductase inhibitors may be used in combination with one of the bile acid-binding resins. These combinations are usually synergistic, allowing the doses of both substances to be lowered. It is thus apparent that drugs utilized in treating dysfunctions of lipid metabolism must be selected and tailored according to the individual condition.

## ACKNOWLEDGEMENTS

Supported by NIH Grants, HL18574 and HL59483.

## REFERENCES

1. Ross, R., Atherosclerosis — an inflammatory disease, *N. Engl. J. Med.*, 340, 115, 1999.
2. Stampfer, M.J. et al., A prospective study of cholesterol, apolipoproteins and the risk of myocardial infarction, *N. Engl. J. Med.*, 325, 373, 1991.
3. Breslow, J.L., Genetics of lipoprotein disorders, *Circulation*, 87 (suppl. III), III-16, 1993.
4. Austin, M.A. et al., Atherogenic lipoprotein phenotype: A proposed genetic marker for coronary heart disease risk, *Circulation*, 82, 495, 1990.
5. Krauss, R.M., Triglycerides and atherogenic lipoproteins: rationale for lipid management, *Am. J. Med.*, 105, 58S, 1998.
6. Buring, J.E. et al., Decreased $HDL_2$ and $HDL_3$ cholesterol, apoA-I and apoA-II, and increased risk of myocardial infarction, *Circulation*, 83, 22, 1992.
7. Rubin, E.M. et al., Inhibition of early atherogenesis in transgenic mice by human apolipoprotein A-I, *Nature*, 353, 265, 1991.
8. Fielding, C.J. and Fielding, P.E., Molecular physiology of reverse cholesterol transport, *J. Lipid Res.*, 36, 211, 1995.
9. Oram, J.F. and Vaughan, A.M., ABCA1-mediated transport of cellular cholesterol and phospholipids to HDL apolipoproteins, *Curr. Opin. Lipidology*, 11, 253, 2000.
10. Bodzioch, M. et al., The gene encoding ATP-binding cassette transporter 1 is mutated in Tangier disease, *Nature Genetics*, 22, 347, 1999.
11. Clee, S.M. et al., Age and residual cholesterol efflux affect HDL cholesterol levels and coronary artery disease in ABCA1 heterozygotes, *J. Clin. Invest.*, 106, 1263, 2000.
12. Glomset, J.A. et al., Lecithin:cholesterol acyltransferase deficiency and fish-eye disease, in *The Metabolic Basis of Inherited Diseases,* Scriver, C.R. et al., Eds., McGraw-Hill, New York, 1995.
13. Warden, C.H. et al., Atherosclerosis in transgenic mice overexpressing apolipoprotein A-II, *Science*, 261, 469, 1993.
14. Castellani, L.W., Goto, A.M., and Lusis, A.J., Studies with apolipoprotein A-II transgenic mice indicate a role of HDLs in adiposity and insulin resistance, *Diabetes*, 50, 643, 2001.
15. Hevonoja, T. et al., Structure of low density lipoprotein (LDL) particles: Basis for understanding molecular changes in modified LDL, *Biochim. Biophys. Acta*, 1488, 189, 2000.
16. Young, S.G., Recent progress in understanding apolipoprotein B, *Circulation*, 82, 1574, 1990.
17. Berg, K., A new serum type system in man: The LP system, *Acta Pathol. Microbiol. Scand.*, 59, 369, 1963.
18. Scanu, A.M., Nakajima, K., and Edelstein, C., Apolipoprotein(a): Structure and biology, *Frontiers in Bioscience*, 6, D546, 2001.
19. Zampoulakis, J.D. et al., Lipoprotein (a) is related to the extent of lesions in the coronary vasculature and to unstable coronary syndromes, *Clin. Cardiol.*, 23, 895, 2000.
20. Corder, E.J. et al., Gene dose of apolipoprotein E type 4 allele and the risk of Alzheimer's disease in late onset families, *Science*, 261, 921, 1993.
21. Brown, M.S. and Goldstein, J.S., How LDL receptors influence cholesterol and atherosclerosis, *Sci. Amer.*, 251, 58, 1984.
22. Steinberg, D., Low density lipoprotein oxidation and its pathobiological significance, *J. Biol. Chem.*, 272, 20963, 1997.
23. Levy, R.L., Troendle, A.J., and Fattu, J.M., A quarter century of drug treatment of dyslipoproteinemia with a focus on the new HMG CoA reductase inhibitor, fluvastatin, *Circulation*, 87(Suppl. III), III-45, 1993.

# 18 The Pulmonary Respiration, Hematopoiesis and Erythrocytes*

*Massimo De Martinis*
University of L'Aquila

*Paola S. Timiras*
University of California, Berkeley

## CONTENTS

## I. THE PULMONARY RESPIRATORY SYSTEM

*PAOLA S. TIMIRAS*

Respiratory function includes *an external process* — oxygen ($O_2$) absorption from atmospheric air in the lungs, and carbon dioxide ($CO_2$) removal from tissues and organs — and an *internal metabolic process* — gaseous exchanges at the cellular level. Traditionally, respiration was assessed essentially by physiologic tests. In the last decade, advances in cellular and molecular biology have facilitated diagnostic evaluation of normal function, have provided a better understanding of disease processes, and have suggested new and more effective approaches to therapy.[1,2] This chapter is concerned with the functions of *external respiration*. In Part I, changes that may occur with old age in the structure and function of the lungs are discussed as a consequence of environmental insults and as contribu-

---

* Illustrations by Dr. S. Oklund.

tory factors in alterations of $O_2$ uptake and $CO_2$ excretion as well as causes of some of the respiratory diseases prevalent in the elderly. In Part II, the aging of the hematopoietic system focuses on the study of the erythrocytes (red blood cells) and their function in carrying $O_2$ and $CO_2$ to and from the lungs and tissues. Human erythrocytes have a relatively short life span of about 120 days and have been extensively used as a model for the study of cellular aging.

## A. THE LUNG: A "BATTERED" ORGAN FROM WITHIN AND WITHOUT

The major function of the lungs is to ensure the efficient exchange of air (oxygen) from the environment to the blood and from the blood to the cells of the body (Box 18.1). Yet, this very function is performed at the peril of contamination and damage from the many toxic substances transported in the air. The degree and the rate of age-related changes in structure and function of the lungs are variable and depend on the habits of the individual (particularly nutrition, physical exercise, and smoking), on the individual's environment (urban versus rural), and

---

**Box 18.1**

**Structure and Functions of the Pulmonary Respiratory System**

In the respiratory system:

The gas-exchange organ is comprised of the two lungs

The pump that ventilates the lungs is comprised of:
  The chest wall
    The respiratory muscles (that increase or decrease the size of the thoracic cavity)
    The brain centers and nerve tracts (that control the muscles)

In addition to regulating gaseous exchange, the lungs perform other functions. They participate in:

Immunologic defenses of the body (by phagocytizing particles from the inspired air and from the blood)
Metabolic functions (by synthesizing, storing, or releasing into the blood such substances as surfactant and prostaglandins)
Endocrine functions (by transforming angiotensin I into angiotensin II, a powerful vasoconstrictor and stimulus for aldosterone secretion, Chapter 10)
The actions of a few biologically active peptides, some with pressor activity (e.g., VIP, vasoactive intestinal peptide) and some with neuronal (e.g., opioid peptides) activity

---

**TABLE 18.1**

**Signs of Impaired Pulmonary Respiration with Aging**

- Reduced maximum breathing capacity
- Progressive reduction in arterial partial pressure of oxygen ($PO_2$) and in $PO_2$ alveolar to arterial differences due to premature airway closure
- Loss of elastic recoil (i.e., springing back of elastic fibers after stretching)
- Weakening of respiratory muscles
- Decreased elasticity of thorax cage and chest wall
- Increased rigidity of internal lung structure
- Less efficient emptying of the lungs
- Earlier and easier fatigability

---

the concomitant occurrence of diseases (infections, industrial diseases).[3,4] The respiratory system in humans is mature, that is, reaches optimal adult function, by age 20.[5] Pulmonary function begins to gradually decline in healthy subjects after the age of 25.[6] This decline is linked to progressive deleterious changes that occur with aging in respiratory structures including the lung, the thoracic cage, and respiratory muscles (Table 18.1), as well as in the respiratory centers in the central nervous system (CNS). These changes, however, are minor compared to the constant effects of the environment and other insults to the respiratory system — infections, pollution, cigarette smoking, disordered immune responses, unfavorable working conditions — to which the organism is exposed throughout the life span.[7–11]

The lungs are not only "battered" by external insults but also by formation of oxygen radicals, so deleterious to tissues, in the pulmonary cells in immediate contact with the gaseous environment rich in $O_2$. Toxic effects of $O_2$ in the lungs and the causative role of cigarette smoking are discussed in Chapter 5.

### OXYGEN TOXICITY IN LUNGS

In the lungs, accumulation of free radicals may contribute to the acute oxygen toxicity that occurs when individuals breathe higher than normal concentrations of oxygen. This acute toxicity is of particular significance for critically ill patients who require respirators (i.e., devices worn over the mouth or nose protecting the respiratory tract or allowing for the administration of $O_2$). Although the lowest possible doses of oxygen are given, the potential damage to the lung must always be considered. Sometimes it may be necessary to choose between "allowing a patient to die immediately or giving pure oxygen which may kill in days."[12] Experiments in rats show that breathing oxygen increases the production of free radicals in pulmonary epithelial cells and macrophages and causes the consequent death of the animals within 3 days. If animals are protected by the administration of the enzymes, superoxide dismutase and catalase, death of the animal is prevented (Chapter 5).

The chronic damage to lung tissue by oxygen results from the inflammatory reaction that is induced by cell damage and death. The responses to the phagocytic cells that infiltrate the inflamed tissue liberate free radicals, induce enzymes such as elastase which breaks down the elastic tissue of the lungs, and reduce elastase inhibitors (see below).

In the absence of disease, none of these functional decrements, singly or in combination, is sufficient to severely incapacitate the old individual. The majority of the elderly are capable of maintaining their lifestyle and a satisfactory respiratory function under resting (steady state) conditions. Some of the impairments become manifest when ambient conditions worsen and may lead to increased pathology and mortality.[13]

Given continuing exposure to external and internal insults, respiratory diseases are more prevalent in older individuals than in the general population. Among these diseases, incidence and severity of infections, chronic obstructive disease, and cancer increase with aging.

## B. AGING-ASSOCIATED CHANGES IN THE LUNG

Air passes in the following manner:

1. It passes first through the nasal passages (nares), where it is filtered of the larger contaminants.
2. It enters the pharynx, where it is warmed and absorbs water vapor.
3. It flows down the trachea and through the bronchi and bronchioles.
4. It proceeds through the respiratory bronchioles and the aveolar ducts to the alveoli. *The alveoli are the functional units of the lungs.*

### ALVEOLAR STRUCTURE AND FUNCTION

The major functional asset of the alveolus is its structure, which provides close proximity of the capillary blood and the alveolar air. Air in the alveolus and blood in the pulmonary capillaries are separated only by the capillary endothelium and the thin basement membrane supporting the alveolar cells. This arrangement facilitates gas exchanges between blood and air.

Two types of epithelial cells (pneumocytes) line the alveolus. Type I cells are extremely thin with few intracellular organelles and are designated agranular pneumocytes. Type II cells contain many organelles and lipid droplets and are designated granular pneumocytes. These type II cells produce surfactant, a proteolipid that coats the alveolar cells and lowers the surface tension at the air–fluid interface. Lower surface tension (a) keeps the alveoli from collapsing and (b) reduces muscular work required to ventilate the lungs. Another and less frequent cell type in the lung is the alveolar macrophage. These cells are loaded with digestive enzymes and can ingest foreign materials, as do white blood cells.

They migrate from the bloodstream and patrol the tissues of the lung on the alveolar side, gliding on the surfactant.

## 1. Structural Changes

The architecture of the lung is altered in aging. Structural changes are associated with impairment of function. The lungs become more voluminous, the alveolar ducts and respiratory bronchioles are enlarged, while the alveoli become shallower and more flat with loss of septal tissue (dividing walls between alveoli). These changes do not appreciably affect total lung capacity (the maximum volume of air in the lungs and airways) when this volume is corrected for the age-related decrease in height: the decrease between the ages of 20 to 60 years is less than 10% in both men and women. However, air distribution is altered with an *increase in alveolar duct air* but *a decrease in alveolar air,* alveolar surface area, which is 75 m² at 30 years decreasing by 4% per decade thereafter. Given the fact that $O_2$ transport is most efficient in the alveoli (and much less efficient in the alveolar ducts), a decrease in alveolar air space will impair optimal $O_2$ diffusion from alveolar air into pulmonary capillaries.

The amount of *elastic tissue*, abundant in the lung and partly responsible for the stretchability of this organ, *is decreased* with age, while fibrous tissue is increased. The importance of lung elasticity is illustrated by the condition of emphysema (see below), in which loss of lung elasticity, due to disruption of elastic tissue, is associated with impaired ventilation (see below). The nature of the exact changes in the elastic fibers during aging is unclear, but it appears that alterations in the distribution of elastic tissue are functionally as significant as changes in amounts. With aging, abnormal location or structure of the fibers may contribute to impairments of ventilation and perfusion of the lungs.

The dome-shaped *diaphragm* that separates the thoracic from the abdominal cavity and constitutes the floor of the former, is the major muscle involved in pulmonary respiration. Contraction of the diaphragm, by lowering its central portion (the diaphragm is anchored around the perimeter of the lower thorax) accounts for 75% of the increase in thoracic volume during quiet inspiration. During inspiration, contraction of the external intercostal muscles further increases volume by elevating and pulling outward the anterior ends of the ribs. Increased thoracic volume and consequent decreased pressure allow for expansion of the lungs and $O_2$ to flow into the alveoli. Other auxiliary muscles from the abdomen and the shoulders are involved in inspiratory processes. Normal expiration is then mostly a passive process attributable to recoil of the elastic tissue of the stretched lungs and thorax. Also involved are the internal intercostal muscles, which, when they contract, lower the ribs and move them inwards, thereby further decreasing thoracic volume. Some abdominal and shoulder muscles

**Systemic and Organismic Aging**

also participate in expiration. During increasing exercise, the abdominal and rib-cage muscles assume a larger role in augmenting ventilation rates. Because of the increased stiffness of the rib cage with aging, the diaphragm takes over a higher proportion of the mechanical effort needed for increasing ventilation.[8,14]

The structure, biochemical properties, and contractile function of respiratory muscles, like those of all skeletal muscles, change in response to variations (a) in the pattern of use (sedentary versus physical exercise habits), (b) in the nutritional state, and (c) during growth and development, in response to the influence of hormones (growth hormone, insulin-like growth factor I, thyroid and gonadal hormones). The diaphragm is a muscle not easily fatigued; further investigation is needed to discover if, during increasing ventilatory activity, it becomes fatigued in old age to the same degree as the other (abdominal and intercostal) auxiliary respiratory muscles.

Exertional dyspnea (i.e., shortness of breath after exercise) is common among the elderly.[9] Yet the contributing roles of decrements in muscle strength, increase in thoracic stiffness, and loss of lung compliance remain to be elucidated.[8,15,16] Inasmuch as fatigue after muscular exercise has a strong central nervous system (CNS) component, factors other than muscular decrements such as impaired coordination, insufficient motivation, arthritic involvement of the joints, and others may also play an important role in limiting the adaptive competence of the elderly in undertaking physical exercise (Chapters 21 and 24).

With aging, it is well accepted that most muscles of the body undergo a certain degree of *sarcopenia* (loss of tissue mass), which may be due, in part, to progressive disuse with advancing old age (Chapters 21 and 24). With respect to the respiratory muscles, it has been suggested that inasmuch as they remain *continuously active* throughout life, they may be spared aging-associated sarcopenia. However, that does not seem to be the case.[8,17,18] The ability of the lungs to shift from resting to maximal function is impaired in some aged individuals, and this impairment depends on the decline in the strength and endurance of the respiratory muscles (Chapter 24). Observed changes in respiratory muscles in older humans and animal models are summarized in Table 18.2. The configuration and mechanical properties of the *chest wall* also change with aging; they follow the changes in bone structure and function with old age as discussed in Chapter 21. Major aging-associated signs are increased curvature of the spine and calcification of the intercostal cartilage.

## 2. Changes in Lung Volumes

Lung volumes and pressures change dramatically from birth to death, with major and rapid changes during childhood and adolescence and slower but progressive

---

**TABLE 18.2**
**Changes with Aging in Respiratory Muscles**

- Muscle strength is decreased
- Muscles of older individuals are more prone than those of adults to fatigue when the work of breathing is increased (as during physical exercise)
- Atrophy of some respiratory muscles (primarily, Type I muscle fibers of slow, red muscles as in long muscles of back and shoulders)
- Ratio of glycolytic (anaerobic) to oxidative (aerobic) metabolism is increased
- Blood supply to muscle is decreased

---

alterations with increasing age. These changes involve the lungs and chest wall and are often divergent (Table 18.3).[6] The chest becomes stiffer because of calcification of rib cartilage, while the lungs become more distended due to a slightly increased compliance (i.e., stretchability) and decreased recoil (i.e., ability of elastic fibers to spring back after stretching). Thus, lung volumes and ventilation rate are decreased at rest and especially, during exercise. Lung volumes and measurements are summarized in Box 18.2.

Vital capacity, in some elderly in the 7th decade of life, may decrease to approximately 75% of its value at age 17 years (taken as men and women average of 4.8 L). During this time, residual volume increases nearly 50%, but *total lung capacity* remains unchanged (Figure 18.1). Other measures of ventilatory mechanics decline as well, including forced expiratory volume/second (32 mL/year decrease in males and 25 mL/year decrease in females starting from the age of 25 years), and airway conductance which also progressively slows with aging.[18] Both of these changes underline the greater difficulty in old age to empty the lung adequately with each expiration, and the consequent increased residual volume and impaired ability of the lungs to promote adequate air diffusion to the alveoli and, therefore, less blood oxygenation.

Expiratory flow rates begin to decline at an age when vital capacity is still intact,[19] perhaps reflecting a reduced elastic recoil in the elderly which, during expiration, may cause a premature closure of some regions of the lungs.[20] Air is trapped in sites distal to the closure, thereby resulting in inadequate blood oxygenation.[21] In contrast, $PCO_2$ remains remarkably constant throughout life.

## C. Control of Ventilation

Control of ventilation by *brain centers in the medulla and pons* and by the peripheral *carotid and aortic body chemoreceptors* is markedly altered in the elderly.[10] It is unclear whether altered ventilation may be due to one or several of the following, and more research is needed in this regard:[20]

**TABLE 18.3**
**Morphologic Changes in the Thorax and Lung with Age**

| Morphological Change | Functional Significance |
|---|---|
| **Thorax** | |
| Calcification of bronchial and costal cartilage | ↑ in resistance to deformation of chest wall |
| ↑ in costovertebral stiffness | ↑ in use of diaphragm in ventilation |
| ↑ in rigidity of chest wall | ↓ in tidal volume |
| ↑ in anterior-posterior diameter | ↓ in response to exercise hyperapnea |
| Wasting of respiratory muscles | ↓ in maximal voluntary ventilation |
| **Lung** | |
| Enlarged alveolar ducts | ↓ in surface area for gas exchange |
| ↓ in supporting duct framework | ↓ in stretchability |
| Alveoli shallow, flatter | ↑ in physiologic dead space (40%) |
| Thinning, separation of alveolar membrane | ↓ in lung elastic recoil |
| ↑ in mucous gland | Vital Capacity ↓ 15–20% |
| ↓ in number, thickness of elastic fibers | RV/TLC ↓ 35–40% |
| ↓ in tissue extensibility (alveolar wall) | ↓ in ventilatory flow rate |
| | ↓ in ventilation distribution |
| ↓ in pulmonary capillary network | ↑ in resistance to flow in small airways |
| ↑ in fibrosis of pulmonary capillary intima | ↓ in ventilation |

*Note:* RV = residual volume; TLC = total lung capacity.

*Source:* Modified with permission from Reddan, W.G., *Exercise and Aging*, Smith, E.L., and Serface, R.C., Eds., Enslow Publishing, Hillside, New Jersey, 1981.

- Intrinsic alteration of neural control such as decreased sensory perception of $PCO_2$, pH, and $PO_2$
- Loss of synchrony of higher CNS inputs
- Alterations of mechanical factors such as stiffness of chest wall
- Reduced neuromuscular competence and responsiveness to neural inputs

Responses to hypercapnia (increased $PCO_2$) and hypoxia (reduced $PO_2$) are reduced by 50% in some aged as compared to young individuals.[21]

**RESPONSES TO EXERCISE**

Physical exercise stimulates the active tissue to utilize more $O_2$ and to eliminate more $CO_2$ through coordinated cardio-

---

**Box 18.2**
**Lung Volumes and Measurements**

Total lung capacity (TLC) of 6 L in men and 4.2 L in women is comprised of the following values (in men):

- **Tidal volume:** amount of air that moves into and out of the lungs with each quiet inspiration and expiration (= 0.5 L)
- **Inspiratory reserve volume** (= 3.3 L) and **expiratory reserve volume** (= 1 L): amount of air that moves into or out of the lungs following maximal inspiration or expiration
- **Residual volume:** air left in the lungs after maximal expiration (= 1.2 L)
- **Dead space:** air in the airways (= 150 mL)
- **Vital capacity:** the greatest amount of air that can be expired after a maximal inspiration. Vital capacity is frequently taken as an index of pulmonary function and ranges from 3 to 4 L in adult females to 4.5 to 5.5 L in adult males. A more precise index is gained by measuring vital capacity per unit of time; for example, in asthma, vital capacity appears normal but when timed, it shows significantly prolonged time because of bronchial constriction
- **Pulmonary ventilation rate** (or respiratory minute volume): is normally 6 L/min
- **Maximal voluntary ventilation:** is the largest volume of air that can be moved into and out of the lungs in 1 min by voluntary effort (it may be as high as 125 to 180 L/min)

**Systemic and Organismic Aging**

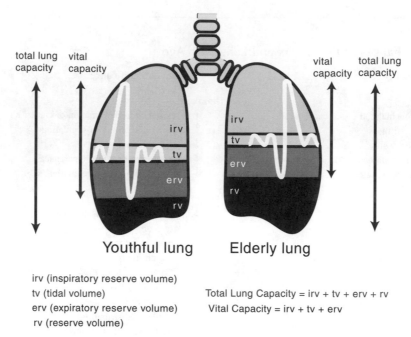

irv (inspiratory reserve volume)
tv (tidal volume)
erv (expiratory reserve volume)
rv (reserve volume)

Total Lung Capacity = irv + tv + erv + rv
Vital Capacity = irv + tv + erv

**FIGURE 18.1** Exchanges in lung volumes with aging. With aging, note particularly the decrease in VC (vital capacity) and the increase in RV (residual volume).

vascular and respiratory adjustments. Circulatory changes increase blood flow to the exercising muscle (Chapter 21). Respiratory adjustments include:

- Increased ventilation, to provide more $O_2$, eliminate more $CO_2$, and dissipate heat
- Increased extraction of $O_2$ from the blood by the exercising muscle
- Increased blood flow and shift in the $O_2$ dissociation curve

With aging, lung ventilation, if already impaired under quiet conditions, is further altered during exercise:

- Ventilation becomes inadequate to muster the necessary adjustments to meet the increased demands of exercise
- Loss of elastic recoil and decreased functional residual capacity (i.e., gas volume remaining in lungs at the end of quiet expiration) inhibit the effective range of tidal volume
- Early closure of the airways similarly inhibits the expiratory flow
- Dyspnea (shortness of breath) ensues and necessitates the early cessation of exercise

The reduction of vital capacity with aging restricts the potential tidal volume that may be reached during maximal exercise. At rest, only a minor fraction of the potential lung volume and flow changes is used, and even during maximum exercise, the maximal inspiratory and expiratory flow rates and volumes are not usually reached. With increasing age, the ability to reach maximal rates and volumes is severely curtailed during moderate to heavy exercise. Therefore, ventilation, in many elderly, cannot increase sufficiently to provide for the increased metabolic demands, keeping in mind significant variations in response to gender and ethnicity.[22,23]

In addition to difficulty in adjusting ventilatory responses to exercise, the elderly have an earlier onset of the shift from aerobic (requiring $O_2$) to anaerobic (independent of $O_2$) metabolism. Both the time required to reach steady state level at the onset of exercise and the time required to return to pre-exercise resting levels is prolonged with age after moderate to heavy exercise. Alveolar-capillary gas exchange is reduced, and alveolar-arterial $PO_2$ differences are increased with exercise in middle-aged men. Nevertheless, at 65 years and older, gas diffusion capacity (that is, the movement of $O_2$ "downhill" from the air through the alveoli and blood into the tissues) is comparable to that at younger ages.[24]

Despite an overall decline in performance during exercise, most elderly can and should undertake a regimen of physical exercise adequate to their capabilities and needs. Such regimens are briefly surveyed in Chapter 24. Other factors influencing performance of physical exercise are discussed in Chapter 21 in relation to skeletal and muscular changes with aging. Age-related respiratory changes during sleep are discussed in Chapter 8.

## D. SOME RESPIRATORY DISEASES RELATED TO AGING OF THE LUNGS

Respiratory diseases are the cause of potentially preventable morbidity and mortality in elderly people. As mentioned earlier, no internal organ other than the lung is so directly exposed to external environmental influences. Thus, decline in physiological competence with aging and the environmental toll exacted by a number of conditions and agents (e.g., smoking-related airway obstruction, asthma, pulmonary tuberculosis) combine to induce multiple pathologies.[10,25,26] One of the most frequent respiratory disorders of the elderly is *chronic obstructive pulmonary disease*, which refers to a spectrum of chronic respiratory diseases; other diseases include lung neoplasia and lung infections, particularly pneumonia (Chapter 15). While some disorders, e.g., bronchitis and emphysema, are not necessarily life-threatening, they represent a considerable "burden" for the well being of the elderly in terms of days of hospitalization, of physician consultation, days of sickness, and almost continuous discomfort. From the perspective of the entire life span, the proportion of deaths due to respiratory diseases (in Western countries) is highest (approximately 30%) in the first year of life, falling to lower values (approximately 5%) in late adolescent and early adulthood. From the fifth decade on, the incidence of respiratory disease rises steadily, and in those over 85 years of age, accounts for 25% of all deaths. In the last 20 years, the death rate from respiratory disease has fallen, except for bronchiogenic carcinoma (still rising, especially in young women) and pneumonia (remaining constantly high). Within the confines of the present chapter, chronic obstructive pulmonary will be the representative disease briefly reviewed.

## E. CHRONIC OBSTRUCTIVE PULMONARY DISEASE (COPD)

This disease refers to a *group of disorders characterized by airflow limitation and persistently impaired gas exchange*, often associated with an inflammatory condition of the lungs.[10,25] In the United States, COPD is ranked among the leading causes of death, with mortality rates on the rise, especially in women. It is often the result of cigarette smoking. It is predicted that by the year 2020, worldwide, mortality from COPD will rise from its current sixth most common cause of death to the third.[26] Reasons for this dramatic increase in COPD include reduced mortality from other causes (e.g., cardiovascular diseases in industrialized countries and infectious diseases in developing countries), and the marked increase in cigarette smoking and air pollution in the developing countries.

COPD is comprised of at least three distinct pathologic processes that may occur separately or concurrently. They are as follows:

- Chronic bronchitis (inflammation of the bronchi), accompanied by hypersecretion of mucus and cough
- Emphysema, characterized by enlargement of air spaces, destruction of lung parenchyma, loss of lung elasticity, and closure of small airways
- Chronic asthma (constriction of the bronchi)

### 1. Genetics and Environmental Risk Factors in COPD

Although cigarette smoking is a major risk factor in the etiology of COPD, the disease occurs only in a small percentage (15 to 20%) of smokers. This suggests that genetic factors may determine which smokers will develop the disease.[27] Other indications for genetic involvement include:

1. The familial clustering of early onset COPD[28]
2. Ethnic differences,[29] for example, the higher than normal risk of COPD in a Taiwanese population associated with increased production of tumor necrosis factor $\alpha$(TNF$\alpha$), but not in a British population with the same increased TNF$\alpha$[30,31]
3. Lower levels of the proteinase inhibitor $\alpha_1$-antitrypsin, leading to early development of emphysema[32] (An hereditary association has been noted among individuals affected by an $\alpha_1$-antitrypsin deficiency; in the homozygous individual, emphysema develops early in life (about 20 years of age) even in the absence of cigarette smoking
4. A polymorph variant of microsomal epoxide hydrolase, an enzyme involved in the metabolism of epoxides (chromosome breaking agents) that may be generated in tobacco smoking, has been associated with a significant increase in COPD incidence[33]

In addition to cigarette smoking, other environmental factors represent major risks for COPD (Table 18.4). In industrialized countries, environmental pollutants — particulates associated with cooking, sulfur dioxide, cadmium, passive smoking — are some among many environmental conditions that can predispose or aggravate COPD. Chronic infections of the respiratory tract and the inflammatory reactions they induce, significantly contribute to the incidence and severity of COPD (Table 18.4).

### 2. COPD and Its Pathophysiology

Causes of COPD range from an inherent defect of elastic tissue to association with fibrotic pulmonary diseases such

---

**TABLE 18.4**
**Major Risk Factors for Chronic Obstructive Pulmonary Disease (COPD)**

- Cigarette smoking
- Air pollution
- Genetic factors
- Bronchial inflammation
- Chronic respiratory tract infections
- Old age
- Family history of COPD
- Male sex

---

**Systemic and Organismic Aging**

## TABLE 18.5
## Major Signs of Chronic Obstructive Pulmonary Disease (COPD)

| Structural | Pathophysiologic |
|---|---|
| • Diffuse distention and overaeration of alveoli<br>• Disruption of interalveolar septa<br>• Loss of pulmonary elasticity<br>• Restructuring of alveoli into large air sacs resulting in poor, uneven alveolar ventilation and inadequate perfusion of under-ventilated alveoli<br>• Increased lung volume<br>• Barrel-shaped chest as the chest wall expands (increased lung volume and increased use of accessory shoulder and abdominal muscles) | • Disturbed ventilation<br>• Altered airflow and blood flow<br>• Frequently partial obstruction of bronchi (hence, the often used name of obstructive disease)<br>• Inspiration and expiration are labored (wheezing), and more work is required for breathing<br>• Resulting hypoxia (low $O_2$ levels) and hypercapnia (high $CO_2$ levels)<br>• Chronic productive cough with mucus<br>• Minor respiratory infections of no consequence to young individuals with normal lungs are fatal or near fatal for the elderly |

as silicosis, to consequence of chronic diffused bronchitis due to age and aggravated by cigarette smoking.

Its major structural and pathophysiologic signs are summarized in Table 18.5.

Chronic inflammation of the airways has a critical role in producing symptoms of emphysema and asthma, hence, the use in the treatment of COPD of anti-inflammatory treatments such as inhaled corticosteroid hormones (one of the commonly used treatments of asthma).[34–38] Smoking increases the number of pulmonary alveolar macrophages, which release a chemical substance that attracts leukocytes to the lungs. Leukocytes, in turn, release proteases such as elastase, which attacks the elastic tissue in the lungs. Normally, the plasma protein $\alpha_1$-antitrypsin inactivates elastase and other proteases. In emphysema, the activity of this enzyme is decreased. This inactivation of the enzyme may be promoted by oxygen radicals that are released by the leukocytes (Chapters 5 and 15). Thus, with smoking, there may be both an increased production of elastase and a decreased activity of the inactivating enzyme with the resulting destruction of elastic fibers. The inflammatory changes and protease imbalance that occur in COPD either as a consequence of infections or of irritants such as tobaccco smoke, would accelerate the functional decline in pulmonary respiratory function that occurs in old age.[25] Latent viral infections have also been suggested as possible causative [39] or contributing factors, perhaps by amplifying the inflammatory responses.[40]

Among the major metabolic disorders associated with emphysema, we must note hypercapnia, which induces acidosis, at first compensated by urinary retention of bicarbonate. When this compensatory mechanism fails, especially in aged individuals in whom renal competence may be diminished (Chapter 19), the ensuing *respiratory acidosis* represents a medical emergency and must be treated accordingly. A consequence of hypoxia is stimulation of the production of red blood cells which are increased in number (polycythemia). This contributes to *hypertension*, which causes the right side of the heart to enlarge and

then leads to cardiac failure (the so-called "cor pulmonale" or congestive right heart failure).[23]

### 3.   COPD Management and Treatment

Given the burden inflicted by the disease and its frequency, treatment and prevention are still inadequate. Although cigarette smoking is the major cause of COPD, quitting smoking does not appear to resolve the inflammatory response in the lung airways, and the first symptoms of the disease may become manifest several years after smoking was stopped.[41] Treatment still remains for the most part symptomatic, despite significant advances.[25,42,43] Major goals of management and available therapeutic strategies are listed in Table 18.6.

## TABLE 18.6
## Treatment of Chronic Obstructive Pulmonary Disease (COPD)

Goals of Management:
• To retard the progression of the disease
• To educate patients
• To improve airflow
• To optimize functional capabilities

Therapeutic Strategies:
1. Administration of pharmacological agents such as:
   • Bronchodilators (to relieve bronchial spasm)
   • Mucus liquefiers (to thin the mucous secretions)
   • Anti-inflammatory agents such as steroids (see above)
   • Protease inhibitors, inhibitors of receptors or enzymes involved in immune responses
   • Antibiotics (to control potential infections)
2. Administration of $O_2$ to be used cautiously to prevent acidosis
3. Optimization of function by:
   • Promoting physical exercise to strengthen abdominal muscles and diaphragm to aid in lung ventilation
   • Meeting social, emotional, and vocational needs
   • Using respiratory aids in the form of aerosols, sprays, etc.

## F. PNEUMONIA

Pneumonia is an inflammatory process of the lung parenchyma most commonly caused by infection. The infectious agent (often the *Bacillus pneumococcus*) is frequently present among the normal flora of the respiratory tract. The development of pneumonia must, therefore, be usually attributed to an impairment of natural resistance. Indeed, the decline of immune competence with aging (Chapter 15) may explain why pneumonia remains, despite the availability of antibiotics, a serious, life-threatening problem for the elderly.

In most cases of community-based epidemics of pneumonia, the elderly are more susceptible than the young with respect to severity and complications (e.g., lung abscess, bacteremia) as well as mortality (as high as 80% in those aged 60 and older.)[44] The rise in pneumonia after the seventh decade registered in several Western countries including the United States, may simply reflect the type of death certification, pneumonia being the terminal expression of other diseases. However, studies of hospitalized elderly suggest a true increase.[45,46] While the true cause may be failure of immunologic competence, the immediate cause has been ascribed to aspiration in the lungs of oropharyngeal flora during sleep or aspiration of food in the trachea during hand-feeding (as in severe dementia).[47] These are common occurrences in the elderly, in whom the mucobronchial defense barrier is impaired due to deficient ciliary activity, mucus production, and mechanical reflexes.

### PRESENTATION AND MANAGEMENT

The diagnosis of pneumonia in the elderly is more difficult than in the young because of the atypical presentation. The classic features of chest pain, cough, and purulent or blood-stained sputum (spit) are uncommon. There is a lack of cough, toxic confusion predominates, and dehydration occurs early. Progression of the disease may produce further lung damage (e.g., abscess) or aggravate extrapulmonary manifestations such as the confusional state or induce additional damage such as pericarditis (inflammation of the pericardium, the sac surrounding the heart), ischemic heart disease, and meningitis (infection of the membranes surrounding the brain). Treatment involves the use of antimicrobial agents, primarily antibiotics, the correction of dehydration, and ultimately the treatment of the underlying disease. Prevention includes the use of anti-influenza vaccination and perhaps pneumococcal vaccination (still under investigation). Rehabilitation after recovery is important physiologically and psychologically and requires the use of an individually tailored program, based on exercise, oxygen therapy, and coordinated support of family and health providers.

## G. TUBERCULOSIS REVISITED

Tuberculosis, induced by *Mycobacterium tuberculosis*, was considered by the end of 19th century an infectious disease definitively eradicated by chemotherapy and improved socioeconomic and hygienic conditions. However, it remains, even today, a health problem, worldwide, primarily among individuals with acquired immunodeficiency syndrome (AIDS) but also among the elderly.[48,49] The disease is chronic in nature, and the agent that causes it may remain dormant for many years but may be reactivated as immune defenses are reduced in old age (Chapter 15). The disease may be more widespread and severe in the elderly due to unfavorable conditions such as malnutrition, alcoholism, and superimposed diseases. Tuberculosis is generally viewed as one of the most easily treatable serious infectious diseases likely to occur in adults, including old adults. Its current reemergence has been ascribed, in part, to an increase in drug-resistant bacterial strains.[50,51]

In the late 1980s, several outbreaks of tuberculosis occurred in nursing homes, where reactivation and primary contact infection may occur. Elderly persons in nursing homes are at greater risk for tuberculosis than those living in the community, because crowded living conditions facilitate transmission of the disease. It is important that all nursing home residents be tested for tuberculosis upon admission to the facility and that vigorous preventive measures be taken to stop the spread of the infection. Even in the elderly living in the community, the geriatrician must be alert to the possibility of the infection, often masked by atypic (simulating cold, influenza, pneumonia) symptoms.

## II. HEMATOPOIESIS AND ERYTHROCYTES

*MASSIMO DE MARTINIS*

The maintenance of a functioning blood-forming system is a requirement for survival. A common view of hematopoiesis holds that a pluripotent hematopoietic stem cell (PHSC) gives origin to committed progenitors of myeloid, erythroid, and megakaryocytic lineages, and these, in turn, give origin to the recognizable hematopoietic precursors of the bone marrow. A complex web of growth factors — colony-stimulating factors, stem-cell factors, some interleukins, erythropoietin, thrombopoietin — as well as specific conditions of the hematopoietic microenvironment — act in concert to achieve a normal blood picture.[52] The present section is concerned with red blood cells (RBC, erythrocytes) (Box 18.3). White blood cells are discussed with the immune system in Chapter 15, and platelets, with the cardiovascular system in Chapter 16.

Hematopoiesis is the study of the origin and development of blood cells, all derived from a pluripotent stem cell, but with distinct trajectories in their developmental potential, depending on the different tissues of origin. Hematopoiesis develops early in fetal life, and the primary source of blood cells changes several times during development, from the yolk sac to the liver and spleen, prenatally, and to the bone marrow, postnatally.

**Systemic and Organismic Aging**

---

**Box 18.3**

**Function and Origin of Red Blood Cells (RBC)**

The bone marrow located in the cavities of the long bones remains a principal center of activity throughout life; it contains the stem cells from which at least eight lineages of mature blood cells, including RBCs, are ultimately derived through a series of steps involving differentiation and proliferation (Table 18.7).

RBCs are biconcave disks. In mammals, they lose their nuclei before entering the circulation. Their function as carriers of oxygen depends on their content in hemoglobin, an iron-containing protein that binds strongly to oxygen and transports it from lungs to cells, where oxygen is needed for metabolic processes. Hemoglobin also plays a minor role in the transport of carbon dioxide from the tissues to the lungs. The average RBC counts (in humans) are 5.4 million/$\mu$L in men and 4.5 million/$\mu$L in women. Human RBCs survive in the circulation for an average of 120 days. When RBCs are no longer flexible enough to squeeze between the endothelial cells that line the blood sinuses of the spleen, they are removed from the circulation, die, and are phagocytized in the spleen.

**Erythropoiesis**, that is, RBCs' formation, is subject to feedback control. It is inhibited by a rise (above normal values) in RBC's number, and it is stimulated by their reduction below normal values, as in anemia, or by a decrease in blood oxygen content, as in hypoxia. Control of erythropoiesis is regulated by several growth factors among which the circulating protein hormone, erythropoietin, secreted from the kidney (85%) and liver (15%). Erythropoietin production is stimulated primarily by hypoxia, and to a lesser extent by androgens and by cobalt salts (cobalt is part of vitamin $B_{12}$, necessary for normal blood cell formation). Red blood cells represent a cell population with unique characteristics, some of them listed in Table 18.8.

Reference will be made here, whenever possible, to studies of hematopoiesis in healthy centenarians, taken as examples of successful aging. Comparison of function at progressive ages may improve our understanding of developmental and aging process in these cells and extend this knowledge to other cell types.

---

**TABLE 18.7**

**Lineages of Mature Blood Cells Derived from Bone Marrow Stem Cells**

- White blood cells
  - Granulocytes
    - Neutrophils
    - Eosinophils
    - Basophils
  - Lymphocytes
    - B cells
    - T cells
- Monocytes
- Erythrocytes (Red blood cells, RBCs)
- Platelets

**TABLE 18.8**

**Some Unique Characteristics of Red Blood Cells**

Red blood cells represent a cell population that:
- Is easily accessible
- Is in continuing renewal
- Has a well-defined life span
- Has become a popular model for the study of cell function at all ages, including old age

**TABLE 18.9**

**Major Functions of Blood Cells**

- Efficient oxygen delivery to tissues and cells (erythrocytes, RBC)
- Hemostasis: prevention of blood loss (e.g., through blood clotting, platelets)
- Immune response, primarily white blood cells (Chapter 15)
- Responsiveness to environmental stimuli (e.g., increase in cellular response to hypoxia)
- Specificity of responses to demands (only relevant lineage is stimulated without expansion of irrelevant ones; e.g., hypoxia selectively stimulates erythroid bone marrow and subsequent erythropoiesis)

The hematopoietic system, responsible for efficient oxygen delivery, hemostasis (i.e., stoppage of bleeding through blood clotting and vascular contraction), and all phases of the inflammatory response, has an astonishing capacity to respond to environmental stimuli (Table 18.9). The responsiveness of each hematopoietic lineage results from coordinate increases in the production and functional activity of appropriate hematopoietic cell types, without expansion of irrelevant ones. An hypoxic stimulus, for instance, will develop a specific expansion of the erythroid bone marrow and a subsequent increase in red blood cells number (erythrocytosis), as occurs in populations living at high altitudes. However, the bone marrow, in response to hypoxia, will not increase production of neutrophils, monocytes, eosinophils, mast cells, or T or B lymphocytes (Table 18.9).

## A. The Stem Cell Pool in Aging

During a normal life span, humans will have produced, cumulatively, red and white blood cells with a mass many times larger than their bodies, but in such a manner that, at any given time, blood counts are maintained within narrow limits. In fact, the short life span of many of the mature blood cell types, as compared to the life span of whole organisms, requires rapid, continuous, and precise recruitment of new cells to maintain each population at its appropriate size.[53] It is now well established that a replenishment of relatively short-lived blood cells is maintained by a small population of primitive, self-renewing, *stem cells*. The question of whether hematopoietic stem cells are altered in aging has been the subject of considerable controversy. Earlier studies in old individuals focused on changes in the capacity of these cells to provide hematological rescue from lethal doses of total body irradiation as well as on their potential for sequential replications.[54]

There are considerable differences in hematopoietic activity in young children and adults on the one hand, and in adults and the elderly on the other. Studies on the frequency of proliferative capacity of a colony-forming unit of granulocytes and macrophages (CFU-GM) in bone marrow and blood from donors of different ages indicate that a functional decline in hematopoietic progenitor (stem cell precursor) cells begins at birth and continues throughout life.[55] The decrease with old age in the number of peripheral blood-CD34+ cells and of peripheral blood CFU-GM suggests a decrease of pluripotent blood stem cells with aging.[56] This is consistent with the reduction in the number of myeloid progenitor cells in the bone marrow of 88-year-old apparently healthy individuals, compared to young controls.[57]

The bone marrow in newborns is very active and fills most of the cavities of the skeleton, whereas in adults, it is mainly found in the pelvis and sternum. In older people, the number and distribution of blood cells in bone marrow decline together with the marrow reserve. Evidence that stem and progenitor cell function may decline with age is further supported by observations (summarized in Table 18.10).

---

**TABLE 18.10**
**Evidence for the Decline in Function of Progenitor Cells with Old Age**

- Decline begins at birth and continues throughout life
- Bone marrow from old donors is more likely to fail when transplanted than bone marrow from young donors
- Ability of old people to respond to hematologic stress is decreased
- Recovery after inhibition of proliferation of bone marrow cells (myelosuppression) is slower in older persons
- Reduction of telomere length

---

## B. Hematopoietic Potential and Remodeling

Whereas hematopoietic potential is maintained in aging, under basal (steady-state) conditions, the capacity of stem cells for self-renewal and recovery after hematologic stress appears to decline gradually with advancing age. Therefore, the developmental and aging potential of stem cells under basal conditions must be differentiated from that under stress conditions.[54]

Basal hematologic parameters show little change with aging, although some aging-associated changes have been reported in a few subpopulations of blood cell groups. Coversely, under conditions of severe hematopoietic demands and in response to stress, the reserve capacity of the bone marrow for recovery may become significantly curtailed in old age.

### 1. Changes with Old Age under Steady-State Conditions

The ability of stem cells derived from old donors to reestablish erythropoiesis as measured in competitive repopulation assays of lethally irradiated young mice, is equally, or even more active, than that from young donors. However, in long-term studies, stem cell function is less efficient in old than young animals. Quantitative analysis of the data suggest that the hematopoietic compartment of aging mice contains a higher proportion of precursor, also called "progenitor," cells, that is, cells preceding the stage of stem cells. Proliferation of precursor cells in old people may represent an attempt to reestablish normal cell number, although these precursor cells are less efficient in doing so than the succeeding stem cells.

Despite the general assumption that important functions may deteriorate with old age, careful examination of the healthy elderly suggests that many physiologic parameters may be preserved throughout life. In particular, most hematological parameters, such as absolute number and percentage of peripheral blood cells (e.g., erythrocytes, with normal hemoglobin levels, neutrophils, monocytes, eosinophils, basophils, and platelets), are well preserved in healthy elderly, including centenarians, and remain quite similar to those of young normal subjects. Thus, the mechanisms responsible for steady-state hematopoietic functions appear to be well conserved in the later decades of human life.

Nevertheless, some changes occur in peripheral blood. For example, the absolute number of the CD34+ hemopoietic progenitor cells decreases with age, as does their capability to form erythroid, granulocyte-macrophage GM), and mixed colonies in semisolid cultures. Such a decrease suggests that hemopoiesis undergoes a complex remodeling with age, although it is a relatively early age phenomenon and does not extend to very old age.

**Systemic and Organismic Aging**

Hematopoiesis is finely controlled by a complex *cytokine network*, cytokines, being hormone-like chemical messengers involved in immune cells communication (Chapter 15). Changes in cytokine production and response occur with aging, as illustrated by the decrease in the production of granulocyte-macrophage colony-stimulating factor (GM-CSF) and of interleukin-3 (IL-3) by phytohemagglutinin (PHA)-stimulated peripheral blood mononuclear cells (PBMC), but an increase in serum levels of stem cell factor (SCF) (Table 18.11). Yet, despite significant changes in the *in vitro* production of hematopoietic cytokines (IL-3 and GM-CSF) and in serum level of hematopoietic growth factors (SCF), stem cells from old subjects remain responsive to hematopoietic cytokines *in vitro*, and are able to form different types of colonies in a way indistinguishable from that of young subjects.

It is possible that the overall mechanisms responsible for the complex remodeling of hematopoiesis depend on modifications of the network of hematopoietic cytokines, rather than on unresponsiveness of hematopoietic progenitors to these growth factors. These considerations are supported by previous studies showing that the *in vitro* production, as well as the plasma levels of cytokines involved in hematopoiesis, such as interleukin-6 (IL-6), increase with age in healthy donors and centenarians. Indeed, the increase of SCF, as well as that of IL-6, could be interpreted as a compensatory mechanism with which to maintain the CD34+ cell pool and to stimulate erythroid cell differentiation. In healthy old people and centenarians, the contrast between the proliferative responsiveness to PHA of peripheral blood mononuclear cells, and the decrease of IL-3 production by PHA-stimulated PBMC, suggest that the production of several hematopoietic growth factors is heterogeneously affected by aging. However, this suggestion may not be applicable to the *in vivo* situation, inasmuch as GM-CSF is also produced by other cell types, such as endothelial cells and fibroblasts. Keeping in mind eventual *in vivo* and *in vitro* differences, the main findings in old people and centenarians compared to young subjects are summarized in Table 18.11.

## 2. Changes in Old Age under Stress Conditions

Hematopoietic competence decreases in old age under various experimental conditions, including stress and stem cell transplantation. In both cases, it is important to distinguish between long-term versus short-term effects on the aging-associated decline in recovery.[57] The reduced potential for sequential replication is not readily expressed in the aged individual, but it may play a role under conditions where the stem cells are subject to replicative stress (that is, stress stimulating cell replication). Therefore, proliferative ability of stem cells must be evaluated before a decision is reached about bone marrow transplantation, in which case, the stem

**TABLE 18.11**
**Changes of Hematopoietic Remodeling in Elderly Subjects**

- Decreased absolute number of CD34+ progenitor cells in peripheral blood
- Well-preserved competence of CD34+ cells to respond to cytokines and to form erythroid, granulocyte-macrophagic, and mixed colonies indistinguishable in number, size, and morphology from those formed from similar cells of young subjects
- Decreased *in vitro* production of granulocyte-macrophagic colony-stimulating factor (GM-CSF) and IL-3
- Increased serum levels of stem cell growth factor (SCF)
- Delayed recovery from hematological stress
- Centenarians exhibit few hematological changes

cells must be able to replicate extensively. In addition, replicative stress may render the cells susceptible to mutations and thereby represent a possible cause of leukemia, secondary to bone marrow transplantation.

While basal hematopoietic potential is well preserved in healthy centenarians, the hematopoietic cytokine network may undergo a complex remodeling with aging. Although significant changes do not occur under steady-state conditions, under pathological conditions (e.g., bacterial infections), or during periods of increased hematopoietic demand (e.g., hemorrhage, hypoxia), an impairment of hematopoietic response may emerge even in healthy elderly and centenarians. Indeed, the well-preserved basal CD34+ cell function in centenarians probably arises from a new equilibrium between hematopoietic cytokines and their hematopoietic target cells, as a result of continuing adaptive processes.[54]

The hypothesis that the aging process is delayed in hematopoietic stem cells, as compared to other cells, is strengthened by the recent observation that telomerase activity is present in hematopoietic stem cells from adult human bone marrow (see also Chapter 4). It can be speculated that the aging process follows diverse timetables in the various cells of the body and that the cell type responsible for maintaining a reservoir of pluripotent cells can be spared, at least in part, from the aging process, thus contributing to individual longevity (Chapter 2).

## C. STEM CELL PROLIFERATIVE CAPACITY

Stem cells are quiescent (S/G2/M) generally, and, at any time, very few are cycling, perhaps as protection against exhaustion of the pool and against mutations. The bone marrow of aged mice contains stem cells that are ready to cycle and to replicate upon incubation with cytokines or seeding onto a thymic stroma; however, replications cease soon after termination of the experimental intervention. In the elderly, induction of stem cell proliferation would depend on the action of cytokines in the bone marrow, as part of normal stem cells replication. With advancing age, once the cells have entered the division cycle, they may become more susceptible to mutations and undergo neoplastic processes, particularly in view of the reduced fidelity of DNA repair.

Analysis of isolated long-term reconstituting stem cells shows a high frequency of the cells in the growth phase.[57] With aging, the increased proportion of the stem cells about to enter the cell division cycle may represent a compensatory process consequent to the reduction in number of active stem cells. A quantitative analysis of the frequency and proliferation of five subsets of primitive hematopoietic cells in the bone marrow[58] revealed that the relative and absolute numbers of the most primitive stem cell subsets is three- to fourfold higher in old than in young mice.[57] The negative correlation between maximum life span and the capacity for proliferation manifested by changes in the frequency and cell-cycle kinetics of primitive hematopoietic stem cells reflects effects that may be both age- and strain-dependent.

The findings that somatic cells have a limited capacity for replication have raised the question of whether the potential for self-renewal is gradually reduced with age, due to a possible finite-cell replication programming (Chapter 4). Serial transplantation of hematopoietic cells from young mice in healthy irradiated recipients showed that cells originating from the transplanted bone marrow donor could not be recovered from the recipients after a limited number of generations. The stem cells have a limited capacity for replication, and the potential for self-renewal would be exhausted in old age (Table 18.12).

## 1.   Telomeres and Telomerase

*Length of the telomeres,* the terminal section of chromosome involved in chromosomal replication and stability, correlates with replication of somatic cells and residual replication potential (Chapter 4). Thus, the sequential loss of telomeric DNA from the ends of the chromosome with each somatic cell division would eventually reach a critical point that may trigger cellular senescence, and may influence the balance between stem cell renewal and proliferation.[59] The loss of telomeric repeats in hematopoietic cells is a dynamic process that is differentially regulated in young children and adults. Human stem cells with a CD34+CD38low phenotype purified from adult bone mar-

row have shorter telomeres than cells from fetal liver or umbilical cord blood. Cells produced in cytokine-supplemented cultures of purified precursor cells show a proliferation-associated loss of telomeric DNA. These findings strongly suggest that the proliferative potential of most, if not all, hematopoietic stem cells is limited and decreases with aging (Table 18.12).

*Telomerase,* the key enzyme that catalyzes the addition of telomeric sequences at the end of chromosomes, is not expressed in somatic cells, but is expressed in germline cells, where telomere length is maintained so that viable chromosomes can be transmitted to the following generation (Chapter 4). If stem cells are to maintain their integrity during replication, for the progeny to manifest the same properties as the parental cells, their telomere length must remain intact through self-renewal. However, experimental evidence shows a decline with aging in the capacity of the stem cells to replicate (replicative senescence). The highest telomerase activity can be detected in immature bone marrow hematopoietic progenitors, and telomerase is downregulated with cellular maturation.

Current data show that telomere loss in normal tissues begins in early adult life and progresses gradually with advancing age. However, rates of telomere loss, surprisingly rapid in young children, are highly variable at subsequent ages. The telomere loss, differentially regulated in leukocytes from young children and adults, may serve as a model for telomere dynamics of other somatic cells.[60]

Telomerase, expressed at basal levels in primitive hematopoietic stem cells, is upregulated in response to cytokine-induced proliferation and cell cycle activation, and downregulated with decreased proliferation and transition to subsequent developmental subsets. Telomerase activity in hematopoietic cells would reduce, but not prevent, telomere shortening during proliferation.[61] Recent studies suggest that other factors besides telomerase activity are involved in the regulation of telomere length.

## 2.   The Bone Marrow Environment

Changes in the stem cell compartment may be related to *primary intrinsic processes or to induction by the stroma* — that is, the collagen matrix of the bone that supports the cells concerned with bone formation (osteocytes, osteoblasts, and osteoclasts) and with bone marrow function — and by other neighboring cells in the microenvironment. Indeed, age-related changes have been noted in stroma cells as well as in mature lymphocytes in bone marrow. Capacity of hematopoietic stroma cell lines for replication was reduced unless they were co-cultured with stroma cell lines that promote stem cell maintenance. Stem cells from aged mice do not show any change in their ability to maintain[57] or to form colonies on stroma cultures, indicating that their capacity to interact with stroma elements is not changed with aging. Persistence of normal

---

**TABLE 18.12**
**Changes in Stem Cell Proliferation in Old Age**

- Proliferative potential decreases
- Telomere length shortens despite telomerase presence and upregulation by cytokines

---

**Systemic and Organismic Aging**

function may be related to compensatory changes with aging in the cytokines produced by stroma cells, as well as by neighboring macrophages and mature lymphocytes. The increased proportion of lymphocytes in the bone marrow of aged mice and their altered pattern of function may be of particular importance.

Relevant aging-related changes in cytokines include:

- A reduced availability of stroma-cell-derived IL-7
- A shift in the cytokine profile from Th1 to Th2
- An increased production of prostaglandin (PGE2) by macrophages
- An impaired expression of hematopoietic growth factors in aged humans and mice[62]

The shift toward Th2 type and proinflammatory cytokines in bone marrow cells of aging mice is similar to the shift detected in the peripheral lymphoid tissues.[63] Cytokine production by bone marrow stroma cells (*in vivo* and *in vitro*) is also dependent on estrogen levels: for example, IL-6 secretion is significantly higher in postmenopausal than in nonmenopausal adult women, an increase prevented by estrogen replacement therapy.[64]

## D. ERYTHROPOIESIS AND AGING

The primary function of the end product of erythropoiesis, the mature red blood cell, is to transport oxygen efficiently through the circulation to all cells and tissues of the body and to respond quickly and appropriately to increased oxygen demands, either acute (e.g., rapid and severe blood loss) or chronic (e.g., hypoxia from pulmonary disease). The overall marrow response, however, is complex, and requires the following:

- The participation of erythroid cells responsive to erythropoietin
- A structurally intact microenvironment
- An optimal iron supply within the marrow

### ANEMIA IN THE ELDERLY

Aging may favor the development of anemia, a disease in which the blood is deficient either in the amount of hemoglobin or the number of RBC or both. There are several types of anemias caused by one or a combination of factors, including:

- Reduced pluripotent hematopoietic stem cell reserve
- Reduced production of growth factors
- Reduced sensitivity of stem cells and progenitors to growth factors
- Microenvironmental abnormalities

Anemia is a common medical problem in the elderly. Most anemias, however, have causes other than aging. Indeed, the overall effects of aging on RBCs in terms of cell turnover, cell renewal, and cell responses to increased functional demand such as hypoxia. appear to be minimal, at least under steady-state conditions. It is still not clear if aging affects erythroid homeostasis and is responsible for low hemoglobin levels.[65]

There are at least two reasons for considering anemia in the elderly a sign of disease and not a physiologic parameter. They are as follows: (1) most older people maintain a normal RBC count, normal hemoglobin, and normal hematocrit, and (2) in most elderly subjects, an underlying cause of anemia can be found when hemoglobin levels are decreased below 12 g/dL (normal values 14 to 16 g/dL). Frequently, patients are already affected by a disorder (e.g., congestive heart failure, cognitive impairment, dizziness, and apathy) that may be worsened by the anemia.[66]

Using the World Health Organization definition of anemia — hemoglobin less than 12 g/dL in women and less than 13 g/dL in men — the prevalence of anemia in the elderly ranges from 8 to 44%, with higher frequency in men than women, and the highest prevalence in men 85 years and older.[67] Cohort studies of the elderly have found that the two most common causes of anemia in the elderly are chronic diseases and iron deficiency. In 15 to 25% of elderly patients with anemia, no cause is found.[68] Sex differences in hemoglobin levels tend to narrow significantly in old age and virtually disappear in the very old. It is unclear if the higher anemia incidence in old men is due to more frequent illnesses or to physiologic changes. Race and ethnicity affects the prevalence of anemia, which is significantly higher in elderly blacks.

As individuals approach 100 years of age, the hematological profile may undergo significant changes[69] (Table 18.13). Counts of hematopoietic stem cells in younger and older mice under steady state or under stress (e.g., radiochemotherapy), revealed significant aging-related changes in the older animals: the number of these cells was reduced, and their proliferation slower in response to stress, perhaps the result of a decline in stem cell reserve.[70] These data are contradicted by recent studies in which old age did not impair RBC recovery after low doses of total body irradiation; these data suggest that the poor tolerance of older patients to radiochemotherapy may be ascribed to nonhematopoietic organ toxicity rather than to aging-related changes in hematopoietic stem cell reserve. Aged mice show normal red cell life span, erythroid precursor numbers, and ferrokinetics; hence, the slightly decreased blood counts may be explained as

---

**TABLE 18.13**
**Hematological Profile of Some Older Individuals**

- Decreased hemoglobin
- Decreased hematocrit
- Decreased RBC number
- Delayed onset of erythropoiesis after severe bleeding
- Decreased erythropoietic responses to erythropoietin administration
- Most of these changes are not experienced by centenarians

resulting from plasma volume expansion rather than a decreased RBCs mass.[53,54]

In experimental animals, major alterations of erythropoiesis possibly associated with old age include:

- Delayed onset of erythropoiesis following severe bleeding
- Altered cycling activity of Pluripotent Hematopoietic Stem Cells (PHSC)[58]
- Altered production of cytokines[71] and altered hematopoietic microenvironment
- Reduced RBCs' responsiveness to erythropoietin *in vivo* and *in vitro*[71]
- Decreased response to stress (e.g., decreased erythroid repopulating capacity)[72]

Studies in animals may not reflect the human condition, and aged animals are not subjected to human-style testing for underlying disease. In humans, several observations suggest a progressive exhaustion of PHSC. The hematopoietic tissue of the bone marrow contracts progressively with aging.[73] The concentration of erythroid colonies in the bone marrow of older individuals is decreased (a decrement not associated with clinical anemia). The comparison of concentration of pluripotent hematopoietic stem cells in the peripheral blood of persons older than age 70 and younger than age 30 have similar baseline values, but the response to the administration of growth factors (GM-CSF) is greater in younger individuals. The reduction in the production of burst-promoting activity (BPA) in the bone marrow of older individuals implies that although aging may not be regarded as a cause of anemia, it may increase susceptibility to the disease.

The main *hormonal modulators* of erythropoiesis are *erythropoietin, testosterone, and IL-3* (Table 18.14).[65] Some (but not all) investigators have reported normal testosterone levels and normal cell responsiveness to testosterone in the elderly. The responsiveness of progenitor cells to androgen does not appear to change with old age, and the consistent age-related differences found in IL-3 levels do not appear to change.[56,74]

With respect to erythropoietin, levels in healthy elderly subjects have been described as unchanged,[75] lower, or even slightly higher than in younger controls. Erythropoietin secretion is usually not compromised in elderly individuals, although it may be reduced in acute or chronic anemia as the result of concomitant disease or exhaustion of its production.[76]

## TABLE 18.14
## Hormonal Regulators of Erythropoiesis

- Erythropoietin
- Testosterone
- Interleukin-3

## TABLE 18.15
## Factors Involved in Earlier and Faster (<120 days) Removal of RBCs from Circulation in Old Subjects

| Causes | Mechanisms |
| --- | --- |
| • Earlier and greater fragility | • Alteration in membrane lipids increases cell fragility as well as decreases glucose transport and utilization |
| • Greater tendency to aggregate | • Accelerated disialyation of membrane glycoconjugates may promote RBCs aggregation |
| • Decreased availability of energy for metabolism | • Decreased activity of $NA^+K^+$-ATPase reduces availability of energy (ATP for cell metabolism) |
| • Altered ionic balance, especially in aging-associated diseases | • Increased cytosolic $Ca^{++}$ and decreased $Mg^{++}$ change ionic balance (risk factors: hypertension or diabetes) |

## E.  RED BLOOD CELLS AND AGING

Given the short, 120-day, life span of the RBCs, the continuous renewal of the red blood cell population from the bone marrow should require their functional and structural integrity even at very old ages. However, unsuccessful aging accompanied by either physiologic impairment or degenerative disorders, such as diabetes mellitus and hypertension, may cause alterations in circulating erythrocytes that may shorten their life (Table 18.15). Red blood cell removal from the circulation may be accelerated by changes in their membrane composition associated with old age and capable of altering the rheological properties of these cells.

From studies of erythrocyte turnover in young hosts, several hypotheses have been proposed concerning the mechanisms involved in the generation of senescence signals and the removal of aged red blood cells by splenic macrophages. Among these, changes in membrane components, such as indicated below, have been given considerable attention. They are as follows:

- Desialylation of glycoproteins[77] (i.e., breakdown of sialic or neuramic acid, a major building block of cell membrane)
- Decreased phospholipid asymmetry
- Alterations in membrane proteins with the appearance of senescence surface antigens

**Systemic and Organismic Aging**

*Conflicting results* are reported about RBC membrane lipids:

- Unchanged cholesterol:phospholipid molar ratio despite increased cholesterol and phospholipids[78]
- Decreased cholesterol:phospholipid ratio with decreased membrane cholesterol
- Increased cholesterol:phospholipid molar ratio in healthy centenarians when compared with old subjects, due to a marked decrease in membrane phospholipids accompanied by a smaller decrease in cholesterol

The observed increase in the cholesterol:phospholipid molar ratio, the decrease in membrane cholesterol, together with the increased polyunsaturated:saturated fatty acid ratio in centenarians are likely to provide the RBC membrane with a better fluidity and, thereby, a greater flexibility that would allow RBCs to escape rapid destruction in the spleen.[68] Free radicals accumulate in RBCs as in other tissues, and oxidative damage may be responsible for some of the alterations in RBCs in old age[79,80] (Chapter 5).

The protein composition of the erythrocyte membrane in older healthy individuals (including centenerians) remains essentially unchanged except for a marked increase in content of the microtubular, contractile protein *actin*, the most abundant protein in mammalian cells. The structural organization of the membrane, responsible for the RBC shape and flexibility, is known to be susceptible to the attack of oxidant agents.[81] An elevated actin content in the membrane skeleton might strengthen the spectrin-4.1-actin junctional sites and, consequently, increase the resistance to alterations in shape and to increased fragility by mechanical stress.[68] Longevity is thought to be associated with a well-preserved membrane structure, and a more viable RBC may contribute to a longer life span. It remains to be clarified whether the increase in actin is a change brought about by the aging process or whether it is a characteristic of individuals genetically predisposed to become centenarian. In the latter case, some biochemical characteristics of the RBC membrane may be taken as an index of individual life expectancy (Chapter 2).

With respect to RBCs' survival in the old individual, a number of changes have been reported that may be responsible for an earlier and greater fragility in old age (but with a certain degree of persistence of normal flexibility in centenerians):

1. Degradation of band-3 protein calpain, a calcium-dependent protease involved in glucose transport and utilization, is enhanced in RBCs of old people; reduced band-3 protein may be responsible for decreased glucose transport and utilization in old people (while oxidative metabolism would be spared)[82]

2. Accelerated desialylation of membrane glycoconjugates may promote RBCs' aggregation (by reducing electrostatic repulsion among cells) and trigger the clearance of senescent cells from circulation.[83,84] The maintenance of sialic acid membrane content in centenerians would prevent RBCs from aggregating, preserve their flexibity, and prolong their survival

3. The activity of the enzyme Na,$^+$K$^+$-ATPase that regulates transport of sodium and potassium through the membrane, decreases progressively with aging in older individuals, including centenerians. The decrease of this enzyme, responsible for one third of the total intracellular energy consumption at all ages, may lead to decreased intracellular ATP content during senescence[85]

4. The progressive increase in cytosolic free calcium (Ca) and the decrease in magnesium (Mg), two divalent cations vital to cellular homeostasis, have been correlated with pathological conditions such as hypertension (Chapter 16) and diabetes (Chapter 14). Thus, these ionic changes may be clinically significant, and underlie the predisposition of older subjects to cardiovascular and metabolic diseases.[86]

## REFERENCES

1. Weinberger, S.E., Recent advances in pulmonary medicine, *N. Engl. J. Med.*, Part I, 328, 1389, 1993; Weinberger, S.E., Recent advances in pulmonary medicine, *N. Engl. J. Med.*, Part II, 328, 1462, 1993.
2. Janssens, J.P., Pache, J.C., and Nicod, L.P., Physiological changes in respiratory function associated with ageing, *Eur. Respir. J.*, 13, 197, 1999.
3. Davies, B.H., The respiratory system, in *Principles and Practice of Geriatric Medicine*, Pathy, M.S.J., Ed., John Wiley and Sons, New York, 1990.
4. Mahler, D.A., Ed., *Pulmonary Disease in the Elderly Patient*, Marcel Dekker, New York, 1993.
5. Meisami, E. and Timiras, P.S., Eds., Respiratory development, in *Handbook of Human Growth and Developmental Biology*, Vol. 3, Part B, CRC Press, Boca Raton, FL, 1990.
6. Masoro, E.J., Ed., *CRC Handbook of Physiology in Aging*, CRC Press, Boca Raton, FL, 1981.
7. Mylotte, J.M., Epidemiology of respiratory infections, in *Principles and Practice of Geriatric Medicine*, Vol. 1, Pathy, M.S.J., Ed., John Wiley and Sons, New York, 1998.
8. Tolep, K. and Kelsen, S.G., The effect of ageing on the respiratory skeletal muscles, in *Principles and Practice of Geriatric Medicine*, Vol. 1, Pathy, M.S. J., Ed., John Wiley and Sons, New York, 1998.
9. Mahler, D.A. and Ramirez-Venegas, A., Dyspnoea, in *Principles and Practice of Geriatric Medicine*, Vol. 1, Pathy, M.S.J., Ed., John Wiley and Sons, New York, 1998.

10. Connolly, M.J., Respiratory diseases, in *Principles and Practice of Geriatric Medicine*, Vol. 1, Pathy, M.S.J., Ed., John Wiley and Sons, New York, 1998.

11. Ilowite, J.S. and Rodrigues, J.C., Pulmonary rehabilitation, in *Principles and Practice of Geriatric Medicine*, Vol. 1, Pathy, M.S.J., Ed., John Wiley and Sons, New York, 1998.

12. Marx, J.L., Oxygen free radicals linked to many diseases, *Science*, 235, 529, 1987.

13. Dockery, D.W. et al., An association between air pollution and mortality in six U.S. cities, *N. Engl. J. Med.*, 329, 1753, 1993.

14. Teramoto, S. et al., A comparison of ventilation components in young and elderly men during exercise, *J. Gerontol.*, 50A, B34, 1995.

15. Russous, C.S. and MacLin, P.T., Diaphragmatic fatigue in man, *J. Appl. Physiol.*, 43, 189, 1977.

16. Zhang, Y. and Kelsen, S.G., Effect of aging on diaphragm contractile function in golden hamsters, *Am. Rev. Respir. Dis.*, 142, 1396, 1990.

17. Powers, S.K. et al., Influence of exercise and fiber type on antioxidant enzyme activity in rat skeletal muscle, *Am. J. Physiol.*, 266, R375, 1994.

18. Reddan, W.G., Respiratory system and aging, in *Exercise and Aging*, Smith, E.L. and Serface, R.C., Eds., Enslow Publishing, Hillside, NJ, 1981.

19. Hurwitz, S.A. et al., Lung function in young adults — evidence for differences in the chronological age at which various functions start to decline, *Thorax*, 35, 615, 1980.

20. Davis, C. et al., Importance of airway closure in limiting maximal expiration in normal man, *J. Appl. Physiol.*, 48, 695, 1980.

21. Astrand, I. et al., Reduction in maximal oxygen uptake with age, *J. Appl. Physiol.*, 35, 649, 1973.

22. Shephard, R.J., *Aging, Physical Activity, and Health*, Vol. 8, Human Kinetics, Champaign, IL, 1997.

23. Shephard, R.J., *Gender, Physical Activity, and Aging*, CRC Press, Boca Raton, FL, 2001.

24. McConnell, A.K., Semple, E.S., and Davies, C.T., Ventilatory responses to exercise and carbon dioxide in elderly and younger humans, *Eur. J. Appl. Physiol. Occup. Physiol.*, 66, 332, 1993.

25. Barnes, P.J., Chronic obstructive pulmonary disease, *N. Engl. J. Med.*, 343, 269, 2000.

26. Lopez, A.D. and Murray, C.C., The global burden of disease, 1990–2020, *Nat. Med.*, 4, 1241, 1998.

27. Barnes, P.J., Genetics and pulmonary medicine, 9, Molecular genetics of chronic obstructive pulmonary disease, *Thorax*, 54, 245, 1999.

28. Silverman, E.K. et al., Genetic epidemiology of severe, early-onset chronic obstructive pulmonary disease: Risk to relatives for airflow obstruction and chronic bronchitis, *Am. J. Respir. Crit. Care Med.*, 157, 1770, 1998.

29. Sandford, A.J., Weir, T.D., and Pare, P.D., Genetic risk factors for chronic obstructive pulmonary disease, *Eur. Respir. J.*, 10, 1380, 1997.

30. Huang, S.L., Su, C.H., and Chang, S.C., Tumor necrosis factor-alpha gene polymorphism in chronic bronchitis, *Am. J. Respir. Crit. Care Med.*, 156, 1436, 1997.

31. Higham, M.A. et al., Tumor necrosis factor-α gene promoter polymorphism in chronic obstructive pulmonary disease, *Eur. Respir. J.*, 15, 281, 2000.

32. Mahadeva, R. and Lomas, D.A., Genetics and respiratory disease, 2, Alpha 1-antitrypsin deficiency, cirrhosis and emphysema, *Thorax*, 53, 501, 1998.

33. Smith, C.A.D. and Harrison, D.J., Association between polymorphism in gene for microsomal epoxide hydrolase and susceptibility to emphysema, *Lancet*, 350, 630, 1997.

34. Finkelstein, R. et al., Alveolar inflammation and its relation to emphysema in smokers, *Am. J. Respir. Crit. Care Med.*, 152, 1666, 1995.

35. Pauwels, R.A. et al., Long-term treatment with inhaled budesonide in persons with mild chronic obstructive pulmonary disease who continue smoking, *N. Engl. J. Med.*, 340, 1948, 1999.

36. Niewoehner, D.E. et al., Effect of systemic glucocorticoids on exacerbations of chronic obstructive pulmonary disease, *N. Engl. J. Med.*, 340, 1941, 1999.

37. Wise, R. et al., Effect of inhaled triamcinolone on the decline in pulmonary function in chronic obstructive pulmonary disease, *N. Engl. J. Med.*, 343, 1902, 2000.

38. Mapp, C.E., Inhaled glucocorticoids in chronic obstructive pulmonary disease, *N. Engl. J. Med.*, 343, 1960, 2000.

39. Matsuse, T. et al., Latent adenoviral infection in the pathogenesis of chronic airways obstruction, *Am. Rev. Respir. Dis.*, 146, 177, 1992.

40. Keicho, N. et al., Endotoxin-specific NF-KB activation in pulmonary epithelial cells harboring adenovirus E1A, *Am. J. Physiol.*, 277, L523, 1999.

41. Rutgers, S.R. et al., Ongoing airway inflammation in patients with COPD who do not currently smoke, *Thorax*, 55, 12, 2000.

42. Ferguson, G.T. and Cherniack, R.M., Management of chronic obstructive pulmonary disease, *N. Engl. J. Med.*, 328, 1017, 1993.

43. Ferguson, G.T., Update on pharmacologic therapy for chronic obstructive pulmonary disease, *Clin. Chest Med.*, 21, 723, 2000.

44. Naughton, B.J., Mylotte, J.M., and Tayara, A., Outcome of nursing home-acquired pneumonia: Derivation and application of a practical model to predict 30 day mortality, *J. Am. Geriatr. Soc.*, 48, 1292, 2000.

45. Callahan, C.M. and Wolinsky, F.D., Hospitalization for pneumonia among older adults, *J. Gerontol.*, 51, M276, 1996.

46. Janssens, J.P. et al., Community-acquired pneumonia in older patients, *J. Am. Geriatr. Soc.*, 44, 539, 1996.

47. Johnson, J.C. et al., Nonspecific presentation of pneumonia in hospitalized older people: Age effect or dementia?, *J. Am. Geriatr. Soc.*, 48, 1316, 2000.

48. Liaw, Y. et al., Clinical spectrum of tuberculosis in older patients, *J. Am. Geriatr. Soc.*, 43, 256, 1995.

49. Blower, S.M., Small, P.M., and Hopewell, P.C., Control strategies for tuberculosis epidemics: New models for old problems, *Science*, 273, 497, 1996.

50. Snider, D.E. and Castro, K.G., The global threat of drug-resistant tuberculosis, *N. Engl. J. Med.*, 338, 1689, 1998.

**Systemic and Organismic Aging**

51. Stokstad, E., Drug-resistant TB on the rise, *Science*, 287, 2391, 2000.

52. Saitoh, T. et al., Comparison of erythropoietic response to androgen in young and old senescence accelerated mice, *Mech. Ageing Dev.*, 109, 125, 1999.

53. Van Zant, G., Commentary on "Hemopoiesis in healthy old people and centenarians: Well-maintained responsiveness of CD34+ cells to hemopoietic growth factors and remodeling of cytokine network," *J. Gerontol. Biol. Sci.*, 55A, B67, 2000.

54. Globerson, A., Hematopoietic stem cells and aging, *Exp. Gerontol.*, 34, 137, 1999.

55. Marley S.B. et al., Evidence for a continuous decline in haemopoietic cell function from birth: Application to evaluating bone marrow failure in children, *Br. J. Haematol.*, 106, 162, 1999.

56. Bagnara, G.P. et al., Hemopoiesis in healthy old people and centenarians: Well-maintained responsiveness of CD34+ cells to hemopoietic growth factors and remodeling of cytokine network, *J. Gerontol.*, 55, B61; discussion B67, 2000.

57. Morrison, S.J. et al., The aging of hematopoietic stem cells, *Nat. Med.*, 2, 1011, 1996.

58. De Haan, G., Nijhof, W., and Van-Zant, G., Mouse strain-dependent changes in frequency and proliferation of hematopoietic stem cells during aging: Correlation between life span and cycling activity, *Blood*, 89, 1543, 1997.

59. Robertson, J.D. et al., Dynamics of telomere shortening in neutropholis and T lymphocytes during ageing and the relationship to skewed X chromosome inactivation patterns, *Br. J. Haematol.*, 109, 272, 2000.

60. Frenck, R.W., Blackburn, E.H., and Shannon, K.M., The rate of telomere sequence loss in human leukocytes varies with age, *Cell Biology*, 95, 5607, 1998.

61. Engelhardt, M. et al., Telomerase regulation, cell cycle, and telomerase stability in primitive hematopoietic cells, *Blood*, 90, 182, 1997.

62. Buchanan, J.P. et al., Impaired expression of hematopoietic growth factors: A candidate mechanism for the hematopoietic defect of aging, *Exp. Gerontol.*, 31, 135, 1996.

63. Segal, R. et al., Effect of aging on cytokine production in normal and experimental systemic lupus erythematosus afflicted mice, *Mech. Ageing Dev.*, 96, 47, 1997.

64. Cheleuitte, D., Mizuno, S., and Glowacki, J., *In vitro* secretion of cytokines by human bone marrow: Effects of age and estrogen status, *J. Clin. Endocrinol. Metab.*, 83, 2043, 1998.

65. Carmel, R., Anemia and aging: An overview of clinical and biological issues, *Blood Rev.*, 15, 9, 2001.

66. Smith, D.L., Anemia in the elderly, *Am. Fam. Physician*, 62, 1565, 2000.

67. Nilsson-Ehle, H. et al., Blood haemoglobin declines in the elderly: Implications for reference intervals from age 70 to 88, *Eur. J. Haematol.* 65, 297, 2000.

68. Fredman, M.L. and Sutin, D.G., Blood disorders and their management in old age, in *Brocklehurst's Textbook of Geriatric Medicine and Gerontology*, 5th ed., Churchill Livingstone, New York, 1998, p. 1247.

69. Caprari, P. et al., Aging and red cell membrane: A study of centenarians, *Exp. Gerontol.*, 34, 47, 1999.

70. Zaucha, J.M. et al., Hematopoietic responses to stress conditions in young dogs compared with elderly dogs, *Blood*, 98, 322, 2001.

71. Shearer, G.M., Th1/Th2 changes in aging, *Mech. Ageing Dev.*, 94, 1, 1997.

72. Baraldi-Junkins, C.A., Beck A.C., and Rothstein, G., Hematopoiesis and cytokines. Relevance to cancer and aging, *Hematol. Oncol. Clin. N. Amer.*, 14, 45, 2000.

73. Moscinski, L., The aging bone marrow, in *Comprehensive Geriatric Oncology*, Balducci, L., Lyman, G.H., and Ershler, W.B., Eds., Harwood Academic Publishers, Amsterdam, 1998.

74. Kamenets Y. et al., Relationship between routine hematological parameters, serum IL-3, IL-6 and erythropoietin and mild anemia and degree of function in the elderly, *Aging Clin. Exp. Res.*, 10, 32, 1998.

75. Matsuo, T. et al., An inappropriate erythropoietin responsiveness to iron deficiency anemia in the elderly, *Clin. Lab Hematol.*, 17, 172, 1995.

76. Baraldi-Junkins, C.A., Beck, A.C., and Rothstein, G., Hematopoiesis and cytokines, relevance to cancer and aging, *Hematol. Oncol. Clin. N. Amer.*, 14, 45, 2000.

77. Aminoff, D., The role of sialoglycoconjugates in the aging and sequestration of red cell from circulation, *Blood Cells*, 14, 229, 1988.

78. Solichova, D. et al., Bioanalysis of age related changes of lipid metabolism in nonagenarians, *J. Pharm. Biomed. Anal.*, 24, 1157, 2001.

79. Yanagawa, K. et al., Age-related changes in alpha-tocopherol dynamics with relation to lipid hydroperoxide content and fluidity of rat erythrocyte membrane, *J. Gerontol.*, 54, B379, 1999.

80. Inal, M.E., Kanbak, G., and Sunal, E., Antioxidant enzyme activities and malondialdehyde levels related to aging, *Clin. Chim. Acta*, 305, 75, 2001.

81. Caprari, P. et al., Junctional sites of erythrocyte skeletal proteins are specific targets of tert-butylhydroperoxide oxidative damage, *Chem. Biol. Interact.*, 94, 243, 1995.

82. Guven, M. et al., Age related changes on glucose transport and utilization of human erythrocytes: Effect of oxidative stress, *Gerontology*, 45, 79, 1999.

83. Hadengue, A.L. et al., Erythrocyte disaggregation shear stress, sialic acid and cell aging in humans, *Hypertension*, 32, 324, 1998.

84. Mazzanti, L. et al., Erythrocyte plasma membranes obtained from centenarians show different functional properties, *J. Am. Geriatr. Soc.*, 48, 350, 2000.

85. Rabini, A. et al., Diabetes mellitus and subjects' ageing: A study on the ATP content and ATP-related enzyme activities in human erythrocytes, *Eur. J. Clin. Invest.*, 27, 327, 1997.

86. Barbagallo, M. et al., Cellular ionic alteration with age: Relation to hypertension and diabetes, *J. Am. Geriatr. Soc.*, 48, 1111, 2000.

# 19 The Kidney, the Lower Urinary Tract, Body Fluids and the Prostate*

*Mary Letitia Timiras*
Overlook Hospital and University of Medicine and Dentistry of New Jersey

*Joyce Leary*
University of California, San Francisco

## CONTENTS

## I. THE KIDNEY, THE LOWER URINARY TRACT, AND BODY FLUIDS

*MARY LETITIA TIMIRAS*

As with other systems, the urinary tract is both *directly* and *indirectly affected by aging*.[1] Direct effects are exemplified by intrinsic cellular changes involving the nephron or the muscles of the bladder or the prostate. Indirect effects may be secondary to cardiovascular, endocrine, or metabolic alterations occurring with aging that have vital repercussions on urine formation and excretion.[2,3] Reciprocally, physiologic decrements in renal function and disturbances of the lower urinary tract may not only alter the elimination of end-products of metabolism but also impair other functions, such as regulation of body

---

* Illustrations by Dr. S. Oklund.

0-8493-0948-4/03/$0.00+$1.50
© 2003 by CRC Press LLC

fluids, acid–base balance, and blood pressure. Finally, given the metabolic and excretory functions of the kidney with respect to drugs of environmental, recreational, and therapeutic origin, the decline of renal function with aging may alter the renal handling of drugs and thereby modify drug pharmacokinetics, including toxicity and therapeutic effectiveness (Chapter 23). Such drug-related considerations are important at any age but are particularly so in the elderly, for whom the prescription of multiple drugs, simultaneously, increases the dangers of polypharmacy and toxicity or illness induced iatrogenically (inadvertently by a physician).

In the present chapter, the following aging-related changes in the urinary system will be grouped as follows:

1. Changes in renal function, including considerations of drug excretion
2. Changes in the lower urinary tract, including the problems of urinary incontinence
3. Changes in body fluids and acid–base balance
4. Aging of the prostate

## A. AGING-RELATED CHANGES IN RENAL FUNCTION

All parameters of renal function may be affected by aging, but the age of onset, rate and course of changes, and consequences vary. Thus, glomerular filtration, closely dependent on the efficiency of the renal blood flow and the integrity of the glomerular basement membrane, appears affected at an earlier age and more severely than tubular reabsorption and secretion. After the age of 30, renal function in humans, as measured by several tests (see below), gradually decreases and by 85 years has been reduced by half. This decline has been ascribed to a gradual loss of nephrons and diminished enzymatic and metabolic activity of tubular cells as well as an increased incidence of pathologic processes, primarly atherosclerosis, which affects the renal blood circulation (Chapter 16), an essential factor in determining renal competence. Likewise, a number of experimental animal studies show an overall decline of renal competence involving glomerular and tubular functions. A discussion of renal function in elderly subjects is preceeded by a brief summary of the structure and function of the kidney in Box 19.1.

---

**Box 19.1**

**Major Structural Characteristics and Functions of the Kidney**

The kidneys are paired retroperitoneal organs that lie on either side of the descending (abdominal) aorta and receive their blood via the renal arteries, two main branches of the aorta. A smooth outer capsule covers the cortex, which contains the majority of the glomerular portion of the individual nephron (Figure 19.1). The nephron represents the functional unit and is formed of the glomerulus, in the cortex, and the renal tubule which dips down and occupies the medulla, the inner portion of the renal parenchyma. Running parallel to the tubules in the medulla is a network of blood vessels, the vasa recta, which participate, with the tubular loop of Henle, in controlling the osmolality of the medulla through a countercurrent mechanism. The urine formed in the nephron flows into collecting ducts and out through the calyces and the pelvis to the ureter and from there to the bladder.

The **glomerulus** is formed by a tuft of capillaries between the entering, afferent, and the exiting, efferent arterioles. Filtration is through a fenestrated glomerular endothelium that is separated by a basal lamina from the interdigitated epithelial cells, the so-called podocytes, of the tubular epithelium. The so-called ultrafiltrate has the same composition as plasma, except for the absence of protein. The epithelial cells lining the various segments (proximal tubule, loop of Henle, distal tubule, collecting duct) of the **renal tubule** have been divided into types and subtypes on the basis of minor differences in histologic structure; there is some evidence that these differences correlate with differences in function, which is either reabsorption or secretion. Cells of the afferent arteriolar wall and the abutting distal tubule form the juxtaglomerular apparatus, site of the formation and release of the enzyme renin (see below). Cells of the distal convoluted tubule and the collecting duct are sensitive to hormones, aldosterone for sodium reabsorption and antidiuretic hormone (ADH) (also called vasopressin) for water reabsorption (Chapter 10).

The kidneys adjust the amount of secreted water and electrolytes, including H+ in such a way that volume and composition of body fluid, including acid–base balance and blood pH are maintained in homeostasis. Kidneys also excrete end-products of metabolism. They have certain endocrine functions: they secrete to the blood renin that regulates the production of angiotensin and, hence, indirectly influence blood pressure and aldosterone secretion; they activate vitamin D and thereby play a role in $Ca^{++}$ metabolism; they secrete erythropoietin and, thus, maintain the hemoglobin level of the blood (Table 19.1).

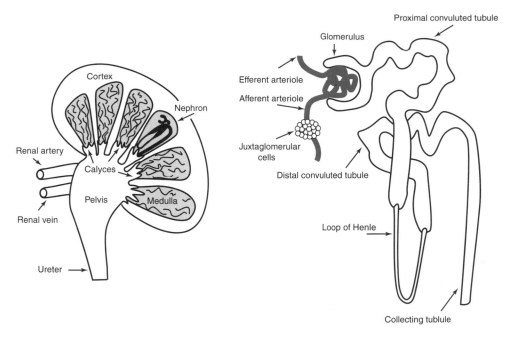

**FIGURE 19.1** Diagram of the kidney (left) and of a nephron (right).

## TABLE 19.1
## Major Functions of the Kidney

Water and electrolyte regulation
Metabolic products excretion
Hydrogen ion excretion and maintenance of blood pH
Endocrine functions:
   Renin-angiotensin secretion (blood pressure)
   Vitamin D activation (Ca++ metabolism)
   Erythropoietin secretion (hematopoiesis)

### TESTS OF RENAL FUNCTION

At all ages, the competence of renal function is assessed by a number of tests, among which the most routinely used are urine volume (per 24 hours), analysis of urine constituents, and urine concentration and dilution tests. Other measures of renal function include clearance tests, that is, clearance of a substance from the plasma during passage through the kidney (for example, inulin to measure glomerular filtration rate and para-aminohippuric acid, PHA, to measure renal blood flow). Radiologic, isotopic, ultrasonic, magnetic resonance spectroscopy and imaging, and tomographic imaging as well as biopsies and other less common tests represent more laborious and expensive means of assessing specific renal conditions.

## 1. Glomerular Function

The function of the glomerulus is to selectively filter plasma to produce a glomerular filtrate which, under normal healthy conditions, is practically free of proteins. Glomerular filtration is determined by measuring the clearance of plasma and the excretion in the urine of a substance such as inulin or creatinine, that is freely filtered through the glomeruli but not secreted or reabsorbed by the tubules. This technique establishes the filtration rate in an average young man as approximately 125 mL/min. According to early and now classic studies, commencing as early as 30 years and continuing into old age, glomerular filtration rate, renal plasma flow, and renal blood flow decrease progressively and significantly.[4,5]

### a. Creatinine Clearance

Both *inulin* (a polymer of fructose), a polysaccharide administered intravenously, and *creatinine*, a normal endogenous product of protein metabolism, are usually utilized to assess glomerular function. Creatinine clearance, which does not involve any drug administration, provides an acceptable index of renal function; it is most frequently used for assessing renal function and determining the dosage of renally eliminated drugs (Chapter 23). Unfortunately, this method requires the collection of a timed urine sample, often difficult to obtain in elderly individuals. To avoid the need for timed urine samples, creatinine clearance is often calculated from easily obtained information on weight, gender, age, and serum creatinine concentration, as shown in the following formula:

$$\text{Creatinine clearance} = \frac{(140 - \text{age}) \times \text{body weight (in Kg)}}{72 \times \text{serum creatinine (in mg\%)}}$$

A 20-year-old individual, with a serum creatinine of 1 mg% and a body weight of 72 kg has a creatinine clearance of 120 mg%, whereas a 90 year old with the

**Systemic and Organismic Aging**

same body weight and serum creatinine has a creatinine clearance of 50 mg%, a greater than 50% reduction. The reduction with aging in the urinary output of creatinine, a muscle-specific metabolite, has been interpreted as reflecting the reduction in lean body mass (demonstrated by decreased radioactivity, in whole body, of labeled potassium, the major intracellular electrolyte) that occurs with old age (Chapter 24). The above equation is valid only with the assumption that renal function decreases with age at a rate of 1 mL/min/year after age 40. Although this may provide a useful index, recent studies have shown great variation in the rate of decline of renal function. In some individuals, no decrease in renal clearance can be detected with advancing age.[6]

Creatinine clearance is often measured in elderly individuals not only as a test of renal function but also as a guide to adjusting dosage of administered drugs to serum creatinine and creatinine clearance. However, a survey of several creatinine-clearance-estimating equations in common clinical use suggests that they are more imprecise and have a narrower range of applicability than hitherto believed.[7]

### b. Changes in Glomerular Morphology and Function

Of the factors that affect glomerular filtration, the most important are characteristics of the glomerular wall and renal blood flow. Studies in humans show that the decrease in glomerular function in old age may be due primarily to a loss or alteration of glomeruli and secondarily to alterations in blood flow, although the reverse sequence may also be true. In humans, the age-related decrease in blood flow that has been frequently reported is often reversible and responsive to improved renal hemodynamics. Irrespective of the primary site of the lesion, studies in rats show a 30% incidence of necrotic glomeruli with age, while, under conditions of restricted food intake (Chapter 24), the incidence is reduced to 2%.[8] In humans, widespread glomerular necrosis is rare: rather, a thickening of the glomerular basement membrane is associated with concomitant biochemical alterations such as a progressive decrease (e.g., 3.7% reduction per decade) in some amino acids, suggesting a diminution of the collagenous component.

Microscopic and ultramicroscopic studies of glomeruli in rats reveal a progressive thickening (from 1300 Å, neonatally to 4800 Å, at the advanced age of 2 years) of the basement membrane in some areas (focal or segmental thickening) and collapse (with loss of distinct layers) in others. Podocytes are lost or undergo swelling or atrophy. Proliferation of interstitial collagen leads to progressive sclerosis.[9] Atherosclerosis of renal blood vessels is frequent, especially in diabetes mellitus, a chronic disease due to insulin insufficiency and insulin resistance; this type of diabetes is associated with severe microangiopathy (abnormality of small vessels) that accelerates the onset and increases the severity of the vascular damage (Chapter 14).

Studies in individuals affected by a congenital loss of urinary proteins have shown that the protein, *nephrin*, is a major component of the *kidney filter* that keeps vital proteins from escaping from the body in the urine.[10] This filter, also known as the "slit diaphragm," is a zipper-like formation of molecules that forms between the podocytes, wraps around the glomerular capillaries, and prevents leakage of proteins from the blood into the glomerular filtrate.[10–12] The gene that encodes nephrin is located on chromosome 19, together with other genes similarly involved in the regulation of the kidney filter. Congenital, adult, or, more frequently, old age mutations of the nephrin gene might predispose to alterations in the structure and function of the kidney filter with consequent alterations of the glomerular capacity to prevent protein loss.[10–12]

### c. Aging-Related Changes in Day/Night Urine Excretion

In the normal adult, urine and electrolyte excretion follow a day/night pattern, with higher levels in the daytime. This pattern may have evolved to permit undisturbed sleep. In the elderly, associated with changes in sleep pattern (Chapter 8), the rhythm of urine excretion is reversed, with increased water and electrolyte excretion during the night. This shift may have multiple renal causes: decrease in renal concentrating capacity, sodium-conserving ability, and secretion of renin-angiotensin-aldosterone. It may also involve hormonal control of urine concentration and may be regulated by extrarenal, neuroendocrine factors,[13] such as alterations of ADH receptor expression,[14] deficiency of ADH production and secretion,[15] and changes in production and function of atrial natriuretic peptide (ANP).[16] Nocturnal polyuria may lead to urinary incontinence and is aggravated in the elderly with Alzheimer's disease. Treatment of nocturnal polyurea with ADH often reduces nocturnal urine production, with improvement in symptoms of frequency, nocturia, and incontinence.[13]

### d. Glomerular Function and Dietary Proteins

Changes with aging in glomerular function do not pose any threat to well-being; likewise, renal function may not be seriously compromised with aging, even in octogenarians and older. If, however, intrinsic renal disease or surgical loss of renal tissue or other factors such as inappropriate diet, add to the glomerular burden of old age, then the course of glomerular sclerosis and impaired glomerular filtration rate may be hastened appreciably.[17] Under normal conditions, little protein is filtered in the glomeruli and excreted in the urine. One manifestation of impairment in glomerular filtration is altered permability with consequent proteinuria, that is, more than the usual trace amounts of protein in the urine. As a consequence of altered glomerular structure, abnormal

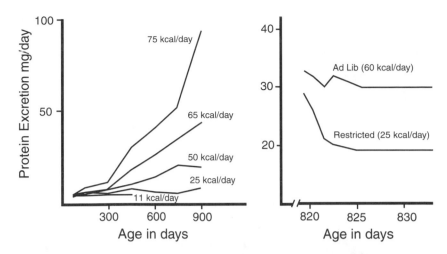

**FIGURE 19.2** Dietary caloric restriction reduces aging-related proteinuria in male rats. (Left) Proteinuria is reduced proportionately to the severity of the caloric restriction: severe food restriction (12.5 kcal/day) initiated at a young age (70 days) abolishes the steep rise in protein excretion with aging (but is not compatible with long life), and a less severe restriction (25 or 50 kcal/day) markedly inhibits it. (Right) This effect of caloric restriction is also observed in older (820 days) rats in which about 1 week of restriction was sufficient to reduce proteinuria significantly.

proteinuria appears in about 25% of male rats at a young age, and its incidence and severity increases progressively with age, although less so in female rats. [18]

High protein diets make it more difficult to prevent filtration of protein. Diets high in proteins may induce glomerular damage, and conversely, reduction of dietary protein intake reduces this damage. For example, feeding rats a diet low in protein or restricted in total calories reduces age-related proteinuria. Dietary restriction extends the life of laboratory animals, and such extension may involve the beneficial effects of a low-protein diet on renal function[19] (Chapter 24). Thus, rats consuming 75 kcal of food per day, a really large quantity, show high protein excretion throughout life from 4 mg/day at 70 days, just after sexual maturation, to 88 mg/day at the old age of 900 days. Normal food consumption is more in the range of 50 to 65 Kcal/day and, with this diet, compared to values of 75 Kcal consumption, protein excretion is halved. If food is severely restricted to 12.5 Kcal/day, the rise in proteinuria with age is abolished.[19] The same modification is apparent even with short-time food restriction in old animals. When the diet is reduced from 60–25 Kcal/day, protein excretion decreases by 40% in one week and continues at this level for the next week, even after discontinuation of the restricted diet (Figure 19.2).

In humans, the usual diet offers sustained (three meals a day) rather than intermittent (feast or famine of animals in the wild) intake of protein. Such sustained "excess" would impose rigorous demands on glomerular filtration rate and renal blood flow, thereby contributing to the decline of these functions with aging. It may also be responsible for the inexorable progressiveness of renal functional deficits and the greater incidence of renal pathology in the elderly. Calorie or protein restriction fairly early in the course of renal disease slows the rate of decline in glomerular filtration. This is true not only for aging, but also for metabolic and renal disorders; this is the case, for example, of obesity, in which diet modification leading to body weight loss reduces the often present proteinuria associated with glomerular sclerosis.[20]

## 2. Tubular Function

Although tubular function remains quite efficient in most individuals 65 years and older, in some elderly persons, the kidney is unable to concentrate urine as well as it does in younger persons; because of this, maintenance of appropriate water and electrolyte metabolism is potentially critical in the elderly, even in those without overt signs or laboratory tests indicating renal dysfunction.

The ability to concentrate or dilute urine may be lost gradually, with the result that the elderly individual is unable to cope optimally with dehydration or water load.[21] This inability is manifested after stimulation and inhibition of ADH secretion[22] (Chapter 10). Administration of ethanol, a drug known to inhibit ADH release, reduces circulating ADH levels for as long as 120 min after administration in young individuals. In old subjects, ADH levels are decreased immediately after alcohol ingestion but increase thereafter as shown at 120 min (Figure 19.3). In the young individual, low ADH levels are associated with the expected diuresis or increased water clearance. In old individuals, ADH inhibition occurs early after alcohol administration and then disappears; concomitantly, diuresis does not occur or is only minimal.[22]

**Systemic and Organismic Aging**

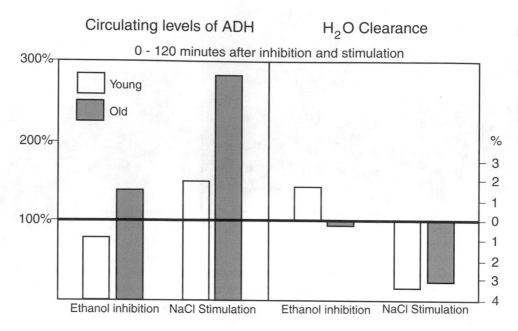

**FIGURE 19.3** Age-related changes in tubular responses to antidiuretic hormone (ADH) and to hypertonic sodium chloride solution. Ethanol inhibits ADH less efficiently in the elderly than in the young; low ADH levels are associated with increased diuresis to ethanol in the young, but diuresis is minimal or absent in the old. Hypertonic sodium chloride increases ADH more in the elderly than in the young, but the expected increase in water retention is less marked in the elderly.

Stimulation or inhibition of ADH secretion to appropriate stimuli differ in the elderly as compared to the young (Chapter 10). In response to a given osmotic challenge (i.e., administration of hypertonic sodium chloride solution), ADH plasma levels increase but more in old subjects than in young. However, the increase in ADH levels is not accompanied in the elderly by the expected increase in water retention observed in the young (Figure 19.3).[22] Attempts to identify the site(s) of the altered function, either renal or extrarenal, have not yet led to definitive conclusions. Administration of a standardized dose of ADH to young and old individuals shows a decline in the ability of the tubules to perform osmotic work and increase water retention in old individuals.[23] However, further studies suggest that decrements in the ability of cellular membranes of the collecting duct to become more water permeable under ADH influence occur primarily in old individuals with renal infections or hypertension; these decrements should not, therefore, be viewed as the usual accompaniment of aging.[24] Experiments in aged rats show that decreased responsiveness of collecting tubular epithelium to ADH is the most likely explanation for the impairment of urine concentrating ability, while reduced secretion and plasma levels of ADH may not play an important role in this impairment.

Whether the defect in renal regulation of water excretion lies in the hypothalamus, the pituitary, or the kidney (Chapter 10), there is no doubt about the relative inability of the aged kidney to concentrate or dilute urine and the consequent ease with which the elderly person may develop acute or chronic renal failure. As long as the total water intake is adequate — 2.5 to 3.0 L/day — renal function will be adequate (Chapter 24). Crises will arise when water intake is diminished due to loneliness, immobility, confusion, fright of incontinence, etc. This may also occur with unnecessary use of diuretics.

Failure to concentrate urine may also be related to sodium loss, particularly in the distal nephron.[19] Elderly persons are more prone to develop hyponatremia (low blood levels of sodium) and hypokalemia (low blood levels of potassium) during diuretic therapy or due to an inadequate diet (Chapter 24). Body potassium content is less in the elderly than young, even when measured in terms of lean (fat-free) body weight. This lower content may be due to renal loss as well as poor intake. Urea, a product of amino acid deamination in the liver, contributes to the establishment of the osmotic gradient in the renal medulla and to the ability to form a concentrated urine in the collecting ducts. Blood urea may be reduced due to diminished protein intake, often present in the elderly, or more rarely, to liver disease and insufficient urea production.

## 3. Renal Disease in the Elderly

Compounding the relatively minor impairment of renal function in the healthy elderly is the increased incidence of renal diseases and the reduced ability to handle the excretion of drugs. The major renal disorders of the young and middle aged — acute nephritis, collagen disorders, and malignant hypertension — are uncommon in old age.

---

**TABLE 19.2**
**Common Renal Problems in the Elderly**

Renal failure
Impaired drug excretion
Urinary tract infections
Hypertension
Miscellaneous disorders:
   Tuberculosis
   Nephritis
   Diabetes, etc.

---

**TABLE 19.3**
**Some Signs of Renal Failure**

Generalized edema
Acidosis
Increased circulating nonprotein nitrogen (urea)
Increased circulating urinary retention products

---

If they occur, they present and are treated, as in the young, with caution to protect water and electrolyte balance.

In the aged population, the common problems affecting renal function are related to damage induced by infections or drugs or by hypertension or miscellaneous disorders such as tuberculosis, nephritis, diabetes mellitus, amyloidoses, and collagen disorders (Table 19.2).[2,3,17] If untreated, these disorders may lead, perhaps more easily than in the young and with more life-threatening consequences, to dysfunction and, finally, to renal failure.

## 4. Failure of Renal Function: Pathogenesis and Management

Renal failure refers to the decline in renal excretion of sufficient severity to result in retention (in blood and extracellular spaces) of nitrogenous waste products, acids, potassium, sodium, and water. Clinically, it is designated as either acute or chronic. Normally, one third of the nephrons can eliminate all normal waste products from the body and prevent their accumulation in body fluids. When the number of active nephrons falls to 10 to 20% of normal, urinary retention and death follow.

### a. Acute Renal Failure

The acute form, frequent in the elderly, occurs in a matter of hours or days. Its consequences depend to a great extent on the food and water intake of the individual. With moderate intake, the most important signs are the following:

1. Generalized edema, resulting from salt and water retention
2. Acidosis, resulting from failure to excrete normal acidic products

3. High concentration of nonprotein nitrogen, especially urea, resulting from failure to excrete metabolic end-products
4. High concentration of other urinary products, including creatinine, uric acid, phenols, etc.

This condition of failing excretion is called *uremia* because of the high concentrations of normal urinary excretory products that collect in the body fluids (Table 19.3).

The causes of acute renal failure are numerous and are divided into prerenal, renal, and postrenal depending on whether they are traced to alterations of the kidney or to extrarenal alterations (Table 19.4). Prerenal causes, depending on factors unrelated to renal pathology and most likely to be found in the aged, include loss of body fluids (e.g., vomiting, diarrhea); inadequate fluid intake (often associated with the overuse of diuretics and laxatives); and surgical shock or myocardial infarction.

Renal causes, depending on renal pathology and relatively rare in the elderly, include drug toxicity due to certain antibiotics (sulfonamides, aminoglycosides, amphoteracin B); x-ray contrast materials (in a rather dehydrated individual); drug-induced immunologic reactions; infectious diseases: gram-negative bacteremia with shock or peritonitis; thrombosis and other circulatory alterations due to atherosclerosis; and intravascular hemolysis, such as may follow transurethral resection of the prostate.

Postrenal causes, depending on factors occurring after normal urine formation, are due primarily to urinary tract obstructions, such as occur with stones, tumors, or prostatic enlargement.

While the outcome of acute renal failure is generally poor in the elderly, a decrease in mortality from 70 to 50% has been reported in the years 1975 to 1990 and continues to date. A similar trend has been observed in young individuals.[26] Prevention is of paramount importance: maintenance of an adequate extracellular volume and drug dosages tailored to the degree of glomerular filtration efficiency are essential.[26]

---

**TABLE 19.4**
**Selected Causes of Acute Renal Failure**

Prerenal:
   Loss of body fluids
   Inadequate fluid intake
   Surgical shock or myocardial infarction
Renal:
   Drug toxicity
   Immune reactions
   Infectious diseases
   Thrombosis
Postrenal:
   Urinary tract obstruction

---

**Systemic and Organismic Aging**

### b.  Chronic Renal Failure

Chronic renal failure results from slow progressive and generally irreversible deterioration of renal function due to destruction of nephrons. It is more often a disease of young and middle-aged individuals, as survival is reduced in the major instances, such as in glomerulonephritis, and polycystic disease. In the elderly, causes of the disease differ from those in younger subjects,[27] they are progressive renal sclerosis (due to atherosclerosis); chronic pyelonephritis (i.e., inflammation of the kidney and pelvis); and obstructive uropathy due to slow but progressive enlargement of the prostate.

The pathology includes renal tubular necrosis and scattered basement membrane disruption. Urine output is acutely reduced to 20 to 200 mL/day (oliguria); the urine contains protein, red cells, epithelial cells, and characteristic dirty brown casts, and the signs of uremia appear with nausea, vomiting, diarrhea, lethargy, and hypertension.

Management depends on the mechanism that is responsible for the failure. Therefore, the initial step involves the differentiation between a prerenal oliguric state versus an intrinsic renal state. This is important, because the treatment for a hypovolemic (reduced blood volume) prerenal state consists of fluid and intravascular volume replacement; this treatment can lead to a dangerous fluid overload in a patient with intrinsic renal disease, such as tubular necrosis. Examination of the urine for the fractional excretion of sodium is the least invasive and most reliable method for making this differentiation (Table 19.5). Pivotal measures of management include: (1) treatment of the underlying cause; (2) monitoring of fluid, electrolytes, and acid–base balance; (3) prevention of infections, and (4) alterations of the diet: not more than 40 g/day protein with sufficient (at least 3000) calories to

---

**TABLE 19.5**
**Kidney Capacity to Conserve Sodium (Na) and Excrete Creatinine (Cr) Estimated by the Fractional Sodium Excretion (FENa) Test**

$<1 \rightarrow$ prerenal

$$FE_{Na} = \frac{U_{Na} \div P_{Na}}{U_{Cr} \div P_{Cr}} \times 100$$

$>1 \rightarrow$ intrinsic renal

The urine sodium concentration is divided by the plasma sodium concentration. This is then divided by the urine creatinine concentration over the plasma creatinine concentration. When this value multiplied by 100 is less than 1, then prenatal azotemia (high blood nitrogen due to altered metabolism) is responsible for renal failure; when FENa is greater than 1, this indicates intrinsic renal failure. Other methods to locate the cause of renal failure are available, but they are potentially more dangerous, especially for the elderly.

---

prevent endogenous catabolism. Other interventions include dialysis treatment and kidney transplantation.

#### DIALYSIS TREATMENT AND KIDNEY TRANSPLANTATION IN THE ELDERLY

The question often arises as to what extent "heroic interventions" are justifiable and advisable in the elderly. In the case of renal failure, dialysis and kidney transplantation represent interventive measures, which are widely utilized with considerable success in the young and the adult. Evidence for their rational use with a favorable outcome in the elderly has emerged from the cases treated so far and from animal experimentation. Clearly, while all contraindications and immediate and long-term risks of these measures must be taken carefully into consideration, as they may be magnified in the elderly, age alone should not deter their appropriate use. Today, up to one third of new patients entering dialysis throughout the world are older than 65 years.[27,28] Major decisions in management, such as the use of dialysis, should be made initially with the assumption that a positive outcome will occur. Thus, aggressive dialysis in acute renal failure should be the rule rather than the exception.[29]

Indication and success of transplantation of organs such as the kidney depend on meeting several criteria, such as normal function and competency of the organ to be transplanted and the age (young) and health (good) of the donor. With respect to age, although the use of kidneys from older donors is controversial, a number of data suggest that age alone should not eliminate using older kidney donors when their renal function and tissue matches are good.[25] Pregancy puts a strain on kidney function, but a report of a successful pregnancy in a renal transplant recipient with a kidney from a 75-year-old donor supports the view that an old kidney may function normally.[30]

Reciprocally, kidney transplants in older individuals (with young kidney donors) seem to fare as well as in the young, once the surgical and pharmacologic measures appropriate to the age of the individuals are taken into account. Until the beginning of the 1980s, older patients were considered a high-risk group.[31] Even today, the reality is that only 4% of patients aged 65 to 74 years under treatment for end-stage renal disease receive a renal transplant[32] and that the proportion of transplantation recipients varies with the country (e.g., from 19% in Switzerland to 2% in Italy).[33]

### 5.  Kidney Susceptibility to Drugs

The kidneys are particularly susceptible to the toxic effects of drugs and other chemical agents because:

1. Blood flow (carrying drugs) to the kidney is high, 20% of cardiac output passes through the kidneys
2. Drugs tend to accumulate in the renal medulla as water is removed from the glomerular filtrate; drug accumulation increases when renal function is impaired
3. Reduced hepatic enzyme activity in the elderly increases circulating drug levels and renal toxicity

**TABLE 19.6**
**Drugs and the Aging Kidneys**

**Questions**

Is the drug excreted primarily by the kidney?

How competent are the kidneys?

What are the side effects?

What are the consequences of drug toxicity when the kidney is impaired?

**Etiopathology of Renal Drug Toxicity**

High renal blood flow

Increased drug concentration and accumulation in kidney

Increased hepatic enzyme inhibition in the elderly

Increased autoimmune disorders in the elderly

4. Incidence of autoimmune disorders increases with aging, with consequent hypersensitivity reactions in the kidney (Table 19.6)

The manifestations of renal drug intoxication are not unique to the elderly; they are, however, more frequent and, when they occur, they may be more severe than in the adult.[34,35] In the kidney of old individuals, toxic effects are seen with lower doses than in the adult, and the consequences are more dangerous, taking into account the multiplicity of drugs taken by the elderly, the generally long duration of treatment, and the often impaired conditions of the kidney (Chapter 23). Examples of drugs and the mechanisms by which they induce renal damage include:

1. Dehydration-induced uremia due to use of diuretics and laxatives
2. Obstruction of urinary tract due to deposition of crystalline matters such as calcium from excessive administration of vitamin D
3. Vascular lesions produced by thiazide diuretics
4. Glomerular damage by penicillin-like antibiotics
5. Interstitial and tubular damage from radiological contrast media
6. Necrosis due to analgesics

## B. AGING-RELATED CHANGES IN THE FUNCTION OF THE LOWER URINARY TRACT

The major health problems of the elderly can be easily recalled by listing them under five words — *I*nstability, *I*mmobility, *I*ncontinence, *I*mpaired cognition, and *I*atrogenic diseases — all starting with the letter "I."[36] Of these, one of the most embarassing and distressing is urinary incontinence. Failure of urinary continence often results from alterations in the function of some of the structures of the lower urinary tract, and in fact, it may sometimes involve all of these structures.[37] Incontinence, therefore, will be the major topic discussed in this section. Major

structures and functions of the lower urinary tract are presented in Box 19.2.

Numerous surveys show that incontinence occurs in 10 to 30% of community-dwelling elderly and in 50 to 60% of those living in institutions. However, such statistics are not accurate inasmuch as people often fail to disclose this condition.[38-43] Many consider it inevitable, and many refuse to admit to it. Normal control of bladder and urethral functions are taken for granted by the majority of individuals and, in fact, remain efficient in many individuals well into old age. However, when it occurs, failure of this control is often considered a main threat to the welfare of those affected; it conjures up fears of rejection, which are often real and further restrict the social interactions of the elderly. Urinary incontinence may result not only from aging of the urinary tract but may also reflect disorders of locomotion[44,45] and of cognitive function,[46,47] and as such, may be viewed as an indicator of overall functional decline and harbinger of frailty and death.[43]

### 1. Urinary Continence

In view of the complexity of the mechanisms regulating micturition, the high prevalence of incontinence in the elderly should not be surprising. Continence depends on a long list of physiologic requirements, several of which undergo changes with aging (Table 19,7). These include motivation to be continent, adequate cognitive function to understand the need for continence, and mobility and dexterity to find and utilize the appropriate sanitary facilities. Normal urinary tract function, another necessary requirement for continence, depends on appropriate storage and emptying of the bladder: urine storage in the bladder depends, in turn, on its relaxation (absence of involuntary contractions, accomodation of increased urine volume under low pressure, closure of sphincters, etc.) as well as its contraction (normal contractile mechanisms, absence of anatomic obstruction, sphincter relaxation, etc.). Physical and environmental barriers (e.g., bed rails, distant toilet location) must also be considered possible causes of incontinence. Finally, the administration of drugs for a variety of ills and their effects, particularly on cognitive functions, may also contribute to the incontinence.

### 2. Causes and Course of Urinary Incontinence in the Elderly

Urinary incontinence can be defined as the involuntary loss of urine sufficient in amount and frequency to be a social or health problem. It is most frequent in the elderly; it is stressful not only for those affected but also for their family and caregivers, and can play a decisive role in the decision to place an elderly subject in a nursing home.[38-42]

Although urinary incontinence is not an inevitable consequence of aging, certain aging-related changes may

**Systemic and Organismic Aging**

---

## Box 19.2
## Structures and Functions of the Lower Urinary Tract

The structures involved in transfer, storage, and excretion of urine are the ureter, the bladder, and the urethra (Figure 19.4). These retroperitoneal and pelvic organs are similar in both sexes except for the urethra and associated structures, such as the prostate, which are sex-specific.

All these structures consist essentially of smooth muscle lined by mucosa: their contraction depends on an intact nervous system. The largest is the **bladder**, which is a smooth muscle chamber comprised of a body, formed of the **detrusor muscle** with a web-like structure, a neck, and near the neck, the orifices through which the ureters and urethra pass. When the detrusor muscle contracts, the bladder neck opens, and the bladder empties. Release of urine from the bladder is also controlled by **internal and external sphincters**. The internal sphincter responds to pressure built up in the bladder by stimulation of stretch receptors; the external sphincter, composed of striated muscle, is under voluntary control (Figure 19.5).

The main function of these structures is **micturition**, which is the process by which the bladder empties when it becomes filled with urine. Basically, micturition may be considered a special reflex, stimulated or inhibited by higher brain centers. Like defecation, it is subject to voluntary facilitation or inhibition. The events of micturition consist of progressive filling of the bladder with urine from the kidneys through the ureters. Once the bladder is filled, there is a tension that initiates a reflex to micturate, or, at least, a conscious desire to urinate.

The innervation of the bladder consists of three components, the two branches of the autonomic nervous system and the somatic nerves. The autonomic fibers, both sympathetic and parasympathetic, act on smooth muscles (detrusor, trigone, and internal sphincter), and the somatic fibers travel to the skeletal (striated) muscle (external sphincter). The act of micturition involves a complex coordination of neural and muscular responses leading to the following:

1. Relaxation of the internal sphincter to the urethra under sympathetic control and of the external sphincter, under somatic control
2. Constriction of the sphincters of the two ureters under sympathetic control to prevent urine retrograde flow to the kidney
3. Contraction of the detrusor muscle, under parasympathetic control

The entire sequence is under voluntary control and can be initiated or inhibited at will. In the absence of inputs from the cerebral cortex and other high brain centers, spinal reflexes may take control and induce automatic emptying of the bladder whenever the retained volume of urine reaches a critical level. When the voluntary system is active, the desire to urinate becomes apparent after approximately 150 to 300 mL of fluid have collected.

---

**TABLE 19.7**
**Physiologic Requirements for Continence**

Motivation to be continent
Adequate cognitive function
Adequate mobility and dexterity
Normal lower urinary tract function
Storage:
  No involuntary bladder contractions
  Appropriate bladder sensation
  Closed bladder outlet
  Low pressure accomodation of urine
Emptying:
  Normal bladder contraction
  Lack of anatomic obstruction
  Coordinated sphincter relaxation and bladder contraction
  Absence of environmental or iatrogenic barriers

**TABLE 19.8**
**Age-Related Changes Contributing to Incontinence**

**In Females**

Estrogen deficiency
  Weak pelvic floor and bladder outlet
  Decreased urethral muscle tone
  Atrophic vaginitis

**In Males**

Increased prostatic size
  Impaired urinary flow
  Urinary retention
  Detrusor muscle instability

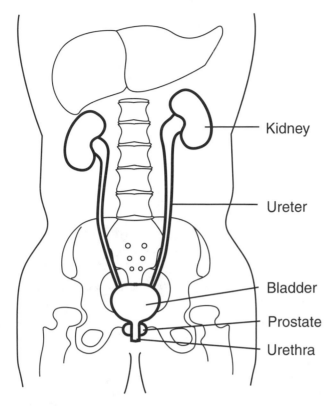

**FIGURE 19.4** Diagram of the major components of the urinary tract in man.

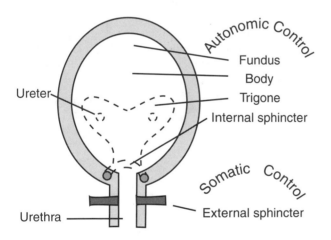

**FIGURE 19.5** Schematic of urinary bladder with structures under autonomic and somatic nervous control.

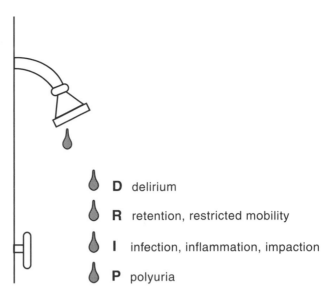

**FIGURE 19.6** Mnemonic device for the major causes of acute urinary incontinence.

contribute to the etiology of incontinence (Table 19.8). For example, in postmenopausal women, the reduction in estrogen levels weakens the tissues around the pelvic floor and bladder outlet, decreases the tone of urethral smooth muscle, and facilitates the occurrence of atrophic vaginitis; all of these contribute to the decrease in urethral contractility and pressure with aging.[48] In elderly men, hypertrophy of the prostate represents one of the major causes of incontinence, as it leads to a decline in urinary outflow rate, increased risk of urine retention, and increased insta-

bility of the detrusor muscle (see below). Other causes include delirium, effects of drugs such as anticholinergics (inducing urinary retention) and diuretics (inducing polyuria), and diseases such as infections and diabetes. A mnemonic device to remember the major causes of acute incontinence is the word DRIP, where *D* stands for delirium, *R* for urinary retention due to anticholinergic drugs and restricted motility, *I* for infection, and *P* for polyuria due to diuretics or diabetes (Figure 19.6).

Many of the causes of acute incontinence may be easily diagnosed and reversed. However, when they have been addressed without any significant improvement, the incontinence is considered persistent, and its management is considerably more difficult than the acute form. Persistent urinary incontinence is categorized into four types to which correspond specific etiologies and interventions:

- Stress incontinence, due to increased intra-abdominal pressure as in coughing and laughing, is caused by weakness of the pelvic floor musculature and is common in multiparous women
- Urge incontinence, due to the inability to delay voiding after perception of bladder fullness, may be caused by mild outflow obstruction or by a central nervous system disorder such as stroke
- Overflow incontinence, due to a leakage of urine resulting from mechanical forces of an overdistended, *a*contractile (i.e., failing to contract) bladder, may be due to prostatic obstruction or to neurogenic disturbances such as a neurogenic acontractile bladder
- Functional incontinence, due to inability or psychologic unwillingness to get to the toilet, is

**Systemic and Organismic Aging**

caused by cognitive deficits, psychologic conditions, or unavailability of caretakers

### DIFFERENTIAL DIAGNOSIS OF INCONTINENCE TYPES

It is based not only on the urinary symptoms but also on a complete history and physical examination, including evaluation of the mental status, physical activity, and ambulation (walking). An abdominal exam should look for signs of bladder distension; a rectal exam for prostate size, rectal sphincter tone, and the presence of fecal impaction; a pelvic exam for uterine prolapse or atrophic vaginitis. Laboratory tests should include a urinalysis, a urine culture, and measurement of postvoid residual bladder urine volume (with the use of an indwelling catheter). A postvoid urine volume greater than 100 mL indicates failure of complete bladder emptying and warrants further urologic or gynecologic evaluation.

### 3. Management of Incontinence

Management varies with the type of incontinence (Table 19.9). Strengthening of abdominal and pelvic muscles by specific exercise, administration of estrogen, surgical bladder neck suspension for stress incontinence; administration of bladder relaxants for urge incontinence; intermittent catheterization or removal of the prostate, for overflow incontinence; and for functional incontinence, habit training, scheduled toileting, undergarment devices, and indwelling catheterization (taking into account the risks involved, such as danger of infection).[49–52]

As indicated above, urinary incontinence is a widespread problem among the elderly. And yet, despite its prevalence and its serious consequences for those affected, it receives little to no attention because of social taboos and ignorance of its physiologic and pathologic causes. Awareness of the condition is important, as even advanced cases may respond well to treatment. Urinary incontinence in the elderly can and must be treated in each and every case. Specific questioning as to urinary continence must be carried out in all routine evaluations of elderly patients and, in this, as in all disorders, an understanding of the

**TABLE 19.9**
**Management of Urinary Incontinence**

| Type | Management |
|---|---|
| Stress | Exercises |
| | Alpha-adrenergic agonists |
| | Estrogen |
| | Surgery |
| Urge | Bladder relaxants |
| | Surgery |
| Overflow | Alpha-adrenergic antagonists |
| | Catheterization |
| Functional | Habit training |
| | Scheduled toileting |
| | Hygienic devices |

physiopathology is essential for a rational diagnostic evaluation and treatment.

### 4. Other Bladder Dysfunctions in Old Age

In addition to urinary incontinence, other major dysfunctions of the lower urinary tract are listed in Table 19.2. Some of these dysfunctions are amenable to pharmacologic interventions, especially infections, efficiently treatable with antimicrobial agents; however, high urinary antibody levels are often associated with decreased survival rates.[53] Urinary retention is responsive to treatment with cholinergic agonists that facilitate bladder emptying by stimulating contraction of the detrusor muscle and relaxation of the internal sphincters. The characteristics of neural control of micturition are presented in Table 19.10. While a number of drugs may act on other parasympathetic targets and, therefore, induce side effects, they are preferred to bladder catheterization with its attendant danger of infection.

In addition to being the target of specific drugs administered for therapeutic effects, the lower urinary tract, particularly the bladder, may bear the consequences or side-effects of many prescribed as well as over-the-counter

**TABLE 19.10**
**Neural Control of Micturition**

| Muscle (Type) | Parasympathetic Nerves (Cholinergic) | Sympathetic Nerves (Adrenergic)[a] | Somatic Nerves |
|---|---|---|---|
| Detrusor (smooth muscle) | Contraction +++ | Relaxation + | No effect |
| Internal sphincter (smooth muscle) | No effect | Contraction ++ | No effect |
| External sphincter (striated muscle) | No effect | No effect | Relaxation ++ |

[a] In the male, adrenergic stimulation of smooth muscle sphincter causes contraction to prevent retrograde flow of semen into the urinary bladder at ejaculation; the same stimulation relaxes the detrusor muscle to prevent coincidental contraction of the muscle during ejaculation.

*Note:* Number of "+"s indicates strength of relaxation or contraction.

drugs. Thus, the bladder may become a victim of polypharmacy (Chapter 23). Drugs in this category include decongestants, antihistamines, antidiarrheals, antipsychotics (such as phenothiazines), and tricyclic antidepressants. Even if taken as prescribed, these drugs may add up to increasing toxicity. Most of these substances have some autonomic activity. The greater number has anticholinergic actions, that is, they block the parasympathetic responses: for example, over-the-counter sleeping, asthma, and antidiarrheal medicines contain the belladonna alkaloids such as atropine or scopolamine or synthetic substitutes. They block the parasympathetic responses, inhibit micturition, and induce a degree of urinary retention.

Sympathomimetics (mimicking the actions of the sympathetic branch of the autonomic nervous system), such as some decongestants and alpha-adrenergic blockers, relax the detrusor muscle, constrict the sphincters, and promote urine retention. Some of the alpha-adrenergic blockers, extensively utilized as antihypertensive drugs, also have some additional side effects, distressing in the older male, for they also prevent contraction of the vas deferens and inhibit ejaculation.

## C. WATER AND ELECTROLYTE DISTRIBUTION AND ACID–BASE BALANCE

The kidney of the normal, healthy, older individual is capable of maintaining water and electrolyte distribution and acid–base balance within homeostatic limits. This is remarkable, as some changes occur in *body composition*. With aging, there is generally a loss of fat-free body weight and tissue mass, a reduction in body mineral, and a gain in body fat.[54] These age differences carry over to sex differences, with women showing a greater increase in total body weight due to increased body fat and men maintaining their body weight. This maintained body weight is due to reciprocal changes in lean body mass (decreased) and body fat (increased). Some gross comparisons in "reference man" between 25 and 70 years show the following percent changes, at age 25: fat, 14%, water, 61%, cell solids, 19%, and bone mineral 6%; and at age 70: 30%, 53%, 12%, and 5%, respectively.

These changes in body composition may vary depending on several factors related either to the individual or the methodology. For example, gain in body fat often increases into the age of 60 years but declines thereafter and, at all ages, depends on a number of variables, among which is the degree of physical activity (Chapter 24). In contrast, lean body mass or body cell mass (as measured by body density through water displacement or helium dilution) continues to decrease with aging, more in males than in females. Some studies show a 3.6% decrease per decade from age 30 to 70 and, thereafter, 9% per decade. If non-fat mass is measured by total body potassium, one observes that potassium decreases with aging as does lean body mass early, but with advancing age, after 70 years, potassium decrease is more pronounced than lean body mass. At older ages, the degenerative loss of lean tissue is in part replaced by other tissue that is low in potassium[54] and has a lower metabolic rate.[55] An alternative explanation with respect to regression of metabolic rate with aging is that the oxygen uptake of cells in old individuals is not significantly different than in young, rather there are fewer functioning cells.[55] While *total body water* is diminished with age, extracellular water remains unchanged, and intracellular water decreases. The reduction in total body and intracellular water in the absence of change in extracellular water can be taken as further support for a loss of functioning cells (lean body mass) with increasing age.

Under normal physiologic conditions and in young adult individuals, *acid–base balance* is maintained by renal excretion of hydrogen ions (eliminated as acids or ammonia); these ions are generated during metabolism of dietary protein and other metabolic processes. When a disturbance in systemic pH (taken as an index of $H^+$ ions in a solution) occurs as a result of an excess or loss of acid or base, shifts in body buffers and adjustments of respiratory excretion of carbon dioxide promptly intervene to correct the pH until it can be stabilized by the appropriate changes in renal excretion of acids. Healthy adults, young and old, manifest a low-grade diet-dependent metabolic acidosis; in some old individuals, the severity of acidosis may increase significantly due to diet, medications, or disease as well as the inability of the kidney to increase the hydrogen ion excretion to match the increased metabolic acidosis.[56,57]

## II. THE PROSTATE

*JOYCE LEARY*

### A. FUNCTION OF THE PROSTATE

The prostate gland functions as a secondary sex organ of the male reproductive system (Chapter 12) (Box 19.3). During ejaculation, the prostate gland expels stored prostatic fluid into the urethra, contributing up to one third of seminal fluid volume. Prostatic fluid consists of water, zinc, citric acid, acid phosphatase, fibrolysin, prostate-specific antigen (PSA), prostaglandins, and proteases. Prostatic enzymes are involved in liquefying ejaculated semen inside the vagina and neutralizing the acidic vaginal environment to facilitate fertilization (Table 19.11).

Approximately the size of an almond at birth, the prostate gland does not resume growth until puberty, when testosterone levels in the male rise dramatically (Box 19.3). Growth, differentiation, and secretion of prostatic fluid depend on the presence of testosterone and dihydrotestosterone (DHT), and, thus, the prostate is said to be androgen-dependent. The amount of testosterone synthesized in

**Systemic and Organismic Aging**

## Box 19.3
## Anatomy and Histology of the Prostate

The prostate is divided into zones by the traversing urethra: the posterior glandular zone and the anterior fibromuscular stroma. The posterior zone is further divided into four zones: the transition zone, central zone, peripheral zone, and periurethral zone. The transition zone surrounds the proximal prostatic urethra, and the central zone surrounds the ejaculatory ducts. The peripheral zone surrounds the distal prostatic urethra, makes up the bulk of prostatic volume, and is the most common site of prostate cancer. Finally, the small periurethral zone is embedded in periurethral smooth muscle and is the most common site of benign prostatic hyperplasia (Figure 19.7). Although these zones are not clearly separated anatomically, they are useful, because pathologic conditions of the prostate have a predilection for occurring in specific zones. The prostate is an encapsulated organ on its posterior and lateral sides, but anteriorly and apically, the anterior fibromuscular stroma makes up the outermost portion of the organ.[58-61]

Histologically, the prostate gland is made up of branched tubulo-alveolar glands arranged concentrically around the urethra. The secretory epithelium of the prostate gland is heterogeneous, exhibiting cuboidal, columnar, and pseudostratified cells, and secretes prostatic fluid. The stroma of the prostate gland is composed mainly of dense collagen, fibroblasts, smooth muscle, and immune cells. Innervation of the prostate is autonomic, with both sympathetic and parasympathetic components. During ejaculation, the stored secretory products of the prostate epithelial cells are expelled into the prostatic urethra.

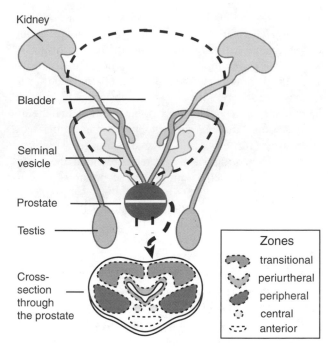

**FIGURE 19.7** Diagrammatic location and structure of the prostate with major zones.

## TABLE 19.11
## Prostatic Fluid

- A slightly alkaline fluid that increases sperm motility and aids in fertilization by neutralizing acidic secretions of the vas deferens and vagina.
- Major components:
  - Water
  - Zinc
  - Citric acid
  - Prostaglandins
  - Acid phosphatase
  - Prostate Specific Antigen (PSA)
  - Other proteases

## TABLE 19.12
## The Prostate and Testosterone

- The healthy prostate is dependent on androgens for growth
- In the prostate: testosterone → dihydrotestosterone (DHT)
- The enzyme catalyzing this reaction is 5-·-reductase
- DHT stimulates growth of the prostate

the testes is regulated via the typical feedback loop with the hypothalamus and anterior pituitary gland (Chapter 12). The hypothalamus releases gonadotropin-releasing hormone (GnRH), which signals the anterior pituitary to release luteinizing hormone (LH) and follicle-stimulating hormone (FSH). In the testes, FSH induces upregulation of LH receptors on testicular Leydig cells. LH then acts on Leydig cells to induce testosterone production. Finally, testosterone negatively feeds back to both the hypothalamus and anterior pituitary to inhibit the release of GnRH, LH, and FSH. DHT is made in the prostate from testosterone by the action of the enzyme 5 α-reductase and promotes epithelial growth, differentiation, and secretion (Table 19.12).

During puberty, the prostate grows approximately to the size of a walnut. Once adulthood is reached, prostate growth ceases, but prostatic fluid continues to be produced. With old age, the outer zones of the prostate progressively atrophy, while, unlike any other organ in the body, the inner zones of the prostate begin to grow again and grow until death[58] (Table 19.13).

**TABLE 19.13**
**Normal Aging of the Prostate**

**After age 40**
Outer regions:
- Atrophy of smooth muscle and proliferation of connective tissue
- Flattening of secretory epithelium

Inner region:
- Increase in the number of cells present (hyperplasia)

**After age 60**
- Slower, but more uniform atrophy of the prostate
- Accumulation of prostate concretions

## B. THE PROSTATE AND URINARY FUNCTION

Because the prostate surrounds the neck of the bladder and the prostatic urethra, any pathology of the prostate gland can contribute to urinary dysfunction. The three most significant pathologic conditions affecting the prostate gland all have the potential to cause urinary dysfunction: prostatitis, benign prostatic hyperplasia (BPH), and prostate carcinoma. Prostatitis, an inflammation or infection of the prostate, is the only of these three conditions that is not associated with aging, occurring most frequently in men between the ages of 20 and 40. In contrast, BPH and prostate cancer occur most often in men over 50. While prostatitis is an inflammatory process, BPH and prostate cancer involve cellular proliferation.

## C. BENIGN PROSTATIC HYPERPLASIA (BPH)

With age (around 40 years), the inner zones of the prostate begin to proliferate. These late changes of the inner prostate transitional and periurethral zones can lead to enlargement of the prostate with or without obstruction of the urethra. While 80% of males have histologic evidence of a hyperplastic prostate by age 80, only up to 25% of these men require treatment for the condition[62] (Table 19.14). Surgery to correct BPH in men over 65 is a leading cause of surgical intervention, second only to cataract surgery.[63] In addition to other symptoms, men with BPH most often complain of increased frequency of urination, difficulty starting and stopping urination, weak urine stream, feeling that the bladder has not emptied completely, urinary retention, or painful urination.[64] In severe cases, the inability to void the bladder can lead to distention and hypertrophy of the bladder, bladder infections, and even kidney infections.[63]

### 1. Etiology

A specific etiology for BPH has been difficult to identify. However, it is clear that two important factors must be present for BPH to develop. First, men with BPH are over age 40. Second, men with BPH always have testosterone-producing testes. Studies have shown that castration results in a decrease in prostate size,[65] and androgen ablation therapy has been shown to cause a decrease in number and a shrinkage of luminal epithelial cells in the prostate.[66] Recall that the prostatic epithelial cells are dependent on the androgen DHT for growth. It is thought that a wide variety of interacting intrinsic and extrinsic factors are responsible for abnormal prostate growth (Table 19.15). Intrinsic factors include proliferation of stromal elements, fibroblasts, smooth muscle cells, and extracellular matrix proteins, while extrinsic factors include hereditary predispositions, dietary factors, environmental toxins, endocrine factors, and more.[67,68] With aging, rising estrogen levels are thought to act synergistically with androgens by inducing transcription of the androgen receptor.[69] Notably, the effects of diet on BPH are evidenced by the fact that BPH is not seen in Asian countries to the same degree as it is seen in the United States. However, on entering the United States, Asian immigrants develop evidence of BPH mirroring that of the American population.[70] In contrast to prostate cancer, the prevalence of BPH does not differ racially within the United States.[62]

**TABLE 19.14**
**Synopsis of Benign Prostatic Hyperplasia (BPH) Characteristics**

- Caused by growth of the prostate from about age 40 until death
- Affects 50% of men >50 years old
- Affects 95% of men >70 years old
- Clinical symptoms due to obstruction of the urethra are present up to 25% of men with histologic evidence of BPH
- BPH tissue resembles normal prostate tissue with increased amounts of smooth muscle, glandular, and stromal components
- An enlarged prostate can strangle the urethra
- BPH is *not* found in men who have been castrated or men who lack 5-α-reductase

**TABLE 19.15**
**Possible Risk Factors for Benign Prostatic Hyperplasia (BPH) and Prostate Cancer**

| Possible Risk Factors for BPH | Possible Risk Factors for Prostate Cancer |
|---|---|
| Aging | Genetic predisposition |
| Use of anabolic steroids | Tobacco exposure |
| Dietary factors | Cadmium exposure |
| Genetic predisposition | Vitamin A deficiency |
| Environmental toxins | Vasectomy |
| No other major risk factors | Sexually transmitted diseases |
| | Mutagenic hormonal factors |
| | Dietary factors (particularly high level of animal fat) |

**Systemic and Organismic Aging**

## 2. Treatment

The first strategy for BPH treatment is watchful waiting, in which the patient is monitored by his physician but receives no other treatment.[71] Watchful waiting is undertaken if the patient's symptoms are minimal. Beyond watchful waiting, 5 α-reductase inhibitors and $\alpha_1$-adrenergic antagonists are drugs of choice for minimizing the symptoms of BPH. Five α-reductase inhibitors act to block the conversion of testosterone to DHT within the prostate, thereby limiting further growth of the gland. Alpha-adrenergic antagonists inhibit the contraction of smooth muscle within the prostate, thereby diminishing the strangulation of the urethra by the prostate. Studies examining these two main pharmacologic therapies for BPH have revealed the heterogeneity of the condition. The variable effectiveness of these drugs in different patients suggests there are two distinct forms of BPH, one dominated by epithelial growth and the other by smooth muscle proliferation. In all cases, 5 α-reductase inhibitors are ideal for slowing growth of the prostate. In patients with a high degree of smooth muscle overgrowth, however, $\alpha_1$-antagonist drugs can further reduce constriction of the urethra through relaxation of prostatic and nonprostatic smooth muscle.[72–76] It is also important to note that BPH patients with a high degree of smooth muscle proliferation may exhibit symptoms of BPH without a significant increase in prostatic volume.[77] This finding can be explained by the fact that smooth muscle proliferation may not add volume but may still have the potential to strangulate the urethra via contraction. Excessive epithelial and stromal proliferation, on the other hand, will inevitably result in a large prostate.

When watchful waiting and drug treatment are not successful, surgical procedures may be required to reduce symptoms. Accessing the prostate via the urethra, incisions can be made in the prostate, or most commonly, the inner portion of the prostate can be removed, reducing urethral constriction. Open prostatectomy and laser prostatectomy are common procedures in which the entire prostate is surgically removed.[71] Further treatments include androgen suppression, aromatase inhibitors, thermotherapy, high-intensity focused ultrasound, and others.

## D. PROSTATE CANCER

The most common form of cancer in men is prostate cancer, and usually it occurs after age 50. As men age, their risk of developing prostate cancer increases: 75% of diagnoses are made in men over 75 years of age.[78] While up to 30% of men have histologic (microscopic) evidence of prostate cancer when they die, only about 10 to 16% are diagnosed with the disease during their lifetimes.[78,79] Moreover, prostate cancer is, in most cases, a slow-growing cancer compared to other forms. For these reasons, it

has been said that men die *with* prostate cancer far more frequently than they die *of* prostate cancer. The lifetime risk of dying from prostate cancer is about 3%.[78] In most cases, prostate cancer is an adenocarcinoma in that it arises from the glandular epithelium, the secretory cells of the gland. Approximately 80% of cases are located in the posterior peripheral zone.

### 1. Ethnic Differences

Notably, unlike BPH, different racial groups within the United States have significantly different incidences of prostate cancer. African Americans show the world's highest incidence in prostate cancer and develop the disease with a startlingly higher incidence than do white Americans.[78] It is much more rare, however, for Asian Americans to develop prostate cancer. These racial differences in prostate cancer incidence have helped motivate research on genetic contributions to its etiology. While Asian immigrants to the United States experience an increase in incidence of prostate cancer, the rate of cancer development in these populations remains lower than that for native Americans. Such evidence suggested not only a genetic influence, but also the influence of environmental factors, especially diet, in the development of cancer. Indeed, studies have linked high consumption of animal fat to prostate cancer.[79]

Genetic predisposition is exhibited by men who have a tenfold increase in cancer risk when three relatives have the disease, and an allele associated with familial clustering is thought to be located on chromosome 1.[80,81] Other risk factors for prostate cancer possibly include vitamin A deficiency, tobacco and cadmium exposure, vasectomy, sexual activity, sexually transmitted disease, and various other mutagenic hormonal factors.[78,79,81]

### 2. Genetic Susceptibility

Prostate cancer pathogenesis is not the same in all individuals. However, studies have shown that certain genetic alterations are common on the path to malignancy. Genetic alterations such as the loss of tumor suppressor genes (including *p53* and *Rb*), and the amplification of oncogenes (including *MYC* seen in a variety of other tumors and *HER2/neu* seen in breast cancer), give prostate cells the ability to proliferate and avoid apoptosis. Racial differences in prostate cancer incidence may be explained by differences in the androgen receptor gene. A region of CAG repeat sequences in the gene varies in number from person to person. On average, African Americans show the fewest number of CAG repeats in the gene, while Asians have the highest number. Activity of the androgen receptor has been shown to be inversely correlated with the number of CAG repeat sequences in the gene. Thus, Asians have more CAG repeats, less androgen receptor activity, and a lower incidence of prostate cancer.[81] These findings suggest that an increase in androgen effect is necessary for the development and progression of prostate cancer.[81] Indeed, the majority of prostate cancers are, like BPH, androgen dependent, or at least, androgen-promoted. In advanced stages, however, genetic alterations allow prostate cancer to become androgen independent. The development of androgen independence appears to be due to an accumulation of mutations in the androgen receptor gene, preventing the receptor from functioning normally within cancer cells.[82,83] Once a state of androgen independence occurs, treatment and management become increasingly difficult.

### 3. Diagnosis and Treatment

Prostate cancer is suspected after a patient's digital rectal exam (DRE) is found to be abnormal and after an elevated serum level of prostate specific antigen (PSA) has been detected (see below). Only 20 to 25%[78] of subsequent prostate biopsies will show unusual glandular tissue: a missing outer basal layer of cells, or large, vacuolated nuclei with one or more nucleolus.[63] Pathologists grade prostate cancer using Gleason's scoring system, wherein the most normal-looking tissue obtains a score of 1, and the most cancerous-looking tissue obtains a score of 5. Because biopsies are rarely uniform in structure, a score is given to the predominant pattern visible, and a second score is given to the second-most predominant pattern.

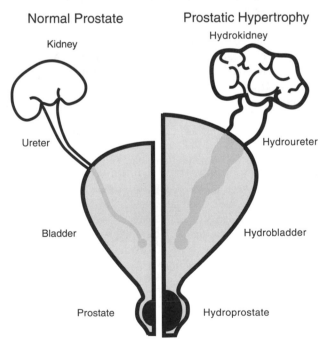

**FIGURE 19.8** Consequences of prostate hypertrophy. The enlarged prostate constricts the urethra with consequent difficulties in bladder emptying and the occurrence of retrograde filling of the ureters and the renal pelvis (hydronephrosis).

These two scores are added together to yield the Gleason score, which can thus be between 2, closest to normal, and 10, most cancerous. Thus far, the Gleason score is the most significant prognostic factor for prostate cancer morbidity and mortality.[79]

Once the diagnosis of prostate cancer has been made, there are multiple treatment options. The appropriate option(s) for an individual depends on his life expectancy, overall health status, personal preferences, size of his prostate, and stage of the disease. Treatments include watchful waiting, surgical procedures to remove prostate tissue, radiation therapy, hormonal therapy, and cryotherapy (freezing cells). Young men are often encouraged to pursue aggressive therapies such as radical prostatectomy, a complete removal of the prostate, so as to guarantee a normal life span if possible. On the other hand, the slow-growing nature of most prostate cancers allows men over 70 to focus less on long-term effects and outcome, so therapies like watchful waiting may be more realistically considered. If the cancer is fully contained within the prostatic capsule, surgery can often eradicate it in its entirety. However, if the cancer has spread beyond the capsule or metastasized to other areas (most often vertebral bone — the finding of cancer in the vertebral bones of older men is virtually diagnostic of metastatic prostate cancer),[63] surgery is usually not a tenable option, leaving radiation and hormonal therapies as the most effective treatments. Researchers continue looking for new prostate cancer therapies. Gene therapy, vaccinations, and new drugs may be on the horizon.[84]

### 4. Prostate Specific Antigen: The Controversy over Screening for Prostate Cancer

Prostate specific antigen (PSA) is one of the products of the prostate gland that contributes to seminal fluid. PSA is an enzyme that helps anticoagulate semen following ejaculation. Normally, a small amount of PSA is found in the blood, usually less than 4ng/mL.[85] However, prostatitis, BPH, and especially prostate cancer can lead to elevations in serum PSA levels. In addition, serum PSA levels increase with age, so a normal PSA level for a 70-year-old man might be a high PSA level for a 50-year-old man. Some studies have suggested that PSA may vary racially, African American men showing significantly higher levels than white men.[86] Other factors that can contribute to an elevated serum PSA include recent ejaculation, finasteride medication (a 5 α-reductase inhibitor usually used for BPH), urinary tract infection, or recent prostate procedure (not including the digital rectal exam[87,88]).

While PSA levels can be elevated in this wide variety of conditions, none causes as dramatic a rise as does prostate cancer (Box 19.4). It is suspected that blood levels of PSA rise more dramatically in cancer than in other conditions, because in cancer, barriers to the

**Systemic and Organismic Aging**

---

**Box 19.4**

**Specificity of PSA Test**

In an attempt to improve the specificity of the PSA test, researchers are experimenting with variations on the basic PSA test. There are two types of PSA in the bloodstream, protein-bound and unbound (free). In cancer, there is a higher percentage of bound PSA, whereas in BPH and other conditions, there is a higher percentage of free PSA. Thus, a free-PSA test would show a reduction in percentage of free PSA with prostate cancer. A more direct measurement of bound PSA is also under investigation. Furthermore, the PSA velocity test will measure the rate of increase with time, while the PSA density test will measure PSA in relation to overall prostate volume. None of these tests is currently in use.[85,92]

**One of the most controversial topics in medicine surrounds the use of the PSA test to screen for prostate cancer.** Since its advent in the early 1980s, a surge in prostate cancer screening has led to an increase in diagnosis of the disease, most notably in smaller, earlier-staged tumors.[93,94] The PSA test detects more tumors while still confined to the gland than does the digital rectal exam.[95] Proponents of prostate cancer screening argue that finding cancer early allows a wider variety of treatment options and, therefore, will decrease mortality from prostate cancer. Unfortunately, because it is still soon after the onset of PSA testing, it is too early to tell if detection of prostate cancer early does, indeed, increase overall survival. This being the case, opponents of widespread screening fear the increase in diagnosis may simply cause unnecessary, anxiety-provoking, and possibly life-altering procedures for men who otherwise might not suffer from their slow-growing, clinically insignificant prostate tumors. In other words, they point out that the rising number of diagnoses of early prostate cancers may be slow-growing cancers that will never progress to cause problems for patients, that wide-span PSA testing leads to the labeling of a problem that does not exist. In the next few years, some key studies should begin to provide definitive evidence whether a possible decrease in mortality from prostate cancer has occurred since the advent of the PSA test.

**Various organizations have issued guidelines for the screening of prostate cancer.** Some, such as the American Cancer Society and American Urologic Association, recommend the PSA test be offered to men over age 50 on a yearly basis, and to men over age 40 who are African American or who have a positive family history for the condition. On the other hand, the National Cancer Institute and American College of Preventive Medicine are not yet recommending annual PSA screening for prostate cancer.

---

bloodstream are often damaged, allowing PSA to escape the prostate.[89] In 80% of cancer cases, a highly elevated PSA will be found (>4 ng/mL in a man over 60).[90,91] However, the inverse is also true: 65 to 80% of the time the PSA is elevated, no cancer will be found.[92] Thus, PSA screening allows for a high sensitivity for cancer, but a low specificity.

# REFERENCES

1. Lindeman, R.D., Overview: Renal physiology and pathophysiology of aging, *Am. J. Kidney Dis.*, 16, 275, 1990.
2. Cameron, J.S., Renal disease in the elderly: Particular problems, in *Issues in Nephrosciences,* D'Amico, G. and Colasanti, G., Eds., Wichting, Milano, 1995.
3. Marino, A.G. and Macias-Nunez, J.F., Renal Disease, in *Principles and Practice of Geriatric Medicine,* 3rd ed., Vol. 2, Pathy, J.M.S., Ed., Wiley, New York, 1998.
4. Davies, D.F. and Shock, N.W., Age changes in glomerular filtration rate, effective renal plasma flow and tubular excretory capacity in adult males, *J. Clin. Invest.*, 29, 496, 1950.
5. Rowe, J.W. et al., The effect of age on creatinine clearance in men: A cross-sectional and longitudinal study, *J. Gerontol.*, 31, 155, 1976.
6. Lindeman, R.D., Tobin, J., and Shock, N.W., Longitudinal studies on the rate of decline in renal function with age, *J. Am. Geriatr. Soc.*, 33, 278, 1985.
7. Malmrose, L.C. et al., Measured versus estimated creatinine clearance in a high-functioning elderly sample: MacArthur foundation study of successful aging, *J. Am. Geriatr. Soc.*, 41, 715, 1993.
8. Durakovic, Z. and Mimica, M., Proteinuria in the elderly, *Gerontology,* 29, 121, 1983.
9. McDermott, G.F. et al., Glomerular dysfunction in the aging Fischer 344 rat is associated with excessive growth and normal mesangial cell function, *J. Gerontol.*, 51, M80, 1996.
10. Wickelgren, I., First components found for key kidney filter, *Science,* 286, 225, 1999.
11. Ruotsalainen, V. et al., Nephrin is specifically located at the slit diaphragm of glomerular podocytes, *Proc. Natl. Acad. Sci. USA*, 96, 7962, 1999.
12. Li, C. et al., CD2AP is expressed with nephrin in developing podocytes and is found widely in mature kidney and elsewhere, *Am. J. Physiol. Renal Physiol.*, 279, F785, 2000.
13. Miller, M., Nocturnal polyuria in older people: Pathophysiology and clinical implications, *J. Am. Geriatr. Soc.*, 48, 1321, 2000.
14. Wenkert, D. et al., Functional characterization of five V2 vasopressin receptor gene mutations, *Mol. Cell. Endocrinol.*, 124, 43, 1996.

15. Ouslander, J.G. et al., Arginine vasopressin levels in nursing home residents with nighttime urinary incontinence, *J. Am. Geriatr. Soc.,* 46, 1274, 1998.

16. Ouslander, J.G. et al., Atrial natriuretic peptide levels in geriatric patients with nocturia and nursing home residents with nighttime incontinence, *J. Am. Geriatr. Soc.,* 47, 1439, 1999.

17. Remuzzi, G. and Bertani, T., Pathophysiology of progressive nephropathies, *N. Engl. J. Med.,* 339, 1148, 1998.

18. Bolton, W.K. and Sturgill, B.C., Ultrastructure of the aging kidney, in *Aging and Cell Structure,* Vol. 1, Johnson, Jr., J.E., Ed., Plenum Press, New York, 1981.

19. Everitt, A.V., Porter, B.D., and Wyndham, J.R., Effects of caloric intake and dietary composition on the development of proteinuria, age-associated renal disease longevity in the male rat, *Gerontology,* 28, 168, 1982.

20. Brenner, B.M., Meyer, T.W., and Hostetter, T.H., Dietary protein intake and the progressive nature of kidney disease: The role of hemodynamically mediated glomerular injury in the pathology of progressive glomerular sclerosis in aging, renal ablation and intrinsic renal disease, *N. Engl. J. Med.,* 307, 652, 1982.

21. Phillips, T.L. et al., Reduced thirst after water deprivation in healthy elderly men, *N. Engl. J. Med.,* 311, 753, 1984.

22. Helderman, J.H. et al., The response of arginine vasopressin to intravenous ethanol and hypertonic saline in man: The impact of aging, *J. Gerontol.,* 33, 39, 1978.

23. Miller, J.H. and Shock, N.W., Age differences in renal tubular response to antidiuretic hormone, *J. Gerontol.,* 9, 446, 1953.

24. Lindeman, R.D. et al., Influence of age, renal diseases, hypertension, diuretics and calcium on antidiuretic responses to suboptimal infusion of vasopressin, *J. Lab. Clin. Med.,* 68, 206, 1966.

25. Modesto-Segonds, A. et al., Renal biopsy in the elderly, *Am. J. Nephrol.,* 13, 27, 1993.

26. Druml, W. et al., Acute renal failure in the elderly 1975–1990, *Clin. Nephrol.,* 41, 342, 1994.

27. Piccoli, G. et al., Death in conditions of cachexia: The price for the dialysis treatment of the elderly? *Kidney Int.,* 43, s282, 1993.

28. D'Amico, G., Comparability of the different registries on Renal Replacement Therapy, *Am. J. Kidney. Dis.,* 25, 113, 1995.

29. Agadoa, L.Y. and Eggers, P.W., Renal replacement therapy in the United States: Data from the United States Renal Data System, *Am. J. Kidney. Dis.,* 25, 119, 1995.

30. Coulam, C.B. and Zincke, H., Successful pregnancy in a renal transplant patient with a 75-year-old kidney, *Surg. Forum,* 32, 457, 1981.

31. Dossetor, J.B., Selection of elderly patients for transplantation, in *Principles and Practice of Geriatric Surgery,* Rosenthal, R.A., Zenilman, M.E., and Katlic, M.R., Eds., Springer-Verlag, New York, 2001.

32. Cameron, J.S. et al., Transplantation in elderly recipients: Is it worthwhile?, *Geriatr. Nephrol. Urol.,* 4, 93, 1994.

33. USRDS (United States Renal Data System). Annual Data Report, *Am. J. Kidney. Dis.,* 18 (suppl.), 38, 1991.

34. Perry, H.M. et al., The effects of thiazide diuretics on calcium metabolism in the aged, *J. Am. Geriatr. Soc.,* 41, 818, 1993.

35. Field, T.S. et al., The renal effects of non-steroidal anti-inflammatory drugs in older people: Findings from the Established Populations for Epidemiologic Studies of the Elderly, *J. Am. Geriatr. Soc.,* 47, 507, 1999.

36. Feigelbaum, L., Geriatric medicine and the elderly patient, in *Current Medical Diagnosis and Treatment,* Schroeder, S.A. and Krupp, M.A., Eds., Lange, Los Altos, CA, 1991.

37. Farrar, D.J. and Webster, G.M., The bladder and urethra, in *Principles and Practice of Geriatric Medicine,* 3rd ed., Vol. 2, Pathy, J.M.S., Ed., Wiley, New York, 1998.

38. O'Donnell, P., Urinary incontinence in the elderly, in *Principles and Practice of Geriatric Surgery,* Rosenthal, R.A., Zenilman, M.E., and Katlic, M.R., Eds., Springer-Verlag, New York, 2001.

39. Roberts, R.O. et al., Urinary incontinence in a community-based cohort: Prevalence and healthcare seeking, *J. Am. Geriatr. Soc.,* 46, 467, 1998.

40. Thom, D., Variation in estimates of urinary incontinence prevalence in the community: Effects of differences in definition, population characteristics and study type, *J. Am. Geriatr. Soc.,* 46, 473, 1998.

41. Ouslander, J.G. et al., Nighttime urinary incontinence and sleep disruption among nursing home residents, *J. Am. Geriatr. Soc.,* 46, 463, 1998.

42. Roberts, R.O. et al., Prevalence of combined fecal and urinary incontinence: A community-based study, *J. Am. Geriatr. Soc.,* 47, 837, 1999.

43. Johnson, T.M. et al., 2nd ed., Urinary incontinence and risk of death among community-living elderly people: Results from the National Survey on Self-Care and Aging, *J. Aging Health,* 12, 25, 2000.

44. Tinetti, M.E. et al., Shared risk factors for falls, incontinence and functional dependence. Unifying the approach to geriatric syndrome, *J. Am. Med. Assoc.,* 237, 1348, 1995.

45. Brown, J.S. et al., Urinary incontinence: Does it increase risk for falls and fractures? Study of Osteoporotic Fractures Research Group, *J. Am. Geriatr. Soc.,* 48, 721, 2000.

46. Skelly, J. and Flint, A.J., Urinary incontinence associated with dementia, *J. Am. Geriatr. Soc.,* 43, 286, 1995.

47. Dugan, E. et al., The association of depressive symptoms and urinary incontinence among older adults, *J. Am. Geriatr. Soc.,* 48, 413, 2000.

48. Thom, D.H. and Brown, J.S., Reproductive and hormonal risk factors for urinary incontinence in later life: A review of the clinical and epidemiologic literature, *J. Am. Geriatr. Soc.,* 46, 1411, 1998.

49. Weinberger, M.W., Goodman, B.M., and Carnes, M., Long-term efficacy of non-surgical urinary incontinence treatment in elderly women, *J. Gerontol.,* 54, M117, 1999.

**Systemic and Organismic Aging**

50. Brown, J.S., Posner, S.F., and Stewart, A.L., Urge incontinence: New health-related quality of life measures, *J. Am. Geriatr. Soc.*, 47, 980, 1999.

51. Chiang, L. et al., Dually incontinent nursing home residents: Clinical characteristics and treatment differences, *J. Am. Geriatr. Soc.*, 48, 673, 2000.

52. Johnson, T.M. et al., Self-care practices used by older men and women to manage urinary incontinence: Results from the national follow-up survey on self-care and aging, *J. Am. Geriatr. Soc.*, 48, 894, 2000.

53. Nicolle, L.E. et al., Urinary antibody level and survival in bacteriuric institutionalized older subjects, *J. Am. Geriatr. Soc.*, 46, 947, 1998.

54. Behnke, A.R. and Myhre, L.G., Body composition and aging: A longitudinal study spanning four decades, in *Oxygen Transport to Human Tissues*, Loeppky, J.A. and Riedesel, M.L., Eds., Elsevier, New York, 1982.

55. Fukagawa, N.K. et al., Effect of age on body water and resting metabolic rate, *J. Gerontol.*, 51, M71, 1996.

56. Ryan, J.J. and Zawada, Jr., E.T., Renal function and fluid and electrolyte balance, in *Principles and Practice of Geriatric Surgery*, Rosenthal, R.A., Zenilman, M.E., and Katlic, M.R., Eds., Springer-Verlag, New York, 2001.

57. Frassetto, L.A. and Sebastian, A., Age and systemic acid-base equilibrium: Analysis of published data, *J. Gerontol.*, 51, B91, 1996.

58. Shapiro, E. and Steiner, M.S., The embryology and development of the prostate, in *Prostatic Diseases I*, 1st Ed., Lepor, H., Ed., W.B. Saunders Co., Philadelphia, PA, 2000.

59. Junqueira, L.C., Carneiro, J., and Kelley R.O., *Basic Histology I*, 9th ed., Appleton & Lange, Stamford, CT, 1998.

60. Young, B., Heath, J.W., *Wheater's Functional Histology: A Text and Colour Atlas*, 4th ed., Churchill Livingstone, New York, 2000.

61. Kerr, J.B., *Atlas of Functional Histology*. Mosby, London, 1999.

62. Isaacs, J.T. and Coffey, D.S., Etiology and disease process of benign prostatic hyperplasia, *Prostate Suppl.*, 2, 33, 1989.

63. Cotran, R.S., Kumar, V., and Collins, T., Eds., *Robbins Pathologic Basis of Disease*, 6th ed., W.B. Saunders, Philadelphia, PA, 1999.

64. Inlander, C.B, and Norwood, J.W., *Understanding Prostate Disease I*, Macmillan, New York, 1999.

65. White, J.W., The results of double castration in hypertrophy of the prostate, *Ann. Surg.*, 22, 1, 1895.

66. Peters, C.A. and Walsh, P.C., The Effect of nafarelin acetate, a luteinizing hormone-releasing hormone agonist, on benign prostatic hyperplasia, *N. Engl. J. Med.*, 317, 599, 1987.

67. Lee, C., Kozlowski, J.M., and Grayhack, J.T., Intrinsic and extrinsic factors controlling benign prostatic growth, *Prostate*, 31, 131, 1997.

68. Partin, A., Etiology of Benign Prostatic Hyperplasia, in *Principles and Practice of Geriatric Surgery*, Rosenthal, R.A., Zenilman, M.E., and Katlic, M.R., Eds., Springer-Verlag, New York, 2001.

69. Moore, R.G., Gazak, J.M. and Wilson, J.D., Regulation of cytoplasmic DHT binding in dog prostate by 17 beta estradiol, *J. Invest.*, 63, 351, 1979.

70. Liao, S. and Hiipakka, R.A., Selective inhibition of steroid 5α-reductase isozymes by tea epicatechin-3-gallate and epigallocatechin-3-gallate. *Biochem. Biophys. Res. Commun.*, 214, 833, 1995.

71. Roehrborn, C.G., The role of guidelines in the diagnosis and treatment of benign prostatic hyperplasia, in *Principles and Practice of Geriatric Surgery*, Rosenthal, R.A., Zenilman, M.E., and Katlic, M.R., Eds., Springer-Verlag, New York, 2001.

72. Walsh, P.C., Treatment of benign prostatic hyperplasia, *N. Engl. J. Med.*, 335, 586, 1996.

73. Lepor, H. et al., The efficacy of terazosin, finasteride, or both in benign prostatic hyperplasia, *N. Engl. J. Med.*, 335, 533, 1996.

74. Gormley, G.J. et al., The effect of finasteride in men with benign prostatic hyperplasia, *N. Engl. J. Med.*, 327, 1185, 1992.

75. The Finasteride Study Group, Finasteride (MK-906) in the treatment of benign prostatic hyperplasia, *Prostate*, 22, 291, 1993.

76. Lepor, H., Alpha-adrenergic blockers for the treatment of benign prostatic hyperplasia, in *Principles and Practice of Geriatric Surgery*, Rosenthal, R.A., Zenilman, M.E., and Katlic, M.R., Eds., Springer-Verlag, New York, 2001.

77. Price, H., McNeal, J.E., and Stamey, T.A., Evolving patterns of tissue composition in benign prostatic hyperplasia as a function of specimen size, *Hum. Pathol.*, 21, 578, 1990.

78. National Institutes of Heath, National Cancer Institute, *Understanding Prostate Changes: A Health Guide for All Men*, 1999, p. 19.

79. Lin, D.W. and Lange, P.H., The epidemiology and natural history of prostate cancer, in *Principles and Practice of Geriatric Surgery*, Rosenthal, R.A., Zenilman, M.E., and Katlic, M.R., Eds., Springer-Verlag, New York, 2001.

80. Smith, J.R. et al., Major susceptibility locus for prostate cancer on chromosome 1 suggested by a genome-wide search, *Science*, 274, 1371, 1996.

81. Lara, P.N. et al., Molecular biology of prostate carcinogenesis, *Oncology/Hematology*, 32, 197, 1999.

82. Taplin, M.E. et al., Mutation of the androgen-receptor gene in metastatic androgen-independent prostate cancer, *N. Engl. J. Med.*, 332, 1393, 1995.

83. Janulis, L., Grayhack, J.T., and Lee, C., Endocrinology of the prostate, in *Principles and Practice of Geriatric Surgery*, Rosenthal, R.A., Zenilman, M.E., and Katlic, M.R., Eds., Springer-Verlag, New York, 2001.

84. Garnick, M.B. and Fair, W.R., Combating prostate cancer, *Sci. Am.*, 279, 74, 1998.

85. Chung, L.W.K., Isaacs, W.B., and Simons, J.W., Eds., *Prostate Cancer: Biology, Genetics, and the New Therapeutics*, Humana Press, Totowa, NJ, 2001.

86. Morgan, T.O. et al., Age-specific reference ranges for serum prostate-specific antigen in black men, *N. Engl. J. Med.*, 335, 304, 1996.

87. Yuan, J.J. et al., Effects of rectal examination, prostatic massage, ultrasonography, and needle biopsy on serum PSA levels, *J. Urol.*, 147, 810, 1992.

88. Brawer, M.K., Ed., *Prostate Specific Antigen*, Marcel Dekker, New York, 2001.

89. Brawer, M.K. et al., Serum PSA and prostate pathology in men having simple prostatectomy, *Am. J. Clin. Pathol.*, 82, 760, 1989.

90. Stamey, T.A. and Kabalin, J.N., Prostate specific antigen in the diagnosis and treatment of adenocarcinoma of the prostate. I. Untreated patients, *J. Urol.*, 141, 1070, 1989.

91. Hudson, M.A., Bahnson, R.B., and Catalona, W.J., Clinical use of prostate-specific antigen in patients with prostate cancer, *J. Urol.*, 142, 1011, 1989.

92. Barrett, D.M., Ed., *Mayo Clinic on Prostate Health*, Kensington Publishing Corp., New York, 2000.

93. Newcomer, L.M. et al., Temporal trends in rates of prostate cancer: Declining incidence of advanced stage disease, 1974 to 1994, *J. Urol.*, 158, 1427, 1997.

94. Stephenson, R., Population based prostate cancer trends in the PSA era: Data from the Surveillance, Epidemiology, and End Results (SEER) program, in *Monographs in Urology*, Vol. 19, Stamey, T.A., Ed., Medical Directions, Montverde, FL, 1998.

95. Paulson, D.F., Impact of radical prostatectomy in the management of clinically localized disease, *J. Urol.*, 152, 1826, 1994.

# 20 The Gastrointestinal Tract and the Liver*

*Paola S. Timiras*
University of California, Berkeley

## CONTENTS

## I.   INTRODUCTION

The study of the aging of the gastrointestinal system discloses few alterations in gastrointestinal function, although cellular changes occur and involve secretory activity and motility of the major structures. These cellular changes are similar to those elsewhere in the body, but in the absence of localized disease, function is usually maintained in line with requirements. Disorders and diseases, however, become more common with advancing age and involve all levels of the gastrointestinal (GI) tract, starting with the mouth and extending to the rectum, anus, and pelvic floor musculature (Figure 20.1). In geriatric clinics, about 20% of all patients have significant gastrointestinal symptoms and morbidity from gastrointestinal diseases, such as cancer of the colon, second only in incidence and mortality to lung cancer. In this chapter, aging of the gastrointestinal tract will be presented first, followed by a brief discussion of the senses of smell and taste because of their close association with gastrointestinal function (Part II); a synopsis of the aging exocrine pancreas will be presented in Part III and of the liver in Part IV.

---

* Illustrations by Dr. S. Oklund.

## II. THE TEETH, THE STOMACH, THE INTESTINES, AND THE SENSES OF SMELL AND TASTE

### A. HETEROGENEITY OF GASTROINTESTINAL FUNCTION AND REGULATION

While, in this chapter, the focus will be on how late and how well digestion and absorption are retained, it must be kept in mind that limiting food intake may have beneficial effects. Since early studies, more than 50 years ago, evidence has been accumulating steadily that a dietary reduction of caloric intake or of specific food constituents will prolong life significantly in rodents, and studies are in progress in primates (Chapter 24). The earlier the restriction begins, the more successful is the subsequent retardation of growth and prolongation of life. Such dietary effects have also been demonstrated in other species. It will always remain unthinkable to perform such restrictive experiments in humans. However, encouragement to limit caloric intake and to prevent or treat obesity is easily justified.[1] Not only is life prolonged, but frequency and severity of a variety of pathological changes are reduced in experimental animals (Chapter 24).

The major function of the gastrointestinal system is to provide the organism with nutritive substances, vitamins, minerals, and fluids. This function is achieved by a series of chemical and mechanical processes involving digestion, storage or propulsion of food, absorption of water and nutrients, and their transfer to blood and tissues or excretion of unabsorbed components (Table 20.1). Secretion of hormones and immune activity represent other key functions.

The elderly are an heterogeneous group of individuals in many important aspects, including nutrition. This is true in relation not only to age but also to degree of health and disease, degree of physical activity, as well as psychological and socioeconomic characteristics. Thus, "young elderly" (aged 65 to 75 years) should be distinguished from "old elderly" (over 75 years) with respect to nutritional needs. However, the latest edition of the American Recommended Dietary Allowances (Food and Nutrition Board, 1985) classifies together persons "aged 51 years and older," disregarding the many individual differences as aging progresses. Other recommendations for "intakes of essential dietary constituents" have begun to take into consideration ages and sex differences: they distinguish males from females and separate 51 to 70 year olds from the 70 years and older age group.[2]

Of the major appendices of the gastrointestinal tract, the liver, the largest gland in the body, and the pancreas, with both endocrine and exocrine functions, the liver appears to be more severely affected by aging.[3–6] Given the numerous liver functions, alterations with aging have widespread repercussions for the well-being of the entire

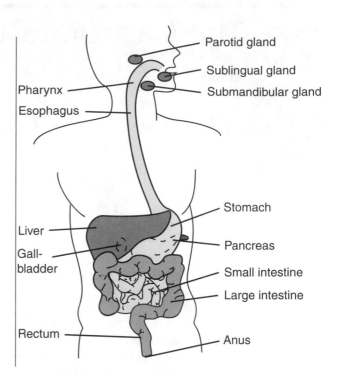

**FIGURE 20.1** Schematic representation of the gastrointestinal tract, with the liver and pancreas.

---

### TABLE 20.1
### Major Functions of the Gastrointestinal System

The major goal of the gastrointestinal system is to provide the organism with nutritive substances, vitamins, minerals, and fluids. This is achieved through the following four key processes:

**Digestion:** chemical (enzymes) and mechanical (teeth, muscles) breakdown of foods in small units that can be absorbed through the intestinal epithelium

**Absorption:** active or passive transfer of substances from the gastrointestinal tract to blood and extracellular spaces

**Motility:** smooth muscle contraction and relaxation regulate digestion and movement of gastrointestinal content along tract

**Secretion:** synthesis and release of hormones, enzymes, chemical mediators, mucus, intrinsic factor

---

organism. However, in the liver as in the whole organism, not all functions are affected simultaneously or with equal severity. The bile, produced in the liver for optimal digestion and absorption of lipids, is important at all ages and is particularly crucial in the elderly whose poor diet may be deficient in several essential elements such as lipid-soluble vitamins (Chapter 24); bile formation remains quite stable in healthy individuals well into old age.[5] Another important hepatic function, the detoxification of many (therapeutic and recreational) drugs, is progressively restricted with advancing age, and this restriction contributes, with the increased use of drugs (polypharmacy), to the greater susceptibility of the elderly to the potential toxicity of excessive or incorrect medication (Chapter 23). In the pancreas, the changes with aging of endocrine functions have been discussed in Chapter 14. Changes with aging of the exocrine functions that involve powerful protein-splitting (proteolytic) enzymes may lead to impaired digestion and absorption.[6]

Regulation of gastrointestinal function with aging depends not only on the type of diet, but also on intrinsic factors, such as the nervous system, hormones, and local chemical mediators that influence gastrointestinal growth, secretory activity, and motility (Chapter 14). The significant role of the immune system in preventing infections to which the gastrointestinal tract is easily exposed throughout its length, was overlooked until the recent discovery of the role of the bacillus, *Helicobacter pylori*, in diseases of the stomach.[7–9] Until then, the gastrointestinal tract was considered practically sterile. However, recent studies show that of the 30,000 genes encoded in the human genome, more than 1000 derive from bacterial species living in the gastrointestinal tract.[10] For example, one-third to one-half of the human population carries *H. pylori,* and once infected, most persons remain infected for decades, if not for life.[11] *H. pylori* infection is associated with two major inflammatory processes, gastritis and peptic ulcer disease. Gastritis is so common, especially among the elderly, that it was thought to represent a characteristic of the aging stomach. We now know that eradication of *H. pylori* from the stomach results in clearance of the gastritis.[7] Not all investigators agree on the beneficial effects of *H. pylori* eradication, and further research is needed to unravel the complexity of immuno-gastrointestinal relations.[8,9]

With respect to the influence of the diet, changes in sensory function with aging (Chapter 9) may result in depressed taste and smell. The decline in olfactory and gustatory sensory functions[12] negatively affects eating behaviors, resulting in less consumption of food, overall decline in health state, and decreased enjoyment of life.[13]

## B. TEETH, GUMS, AND ORAL MUCOSA

In the mouth, food is mixed with saliva, chewed, and propelled into the esophagus. Chewing (mastication), a matter of breaking down large food particles, is a function of the teeth. The saliva contains the digestive enzyme ptyalin, which plays a minor role in starch digestion. It also contains the glycoprotein, mucin, that lubricates the food and facilitates its passage through the esophagus on the way to the stomach.

With aging, the *teeth* undergo characteristic changes:[14–16]

- They acquire a yellowish-brown discoloration from staining by extrinsic pigments from beverages, tobacco, and oral bacteria
- The pulp recedes from the crown, and the root canal becomes narrow and thread-like

- The roots become brittle and fracture easily during extractions
- The layer of odontoblasts (cells secreting dentin) lining the pulp chamber becomes irregular and discontinuous
- The pulp undergoes fibrosis and calcification
- Concomitantly, with faster destruction than reconstruction of the dentin, the mandibular and maxillary bones in which the teeth are imbedded undergo the same aging processes as all other bones
- Osteoporotic changes (i.e., increased bone loss, Chapter 21) result in looser teeth, contributing ultimately to tooth loss

The surfaces of the teeth involved in chewing become progressively worn down throughout life. This attrition is not only a consequence of chewing, but also, in some individuals, of the habit of grinding or clenching the teeth together (so-called bruxism), often during REM sleep (Chapter 8). Abrasion (sometimes due to improper brushing) and erosion (often aggravated by the demineralizing action of soft drinks) are frequent. Although new caries (i.e., cavities with decay) are uncommon in the elderly, loss of interest in dental hygiene and a decline in dexterity needed for tooth brushing may lead to plaque accumulation and caries.[16] About 50% of elderly in the United States have lost the majority of their teeth by age 65 and about 75% by age 75. Teeth loss can be prevented by restoring appropriate hygienic measures self-managed or provided by a dental hygienist.[17]

Recession of the gingivae (gums) occurs in all elderly. The epithelial attachment that forms a cuff around the tooth at the interface with the gums recedes and opens the way to accumulation of particulate material with bacteria (i.e., plaque), swelling, inflammatory hyperplasia, or low-grade infection. Whether such a gum recession is a physiologic process or the result of chronic peridontitis (i.e., inflammation of the peridontal membrane) due to a variety of local irritative factors (e.g., ill-fitting bridges or partial dentures) remains to be clarified. Indeed, after the age of 40, chronic peridontitis is the major cause of tooth loss.[17–19] Peridontal disease is not only common in aging humans, but also, it is usually found in aging experimental animals (e.g., mice, rats, dogs, monkeys, baboons). In humans and experimental animals, the disease is due to local factors as well as some systemic predisposing disease (e.g., diabetes) or stress (e.g., cold exposure).[18] Indeed, in germ-free animals, recession of gums does not occur with age.

With aging, the epithelium of the oral mucosa becomes relatively thin and atrophic. Specialized structures, such as the papillae of the tongue, also become atrophic, and this atrophy is associated with loss of taste. Other structures, like the palatal mucosa, undergo edema

**Systemic and Organismic Aging**

and keratinization (i.e., accumulation of a highly insoluble protein, keratin), a condition which seems to be delayed or prevented by the wearing of dentures. Keratin, also a normal component of the skin, is the product of epidermal cells and undergoes characteristic changes with development as well as extensive cross-linking with age and a number of conditions. The oral epithelium exhibits increasing amounts of glycogen and alterations in collagen.[17]

### Oral Diseases in the Elderly

While the above changes in oral structures with aging are gradual and relatively benign, they may predispose the involved tissues to a variety of pathologic conditions. Although very few oral diseases are characteristic of old age, many pathologic states are seen with greater frequency in older than in younger individuals.[18,19] Among these, chronic periodontal disease (discussed above), xerostomia, mucositis and mucosal atrophy, leukoplakia, and malignant neoplasia (these latter beyond the scope of the present discussion) are the most common.

*Xerostomia* or dry mouth may be due to a large variety of etiologic factors. In the elderly, the major causes are:

Primarily, atrophy of the salivary glands

Decline of salivary secretion

Systemic disease (e.g., diabetes)

Heavy cigarette smoking

Anxiety and depression

Several medications (antihypertensives, antidepressants, antihistamines) depending on the dose (Chapter 23)

In this condition, the oral cavity is extremely dry, the mucosa appears red, fissured, and often coated with food particles and sloughed-off cells. Therapy involves cessation of the underlying cause (e.g., smoking) or treatment of the underlying systemic disease (diabetes) or the use of artificial saliva. The decrease in salivary volume is associated with enzymatic changes, such as reduction in amylase activity and in electrolytes.[14–19]

*Mucositis and mucosal atrophy* are frequent occurrences in elderly individuals in whom the oral mucosa has become atrophic and less resistant to irritation by oral noxious stimuli such as trauma, hot foods, and smoking, to infections and to chemotherapeutic agents or radiation therapy. These stimuli result in a chronic inflammatory process (mucositis) and, in more severe cases, in ulceration with pain.

*Leukoplakia*, or keratosis, represents a hyperplasia of the mucosa with accumulation of keratin, hence, the name of "white patch." It is rarely seen in young individuals but is frequent after 60 years. It may be caused by pipe or cigarette smoking, by ill-fitting dentures, or by infections (e.g., candidiasis). It is associated with precancerous histologic alterations, and in this case, it must be treated as if it were a carcinoma.[19]

### C. Swallowing and Pharyngo-Esophageal Function in the Elderly

Dysphagia, or difficulty in swallowing, is a common complaint of elderly individuals. It can result from alteration of any of the components of deglutition, a complex motor activity, involving the mouth, the esophagus, and several levels of nervous control.

Deglutition or swallowing is a reflex response that pushes the contents of the mouth into the esophagus. The afferent stimuli are generated by the voluntary collection of the oral contents on the tongue and their propulsion backward into the pharynx, and they are carried by several nerves to the brain in the medulla oblongata, where they are integrated. The efferent fibers are carried, also by several nerves, to the pharyngeal musculature and the tongue. Inhibition of respiration and closure of the glottis are part of the reflex. Swallowing is impossible when the mouth is open; it is rapid during eating but continues at a slower rate between meals. Upon swallowing, the upper portion of the esophagus relaxes to permit entrance of the swallowed material, which then progresses through the esophagus to the stomach by peristaltic movements (circular waves). In the standing position, liquids and semisolid foods may fall by gravity to the lower esophagus, where the musculature relaxes upon swallowing and permits the passage of food into the stomach.

The act of swallowing is divided into three stages, all of them affected by aging.[20,21] The first stage, in which the material to be swallowed is passed from the mouth to the pharynx, is a voluntary act, mediated through stimulation of skeletal muscles. These undergo aging-related changes with atrophy and increasing weakness common to all skeletal muscles (Chapter 21). The second stage, reflex in nature is short in duration but complex in its neural control and involves the relaxation of the sphincter between pharynx and esophagus. In the third stage, reflex transport sweeps the contents onward through smooth muscle peristalsis. All of these stages that require a precisely timed contraction and relaxation sequence are affected; they may become desynchronized and result in less efficient deglutition. Dysphagia of varying degrees of severity is a common complaint. In the most severe cases, dysphagia is associated with symptoms of choking and drowning with aspiration or regurgitation of food, while mild dysphagia may be found in otherwise healthy elderly. Severe dysphagia is always a symptom of a systemic disease, either of the muscle or the nervous system. In addition to being an uncomfortable and unpleasant symptom, dysphagia leads to reduced and altered nutritional intake, particularly in the elderly. Malnutrition, weight loss, and dehydration are common features of this condition.[22]

Presbyesophagus or old esophagus is the most common disorder of the esophageal motility associated with old age.[23–27] Radiologic and manometric studies reveal:

- An increased incidence of nonperistaltic contractions (with failure of the lower esophageal sphincter to relax)
- A reduced amplitude of peristaltic contractions
- A decreased responsiveness to cholinergic stimulation

Other motor disorders of the esophagus in addition to those mentioned above and occurring at all ages, including old age, are *achalasia*, in which food accumulates in the esophagus and the organ becomes extremely dilated; *sphincter incompetence*, which permits reflux of acid from the stomach; *aerophagia*, or ingestion of air, which can be regurgitated or absorbed in the intestine or expelled as flatus.[25]

## D. THE STOMACH AND DUODENUM: PHYSIOLOGIC AND PATHOLOGIC CHANGES WITH AGING

The stomach:

- Serves as a food reservoir
- Breaks down ingested food by its churning movements (due to the presence of three smooth muscle layers)
- Secretes the gastric juice which contains a variety of substances: digestive *enzymes* (e.g., pepsin), *mucus* (which lubricates the food), *hydrochloric acid* (which destroys the ingested bacteria, aids in protein digestion, and is necessary for the transformation of iron for the synthesis of hemoglobin)
- Secretes and is responsive to hormones: *gastrin, glucagon, somatostatin* (Chapters 10 and 14), and to special peptides also found in the brain:

*vasoactive intestinal polypeptide* (VIP) that (a) stimulates intestinal secretion of water and electrolytes; (b) relaxes intestinal smooth muscle, including sphincters; (c) dilates peripheral blood vessels; and (d) inhibits gastric acid secretion; it would also produce and possibly release in the circulation substance P, also a peptide, with a stimulatory action on small intestine motility
- Secretes an intrinsic factor (a glycoprotein) necessary for the absorption of Vitamin $B_{12}$ from the small intestine; vitamin $B_{12}$ is necessary for the maturation of red blood cells, and deficiency of the vitamin or intrinsic factor results in a severe type of anemia (Chapter 18)

With aging, changes in the stomach and neighboring duodenum involve the mucosa cells and the hydrochloric acid and pepsin secretions; under basal conditions and in the "healthy" elderly, both are decreased (Figure 20.2). This decrease may contribute to some of the difficulty of digestion (involving foods rich in protein such as meat) that affects a large proportion of the elderly (Chapter 24). However, more threatening to the health status of the elderly is the possible disruption of the so-called "gastric mucosal barrier" that protects the mucosa from attack by acid and pepsin; failure of this protection results in injury and death of mucosal cells.

The protective barrier is formed of tight junctions between the cells and of a thick layer of mucus. Because a pH of 7 to 8 is present at the mucosal cell surface, and a pH of 2 is present in the gastric lumen, a gradient from mucosa to gastric lumen is established, under normal conditions, to prevent back-diffusion of hydrogen ions to the gastric cells. Bile acids, nonsteroidal anti-inflammatory

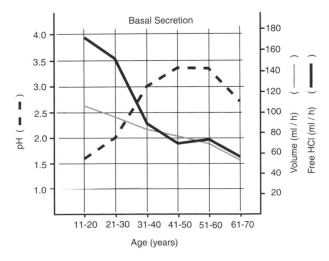

**FIGURE 20.2** Changes with age, under basal (preprandial) conditions of gastric secretions. Note, with advancing years, the moderate decrease in total volume and the greater decrease in free hydrochloric acid (HCl). As the free HCl decreases, pH increases.

**Systemic and Organismic Aging**

drugs (e.g., aspirin), other substances (ethanol, caffeine), and bacteria may disrupt this mucous protective layer and allow acid and pepsin to damage and destroy mucosal cells.[28–30] Prostaglandins, bioactive lipids present in the stomach and duodenum (as well as in many other tissues), stimulate bicarbonate secretion and thus play a role in gastric mucosa defense. However, prostaglandin secretion decreases with aging. Gastrin, one of the gastrointestinal hormones, secreted by the gastric mucosal cells as well as the pituitary, pancreas (fetal), and brain, stimulates gastric acid and pepsin secretion. Gastrin may also participate to some degree in regenerating processes by promoting mucosal growth; this growth-promoting action, beneficial in cases of mucosal damage, is mediated through the release of epidermal growth factor (EGF) from the duodenal Brunner's glands.

Changes with aging involve some of the digestive enzymes as well; they may be consequent directly to changes in the enzyme secreting cells and organs or indirectly to hormonal and neural regulatory alterations (Table 20.2). The direct changes involve cellular or enzymatic reduction. Failure of enzymatic activity may also depend on alterations of the regulatory mechanism of enzyme synthesis and release due either to a change in the hormonal or neural stimuli or to a change in response (altered receptors).

In addition to physiologic changes with aging, the stomach undergoes pathologic changes,[28–30] among which the most common are gastritis and peptic ulcer disease.[8,9]

## 1. Gastritis and Peptic Ulcer Disease

The incidence and prevalence of gastritis, an inflammatory process of the gastric mucosa increase with advancing aging. This is also the case of peptic ulcer disease; in this condition, one or several ulcers (i.e., break in the mucosa with loss of cells) may be situated in the stomach (usually near the pylorus, the region close to the duodenum) or the duodenum and will then be designated as gastric and duodenal ulcers. The incidence of peptic ulcer is on the rise in developing countries, particularly in India and Africa. In Western countries, including the United States, incidence is decreasing in the overall population (perhaps reflecting advances in treatment) but increasing in the elderly group.[11] In this group, not only is the incidence on the rise, but the severity of the disease and its consequences are also greater than in younger individuals.[31] Elderly patients tend to predominate among those admitted to hospitals for peptic disease; patients over 60 years of age account for nearly 50% of those with gastric and 40% of those with duodenal ulcers. In these individuals, age is the major mortality risk.

Excessive acid production was believed to be the major cause of gastritis and peptic ulcers until two decades ago, when infection with the common microorganism, *Helicobacter pylori*, usually attached to mucosal cells, but

**TABLE 20.2**
**Possible Mechanisms of Aging-Related Changes in Digestive Enzyme Secretion**

**Enzyme Secreting Organs**
Reduction in:
  Number of cells
  Enzyme concentration
  Enzyme synthesis and release

**Hormonal and Neural Regulation of Enzyme-Secreting Organs**
Reduction in:
  Number of gastrointestinal endocrine cells
  Hormone concentration
Impairment of:
  Sensitivity of endocrine cells to digestive stimuli
Alteration of:
  Distribution and metabolism of gastrointestinal hormones
  Number and affinity of endocrine or neural receptors

rarely located intracellularly, was established as a major cause of gastritis and ulcer. The presence of *H. pylori* had been noted in the stomach since the end of the 19th century, but it is only in recent years that its association with gastritis and peptic ulcer led to one of the major medical revolutions of the 20th century.[9] One important aspect of the infection with *H. pylori* is the diversity of its consequences depending on:

1. The virulence of the infection
2. The genetic characteristics of the host: some have the infection but never develop the gastrointestnal problems
3. The presence of environmental cofactors (e.g., nonsteroidal anti-inflammatory drugs, smoking)
4. The age of the affected individual, and the age when the infection was acquired[8, 31–34]

Current studies of the molecular and cellular biology of *H. pylori* have provided two complete genome sequences, population genetics have been initiated, and potential therapeutic targets with novel antibacterial drugs or vaccines are being developed.[9]

Symptoms for gastritis and peptic ulcer are often indistinguishable, but generally, always more time is allotted for severe peptic ulcer. They consist primarily of epigastric pain, (over the region of the stomach) occurring periodically (more often during the night) and gastrointestinal bleeding. Symptoms may be atypical in the elderly with vague abdominal pain and weight loss. Major life-threatening consequences are perforation of the gastric/duodenal wall and hemorrhage: these can occur, suddenly, even without previous symptoms. A summary of peptic ulcer management is presented in Box 20.1.

**Box 20.1**

**Management of Peptic Ulcer**

Procedures used in the treatment of ulcers include the following interventions:

1. Dietary
2. Pharmacologic
3. Surgical

Pharmacologic interventions were originally aimed at inhibiting acid secretion and enhancing mucosal resistance to acid. Currently, the use of antimicrobial agents, primarily **antibiotics**, (e.g., tetracycline, amoxicillin) is favored, eventually in association with **antacids** and antisecretory agents.

A variety of antacids, most of which contain aluminum and magnesium hydroxide or calcium carbonate are available. Inhibition of parasympathetic inputs (which stimulate acid secretion) by **atropin** gives variable responses and has many undesirable side effects. **Receptor blockers of histamine** (an amine derived from the amino acid histidine and a powerful stimulator of gastric secretion), such as cimetidine and ranitidine (some of the most commonly prescribed drug in the United States), are also often used. **Drugs capable of inhibiting H⁺K⁺ATPase** are also widely used. **Epidermal growth factor** is still at the experimental stage. A number of substances increase the resistance of mucosal cells to acid by forming adherent protein complexes at the ulcer site. Usually, the pharmacologic treatment is associated with special diets and **cessation of cigarette smoking,** inasmuch as healing rates of duodenal ulcer are probably adversely affected by cigarette smoking, and the incidence of ulcers is higher in smokers than in nonsmokers.

If pharmacologic, dietary, and hygienic measures fail, **surgery** is advisable, with the caution necessary for the older patient.[33,34]

**Box 20.2**

**Structure of the Small Intestine Mucosa**

Throughout the length of the small intestine, the mucosa displays many folds and is covered with villi formed of a single layer of columnar epithelium and containing a network of capillaries and one lymphatic vessel. The free edges of the mucosal cells are divided into microvilli which form a "brush border." Folds, villi, and microvilli augment considerably the absorptive surface. The mucosal cells are formed from mitotically active undifferentiated cells located at the bottom of the villi (in the so-called crypts of Lieberkhun). They migrate up to the tips of the villi, where they are sloughed into the intestinal lumen. The average life of these cells lasts 2 to 5 days, thereby representing a potential model (still little utilized) for the study of cellular aging. The crypts are also the site of active secretion of water and electrolytes. Studies in humans have reported, with aging, a reduced height and a frequent convoluted pattern of villi; changes were not observed in villus width, cell height, or mucosal thickness.[39]

Circulatory alterations are not confined to bleeding but also encompass atherosclerotic lesions (Chapter 16) that may result in ischemia (i.e., local and temporary deficiency of blood), that may lead to mesenteric infarction. Ischemia produces symptoms of varying severity from transient intestinal discomfort, to abdominal angina, infarction, or in cases of segmental ischemia, to ischemic colitis.[38]

### CARCINOMA OF THE STOMACH

For unknown reasons (perhaps a change of dietary and other habits?), the incidence of carcinoma of the stomach, once one of the most frequent cancers in men, has been declining in the last 30 years. It is still relatively frequent and is situated primarily in the lower regions of the stomach (antrum and pylorus). It has a very unfavorable outcome (prognosis). Peak incidence is reached in 80- to 90-year-old men and may have a familial occurrence. Diagnosis is made on the basis of gastroscopy with biopsy. Unfortunately, treatment, either by surgery, radiation or chemotherapy shows a low (5 to 10%) 5-year survival rate.

### E. THE SMALL AND LARGE INTESTINES

### 1. Changes in Intestinal Absorption

Intestinal absorption depends on the structural and functional integrity of the intestinal mucosa (Box 20.2). In a number of animal species, advancing age is accompanied by alterations involving one or several of the following changes:[39]

## 2. Vascular Alterations

As indicated above, *bleeding* in the elderly presents a special problem — incidence and mortality are high after the age of 60 years.[35,36] Another cause of bleeding in the elderly is *vascular malformations, or ectasias,* of the intestinal vessels which are easily subject to bleeding. These ectasias occur throughout the gastrointestinal tract, particularly the small bowel, stomach, and colon. They are associated with aortic stenosis, age-related degenerative changes in tissues, inherited collagen, or ground substance defects.[37]

**Systemic and Organismic Aging**

- Overall shape of the villus
- Increase in collagen
- Mitochondrial changes
- Lengthening of the crypts
- Prolonged replication time of the crypt stem cells

All of these changes are minor, and none appears sufficient to explain the impaired absorption often found in the elderly. Other factors may intervene:

- Altered villus motility limiting functional surface area
- Inadequate intestinal blood supply (due to atherosclerotic involvement of major intestinal vessels)
- Impaired water "barrier" restricting diffusion and transport, and changes in small intestine permeability[40]

Research concerning absorption of nutrients and vitamins is sparse in humans and even in laboratory animals. With increasing age, however, absorption of several substances (e.g., sugar, calcium, iron) is reduced, whereas digestion and motility remain relatively unchanged. Among the substances for which absorption has been studied in the elderly, calcium probably presents the best evidence for a gradual reduction with increasing age (Table 20.3). In addition to changes in bone calcium with aging, calcium absorption and transport are significantly reduced starting at about age 60 (Chapter 24). When comparing young and old subjects, the young individuals are capable of responding to a low calcium diet by increasing intestinal calcium absorption; a response no longer found in the elderly. Mechanisms responsible for reduced calcium absorption are listed in Table 20.3. There is conflicting evidence for other substances such as dextrose, xylose, iron, vitamin $B_{12}$, fats, which may be reduced or may be absorbed normally at all ages.

Increased general frailty and weight loss may occur in the old, without evidence of any specific underlying cause, in the presence of a well-maintained appetite and a balanced diet. However, a relatively high percentage of old individuals suffer from malabsorption; at least 7% of residents in nursing homes are likely to have impaired absorptive ability.

## 2. Absorption of Nutrients and Malabsorption

Adequate nutrition is indispensable at all ages but especially in the young, who must provide extra calories for growth, and in the old, whose gastrointestinal function is only marginal. Indeed, dietary interventions to ensure a long and vigorous life have been popular for many centuries. Once achieved, old age has been recognized as a period requiring special attention to dietary habits

---

**TABLE 20.3**

**Mechanisms of Decreased Intestinal Calcium Absorption with Aging**

↓ intake of vitamin D (poor nutrition)
↓ vitamin D conversion in skin (reduced sunlight exposure)
↓ intestinal absorption
↓ vitamin D metabolism (hepatic) and activation (renal)
↓ cellular calcium binding (decreased receptors)

---

**TABLE 20.4**

**Important Factors for the Maintenance of Optimal Small Intestinal Function**

1. Anatomic integrity of the absorbing small intestine and normal cell replication of intestinal mucosal cells
2. Normal gastrointestinal secretions, including basal and postprandial secretions from salivary, gastric, pancreatic, and hepatic cells
3. Coordinated gastrointestinal motility
4. Normal intestinal uptake and transintestinal transport
5. Adequate intestinal blood supply to maintain cell oxygenation and cell nutrient supply
6. Normal defense mechanisms against toxic injurious agents from the intestinal lumen:
    normal clearance of bacteria in intestinal lumen
    immunologic response to injury
    mucosal cell wall integrity to prevent macromolecular uptake
    mucosal cell detoxification of toxic absorbed materials

---

In the intestine, primarily the small intestine, the intestinal contents are mixed with mucus, pancreatic juice, and bile. Digestion, which begins in the mouth and stomach, is completed in the lumen and mucosal cells of the small intestine. Digestion depends upon a number of enzymatic processes under neural and hormonal stimuli which can be affected in the elderly (Table 20.2). The products of digestion are then absorbed, along with vitamins and fluid in the small intestine (water also in the colon) and carried to the liver by the portal blood. The small intestine, with its many folds, finger-shaped villi, and array of microvilli on the luminal side of the cells, is particularly well designed for this function of absorption. The rate of nutrient transport by the intestine is related primarily to the surface area that is functionally exposed to the luminal contents. Several factors are important for the maintenance of optimal function of the small intestine (Table 20,4) and may be affected by aging. Thus, any change in intestinal architecture or diffusion barrier of the mucosal cells greatly affects transport.[40]

### MALABSORPTION AND DISEASE

Many generalized conditions, such as rheumatoid arthritis (Chapter 21), afflicting the elderly may have some detrimental affects on absorption. A number of small bowel disorders may cause malabsorption, but their incidence is low and symptomatology vague.[41] Other frequent causes of malabsorption include:

Infections (e.g., after gastrointestinal surgery, diarrhea)

Small intestine diverticula (i.e., small dilations or pockets leading off from the intestinal tube)

Pancreatic insufficiency (rare)

Celiac disease (alterations of mucosa and cell transport)

Mental disorders (e.g., dementia)

Many old individuals with malabsorption are severely undernourished, weak, and debilitated. Management, therefore, includes treatment of the specific underlying disease and appropriate diet. With energetic treatment, even severely ill patients have a good prognosis.

## 3. Aging of the Large Intestine

Disorders of the large bowel are almost exclusive to the elderly.[42,43] The physiopathology of such disorders is still little known, but given the multiple clinical problems, more attention is currently being dedicated to the study of physiologic changes with aging in this intestinal segment.

Anatomical changes are similar to those of the small intestine and include:

1. Atrophy of the mucosa
2. Proliferation of connective tissue
3. Vascular changes, mostly of an atherosclerotic nature

The large intestine major functions are storage, propulsion, and evacuation of the intestinal content (feces). Especially important are those conditions associated with bowel motility, such as constipation or diarrhea.[43]

In the colon (the last portion of the large intestine before the rectum), the most obvious aging-related change is the increased prevalence of *diverticula*, small, pocket-like mucosal herniations through the muscular wall. They vary in diameter from 3 mm to more than 3 cm and are present in 30 to 40% of persons over the age of 50 and with increasing incidence thereafter. Diverticula are often responsible for severe bleeding from the rectum and often become inflamed, causing diverticulitis.[44–46] A highly refined, low residue, diet as often consumed today, may be responsible for the formation of diverticula. The lack of dietary fiber and bulk is associated with spasm of the colon. The intraluminal pressure builds up, and the mucosa eventually pushes through the muscular coat at weak points, usually where the colonic blood vessels pierce the muscle to supply the mucosa. Diverticula become filled with packed feces and may ulcerate into the thinned mucosa, causing infection and inflammation.

The presence of diverticula may induce nonspecific abdominal pain, diarrhea, or constipation. A diet to increase the fiber content may alleviate these symptoms. Major complications include diverticulitis, hemorrhage, and colonic obstruction or perforation requiring surgery and rigorous pharmacologic and dietary treatment.[46]

## 4. Constipation

The major cause of constipation is decreased motility of the large intestine, but diet (unbalanced with respect to bulk) and lack of exercise may also be implicated in its etiology.[43,45,47,48] Constipation is considered one of the most common gastrointestinal complaints of the elderly. Its prevalence seems to be greater in women, although this sex difference may not be real but rather due to the larger number of old women than old men and the overall degree of old women's disabilities (Chapter 2). Treatment involves increasing the bulk in the diet and increasing physical activity (Chapter 24).

### PELVIC STRUCTURES AND DEFECATION

Distension of the rectum, the last portion of the intestine, initiates reflex contractions of its musculature. In humans, the internal involuntary sphincter is excited by the sympathetic but inhibited by the parasympathetic nerve supply. The external sphincter (voluntary) is innervated by somatic nerves. The urge to defecate begins when the rectal pressure increases to a certain level (about 55 mm Hg), at which time both sphincters relax and the rectal contents are expelled. At a lower rectal pressure, defecation can be initiated by voluntary relaxation of the external sphincter and contraction of the abdominal muscles. Thus, defecation is a spinal reflex that can be voluntarily inhibited by contraction of the external sphincter or facilitated by its relaxation.

## 5. Incontinence

As discussed in Chapter 19, urinary incontinence is one of the major afflictions of old age. The same tragic consideration applies to fecal incontinence (Figure 20.3).[49] The maintenance of normal control on fecal evacuation is regulated by complex neuromuscular functions (Figure 20.4). Should any physiologic decrement occur in the activity of the intestine, or in the muscles of the pelvic floor, or in the neural inputs, then the control of defecation may break down. A person with efficient sphincters may find control impossible during an attack of severe diarrhea. In the elderly, loss of sphincter muscle strength (Figure 20.5), merely as a consequence of aging, creates a more difficult problem when confronted with diarrhea. Certain neurologic conditions affect the pelvic floor muscles, and these may be so severe that, even with a normal stool, continence cannot be preserved. At all ages, there may be organic deficiencies in the muscle ring due to trauma, and such deficiencies are more likely to occur in the elderly person.[50]

**Systemic and Organismic Aging**

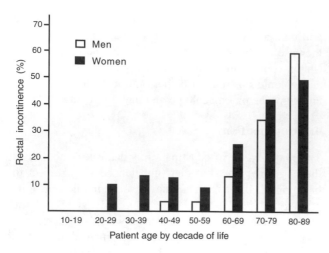

**FIGURE 20.3** Relationship of age and sex to fecal incontinence. (From Stewart, E.T. and Dodds, W.J., *Am. J. Roentgenol.*, 132, 197, 1979. With permission.)

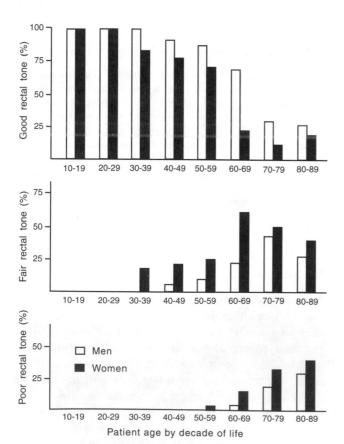

**FIGURE 20.4** Relationship of sphincter tone to age. (From Stewart, E.T. and Dodds, W.J., *Am. J. Roentgenol.*, 132, 197, 1979. With permission.)

With aging, the rectal muscle mass is decreased in size, and the sphincter is weakened. The external sphincter is always the most affected of the pelvic floor muscles. The high incidence of incontinence in the elderly makes it essential to exclude any possible underlying gastrointes-

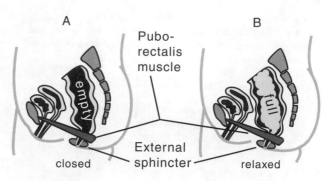

**FIGURE 20.5** Diagrammatic representation of some of the structures involved in defecation (here, in women). Note that the external sphincter and the puborectalis muscle play an important role in maintaining normal fecal continence: (A) when the rectum is empty and the sphincter is closed, and (B) when it is full, and the sphincter is relaxed. Note that one important action of the puborectalis muscle is to maintain angulation (bend) between the lower rectum and anal canal, upon which continence is largely dependent. Also note that with increased rectal pressure, the activity of the sphincter increases, thereby protecting the individual from involuntary defecation. However, above a certain pressure (fecal volume) this protection is lost.

tinal infection or systemic disease. One classification is shown in Table 20.5 and may be compared with the causes of urinary incontinence (Chapter 19).

Fecal incontinence may be caused by:

1. Neurogenic alterations involving the cortex (e.g., dementia) or the spinal cord (e.g., failure to inhibit defecation upon entrance of feces into rectum)
2. Muscle atrophy (due to trauma as in prolonged and difficult labor), a direct analogy to stress incontinence of urine (Chapter 19)
3. Constipation (due to immobility, poor reflexes, difficulty in reaching the toilet)
4. Diarrhea which may cause incontinence at all ages but more frequently at older ages

Knowledge of the etiology of incontinence leads to some practical interventions. A first step is to rule out

---

**TABLE 20.5**
**Classification of Fecal Incontinence**

| Cause | Consequence |
| --- | --- |
| Neurogenic | Loss of inhibitory control |
|   Cerebral cortex | Reduced reflex activity |
|   Spinal cord | "Stress" incontinence |
| Muscle atrophy | Constipation |
| Retention | Diarrhea |
| Overflow (bacteria, virus, allergies) | |

constipation or diarrhea and, if present, to treat them.[51] If fecal incontinence persists, then other causes must be sought and appropriate treatment established.

### CARCINOMA OF THE LARGE INTESTINE

Carcinoma of the large bowel, colorectal cancer, is the second (after lung carcinoma) most common malignancy in individuals over 70 years of age. Cancer of the colon would be more frequent in women and cancer of the rectum in men.[42,52–54] Polyps (small tissue mass) resulting from hypertrophy of the intestinal mucosa and extending into the intestinal cavity are also frequent. They may be benign tumors or possible precursors of carcinoma.[52,53]

## F. THE SENSES OF SMELL AND TASTE

### 1. Smell

The sense of smell or olfaction is important in food selection and nutrition, in social interaction, and enjoyment of life.[13,54] Olfaction helps in the promotion of health by helping us to avoid the putrid smell of rotten matter. In addition to these values, olfaction warns against certain dangers posed by modern living, such as odiferous atmospheric pollutants. The olfactory system is comprised of peripheral chemoreceptor cells situated in the nasal mucosa and stimulated by molecules in solution in mucus, in the nose (and saliva in the mouth for taste). The central pathways include the olfactory bulbs, the olfactory cortex (in humans, the piriform cortex), and the limbic system, where, presumably, olfactory discrimination and conscious perception of odors are mediated.

With aging, olfactory ability declines (hyposmia) or may be completely lost (anosmia). Decline in olfactory function can be demonstrated by two kinds of evaluations:

- Decreased sensitivity to odor thresholds, that is, decreased ability to identify various odors
- Decreased discriminatory ability to identify odor constituents in a mixture

The decrease in olfaction worsens with progressing age and may culminate in failure of detection by age 80 years and older; it is greater for some odors than others, and, overall, females score better than males.[12,55] The decline in the ability to recognize odors has been related to alterations of peripheral and central nervous control.[12] Peripherally, in the nasal cavity, aging changes begin to occur at a relatively early adult age and steadily progress with advancing age. They include:

- Loss of cilia from cells of the nasal mucosa, followed by loss of cells, and by slower replacement of lost cells
- Loss of neurons in brain olfactory centers such as the olfactory bulb. Neuronal losses may be

secondary to the loss of the sensory cells from the nasal mucosa (and hence, to loss of sensory stimulation) or to cerebral degenerative changes (e.g., in Parkinson's and Alzheimer's diseases).

Deficiency of olfactory function in old age may be one of the causes of poor appetite and irregular eating habits[13,56–58] and of some oral and pharyngeal diseases.[55] Adding flavor to foods enhances meal satisfaction and improves dietary intake and body weight in the elderly.[58,59]

### 2. Taste

Taste or gustatory sense is served by special taste buds located in the tongue and other regions of the oral cavity (mouth and pharynx). Taste buds contain taste cells, chemoreceptors similar to the olfactory cells and like them, in constant renewal (i.e., they have a 10-day cycle). Taste cells are in synaptic contact with the gustatory sensory neurons carrying, via the taste nerves, the information to the taste center in the medulla, and from there, gustatory stimuli are relayed to other sensory centers in the brain.

Classically, there are four primary tastes: sour, sweet, salty, and bitter, localized in different regions of the tongue: sweet at the tip, bitter at the back, and sour and salty on the sides. With aging, a spectrum of changes has been described in the sense of taste, although it is generally accepted that, compared to olfaction, taste is less affected. As for smell, threshold sensitivity to specific food stimuli and discriminatory ability to identify flavors in a mixture decline with aging.[12,54–59] As for smell, this decline may be ascribed to peripheral (e.g., loss of taste buds) and, possibly, central degeneration of neural centers of the gustatory system.

Smell and taste losses in the elderly can reduce appetite and food intake, they may fail to detect noxious substances, and may alter neural-immune connections as well.[60] Stimulation of taste and smell by increasing salivary flow and excitability of olfactory cells may improve eating habits and satisfaction of life. They may also stimulate immune responses and thereby may help to remedy the immune deficiencies and the dry mouth frequently affecting the elderly.[59,58,60,61]

## III. AGING OF EXOCRINE PANCREAS

Besides its hormonal secretions, the pancreas produces a pancreatic juice containing enzymes (amylase, lipase, proteases) important for digestion. The enzymes are discharged by exocytosis, and their secretion is controlled by a reflex mechanism and by the hormones, secretin and cholecystokinin. The major enzyme is trypsin, which is secreted as an inactive proenzyme, trypsinogen. The active form, trypsin, has a proteolytic action (i.e., catalyzes the hydrolysis of peptide bonds in the basic amino acids arginine and lysine). Some uncertainty exists regarding the effects of advancing age upon pancreatic secretion.[6,62] The senile gland is smaller, harder than normal (due to increasing fibrosis), and yellow-brown (due to accumulation of lipofuscin). Of the major enzymes, some (amylase) remain constant, whereas others (lipase, trypsin) decrease dramatically. Secretin-stimulated pancreatic juice and bicarbonate concentration remain unchanged. Little is known so far about age-related changes in the hormones that regulate pancreatic function. While a decline in some functions of the pancreas occurs with aging, the genesis of this decline is unknown but has been related to:

1. Diet
2. Drugs (e.g., alcoholism)
3. Vascular sclerosis
4. General fibrosis
5. Lack of cell regeneration

However, as only one tenth of pancreatic secretion is needed for normal digestion, it is not probable that only age is responsible for a significant pancreatic insufficiency capable of inducing severe digestive disorders. In general, these age-related changes do not seriously compromise pancreatic function, but their presence may increase the incidence of pancreatic disease (acute and chronic pancreatitis, cancer) in the elderly.[6, 62–65]

## IV.  AGING OF THE LIVER

The liver is an organ with many functions, to mention only a few:

- Bile formation
- Carbohydrate storage and metabolism
- Ketone body formation
- Reduction and conjugation of steroid hormones
- Inactivation of polypeptide hormones
- Detoxification of many drugs and toxins
- Manufacture of plasma proteins
- Urea formation
- Regulation of lipid metabolism

As is the case with other multifunctional organs, not all liver functions age at the same pace.[4,66,67] This section will focus on changes in morphology and function, particularly, of bile formation and excretion.[5] Changes in enzyme activity with aging and hepatotoxic effects of various drugs are considered in Chapter 23.

### A.  STRUCTURAL CHANGES

Major changes with aging in liver size and liver cells are listed in Table 20.6.

---

**TABLE 20.6**
**Some Structural and Function Alterations Associated with Aging in the Liver**

1. Atrophy and ↓ weight (cell loss beginning around 60 years and accelerating into the eighties and older)
2. Cell size remains unchanged or may ↑ (in contrast to malnutrition, when cell size ↓)
3. ↑ collagen with aging associated changes (Chapter 22)
4. Alterations of the usual cycle of hepatic cell and degeneration/regeneration
   a. ↑ binucleate cells
   b. ↑ ploidy (more than two full sets of homologous chromosomes)
   c. ↓ regeneration (perhaps due to growth inhibition)
   d. ↑ number of degenerating cells
   e. ↑ compensatory hypertrophy of remaining cells
5. ↓ number but ↑ size of mitochondria (suggesting compensatory attempts to maintain function)
6. ↓ size of endoplasmic reticulum and ↓ ability to metabolize drugs (Chapter 23)
7. Unchanged liver function tests involving metabolism and elimination of specific dyes and radioisotopes, protein synthesis, and metabolic functions

---

### HEPATIC CELLS

The liver is organized in lobules formed of hepatic cells, hepatocytes, lined up in rows irradiating from the center of the lobule to the periphery. The hepatic veins are situated in the center of the lobule and the biliary ducts, the portal veins and hepatic arteries are at the periphery. The specialized capillaries, the sinusoids, are lined with phagocytic cells (Kupffer's cells) that engulf bacteria or other foreign particles (see Chapter 15). The lobules are separated by a small amount of interlobular connective tissue in which are found the blood vessels and the beginnings of the biliary ducts and lymphatic vessels (Figure 20.6).

Changes with aging may reflect degenerative processes (e.g., reduced liver weight, cell loss, decreased mitochondrial number) or compensatory processes (e.g., increased cellular and mitochondrial size). That this is the case and that aged hepatic cells are active is supported by the increased activity of some enzymes (e.g., succinic dehydrogenase).

Hepatic cells regenerate throughout life, but their turnover slows with aging, perhaps due to the absence or deficit of growth factors for cell replication or to the excess of growth inhibitory factors.[68] The decelerated regeneration may also be due to cellular alterations such as binucleation (two nuclei in a cell) and polyploidization (increased nucleus size with more than two full sets of homologous chromosomes). Administration of growth-promoting hormones, such as thyroxine, does not appear to normalize regenerative processes. Indeed, the compensatory cell hypertrophy when regenerative power

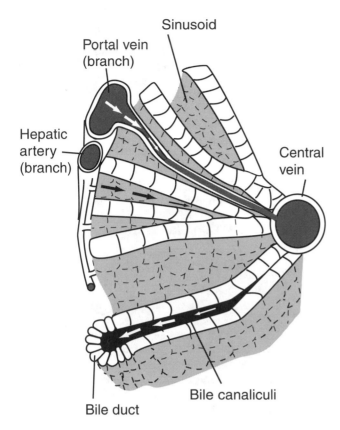

**FIGURE 20.6** Diagrammatic representation of the liver, which is organized in lobules formed of hepatic cells, hepatocytes, lined up in rows irradiating from the center of the lobule to the periphery. The hepatic veins are situated in the center of the lobules and the biliary ducts, the portal veins, and the hepatic arteries at the periphery. The specialized capillaries, the sinusoids, are lined with phagocytic cells (Kupffer's cells) that engulf bacteria or other foreign particles. The radial disposition of the liver cell plates and sinusoids around the terminal hepatic venule or central vein show the centripetal flow of blood from branches of the hepatic artery and portal vein and the centrifugal flow of bile to the small bile duct.

is more limited, would accelerate cell loss and, thereby, result in a vicious cycle of cell destruction and compensatory hypertrophy.

Smooth endoplasmic reticulum is decreased in the rat hepatocytes, a morphologic correlate underlining the age-related reduction in the hepatic capacity to metabolize drugs (Chapter 23). Other aspects of hepatic structure are not markedly altered, and maintenance of normal morphology agrees with minimal or no alteration of hepatic metabolic functions such as protein synthesis.

## B. FUNCTIONAL CHANGES

Alteration of hepatic structure and enzymatic functions with aging is moderate. In the healthy elderly, routine tests of liver function involving the metabolism and elimination of specific dyes (e.g., bromsulphalein) and radio-

isotopes as a test of hepatic clearance, and protein synthesis (e.g., plasma albumin levels), do not show significant differences between individuals aged 50 to 69 and 70 to 89 years.[66] Similarly, tests of liver dysfunction, including cholestasis (blockage of bile flow), cell necrosis, inflammation, and impaired detoxification are usually negative in the absence of specific liver damage and systemic disease.[69]

In contrast to humans, studies in experimental animals (rats) show consistent reduced removal from the blood of many of the dyes used to measure hepatic function, storage, and transport.[70] These changes have been ascribed to both circulatory alterations and increased collagen deposition (and hence, impaired transport) in the liver.

## C. BILE FORMATION

Bile production is a major function of the liver. Bile, considered as both a secretion and an excretion, is formed continuously by the hepatic cells. It is a greenish-yellow fluid composed of water, bile salts, bilirubin, cholesterol, and various inorganic salts; its many important functions are listed in Table 20.7.

### THE BILIARY SYSTEM

Bile formed of the hepatic cells is carried through several bile caniculi to the right and left hepatic ducts that join to form, outside the liver, the common hepatic duct. This duct joins the cystic duct that drains in the gallbladder, the reservoir for bile located on the undersurface of the liver. The hepatic duct unites with the cystic duct to form the common bile duct which enters the duodenum, usually united with the pancreatic duct, and pours the bile into the duodenum.

The bile, primarily through the action of its salts, has several important functions in the digestion by emulsifying lipids and activating the lipid enzymes, lipases. Bilirubin, the major pigment of the bile, results from the breakdown of hemoglobin, myoglobin, and respiratory enzymes in the reticuloendothelial system of the liver and spleen. Unconjugated bilirubin thus formed is carried to the liver cell, where it is conjugated and bound to proteins; this conjugated form is water soluble. In the colon, conjugated bilirubin is hydrolyzed to urobilin and excreted.

---

**TABLE 20.7**
**Major Functions of the Bile**

- Emulsification of lipids
- Activation of enzymes for digestion of lipids
- Conjugation of bilirubin (derived from hemoglobin breakdown) to form a water-soluble product for excretion
- Excretion of cholesterol (Chapter 17)
- Neutralization (by $HCO_3^-$) of acid delivered to duodenum from stomach
- Excretion of drugs, heavy metals, and environmental toxins

---

**Systemic and Organismic Aging**

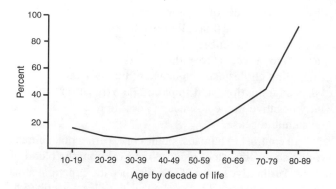

**FIGURE 20.7** Incidence of bile duct stones at cholecystectomy in relation to age. Note the sharp increase from 60 years on.

Little is known of changes with aging in bilirubin metabolism. However, biliary disease is common, and its incidence increases steadily with age (Figure 20.7).[5] Biliary disease (gallstones), a common problem of most Western societies, affects 15 to 20% of adults of all ages and 30 to 50% of elderly persons by age 75, with a ratio of 2:1 for females to males.[71] As people live longer, the incidence of biliary disease has increased in the last 30 years. Forty to sixty of patients with gallstones show no symptoms, and it is possible that the condition starts this way in most individuals.

In the gallbladder, the bile is concentrated by absorption of water (water is 97% in the liver bile and 89% in the gallbladder bile). Stones form in the gallbladder or bile ducts when a substance that is not normally present appears in the bile or the composition of the bile changes so that a normal constituent precipitates. For example, cholesterol stones form when the proportion of cholesterol, lecithin, and bile salts in the bile are altered.

Several aspects of biliary disease are characteristic of the elderly. These include:

1. A greater incidence of acute versus chronic cholicystitis (i.e., inflammation of the gallbladder)
2. The presence of stones in the bile duct
3. The recurrence of the disease after a previous operation
4. The greater severity of the disease and the higher mortality

An increasing incidence of stones in bile ducts and gallbladder with aging has been well documented.[72,73] With this increase, the incidence of related complications (e.g., jaundice, pancreatitis, cholecystitis, liver abscesses, and systemic sepsis) also rises. The treatment of choice is surgical and consists in the removal of the stone(s); it entails, in the elderly, more complicated operations to control or correct the disease.[72–74]

## REFERENCES

1. Jung, R.T., Obesity, in *Principles and Practice of Geriatric Medicine*, Pathy, M.S.J., Ed., Vol. 1, 3rd ed., John Wiley & Sons, New York, 1998.
2. FAO/WHO/UNA (Food and Agriculture Organization of the United Nations/World Health Organization/United Nations Association), *Energy and Protein Requirements Report of a Joint FAO/WHO/UNA Expert Consultation*, Geneva, WHO, Technical Report Series No. 724, 1985.
3. Levine, J.L. and Zenilman, M.E., Age-related physiologic changes in the gastrointestinal tract, in *Principles and Practice of Geriatric Surgery*, Rosenthal, R.A., Zenilman, M.E., and Katlic, M.R., Eds., Springer, New York, 2001.
4. Morris, J.S., Diseases of the liver, in *Principles and Practice of Geriatric Medicine*, Pathy, M.S.J., Ed., Vol. 1, 3rd ed., John Wiley & Sons, New York, 1998.
5. Morris, J.S., Diseases of the gall-bladder and bile ducts, in *Principles and Practice of Geriatric Medicine*, Pathy, M.S.J., Ed., Vol. 1, 3rd ed., John Wiley & Sons, New York, 1998.
6. Morris, J.S., Diseases of the pancreas, in *Principles and Practice of Geriatric Medicine*, Pathy, M.S.J., Ed., Vol. 1, 3rd ed., John Wiley & Sons, New York, 1998.
7. Blaser, M.J., *Helicobacter pylori* and the pathogenesis of gastroduodenal inflammation, *J. Infect. Dis.*, 161, 626, 1990.
8. Clayton, C.L. and Mobley, L.T., Eds., *Helicobacter Pylori Protocols*, Humana Press, Totowa, NJ, 1997.
9. Achtman, M. and Suerbaum, S., *Helicobacter Pylori: Molecular and Cellular Biology*, Horizon Scientific Press, Norfolk, England, 2001.
10. Jasney, B.R. and Kennedy, D., The Human Genome, *Science*, 291, 1153, 2001.
11. Pounder, R.E. and Ng, D., The prevalence of *Helicobacter pylori* infection in different countries, *Aliment. Pharmacol. Ther.*, 9, S33, 1995.
12. Winkler, S. et al., Depressed taste and smell in geriatric patients, *J. Am. Dent. Assoc.*, 130, 1759, 1999.
13. Drewnowski, A. and Schultz, J.M., Impact of aging on eating behaviors, food choices, nutrition, and health status, *J. Nutr. Health Aging*, 5, 75, 2001.
14. Toga, C.J., Nandy, K., and Chauncey, H.H., *Geriatric Dentistry*, Lexington Books, Lexington, MA, 1979.
15. Ship, J.A. et al., Geriatric oral health and its impact on eating, *J. Am. Geriatr. Soc.*, 44, 456, 1996.
16. Shay, K. and Ship, J.A., Importance of oral health in the older patient, *J. Am. Geriatr. Soc.*, 43, 1414, 1995.
17. Griffiths, J.E., Oral health, in *Principles and Practice of Geriatric Medicine*, Pathy, M.S.J., Ed., Vol. 1, 3rd ed., John Wiley & Sons, New York, 1998.
18. Jette, A.M. et al., Oral disease and physical disability in community-dwelling older persons, *J. Am. Geriatr. Soc.*, 41, 1102, 1993.
19. Walker, D.M., Oral disease, in *Principles and Practice of Geriatric Medicine*, Pathy, M.S.J., Ed., Vol. 1, 3rd ed., John Wiley & Sons, New York, 1998.

20. Patel, G.K., Diner, W.C., and Texter, E.C., Swallowing and pharyngoesophageal function in the aging patient, in *The Aging Gut*, Texter, E.C., Ed., Masson Publ., New York, 1983.

21. Sonies, B.C. et al., Durational aspects of the oral-pharyngeal phase of swallow in normal adults, *Dysphagia*, 3, 1, 1988.

22. Hellemans, J., Vantrappen, G., and Pelemans, W., Oesophageal problems, in *Gastrointestinal Tract Disorders in the Elderly*, Hellemans, J. and Vantrappen, G., Eds., Churchill Livingstone, London, England, 1984.

23. Soergel, K.H., Zboralske, F.A., and Amberg, J.R., Presbyesophagus, Esophageal motility in nonagenarians, *J. Clin. Invest.*, 43, 1472, 1964.

24. Hollis, J.B., and Castell, D.O., Esophageal function in elderly men—A new look at "presbyesophagus," *Ann. Intern. Med.*, 80, 371, 1974.

25. Morris, J.S., Diseases of the oesophagus, in *Principles and Practice of Geriatric Medicine*, Pathy, M.S.J., Ed., Vol. 1, 3rd ed., John Wiley & Sons, New York, 1998.

26. Spector, S.A. and Seymour, N.E., Benign esophageal disease in the elderly, in *Principles and Practice of Geriatric Surgery*, Rosenthal, R.A., Zenilman, M.E., and Katlic, M.R., Eds., Springer, New York, 2001.

27. Freston, J.W., Ed., *Diseases of the Gastroesophageal Mucosa*, Humana Press, Totowa, NJ, 2001.

28. Batty, G.M. and Gelb, A.M., Diseases of the stomach and duodenum, in *Principles and Practice of Geriatric Medicine*, Pathy, M.S.J., Ed., Vol. 1, 3rd ed., John Wiley & Sons, New York, 1998.

29. Sonnenberg, A., Peptic ulcer, in *Digestive Disease in the United States: Epidemiology and Impact*, Everhart, J.E., Ed., U.S. Department of Health and Human Services, NIH no. 94–1447, Washington, DC, 1994.

30. Cima, R.R. and Soybel, D.I., Pathophysiology and treatment of benign diseases of the stomach and duodenum, in *Principles and Practice of Geriatric Surgery*, Rosenthal, R.A., Zenilman, M.E., and Katlic, M.R., Eds., Springer, New York, 2001.

31. Cullen, D.J.E. et al., When is *Helicobacter pylori* infection acquired?, *Gut, 34, 1681, 1993.*

32. Schubert, T. et al., Ulcer risk factors: Interactions between *Helicobacter pylori* infection, nonsteroidal use, and age, *Am. J. Med.*, 94, 413, 1993.

33. Veldhuyzen van Zanten, S.J.O. et al., Increasing prevalence of *Helicobacter pylori* infection with age: Continuous risk of infection in adults rather than cohort effect, *J. Infect. Dis.*, 169, 41, 1994.

34. Walsh, J.H. and Peterson, W.L., Treatment of *Helicobacter pylori* infection in the management of peptic ulcer disease, *N. Engl. J. Med.*, 333, 984, 1995.

35. Kumpuris, D., Gastrointestinal bleeding in the older patient, in *The Aging Gut*, Texter, E.C., Ed., Masson Publ., New York, 1983.

36. Kaplan, R.C. et al., Risk factors for hospitalized gastrointestinal bleeding among older persons, *J. Am. Geriatr. Soc.*, 49, 126, 2001.

37. McGinty, D.P. et al., Vascular ectasias of the gastrointestinal tract: a new problem in our aging population, in *The Aging Gut*, Texter, E.C., Ed., Masson Publ., New York, 1983.

38. Reilly, J.M. and Sicard, G.A., Ischemic disorders in the large and small bowel, in *Principles and Practice of Geriatric Surgery*, Rosenthal, R.A., Zenilman, M.E., and Katlic, M.R., Eds., Springer, New York, 2001.

39. Webster, S.G.P., Small bowel morphology and function, in *Gastrointestinal Tract Disorders in the Elderly*, Hellemans, J. and Vantrappen, G., Eds., Churchill Livingstone, London, England, 1984.

40. Saltzmann, J.R. et al., Changes in small-intestine permeability with aging, *J. Am. Geriatr. Soc.*, 43, 160, 1995.

41. Tai, V. et al., Celiac disease in older people, *J. Am. Geriatr. Soc.*, 48, 1690, 2000.

42. Dew, M.J., Diseases of the colon and rectum, in *Principles and Practice of Geriatric Medicine*, Pathy, M.S.J., Ed., Vol. 1, 3rd ed., John Wiley & Sons, New York, 1998.

43. Evans, J.M. et al., Relation of colonic transit to functional bowel disease in older people: A population-based study, *J. Am. Geriatr. Soc.*, 46, 83, 1998.

44. 44. Deckmann, R.C. and Cheskin, L.J., Diverticular disease in the elderly, *J. Am. Geriatr. Soc.*, 40, 986, 1993.

45. Camilleri, M. et al., Insights into the pathophysiology and mechanisms of constipation, irritable bowel syndrome, and diverticulosis in older people, *J. Am. Geriatr. Soc.*, 48, 1142, 2000.

46. Thornton, S.C., Diverticulitis and appendicitis in the elderly, in *Principles and Practice of Geriatric Surgery*, Rosenthal, R.A., Zenilman, M.E., and Katlic, M.R., Eds., Springer, New York, 2001.

47. Harari, D. et al., Constipation in the elderly, *J. Am. Geriatr. Soc.*, 41, 1130, 1993.

48. Clinch, D.P. and Hilton, D.A., Constipation in *Principles and Practice of Geriatric Medicine*, Pathy, M.S.J., Ed., Vol. 1, 3rd ed., John Wiley & Sons, New York, 1998.

49. Brocklehurst, J.C., The problem of faecal incontinence, in *Gastrointestinal Tract Disorders in the Elderly*, Hellemans, J. and Vantrappen, G., Eds., Churchill Livingstone Press, London, England, 1984.

50. Swash, M., Physiology and pathophysiology of sphincter function, in *Principles and Practice of Geriatric Medicine*, Pathy, M.S.J., Ed., 3rd ed., John Wiley & Sons, New York, 1998.

51. Muir Gray, J.A. and Tinker, G.M., Incontinence in the community, in *Principles and Practice of Geriatric Medicine*, Pathy, M.S.J., Ed., 3rd ed., John Wiley & Sons, New York, 1998.

52. Winawer, S.J., Colorectal cancer screening comes of age, *N. Eng. J. Med.*, 328, 1416, 1993.

53. Jasleen, J. et al., Neoplastic diseases of the colon and rectum, in *Principles and Practice of Geriatric Surgery*, Rosenthal, R.A., Zenilman, M.E., and Katlic, M.R., Eds., Springer, New York, 2001.

54. Shiffman, S.S., Taste and smell losses in normal aging and disease, *J. Am. Med. Assoc.*, 278, 1357, 1997.

**Systemic and Organismic Aging**

55. Doty, R.I. and Zrada, S.E., Smell and taste, in *Principles and Practice of Geriatric Medicine,* Pathy, M.S.J., Ed., Vol. 1, 3rd ed., John Wiley & Sons, New York, 1998.

56. Ship, J.A., The influence of aging on oral health and consequences for taste and smell, *Physiol. Behav.,* 66, 209, 1999.

57. De Jong, N. et al., Impaired sensory functioning in elders: The relation with its potential determinants and nutritional intake, *J. Gerontol.,* 54, B24, 1999.

58. Rolls, B.J., Do chemosensory changes influence food intake in the elderly? *Physiol. Behav.,* 66, 193, 1999.

59. Schiffmann, S.S., Intensification of sensory properties of foods for the elderly, *J. Nutr.,* 130, 927S, 2000.

60. Mathey, M.F. et al., Flavor enhancement of food improves dietary intake and nutritional status of elderly nursing home residents, *J. Gerontol.,* 56, M200, 2001.

61. Schiffman, S.S., Effect of taste and smell on secretion rate of salivary IgA in elderly and young persons, *J. Nutr. Health Aging,* 3, 158, 1999.

62. Laugier, R. and Sarles, H., Pancreatic function and diseases, in *Gastrointestinal Tract Disorders in the Elderly,* Hellemans, J., and Vantrappen, G., Eds., Churchill Livingstone, London, England, 1984.

63. Lillemoe, K.D., Pancreatic disease in the elderly patient, *Surg. Clin. N. Am.,* 74, 317, 1994.

64. Banks, P.C., Pancreatic disease in the elderly, *Sem. Gastrointest. Dis.,* 5, 189, 1994.

65. Zdankiewicz, P.D. and Andersen, D.K., Pancreatitis in the elderly, in *Principles and Practice of Geriatric Surgery,* Rosenthal, R.A., Zenilman, M.E., and Katlic, M.R., Eds., Springer, New York, 2001.

66. Schmucker, D.L., Aging and the liver: An update, *J. Gerontol.,* 53, B315, 1998.

67. Hagmann, M., New genetic tricks to rejuvenate ailing livers, *Science,* 287, 1185, 2000.

68. Schmucker, D.L., A quantitative morphological evaluation of hepatocytes in young, mature and senescent Fischer 344 male rats, in *Liver and Aging,* Kitani, K., Ed., Elsevier/North Holland Biomedical Press, Amsterdam, 1978.

69. Trauner, M. et al., Molecular pathogenesis of cholestasis, *N. Engl. J. Med.,* 339, 1217, 1998.

70. Skaunic, V., Hulek, P., and Martinkova, J., Changes in kinetics of exogenous dyes in the ageing process, in *Liver and Aging,* Kitani, K., Ed., Elsevier/North Holland Biomedical Press, Amsterdam, 1978.

71. Hermann, R.E., Biliary disease in the aging patient, in *The Aging Gut,* Texter, E.C., Ed., Masson Publ., New York, 1983.

72. Mason, D.L., and Brunicardi, C.F., Hepatobiliary and pancreatic function: Physiologic changes, in *Principles and Practice of Geriatric Surgery,* Rosenthal, R.A., Zenilman, M.E., and Katlic, M.R., Eds., Springer, New York, 2001.

73. Kahng, K.U. and Wargo, J.A., Gallstone disease in the elderly, in *Principles and Practice of Geriatric Surgery,* Rosenthal, R.A., Zenilman, M.E., and Katlic, M.R., Eds., Springer, New York, 2001.

74. Johnston, D.E. and Kaplan, M.M., Pathogenesis and treatment of gallstones, *N. Eng. J. Med.,* 328, 412, 1993.

75. Bender, J.S. and Zenilman, M.E., Laparoscopic cholecystectomy in the nonagenarian, *J. Am. Geriatr. Soc.,* 41, 757, 1993.

# 21 The Skeleton, Joints, and Skeletal and Cardiac Muscles*

*Paola S. Timiras*
University of California, Berkeley

## CONTENTS

## I.  INTRODUCTION

As with other systems of the body, the musculoskeletal system, including the skeleton, joints, and muscles, passes through phases of growth, maturation, and decline. The first two phases have been studied extensively. The post-mature phase, often associated with involutional processes, is less well understood. It is this latter phase that is addressed in the present chapter.

Compared with other systems, the skeleton is rugged and durable; it usually carries on its tasks into advanced years, resists damage well, and has an efficient self-repair capability; indeed, normal or abnormal aging of the skeleton is seldom the cause of death. However, bones, the skeleton's major components, are subject, as are other parts of the body, to various hazards, primarily trauma, deficient metabolism and nutrition, and multiple degenerative changes.

---

* Illustrations from Dr. S. Oklund.

While aging of the skeleton usually occurs without conscious awareness on the part of the individual, aging of the articulations (joints) induces considerable physical pain and causes severe disability. Arthritic diseases, some of the most common expressions of joint aging, are among the most frequent and debilitating diseases of old age. The functional impairment and pain resulting from normal or pathologic aging of the joints limit the movement of elderly individuals, thus hindering their ability to care for themselves, eroding their independence, and, by forcing them to varying degrees of immobility, contributing to the decline in competence of other systems (such as the circulatory system). Some pathology of the joints begins at a relatively young adult age (rheumatoid arthritis may begin in adolescence) but shows increasing prevalence with advancing age. Thus, prevalence of rheumatoid arthritis is less than 1% before the age of 30 years and thereafter rises in each decade to 1 to 3% in the late fifties and 8 to 11% in the late sixties. Other manifestations of articular aging occur at later ages; osteoarthritis that affects 85% of persons 70 to 79 years of age is one of the major causes of invalidism confining the affected individuals to bed or to the wheelchair. Aging of the skeleton and joints will be considered in the first part of this chapter.

The study of aging processes in muscle is greatly complicated by the fact that muscle fibers do not constitute an homogenous tissue (e.g., differences between skeletal and cardiac muscle, striated and smooth muscle). Furthermore, the state of the tissue at any time depends on the extent and nature of many influences (e.g., nutrition, neural control, hormones), some of which are specific to muscle (i.e., physical exercise). Thus, the characteristic decline of muscular performance associated with advancing age is variable and may be caused not only by primary aging changes in the muscle fibers but also by aging of other body systems, nervous, vascular, and endocrine. Aging of the skeletal muscle is considered in the second part of this chapter, and the influence of nutrition and exercise thereon is discussed in Chapter 24.

The function of the heart is a major determinant of blood circulatory competence, and any changes with aging in this function are likely to play a central role in aging of the organism. The debate of whether atherosclerosis (Chapters 16 and 17) and its cardiovascular consequences of cardiac infarct, stroke, aneurysm, and gangrene are superimposed on the aging process or merely occur in older persons due to their longer exposure to risk factors, continues to generate considerable controversy.

With the current "reductionist approach," cardiovascular pathology research has shifted from the heart–lung preparation to the genes that encode the contractile proteins that regulate the cardiac pump. Cellular and molecular changes with aging in cardiac function have been studied primarily in experimental animals (rodents). There is good reason, however, by incorporating basic knowledge with new technology, to look optimistically to progress in our understanding of cardiac function in old humans as discussed in the third section of this chapter.

## II.  AGING OF THE SKELETON

The skeleton, the heaviest and most durable part of the body, provides the body framework and derives its properties from the unique characteristics of bone. Major functions of the skeleton provide:

- Body support
- Body movement
- Storage of calcium and other minerals, thereby aiding in mineral homeostasis: storage of calcium provides, upon request, calcium to blood, and blood calcium is important for regulating important cellular activities
- Maintenance of acid–base balance in association with the lungs and kidneys
- Regulation of phosphate and carbonate for buffers
- Storage for bone marrow, where white and red blood cells and platelets are produced; the important functions and the changes with aging of these cells are discussed in Chapter 15 (the immune system), in Chapter 20 (hematopoiesis and red blood cells), and in Chapter 16 (platelets)

### A.  PATTERNS OF BONE REMODELING

Maintenance of bone structure and function is a dynamic two-phase process by which bone mass is regulated throughout adult life, involving bone resorption by osteoclasts and bone formation by osteoblasts (Box 21.1). These processes are influenced by a number of factors, as illustrated in Figure 21.1. Calcium of bone is derived from circulating free calcium which is dependent upon calcium absorption, mainly through the gastrointestinal tract (Chapter 20). Concentration of circulating calcium also depends on its excretion, mainly through the kidneys (Chapter 19). A number of osteogenic factors promote bone formation through stimulation of osteoblastic activity or bone resorption through osteoclastic activity.

Bone is in a continuous state of flux. Although net bone mass does not change throughout much of adult life, bone is never metabolically at rest. It constantly remodels and reapportions its mineral stores along lines of mechanical stress. The major site of remodeling is the cancellous bone, lining the bone marrow cavities. Thus, bone maintenance includes:

- Formation of new bone by the osteoblasts
- Resorption of old bone by osteoclasts
- Carrying out of mature bone functions by osteocytes

## Box 21.1
## Bone Structure and Histology

Bone is a hard form of connective tissue (Chapter 22) consisting of bone cells or osteocytes imbedded in a collagenous protein matrix that has been impregnated with mineral salts, especially phosphates of calcium. The matrix consists of two phases: an organic one that comprises collagen, proteins, and glucosaminoglycans (polysaccharide and protein complexes) (Chapter 22) and an inorganic one that contains mainly hydroxyapatites (calcium phosphate) and minor amounts of other minerals. The collagen fibers provide resilience, and the minerals hardness. Given individual variations, it can be stated that, in childhood, about two thirds of the bone substance is composed of connective tissue, whereas in aged individuals, about two thirds consists of minerals. This transposition in content results in decreased flexibility and increased brittleness with advancing age.[1]

Histologically, bone is distinguished, as (1) "compact bone" found in shafts of long bones and outer surfaces (periosteum) of flat bones; (2) "cancellous bone" making up the trabecular space containing the bone marrow; and (3) "woven bone," which is an immature form of bone also involved in fracture repair. Bone cells are primarily concerned with bone formation and resorption; **osteocytes**, mainly in cancellous bone, maintain bone structure; **osteoblasts**, cells in the periosteum and endosteum (tissue surrounding the inner bone cavity) are bone-forming cells; and **osteoclasts**, found in the same regions as the osteoblasts, resorb bone by phagocytosis and digestion in their cytoplasm. Bone is well vascularized and contains an abundant nervous network.

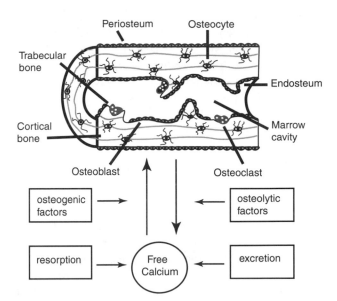

**FIGURE 21.1** Factors responsible for maintenance of bone structure. Calcium in bone is derived from circulating free calcium which depends on calcium absorption and excretion. Formation of new bone (osteogenesis) depends on osteoblasts, and destruction of all bone (osteolysis) depends on osteoclasts. New bone is continuously formed in the periosteum and endosteum. Active bone is formed of osteocytes, which are part of the cortical and trabecular bone. Blood cells are continuously produced in the bone marrow.

The mechanisms responsible for the calcification of newly formed bone matrix are incompletely understood despite intensive investigation: precipitation of calcium phosphate in bone may depend on some critical calcium concentration and is associated with the activity of certain enzymes (e.g., alkaline phosphatase) and specific proteins capable of binding calcium.

With aging, the balance between rates of bone formation and of bone resorption (i.e., the process of breaking down differentiated tissue and assimilating resulting particles) is disturbed, and the ensuing changes lead to a decrease in bone mass. After age 40, formation rates remain constant, while resorption rates increase. Over several decades, the skeletal mass may be reduced to half the value it had at 30 years. Progression of this loss may be measured by counting the number of Haversian canals or osteons that refer to cavities represented by cylinders of

bone containing a central blood vessel and nerve fibers. With advancing age, there is an increase in osteons at the bone shaft region, beginning at the ends of long bones. Between the ages of 42 to 52 years, this medullary cavity extends to the neck of the femur; between 61 and 74 years, it reaches the epiphyseal line close to the articular end. Changes that occur in the humerus may serve as an example: after 61 years, the outer bone surface becomes rough and the cortex thinner; the medullary cavity reaches the epiphyseal line; after 75 years, there is little spongy tissue left, and the cortex is very thin. Articular surfaces become very thin and may collapse. A major factor in these age-related changes is a loss of bone matrix that is confined primarily to the inner bone core. In the elderly, the periosteal tissue at the outer surface of the bones tends to remain constant or even increases in some bones (e.g., metacarpal bone of the hand),[2] but the endosteal tissue at the interior of bones is increasingly resorbed.[3]

### B. CELLULAR ELEMENTS DETERMINING BONE REMODELING

In the remodeling process of bone, the functions of *osteoblasts* and *osteoclasts* are intimately linked: osteoblasts synthesize and secrete molecules[4] that, in turn, initiate and control osteoclast differentiation.[5] The functions of both groups of cells are, in turn, correlated to

**Systemic and Organismic Aging**

the function of the *extracellular matrix*, composed primarily of collagen, proteoglycans, and various proteins (i.e., glycoproteins).[6] Osteoclasts are specialized macrophages; their differentiation is regulated by growth factors, such as granulocyte/macrophage colony-stimulating factor (GM-CSF), a cytokine (Chapter 15), by a nuclear factor kappa B (RANK) and its ligand (RANKL), and by the protein osteoprotegerin.[7–10] Binding of nuclear factor RANK to its ligand (RANKL) provides signals for survival and proliferation of osteoclasts, whereas overexpression or administration of osteoprotegerin reduces osteoclast formation.[7–10] RANK and RANKL are members of the tumor necrosis factor (TNF) and TNF receptor families, respectively.[11] Other molecules important in the resorption process, and thus in bone remodeling, are the integrins, a superfamily of adhesion molecules involved in the adhesion of cells to extracellular matrix: for the resorption process to take place, recognition and a physical intimacy are essential between osteoclasts and the surrounding matrix, and these are controlled by integrins.[12,13]

The realization that the osteoclast is part of the monocyte-macrophage family (Chapter 15) has prompted the development of techniques to stimulate bone formation in which macrophages may be made to differentiate into osteoblasts or, conversely, differentation into osteoclasts may be inhibited or slowed.[14] For example, osteoporosis (i.e., reduction of bone mass), frequent in women after menopause[15,16] is associated with a rise in osteoclast number, driven by increases in cytokines that induce osteoclast generation;[5,17–19] replacement therapy with estrogens or SERMs (Selective Estrogen Receptor Modulators) will block this increase and slow the osteoporotic process[20,21] and its consequences (falls and fractures) (Chapter 11).[22] A better understanding of the biology of osteoblasts and osteoclasts is providing new opportunities for developing therapies to act, singly, or in synergism, with other substances (e.g., growth factors, hormones), to treat diseases of bones.[14,21]

## C. BONE STRENGTH

Bone strength is an important property that allows bones to withstand the forces applied in the various movements of daily life. Several tests of bone strength have been studied, and the results show life cycle trends from youth through maturity into old age (Figure 21.2) and disclose a consistent decline of strength with aging.[23] Comparison of the change in strength of several musculoskeletal components with aging indicates that the fastest decrease in strength involves the cartilage followed by muscle, bone, and finally, tendon. Also, the comparison of bone with other tissues shows a much slower rate of decline for bone than for intestine and muscle but faster than for kidney.[24]

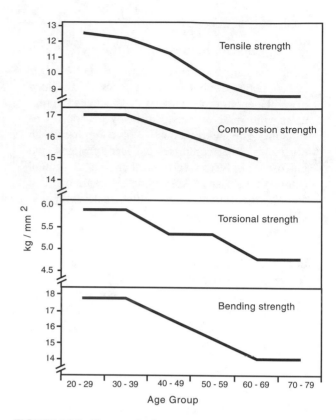

**FIGURE 21.2** Changes in bone strength with aging. Bone strength provides the ability of the bone to undergo force applied in (bottom to top) bending, twisting (torsional strength), compressing, and stretching (tensile strength).

## D. FACTORS AFFECTING BONE AGING

### 1. Role of Calcium in Bone Metabolism

Adequate circulating calcium is vital in maintaining bone mass. Indeed, the body regulates few parameters with greater fidelity than the concentration of extracellular and intracellular calcium. The constancy of extracellular calcium levels depends in part on its absorption in the intestinal mucosa and its excretion from the kidneys (Figure 21.3).

Calcium of the human body comprises 1.5% of body weight (1100 g, 99% of which is in the skeleton).[25] The need for calcium increases with aging. For example, while for young adults the recommended daily allowance is 600–800 mg, this value is increased to more than 1500 mg for women over 50 years of age. This increased need for calcium may be explained by a progressively less efficient absorption from the upper intestinal tract (Chapter 20).[26] Thus, in the elderly, more dietary calcium is needed to maintain an adequate calcium balance (Chapter 24). In addition to dietary calcium, other dietary components, physical exercise, and gender influence bone growth and aging.[27–29]

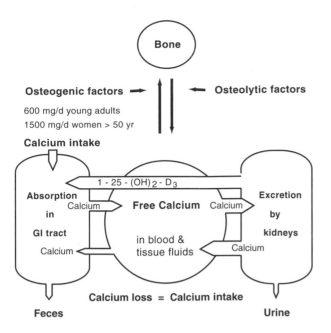

**FIGURE 21.3** Calcium metabolism. Note the different daily requirements of calcium with age; they almost double in women after 50, as compared to young adults.

## CALCIUM METABOLISM

Three hormones, parathyroid hormone, calcitonin (Chapter 13), and calcitriol are primarily concerned with calcium metabolism (Figure 21.4):

- Parathyroid hormone directly increases bone resorption and mobilizes calcium from the bone into the blood, thereby elevating plasma calcium. It also depresses plasma phosphate by increasing its urinary excretion
- Calcitonin lowers body calcium by inhibiting bone resorption
- Calcitriol (dehydroxycholecalciferol, vitamin $D_3$), a sterol derivative, increases calcium and phosphate absorption from the intestine and decreases their renal excretion; it also enhances bone resorption

Calcitriol derives through a series of steps either from the diet through absorption in the intestine, or through transformation of cholesterol in the skin by the action of sunlight. In the skin, the first step is the formation of previtamin $D_3$ which is converted to vitamin $D_3$ (cholecalciferol). Vitamin $D_3$ from the skin and intestine is carried to the liver, where it is hydroxylated to calcidiol, which, in turn, is carried to the kidney, where it is again hydroxylated to form the active product calcitriol.

The regulation of the relationship between calcium in blood and in bone depends primarily on parathyroid hormone but also involves other factors acting on calcium in tissues and cells. Indeed, calcium has many functions and is known to be a "universal regulator."[25,26] The many actions of calcium are summarized into six categories (Table 21.1): cell movement, cell excitability, cell secretion, phagocytosis, intermediary metabolism and respiration, and cell reproduction.

Despite the universal role of calcium in cell function and metabolism, few studies have inquired into the potential

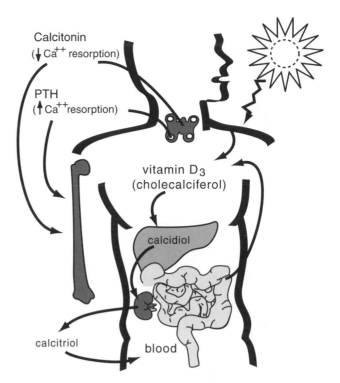

**FIGURE 21.4** Major hormones that regulate calcium metabolism. The parathyroid gland secretes the parathyroid homeone (PTH) which acts directly on bone to increase its resorption to mobilize calcium, thereby increasing (+) plasma calcium. Calcitonin from the thyroid gland inhibits bone resorption and lowers (-) plasma calcium. Calcitrol, a sterol derivative, increases calcium absorption from the intestine and decreases renal excretion. It also enhances bone resorption. Vitamin $D_3$ derives from dietary sources and is absorbed in the intestine or is formed in the skin under the influence of ultraviolet radiation. In the liver, vitamin $D_3$ is converted to calcidiol which, in turn, is activated in the kidney to calcitriol, the most active form in calcium regulation.

changes in calcium with aging. The most prominent actions of calcium related to aging are listed in Table 21.2. Considering the importance of calcium in regulation of cell metabolism and function, it seems reasonable to hypothesize that changes in calcium concentration and the consequent alterations in cell function would have an important bearing on the causes and course of the aging process. Research in this direction is still limited and should be pursued actively.

## 2. Hormonal Regulation of Bone Metabolism

Bone remodeling is accomplished by the interaction of several growth factors and locally acting cytokines operating on diverse cell populations in a highly coordinated manner. Hormones may affect bone function either by stimulating resorption or formation. Among those that affect resorption is the hormone from the parathyroid gland (as already considered in Chapter 13). Bone changes with aging apparently occur without marked alterations in the parathyroid gland or in the levels of parathyroid hormone, PTH (Table 21.3). However, in some old women, parathyroid hormone levels

**Systemic and Organismic Aging**

**TABLE 21.1**
**Intracellular Calcium Actions**

| Action | Examples of Change of State |
|---|---|
| 1. Cell movement | Muscle contraction |
| | Ciliate and flagellate movement |
| | Chemotaxis |
| 2. Cell excitability | Muscle action potential |
| | Myocardial action potential |
| | Response of the eye and other photoreceptor types to light |
| 3. Cell secretion | Neurotransmitter and hormone secretion |
| 4. Cell phagocytosis | Vesiculation of particles or soluble substances |
| 5. Intermediary metabolism and respiration | Glucose production |
| | Lipolysis |
| | Blood coagulation |
| 6. Cell reproduction | Lymphocyte transformation |
| | Ovum fertilization and sperm capacitation |
| | Ovum maturation and meiosis |

*Note:* In each category, changes in intracellular calcium translate into specific actions. For example, in (1) by promoting cell movement, calcium causes muscle contraction, ciliate and flagellate movement, and chemotaxis. In (2), the production of an action potential in appropriate cells causes muscle contraction or photoreceptor response. In (3), another pertinent example related to neural and endocrine functions is the role of calcium in cell secretion of neurotransmitters and hormones. In (4), calcium is involved in the uptake of particles by phagocytosis or uptake of soluble substances by vesiculation. (5) refers to the role of calcium in glucogenesis, lipolysis, and blood coagulation. And, finally, (6) intracellular calcium acts in the regulation of various phases of cell reproduction.

**TABLE 21.2**
**Intracellular Calcium-Dependent Changes Relating to Aging**

| Actions | Relation to Aging |
|---|---|
| 1. Intermediary metabolism and respiration | Activation of oxygen radicals |
| | Obesity and diabetes |
| | Arthritis and other diseases |
| 2. Tissue calcification | Loss of bone calcium (osteoporosis) |
| | Abnormal deposition on normal and injured tissues (e.g., atherosclerotic lesions) |
| 3. Cell excitability | Changes in cell potentials |
| | Changes in response to drugs |
| 4. Cell secretion | Alterations in neurotransmitter and hormone production |
| 5. Phagocytosis | Alterations in immune responses |

*Note:* The most prominent involvements of calcium related to aging appear to be: (1) to trigger the intracellular production of oxygen radicals which, as discussed in Chapter 5, lead to the accumulation of potentially toxic substances; (2) to affect normal and abnormal bone metabolism; a possible role of calcium in the pathogenesis of a number of diseases such as arthritis, obesity, diabetes, cystic fibrosis, etc.; calcification of injured and necrotic tissues such as occurs in atherosclerotic lesions (Chapter 16); (3) to induce age-related changes in cell excitability and in the action of several classes of drugs such as anesthetics, analgesics, and cardiovascular drugs; (4) to be involved in ion movement and to affect neurotransmitter and hormone production and secretion; (5) to participate in phagocytosis, with relation to aging of the immune system (Chapter 15).

increase with aging, but the contribution of this increase to bone loss may be minimal.[30,31] Some racial and gender differences have been reported: black and Asian postmenopausal women, have lower levels of parathyroid hormone and higher levels of calcium well into advanced age than white women, and men maintain lower levels coincident with a lower incidence of osteoporosis than women.

The increase in parathyroid hormone levels with aging may represent a compensatory response to reduced intestinal absorption of calcium, and hence, lower plasma calcium levels. Reduced exposure to sunlight in the immobile and house-bound elderly will also impair vitamin D manufacture in the skin and accentuate any dietary deficiencies of calcium. Reduced renal degradation and excretion of

## TABLE 21.3
## Parathyroid Hormone Changes with Aging

### Increased Parathyroid Hormone Plasma Levels

Causes:

    Decreased calcium intestinal absorption

    Decreased production of vitamin D (skin)

    Parathyroid tumors

Consequences:

    Increased bone resorption (osteoporosis)

    Symptoms resembling cognitive disorders (e.g., Alzheimer's Disease)

    Experimental progeria-like syndrome (rats)

### Lesser Increase of Parathyroid Hormone Levels

Men

Black and Asian women: lower incidence and severity of osteoporosis

parathyroid hormone may be another important factor, although it is still controversial whether or not such reductions occur and, if they do, how important they are.

Although best recognized for promoting bone resorption and elevating blood calcium levels, parathyroid hormone can also stimulate bone formation.[32–34] There is some evidence that intermittent parathyroid hormone administration increases mechanical strength and mass of intratrabecular bone, and that this action is mediated by transformation of the precursor cells into osteoblasts.[35] Parathyroid hormone may also increase bone formation by preventing apoptosis of osteoblasts.[36]

### DISEASES OF THE PARATHYROID GLAND

Structure, hormone, and hormone regulation of the parathyroid gland are presented in Chapter 13. Diseases of this gland are infrequent at all ages; however, they bear mentioning here, for the symptoms associated with aging may mask parathyroid pathology.[37] Indeed, if the disorder is recognized as a possible parathyroid dysfunction and is corrected, the symptoms described as aging changes may be ameliorated. Hyperparathyroidism presents with a variety of symptoms, such as increased plasma calcium (hypercalcemia), renal calculi, peptic ulceration, and, in a few individuals, mental aberrations with psychotic components. The latter have been sometimes erroneously identified as senile dementia of the Alzheimer type (Chapter 8). In advanced stages, characteristic bone lesions are present. Hypoparathyroidism is quite rare at all ages and is easily recognizable, for it generally follows ablation of the glands during thyroid gland surgery.

Other hormones that affect bone formation are briefly discussed below. Additional information is available in Chapters 10 through 14, where the corresponding endocrine gland is presented in more detail. The effects of gonadal hormones during the period of growth in childhood and adolescence are well known; those occurring during aging have been considered in Chapters 11 and

12. Particularly relevant to bone changes with old age, are those occurring in women in whom the decrease in estrogen levels at menopause plays a key role in accelerating osteoporosis. Also noteworthy, is the efficacy of replacement therapy with estrogens in delaying the progression of bone loss and in reducing the risk of bone fractures (Chapter 11). Following are hormones that affect bone formation:

1. *Glucocorticoids* lower plasma calcium levels, and over long periods of time, may cause osteoporosis by decreasing bone formation (due to inhibition of cellular replication, protein synthesis, vitamin D, absorption of calcium from the intestine, and reduced function of osteoblasts) and increasing bone resorption (due to stimulation of parathyroid hormone secretion) (Chapter 10).[38–40]

2. *Growth hormone (GH)* increases calcium excretion in urine, but because it has a greater effect in increasing intestinal absorption of calcium, overall it produces a positive calcium balance.[14–16,38] Somatomedins (IGFs-I and II), induced by growth hormone, also stimulate protein synthesis in bone. Risks and benefits of GH and IGF administration to increase bone mass in elderly men and women are discussed in Chapter 10.

3. *Thyroid hormones* may induce hypercalcemia and hypercalciuria and, in some cases, may also produce osteoporosis, but the mechanisms of these actions are unclear (Chapter 13).

4. *Insulin* promotes bone formation, and insulin deficiency (diabetes) is associated with significant bone loss (Chapter 14).

5. *Calcitonin* inhibits bone resorption by blocking osteoclastic activity (Chapter 13). The major physiological role of this hormone (secreted by the C cells of the thyroid gland) is the fine-tuning of extracellular calcium regulation, especially under specific conditions, such as growth, pregnancy, lactation, stress, etc. However, over time, a resistance to the affect of the hormone can develop, due to loss of receptors.[41,42] Development of resistance may restrict the therapeutic use of the hormone.

6. *Local growth promoting or inhibitory factors* may also influence bone formation and resorption, thereby participating in continuing bone remodeling. Most of these factors (e.g., epidermal growth factor, EGF; fibroblast growth factor, FGF; platelet-derived growth factor, PDGF; transforming growth factor, TGF) promote bone growth, while some inhibit growth (e.g., tumor necrosis factor, TNF).[43] Prostaglandins E

**Systemic and Organismic Aging**

secreted by certain tumors increase plasma calcium levels. Osteoclast activating factor, produced by lymphocytes, induces bone loss in tumors of the bone marrow. Cytokines may play a key role in bone metabolism by their ability to promote differentiation and maturation of osteoblasts or osteoclasts.

Physical activity also affects bone in several ways; it increases stress and strain on the skeleton due to muscular contraction and gravity; it improves blood flow to exercising muscles and, indirectly, increases venous return; and it stimulates bone mineral accretion. These effects are most marked in the young but are also present, although to a lesser degree, in the old (Chapter 24).

### E.   Aging-Related Fractures

Fracture patterns in the elderly differ markedly from those in the younger adult. Whereas in the younger adult (20 to 50 years) considerable violence is required to break the bones, in the elderly, fractures result from minimal or moderate trauma. This is due to the progressive loss of absolute bone volume, both compact and spongy bone, with aging. While the consequences of bone loss become manifest at 40 to 45 years of age in women and 50 to 60 years in men, bone loss may in fact begin much earlier, at the end of the growth period.

In younger adults, fracture incidence is higher in males than in females, perhaps due to the greater physical activity and exposure to accidental falls in the former. In the elderly, however, the reverse is true, especially in the case of fracture of the vertebral bodies, the lower end of the forearm, and the proximal femur. The relation between menopause and accelerated bone loss (osteoporosis) in females is discussed in Chapter 11.

Not only is the incidence of fractures higher in the elderly than young, but also, the sites of fracture are often different.[44] In the elderly, fractures occur through cancellous bone, usually next to a joint, rarely through the shaft of the bones, the most frequent site of fracture in the young. Orthopedic interventions to prevent or repair fractures are more difficult, and recovery is slower in the elderly than the young, due to bone fragility and overall less robust condition of the elderly.[44,45]

In addition to osteoporosis, other age-related bone disorders are associated with fractures. One of the most frequent of these is Paget's disease or Osteitis Deformans characterized by pain, deformity, and fractures of the bones. The incidence is higher in men than women. The early lesions are localized osteolytic lesions followed by accelerated bone turnover, that is, rapid remodeling leading to excessive and somewhat disorganized ("mosaic" or "woven" aspect) bone tissue balance. The resulting bone, although more dense than normal bone, has lost some of its elastic properties and is more susceptible to fractures. Complications include neoplastic transformation, neurologic signs due to compression of CNS structures, and heart failure. Management is based on symptomatic administration of aspirin for pain relief and of calcitonin (Chapter 13) and diphosphonates (compounds that slow the formation and dissolution of hydroxyapatite) to decrease bone resorption and bone formation.

### Ethnic Differences in Bone Mass and Consequences in Aging

The rates of fractures due to osteoporosis and other bones diseases associated with old age are substantially lower among blacks than white persons (three times lower among black women and five times lower among black men).[45] These differences are generally attributed to the 10 to 20% greater mass and density of adult bone in blacks. This greater bone mass and density are apparent before birth.[46] In blacks, bone remodeling also proceeds at a lower rate than in whites. Such observations suggest that hormonal, metabolic, and genetic factors play a role in these black/white differences. Age-matched bone density and mass among other ethnic groups in the U.S. population, although less studied than those among blacks and whites, show significant differences: for example, a survey of proximal femur fracture rates reveal that incidence rates are higher among Asian than white women.

### F.   Aging of the Joints

Joints or articulations comprise the junction between two or more bones (Box 21.2). They contribute to the function of the skeleton in promoting movement of body parts and

---

**Box 21.2**
**Structure of the Joints**

The articular system is comprised of simple and complex joints associated with the skeletal system and blood and nerve supplies to the joints. They represent junctions between two or more bones or cartilages (Figure 21.5).

In the skull, the joints (synarthroses) are immovable, and the connected bones are separated by a thin layer of connective tissue.

The articulations between the vertebrae (amphiarthroses) are somewhat more movable, and the bones are united by dense fibrous tissue and intervening cartilage.

Most bones are freely movable (diarthroses), as the adjoining ends are coated with smooth cartilage separated by a short tube of strong fibrous tissue containing synovial fluid.

Joint cartilage is a special type of connective tissue with an extracellular matrix of proteoglycans and collagen, synthesized by the cartilage cells or chondrocytes.

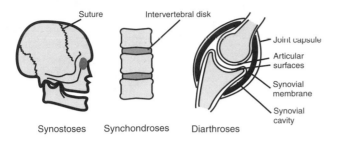

**FIGURE 21.5** The three major types of joints.

locomotion. The pattern and prevalence of joint changes in the elderly reflect that certain disorders arise more frequently with increasing age, and there is a steady cumulative effect with age inasmuch as many joint disorders are chronic.[46–48] By age 65, 80% of the population has some articular disorder. Beyond the aging changes affecting the articulations themselves, the problems of added illness, frailty, diminished motivation, and social isolation affect the problems of management and outcome of the disorder (Chapter 3).

Osteoarthritis and rheumatoid arthritis are examples of articular disorders occurring with increasing frequency from middle to old age. In fact, osteoarthritis is so common among the elderly that it is often assumed to be a normal accompaniment of aging rather than a disease. Radiological signs of osteoarthritis are universal in the later decades of life, but the changes are not always associated with significant pain or disability. It is particularly difficult to draw a dividing line between disease and normal aging with respect to degenerative changes of the cartilage.[46] With aging, decreased proteoglycan synthesis, loss of chondroitin sulfate and collagen glycation (i.e., covalent attachment of a carbohydrate to a polypeptide or polynucleotide), render the collagen network of the cartilage stiffer, more cross-linked, and more prone to fatigue.[49] It has been suggested that these alterations of the cartilage may be extremely important in the etiology of osteoarthritis.[49]

## 1. Aging-Related Changes in Joints

With aging, bony excrescences or osseous outgrowths (called osteophytes or bone spurs) occur on the heads of long bones, for example, on the head of the femur in 33% of individuals beyond 50 years of age; they are viewed as a purely age-related phenomenon. Areas where smooth cartilage has been replaced by a rough surface occur in all joints and are often seen as early as the second decade of life and spread to the periphery of the joint with age.

The undulations and hollows that can be seen on the cartilage surface by electron microscopy increase both in depth and diameter with increasing age, and the outer surface irregularities become more common. Whether these consistent surface changes may be related to the development of osteoarthritis is not known. Osteoarthritis is associated with cartilage thinning.

Such findings are contrary to the observation that the cartilage, in the absence of pathology, actually becomes thicker with increasing age,[48] except in the case of the patella (kneecap), which becomes thinner, especially in women. As cartilage ages, it loses some of its elasticity and becomes more easily stretched. Such changes lead to easier fatigability and higher susceptibility to osteoarthritis. There is a progressive pigmentation of the cartilage cells due to deposition of amino acid derivatives (probably from the protein of the matrix), and a gradual reduction in collagen, but no change in water content; this latter change is invariably associated with the early stages of osteoarthritis.

## 2. Pathologic and Clinical Aspects of Aging of Bones and Joints

Normal and pathological changes of skeleton in the elderly severely curtail the well-being of this population and contribute significantly to immobility. It is difficult at best to encourage an elderly individual to maintain a program of physical fitness or activity if mobility is limited by joint pain. Chronic pain and disability of whatever cause or nature induce anxiety and dampen the morale of any individual, especially the elderly, who are already prone to depression. For the elderly person who may be suffering from other concomitant losses, the inability to participate in certain physical activities can be quite devastating.

The list of musculoskeletal disorders of the elderly is quite long, and only an abbreviated one is presented here (Table 21.4); one of these disorders, osteoarthritis, will be briefly compared with "normal" aging changes. Joint disorders are essentially divided into two major types (Table 21.4):

- Those which affect joints without involvement of other organ systems
- Those which affect the skeleton as a manifestation of systemic disease involving several organ systems

Most of the systemic disorders are often referred to as "collagen-vascular," as they are frequently manifested by

**TABLE 21.4**
**Types of Musculoskeletal Disorders**

| Nonsystemic | Systemic |
|---|---|
| Noninflammatory | Non-autoimmune |
|   Osteoarthritis |   Polymyalgia Reumatica |
| Inflammatory | Autoimmune |
|   Gout |   Rheumatoid arthritis |
| |   Systemic Lupus Erythematosus |

**Systemic and Organismic Aging**

changes in connective or vascular tissues. Others are referred to as "autoimmune," because they are associated with the presence of antibodies that attack and damage the host's own tissue (Chapter 15). Among the nonsystemic diseases, osteoarthiritis and gout are the most frequent.

## 3.  Osteoarthritis

As already indicated, the most common articular disorder of the elderly is osteoarthritis or, less frequently, degenerative joint disease. Although it affects over three-quarters of the elderly, only 30% have any symptoms, and only 10% suffer significant disability. In a large proportion of the elderly, osteoarthritic lesions are often discovered incidentally (for example, osteoarthritis of the spine at the occasion of a chest x-ray). Risk factors include heredity as well as a history of trauma or injury to a certain body part or bone. Little is known of the pathogenesis of osteoarthritis. Why the process occurs at an accelerated pace in some individuals and why it affects certain joints and not others remain a mystery.

Loss of proteoglycans represents the cause of early changes that occur in the articular cartilage that lines the joint surfaces of the bones. Subsequent changes involve stiffening of collagen. Eventually, the joint space narrows. The bone underlying the articular cartilage is undergoing aging changes and becomes more susceptible to damage and microfractures. In the attempt to repair the damage, cysts and osteophytes form, superimposed on sclerosis.

Osteoarthritis is not a systemic disease, it affects only individual joints. Susceptible locations include the hands, hips, knees, feet, and spine. Symptoms usually begin with aches and pain in the involved joints. Initially, pain occurs with motion or weight bearing. In later stages, pain can occur at rest. Hand involvement occurs more frequently in women than in men and usually affects the distal interphalangeal joints. Small paired dorsal cysts of these affected joints are called Heberden's nodes. The first carpometacarpal joint at the base of the thumb is also commonly affected and gives the thumb a squared-off appearance. Hip involvement occurs more frequently in men and can present with knee or groin pain.

Diagnosis is made from clinical presentation and x-rays. It is important to differentiate the disease from other much more serious and life-threatening illnesses. For example, an elderly person with back pain should be carefully evaluated to ensure that no serious pathology such as metastatic malignancy exists before attributing the symptoms to osteoarthritis.

## 4.  Management of Osteoarthritis: Nonsteroidal Anti-Inflammatory Drugs Including COX-2 Inhibitors

In contrast to the significant advances that have been made in the treatment of some old age-associated diseases (e.g.,

atherosclerosis, Chapter 16), less progress has occurred with respect to osteoarthritis. There are no measures known that prevent or reverse the progressive joint damage. Therefore, management involves symptomatic relief of pain (analgesia).[50] In view of the lack of direct inflammatory activity in the affected joints, some argue that acetaminophen (Tylenol®), which acts primarily as an analgesic (i.e., to reduce pain), is a good first-line medication, particularly in individuals susceptible to the gastrointestinal side effects of aspirin. The nonsteroidal anti-inflammatory agents, NSAIDs (such as aspirin, ibuprofen) and the more recently discovered class of cyclooxigenase inhibitors are more potent than Tylenol but potentially have more adverse effects.

NSAIDs are among the most widely prescribed drugs in the world.[51,52] It is estimated that 15 to 25% of the ambulatory elderly use NSAIDs; this usage may be higher in view of the fact that these drugs are sold over-the-counter.[51–53] The widespread use of NSAIDs among the elderly is due to their unique features:

- They provide rapid and prolonged relief of pain
- They are without risk of addiction or tolerance
- They reduce swelling, tenderness, and stiffness of joints
- The more than 20 marketed medications (with different chemical structure, potency, pharmacokinetics properties, Chapter 23) provide great flexibility of choice

NSAIDs used in the treatment of osteoarthritis and other joint diseases inhibit the enzyme cyclooxigenase that catalyzes the first reaction in the conversion of arachidonic acid to prostaglandins. These are ubiquitous molecules with diverse and numerous actions (Table 21.5)[54] and, in the case of osteoarthritis, decreased prostaglandin production or release is manifested by elimination of pain (analgesic action), reduction of fever (antipyretic action), and reduction of inflammation that accompanies articular damage (anti-inflammatory action) (Table 21.6). Unfortunately, NSAIDs induce severe side effects, particularly in the gastrointestinal tract, where they may be responsible for impaired digestion (dyspepsia), erosion of the gastric mucosa, and gastric and duodenal ulcer (Chapter 20);[55] they may also have serious toxic consequences for the liver,[56] where they are metabolized (Chapter 20) and the kidneys,[57] from where they are excreted (Chapter 19) (Table 21.6).

The identification of two isoforms of the enzyme cyclooxygenase, COX-1 and COX-2, and of two inhibitors of COX-2, celecoxib and rofecoxib, now offers patients:

- A safer alternative to standard NSAIDs
- Comparable analgesic and anti-inflammatory efficacy
- Significantly reduced gastrointestinal toxicity[58,59]

## TABLE 21.5
### Some Characteristics and Actions of Prostaglandins

- Bioactive lipids generated by the action of cyclooxygenase from arachidonic acid
- Produced by virtually all tissues, but effects depend on specific prostaglandin and tissue or organ where produced
- Released when cells are damaged
- Detected in increased concentrations in inflammatory exudates
- Mediate increased body temperature (fever) due to infection, tissue damage, inflammation, others
- Contribute to pain and inflammation
- Sensitize pain receptors, lower threshold of pain responses, potentiate transmission of pain signals
- Protect gastric mucosal barrier
- Regulate intestinal motility
- Regulate renal blood flow and sodium reabsorption
- Stimulate $HCO_3^-$ secretion from the duodenum (to neutralize acid stomach contents)
- Potent vasodilators and synergistic to other vasodilators (e.g., histamine)
- Promote platelet agglutination

COX-1 is found in many tissues, including the stomach, where it provides some cytoprotection. COX-2 is also found in many tissues; it increases substantially during inflammation induced by a variety of stimuli (e.g., endotoxins, cytokines).

In addition to NSAID and COX-2 inhibitor administration, other helpful interventions include the following (Table 21.7):

- Avoidance of heavy weight-bearing (especially in osteoathritis of the knee) as well as weight loss
- Use of a cane and orthotics (mechanical devices to support and brace weak joints and muscles)

- Muscle strengthening by physical exercise
- Intra-articular injection of glucocorticoid hormones for temporary relief due to the analgesic and anti-inflammatory actions of these hormones (Chapter 10)
- Surgical interventions (orthoplasty), especially for hip or knee replacement

## III. AGING OF MUSCLE

The major function of muscle is the generation of force and performance of work through the conversion of chemical to mechanical energy (Box 21.3). Muscle force and work are necessary to maintain structural integrity, to maintain posture, for locomotion, for breathing, for digestion, and for almost all functions of the body. Muscle strength reaches a peak between 20 and 30 years of age. It declines beginning in middle age, and continues to decline at an approximately constant rate with increasing age, irrespective of the muscle group considered. There is, however, great variability among muscle groups in the rate of decline with aging. For example, the diaphragm remains active throughout life and undergoes little change with aging (Chapter 18). In contrast, the soleus muscle of the leg, relatively inactive in the less mobile elderly, shows decreased strength with aging.[60] While the focus here is on skeletal muscle, many of the age-related changes identified in this type of muscle are similar to those of cardiac (see below) and smooth muscle. Those occurring in skeletal muscle are particularly prominent because of the proportionally high distribution of this type of muscle throughout the body and its relationship to lean body mass (body mass minus bone, mineral, fat, and water). Additionally, it is possible to increase skeletal muscle power (i.e., speed and force of contraction) by physical training, even in old age.[61,62]

## TABLE 21.6
### Actions of Nonsteroidal Anti-Inflammatory Drugs (NSAIDs) and COX-2 Inhibitors

**Mechanism of Action:**
Inhibition of the Enzyme Cyclooxygenase that Catalyzes the First Step in the Conversion of Arachidonic Acid to Prostaglandins

| Aspirin, Tylenol (Acetaminophen), Ibuprofen (Motrin, Other Propionic Acid Derivatives) | | Celecoxib Rofecoxib |
|---|---|---|
| **Beneficial Effects** | **Adverse Effects** | **Beneficial/Adverse Effects** |
| - Reduction of joint swelling, tenderness, stiffness<br>- Analgesia<br>- Fever reduction<br>- Anti-inflammatory action<br>- Others[a] | - Gastrointestinal tract: perforation, ulceration, bleeding<br>- Liver and kidney toxicity | - Same overall effects as standard NSAIDs but longer duration and less adverse effects |

[a] Anti-inflammatory and antiplatelet agglutination actions have been described as beneficial in atherosclerosis (Chapter 16) and in cognitive disorders such as Alzheimer's Disease (Chapter 8).

**Systemic and Organismic Aging**

## TABLE 21.7
## Management of Osteoarthritis

| Physical Interventions | Medical Treatment |
| --- | --- |
| Weight loss | Acetaminophen (Tylenol) |
| Use of cane | Aspirin and nonsteroidal anti- |
| Exercise to increase muscle | inflammatory drugs |
| strength | Intra-articular steroid injection |
| Orthotics | |
| Intra-articular lavage | |
| Arthroplasty (joint replacement) | |

## A. AGING-RELATED CHANGES IN SKELETAL MUSCLE

The major aspect of muscle aging is the greatly *reduced muscle mass*, also called *"sarcopenia"* (or muscular wasting or cachexia) with consequent muscle weakness.[63–65] In addition to muscle fiber loss, the spinal cells of the anterior horn regulating muscular activity also are reduced in number and show increased size variability (suggesting unsuccessful compensatory attempts to restore appropriate nervous control).[66–68]

Macroscopically, as muscles age:

- They become smaller in size, due to loss of motor units (sarcopenia)
- They lose the usual red-brown color of normal muscle due to loss of myoglobin pigment
- They become yellow due to the deposition of lipofuscin pigment and increased fat cells or grey due to increased amounts of connective (fibrous) tissue

Microscopically, muscle fibers are subjected to the following:

- Decrease in number, especially of fast-contracting Type II fibers
- Increase in size variability due to selective loss of fast Type II fibers
- Proliferation of the T system and sarcoplasmic reticulum
- Decrease in synthesis of contractile protein,
- Reduction of mitochondrial mass, with consequent decrease of muscle oxidative capacity

It is uncertain which of these changes represents the primary aging phenomenon. Together with muscle changes, the myoneural junction undergoes several changes:

- The capability to sustain transmission of the nerve impulse from the neuronal axon to the

## Box 21.3
## Skeletal Muscle Structure and Function

The major function of muscle is to contract with utilization of energy and production of work and heat. Skeletal muscle is formed of individual muscle fibers containing fibrils, which are divisible in filaments made up of contractile proteins, **myosin**, **actin, tropomyosin,** and **troponin**. Myosin forms the thick muscle filaments, and the three other proteins form the thin filaments. Cross-linkages are formed between myosin and actin molecules. During contraction, by breaking and reforming cross-linkages between myosin and actin, the thin filaments slide over the thick filaments, thereby shortening muscle length and utilizing ATP for energy. Muscle fibrils are surrounded by the **sarcotubular system,** formed of vesicles and tubules distinguished into a T system, and a **sarcoplasmic reticulum**. The function of the T system is the rapid transmission of the action potential from the cell membrane to all myofibrils. The sarcoplasmic reticulum regulates calcium movement and muscle metabolism. Muscle fibers are distinguished into two major types: the slower contracting Type I fiber (e.g., long muscles of the back) and the fast-contracting Type II fiber (e.g., muscles of the hand). Of the two types, the most affected with aging are the Type II fibers.

Muscles are innervated by somatic nerves (skeletal, voluntary muscles) or autonomic nerves (smooth, visceral muscles) or have their own, specialized nervous conduction system (cardiac muscle). In skeletal muscle, nerve endings, rich in the neurotransmitter **acetylcholine** (Chapter 7) fit into a motor-end-plate, a thickened portion of muscle membrane, to form the **neuromuscular or myoneural junction**. In smooth and cardiac muscle, there are no recognizable end-plates, and the neurotransmitter may be acetylcholine or **norepinephrine**. Muscle contraction is initiated by calcium release through depolarization at the neuromuscular junction; the action potential is transmitted to all fibrils via the T system and triggers the release of calcium from the sarcoplasmic reticulum. This process is called excitation–contraction coupling.

Movements of skeletal muscle are voluntary; they are regulated by inputs from the cerebral cortex (motor cortex, area 4), the basal ganglia, and the cerebellum, that are relayed to the spinal cord (by the corticospinal and corticobulbar tracts) and from there to the muscles.

muscle fiber decreases, as muscle and nerve fibers show increased refractoriness

- The amount of acetylcholine (ACh), the neurotransmitter at the neuromuscular junction, declines
- The motor nerve conduction velocity is reduced
- The balance between nerve terminal growth and degeneration becomes less stable,
- The membrane alterations include membrane potentials, lowered uptake of choline (the ACh precursor), and less uniform distribution of ACh receptors

These changes, characterized by their high variability (i.e., some neuromuscular junctions remain unaffected), may be due to the following:

- Instrinsic muscle changes
- Alterations of the neuromuscular junctions
- Impairment of CNS control, such as mental impairment, observed in demented individuals (Chapter 8)

## B. MUSCLE ENERGY SOURCES AND METABOLISM

Muscle activity requires energy; the muscle has been called a machine for converting chemical to mechanical energy.[68] In the case of the heart, the requirement is for continued sustained work with limited resting time, when compared to skeletal muscle activity. Activity in skeletal muscle is "on demand," with intervening periods of contractile inactivity. To provide for the energy requirements, muscle has an abundant blood supply, numerous mitochondria, and a high content of myoglobin, the latter an iron-containing pigment, resembling hemoglobin. Myoglobin serves as an oxygen supply during contractions, which cut off blood flow, and also facilitates the diffusion of oxygen to mitochondria where the oxidative reactions occur.

Muscles at rest derive most of their energy from fats (free fatty acids) and less from carbohydrates; when activity increases, carbohydrates become the major source of energy, while utilization of fats decreases (Chapter 24). When activity is extreme or once activity ceases, fats again become the primary energy source. With aging, the metabolic pathways and energy resources are the same as in young individuals, but metabolic adjustments are slower. Further, the greater difficulty involved in activating bones and joints renders the older individual less capable of meeting the increasing energy requirements of the exercising muscle (Chapter 24). The safest approach for measuring energy requirements in the elderly is to examine metabolic activity in the context of daily routines, such as walking (which normally demands 35 to 40% of maximum oxygen intake) or after muscle activity of such intensity to induce difficulty in speaking due to breathlessness (usually the anaerobic threshold, 60 to 70%

of maximum oxygen intake). Under both conditions, metabolic activity is decreased, while cardiac rate increases, thereby reducing muscle efficiency and accelerating fatigability even in highly selected healthy 65- to 89-year-old men and women.[69]

However, even in very old individuals, physical activity performance may be considerably improved with exercise (Chapter 24). In addition, physical training may increase the ability to utilize lipids as an energy source for exercise. Long-term heavy exercise may decrease LDL and have a beneficial effect on cardiovascular function (Chapter 24), but it may also increase the production of free radicals with unfavorable consequences on the same cardiovascular function (Chapter 5).

## C. RESPONSIVENESS OF AGING SKELETAL MUSCLE TO EXERCISE

Muscle responsiveness to appropriate stimuli is well demonstrated by the improvement induced by physical exercise in muscle strength, ambulatory ability, and endurance in the young and adult as well as in the elderly, including the "oldest old" (80 years and older).[62] For example, high-resistance weight training in nonagenarians leads to significant gains in muscle strength, size, and motility.[70,71] Exercise training in 60- to 70-year-olds increases bone density as well. The benefits and risks of exercise on muscle and other organs and functions[72] and the most appropriate regimens of physical exercise for the elderly are considered in Chapter 24.

Notwithstanding the alterations of the muscle fibers at the myoneural junction, reinnervation by motor nerve sprouting may occur.[73] Such "sprouting effects" have provided evidence for the concept of a continual, dynamic turnover of synaptic sites in muscle. Despite some conflicting reports, the prevalent view is that remodeling of aging neuromuscular junctions would replace those that degenerate after a limited lifetime.[74] Thus, adaptation to training may remain operative in some motor units well into old age.[75] In old rats, treadmill running and ablation of one of two synergistic muscles significantly increases (by 45 to 75%) muscle mass and force-generating capacity. The increase in old animals is comparable to that occurring in adults and young rats. Exercise may also stimulate neuronal regeneration not only in the spinal cord but also in the brain, as suggested in a number of recent studies presented in Chapter 8.

### MUSCLE AND GROWTH HORMONE

The considerable interest in the possible ability of growth hormone and insulin-like growth factor I to increase muscle mass and strength in old age has generated a plethora of over-the-counter (dietary) supplements with claims to improve and rejuvenate a large number of functions (e.g., "feel 10 years younger with oral growth hormone spray!"). As discussed in

**Systemic and Organismic Aging**

Chapter 10, administration of recombinant human growth hormone (available only with physician's prescription and quite expensive) results in modest beneficial effects on muscle and bone that are, at best, temporary, and, at worst, associated with adverse side-effects that represent a real danger of long-term treatment.

## IV.  AGING OF CARDIAC MUSCLE

The major function of the heart is to pump blood through the systemic and pulmonary circulations and thereby transport respiratory gases, nutrients, and metabolic products to and from tissues. It is not surprising, therefore, that most alterations in cardiac function have life-endangering consequences. Structural changes with aging (Figure 21.6) involve all components of the heart, as listed in Table 21.8.

Heart disease, associated with advancing age and atherosclerosis, remains the most important single cause of death, worldwide, in individuals 65-years-old and older, and in both sexes (Chapter 16). While the mortality from cardiac diseases due to atherosclerosis (e.g., coronary heart disease) has progressively declined since the 1970s (Chapter 2), two other types of cardiac pathology are emerging and reaching epidemic proportions: they are heart failure and atrial fibrillation.[76,77] Both conditions, generated by intrinsic alterations of cardiac muscle (heart failure) or of cardiac rhythm (atrial fibrillation), result in the inability of the heart to maintain blood flow to body tissues. Heart failure is a sudden fatal cessation of the heart pumping function due to inability of the ventricles to contract and maintain an adequate blood ejection and flow. It is usually associated with hypertension and has been related to a loss of cardiac muscle cells below a critical threshold required for myocardial contraction; attempts to compensate the loss of cells by hyperplasia of remaining cells will generate cardiac hypertrophy and dysfunction manifested in cardiac failure and arrhythmias. Atrial fibrillation, the most common form of chronic arrhythmia, is responsible for abnormal cardiac rhythm generated by loss or damage of pacemaker cells that drive the cardiac rhythm. As the cardiac rhythm is disrupted, the rate of ventricular contraction becomes rapid and irregular, and blood flow is interrupted.

### A.  AGING-RELATED CHANGES IN CARDIAC STRUCTURE

Cardiac structural changes may be caused by the aging process, or may be secondary to disease. They may be:

- Primary to the heart
- Secondary to vascular lesions or to pulmonary disease

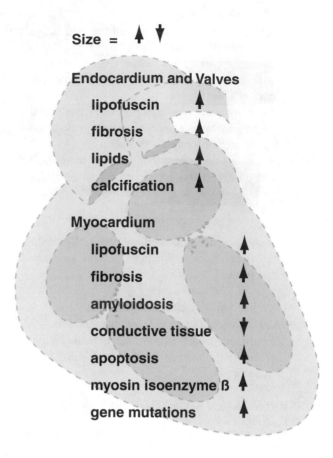

**FIGURE 21.6** Some age-related changes in the heart.

## TABLE 21.8
## Major Cardiac Structures

- The *myocardium,* the cardiac muscle, closely resembling the skeletal muscle with the difference that myocardial fibers act as a syncytium
- The *cardiac conduction system,* the specialized pacemaker tissue that can initiate repetitive action potentials and regulates autonomous, rhythmic contractions
- The *endocardium,* the endothelial layer lining the internal surface of the cardiac cavities
- The *pericardium,* the serous membranous sac that encloses the heart

- So severe as to be present at rest
- Manifest only under conditions of increased demand, such as physical exercise

They involve, in varying degrees, all cardiac elements:

- Muscle
- Endocardium
- Connective tissue
- Valves
- Conduction tissue
- Cardiac vasculature

Decline in cardiac function with aging is often associated with concomitant atherosclerosis of the coronary arteries, with consequent severe pathology such as coronary heart disease, the major cause of cardiovascular disease in the elderly (Chapter 16). Another consequence of age-related changes in heart and vasculature is hypertension, which today is amenable to successful treatment. Cardiac size and weight remain essentially unchanged with aging, although some studies suggest enlargement (particularly of the left ventricle) due to muscle hypertrophy responding to increased contractile effort in hypertension and atherosclerosis, or, vice versa, cardiac atrophy, due to loss of muscle mass. Normally, a steady drop-out of cardiac cells occurs during life.[78-83] With aging and, especially, when complicated by hypertension, *apoptosis* or cell programmed death (Chapter 4) of myocardial cells becomes more rapid and extensive, and the remaining cells are insufficient to maintain adequate contraction with consequent cardiac failure.[78-83] In addition to loss of muscle cells by apoptosis, the surviving myocardial cells show an altered phenotype that has been linked to altered function and is associated with the secretion of molecules, such as natriuretic and opioid peptides, usually released in response to stress.[84] In rodents, molecular shifts with aging occur in the *myosin heavy chain isoforms:* these isoforms are present as α and β isoforms and, with aging, the β isoform becomes predominant. This predominance results in a contraction that exhibits reduced velocity and prolonged time course. Inasmuch as reduced velocity conserves energy and longer contraction extends blood ejection time, these changes may be interpreted as adaptive responses to the stiffness of the atherosclerotic vasculature.[85] Other mutations that affect contractile proteins,[86-88] cytoskeletal proteins,[89,90] structural proteins,[91-93] or calcium-sensitive proteins,[94] and their consequences for cardiac muscle contractility, are summarized in Table 21.9.

The normal contractile function of cardiac cells is further endangered by intracellular — accumulation of lipofuscin — and extracelluar changes — increased connective tissue and accumulation of fat deposits in the ventricles and interatrial septum. In the latter, such deposits may displace conduction tissue in the sino-atrial node and lead to disturbances of conduction in some severe cases (as in atrial fibrillation). Other extracellular deposits that may reduce the contractile efficiency of cardiac muscle include some degree of amyloidosis (Chapter 8) and fibrosis, as occurs in myocardial infarction (Chapter 16). Mitochondrial DNA damage, perhaps due to free radical damage and to lipid alterations, may play a role in decreased metabolic activity of human hearts, cell loss, and deterioration of function (Chapter 5).

*Specialized conducting cells* (forming the sino-atrial and atrioventricular nodes and the intraventricular bundle of His) may be lost with aging, although this loss is usually moderate and may be associated with an increase in con-nective tissue elements, particularly collagen and elastin. It is not clear, however, whether and to what extent these changes interfere with conduction.

Endocardial changes — lipofuscin deposition and varying degrees of fibrosis — are probably influenced by mechanical factors such as blood turbulence (Chapter 16). These mechanical factors may also be responsible for thickening of the atria and valves, which also show lipid deposition and calcification. The timetable of these valvular lesions varies with each valve.

### EXCITATION–CONTRACTION COUPLING

In skeletal muscle, the nerve impulse triggers release of calcium from its stores in the sarcoplasmic reticulum and initiates contraction. Calcium binds to troponin C, uncovering several myosin binding sites on actin. These molecular changes result in a decrease in the number of cross-linkages that bind myosin to actin and facilitate the sliding of the actin and myosin filaments along each other, inducing shortening of the muscle during contraction. In cardiac muscle during the action potential, calcium ions diffuse into the myofibrils from stores in the sarcoplasmic reticulum and also from the T-system. Cardiac T-tubules are much larger and contain many times more calcium than those in skeletal muscle. This extra supply of calcium from the T-tubules is at least one factor responsible for prolonging the cardiac muscle action potential for as long as one third of a second, ten times longer than in skeletal muscle. Then, when calcium concentration is lowered, chemical interaction between myosin and actin ceases, and the muscle relaxes. ATP provides the energy necessary for the active transport of calcium.

With aging, the excitation–contraction coupling in the heart appears to be markedly altered in some elderly persons. In experimental animals or with isolated heart models, excitation appears insufficient — or less sufficient than in the young — to trigger the release of calcium from its stores; the calcium stores would be reduced, and the accumulation of calcium in the sarcoplasmic reticulum would be slowed.[95,96] Mutations of the contractile proteins,[85,86] of calcium-sensitive proteins,[94] and of microtubule assembly proteins,[89,90] would inhibit active transport of calcium and prolong duration of contraction and relaxation (Table 21.7). Alterations with aging in calcium stores and movements are also reflected in diminished cardiac responses to inotropic (promoting muscular contraction) agents such as catecholamines. These agents exert their stimulatory action on cardiac contraction by loading calcium stores and by facilitating calcium transport into the cells. With aging, calcium storage and movements may be altered, hence, the reduced effectiveness of inotropic factors in activating cardiac contraction.

### B. AGING-RELATED CHANGES IN CARDIAC OUTPUT

The overall expression of cardiac function is cardiac output, that is, the amount of blood pumped by the heart into the circulation per unit of time (in the healthy, resting, adult man, cardiac output is 5.5 L/min). Cardiac output depends on stroke volume and cardiac rate as well as on venous return (Figure 21.7). Stroke volume depends on strength of muscle contractility, cardiac rate, autonomic

**Systemic and Organismic Aging**

**TABLE 21.9**
**Some Molecular Changes in Cardiac Muscle Cells with Aging[a]**

| Contractile Proteins | | Cytoskeletal Proteins | | Ca++ Sensitive Proteins | |
|---|---|---|---|---|---|
| **Changes** | **Consequences** | **Changes** | **Consequences** | **Changes** | **Consequences** |
| • Mutant gene encoding heavy chain of myosin, with preponderance for β isoenzyme <br> • Mutant genes encode six proteins that regulate muscle contraction (two myosin light chains, tropo-myosin, troponins T and I myosin binding protein C) <br> • Gene mutations that encode structural proteins like dystrophin, LIM, others | • ↓ velocity of contraction saves energy; longer period of contraction permits blood ejection to last longer <br> • ↑ hypertrophic cardiopathy in young and older people with stretched-out thin-walled heart that fills with excess blood without efficiently pumping it out <br> • May impede smooth sliding of myosin over actin, slowing down contraction <br> • Conversely, Troponin T mutations accelerate sliding of actin past myosin, consuming more energy and producing local energy shortage <br> • Failure of linking muscle cytoskeleton to extracellular matrix and therefore failure to anchor myocytes to extracellular milieu | • ↑ microtubule component of cytoskeleton of cardiac muscle cells, perhaps due to posttranslational modifications of tubulin or ↑ microtubule assembly protein (MAPs) | • Cellular contraction dysfunction | • ↑ Ca++ enzyme, calcineurin, activates the NF-AT proteins by removing the phosphate group and allowing them entrance into muscles, linking to GATA4 protein and turning on the genes for hypertrophy | • NA-AT – GATA4 system induces cardiac cell hypertrophy and dysfunction *in vitro*; cyclosporin-A inhibits calcineurin and prevents cardiac hypertrophy; in transgenic mice, similar effects are seen |

**Apoptosis**

| Changes | Consequences |
|---|---|
| • ↑ apoptosis of cardiocytes (cardiac muscle cells) | • Cardiac atrophy, thin myocardial wall, decreased strength of contraction, cardiac failure <br> or <br> • Compensatory responses to loss of cells induces hyperplasia of remaining cells with enlargement of cardiac muscle and abnormal cardiac function resulting in heart failure |

[a] For more information, refer to the text and the references in the text.

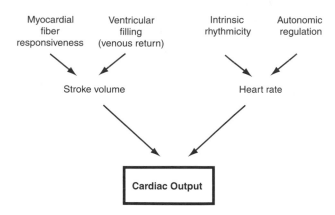

**FIGURE 21.7** Factors regulating cardiac output.

---

**TABLE 21.10**
**Effects of Aging on Venous Return**

Venous competence depends on:
    Venous diameter
    Intrathoracic pressure
    Total blood volume
With aging:
    Altered venous smooth muscle and elastin (varicose veins, hemorrhoids)
    Thrombophlebitis
    Orthostatic hypotension ($\downarrow$ sympathetic tone, $\downarrow$ baroreceptor activity, $\uparrow$ polypharmacy)

---

**TABLE 21.11**
**Autonomic Cardiac Regulation**

Parasympathetic Stimulation
    Decreased heart rate
    Decreased conduction
    Increased refractory period
Sympathetic Stimulation
    Increased heart rate
    Increased strength of contraction

---

innervation, venous return (amount of blood returning to the heart), and competence of the veins. Some aspects of intrinsic cardiac conduction and extrinsic autonomic innervation, in health and disease, are summarized in Box 21.4. Cardiac output may be reduced in a large number of elderly, although in others, it remains efficient well into old age.

**1. Venous Return**

Changes with aging in the venous compartment occur at all ages, but are more frequent at older ages. As summarized in Table 21.10, they involve the following:

---

**Box 21.4**
**Cardiac Innervation and Some Disorders with Aging**

Instrinsic conduction tissue is characterized by an unstable potential related to an unstable permeability to potassium. With aging, intracellular accumulation of lipofuscin, extracellular amyloidosis, loss of conduction cells, and fatty and fibrotic infiltrates lead to a decline in function and an increase in instability. Thus, **cardiac dysrhythmias** or **arrhythmias, (alterations of cardiac rhythmic contractions)** are frequent in elderly subjects whose hearts are more vulnerable to such biochemical insults as hypoxia (decreased oxygen), hypercapnia (increased carbon dioxide), acidosis (decreased pH), and hypokalemia (decreased blood potassium). All these conditions increase cardiac irritability, particularly of the atria. This, coupled with reduced cardiac output and coronary blood flow, leads, in the elderly more than in young adults, to cardiac failure and death.

Another conduction defect in the elderly is the **interruption or block of conduction**, most often between atria and ventricles. Cardiac blocks reflect potential cardiac disorders but are usually asymptomatic and non-life-threatening.

The **extrinsic innervation of the heart involves the two branches of the autonomic nervous system**: the **parasympathetic** vagal innervation slows heart rate, and the **sympathetic** increases cardiac rate and strength of contraction (Table 21.11). Both undergo changes with aging. Some consistent findings are (a) decreased inotropic (force of cardiac contraction) responses to catecholamines (primarily, epinephrine and norepinephrine), and (b) decreased sensitivity to a variety of drugs (e.g., sympathetic agonists or antagonists) or hormones (e.g., thyroid hormones). Decreased sympathetic responses may be due to loss of neural cells resulting in reduced catecholamine content. In the rat, for example, cardiac norepinephrine is halved between the age of 1 month (young) and 28 months (old). Another contributing factor may be the reduction in number or affinity of , receptors. Although little is known in man, in experimental animals, the number of receptors appears to be markedly decreased.

---

- Slower peripheral venous circulation due to impairment of elastic and smooth muscle components, as in *varicose veins* (i.e., tortuous and enlarged veins, most frequently found in the lower limbs) or in *hemorrhoids* (i.e., enlargement of the rectal veins)
- Inflammatory processes, as in *thrombophlebitis,* an inflammation of the veins due to the

**Systemic and Organismic Aging**

presence of a thrombus induced by slow blood flow in the unusually dilated veins

- *Postural hypotension,* common in the elderly, and has increasing incidence with aging; this condition, characterized by a fall in blood pressure rising from the supine to the standing position, has been ascribed to autonomic, particularly sympathetic, insufficiency, to low circulating levels of epinephrine, or to decreased activity of the baroreceptors, which sense blood pressure

## 2. Stroke Volume

Strength of ventricular contraction may be altered by a number of factors, some of which are illustrated in Figure 21.8; some are related to changes in the muscle (loss of myocardial tissue, as mentioned above), others to alterations in autonomic nervous control, particularly with respect to levels of catecholamines, or to effects of hormones and selected drugs, and alterations in blood oxygen, electrolytes, and pH.

## 3. Heart Rate

The decreasing cardiac output with age may not have significant impact on the circulation at rest, but it severely curtails the ability of the heart to respond to increased demand with acceleration of heart rate, as in physical exercise. In the healthy young, exercise provokes a rise in cardiac rate and a consequent increase in cardiac output, providing for augmented blood supply and oxygen consumption in muscle; this expected increase is considerably blunted in the elderly. During exercise, the expected rise in cardiac rate necessary to increase cardiac output in order to provide for the increased oxygen consumption by the exercising muscles is much lower in the elderly than in the young. The maximum achievable heart rate with exercise decreases linearly with age and may be calculated empirically, taking 220 beats/min as the maximum in the adult and subtracting the age of the individual from the 220 value; for example, in an 80-year-old, the maximal heart rate that can be achieved while exercising is 220 – 80 = 140 beats/min. Although the decline is progressive, the change is more steep after 50 years of age.

During exercise in the young adult, cardiac output is also increased by increasing stroke volume. Here, the most important factor is the vasodilation in exercising muscles; vasodilation leads to a fall in vascular resistance, thereby augmenting venous return to the heart, increasing ventricular blood-filling in diastole, with a resulting stronger stroke volume. With aging, peripheral vasodilation is less efficient or absent, and muscle mass is decreased; vascular resistance is increased due to atherosclerosis and cannot be overcome by the strength of cardiac contraction.

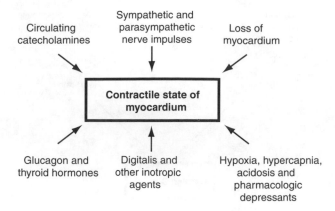

**FIGURE 21.8** Factors that influence the contractile state of the myocardium.

The heart is highly dependent on aerobic processes, and, normally, less than 1% of the energy liberated in cardiac tissue is due to anaerobic metabolism. Under basal (at rest) conditions, 35% of the caloric needs of the heart are provided by carbohydrates, 5% by ketones and amino acids, and 60% by fat. The proportion of these substrates varies with the nutritional state. Oxygen consumption of the heart is primarily dependent, among other factors, on the heart rate and the contractile state of the myocardium. Increasing the heart rate increases oxygen consumption, at least temporarily until the onset of appropriate compensatory adjustments, and so, too, does increased peripheral resistance, which causes the heart to work harder. This explains why in angina pectoris (Chapter 16), there is a relative deficiency of oxygen due to the greater oxygen demand to expel blood against increased peripheral pressure.

## C. Adaptive Adjustments in the Aging Heart

Despite some of the cardiovascular alterations described with aging in this and previous chapters (Chapters 16 and 17), the heart is capable of mustering those compensatory responses necessary to maintain adequate function. For example, while the rate at which the left ventricule fills with blood during early diastole declines markedly between the ages of 20 and 80, the enhaced filling in later diastole keeps ventricular filling adequate in elderly individuals. The myocardium remains capable of adaptive hypertrophy to maintain normal heart volume and pump function in the presence of a moderate increase in systolic pressure. Such hypertrophy may be mediated through hyperplasia of myocardial cells,[97] and may be due to persistent activation of cardiac gene expression and its hormonal regulation.[98]

## D. Hormonal and Chemical Actions on the Aging Heart: The Emerging of New Therapies

Although several hormones may indirectly affect the heart through their generalized metabolic and physiologic actions, glucagon and thyroid hormones directly influence cardiac contractility. Glucagon (Chapter 14) increases the formation of cAMP through its binding to receptors other than the adrenergic β-receptors. Its inotropic action, therefore, can be beneficial to individuals suffering from toxicity due to the administration of adrenergic blockers. Thyroid hormones increase the number of β-receptors in the heart, and their effects resemble those of sympathetic stimulation (Chapter 13).

Various drugs also modify myocardial contractility and rhythmicity. Xanthines, such as caffeine, exert their inotropic effect by inhibiting the breakdown of cAMP (Chapter 24). The inotropic effect of digitalis and related drugs is due to their inhibitory effects on $Na^+K^+ATPase$; this inhibition increases intracellular sodium which, in turn, increases calcium availability to the cell and initiates contraction. Depressants such as barbiturates depress myocardial contractility. Evaluation of therapy should take into consideration that cardiac responses depend in large measure on the condition of the vascular tree, and that the agents considered exert cardiac and extracardiac effects, the latter which may, indirectly, influence the myocardium.

Current molecular advances provide a better understanding of the pathogenesis of alterations of cardiac muscle contractility and excitability and open the way for better diagnosis and better strategies to cure the life-threatening disorders of the heart. By correlating clinical outcome with genetic susceptibility profiles, it will be possible to implement a more efficient therapy if the cardiopathy is diagnosed at young ages. By identifying genes that initiate or aggravate cardiac myopathy, it will be possible to target new drugs and gene therapies directly to the causative molecule(s). The improvements in the construction of artifical hearts and the availability of many models (e.g., transgenic animals, cultured cells, stem cells, cell tissue, and organ transplants) will make it possible to test new preventive and therapeutic interventions. In the current period, when scientists are eager for better drugs and technologies, the treatment of cardiac failure represents a worthy challenge that can save lives and improve the quality of life of many elderly persons.

# REFERENCES

1. Antich, P.P. et al., Measurement of intrinsic bone quality *in vivo* by reflection ultrasound, correction of impaired quality with slow-release sodium fluoride and calcium citrate, *J. Bone Mineral. Res.*, 8, 301, 1993.
2. Canalis, E., Ed., *Skeletal Growth Factors*, Lippincott Williams & Wilkins, Philadelphia, PA, 2000.
3. Fedarko, N.S. and Shapiro, J.R., Physiologic changes in soft tissue and bone as a function of age, in *Principles and Practice of Geriatric Surgery*, Rosenthal, R.A., Zenilman, M.E., and Katlic, M.R., Eds., Springer, New York, 2001.
4. Ducy, P., Schinke, T., and Karsenty, G., The osteoblast: A sophisticated fibroblast under central surveillance, *Science*, 289, 1501, 2000.
5. Teitelbaum, S.L., Bone resorption by osteoclasts, *Science*, 289, 1504, 2000.
6. Scott, J.E., Extracellular matrix, supramolecular organization and shape, *J. Anat.*, 187, 259, 1995.
7. Udagawa, N. et al., Origin of osteoclasts: Mature monocytes and macrophages are capable of differentiating into osteoclasts under a suitable microenvironment prepared by bone marrow-derived stromal cells, *Proc. Natl. Acad. Sci. U.S.A.*, 87, 7260, 1990.
8. Dougall, W.C. et al., RANK is essential for osteoclast and lymph node development, *Genes Dev.*, 13, 2412, 1999.
9. Simonet, W.S. et al., Osteoprotegerin: A novel secreted protein involved in the regulation of bone density, *Cell*, 89, 309, 1997.
10. Lacey, D.L. et al., Osteoprotegerin ligand is a cytokine that regulates osteoclast differentiation and activation, *Cell*, 93, 165, 1998.
11. Merkel, K.D. et al., Tumor necrosis factor-α mediates orthopedic implant osteolysis, *Am. J. Pathol.*, 154, 203, 1999.
12. Engelman, V.W. et al., A peptidomimetic antagonist of the $\alpha_v\beta_3$ integrin inhibits bone resorption *in vitro* and prevents osteoporosis *in vivo*, *J. Clin. Invest.*, 99, 2284, 1997.
13. McHugh, K.P. et al., Mice lacking $\beta_3$ integrins are osteosclerotic because of dysfunctional osteoclasts, *J. Clin. Invest.*, 105, 433, 2000.
14. Rodan, G.A. and Martin, T.J., Therapeutic approaches to bone disease, *Science*, 289, 1508, 2000.
15. Marcus, R., Feldman, D., and Kelsey, J., Eds., *Osteoporosis*, Academic Press, New York, 1996.
16. Bialas, M. and Stone, M., Osteoprosis, in *Principles and Practice of Geriatric Medicine*, Pathy, M.S.J., Ed., John Wiley & Sons, New York, 1998.
17. Jilka, R.L. et al., Increased osteoclast development after estrogen loss: Mediation by interleukin-6, *Science*, 257, 88, 1992.
18. Manolagas, S.C. and Jilka, R.L., Bone marrow, cytokines, and bone remodeling — emerging insights into the pathophysiology of osteoporosis, *N. Engl. J. Med.*, 332, 305, 1995.
19. Michaelsson, K. et al., Hormone replacement therapy and risk of hip fracture: Population based case-control study, *Br. Med. J.*, 316, 1858, 1998.
20. Love, R.R. et al., Effects of tamoxifen on bone mineral density in postmenopausal women with breast cancer, *N. Engl. J. Med.*, 326, 852, 1992.
21. Sato, M. et al., Emerging therapies for the prevention or treatment of postmenopausal osteoporosis, *J. Med. Chem.*, 42, 1, 1999.
22. Richmond, J.H., Koval, K.J., and Zuckerman, J.D., Orthopedic injuries, in *Principles and Practice of Geriatric Surgery*, Rosenthal, R.A., Zenilman, M.E., and Katlic, M.R., Eds., Springer, New York, 2001.
23. Smith, E.L., Sempos, C.T., and Purvis, R.W., Bone mass and strength decline with age, in *Exercise and Aging, The Scientific Basis*, Smith, E.L. and Serfass, R.C., Eds., Enslow Publishers, Hillside, NJ, 1981.

24. Yamada, H., *Strength of Biological Materials,* Williams & Wilkins, Co., Baltimore, MD, 1970.

25. Anghileri, L.J. and Tuffet-Anghileri, A.M., *The Role of Calcium in Biological Systems*, CRC Press, Boca Raton, FL, 1982.

26. Campbell, A.K., *Intracellular Calcium,* John Wiley & Sons, London, England, 1983.

27. Grympas, M.D. et al., The effects of diet, age and sex on the mineral content of primate bones, *Calcif. Tissue Int.,* 52, 399, 1993.

28. Nilas, L., Nutrition and fitness in the prophylaxis for age-related bone loss in women, *World Rev. Nutr. Diet.,* 72, 102, 1993.

29. Holloszy, J.O., Exercise, health and aging, a need for more information, *Med. Sci. Sports Exerc.,* 25, 538, 1993.

30. Flicker, L. et al., The effect of aging on intact PTH and bone density in women, *J. Am. Geriatr. Soc.,* 40, 1135, 1992.

31. Lindsay, R. et al., Randomized controlled study of effect of parathyroid hormone on vertebral-bone mass and fracture incidence among postmenopausal women on oestrogen with osteoporosis, *Lancet,* 350, 550, 1997.

32. Mosekilde, L. et al., The anabolic effects of parathyroid-hormone on cortical bone mass, dimensions and strength-assessed in a sexually mature, ovariectomized rat model, *Bone,* 16, 223, 1995.

33. Lane, N.E. et al., Parathyroid hormone treatment can reverse corticosteroid-induced osteoporosis, *J. Clin. Invest.,* 102, 1627, 1998.

34. Dobnig, H. and Turner, R.T., Evidence that intermittent treatment with parathyroid hormone increases bone formation in adult rats by activation of bone lining cells, *Endocrinology,* 136, 3632, 1995.

35. Onyia, J. E. et al., *In-vivo,* human parathyroid-hormone fragment (HPTH 1–34) transiently stimulates immediate-early response gene-expression, but not proliferation, in trabecular bone-cells of young-rats, *Bone,* 17, 479, 1995.

36. Jilka, R.L. et al., Increased bone formation by prevention of osteoblast apoptosis with parathyroid hormone, *J. Clin. Invest.,* 104, 439, 1999.

37. Marx, S.J., Hyperparathyroid and hypoparathyroid disorders, *N. Engl. J. Med.,* 343, 1863, 2000.

38. Rodan, G.A., Use of growth factors for osteoporotic therapy, in *Skeletal Growth Factors,* Canalis, E., Ed., Lippincott Williams & Wilkins, Philadelphia, PA, 2000.

39. Reid, I., Veale, A.G., and France, J.T., Glucocorticoid osteoporosis, *J. Asthma,* 31, 7, 1994.

40. Eastell, R., Management of corticosteroid-induced osteoporosis, *J. Intern. Med.,* 237, 439, 1995.

41. Rakopoulos, M. et al., Short treatment of osteoclasts in bone-marrow culture with calcitonin causes prolonged suppression of calcitonin receptor messenger-RNA, *Bone,* 17, 447, 1995.

42. Wada, S. et al., Physiological levels of calcitonin regulate the mouse osteoclast calcitonin receptor by a protein kinase Alpha-mediated mechanism, *Endocrinology,* 137, 312, 1996.

43. Groeneveld, E.H. and Burger, E.H., Bone morphogenetic proteins in human bone regeneration, *Eur. J. Endocrinol.,* 142, 9, 2000.

44. Pollitzer, W.S., and Anderson, J.B., Ethnic and genetic differences in bone mass, a review with a hereditary vs. environmental perspective, *Am. J. Clin. Nutr.,* 50, 1244, 1989.

45. Gilsanz, V. et al., Changes in vertebral bone density in black girls and white girls during childhood and puberty, *N. Engl. J. Med.,* 325, 1597, 1991.

46. Wright, V., Diseases of the joints, in *Principles and Practice of Geriatric Medicine,* Pathy, M.S.J., Ed., John Wiley & Sons, New York, 1998.

47. Roth, R.D., Joint diseases associated with aging, *Clin. Podiat. Med. Surg.,* 10, 137, 1993.

48. Armstrong, G.G. and Gardner, D.L., The thickness and distribution of human femoral head articular cartilage changes with age, *Ann. Rheum. Dis.,* 36, 407, 1977.

49. DeGroot, J. et al., Age-related decrease in proteoglycan synthesis of human articular chondrocytes: The role of non-enzymatic gycation, *Arthritis Rheum.,* 42, 1003, 1999.

50. Bell, G.M. and Schnitzer, T.J., COX-2 inhibitors and other nonsteroidal anti-inflammatory drugs in the treatment of pain in the elderly, *Clinics in Geriatric Medicine,* 17, 489, 2001.

51. 51, Woodhouse, K.W. and Wynne, H., The pharmacokinetics of non-steroidal anti-inflammatory drugs in the elderly, *Clin. Pharmacokinet.,* 12, 111, 1987.

52. Phillips, A., Polisson, R., and Simon, L., NSAIDs and the elderly. Toxicity and economic implications, *Drugs Aging,* 10, 119, 1997.

53. Elseviers, M. and De Broe, M., Analgesic abuse in the elderly, *Drugs Aging,* 12, 391, 1998.

54. Crofford, L. et al., Basic biology and clinical application of specific cyclooxygenase-2 inhibitors, *Arthritis Rheum.,* 43, 4, 2000.

55. Wolfe, M., Lichtenstein, D., and Singh, G., Gastrointestinal toxicity of nonsteroidal anti-inflammatory drugs, *N. Engl. J. Med.,* 34, 1888, 1999.

56. Bjorkman, D., Current status of nonsteroidal anti-inflammatory drug (NSAID) use in the United States: Risk factors and frequency of complications, *Am. J. Med.,* 107 (Suppl. 6A), 3S, 1999.

57. Boyce, E. and Breen, G., Celecoxib: A COX-2 inhibitor in the treatment of osteoarthritis and rheumatoid arthritis, *Formulary,* 34, 405, 1999.

58. Furst, D., Pharmacology and efficacy of cyclooxygenase inhibitors, *Am. J. Med.,* 107 (Suppl. 6A), 18S, 1999.

59. Hawkey, C. et al., Comparison of the effect of rofecoxib (a cyclo-oxygenase-2 inhibitor), ibuprofen and placebo on the gastroduodenal mucosa of patients with osteoarthritis, *Arthritis Rheum.,* 43, 370, 2000.

60. Pearson, M.B., Bassey, E.J., and Bendall, M.J., Muscle strength and anthropometric indices in elderly men and women, *Age and Ageing,* 14, 49, 1985.

61. Cunningham, D.A. et al., Exercise training of men at retirement, a clinical trial, *J. Gerontol.,* 42, 17, 1987.

62. Fiatarone, M.A. et al., High-intensity strength training in nonagenarians, Effects on skeletal muscle, *J. Am. Med. Assoc.*, 263, 3029, 1990.

63. Harridge, S.D.R. and Young, A., Skeletal muscle, in *Principles and Practice of Geriatric Medicine*, Vol. 2, Pathy, M.S.J., Ed., John Wiley & Sons, New York, 1998.

64. Shephard, R.J., Ed., *Gender, Physical Activity and Aging*, CRC Press, Boca Raton, FL, 2002.

65. Brooks, G.A. et al., *Exercise Physiology*, 3rd ed., Mayfield Publ. Co., Mountain View, CA, 1999.

66. Faulkner, J.A., Brooks, S.V., and Zerba, E., Skeletal muscle weakness in old age: Underlying mechanisms, *Ann. Rev. Gerontol. Geriatr.*, 10, 147, 1990.

67. Doherty, T.J. and Brown, W.F., The estimated number and relative sizes of thenar motor units as selected by multiple point stimulation in young and older adults, *Muscle Nerve*, 16, 355, 1993.

68. Doherty, T.J. et al., Effect of motor unit losses on strength in older men and women, *J. Appl. Physiol.*, 74, 868, 1993.

69. 69. Skelton, D.A. et al., Strength, power and related functional ability of healthy people aged 65 to 89 years, *Age Ageing*, 23, 371, 1994.

70. Fiatarone, M.A. and Evans, W.J., Exercise in the oldest old, *Geriatr. Rehabil.*, 5, 63, 1990.

71. Fiatarone, M.A. et al., The Boston FICSIT study: The effects of resistance training and nutritional supplementation on physical frailty in the oldest old, *J. Am. Geriatr. Soc.*, 41, 333, 1993.

72. Fujita, Y., Avoidance of drug therapy in the elderly: Exercise as a preventative prescription, in *Exercise for Health*, Shanahan, J., Ed., Adis International, Auckland, 2000.

73. Andonian, M.H. and Fahim, M.A., Effects of endurance exercise on the morphology of mouse neuromuscular junctions during ageing, *J. Neurocytol.*, 16, 589, 1987.

74. Wernig, A. and Herrera, A.A., Sprouting and remodeling at the nerve-muscle junction, *Prog. Neurobiol.*, 27, 251, 1986.

75. Stanley, S.N. and Taylor, N.A., Isokinematic muscle mechanics in four groups of women of increasing age, *Eur. J. Appl. Physiol. Occup. Physiol.*, 66, 178, 1993.

76. 76. Braunwald, E., Shattuck Lecture — Cardiovascular medicine at the turn of the millennium: Triumphs, concerns and opportunities, *N. Engl. J. Med.*, 337, 1360, 1997.

77. Rich, M.W., Heart failure in the 21st century: A cardiogeriatric syndrome, *J. Gerontol.*, 56, M88, 2000.

78. Svanborg, A., Age-related changes in cardiac physiology: Can they be postponed or treated by drugs?, in *Cardiovascular Diseases in the Elderly*, Mallarkey, G., Ed., Adis International, Auckland, 2000.

79. Kajstura, J. et al., Necrotic and apoptotic myocyte cell death in the aging heart of Fisher 344 rats, *Am. J. Physiol.*, 271, H1215, 1996.

80. Mallat, Z. et al., Evidence of apoptosis in arrhythmogenic right ventricular dysplasia, *N. Engl. J. Med.*, 335, 1190, 1996.

81. Colucci, W.S., Apoptosis in the heart, *N. Engl. J. Med.*, 335, 1224, 1996.

82. Olivetti, G. et al., Apoptosis in the failing human heart, *N. Engl. J. Med.*, 336, 1131, 1997.

83. Williams, R.S., Apoptosis and heart failure, *N. Engl. J. Med.*, 341, 759, 1999.

84. Lakatta, E.G., Deficient neuroendocrine regulation of the cardiovascular system with advancing age in healthy humans (point of view), *Circulation*, 87, 631, 1993.

85. Effron, M.B. et al., Changes in myosin isoenzymes, ATPase activity and contraction duration in rat cardiac muscle with aging can be modulated by thyroxine, *Circ. Res.*, 60, 238, 1987.

86. O'Neill, I. et al., Progressive changes from young adult age to senescence in mRNA for rat cardiac myosin heavy chain genes, *Cardioscience*, 2, 1, 1991.

87. Marian, A.J. and Roberts, R., Molecular genetic basis of hypertrophic cardiomyopathy, *J. Cardiovasc. Electrophysiol.*, 9, 88, 1998.

88. Barinaga, M., Tracking down mutations that can stop the heart, *Science*, 281, 32, 1998.

89. Tsutsui, H., Ishihara, K., and Cooper, G., Cytoskeletal role in the contractile dysfunction of hypertrophied myocardium, *Science*, 260, 682, 1993.

90. Towbin, J.A., The role of cytoskeletal proteins in cardiomyopathies, *Curr. Opin. Cell Biol.*, 10, 131, 1998.

91. Leiden, J.M., The genetics of dilated cardiomyopathy — Emerging clues to the puzzle, *N. Engl. J. Med.*, 337, 1080, 1997.

92. Hirota, H. et al., Loss of a gp130 cardiac muscle cell survival pathway is a critical event in the onset of heart failure during biochemical stress, *Cell*, 97, 189, 1999.

93. Olson, E.N., Molecular pathways controlling cardiac myogenesis and morphogenesis, *Molecular Biology of the Cell*, 7 (Suppl.), 504A, 1996.

94. Olson, E.N. and Williams, R.S., Calcineurin signaling and muscle remodeling, *Cell*, 101, 689, 2000.

95. Lakatta, E.G. and Yin, F.C., Myocardial aging, functional alterations related to cellular mechanisms, *Am. J. Physiol.*, 242, H 927, 1982.

96. Lakatta, E.G., Cardiovascular aging research: The next horizons, *J. Am. Geriatr. Soc.*, 47, 613, 1999.

97. Anversa, P. et al., Myocyte cell loss and myocyte cellular hyperplasia in the hypertrophied aging heart, *Circulation Res.*, 67, 871, 1990.

98. Chien, K.R. et al., Regulation of cardiac gene expression during myocardial growth and hypertrophy, molecular studies of an adaptive physiologic response, *FASEB J*, 5, 3037, 1991.

**Systemic and Organismic Aging**

# 22 The Skin*

*Mary Letitia Timiras*
Overlook Hospital and University of Medicine and Dentistry of New Jersey

## CONTENTS

## I.   INTRODUCTION

Perhaps no aspect of aging is as dramatic or readily obvious as that which occurs in the skin and its appendages. The development of either gray hair or facial and body wrinkles represents irrefutable evidence of the passage of time and the aging process. Many descriptions of aging skin fail to distinguish between intrinsic aging changes and changes caused by environmental insults that accumulate with exposure time. For example, chronic solar damage changes are far more prevalent in the elderly because they result from cumulative exposure over time.[1–4] These changes involve cells,[5,6] glands,[7] and connective tissue (particularly, collagen)[8] of the skin. Intrinsic changes in the structure and function of skin that occur with aging also make the skin more vulnerable to external insults. It is not surprising then, that dermatologic problems are very common in the elderly.[9–11]

Several studies describe how almost one half of persons over 65 years of age have at least one dermatologic disease requiring medical attention. About one third of these have more than one skin problem. Multiple skin conditions are characteristic of the very old, and the common ones are different from those affecting the young.[12]

In order to provide an understanding of the nature and origin of age-related changes in the skin and of how these changes lead to dermatologic problems in the aged,

this chapter will review some of the anatomic and physiologic changes that occur in the skin and collagen with normal senescence (Part II). Pressure sores will be discussed as a clinical example of skin dysfunction in the elderly (Part III).

## II.   AGING OF THE SKIN

The skin is one of the largest organs of the body and accounts for approximately 16% of the total body weight. It is part of the integument, a covering of the entire body, which, in addition to the skin, includes the nails, hair, and various types of glands, all accessory organs derived from the skin. The skin has several functions, the most important of which are as follows:

1. It provides a barrier to exclude harmful substances
2. It prevents water loss
3. It plays a role in the control of body temperature
4. It readily repairs itself
5. It receives sensory stimuli: touch, pressure, temperature, and pain
6. It excretes waste products through sweat glands
7. It secretes special products, such as milk from mammary glands

---

* Illustrations by Dr. S. Oklund.

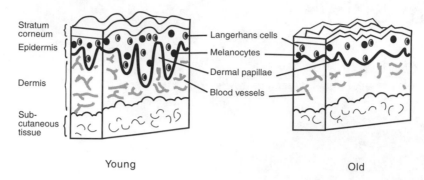

**FIGURE 22.1** Histologic changes in normal aging skin. Changes from top to bottom include (1) rougher stratum corneum, (2) fewer melanocytes and Langerhans cells, (3) flattening of the dermo-epidermal junction, (4) fewer blood vessels in the dermis, and (5) less subcutaneous tissue.

The skin consists of an epithelium, the epidermis, and the dermis considered as part of the connective tissue (Figure 22.1).

## A. EPIDERMIS AND DERMIS

The main layers of the skin consist of the surface layer known as the epidermis, and a deeper connective tissue layer known as the dermis. The interface between the epidermis and dermis is normally uneven and forms wavy interdigitations. The projections of the dermis into the epidermal layer are called dermal papillae.

The epidermal cells produce *keratin,* an insoluble, fibrous protein that forms the dead superficial layers of the skin (stratum corneum) and is essential for the protection of the outer body surface. The epidermis is locally specialized into various skin appendages, such as the hair, nails, sweat glands, and oil or sebaceous glands. Nails are formed of α-keratin, a protein rich in cysteine. The superficial keratinized skin cells are continuously exfoliated and are replaced by cells that arise from the basal layer of the epidermis. The sequence of changes of an epidermal cell from the basal layer of the epidermis to keratinization and exfoliation normally takes from 15 to 30 days, depending on the region of the body.[5]

Other types of cells that are found in the epidermis include *melanocytes*,[5] which produce the pigment melanin that protects against ultraviolet irradiation; melanocytes have been proposed as a model for studying the complexity of cellular changes with aging.[13] The *Langerhans cells*, also present in the epidermis, are responsible for recognition of foreign antigens, and they participate in the protective function of the skin.

Because the epidermis has *no blood vessels*, the dermis, which has abundant blood vessels, plays an important role in providing the epidermis with nutrients and in contributing to thermoregulation. The amount of heat lost from the body is regulated to a large extent by blood flowing through the skin. The presence of direct connections between arterioles and venules bypassing the capillaries (arteriovenosus anasto-

moses) in the skin of fingers, palms, toes, and earlobes coupled with a rich autonomic innervation support a blood circulation that varies greatly (from 1 to as much as 150 mL/100 gm of skin/min) in response to thermoregulatory stimuli.

Although the epidermis contains *nerve endings*, there is a richer *nerve supply* in the dermis, because it also contains the nerves that lead to the sensory nerve endings. In addition, epidermal appendages such as the hair follicles and sebaceous and sweat glands extend down into the dermis. Below the dermis is looser connective tissue that usually consists of subcutaneous adipose tissue which plays a role in fat metabolism.

## B. EPIDERMAL CHANGES WITH AGING

Dryness and roughness of the skin are two of the most readily appreciated changes that occur with aging. This could be due to a decrease in the moisture content of the stratum corneum or to a decrease in vertical height and increase in overall surface area of epidermal cells. The increased dryness results in a surface likened to "shingles on an old roof" (Figure 22.1).[14] The exact pathogenesis of the drying process is still unclear but probably represents a combination of several contributing factors, primarily, the aging of collagen and matrix.

Another well-recognized change that occurs with aging is the increased incidence of benign and malignant epidermal neoplasms.[16] Although there is no question that chronic ultraviolet light exposure contributes to this problem, some intrinsic cellular alterations of basal cells may also contribute to the increased incidence of skin cancer in the elderly.

### EFFECT OF CUMULATIVE ENVIRONMENTAL INJURY ON THE SKIN

Several environment hazards cause injury to the skin, but the most widespread and damaging is solar ultraviolet radiation (UVR), which has a wavelength of 200 to 400 nm. Of the three (A,B,and C) types of UVR, the UVB appears to be the most significant biologically. The degree of skin pigmentation is of critical importance in modulating the damaging effects of UVR: the darker the skin, the lesser the damage. UVR chronic exposure first causes hyperplasia (thickening) of the epidermis, followed by atrophy. The major alterations occur in the dermis, which undergoes degenerative changes affecting the connective tissue of upper and middle layers and causing wrinkling (characteristic of old age). Prevention of solar damage may be obtained — and in fact, is strongly recommended — by reducing exposure to the sun and by using special compounds (applied topically to the exposed body parts) that may act as sunscreens.[16]

One of the histologic changes that occurs most consistently in the elderly is the flattening of the dermo-epidermal interface and effacement of the dermal papillae. This reduces the total area of dermo-epidermal junction per area of external body surface area. This change predisposes older persons to blister formation and shear-type injuries and easy abrasions. Another observation is an age-associated decrease in epidermal turnover rate of about 50% between the third and seventh decades of life. This means fewer basal cells are replaced, and it takes longer for a basal cell to reach the stratum corneum and be exfoliated. The slower movement prolongs the exposure of epidermal cells to potential carcinogens and contributes to an increased incidence of skin cancer; in addition, it is responsible for the slowing of wound healing.

The number of melanocytes decreases with old age by approximately 8 to 20% per decade after the age of 30 in exposed and unexposed areas.[17] This reduction leads to irregular pigmentation, especially in sun-exposed areas (hence, the "age spots" frequently seen on the back of the hands), and to the inability to tan as deeply as when younger. However, other studies seem to negate or at least to minimize this reduction in melanocytes.[5] The number of Langherhans cells decreases as well. This change is expected to contribute to a decline in cell-mediated immune responses in the skin. The reduction of melanin and its protective action, the reduced inflammatory warning signs, and the reduced immune capacity combine to increase the risk for tumorigenesis (Chapter 15). Elderly patients require longer ultraviolet exposure to develop erythema and edema (sunburn), than younger patients. Thus, the body's warning system as well as its defense system in relation to skin cancer become blunted with age. In addition, ultraviolet light further reduces the quantity of Langerhans cells. These changes, together with the increase in cumulative irradiation exposure, possibly explain why skin cancers (nonmelanomatous) are prone to occur on sun-exposed skin.

## C. Dermal Changes

The dermis in elderly individuals has a decreased density, fewer cells, and fewer blood vessels. The total amount of collagen decreases 1% per year in adulthood, therefore, skin thickness decreases linearly with age after 20. Collagen becomes thicker, less soluble, and more resistant to digestion by enzyme collagenase with age. Changes in the number and types of cohesive bonds make collagen stronger and more stable. Collagen alteration predisposes the dermis to tear-type injury because there is less "give" in the tissue. Architectural rearrangements of collagen fibers may also be responsible for changes in dermal tissue properties.

The total amount of hyaluronic acid and dermatan sulfate, both components of the extracellular matrix, decreases in the dermis with aging, affecting the viscosity of the dermis which, in turn, may alter the rate of dermal clearance of substances. Changes in the elastic fibers of the dermis also result in loss of stretch and resilience. A consequence is skin sagging and wrinkling[18] and predisposition to injury of the underlying tissues following trauma. While even the very old (beyond 85 years) can effectively repair extensive wounds, elderly individuals, in general, lag behind younger controls at every stage of wound healing.[19,20]

Pale skin results from the decrease in dermal blood vessels. Skin surface temperature is also decreased due to the diminished vascularity. These changes, together with a decrease in the thickness of the subcutaneous tissue, make thermoregulation more difficult in the elderly. The vascular changes described above also result in a decrease in dermal clearance of foreign materials with consequent prolonging or exacerbating cases of contact dermatitis.

## D. Aging of Skin Appendages

Older individuals produce less sweat, because sweat glands decrease in number or functional efficiency, a decrease that interferes further with thermoregulation. Although the number of sebaceous glands remains constant with age, their size increases, while the sebum output as well as wax production decline with age.[7] The diminished sweat and oil production no doubt contribute to skin dryness and roughness in the elderly.

The rate of linear *nail growth* decreases with aging. Nail plates usually become thinner, more brittle, and fragile.[21,22] *Hair graying* occurs because of a progressive loss of functional melanocytes from hair bulbs.[23] By the age of 50, it is said that 50% or more of the population have at least 50% of their body hair gray, regardless of sex or hair color. Heredity plays an important role in hair graying as well. A decrease in the number of hair follicles in the scalp and consequent increased balding have also been described.[23]

Although no changes in the free nerve endings in aged skin have been found, the number of *Pacini's and Meissner's corpuscles* (end organs responsible for the sensation of pressure and light touch) decrease with age. This results in decreased sensation, which predisposes the elderly to injury, and in decreased ability to perform fine maneuvers with the hands. A summary of structural and functional changes with aging in the main skin components, epidermis, dermis, and appendages, is presented in Tables 22.1 and 22.2, respectively; considerations on aging of connective tissue and collagen are presented in Box 22.1.

**Systemic and Organismic Aging**

---

**Box 22.1**

**Aging of Connective Tissue and Collagen**

Aging of the connective tissue, a major constituent of many organs and systems, involves several functions, of which the most important include the following:

1. Mechanical support as provided by bones and joints
2. Exchange of metabolites between blood and tissues as provided by blood circulation
3. Storage of fuel in its adipose cells
4. Protection against infection, and repair of injury

In the skin, connective tissue is represented by the ground substance (matrix) in which are embedded collagen and elastic fibers. Because it is ubiquitous throughout the body and undergoes identifiable changes with age, collagen has been considered a possible primary source for the onset of the aging processes.[24] The striking changes that take place with age in the structure and chemistry of collagen fibers and the surrounding extracellular matrix have been ascribed to metabolic alterations in tissues; these alterations would lead to the formation of covalent bonds among polymeric chains or cross-linkages.[25] Cross-linking, therefore, would represent a fundamental mechanism by which overall functional impairment is induced in the aged.[26]

Collagen, produced by the fibrocytes present in the matrix, represents the major fibrous protein of connective tissue and forms a large portion (30 to 40%) of all proteins of the body. Collagen contains two specific amino acids — hydroxyproline and hydroxylysine — that do not occur in significant amounts in other animal proteins, and their content in a tissue, can be taken as an index of its collagen content. The structure of collagen is illustrated in Figures 22.2 and 22.3, and major changes with aging are summarized in Table 22.3.

Changes with aging involve other cutaneous components of connective tissue, including elastin and the extracellular matrix. With aging, matrix size is progressively reduced; this is associated with a corresponding decrease in water content — a water loss consistently observed throughout the life span in many organs and species. The loss of extracellular matrix has been ascribed to a slower turnover in aging compared to that of collagen. Extracellular matrix must be renewed within days or weeks, whereas the more inert collagen persists for a considerably longer period of time. Changes in the composition of the extracellular matrix with age have also been reported; they appear to differ from tissue to tissue with a progressively increasing occurrence of cross-links. In addition, disregulation of the matrix metalloproteinases with aging may result in excessive proteolytic activity and tissue damage.[27] Changes with aging in elastic fibers have been discussed in Chapter16 in relation to the arterial vascular wall and in Chapter 18 in relation to the lungs.

---

**TABLE 22.1**
**Changes in Normal Aged Skin**

| Decrease | Increase or Other Changes |
|---|---|
| **Epidermis** | |
| Epidermal turnover rate | Severe dryness and roughness |
| Number of melanocytes | Flattening of dermoepidermal |
| Number of Langerhans cells | junction |
| **Dermis** | |
| Density | Stiffer collagen |
| Cells | Stiffer elastic fibers |
| Blood vessels | |
| Clearance of foreign substances | |
| **Appendages** | |
| Sweat production | Gray hair |
| Sebaceous glands | Thinner nails |
| Hair follicles | |
| Rate of nail growth | |
| Sensory end-organs | |

**TABLE 22.2**
**Functional Changes in Aging Skin**

| Decreased Function | Increased Pathology |
|---|---|
| Wound-healing capability | Blister formation |
| Cell-mediated immune response | Incidence of infection |
| Thermoregulation | Incidence of cancer |
| Clearance of foreign substances | Dryness |
| Tanning | Roughness |
| Elasticity | Fragility |
| Sweat and oil production | Sensory deprivation |
| Thickness | |

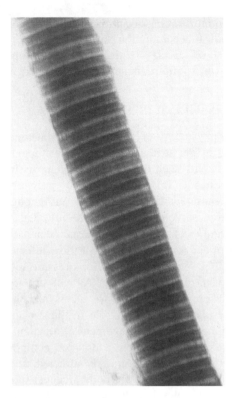

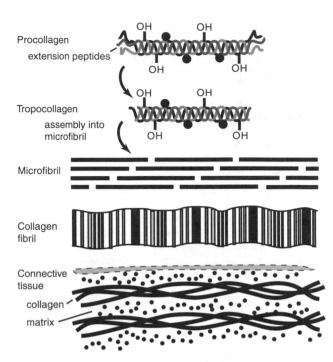

**FIGURE 22.2** Electron micrograph of a negatively stained microfibril of collagen isolated from rat tendon. (Original magnification ×96, 541.) (Courtesy of Dr. N.B. Gilula.) One dark and one light segment represent a period produced by the arrangement of tropocollagen molecules that would come together in a parallel arrangement and overlap by about a quarter of their length to produce a staggered array resulting in cross striation (refer to Figure 22.3). This structural arrangement makes the collagen fibers flexible and highly resistant (e.g., capable of withstanding several hundred kg/cm$^2$ from a pulling force). These properties of collagen resemble those of a cable that ties a ship to shore: it is sufficiently flexible to be curled when not in use, but strong enough not to allow movement of a ship at anchor.

**FIGURE 22.3** Schematic representation of the composition of a collagen fibril and the distribution of tropocollagen molecules in the connective tissue. Collagen fibrils derive from tropocollagen molecules formed from larger procollagen precursor molecules; the latter contain the amino acids that will give rise to the tropocollagen as well as extra amino acids called "extension peptides" that prevent the formation of large collagen fibrils, damaging to the cell. Collagen fibers are embedded in a matrix (or ground substance) formed mainly of protoglycans (high molecular weight complexes of proteins and polysaccharides, serving as lubricants and support elements), some proteins and interstitial fluid filtered from the capillaries.

## E. LOOKING YOUNG FOREVER!

In one of his tragedies, the 17th century, French playwriter, Jean Racine, wrote about the struggle of his heroine, Athalie, queen of Judea, to "repair the irreparable outrage inflicted by the passing years" on her beauty. The enormous amount of money currently spent on cosmetics and comestic interventions throughout the world attests to the continuing human devotion to a beautiful appearance. Some cosmetics have a physiologic justification, and some have well-proven therapeutic actions. Estrogens and retinoic acid will be considered here, and a brief listing of other types of interventions is included.

### 1. Estrogens

Both male and female sex hormones change with age, but the female sex hormones, *estrogens*, decrease at meno-

pause more dramatically than any other hormone (Chapter 11). As the population of postmenopausal women continues to grow, interest in the effectiveness of replacement therapy with these hormones on various systems and functions is also increasing.[28] Several of these effects have been discussed in Chapter 11.

There are many ways by which steroidal and nonsteroidal estrogens may influence skin aging;[28–30] they include:

- Maintenance of skin thickness by preventing skin collagen decrease
- Maintenance of skin moisture by increasing mucopolysaccharides and hyaluronic acid in skin matrix
- Maintenance of the "barrier function" of stratum corneum
- Increase in sebum levels
- Maintenance of skin elastic fibers and collagen and, therefore, reduction or delayed appearance of wrinkles

**Systemic and Organismic Aging**

**TABLE 22.3**
**Major Changes with Aging in Collagen Structure and Function**

↓ Chemical Excitability
- Fibers become tougher
- Tensile strength is reduced
- Plasticizing function is impaired or lost

↓ Effectiveness of Enzymatic Degradation
- Turnover of fibers is slowed due to:
  - ↓ amount of collagenase (the major degradating enzyme) and/or
  - increased resistance of fibers to degradation by collagenase
  - ↓ proportion of labile — easily degraded to stable — less degradable collagen
  - ↓ ability of cortisol and other hormones to stimulate collagenase activity

↓ Gluconeogenic action (?)
- Collagen would release amino acids that convert to glucose
- Under stress, this release of energy conversion is decreased or prevented

↑ Cross-linking
- ↑ intramolecular binding of ester bonds
- ↑ rigidity of fibers
- ↑ time of contraction and relaxation

---

- Improvement and acceleration of wound healing by regulating cytokine levels (in elderly women and men)
- Some still controversial effects on skin scarring

## 2. Retinoids

These, either naturally occurring or synthetic substances, include primarily, retinol, a synonym of vitamin A (Chapter 24), and retinoid acid, a derivative of vitamin A. One of their characteristics is the ability to bind nuclear receptors of the steroid/thyroid hormones superfamily (Chapters 10 and 13). They regulate cellular proliferation and differentiation.[31–35] Retinol is a coenzyme necessary for the function of the retina, bone growth, and differentiation of epithelial tissue. Some retinoids, with topical action, like tretinoin, appear to prevent and repair skin photoaging damage by:

- Preventing loss of collagen
- Stimulating new collagen formation

These actions are also applicable to the aged skin, which is, apparently, highly sensitive to retinoid action. Retinoids must be used with caution, especially when administered orally.[34] The use of topical or oral retinoids continues to expand within and beyond the field of dermatology.

Other frequent interventions for cosmetic use to flatten wrinkles and improve skin condition, include the following:

- Administration of antioxidants to prevent free radical accumulation during skin aging[36–39]
- Use of lasers,[40,41] injections of the botulinum toxin,[42] radiofrequency,[43] and dermabrasion[44]

## III. PRESSURE SORES

The pathophysiology of pressure sores illustrates several principles in geriatric management.[45] Although not the most commonly seen dermatologic diagnosis in the elderly (skin cancer is), the problem of pressure sores is frequent enough and produces serious enough consequences to deserve discussion.[46] The prevalence of pressure ulcerations increases with age such that patients over 70 years of age account for 70% of those affected. In this age group, 70% of patients develop pressure sores within 2 weeks of hospital admission.[47] As with urinary incontinence, the presence of pressure ulcers and their status frequently play a prominent role in the decisions made regarding the ultimate management of a patient.

Pressure sores represent a dreaded complication of immobility. The most commonly affected sites are the sacrum, ischial, trochanteric and calcaneal tuberosities, as well as the lateral malleolus. They arise from four different mechanisms: pressure, shear, friction, and moisture leading to maceration (softening of tissues). The role of pressure is the most critical in the development of the pressure ulcer. The average period of time necessary to produce pressure-associated changes varies but can be as little as 2 h. Hence, a patient who does not move every 2 h, for whatever reason, is at risk for developing sores. Shearing forces are produced by an improper position, friction occurs with improper handling of the patients, and moisture is due to perspiration or urinary and fecal incontinence. Due to all of these factors, the skin changes previously described increase the risk of developing pressure sores.

A uniform classification system of pressure sores has been proposed and is routinely utilized in their management.[48] Stage I occurs when there is redness not reversible with pressure (called "nonblanchable erythema") of intact skin. Stage II occurs when there is partial thickness skin loss involving epidermis or dermis. The ulcer is superficial and presents clinically as an abrasion, blister, or shallow crater. Full thickness skin loss involving damage or necrosis of subcutaneous tissue, which may extend down to, but not through, underlying fascia, represents Stage III. Finally, Stage IV occurs when there is full thickness skin loss with extensive destruction, tissue necrosis, or damage to muscle, bone, or supporting structures (i.e., tendon).

There is no cure for a pressure sore once it develops. Even if an ulcer heals, there is always a significant chance that it will recur.[47] Therefore, prevention of the development of pressure sores, as with many other geriatric problems, is the most important aspect of management. Risk factors for the development of pressure sores include: incontinence, edema, obesity, diabetes with neuropathy, sepsis, vascular disease, immobility due to fractures, dementia or restraints, and finally, systemic factors related to malnutrition such as hypoalbuminemia, anemia, and vitamin deficiency. Once patients at risk are identified, specific measures need to be taken to avoid the causative factors mentioned earlier. This is best accomplished through the use of a multidisciplinary team approach. The medical primary care provider optimizes physiologic function and treats any underlying illnesses. The nursing staff ensures feeding, turning, positioning, and initiates a bowel and bladder program. The nutritionist develops an aggressive nutritional strategy, and physical therapy can institute a mobilization plan, encourage strengthening exercises, and provide special pressure-sparing equipment.

Relief from pressure is the most important factor in preventing as well as treating pressure sores. In patients at risk, this may necessitate turning the patient in bed every 2 h. Padding for the bed or chair as well as special air and water matresses have also been recommended.[49] Providing a clean and moist environment for tissue to heal is important for Stages II–IV. This is usually accomplished with sterile gauze moistened with normal saline or synthetic colloid dressings. In addition, for Stages III and IV, it is necessary to remove necrotic debris. This is accomplished through surgical or mechanical dressing change debridement, as well as enzymatic products. Avoiding irritating substances such as betadine or hydrogen peroxide also helps with wound healing. Finally, treating local infection with frequent dressing changes or debridement, and systemic infections with empirical antibiotics will also help wounds heal faster. If the wound is large enough, plastic surgery utilizing tissue flaps may be necessary to fill in large gaps. The management of pressure sores is much more successful and satisfying when underlying pathophysiologic principles are utilized in conjunction with a comprehensive and multidisciplinary approach.

# REFERENCES

1. Kang, S., Fisher, G.J., Voorhees, J.J., Photoaging: Pathogenesis, prevention and treatment, *Clin. Geriatr. Med.*, 17, 643, 2001.
2. Trautinger, F., Mechanisms of photodamage of the skin and its functional consequences for skin ageing, *Clin. Exper. Dermatol.* 26, 573, 2001.
3. Jackson, R., Elderly and sun-affected skin. Distinguishing between changes caused by aging and changes caused by habitual exposure to sun, *Can. Fam. Physician*, 47, 1236, 2001.
4. Yaar, M. and Gilchrest, B.A., Skin Aging: Postulated mechanisms and consequent changes in structure and function, *Clin. Geriatr. Med.*, 17, 617, 2001.
5. Yaar, M. and Gilchrest, B.A., Ageing and photoageing of keratinocytes and melanocytes, *Clin. Exper. Dermatol.*, 26, 583, 2001.
6. Grewe, M., Chronological ageing and photoageing of dendritic cells, *Clin. Exper. Dermatol.*, 26, 608, 2001.
7. Zouboulis, C.C. and Boschnakow, A., Chronological ageing and photoageing of the human sebaceous gland, *Clin. Exper. Dermatol.*, 26, 600, 2001.
8. Ma, W. et al., Chronological ageing and photoageing of the fibroblasts and the dermal connective tissue, *Clin. Exper. Dermatol.*, 26, 592, 2001.
9. Shenefelt, P.D. and Fenske, N.A., Aging and the skin, Recognizing and managing common disorders, *Geriatrics*, 45, 57, 1990.
10. Kurwa, H.A. and Marks, R., Skin disorders, in *Principles and Practice of Geriatric Medicine,* 3rd ed., Pathy, M.S.J., Ed., John Wiley & Sons, New York, 1998.
11. Smith, E.S., Fleischer, A.B., and Feldman, S.R., Demographics of aging and skin disease, *Clin. Geriatr. Med.*, 17, 631, 2001.
12. Millard, T.P. and Hawak, J.L., Photodermatoses in the elderly, *Clin. Geriatr. Med.*, 17, 691, 2001.
13. Bandyopadhyay, D. et al., The human melanocyte: A model system to study the complexity of cellular aging and transformation in non-fibroblastic cells, *Exper. Gerontol.*, 36, 1265, 2001.
14. Fenske, N.A. and Lober, C.W., Structural and functional changes of normal aging skin, *J. Am. Acad. Dermatol.*, 15, 571, 1986.
15. Sachs, D.L., Marghoob, A.A., and Halpern, A., Skin cancer in the elderly, *Clin. Geriatr. Med*, 17, 715, 2001.
16. Scherschum, L. and Lim, H.W., Photoprotection by sunscreens, *Am. J. Clin. Dermatol.*, 2, 131, 2001.
17. Quevedo, W.C., Szabo, G., and Vicks, J., Influence of age and ultraviolet on the populations of dopa-positive melanocytes in human skin, *J. Infect. Dis.*, 52, 287, 1969.
18. Akazaki, S. and Imokawa, G., Mechanical methods for evaluating skin surface architecture in relation to wrinkling, *J. Dermatol. Sci.*, 27 (Suppl. 1), S-5, 2001.
19. Grove, G.L., Age-related differences in healing of superficial skin wounds in humans, *Arch. Dermatol. Res.*, 272, 381, 1982.
20. Grove, G.L. and Kligman, A.M., Age-associated changes in human epidermal cell renewal, *J. Gerontol.*, 38, 137, 1983.
21. Selmanowitz, V.J., Rizer, R.L., and Orentreich, N., Aging of the skin and its appendages, in *Handbook of the Biology of Aging*, Finch, C.E. and Hayflick, L., Eds., Van Nostrand Reinhold Co., New York, 1977.
22. Bolognia, J.L., Aging skin, *Am. J. Med.*, 98, 99S, 1995.
23. Tobin, D.J. and Paus, R., Graying: Gerontobiology of the hair follicle pigmentary unit, *Exper. Gerontol.*, 36, 29, 2001.
24. Verzar, F., Intrinsic and extrinsic factors of molecular aging, *Exp. Gerontol.*, 3, 69, 1968.
25. Sinex, F., Cross-linkage and aging, in *Advances in Gerontological Research*, Vol. 1, Strehler, B.L., Ed., Academic Press, New York, 1964.

**Systemic and Organismic Aging**

26. Sobel, H., Aging of ground substance in connective tissue, in *Advances in Gerontological Research,* Vol. 2, Strehler, B.L., Ed., Academic Press, New York, 1967.

27. Herouy, Y., Matrix metalloproteinases in skin pathology, *Int. J. Mol. Med.*, 7, 3, 2001.

28. Wines, N. and Willsteed, E., Menopause and the skin, *Australas. J. Dermatol.*, 42, 149, 2001.

29. Phillips, T.J., Demircay, Z., and Saliu, M., Hormonal effects on skin aging, *Clin. Geriatr. Med.*, 17, 661, 2001.

30. Shah, M.G. and Maibach, H.I., Estrogen and skin, *Am. J. Clin. Dermatol.*, 2, 143, 2001.

31. Fisher, G.J. and Voorhees, J.J., Molecular mechanisms of photoaging and its prevention by retinoic acid: Ultraviolet irradiation induces MAP kinase signal transduction cascades that induce Ap-1-regulated matrix metalloproteinases that degrade human skin *in vivo*, *J. Investigative Dermatol. Symp. Proc.*, 3, 61, 1998.

32. Thacher, S.M., Vasudevan, J., and Chandraratna, R.A., Therapeutic applications for ligands of retinoid receptors, *Current Pharmaceut. Des.*, 6, 25, 2000.

33. Griffitths, C.E., The role of retinoids in the prevention and repair of aged and photoaged skin, *Clin. Exp. Dermatol.*, 26, 613, 2001.

34. Ellis, C.N. and Krach, K.J., Uses and complications of isotretinoin therapy, *J. Am. Acad. Dermatol.*, 45, S150, 2001.

35. Krutman, J., New developments in photoprotection of human skin, *Skin Pharmacol. Appl. Skin Physiol.*, 14, 401, 2001.

36. Wei, Y.H. et al., Oxidative stress in human aging and mitochondrial disease — consequences of defective mitochondrial respiration and impaired antioxidant enzyme system, *Clin. J. Physiol.*, 31, 44, 2001.

37. Podda, M. and Grundmann-Kollmann, M., Low molecular weight antioxidants and their role in skin ageing, *Clin. Exp. Dermatol.*, 26, 578, 2001.

38. Thiele, J.J. et al., The antioxidant network of the stratum corneum, *Curr. Probl. Dermatol.*, 29, 26, 2001.

39. Dreher, F. and Maibach, H., Protective effects of topical antioxidants in humans, *Curr. Probl. Dermatol.*, 29, 157, 2001.

40. Rohrer, T.E., Lasers and dermatologic surgery for aging skin, *Clin. Geriatr. Med.*, 17, 769, 2001.

41. Biiesalski, H.K. and Obermueller-Jevic U.C., UV light, beta-carotene and human skin — beneficial and potentially harmful effects, *Arch. Biochem. Biophys.*, 389, 1, 2001.

42. Carruthers, J. and Carruthers, A., Botulinum toxin (botox) chemo-denervation for facial rejuvenation, *Facial Plast. Surg. Clin. North Am.*, 9, 197, 2001.

43. Carruthers, A., Radiofrequenct resurfacing: Technique and clinical review, *Facial Plast. Surg. Clin. North Am.*, 9, 311, 2001.

44. Hruza, G.J., Dermabrasion, *Facial Plast. Surg. Clin. North Am.*, 9, 267, 2001.

45. Bar, C.A. and Pathy, M.S.J., Pressure sores, in *Principles and Practice of Geriatric Medicine*, Pathy, M.S.J., Ed., 3rd ed., John Wiley & Sons, Ltd., New York, 1998.

46. Allman, R., Epidemiology of pressure sores in different populations, *Decubitus*, 2, 30, 1989.

47. Cooney, T.G. and Reuler, J.B., Pressure sores, *West J. Med.*, 140, 622, 1984.

48. Patterson, J.A. and Bennett, R.G., Prevention and treatment of pressure sores, *J. Am. Geriatr. Soc.*, 43, 919, 1995.

49. Ferrell, B.A., Osterweil, D., and Christenson, P., A randomized trial of low-air-loss beds for treatment of pressure ulcers, *J. Am. Med. Assoc.*, 269, 494, 1993.

# Part III

Prevention and Rehabilitation

# 23 Pharmacology and Drug Management in the Elderly*

*Mary Letitia Timiras*
Overlook Hospital and University of Medicine and Dentistry of New Jersey

*Jay Luxenberg,*
Jewish Home and University of California, San Francisco

## CONTENTS

## I.   INTRODUCTION

In large part, advances in pharmacology throughout the past centuries, and, particularly, the recent and rapid advances witnessed in the 19th and 20th centuries, have contributed to the increase in the number and longevity of elderly persons as described in Chapter 2. In addition to decreasing death from infections and acute medical illnesses in youth and middle age, modern pharmacotherapy has begun to address the causes and treatment of disease and disability in old age, including hypertension, stroke, congestive heart failure, adult-onset diabetes, osteoporosis, and cancer.

Despite the evident benefits of pharmacologic medications, their use by elderly patients has always been a focus of concern for health professionals caring for the aging patient. The elderly make up 13% of the U.S. population, yet it is estimated that they consume 30% of prescribed medications and 40 to 50% of over-the-counter (OTC) medications

(Figure 23.1).[1] Physiologic changes, coupled with increased use of medications, place older patients at risk for adverse effects and drug interactions. The concept of "polypharmacy," originally meaning "many drugs," has acquired a derogatory connotation of excessive and unnecessary use of medication (Table 23.1). Studies have shown that 9 to 31% of hospital admissions in elderly patients may be medication related.[2-4] In addition, the elderly seem to be two to three times as likely to experience adverse drug reactions as compared to younger adults.[5]

Thus, discussion in this chapter will focus on how best to evaluate the utility of a drug for use in the elderly. Ideally, one would consult available data, allowing for assessment of benefits and side-effects in the appropriate age group and in persons with similar medical conditions. However, data for a complete evaluation are rarely available. In general, risk for side-effects increases with age. Expectation of remaining years of life decreases with advancing age, and, therefore, the time for a potentially

---

* Illustrations by Dr. S. Oklund.

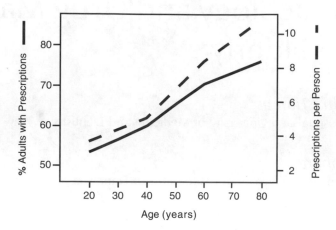

**FIGURE 23.1** Increase with age in the percentage of adults receiving prescriptions and in the number of medications prescribed per person.

beneficial distant effect decreases. Elderly persons are more likely to be taking multiple medications, so the potential for drug–drug interactions increases with age, while the alterations in metabolism associated with age and chronic illness tend to increase risk of adverse events. Conversely, the incidence of many outcomes such as stroke, myocardial infarction, and many cancers increases with age, so the potential benefit of preventative interventions may be higher in the elderly.

There is a large interindividual variation in drug metabolism that contributes to the susceptibility of the elderly to drug–drug interactions.

Advances in pharmacogenomics, that is, the relationship between functional genetics and rational therapy (based on genetic characteristics of each individual) have improved our ability to predict some of this inter-individual variability.[6,7] Knowledge of the phenotype related to the metabolism of a particular drug can predict the risk of drug side-effects and provide an individualized, so-to-speak "customized" treatment.[8,9]

## II. PHYSIOLOGIC CHANGES AFFECTING PHARMACOKINETICS

Pharmacokinetics is defined as the handling of a drug within the body, including its absorption, distribution, metabolism, and elimination. There are various physiologic changes that occur with aging that may affect drug disposition in the body (Table 23.2).

### A. ABSORPTION

While the extent of gastrointestinal (GI) absorption of most drugs is not significantly changed in the elderly, there are several changes in the GI tract that may affect the pattern of absorption (Chapter 20). Gastric emptying time may be prolonged, causing a delay in absorption from the GI tract. This would only be clinically

**TABLE 23.1**
**Features of Polypharmacy**

- Medication not indicated
- Duplicate medications
- Concurrent interacting medications
- Contraindicated medications
- Inappropriate dosage
- Drug treatment of adverse drug reaction
- Improvement following discontinuance.

**TABLE 23.2**
**Age-Related Physiologic Changes Affecting Pharmacokinetics**

Gastrointestinal system — rarely clinically significant
- Acid production generally unchanged
- Drug–drug interactions may alter absorption
- Splanchnic blood flow decreases with no effect on drug absorption

Liver
- Decrease in hepatic blood flow often associated with decreased first-pass metabolism
- Phase I metabolism affected
- Phase II metabolism generally preserved

Fluid and tissue compartments
- Decrease in total body water
- Increase in fat compartment
- Decrease in muscle mass

Plasma drug-binding proteins — rarely clinically significant
- Decrease in serum albumin (primarily disease-related)
- No change in α-acid glycoprotein

Kidneys
- Decrease in renal blood flow
- Decline in creatinine clearance
- Decline in tubular secretion

significant for acutely administered medications, such as analgesics. GI motility may be decreased, thereby prolonging the absorption phase of a medication. Decreased mucosal cells have been noted in the aging GI tract, and intestinal blood flow may be diminished, particularly if the patient has congestive heart failure. Absorption by active transport is reportedly decreased, but because the majority of medications are absorbed by passive diffusion, changes in absorption are not usually clinically significant. Changes in absorption can be clinically significant in the presence of concomitant administration of other medications. An example of this would be the marked decrease in absorption of fluoroquinolone antibiotics with calcium supplements.

Absorption of medications via the intramuscular route (IM) has not been well studied in the elderly, but some predictions about absorption can be made. Absorption of IM injections may be impaired in the elderly because of decreased peripheral blood flow, particularly in patients with peripheral vascular disease. Increased connective tissue in aging muscle (Chapters 21 and 24) can impair permeability, decreasing systemic absorption of an IM injection. Intramuscular injections may also be difficult to administer and painful to the patient, due to the decreased muscle mass in many elderly patients.

Systemic absorption must also be considered when using a transdermal product in older patients. Transdermal absorption of medications has not been extensively studied in the elderly, but there are changes in the aging skin that may affect transdermal absorption (Chapter 22). Older patients tend to have decreased skin hydration and decreased surface lipid content, factors important for transdermal penetration. They may also have increased keratinization, further impairing absorption. Decreased peripheral blood flow and compromised microcirculation may impair systemic absorption from transdermal products.[10] Of the currently marketed transdermal products, most (nitroglycerine, estradiol, clonidine, fentanyl) are commonly prescribed for older patients and should therefore be studied in this population.

## B. DISTRIBUTION

The distribution of medications within the body is dependent on whether the drug is lipid soluble or water soluble and the extent of protein binding. *Age-related changes in body composition* may significantly affect drug distribution. With aging, lean body mass and total body water decrease, while fat content increases (Chapter 21).[11] This can significantly increase the volume of distribution of fatsoluble drugs such as the benzodiazepines, a group of drugs with sedative-hypnotic actions (i.e., capable of inducing drowsiness and sleep). Clinically, it may take longer to reach steady-state levels and for the drug to be eliminated from the body.

The extent to which *drugs bind to proteins* can significantly alter the volume of distribution. Many elderly, particularly those living alone in the community, have inadequate protein intake, and therefore, are hypoproteinemic (Chapter 24). Because only the unbound portion of a drug is pharmacologically active, a reduction of plasma protein, specifically albumin, can result in higher free drug levels and increased effects. This is clinically significant for highly protein bound drugs such as phenytoin (an antiepileptic drug) and warfarin (an anticoagulant drug). Caution should be advised when prescribing two or more highly protein-bound drugs.

Cardiac output may be decreased in the elderly, altering regional blood flow. Decreased splanchnic, renal, and peripheral blood flow affects the distribution, and, therefore, the effectiveness of some medications. Tissue barriers may be altered with age, also affecting the outcome of the drug.

## C. METABOLISM

*The liver* is the main site of metabolism or biotransformation of drugs, and with increased age, there is a decrease in hepatic mass and blood flow (Chapter 20).[12] The first pass metabolism of some medications is decreased with aging, thus increasing their relative bioavailability.

Hepatic metabolism of drugs occurs mainly through a number of chemical reactions that are classified as phase I and phase II reactions:

- Phase I reactions usually convert the drug considered to a more easily excreted metabolite, primarily by oxidation but also by reduction and hydrolysis.
- Phase II reactions, also called synthetic or conjugation reactions, intend to facilitate drug excretion by coupling between the drug or its metabolite with an endogenous substrate such as glucuronic acid, sulfuric acid, acetic acid, or an amino acid.

Various factors besides age, such as gender, genetics, smoking, alcohol, and medications, affect metabolism. With age, phase I reactions are decreased, particularly in older men. Medications affected include theophylline (a compound found in tea with CNS stimulatory, smooth muscle relaxing, and diuretic actions), propranolol (a β-adrenergic blocker), and diazepam (an antianxiety drug), to name a few. Phase II reactions are minimally affected by age, but because polypharmacy occurs commonly, these reactions are affected by concomitant drugs that alter metabolism through these pathways. Examples here are lorazepam, (an antianxiety drug) and acetaminophen (an analgesic/anti-inflammatory drug). The inducibility of hepatic enzymes by smoking, alcohol, and drugs appears to be diminished, though results from various studies are conflicting.[13] Age-related changes in enzyme inhibition induced by cimetidine (a blocker of gastric acid secretion) have not been reported. It appears that enzyme activity is decreased by the same amount in elderly and younger subjects.[13]

## D. ELIMINATION

Renal elimination is considered the most significant pharmacokinetic change in the elderly (Chapter 19). Between the fourth and eighth decades of life, renal mass decreases on average by 20%, and renal blood flow decreases by 10 mL/min per decade after age 30.[14] Glomerular filtration rate, as expressed by the creatinine clearance (CrCl), also declines linearly with age beginning in the fourth decade. While measuring creatinine clearance via a 24-h urine collection is the best way to obtain an accurate level, these collections are often difficult, if not impossible, in the elderly patient, and usually entail inserting a catheter (a tubular device) into the bladder by intubation through the urethra (Chapter 19).

Alternatives to urine collections rely upon empirical estimations of creatinine clearance. A common approach relates renal function to the serum creatinine alone. However, in the elderly, serum creatinine is an inadequate measure of renal function because creatinine production decreases as muscle mass decreases. Therefore, it is more appropriate to estimate creatinine clearance using an equation that takes into account age, weight, serum creatinine (Scr), and gender. A commonly used equation is that developed by Cockcroft and Gault (Equation 23.1), where the weight used is the patient's lean body weight (LBW) and, which provides a formula for calculating LBW, including patients that are overweight (Equation 23.2).[15,16]

$$CrCl_{men} = \frac{(140 - Age)LBW(kg)}{72 \times Scr(mg/dL)} \qquad (23.1)$$

$$CrCl_{women} = CrCl_{men} \times 0.85$$

$$LBW_{men} = 50 \text{ kg} + 2.3 \text{ kg/inch above 5 feet} \qquad (23.2)$$

$$LBW_{women} = 45.5 \text{ kg} + 2.3 \text{ kg/inch above 5 feet}$$

The Cockcroft–Gault equation is useful when calculating doses for drugs that are eliminated by glomerular filtration, such as aminoglycosides and vancomycin (antibiotics), digoxin (for the treatment of cardiac failure), lithium (a mood stabilizing drug), and histamine antagonists.

Drugs that are eliminated by tubular secretion also exhibit decreased excretion with age. This can occasionally be clinically significant for drugs that rely on tubular secretion for elimination.

## III. PHYSIOLOGIC CHANGES AFFECTING PHARMACODYNAMICS

Pharmacodynamics refers to the processes involved in the interaction between a drug and an effector organ that results in a response, either therapeutic or adverse. Pharmacodynamics measures the intensity, peak, and duration of action of a medication. With aging, physiologic changes may affect the body's response to medications. Some changes increase the older patient's sensitivity to a drug, some may decrease the effect of the medication, and some may make the older person more susceptible to the adverse effects of a drug. The predictability of a drug's response is decreased in the elderly, and the practitioner cannot only rely on the pharmacokinetic changes but must include pharmacodynamic factors as well.

Pharmacodynamic changes in the elderly can best be understood by reviewing examples of medications in which pharmacodynamic changes have been described. In the following section, several medications that illustrate pharmacodynamic changes are discussed.

*Postural hypotension* appears to be more of a problem in older patients as compared to younger patients. This is due to a decrease in baroreceptor function and decreased peripheral venous tone.[17,18] When a younger person taking a vasodilator drug stands up abruptly, the body responds to the immediate hypotension with a reflex acceleration of cardiac rate (tachycardia), which helps to restore normal blood pressure. This tachycardia does not always occur in older patients, putting them at risk for dizziness, syncope, and falls. For this reason, vasodilators should be used cautiously in the elderly. Older patients who are treated with hypotensive or vasodilatory drugs should be instructed to change positions slowly when rising from a lying or sitting position. Increased postural (orthostatic) hypotension is seen with nitrates, nifedipine, tricyclic antidepressants, antipsychotics, and diuretics.

The *hyponotic-sedative benzodiazepines* are (unfortunately) widely used in the geriatric population. Based on pharmacokinetic factors, we can choose a benzodiazepine that would be relatively well tolerated in the elderly (short half-life and conjugated metabolism), but pharmacodynamic factors must also be considered. Reidenberg and colleagues studied the relationship between diazepam dose, plasma level, age, and central nervous system (CNS) depression.[19] Their results demonstrated that older patients require a lower dose and plasma level of diazepam than younger patients to reach the same level of sedation. An increased sensitivity was also demonstrated by Castleden and colleagues, who found increased psychomotor impairment in older subjects as compared to younger subjects while on nitrazepam.[20]

In general, the elderly will be more sensitive to the effects of a medication that depresses or excites the CNS. The elderly have less CNS reserve and therefore are more sensitive to an "insult" by a medication (Chapters 7–9).

Though the data are conflicting, most studies on *adrenergic β-receptors* have demonstrated that the elderly have a decreased β-receptor sensitivity as evidenced by a decreased cardiac response to the ß-agonist isoproterenol and the beta-antagonist (blocker) propranolol.[21] It was originally thought that this decreased response was due to a decrease in the number of ß-receptors. More overall decrease was not demonstrated, but rather a decrease in the number of receptors that have a high affinity for these drugs was shown.[22] The elderly have a decreased ability to form high-affinity binding complexes. Desensitization of the receptors would occur because norepinephrine levels increase with age (Chapter 10), yet response to beta agonists decreases.[23] The clinical implications of all of these changes are that the elderly may have a decreased response to ß-blockers and agonists, including the ß-$_2$ bronchodilators.

Elderly patients seem to be more susceptible to the side effects of *antipsychotic medications.* Extrapyramidal symptoms, orthostatic hypotension, and anticholinergic effects occur more frequently than in younger patients and are less well tolerated.[24] Elderly patients on antipsychotics are more likely to experience *parkinsonism,* probably because of an already depleted dopamine reserve (Chapter 7). Adding a dopamine antagonist, such as a phenothiazine or haloperidol, can "tip" the older patient into parkinsonism or unmask latent Parkinson's Disease. In one series of new Parkinson's cases, 51% of the patients had drug-induced parkinsonism.[25] Clinical features clear completely in only 66% of the cases, taking up to 36 weeks to resolve. Elderly patients experience *tardive dyskinesia* (i.e., impairment of voluntary movements) more frequently and earlier in treatment, even when on low doses of neuroleptics.[26,27] This may be due to a "hypersensitivity" of the dopamine receptors in the nigrostriatum. Tardive dyskinesia is more likely to be persistent and severe in the elderly, and women seem to be particularly at risk.

Physiologic changes in the geriatric patient may exaggerate the effects of *anticholinergic agents* (Chapters 7 and 8). Slowed gastrointestinal (GI) motility increases the risk of constipation. Urinary retention is enhanced in patients with an outflow obstruction, such as an enlarged prostate. Anticholinergic-induced CNS effects such as delirium and memory impairment may be more pronounced in the elderly because of decreased CNS reserve. Dry eyes and mouth (signs of parasympathetic block) may be more pronounced in the elderly and are more bothersome than in younger patients.

## IV. ADVERSE DRUG REACTIONS IN THE ELDERLY

As stated earlier, studies have demonstrated that the elderly are two to three times more at risk for adverse drug

---

**TABLE 23.3**

**Factors Associated with Increased Incidence of Adverse Drug Reactions in the Elderly**

1. Reduced (small) stature
2. Reduced renal and hepatic function
3. Cumulative insults to body
   a. Disease
   b. Faulty diet
   c. Drug abuse
4. Medications, multiple and potent
5. Altered pharmacokinetics
6. Noncompliance

---

reactions as compared to younger adults.[5] This is due to a number of factors summarized in Table 23.3, the most important of which include the following:

- Increased number of medications taken by the elderly
- Increased sensitivity to medications related to the pharmacokinetic and pharmacodynamic changes described above

Many adverse reactions are iatrogenic in nature. For example, they may be due to:

- Choice of an inappropriate medication or dosage
- Inadequate monitoring of the patient
- Failure to recognize or let adverse effects go unnoticed by the patient and the prescriber
- Drug–drug interactions, drug–food interactions, or drug–disease interactions
- Noncompliance

Subtle effects such as GI complaints, dizziness, mental status changes, change in libido, instability and falls, and bowel or bladder habits may be attributed to "old age" or be treated as a new disease state. Health professionals should inquire about specific adverse effects and should encourage the patient to report any unusual occurrences while on medication.

Noncompliance is the failure of a patient to follow instructions regarding medications. It represents another factor often contributing to adverse drug reactions. Psychosocial complications such as poverty, dementia, and loneliness exacerbate this problem. Effective communication between the prescriber and patient and caregiver can help eliminate this factor.

### A. Specific Principles

Although a comprehensive review of all drug classes is impractical here, certain groups of medications are problematic enough in the elderly to warrant separate consid-

**TABLE 23.4**
**Commonly Used Medications Best Avoided in the Elderly**

Anticholinergic preparations
    Diphenhydramine (Benadryl)
    Amitriptyline (Elavil)
    Oxybutynin (Ditropan)
    Doxepin (Sinequan)
    Dicyclomine (Bentyl)
Benzodiazepines with active metabolites
    Diazepam (Valium)
    Chlordiazepoxide (Librium)
    Flurazepam (Dalmane)
Central Acting CNS agents
    Alpha-methyldopa (Aldomet)
    Clonidine (Catapres)
Analgesics
    Propoxyphene (Darvon)
    Meperidine (Demerol)
    Indomethacin (Indocin)

*Source:* Adapted from Beers, M.H., *Arch. Intern. Med.*, 135, 703–710, 2001.

eration. A list of undesirable medications has been formulated by a consensus panel (Table 23.4).[28]

## B. Drugs Acting on the Nervous System

*Psychotropic medications* are responsible for the most adverse drug reactions.[29] Such reactions include: worsening mental status, falls, dehydration, orthostatic hypotension, and disorders of movement (e.g., extrapyramidal signs) and tardive dyskinesia. An important disease state – drug interaction common to all psychotropics is the increased sensitivity to side effects exhibited by elderly persons with common brain diseases such as Alzheimer's disease, Parkinson's disease, and stroke. A side effect profile that minimizes anticholinergic effects should be chosen when prescribing psychotropics.[30,31]

*Analgesics* (drugs relieving pain) are another group of medications frequently prescribed and therefore also warrant special consideration. Nonsteroidal anti-inflammatory drugs (NSAIDs) are currently ubiquitous, despite the plethora of adverse drug reactions that may arise from their use.[32] The most common side effects are gastrointestinal symptoms ranging from gastritis to life-threatening hemorrhage.[33,34] Renal effects are usually reversible as well as are photosensitivity and urticaria. Pulmonary edema arises from congestive heart failure due to fluid and sodium retention. NSAIDs also interfere with hypertension treatment. CNS effects range from headache, to altered mental status, to frank psychosis. Finally, genitourinary effects include ejaculatory dysfunction. In view of the aforementioned problems, geriatricians are conservative in the use of

NSAIDs. The nonpharmacologic management of osteoarthritis (i.e., physical therapy) is utilized whenever possible, and finally, acetaminophen is recommended for analgesia (Chapter 21). When NSAIDS are indicated, the newer COX-2 specific drugs may be safer.

## C. Anticoagulant Drugs

Elderly patients seem to be more sensitive to the effects of *anticoagulants* such as warfarin, and decreased doses are needed to adequately anticoagulate the older patient.[35,36] Just why this sensitivity occurs is not entirely clear. There may be a relative deficiency of vitamin K or vitamin K-dependent clotting factors in the elderly (Chapter 24). There may also be increased concentrations of inhibitors of coagulation. In addition, warfarin is strongly bound to albumin, so alterations in binding must be considered. Whether age is an independent risk factor for complications of warfarin therapy has been debated.[37,38] Either way, it is prudent for the practitioner to closely monitor all elderly patients on warfarin therapy.

## D. Cardiovascular Drugs

Management of an elderly patient on *digoxin* illustrates all of the principles already mentioned. Digoxin is one of the most used preparations of digitalis from the leaf of the foxglove plant. The main action of digoxin is its *ability to increase the force of myocardial contraction*. Beneficial consequences of this action include increased cardiac output, decreased cardiac size, venous pressure, and blood volume, slow-down cardiac rate, promotion of diuresis, and relief of edema. Many of these actions will benefit alterations in cardiac functions at all ages, including old age (Chapter 21). Administration of digoxin to an older individual with cardiac alterations depends on several considerations:

- First, the indications for the use of digoxin have been scrutinized, and digoxin is no longer utilized in an indiscriminate manner to everyone with congestive heart failure.[39] Rather, the indications for its use have been narrowed to specific clinical situations.[40,41]
- Second, because of its reduced elimination, it is usually given in lower doses in the elderly. Digoxin serum levels are monitored to prevent toxicity. Even at normal therapeutic levels, it can produce a range of adverse effects including anorexia and altered mental status.

Studies have been performed to examine the effects of withdrawing patients from this drug after chronic administration showing that many do well.[41] There have also been reported some sensitivity changes at the digoxin receptor level in elderly patients showing different endorgan reaction to the drug.[42] Thus, this classic medication serves as a good example of a drug that undergoes altered pharmacokinetics as well as altered pharmacodynamics in the elderly.

## V.  GENERAL GUIDELINES

In the interest of avoiding adverse drug reactions, a number of principles, partly based on physiologic considerations, should be followed:

1. Nonpharmacologic management should be used whenever possible.
2. The number of drugs prescribed should be kept to a minimum.
3. The drug regimen should be simplified to aid in compliance.
4. Treatment should be prescribed only with clear goals or endpoints in mind.
5. Dosage should be adjusted to take into consideration altered physiologic parameters (i.e., start with low dose and increase slowly)
6. Utilize laboratory monitoring when indicated.
7. Reevaluate the medication regimen regularly.

The drug regimen needs to be regularly reviewed and re-assessed for possible changes.[43] This includes an "obsessive" (i.e., as detailed as possible!) drug history that may be facilitated by the patient bringing in all medications as well as over-the-counter drugs for scrutiny by a health professional. Finally, one must suspect a drug reaction when any otherwise unexplained symptoms occur, such as a change in mental status.

*One of the most common interventions a geriatrician makes is to discontinue medications.* It has been shown that, when done judiciously, most patients benefit from this maneuver. Equipped with a basic knowledge of physiologic changes that occur with aging, pharmacologic principles, and common sense, health care professionals should be able to prescribe medications for elderly patients in a safe and effective manner.

## REFERENCES

1. National Medical Expenditure Survey: Prescribed Medicines: A Summary of Use and Expenditures by Medicare Beneficiaries: Research Findings, *U.S. Department of Health and Human Services publication 89–3448*, National Center for Health Services Research and Health Care Technology Assessment, Rockville, MD, 1989.
2. Grymonpre, R.E. et al., Drug-associated hospital admissions in older medical patients, *J. Am. Geriatr. Soc.*, 36, 1092, 1988.
3. Colt, H.G. and Shapiro, A.P., Drug-induced illness as a cause for admission to a community hospital, *J. Am. Geriatr. Soc.*, 37, 323, 1989.
4. Williamson, J. and Chopin, J.M., Adverse reactions to prescribed drugs in the elderly: A multicentre investigation, *Age Ageing*, 9, 73, 1980.
5. Nolan, L. and O'Malley, K., Prescribing for the elderly. Part I: Sensitivity of the elderly to adverse drug reactions, *J. Am. Geriatr. Soc.*, 36, 142, 1988.
6. Evans, W.E. and Relling, M.V., Pharmacogenomics: Translating functional genomics into rational therapeutics, *Science*, 286, 487, 1999.
7. Kalow, W., Pharmacogenetics, pharmacogenomics and pharmacobiology, *Clin. Pharmacol. Ther.,* 70, 1, 2001.
8. Phillips, K.A. et al., Potential role of pharmacogenomics in reducing adverse drug reactions: a systematic review, *J. Am. Med. Assoc.*, 286, 2270, 2001.
9. Pollock, B.G. et al., Prospective cytochrome P450 phenotyping for neuroleptic treatment in dementia, *Psychopharmacol. Bull.*, 31, 327, 1995.
10. Roskos, K.V., Maibach, H.I., and Guy, R.H., The effect of aging on percutaneous absorption in man, *J. Pharmacokinet. Biopharmaceut.*, 17, 617, 1989.
11. Yuen, G.J., Altered pharmacokinetics in the elderly, *Clin. Geriatr. Med.*, 6, 257, 1990.
12. James, O.F.W., Drugs and the ageing liver, *J. Hepatology*, 1, 431, 1985.
13. Durnas, C., Loi, C.M., and Cusack, B.J., Hepatic drug metabolism and aging, *Clin. Pharmacokinet.*, 19, 359, 1990.
14. Bennett, W.M., Geriatric pharmacokinetics and the kidney, *Am. J. Kidney Dis.*, 16, 283, 1990.
15. Cockcroft, D.W. and Gault, M.H., Prediction of creatinine clearance from serum creatinine, *Nephron*, 16, 31, 1976.
16. Lott, R.S. and Hayton, W.L., Estimate of creatinine clearance from serum creatinine concentration — a review, *Drug Intell. Clin. Pharm.*, 12, 140, 1978.
17. Caird F.I., Andrews, G.R., and Kennedy, R.D., Effect of posture on blood pressure in the elderly, *Br. Heart J.*, 35, 527, 1973.
18. Gribbon B. et al., Effect of age and high blood pressure on baroreflex sensitivity in man, *Circ. Res.*, 29, 424, 1971.
19. Reidenberg, M.M. et al., Relationship between diazepam dose, plasma level, age and central nervous system depression, *Clin. Pharmacol. Ther.*, 23, 371, 1978.
20. Castleden, C.M., Kay, C.M., and Parsons, R.L., Increased sensitivity to nitrazepam in old age, *Br. Med. J.*, 1, 10, 1977.
21. Vestal, R.E., Wood, A.J.J., and Shand, D.G., Reduced ß-adrenoreceptor sensitivity in the elderly, *Clin. Pharmacol. Ther.*, 26, 181, 1979.
22. Feldman, R.D. et al., Alterations in leukocyte beta-receptor affinity with aging. A potential explanation for altered ß-adrenergic sensitivity in the elderly, *N. Engl. J. Med.*, 310, 815, 1984.
23. Scarpace, P.J., Decreased receptor activation with age. Can it be explained by desensitization? *J. Am. Geriatr. Soc.*, 36, 1067, 1988.

24. Peabody, C.A. et al., Neuroleptics and the elderly, *J. Am. Geriatr. Soc.*, 35, 233, 1987.

25. Stephen, P.J. and Williamson, J., Drug-induced parkinsonism in the elderly, *Lancet*, 2, 1082, 1984.

26. Caligiuri, M.P. et al., Incidence and risk factors for severe tardive dyskinesia in older patients, *Br. J. Psychiatry,* 171, 148, 1997.

27. Morton, M.R., Tardive dyskinesia: Detection, prevention and treatment, *J. Geriatr. Drug Therapy*, 2, 21, 1987.

28. Beers, M.H., Explicit criteria for determining potentially inappropriate medication use by the elderly, *Arch. Intern. Med.*, 135, 703, 2001.

29. Thompson, T.L., Moran, M.G., and Nies, A.S., Psychotropic drug use in the elderly, *N. Engl. J. Med.*, 308, 194, 1983.

30. Jenike, M.A., Psychoactive drugs in the elderly; Antidepressants, *Geriatrics*, 43, 43, 1988.

31. Jenike, M.A., Psychoactive drugs in the elderly; Antipsychotics and anxiolytics, *Geriatrics*, 43, 53, 1988.

32. Sack, K.E., Update on NSAIDS in the elderly, *Geriatrics*, 44, 71, 1989.

33. Fries, J.F. et al., Toward an epidemiology of gastropathy associated with nonsteroidal anti-inflammatory drug use, *Gastroenterology*, 96, 647, 1989.

34. Griffin, M.R., Ray, W.A., and Schaffner, W., Nonsteroidal anti-inflammatory drug use and death from peptic ulcer in elderly patients, *Annu. Intern. Med.,* 109, 359, 1988.

35. Shepherd, A.M.M. et al., Age as a determinant of sensitivity to warfarin, *Br. J. Clin. Pharmacol.*, 4, 315, 1977.

36. Redwood, M. et al., The association of age with dosage requirement for warfarin, *Age Ageing*, 20, 217, 1991.

37. Gurwitz, J.H. et al., Age-related risks of long-term oral anticoagulant therapy, *Arch. Intern. Med.*, 148, 1733, 1988.

38. Landefeld, C.S. and Goldman, L., Major bleeding in outpatients treated with warfarin: Incidence and prediction by factors known at the start of outpatient therapy, *Am. J. Med.*, 87, 144, 1989.

39. Fleg, J.L., Gottlieb, S.H., and Lakatta, E.G., Is digoxin really important in treatment of compensated heart failure? *Am. J. Med.,* 73, 244, 1982.

40. Papadakis, M.A. and Massie, B.M., Appropriateness of Digoxin use in medical outpatients, *Am. J. Med.*, 85, 365, 1988.

41. Sueta, C.A., Carey, T.S., and Burnett, C.K., Reassessment of indications for digoxin, *Arch. Intern. Med.*, 149, 609, 1989.

42. Kelly, J.G., Copeland, S., and McDevitt, D.G., Erythrocyte cation transport and age: Effects of digoxin and furosemide, *Clin. Pharmacol. Ther.*, 34, 159, 1983.

43. Avorn, J. et al., A randomized trial of a program to reduce the use of psychoactive drugs in nursing homes, *N. Engl. J. Med.*, 327, 168, 1992.

# 24 Healthful Aging: Nutrition and Exercise and Experimental Strategies in Dietary Restriction*

*Franco Navazio and Paola S. Timiras*
University of California, Berkeley

## CONTENTS

---

* Illustrations by Dr. S. Oklund.

## I. TOWARD A NEW IMAGE OF AGING

Epidemiologic studies in humans and experimental data reported in several animal species indicate that good nutrition and regular physical exercise not only contribute significantly to health and well-being but also may prolong average and maximum life span. With scientific validation, these two interventions are now accepted as "natural" ways to deal with a number of disabilities and diseases related to old age. Accordingly, a brief discussion of their impact on functional competence and on longevity will serve as an appropriate conclusion to our study of the physiologic changes that occur with advancing age.

Lengthening the "health span," that part of the life span enjoyed in good health and without disabilities, has become one of the major goals of those researchers and educators working in the field of gerontology and geriatrics. We know already that biomedical progress in the United States and in many other world countries have allowed us to live several years longer than our parents and ancestors (Chapter 2). Similarly, preventive and rehabilitative strategies are allowing individuals to maintain an active lifestyle, and thus, their personal independence, which contributes considerably to preserving intellectual and functional competence into old age. Yet, these benefits, while considerable, fall short of assuring a good quality of life. We are now witnessing rapid biotechnological advances that we must exploit, in full, to treat the diseases of old age and to prolong and improve the health span. Some of the classic and new directions to come from this research are briefly presented in this chapter.

As for the preventive and rehabilitative potential of nutrition, physical exercise, and other interventions in an aging population, it must be kept in mind that: (a) *the duration of the intervention*, short-term or long-term, is likely to significantly influence the outcome; (b) old age may result in or from *differential gene expression*, such that the requirements for nutrition and physical exercise of the elderly may differ from those of the young; (c) *genetic variation among individuals and ethnic groups* may justify an individualized or "customized," approach to treatments involving specific dietary components such as vitamins and trace nutrients; and, finally, (d) any preventive or therapeutic program for the elderly must take into consideration the possible presence of *multiple pathologies of the elderly* (Chapter 3) and the possibility that *polypharmaceutical interventions* (Chapter 23) may interfere with or enhance the regimen chosen for a given individual.

Given the complexity of the molecular, cellular, and organismic mechanisms involved in the aging process, particularly in mammals, attempts have been made to develop experimental models in which the rate of aging can be manipulated in a predictable manner (Chapter 1). In experimental animals, one of the simplest but most effective methods of extending the life span and, by implication, slowing down the rate of aging, is dietary (caloric) restriction. This method consists of limiting access to food until the body weight of the experimental animals is about 50% that of age-matched, fully fed control animals. In addition to prolonging the life span, dietary restriction has the effect of maintaining a number of physiologic and metabolic processes in a youthful state, thus, delaying and reducing the incidence of aging-related pathology. This animal model is now being exploited as an important experimental tool in aging research, as evidenced by the rapid growth, in the last 20 years, of restricted feeding as a way to retard physiological aging in a variety of animal species. However, two questions about the use of caloric restriction remain unanswered: what mechanism is at work in the findings reported? And, will dietary restriction work the same way in humans as it appears to work in other animals?[1]

Although many diseases associated with old age in humans are susceptible to dietary intervention as well as to physical exercise, little is known of their specific effects on the rate of aging. The brain, specifically, appears to be quite sensitive to the type of nutrition and the amount of exercise in a person's regime. This fact was recognized in a recent article, where the author, extolling the neuroprotective actions of these simple lifestyle interventions, stated provocatively: "Take away my food and let me run."[2] The effectiveness of interventions at this level is reviewed in this chapter.

Maintaining of quality of life in old age is not merely a matter of improving physical fitness or eliminating age-related disease. We know that many individuals, including the elderly, do not always see health as the most important issue in their lives. Indeed, social, cultural, economic, and emotional factors are integrally involved in lifestyle choices related to health, and they must be considered. So far, the increase in life expectancy witnessed in the last century has been attributed as much to improvements in hygiene, better nutrition, and the rising standard of living as to the successful eradication of a number of infectious diseases. Further advances in gerontology and geriatrics will come about when we integrate these disparate factors of lifestyle modification and biomedical breakthroughs with socioeconomic realities. In this sense, beyond the role the individual must play (eating properly, exercising, following good hygienic practices, remaining intellectually alive, and so forth), society has a role that goes farther than that of strengthening its health care systems and disseminating public health information: it would do well to devise more creative opportunities for the elderly so that they can retain or embark on productive roles in the community.

Certainly, a myriad of sociocultural variables contribute to a person's overall health and rate of physiologic decline. A simple example is ones' level of education: the better the education, with its attendant nutritional and lifestyle habits, the longer the life span and the fewer the disabilities associated with old age (Chapter 8). On a societal level, effective socioeconomic networks, public health and safety programs, and community support systems play as important a role in determining the quality of life at all ages as does good medical care.

The authors' lack of expertise in these critical areas, along with present limitations of time and space, prevent forays into material that is not biomedically oriented. Nevertheless, we acknowledge the extreme importance of these nonbiomedical factors in any discussion of ways to promote the highest possible quality of life in the elderly. Below is a brief discussion of the myths surrounding old age. As is well known, myths are powerful in any society, dictating attitudes and behaviors to a great extent.

## MYTHS AND STEREOTYPES ABOUT OLD AGE

A number of myths and stereotypes about old age pervade our culture. Not only do they obscure the facts about the later years, they undermine the real potential most of us have today for a vigorous and healthy senescence.

According to Henry Miller, "At eighty I believe I am a far more cheerful person than I was at twenty or thirty. I most definitely would not want to be a teenager again. Youth may be glorious, but it is also painful to endure. Moreover, what is called youth is not youth; it is rather something like premature old age." [3]

For a true picture of physiologic aging to emerge, it is essential that these myths be disavowed. The most common and insidious among them are:

- Old age starts at 65 years (in the words of Picasso, "One starts to get young at the age of sixty, and then it's too late")[4]
- Old age is a disease
- Old age always brings mental impairment
- The elderly are an homogeneous group in terms of their physiologic decline over time
- Old people are poor
- Old people are powerless
- Nothing can be done to modify aging; the elderly of the future will have the same problems as they have today
- The United States has been and will always be a youth-focused culture

With intensive efforts at reeducation, starting at an early age, these negative images can be reversed. The truth is that:

- Good health is important during early ages as a forerunner of good health in old age
- Considerable advances have been achieved in prolonging life expectancy and in preventing and treating many diseases, and similar advances may be expected in the future
- Quality of life in old age depends on biomedical as well as socioeconomic and cultural factors
- Prevailing attitudes about aging and negative images of the elderly can be turned around through education

At all ages, a sense of well being or "wellness," the overworked but useful concept coined in the 1970s, depends on good mental and physical health and a favorable environment. Although elderly persons afflicted by declining function and increasing pathology are more vulnerable to the everyday stresses of life, maintenance or even restoration of wellness can be achieved at all ages. Various common conditions associated with physical, emotional, and social circumstances capable of being addressed by remedial strategies are illustrated in Table 24.1.[5] For each condition, the table indicates the underlying causes in terms of nutrition, exercise, drugs, smoking, alcohol, and injury, and suggests appropriate remedial strategies. It emerges clearly that achieving a state of wellness, especially in later years of life, demands a judicious combination of medical, social, economic, and psychological interventions.[6]

### A. THE ELDERLY AS AN ASSET

As amply illustrated throughout this book, people turning 50 today have nearly half of their adult life ahead of them. What do they have to look forward to? While their quality of *physical* life will be considerably better than that of their predecessors, what about their *political, economic, and cultural* roles so vital to well-being at all ages? And when they retire, will they opt for a lifestyle of leisure and recreation or will they want to continue working or volunteering their skills?

Responding to these questions is not only critical for the individual men and women involved, but it is an important task for society.[7–9] As communities across the nation develop policies and services, they must give considered thought to their older members. Enlightened policies that are inclusive of the elderly will need to rely on a *global assessment* of their need, by age group, by gender, and by health status, for the latter translates to the potential this age group has for continuing to contribute to society. Certainly, there will be variations from one community to another, but only by a thorough demographic survey can appropriate policy decisions be made.

For example, with respect to older citizens, policy makers would be well advised to adopt a dual but synergistic approach that works to:

- Prolong the health span by strengthening medical care and public health education to include

**Prevention and Rehabilitation**

**TABLE 24.1**
**Factors Affecting the Elderly and Amenable to Multifactorial Interventions***

| Conditions | Underlying Cause | Control |
|---|---|---|
| **Issues Pertaining to Nutrition** | | |
| Reduced calcium absorption; loss of lean body mass and bone mass; low levels of vitamins K and D; muscle weakness and cramps | Inadequate caloric intake and food variety; excess processed food; lack of nutritional knowledge; use of alcohol, drugs; low income and social status; illness; physical difficulty in daily activities | Improved economics status and health knowledge aimed at improving diet: Low Na, low saturated fats, high complex CHO diet |
| Hypertension; obesity; increased insulin resistance; high cholesterol; low HDL | Excess salt; excessive food intake; high caloric food intake; inactivity; heredity | Weight loss; regular exercise; antihypertensive drugs; anticholesterol drugs |
| **Issues Pertaining to Exercise** | | |
| Reduced muscle strength and endurance; loss of lean body mass; increased fat | Inactivity due to lack of stimulation in the environment; depression; unplanned retirement | Education regarding need of exercise; improved social contacts |
| Poor neurologic control; impaired coordination, balance, and mental function | House-bound due to fear of crime; lack of knowledge regarding benefits of exercise | Treatment of underlying medical conditions; access social services in community |
| Reduced joint mobility, bone mass | Joint and bone discomfort; arthritis; osteoporosis; cardiovascular disease; respiratory disease | Enrollment in regular exercise or walking program; massage; increased fluid intake |
| **Issues Consequent to Drugs, Smoking, and Alcohol** | | |
| Polypharmacy; improper therapeutic drug dosage; drug addiction | Lack of knowledge of geriatric pharmacology; inaccurate diagnosis; noncompliance; unawareness of side effects; cost of drugs; lack of health insurance | Adequate diagnosis and treatment; surveillance of appropriate intake and education regarding side effects of drugs; economic support; enrollment in rehabilitation program |
| Respiratory insufficiency; lost capacity to perform daily tasks | Mass media influence; peer pressure | Stop smoking; support |
| Impaired hepatic function; neurologic symptoms | Addiction; depression; distorted self-image; peer pressure; socioeconomic conditions; heredity | Stop drinking; peer and family support; community programs |
| **Issues Caused by Injury** | | |
| Cognitive impairment | Osteoporosis | Weight-bearing exercise |
| Balance and gait impairment | Sarcopenia | Active exercise |
| Postmenopausal complaints | Hormonal deficiencies | Hormone replacement therapy |
| General frailty | Depression | Calcium and vitamin D supplements |
| Use of diuretics | Dehydration | Community assistance and programs |
| Use of psychotropic drugs | Lack of social support/finances | |
| Hazards in the home | | |

*Source:* Adapted from Minkler, M. and Pasick, R.J., Health promotion in the elderly: A critical perspective on the past and future, in *Wellness and Health Promotion in the Elderly*, Dychtwald. K., Ed., Aspen Systems Corp., Rockville, MD, 1986.

recommendations about hygiene, nutrition, and physical exercise

• Create opportunities in which the elderly may be encouraged to participate (e.g., working part-time on a paid or volunteer basis in outreach programs, in service jobs, as mentors, and as resource volunteers) so that individuals can retain a strong and productive role in the community in which they live, and, simultaneously, give back vital services that are of benefit to the community at large

The continuing societal involvement of persons of this age group in their communities and beyond is well justified. Indeed, if they are not struggling with disease or poverty, the elderly represent individuals who are variously skilled, literate, energetic, and well informed. They have overcome many obstacles, and their survival suggests a certain resilience. As with children, holding low expectations, both by the elderly and the society in which they live, is not conducive to achievement. Yet, at the moment, we are too willing to make assumptions about this age group that relegate them to lives in which they no longer are contributing members of society, buoyed by a sense of purpose and meaning. It is foolhardy, in fact, to regard older adults as a "throw-away" population whose usefulness is at an end. Rather, they represent a largely untapped resource in this country.[7,8]

To meet the challenge of our growing elderly population, we must endeavor to better understand the nature of the aging process and its physiologic and pathologic consequences. Concurrently, we must hold a positive view of the potential of older adults. If older people are to contribute to society in meaningful ways, the vast network of agencies and programs set up to provide services and programs would do well to maintain *high expectations* of them; they must distinguish between those in need and those able and willing to serve the needs of others. Retirement may be viewed more as a lifestyle transition than a termination of employment. Few in this age group need or want to take to their rockers; it is well documented that pursuing useful activities (old or new) is what enables older adults to stay physically and mentally alert, flexible, and in touch.[9]

In the last analysis, whatever the outward limits of our 21st century vision, we would like to be able to remark, as Leo Tolstoy did in the 19th century in a letter to a friend, "Don't complain about old age. How much good it has brought me that was unexpected and beautiful...[and] the end of old age and of life will be just as unexpectedly beautiful."[10]

## II.  NUTRITIONAL REQUIREMENTS IN OLD AGE

Most old people retain dietary patterns similar to those acquired in their youth. The American Dietetic Association acknowledges this fact in its emphasis on regarding nutrition and dietary changes as part of a continuum of one's health care programs throughout life.[11] Our intent here is to provide useful information about nutrition and diet among the elderly, drawing from selected physiologic and clinical observations.

### A.  NUTRITIONAL RISK FACTORS FOR INDIVIDUALS OVER 60

While optimal nutritional status is critical to good health at any age, to achieve it demands special attention in the older age group. For one thing, adults, 65 years of age and older, for numerous psychosocial reasons, often face obstacles in trying to prepare nutritious meals on a day-to-day basis. In addition, given the increasing number of prescriptions and over-the-counter medications many older people take (Chapter 23), they may suffer drug-associated nutritional deficiencies of which they might not be aware. Indeed, the detection and treatment of nutritional disorders in older persons has emerged as a major public health concern nationally.

A variety of *screening tests* has been recommended by different agencies and organizations concerned with the aging population, but many of these tests are not easily performed or properly interpreted.[12] The most important risk factor(s) associated with poor diet (in this age group) include low income, social isolation and illness.[13] Commonly used screening tests and assessments focus on the following:

### 1.  Anthropometric and Biochemical Parameters

One anthropometric measure frequently employed to assess overall nutritional status of elderly persons is the Body Mass Index (BMI), which relates body weight (in kilograms) to height (in meters squared):

$$(BMI) = \frac{\text{weight (in kg)}}{\text{height (in m}^2)} *$$

in which "normal" individuals register BMI values ranging from =18.5 to ≤25.0.

Unfortunately, in the elderly, in whom intervertebral disc spaces are often narrowed and osteoporotic vertebral compressions are frequently encountered (Chapter 21), the BMI may give inaccurate information. Even alternative measures such as the height to the knee, the length of the arms (the so-called total arm length), and the arm span, are far from ideal measures.[14]

From a practical as well as clinical point of view, an accurate serial recording of body weight may be considered the most used and useful screening method to ascertain nutritional status in all age groups: for example, significant signs of impending malnutrition are the (unintended) loss of 1 kg in 1 month or 3 kg in 6 months.

Assessment of poor nutritional status can be made by laboratory tests, which include the following:

- Low ferritin levels (below 15 µg/dL)
- Low lymphocytic counts (less than 1200/dL)
- Abnormal level of Thyroid Stimulating Hormone (TSH) (Chapter 13)
- A total cholesterol level below 160 mg/dL (Chapter 17)
- Low albumin (below 4.0 g/dL) and prealbumin (below 1.3 g/dL) levels

Currently, the presence of hypoalbuminemia and, particularly, of hypoprealbuminemia is recognized as a strong indicator of undernourishment, a marker for the possible onset of medical complications related to malnutrition and a factor in the prognosis of mortality.[15]

---

* Weight in lbs × 0.45 = kg; Height in inches × 0.354 = m.

## 2. Clinical Assessment

In screening the nutritional status of the elderly, it is important, before any dietary recommendations are made, to assess the person's clinical status, particularly the functional status of the gastrointestinal tract (Chapter 20), and certainly to take a drug and medication history.[16]

In addition, it must be borne in mind that:

- An elderly individual's ability to eat, specifically to chew and swallow, may be impaired by poor teeth and dry mouth (Chapter 20)
- The pleasure of eating may have declined as a result of decreased taste or smell (Chapter 9)
- Urinary frequency or incontinence (Chapter 19) may have led to restriction of fluids and, concomitantly, poor food intake

*Specific recommendations* are in order for a number of conditions. For those with gastroesophageal-regurgitation disorder and with diaphragmatic hernia, sound advice is to maintain elevation of head and chest when sleeping and to avoid alcohol. The presence of diverticular disease of the colon (Chapter 20) may require initiating a low-fiber diet, certainly during flare-ups (associated with abdominal pains and diarrhea). Constipation may be corrected by increasing fluid and fiber intake, exercising (simple walking), reviewing medications for side effects of constipation, and reassuring that a bowel movement every 2 or 3 days is not abnormal. Should rectal bleeding be present, all possible causes (hemorrhoids, ischemic colitis, diverticular disease of the colon, and colon cancer, etc.) should be explored.

Even though digestion in the elderly tends to be slower than in younger adults, actual intestinal absorption appears essentially unchanged; some nutrients' absorption, like vitamin $B_{12}$, may be adversely affected by low gastric hydrochloric acid content. What interferes most with digestion in this age group is the quality of the nutrients ingested and the prolonged use of medications.

### The Obese Elderly Individual

Total body fat increases until about age 40 in men and age 50 in women. It remains unchanged thereafter until age 70, in both sexes, when it tends to decrease. Fats provide some benefits, especially for the elderly, by (1) providing storage for excess calories and protection in acute illnesses, (2) protecting vital organs and bones from injuries due to falls, and (3) maintaining core body temperature. Body fat distribution has been categorized in the so-called low to high "waist-to-hip ratio." The low ratio (a "pear shape" distribution of fat) may protect from some of the complications associated with obesity, whereas the high ratio (an "apple shape" distribution of fat) is associated with an increased risk of hypertension, late onset diabetes mellitus, coronary heart disease, and premature death.[17]

Elderly individuals with pathological obesity (BMI greater than 35) are few, because they die before reaching advanced age.* Less severe obesity is relatively more frequent and associated with less severe complications. In elderly men, weight gain generally comes from a decrease in physical activity, whereas in elderly women, it is associated with a loss of estrogens; in both men and women, it may be associated with a reduced level of growth hormone.[18] Treating obesity is a less important problem in the elderly than in younger individuals. As discussed in the following section, an appropriate exercise program is often all that is needed to maintain a healthy weight. A 1 mile walk burns about 100 kcal, and a walk (even in a mall) of 2 to 3 miles, four times a week, if accompanied by a slight restriction in caloric intake, should result in a gradual weight loss. If osteoarthritis (Chapter 21) is a problem, an upper body exercise is recommended. Elderly patients on weight reduction diets are at a significant risk of developing protein deficiencies, and thus blood albumin and pre-albumin levels should be monitored.[19]

### B. Dietary Recommendations

The daily intake of nutrients as recommended by the Food and Nutrition Board of the National Academy of Science[20] does not differ much between younger and older individuals. The age-associated physiologic and metabolic changes occurring in the elderly from age 51 to over 80 show only minor variations in nutrient requirements. The age groups charted were 51 to 70 years, 71 to 80 years, and over 81 years.[21] For all three groups, the emphasis was on achieving a balance between exercise activity and the intake of calories and nutrients. Since 1990, the so-called Food Guide Pyramid has been regarded as a valid tool. The original pyramid was modified in 1995 to include abundant cereals, vegetables, and fruits together with olive oil (Figure 24.1); it was termed the "Mediterranean diet" by Keys,[22] following his observations of dietary patterns, especially in relation to cardiovascular mortality, in 16 southern European populations from countries bordering the Mediterranean Sea. The low rates of cardiovascular deaths in these populations led Keys to conclude that their typical dietary pattern might explain the health benefits observed. The great emphasis on cereals, fruits, and vegetables is reflected in the present nationwide campaign to promote the daily consumption of at least five servings of fruit and vegetables. Current recommendations emphasize that a proper diet along with exercise are key to controlling weight and maintaining optimal health (Table 24.2).[23]

In terms of energy, the Recommended Dietary Allowance (RDA) lists:

- For individuals aged 51 to 70 years, a 2300 kcal diet for a 75 kg man, and 1900 kcal for a 56 kg woman
- For individuals aged 75 years and older, a 2300 kcal for men and 1700 kcal for women

---

* Obesity = BMI > 30; High-Risk Obesity = BMI >35; Morbid Obesity = BMI >40.

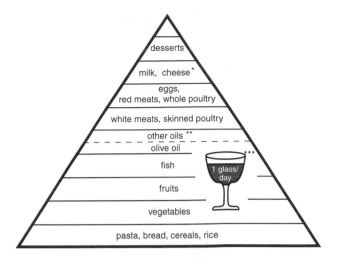

**FIGURE 24.1** Adaptation of the Food-Guide Pyramid for the Mediterranean Diet.[20] Note the two separate tiers for fruits and vegetables; the tier for fats consists primarily of olive oil and other oils. Note as well the presence of one glass of red wine per day. *Low priority due to presence of saturated fats (especially myristic acid); ** Unsaturated fats; *** May interact with a number of medications.

## TABLE 24.2
## Dietary Guidelines

1. Eat a variety of foods, keep foods safe to eat
2. Balance the food you eat with physical activity; achieve and maintain optimal weight
3. Select foods low in fat (especially saturated fats) and cholesterol, moderate in total fat
4. Include plenty of grain products, vegetables, and fruits
5. Choose beverages and foods to moderate intake of sugars
6. Choose and prepare foods with less salt, lower salt intake to less than 5 mg/day
7. If you drink alcoholic beverages, do so in moderation

*Source:* Adapted from Guidelines Advisory Committee, United States Department of Agriculture, *Dietary Guidelines for Americans,* 4th ed., U.S. Departments of Agriculture, Health, and Human Services, Washington, DC, 1995.

These estimates are approximate and do not specify the kind of activity performed by the individuals over 24 h. In view of the most recent Surgeon General's report on physical activity and health,[24] a slight upward revision of the RDA for the calories may be warranted.

### HISTORICAL NOTES

The ancient Egyptians, in addition to the more common kinds of cereals and beers for the masses, used to prepare special diets, set aside exclusively for the high priests and the Pharaohs in the temples of Thebes. These diets contained special ingredients including prepared extracts from leaves of lotus and papaverus, in which modest amounts of opium were present;[25] they were thought to ensure a much longer and happier life.

The word "cereal" is derived from the name of the Roman goddess Ceres, sister of Jupiter and patroness of grains in general. A temple to Ceres was built in Rome in the fourth century B.C. and given in custody to the Ediles, or elected individuals with the mandate to provide Rome with adequate quantities of cereals. In statues and frescos, Ceres is always represented with a crown of wheat on her head and a sheaf of wheat in her arms.

In the fifth century B.C., Hippocrates of Coos, proclaimed the benefits of barley, in any form, as conducive to a healthy and long life.[25] Galenus of Pergamon, personal physician to the emperor Marcus Aurelius in imperial Rome (second century A.D.), was an advocate of the frequent use of onions and garlic, ingredients once again recommended for a longer and happier life.[25] The benefits of garlic are touted even today, although, pharmacologically, no specific effect has ever been verified.

In the middle ages, a host of recipes were prepared by "witches" promising to provide humans with eternal youth — especially the rejuvenation of somnolent sexual organs to restore virility.[26] Even as the Inquisition disposed of more than one million individuals, most of them old women, these special diets and potions for the nobility continued; witchery was apparently well protected. Madame de Montespan, a favorite of King Louis XIV of France, "le roi soleil," used to prepare special dinners for her royal lover, dinners in which "cantaridin" was a prime ingredient.[26] Cantaridin prepared from the drying of the insect cantaris, induces congestion in the genitals. Although it facilitates penile erections, the cost in terms of renal toxicity and gastrointestinal problems is significant. Ingestion of this substance might well explain the chronic gastrointestinal upset of the poor king.

## 1.   The Macronutrients: Proteins, Carbohydrates, Fibers, and Fats

### a.   Proteins

The RDA for protein is about 0.8 g/kg body weight/day, which corresponds to 63 g/day for a 75 kg adult man on 2300 kcal/day, and 50 g/day for a 56 kg adult woman on 1900 kcal/day. In the elderly, protein deficiency may be due to impaired utilization, lower metabolic demand, lower caloric intake, and, possibly, lower protein requirements.[27,28] Protein deficiency is associated with lack of energy, weakness, decreased muscular strength, cognitive dysfunction, and depression (sometimes indicating the onset of anemia and hypothyroidism). Although these changes contribute to the loss of muscle mass and strength in this age group, studies (described below) have found that when the elderly participate in regular resistance training exercise, their protein synthesis rates are restored to a level similar to that found in younger individuals.

In terms of percentages, the amounts of protein present in foods typically consumed in the United States today are: 0.5% in cooked vegetables, 4% in low-fat milk,

**Prevention and Rehabilitation**

10% in uncooked pasta, and 23% in meat, An impressive 40 grams of protein are present in each of the following:

200 g fish or poultry or red meat
400 g cooked rice
150 g tree nuts
470 g cooked beans
130 g peanuts
500 g tofu
200 g cheese
1 L milk
12 egg whites

### b.  Carbohydrates and Fiber

Complex carbohydrates should represent 55 to 60% of the total caloric intake in a well-balanced diet. This amount will also provide about 25 to 30g of daily dietary fiber. Present advice from nutritionists is to limit the intake of refined sugars, but the consumption of starch is still recommended. The portion of starch that may escape complete digestion in the upper intestinal tract, whether as a consequence of insufficient mastication or rapid transit through the small intestine, is converted to butyric acid, which, in fact, seems to improve colon health and may even add some protection against colon cancers.[29]

Dietary fibers are of two classes, soluble and insoluble. The soluble fibers, e.g., pectin, mucillagen, and gums, form a sort of gel matrix that slows intestinal absorption but is later metabolized in the large intestine by bacteria, giving rise to short-chain fatty acids. Fruits, vegetables, and legumes are an excellent source of soluble fiber. The insoluble fibers, e.g., cellulose, hemicellulose, and lignin, have strong water-binding capacity and are found in wheat-bran, whole-grain breads, and cereals and also in the skin of fruits and vegetables. Because of their water-binding characteristic, they tend to facilitate intestinal motility. In elderly individuals, high dietary fiber may interfere with the intestinal absorption of some drugs, such as the cardiac medication, digitalis (Chapter 23).

### c.  Fats

The current recommendation regarding the intake of dietary fats in the diet is that they should be limited to 30% or less of the total energy requirement; the total cholesterol intake should be no more than 300 mg/day. A useful guideline is that a single egg yolk contains about 300 mg of cholesterol. Apart from cholesterol, fats should be essentially unsaturated, such as the monounsaturated fatty acids in olive oil.

In milk and milk derivatives, because most of the fatty acids are saturated (such as the highly represented myristic acid), low-fat or nonfat milk and milk derivatives are generally recommended. Neither milk nor cheese should be banned altogether because they are an invaluable source of calcium, proteins, and vitamins.

Definite benefits seem to be found in the oils of cold-water fish, the omega 3-polyunsaturated fatty acids. Studies have indicated that in high amounts, they decrease triglyceride levels and perhaps increase HDL levels[30] (Chapter 17). In addition, they may have antiarrhythmic properties and may reduce the risk of sudden death after myocardial infarction.[31]

The campaign against saturated fats, specifically against cholesterol, started 20 years ago and has unquestionably been effective. Reduced levels of serum cholesterol have coincided with a 43% decline in death rate from all cardiovascular diseases (Chapters 16 and 17). The current goal of the National Cholesterol Education Program is to bring the mean blood cholesterol concentration to well below 200 mg/dL.[32]

## 2.  The Micronutrients: Vitamins and Minerals

In the most recent RDA,[23] recommended vitamin intake does not differ between younger and older adults, with the exception of vitamin $B_6$ up from 1.3 to 1.7 mg/day. However, mild vitamin deficiency, with or without trace mineral deficiency, is common among the elderly, particularly in frail and institutionalized individuals, and is accompanied by cognitive impairment, poor wound healing, anemia, and increased propensity for developing infections.[33]

The advisability of routine vitamin and mineral supplements in healthy older people remains controversial. A diet including at least five to six servings of fruits and vegetables usually contains sufficient vitamins and minerals, as well as useful phytochemicals, to satisfy the necessary requirements. Supplementation is usually required in cases of less healthful diets, especially to maintain a normal immune status. There is a large and increasing body of literature on the possible benefits of vitamin and mineral supplements, although many of these purported benefits have not been substantiated by rigorous clinical study. For presentation here, we have chosen only those vitamins and minerals with well-established properties.

### a.  The Liposoluble Vitamins

A list of liposoluble vitamins, selected major functions, and effects of their deficiency or excess is presented in Table 24.3.

### b.  The Hydrosoluble Vitamins

The B-complex vitamins (so called because they were originally identified in beer yeast) act as coenzymes for various metabolic reactions and are needed for cell replication and hematopoiesis (Chapter 18). A list of hydrosoluble vitamins, their selected major functions, and effects of their deficiency or excess is presented in Table 24.4.

**TABLE 24.3**
**Selected Characteristics of Liposoluble Vitamins Important for Maintaining Health After 60[a]**

| Vitamin A (retinol) | Vitamin D | Vitamin K | Vitamin E (alpha-tocopherol) |
|---|---|---|---|
| Required for cell development and visual pigment production | Increases intestinal calcium absorption and bone calcium metabolism (Chapter 21) | Essential for blood coagulation (activates prothrombin and VII, IX, X coagulation factors) | Antioxidant (Chapter 5) |
| DRI: 700–900 microgram (retinol equivalents 2500–3000 IU) | May also protect against cancer, especially of prostate[35] | DRI: 90–120 microgram/day | DRI: 15 milligram/day; doses higher than 800 IU in dietary supplements should be avoided (possible anticoagulant effect) |
| True deficiency (rare) may lead to night blindness | DRI: 800 IU (200 IU = 5 microgram) | Abundant in leafy green vegetables | Present in vegetable oils and derived products (margarine, mayonnaise) |
| Abundant in fruits and vegetables | Fortified dairy products rich in vitamin D (30 mL fortified milk contains 100 IU of vitamin D) | Deficiency may occur in the elderly only under specific conditions: poor diet in intensive care units, recent surgery, uremia, and multiple antibiotic therapy | Risk of deficit is rare in elderly but may become real in rigorous low-fat diets |
| Toxicity affects liver and skin | Also produced in skin under ultraviolet irradiation (sun exposure) | | Protection of immune function and against neurodegenerative disease is controversial |
| Not recommended are beta-carotene supplements structurally related to vitamin A, because they can color the skin yellow-orange and have been related to cancer risk[34] | Sun exposure may increase vitamin D in nursing home residents (however, there is a risk of skin photoaging) (Chapter 22) | | |
| | Toxicity after doses higher than 1500 IU due to high blood and urine calcium levels[36] | | |

[a] Superscripts refer to number in References.
[b] DRI = Dietary Reference Intake.

**Prevention and Rehabilitation**

**TABLE 24.4**
**Selected Characteristics of Hydrosoluble Vitamins Important for Maintaining Health After 60[a]**

| Thiamine (Vitamin B$_1$) | Riboflavin (Vitamin B$_2$) | Pyridoxin (Vitamin B$_6$) | Folic Acid (folate) | Vitamin B$_{12}$ | Vitamin C |
|---|---|---|---|---|---|
| Cofactor in decarboxylation (in metabolic reactions) | Constituent of special proteins (flavoproteins) | Cofactor in several neurotransmitter synthesis | Coenzyme in methylating reactions | Coenzyme in amino acid metabolism and stimulates erythropoiesis | Necessary for collagen synthesis, strong reducing agent with antioxidant properties |
| DRI: 1.2 mg/day | DRI: 1.3 mg/day | DRI: 1.7 mg/day | DRI: 400 micrgrams/day | DRI: daily 200 microgram/mL | DRI: 40 mg/day |
| Symptoms are vague (anorexia, weight loss, confusion, apathy, weakness) | Deficiency rare but may occur in the elderly with chronic inflammation of the mouth (typical cracking at the oral commissures) | Deficiency associated with peripheral neuropathies, convulsions, impaired immune function[37] | Decreased intestinal absorption may cause deficiency in the elderly[38] | Deficiency, frequent in elderly vegetarians or with gastric and intestinal disturbances (Chapter 20). Consequences include anemia and neurologic abnormalities[40,41] | For those bruising easily, increase daily intake to 500–1000 mg/day (doses above this level remain controversial[43]) |
| Supplementation (100 mg/day) is recommended in persons affected by chronic alcoholism to prevent neuropsychiatric symptoms | May be indicated in treatment of migraine headaches[36] | | Deficiency as induced by anti-bacterial, anti-inflammatory, antiacid drugs or excessive alcohol and smoking; associated with developmental defects of neural tubes and nervous disorders | Supplementation to restore normal levels (2.4 mg/day by injection or 1000–2000 mg/day orally)[42] | Vit C increases iron absorption and high levels of serum iron may represent a risk factor for cardiovascular diseases, especially in individual carriers of the hemochromatosis gene[36] |
| | | | By association with increased blood levels of homocysteine, a risk factor for coronary heart disease (Chapters 16 and 17) | | |
| | | | Supplements may be used as adjunct treatment for depression and Alzheimer's disease[39] | | |

[a] Superscripts refer to number in References.
[b] DRI = Dietary Reference Index.

**TABLE 24.5**
**Selected Characteristics of Minerals Important for Maintaining Health After 60[a]**

| Calcium | Iron | Selenium | Zinc |
|---|---|---|---|
| Many important functions (Ch. 21) | Necessary for hemoglobin formation, respiration, hematopoiesis | Plasma concentration decreases progressively with age | DRI: 11 mg/day for men, 8 mg/day for women |
| DRI[b] over 51 years: 1200 mg/day<br>DRI over 64 years: 1500 mg/day[44] | DRI: 8 mg/day and does not change with age<br>Moderate deficiency can occur in undernourished individuals | DRI: 55 mg/day | Deficiency implicated in skin ulcers, depressed immune responses, impaired body growth |
| In the elderly, calcium intake should be increased to compensate for reduced intestinal absorption | High levels associated with higher incidence of coronary heart disease and oxidative damage[46,47] (results are still controversial and do not yet warrant changes in current DRI) | As an essential component of the antioxidant defenses (Chapter 5), supplementation with selenium has been recommended for its antioxidant properties[48] | No evidence of true zinc efficiency in the elderly, although some anecdotal reports suggest an association with wound healing, appetite, and possibly CNS and immune function[49] |
| Urinary calcium excretion is increased by high intake of protein, sodium, caffeine, foods rich in phosphorus (Including meat) Example: 6 mg of calcium is lost through urine with each cup of brewed coffee[45] | | | |

[a] Superscripts refer to number in References.
[b] DRI = Dietary Reference Index.

*c. The Minerals*

A list of minerals, their selected major functions, and effects of their deficiency or excess is presented in Table 24.5.

## 3. Electrolytes and Water

### a. Sodium and Potassium

Although sodium is present in a multitude of substances, unquestionably the main source of this important electrolyte is table salt (sodium chloride). A diet completely deprived of sodium would induce insufficiency of the adrenal glands (Chapter 10). Hence, in areas of the world where salt is scarce, it is considered as valuable as money. Indeed, the term "salary" originates from the Latin *salarium*, that is, allowance given to the Roman soldiers for salt, hence any allowance or pay. Salt was frequently used by the soldiers of the Roman legions in Africa or Asia, as barter in transactions with local people.

In the current American diet with its overreliance on canned foods, fast foods, restaurant food, and the frequent "salt shaking habit," researchers have determined that our average intake of salt is 10 mg/day. The National Heart, Lung and Blood Institute as well as the National High Blood Pressure Education Program have recommended that we all, young and old, reduce salt intake to less than 5 mg/day. In those affected by hypertension (especially in sodium-sensitive individuals), salt intake should be further reduced to less than 4 mg/day;[50] fortunately, the taste for sodium diminishes after a few months of restriction, and such a goal is not difficult to attain.

Potassium is abundantly present in fruits and vegetables, especially those with bright colors. In the elderly, the frequent use of diuretics (prescribed for various medical conditions) may lead to a deficit; in these individuals, laboratory tests of serum potassium levels may suggest the need for supplementation. Hyperkalemia (high blood levels of potassium), may also occur in the elderly due to declining levels of aldosterone and impairment of renal function (Chapter 10).

### b. The Need for Water

How much water should we drink daily? Advice continues to be six to eight glasses a day. One check for dehydration is measuring the concentration of sodium in the urine; more than 20 mEq/L is a sign of water depletion. A practical recommendation is to drink a full glass of water any time the urine color is yellow and to maintain almost colorless urine. If following this advice means the annoyance of taking frequent trips to the bathroom, it is an annoyance that must be borne; sufficient intake of water is vital in the young and the old. Elderly people need to pay special attention to their water intake, because, with age, there is a diminished sense of thirst and impaired renal concentrating ability (Chapter 19). Dehydration is one of the common causes of fluid and electrolyte imbalance in the elderly.

## ADDICTING SUBSTANCES

Alcohol has been with us since prehistory. Beer was the first alcoholic beverage ever reported in written documents. Since the year 3000 B.C., the Nile River was literally lined with beer houses. Though alcohol in moderate amounts may indeed be beneficial on an individual basis, those benefits would seem to be outweighed by the statistics that alcohol contributes annually, in the United States alone, to more than 100,000 deaths.[51] To be added are the many other social, economic, and health-related costs associated with alcohol use and abuse.

Given the socioeconomic conditions of many elderly individuals, it is understandable that they may be tempted to fight their loneliness and depression with a "drink or two." The impairment of judgment that accompanies alcohol use is reason enough to discourage the elderly from developing such a habit. On the other hand, if we look at the benefits of moderate drinking, we find evidence that alcohol may offer some protection against diseases of the cardiovascular system. Even patients dying of alcoholic liver cirrhosis may present on postmortem examinations, surprisingly elastic aortas. HDL (the "good cholesterol"), is elevated in alcoholics, and this high value may give patients a false sense of security and lead them to indiscriminate drinking. The debate continues, but recent studies conclude that a modest intake of alcoholic beverages — translated into "no more than one drink a day" and preferably wine — may be of some benefit.[52] Still, the inappropriate interaction between alcohol and medications must always be kept in mind.

*Coffee and its surrogates (chocolate)* — Coffee and its component, caffeine, an alkaloid with a stimulatory action on the brain, make an extremely pleasant beverage but a remarkably mediocre food stuff. In a cup of American coffee, 0.04–0.05 g caffeine is present; in the very small cup of Italian espresso, there is about 0.02g of caffeine. While coffee and caffeine may at times be toxic for children, they are usually safe for adults and elderly individuals. As most are aware, coffee is not recommended in individuals affected by insomnia, but it also should definitely not be drunk by elderly people with a tendency to disturbances of cardiac rhythm (tachycardia).

In those elderly affected by orthostatic hypotension, i.e., a tendency to experience a drop in blood pressure upon moving from supine to erect position, one or, better, two cups of brewed coffee may be an effective aid that they can safely use.[53]

Cocoa, like caffeine, provides a characteristic excitation, which is due to the presence of theobromine, a substance similar to, but less excitatory, weaker than caffeine. Chocolate or cocoa may induce cardiac arrythymias, and its chronic use may cause loss of appetite.

*Tea* — A Chinese text, the *Pent-S-Ag*, dated 2700 B.C. provides a description of tea as a tonic beverage. Its active ingredients are Teine, similar to caffeine, and teophylline, the latter in minimal amounts, fortunately; teophylline has powerful pharmacological effects on both the cardiovascular and the respiratory systems. The presence of tannin, a derivative of tannic acid, in tea is responsible for the astringent effect of the beverage. Coffee stimulates gastrointestinal motility, unlike tea, which, if it acts at all at this level, would have an inhibitory effect.

In summary, certainly caffeine and teine are both "drugs" capable of inducing dependency. Indeed, caffeine is currently the most widespread drug used the world over. However, genetic, social, and neurological factors render caffeine dependency far less onerous than that associated with its more feared cousins: alcohol, nicotine, cocaine, and opium derivatives.[54]

## 4. Dietary Supplementation

Definitive proof is still lacking that dietary supplements of multivitamins and minerals are appropriate or beneficial for the elderly. They are well advised for all those on a limited caloric diet in which key nutrients might be absent or deficient. Calcium and vitamin D supplements may also be advisable.[55] Medical recommendations for specific supplements such as folate in individuals with elevated serum homocysteine levels are to be followed conscientiously. B$_{12}$ supplementation is often recommended, especially in individuals with low gastric hydrochloric acid content. Assuming a healthy and varied diet, most needed supplements are likely to be available in a single daily pill, which should not contain more than 100% of the recommended daily allowance for the respective vitamins and minerals.[55] In general, for the elderly, the number of pills should be kept to the necessary minimum to avoid the confusion created by a multiplicity of pills on multiple schedules.

As difficult as it may be for the elderly, so many of whom live alone, eating with others is highly recommended for its obvious social benefits as well as for the fact that seeing others eat stimulates the appetite and may facilitate digestion. *Companatico e compagnia*, meaning bread and whatever goes with it, including company, is an old Italian saying that remains valid everywhere.

## C. Neuroendocrinological Control of Food Intake

Food intake is regulated not only by the availability of food but also by the factors that regulate appetite and satiety. While the mechanisms of feeding behavior are not completely known, it is accepted that they involve nervous and endocrine controls. The major CNS regulatory centers are in the hypothalamus, and the major hormones involved are those from the gastrointestinal tract and from adipose cells (Table 24.6). In the hypothalamus, the actual drive for food seems to be under the control of a powerful neuro-opioid peptide, dynein (also known as neuropeptide Y, NPY), which stimulates appetite and decreases with age in some individuals. The sense of satiety is stimulated by the polypeptide hormone cholecystokinin secreted by the cells of the upper small intestinal mucosa. The neurotransmitter nitric oxide (NO) is secreted by the gastric mucosa cells and induces relaxation of the stomach fundus; in many elderly persons, its secretion may be deficient, and the

## TABLE 24.6
## Endocrine Control of Food Intake and Utilization

Dynein: neuro-opioid; controls the drive for food

Cholecystokinin: intestinal peptide; increases the sense of satiety

Nitric Oxide (NO): neurotransmitter; increases stomach relaxation

Leptins: peptides produced by adipose cells; inhibit appetite, decrease food intake, and increase energy metabolism

Cytokines (TNF, IL-2, IL-3): proteins produced by lymphocytes (mainly T cells) and other immune cells; decrease food intake

Activins: peptides of gonadal origin with receptors in many tissues including brain stem; decrease appetite (in mice, induce wasting syndrome)

resulting absence of stomach relaxation may induce a false sense of satiety and fullness even after minimal food intake.

The leptins, produced and secreted by fat cells, inhibit food intake and stimulate energy metabolism. High levels of leptin tend to decrease body fat. After age 70, leptins decrease in elderly women; in men, leptins tend to increase slightly in parallel with the decrease in testosterone levels. The cytokines, produced by macrophages and lymphocytes, have important regulatory immune function (Chapter 15) and also tend to decrease food intake. In elderly individuals and in situations of stress, cytokines (especially TNF, IL-2, IL-3) (Chapter 15) are elevated. The peptide activins secreted by the testes and ovaries have receptors in many tissues, including the brain (specifically the midbrain) where they may act as neurotransmitters and modulators of appetite and satiety signals; in mice, they decrease food intake and produce a so-called "wasting syndrome." Activins increase in older men but not in women.[56]

Eating disorders are classically considered illnesses of adolescence and young adulthood. However, a small percentage of cases have a late onset, occurring between the ages of 40 and 75 years (average age at onset 56 years and clinical presentation at 60 years).[57] These cases comprise almost exclusively women of diverse ethnicity. They present all the characteristic symptoms of anorexia nervosa and bulimia nervosa: self-induction of starvation or binge eating, morbid fear of being fat, and the denial of the seriousness of their low body weight. As is common in eating disorders of younger individuals, many of the elderly cases have psychiatric components, especially affective disorders and perfectionism. As these elderly women are postmenopausal, alteration of the hypothalamo–pituitary–ovarian axis is not detectable: increased levels of FSH and LSH characteristic of the postmenopausal profile are present and similar to those of unaffected women (Chapter 11). The treatment of these late-onset cases is similar to that of younger subjects, but increased morbidity and the presence of previous psychological disorders may complicate treatment.[57]

**Prevention and Rehabilitation**

## III. EXERCISE AND PHYSICAL FITNESS FOR THE AGED

Research on physical activity and its relation to physiologic aging remains very active today. Exercise facilitates maintenance of sound cardiovascular function and lessens many risk factors associated with heart disease, diabetes, insulin resistance, and some cancers.[58] In addition, other losses associated with aging and once thought to be inevitable (e.g., loss of muscle mass and strength, bone density, and postural stability) can be effectively counteracted through regular exercise.[58] Concurrent psychological benefits of exercise include preservation of cognitive function and self-sufficiency as well as reduced episodes of depression. A regular program of physical activity, therefore, is extremely beneficial, whether one begins early in life, in adulthood, or after 60; indeed, the consequences of inactivity are being increasingly chronicled (Chapter 3). A powerful endorsement of the role of physical exercise in promoting a healthy life span was provided by the 1996 Surgeon General's Report on Physical Activity and Health.[58]

### HISTORICAL NOTES

The recognition that physical activity can improve one's functional state and sense of well being is neither new nor restricted to the Western hemisphere. As early as the year 3000 B.C., the Chinese classic "Yellow Emperor's Book of Internal Medicine" described the principle that "harmony" with the surrounding world was the key to prevention of disease and aging. This and other principles grew into concepts that became central to sixth-century Chinese philosophy, Taoism, where longevity was to be obtained by living simply. "Tai chi chuan" is a system of meditative movements practiced as a daily exercise. It is deeply embedded in that culture, which teaches graceful movements encoded by Hua T'o as early as 200 B.C. Tai Chi is widely practiced today in this country and elsewhere specifically to ward off the incidence of falls among the elderly.

In Greece, Hippocrates, in his second book on hygiene (400 B.C.) wrote, "Only eating will not keep a man well; he must also take exercise."[1] Hippocrates's influence extended to Rome, where Galen, in his book on hygiene (second century) wrote "The uses of exercises are twofold: one for the evacuation of excrements, the other for the production of good condition of the firm parts of the body."[25]

In the United States, it was President Dwight Eisenhower in 1956 who, to promote national fitness, established the President's Conference on the Fitness of American Youth. Later, renamed by President John F. Kennedy, the intent remained to promote physical activity and sports for Americans of all ages.

Over the past two decades, experts from numerous disciplines have determined that exercise substantially enhances physical performance, and their recommendations are being constantly refined and monitored by organizations such as the American College of Sports Medicine, the American Heart Association, and others.

### A. CHANGES IN MUSCLE STRUCTURE AND FUNCTION WITH OLD AGE

Sarcopenia (from the Greek, meaning "poverty of flesh") denotes the decline in muscle mass and strength that occurs with old age (Chapter 21). Although it occurs universally, the age of onset and severity of decline vary with the individual. This decrease differs from other muscle changes that occur as a result of an individual's lack of physical activity or coexistent with acute or chronic diseases.[59] The exact causes of sarcopenia are still unclear, but its occurrence has been ascribed to a number of factors summarized in Table 24.7.

In healthy young individuals, muscle tissue represents 30 to 40% of body weight; adipose tissue, 20%; and bone, 10%. Muscle mass accounts for 50% of lean body mass and 50% of total body nitrogen. By age 75, muscular mass has decreased to less than 15%, adipose tissue has increased to 40%, and bone has decreased to 8%.[59]

This aging-related decrease in muscular mass is due primarily to actual loss of muscle fibers. Prior to their disappearance by apoptosis, individual fibers in the aged change shape and appear to be more rounded (even banana-shaped) than those of the young. These changes in morphology affect all fibers. Of the two major types, the slower-contracting Type I fiber (e.g., long muscles of the back) and the faster-contracting Type II fiber (e.g., muscles of the hands), the most affected is the Type II fiber, which is significantly reduced in number among the elderly. The mechanism underlying this selective loss has been ascribed to decreased innervation at the myoneural junction, leading to fiber death.[59] The greater atrophy and death of Type II fibers creates a relative predominance of Type I fibers, and explains why a 10-year-old boy will outrun his 70-year-old grandfather in a 100-m race, whereas grandpa might still defeat junior in a 10-km march.

In addition to Type I and Type II fibers, muscle fibers also include "hybrid" fibers, which consist of both fast and slow myosin isoforms. These hybrid fibers predominate in old muscle, and, in the elderly, one-third of the

---

**TABLE 24.7**
**Major Factors Responsible for Sarcopenia in the Elderly**

1. Inactivity
2. Changes in central and peripheral nervous systems
3. Reduced protein synthesis
4. Reduced protein intake
5. Reduced blood supply
6. Reduced mitochondrial mass in muscle

muscle fibers are neither strictly slow nor fast but something in between (hence, the term "hybrid").[60] The preponderance of hybrid fibers together with the decrements in the supply of motor nerves at the myoneural junction explain why aging muscle is less excitable and has a longer refractory period; that is, the muscles demand greater stimulus for contraction and longer time to respond to stimulation. One of the responses to regular exercise training is growth of hybrid fibers.

Another important group of muscle cells is the "satellite cells," so-called because of their peripheral location in muscle. They are considered to be usually dormant "adult" myoblasts that do not participate in contractile activity. However, under conditions of muscle injury or damage, they migrate to the affected areas, undergo maturation, and develop into new muscle fibers. Thus, satellite cells play a critical role in exercise-induced muscle growth and regeneration, and even in old age, they may support muscle growth.[59]

## MUSCLE ENERGY SOURCES AND METABOLISM

Muscle contraction requires energy to convert chemical energy into mechanical work.[61] Overall, the metabolic pathways and energy resources used by muscle tissue for contraction are similar in the old as in the young individual. However, progressive decrements in motor nerve inputs at the myoneural junctions (Chapter 21), less available blood supply due to atherosclerosis (Chapter 16), slower metabolic adjustments to the demand on the muscle fibers for maximal effort, and the greater difficulty involved in activating bones and joints (due to arthritis, osteoporosis, etc., Chapter 21) all contribute to rendering the older individual less capable of meeting the increased energy requirements of the exercising muscle.

Muscles at rest or at minimal level of activity derive most of their energy from fats (free fatty acids) and somewhat less from carbohydrates. As one's exercise level increases in intensity, the maximal energy source shifts from fats to carbohydrates; glucose utilization increases rapidly, while utilization of fats decreases.[61,62] This phenomenon has been described as the cross over concept[59] (Figure 24.2). Once physical effort ceases, fats reestablish their prominent role as the primary source of energy for the muscle.[59] The lactate produced by glycolysis during exercise was formerly considered toxic, and this toxicity was regarded as the main reason for muscle fatigue and pain. Current studies have instead revealed lactate to be an excellent source of energy to be used by the same cells where it is produced or by other cells through the "intra- and extracellular shuttle mechanism."[59] Lactate, in small amounts, may be redirected to replenish impoverished glycogen stores or used as an additional source of energy through oxidation in mitochondria.[59] Only in prolonged and intensive exercise are fatty acids, eventually in combination with proteins, again used as energy sources.[61,62] Currently, this "intracellular and extracellular shuttle mechanism" allows lactate to be used to produce more energy.[59,61,62] Additional information on muscle metabolism is presented in Chapter 21. The role of exercise in muscle metabolism and its relation to insulin action in preventing and treating late onset diabetes appears in Chapter 14 and below.

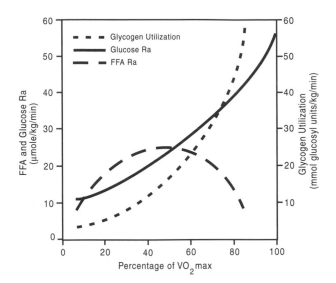

**FIGURE 24.2** *The Cross-Over Concept.* Utilization of free fatty acids (FFA) decreases, and glucose/glycogen increases with increased intensity of exercise. $VO_2$ Max = maximum oxygen consumption taken as an index of energy utilization; Ra = rate of appearance. Adapted from Reference 59. Not shown here: when exercise is prolonged, FFA appearance and utilization increase again due to the progressive depletion of glycogen stores.

## B. MUSCLE RESPONSIVENESS TO EXERCISE IN OLD AGE

Current exercise programs for the elderly involve "endurance" and "resistance" types of activities.[63,64] While *resistance training* may be the most effective in increasing muscle mass and strength, there were, originally, concerns, no longer substantiated, about the danger of a concomitant increase in blood pressure. In contrast, *endurance training*, even when relatively strenuous, yields only a small increase in the percentage of Type II fibers, which is probably associated with increased vascularization from the aerobic activity.[63]

From reports of longitudinal studies of resistance training for the elderly, the consensus is that this form of exercise promotes substantial gains in muscle strength, even among the very old.[64–67] For example, men and women ages 61 to 78 years performing a program of 12 resistance exercises, three times a week for 50 weeks, showed significant increase in strength as early as 8 weeks into their exercise program. In another study, a group of older subjects (72 to 98 years of age) during a 10-week program of progressive resistance training, showed a 113% increase in muscle strength. Other significant gains were also noted in gait velocity (improved by 11.8%) and stair climbing power (improved by 24.8%). Nutritional supplements were provided in some of these studies, but their contribution to these findings is difficult to evaluate.[67]

**Prevention and Rehabilitation**

Even the frail elderly can benefit from endurance and resistance exercise programs, improving their gait, balance, and overall functional ability. If the programs are pursued over a long period, lean muscle mass, bone, and connective tissue may also show some small increments of gain.[68–70]

## C. Metabolic Adaptation to Exercise by Muscle

Significant structural and metabolic responses in skeletal muscle function are associated with regular and protracted exercise training. They include (a) an increase in the number and size of mitochondria with a consequent increase in oxidative enzymatic activity,[62] (b) an increase in myoglobin content of muscle with possible increased oxygen storage in individual muscle fibers,[62,63] (c) an increase in the number of capillaries and muscle blood flow enhancing the oxidative capacity of the exercised and trained muscle, (d) an increased capacity of skeletal muscle to store glycogen, and (e) an increased mobilization of free fatty acids from fat deposits, with consequent increased activity of muscle enzymes responsible for fat oxidation (Table 24.8).[59]

## D. Hormonal Responses to Exercise in Old Age

At any age, exercise sends a number of signals, mostly beneficial ones that mobilize endocrine responses. Of the various hormonal responses to the moderate level of exercise practiced by the elderly, some of the most significant are those that regulate blood glucose (Table 24.8). During exercise, glycemia (blood glucose level) depends on multiple hormonal actions; insulin stimulates the entrance of glucose into most cells, but not in the working muscle, which utilizes glucose independently of insulin. Glucagon and the so-called "counterregulatory hormones" (e.g., gastrointestinal peptides, prostaglandins, growth hormone, epinephrine, cortisol) have an opposite (hyperglycemic) action that prevents the risk of hypoglycemia. In some older individuals, an imbalance of these mechanisms may induce or contribute to the onset of diabetes mellitus (Type II) and, vice versa, exercise may help restore normal glucose metabolism (Chapter 14). When heavy exercise is experienced as a stress, then the hypothalamo–pituitary–adrenocortical axis is likely to be stimulated, and high levels of glucocorticoids (e.g., cortisol) released; given the anti-insulin effects of glucocorticoids, this cortisol excess may contribute to the pathogenesis of Type II diabetes mellitus (Chapter 10).

In evaluating the significance and consequences of hormonal responses to exercise, it has been reported that metabolic changes are few, and endocrine responses are progressively dampened in young and adult individuals who have been appropriately trained. Whether this difference between trained and untrained is also the case in older individuals has not yet been documented.

---

**TABLE 24.8**
**Selected Muscular Responses and Hormonal Effects Induced by Exercise**

**Muscle Responses**
- Increase of mitochondria (number and size)
- Increase in muscle fiber myoglobin
- Increase in capillary network and circulation
- Increase in glycogen storage capacity
- Increase in free fatty acid utilization (only in prolonged, steady exercise)[a]

**Hormonal Effects**
- Insulin blood level decreases, muscle glucose uptake increases, while glucose blood levels remain unchanged
- Glucagon levels increases slightly, freeing glucose from liver (glycogenolysis stimulation)
- Epinephrine increases muscle glycogenolysis (only in intense exercise)[b]
- Growth hormone and Insulin Growth Factor-1 (IGF-1) may increase slightly and induce increased protein synthesis and lipolysis
- Cortisol increases only during intense, prolonged exercise and enhances proteolysis
- Amylin[c] may enhance glycogenolysis and lactate production

[a] Seldom in elderly.
[b] Moderate exercise has little or no effect on catecholamine levels.
[c] Still under study.

---

As presented in Chapter 10, repeated stress may create a serious challenge to the multiple "allostatic" mechanisms necessary for maintaining homeostasis, and thereby contribute to the etiology or aggravate the course of several so-called diseases of adaptation (Chapter 10). Among the beneficial effects noted below, a consistent exercise regimen may serve to reduce the deleterious consequences of stress and thus explain, at least in part, the beneficial effects of exercise as reflected in longevity and good health.

## E. Exercise and the Aged Heart

Overall, cardiac function changes little in old age, even when measured with a number of advanced techniques that test the heart at rest and under exercise (Chapter 21).[71–73] In the healthy elderly, these tests reveal, on average, only a modest increase in *diastolic and systolic stroke volume* in men, and less in women. The most significant cardiovascular change associated with old age is an increase, at rest, of the systolic blood pressure, together with some left ventricular hypertrophy and dilation of the aorta.[73] These changes are most likely due to arterial "stiffness" consequent to atherosclerosis (Chapter 16). During exercise, the fraction of cardiac blood ejected increases compared to the resting state, more in the young (about 90%) than in the elderly (about 55 to 50% and higher).

In fact, any ejection fraction below 55 to 50% at rest or during exercise is to be considered pathological.

In some individuals, sudden standing causes an abrupt fall in blood pressure, dizziness, dimness of vision, and even fainting. This condition, known as *orthostatic (postural) hypotension,* is quite frequent in the elderly (from 10 to 30% of older people show a drop in systolic pressure of 20 mm Hg or more with a rapid change in posture) and represents a strong contributing factor to falls (Chapter 8).

Heart rate is little altered during light exercise in the older person. During maximal effort, however, maximal cardiac output is reached at a lower heart rate than in younger persons. This decline in achievable cardiac rate has been ascribed to stiffness of the cardiac wall and a reduced chronotropic response to catecholamines (Chapter 21).

In some old persons, a decrease in blood oxygen saturation may be due to chronic chest diseases and decreased pulmonary diffusing capacity (Chapter 18). In exercising, the heart may find it difficult to pump the blood through the strongly contracting muscles: vascular resistance may be 30% higher in a 70-year-old individual than in a 35-year-old. Hence, older persons are at particular risk for this type of peripheral blood flow limitation. Insufficient blood flow to the exercising muscle depends, in part, on the greater demand of blood flow by the skin. The sweat rate and heat dissipation at the skin surface slows during exercise in older people with significant subcutaneous fat; at the same time, they experience a reduction in the evaporative heat loss and, hence, lower heat tolerance.

## F. Exercise Programs for the Aged

A large percentage (35 to 45%) of the U.S. elderly population practices some form of physical activity. However, only 20 to 25% undertake regular exercise — 30 min or more three times a week. Women are generally less active in this respect than men. Low income and poor education appear to be predictors of inactivity. It is well recognized that regular exercise not only enhances the quality of life but also decreases mortality rates by 20 to 50% and increases overall life expectancy an average of two years.[58] Well-established and probable benefits brought about by exercise are summarized in Table 24.9.

### 1. Benefits of Exercise

The benefits of exercise by far exceed the risks. Regular physical activity (a) reduces mortality rates even in smokers and obese persons, (b) preserves muscle strength, (c) improves aerobic capacity, (d) slows bone density loss, and (e) may help with weight loss (when combined with reduced caloric intake). Indirect benefits include those associated with social interaction, an enhanced sense of well being, and improved quality of sleep.

---

**TABLE 24.9**
**Major Benefits Obtained with Exercise**

**Definite Benefits**
1. Metabolic: Increase in insulin sensitivity and glucose tolerance
2. Cardiovascular: Stimulation of vasodilation, reduction of blood pressure, and occurrence of arrythmias
3. Cerebral: Increase in blood supply to the brain and reduction in episodes of thrombosis
4. Improvement of balance, hence fewer falls
5. Lower mortality rate
6. Higher HDL/LDL ratio
7. Diminished severity of osteoporosis

**Possible Benefits**
8. Prevention of osteoporosis; gender differences (better prevention in males than in females); better results with associated administration of calcium and vitamin D
9. Prevention of colon cancer (probably due to faster intestinal transit time)
10. Mood changes (variable; more significant in males than in females)

---

In addition to improved physical functioning, regular exercise in the elderly increases insulin sensitivity and glucose tolerance (Chapter 14), reduces blood pressure, and normalizes blood lipid levels by reducing triglycerides and increasing HDL levels (Chapter 17). Exercise helps to prevent cardiovascular diseases, diabetes, osteoporosis, obesity-related diseases, colon cancer, and, possibly, pancreatic cancer. It also lowers the incidence and severity of falls, prevents or relieves depression, and may improve cognition.

### 2. Adverse Effects of Exercise

The few adverse effects include the possibility of musculoskeletal injuries, metabolic abnormalities, and problems often seen in athletes but seldom reported in the elderly (Table 24.10).

#### Sudden Death

In young and middle-age athletes, sudden death is usually the consequence of congenital hypertrophic cardiomyopathy and often secondary to severe arrhythmia. It can be prevented by implantation of a defibrillator. In older runners or joggers who are not physically active in general, sudden death would be ascribed to Coronary Artery Disease (CAD) (Chapter 16); this complication is actually rather rare in the United States — one case per 15,000 joggers per year. Anaphylactic shock should also be suspected if joggers, or even walkers, are allergic to bee poison and collapse when stung.[74]

Despite these few risks, the benefits of exercise in older persons, in the absence of obvious contraindications such as illness or injury, strongly outweigh the possible potential but unlikely hazards. With proper screening,

**Prevention and Rehabilitation**

**TABLE 24.10**
**Major Adverse Effects of Exercise in the Elderly**

1. Musculoskeletal Injuries
   - Dislocation of shoulder
   - Achilles tendon tear
   - Intervertebral disk injury
   - Traumas due to repetitive motion (jogging)
   - Traumas due to falls or collisions
2. Metabolic Abnormalities (Consequent to Severe Exertion)
   - Hyperthermia
   - Electrolyte imbalance
   - Dehydration
   - Hypothermia (especially in water and winter sports)
3. Pathologies Occurring with Exercise (but Seldom Seen in the Elderly)
   - Hemolytic anemia and hemoglobinurias (in distance runners)
   - Hematurias (blood in urine), secondary to urinary bladder traumas
   - Muscle degeneration (rabdomyolysis) in weight lifters lifting very heavy weights

training, and supervision, the risk of injury or sudden death is less in older adults than in younger persons. Therefore, older individuals should strongly be encouraged to participate in regular physical activity.

### 3. Preexercise Screening

Even though exercise is contraindicated only in a relatively small fraction of the elderly population, all elderly individuals starting an exercise program should be screened by interview or questionnaire.[75] However, any individual with a history of cardiac risk factors (hypertension, obesity, diabetes, etc.) should undergo a full medical screening and functional fitness tests; they may also require medical supervision during exercise. For the ambulatory elderly in a walking program, the simple advice to gradually increase walking time is probably safe and does not require comprehensive medical screening. Similarly, a comprehensive medical screening is probably not needed for healthy patients of any age prior to starting an activity program of moderate intensity.

### G. PRACTICAL RECOMMENDATIONS

In general, regardless of age, physical exercise programs need to be followed on a regular basis and work best when the specific activity selected is incorporated into daily routines.[75–77] For the elderly, most senior centers offer invaluable, low-cost physical fitness classes, from mall walking, to tai chi, to Yoga, to low-impact aerobics, and more. In addition to specific information on endurance, resistance, and other types of exercises discussed below, general recommendations follow:

- The choice of exercise should be based more on how regularly it can be repeated than on the actual degree of effort it requires.
- The exercise period should be preceded by a period of warming (e.g., stretching) and followed by a corresponding period of cooling down (to prevent cardiovascular accidents).
- Racing against one's self should be discouraged.
- The success of exercise programs depends on their being enjoyable; if not, they will soon be abandoned.
- Always maintain an appropriate hydration.

### ENDURANCE EXERCISE

Endurance exercises consist of walking, cycling, dancing, swimming and all low-impact aerobics; these provide the best documented health benefits for the elderly. Walking is the most common exercise among the elderly in the United States and the most commonly recommended by physicians. Some statistics suggest that walking an average of 3.2 km/day (2 mi/day) may reduce mortality rate by about 50%.[58] It also reduces the risk of heart disease and helps to prevent falls.[58] Jogging is generally not recommended for the elderly unless they are already accustomed to this type of exercise. The maximum heart rate to be achieved during the exercise period, the so-called "target heart rate," may be monitored during exercise through pulse checks or by wearing a monitor. For a moderate-intensity exercise, the target should reach 60 to 79% of the maximum heart rate, i.e., 220 beats/min minus age in years. However, this formula is less reliable for the elderly; a simpler way is to rely on one's actual "perception of exertion." If you can talk without discomfort while continuing to exercise, you can assume that you are not overexerting. In a clinical setting, the health care worker can refer to a special table where you can identify your "Rate of Perceived Exertion" during exercise.

"Reconditioned" individuals, those who have interrupted their exercise programs with periods of inactivity, especially bed rest, should resume exercising at about half the intensity to which they were previously accustomed.

### RESISTANCE EXERCISE

The consensus today is that elderly people can and should perform some kind of resistance exercise at least twice a week, such as lifting a weight equivalent to 60 to 70% of the individual's maximal capacity, and then executing two sets of 10 repetitions on different exercise machines. With this routine, in 1 year, the gain in strength can reach 100% or more, and muscle hypertrophy may become apparent. Programs of this kind are highly recommended for individuals with clear evidence of sarcopenia. Even moderate-intensity programs using calisthenics or elastic tubing can increase strength by 10 to 20%. After initial training on free weights, these exercises can be performed at home after a period of warm-up. Weights can be gradually increased as strength improves. Starting with weights that are too high, or using them without professional guidance, can lead to injury, and the outstanding benefits to be achieved are lost.[67–76]

### Exercise for Maintaining Balance

Exercises such as tai chi (a sequence of movements originally used in the martial arts) are very effective in balance training. Walking with outstretched arms, crossed arms over the chest, and crossed arms holding weights are all effective for balance training.

### Flexibility Exercises

Stretching exercises are recommended for all age groups, but they are critical for older individuals who have not been on a regular fitness program. Many older individuals report that stretching alone makes them feel appreciably better. These exercises, some done standing, some lying on a mat on the floor, are recommended to precede and follow any endurance and resistance exercises. There is a sequence of stretches most trainers recommend. They are posted in chart form in gyms and fitness facilities. Each stretch is held for 10 to 30 sec and repeated three to five times per session.

## IV. EXPERIMENTAL STRATEGIES IN DIETARY RESTRICTION

The physiologic and pathologic changes associated with old age have been variously ascribed to (1) extrinsic or intrinsic factors that damage intra- or extracellular molecules, (2) programmed changes in gene expression, or (3) an interaction of the two (Chapters 1 through 6). Regardless of which hypothesis is correct, biogerontology research over the past 70 years has used caloric restriction as a key tool for testing causal theories about aging in animals.

As mentioned above, caloric restriction not only prolongs functional competence and postpones aging-related pathology, but it reliably extends the mean and maximum life span in several species, including mammals. Yet, despite years of intense research, the mechanism(s) of action underlying these effects and the applicability of caloric restriction to humans remain enigmatic. A summary of the methods that have been employed in these studies and the results obtained to date, are presented here to serve as a model for the effects of environmental manipulation on aging processes and its implications for humans.

### A. The Dietary Restriction Model

With the exception of selective breeding, used especially in less complex animal species (Chapter 6), dietary restriction has been, so far, the only experimental approach to extending the life span that has been successfully reproduced in a wide range of invertebrate and vertebrate species, and the only one in mammalian species. The majority of studies relying on dietary restriction has been conducted in rodents; indeed, it is in diet-restricted rats that life span extension was first reported in 1935 by McCay and collaborators.[78] A number of strategies have since been used to obtain robust and reproducible results.[79–81] As summarized in Table 24.11, these strategies take into account the following.

### a. Species Selection

Species and strains most used include the rat (primarily Fisher, Brown Norway, and a $F_1$ hybrid of the two), the mouse (primarily, C57BLand DBA or B6), the hamster (Syrian), and some primate species (rhesus monkey, squirrel monkey, and others), although for the latter species, data are still incomplete.[81–83] Also available are animals selectively bred as models of human pathology or short-lived strains carrying a harmful mutation; a number of these symptoms have been reported to be ameliorated by caloric restriction.[83,84]

### b. Health Status

Animals may be kept under isolated, germ-free hysterectomy-derived, barrier–maintained environment or may be kept under conventional conditions without any of these protections. Under all conditions, if the animals remain in good health, caloric restriction will extend mean and maximum survival.

### c. Age of Onset

In early experiments using rodents, caloric restriction was initiated at weaning and continued for 1 year, at the end of which the animals were returned to full feeding. It is now accepted that the greatest effect (30 to 70% life extension) of caloric restriction regimens on subsequent survival occurs in animals underfed throughout most of their postweaning life. Successful caloric restriction, when started in adult ages, depends on gradual adaptation to the reduced rations; the expectation is of a lesser (10 to 20%) shortening of life span extension than that obtained (30 to 70%) when the restricted regimen is started at weaning.[79]

### d. Type of Diet

Restricted feeding regimens capable of successfully extending the life span need to provide all essential nutrients and vitamins, while restricting caloric intake by 30 to 70% of the ad libitum intake. A detailed list of dietary components that assure success in a calorie-restricted diet is presented in the excellent review by Merry.[85] Diets may be "non-purified" (that is, composed mainly of unrefined plant and animal materials with added minerals and vitamins) or "purified" (that is, manufactured from refined components). In general, commercially available diets are designed to maximize growth and fertility in young rodents rather than to promote longevity;[80] further research is required to determine the optimal composition of longevity-promoting diets.

Another type of diet is one in which an essential amino acid has been significantly reduced such that the diet becomes extremely unpalatable and the rats spontaneously reduce their food intake. Rats on a tryptophan-deficient diet, for example, grow significantly less than controls (animals fed the same diet with normal tryptophan levels). When compared to calorie-restricted rats, they present a similar degree of life span extension with prolonged functional competency in old age and delayed appearance of pathology.[86] This approach obviates the need to measure the amount of food allotted daily to each animal.

Changing the temporal pattern of food intake from "nibbling" for the ad libitum fed to "meal eating" for the calorie restricted does not appear to alter the beneficial effects of caloric restriction.[87]

### e. Housing

Individual versus group housing may influence body weight of rats, because group housing may impose an additional stress on the animals and, thereby, affect their growth. On the other hand, some studies report lower body weights in singly housed animals as compared to those housed in groups, where, presumably, the rats can huddle to conserve body heat (Figure 24.3).

## B. Effect on Longevity

That caloric restriction significantly extends survival of rats and mice is shown by the example illustrated in Figure 24.4.[88] As already discussed, the extent of life prolongation depends on species of animal, its strain and sex, the age at which the regimen is instituted, the animal health status, the type, severity, and duration of restriction, and the housing conditions (Table 24.11). A useful comparison of these parameters in selected strains of rats

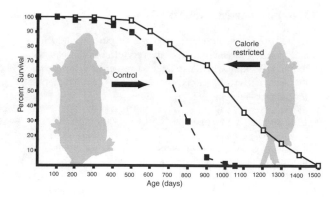

**FIGURE 24.4** The effect of calorie-restricted feeding on lifetime survival for CFY male rats. Control rats were fed ad libitum, while food intake for experimental animals was maintained at approximately 50% that of age-matched control animals from weaning. (Reproduced from Merry, B.J., *Biologist*, 46, 114, 1999. With permission from the Institute of Biology.)

and mice is provided by Weindruch and Walford[79] and by Merry.[85]

## C. Enhancement of Functional Capacity and Delay of Diseases Associated with Old Age

In addition to extending the mean and maximum life span, caloric restriction in rodents affects a number of functional changes and modifies the course of diseases associated with old age. Those most frequently reported are as follows.

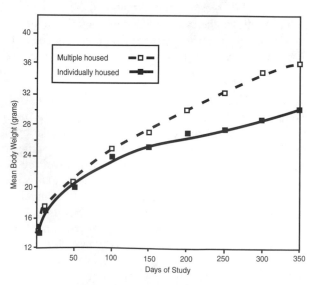

**FIGURE 24.3** Comparison of body growth (body weight in grams) in individually housed and multiple-housed B6D2F1, female mice from weaning until 1 year of age. Multiple-housed mice showed a significantly heavier body weight than those housed individually, starting on day 150 and continuing until the end of the experiment. (Courtesy of L. McCook.)

**TABLE 24.11**
**Schematic Flow Chart of Steps and Selections Needed to Produce Dietary Extended Survival***

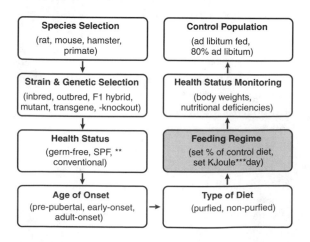

---

\*   Adapted from Reference 85 with permission.

\*\*  hysterectomy-derived and barrier-maintained with routine microbiological screening

\*\*\* Joule, a unit of work or energy equal to 0.239 calories.

## 1. Reproduction

Original studies in this research area linked arrested development with enhanced survival; however, the dietary restriction was so severe and the duration so long (up to 2 1/2 years) that rats lived longer but remained in a pre-pubertal condition.[89] However, re-feeding the animals resulted in rapid sexual maturation and fertility as indicated by the production and survival of litters. When caloric restriction is moderate, puberty is slightly delayed in rats but markedly delayed in mice, and there is no loss of fertility.[90,91]

## 2. Endocrine and Neuroendocrine Responses

The delay in maturation of reproductive function in rats and mice kept on a calorie-restricted diet has been attributed to retarded maturation of the *hypothalamo–pituitary–gonadal axis* and its regulation by central nervous system and hormonal feedbacks rather than to intrinsic alterations of the ovary or testis. Other neuroendocrine axes appear to be similarly modified by caloric restriction. The *hypothalamo–pituitary–adrenocortical (HPA) axis*, for example, follows a triphasic pattern: on initiation of caloric restriction, it is stimulated (underfeeding may be considered a form of stress), but continued underfeeding desynchronizes the circadian corticoid rhythm and depresses responses to stress. When caloric restriction is further prolonged (and the animal has adapted to the stress of being underfed) and when normal feeding has been restored, the HPA's ability to respond to stress is gradually recovered. In other words, the inability of calorie-restricted animals to increase corticosterone in response to the stress of underfeeding may originate at the hypothalamo-pituitary level rather than at the adrenal cortex.

With aging, the ability of the anterior pituitary to secrete *growth hormone* in response to several stimuli (e.g., hypoglycemia, arginine administration) usually capable of triggering such response in young animals, is significantly diminished in rats and humans. This decrease has been ascribed to a reduced secretion from the hypothalamus of Growth Hormone Releasing Hormone (GHRH) (perhaps due to increased secretion of the inhibitory hormone, somatostatin) or to diminished responsiveness of the pituitary to the stimulatory GHRH (Chapter 10). The expected decline in growth hormone levels with old age is associated with an aging-related loss of nocturnal surges of the hormone and a small decrease in plasma *Insulin-like Growth Factor-I (IGF-I)*. Administering these hormones to old animals has shown some beneficial effects, suggesting that their absence may contribute to the aging phenotype.

Caloric restriction would have the effect of increasing growth hormone secretion and growth hormone receptors in muscle and liver as well as stimulating the activity of enzymes involved in protein synthesis.[92] These actions are known to ameliorate the rate of protein synthesis in muscle and liver and to improve cerebral flow and brain function.[93] In contrast, caloric restriction would reduce IGF-I levels, the latter a risk factor for tumorogenesis. The reduction of IGF-I levels would reflect the reduced capacity of growth hormone to induce IGF-I gene expression.[93] Changes in growth hormone/IGF-I levels may have profound effects: an increase in growth hormone because of its anti-insulin effect, may be a contributing factor in late-onset diabetes, whereas a decrease in IGF-I secretion may protect against the risk of tumors.

## 3. Immunologic Responses

A common response of mice, rats, guinea-pigs, and monkeys to moderate but chronic caloric restriction is a depression of antibody production and natural killer cell activity and an enhancement of cell-mediated immunity[94] (Chapter 15). Thymus development is delayed in calorie-restricted animals, but, despite a reduction of cell numbers in immune organs and white cells in blood, resistance to pathology is enhanced, and survival is extended. These responses, markedly influenced by genetic factors, may be mediated through changes in fatty acid composition of phospholipid fractions of cell membranes.[95] While the incidence of autoimmune diseases increases with normal aging (Chapter 15), it is significantly decreased in the presence of caloric restriction, especially in autoimmune-prone mouse strains. The mechanism(s) of this protective action remain(s) to be elucidated.

## 4. Metabolic Responses

Early studies in rodents suggested that caloric restriction reduced *metabolic rate*. Current research, including work with monkeys[96,97] indicates that this is not the case; the animals on calorie-restricted diets adapt by reducing tissue deposition and increasing the efficiency of energy utilization, thereby preventing obesity and maintaining a metabolic rate similar to that of fully fed animals.[98] More controversial are the effects of dietary restriction on *core body temperature:* superimposed on the decrease in core temperature associated with aging in control and restricted animals are temperature increases that may be related to the timing and periodicity of feeding. Other effects of dietary restriction that have been extensively reported are reductions of *mitochondrial damage, free radical accumulation, and lipid peroxidation* (Chapter 5). This reduction would be mediated, at least in part, by the decrease in energy expenditure brought about by the increased activity of those enzyme systems responsible for detoxification of reactive oxygen species.[99]

In rodents, a consistent, but still little understood, metabolic response to dietary restriction is *improved insu-*

*lin sensitivity* and permanently lowered glycemia. Improved insulin sensitivity may be due to tissue-specific effects on cell insulin receptor binding (perhaps related to effects on membrane lipid composition) that do not alter glucose transporter protein content.[100–102] A similar action has been reported in monkeys; insulin sensitivity is increased in parallel with a decrease in fat deposits.

Because experiments in monkeys are still in progress, physiopathologic and metabolic changes with aging and their possible modification by caloric restriction cannot yet be ascertained.[82] Focus on two metabolic outcomes of dietary restriction shows a reduction in oxidative stress and improved glucose regulation.[103]

These actions can be imitated or "mimicked" by interventions other than caloric restriction that have similar effects.[103–105] Research *on caloric restriction "mimetics"* includes improvement of mitochondrial function, use of antioxidants, administration of compounds known to lower blood glucose levels and increase insulin sensitivity, regular physical exercise activity, maintenance of body weight and composition over the life span, and others.[103–105] The antidisease markers observed thus far suggest that animals subjected to caloric restriction mimetics may escape the development of diabetes, cardiovascular disease, obesity, immune dysfunction, and possibly cancer more successfully than their counterparts fed ad libitum. As discussed in Chapter 1, one major challenge in aging studies is the choice of biomarkers reflecting the rate of aging and concomitant functional capacity. This choice is especially critical when working with long-lived species such as monkeys and, eventually, humans. Interspecies comparisons as well as the use of several markers that can be measured routinely (and are minimally invasive) in biochemical and hematologic profiles show considerable promise not only for improving our understanding of the effects of caloric restriction in monkeys but also for their possible applicability to humans.[105] So far, several blood chemistry measures (e.g., total protein, several enzymes, creatinine, dehydroepiandrosterone, T-cell subsets) show a trend toward slower age-related change in calorie-restricted than ad libitum fed monkeys.[83,107,108]

## 5. Collagen Responses

The rate of cross-linkage between the fibers of the structural protein collagen, a component of connective tissue, is usually taken as a biomarker for the rate of aging. In old animals, diminished gene expression for collagens and proteinases may slow collagen turnover and increase the cross-linking of collagen molecules[106] (Chapter 22). Dietary restriction reduces the amount of total collagen, the amount and type of cross-links, and the accumulation of advanced glycosylation products in several tissues (e.g., kidneys, lung, liver).[109] While this retardation may have a beneficial effect on cataract development (Chapter 9), it may have adverse effects on wound healing,[107] despite a trend toward faster wound closure.[110,111]

## 6. Tumor Incidence

With respect to tumors, the effects of caloric restriction vary depending on the type of tissue or organ and the animal species or strain. In both rats and mice, the frequency of tumors is reduced in pituitary gland, lung, pancreatic islet cells, liver, mammary gland, and skin; increased in epithelial tissue and adrenal and parathyroid glands, and reticulum cells of lymphoid glands; and unaffected in soft tissues, thyroid gland, and bladder.

In addition to its effects on spontaneous tumors, caloric restriction also confers a certain degree of protection against tumors (primarily skin and intestine) that are induced by highly mutagenic exogenous carcinogens. This protective action has been attributed to the animal's reduced body fat, which, in turn, limits the storage space for carcinogenic substances and increases their metabolism.

## D. Effect on Gene Expression Profile

Studies in skeletal muscle (which is composed of long-lived, high-oxygen-consuming postmitotic cells) of C57BL/6 mice have revealed that aging results in a differential gene expression pattern indicative of a marked stress response (characterized by cell energetic deficit, mitochondrial dysfunction, and damaged proteins) and lower expression of metabolic and biosynthetic genes.[112] Of the 6347 genes surveyed, 58 (0.9%) displayed a greater than twofold increase in expression levels and 55 (0.9%) a greater than twofold decrease in gene expression; thus, the aging process does not appear to be due to a large, widespread alteration in gene expression. Of the 58 genes that increased in expression with age, 16% were mediators of stress responses, in particular, genes encoding enzymes involved in free radical inactivation, neuronal growth, and neurite extension. Of the genes showing decreased expression with age, the majority were involved in energy metabolism (especially mitochondrial enzymes), carbohydrate metabolism, and biosynthetic enzymes (for protein and lipid synthesis and turnover).[112]

After chronic (28 months) reduction of calorie intake (76%), aging-related changes that occur in mice fed ad libitum, were markedly (84%) attenuated: of those characterized by increased gene expression, 29% were completely prevented and 34% greatly diminished. Caloric restriction may act through metabolic reprogramming with a transcriptional shift (in flies, perhaps triggered by insulin[113]), toward reduced energy metabolism and increased biosynthesis and protein turnover.[114] This modification of the metabolic profile of old mice may vary, depending not only on caloric restriction but also on strain, sex, and age. As mentioned above, caloric restriction markedly influences the expression of pathologic genotypes in rodent species that have been selectively bred as models of human pathology.[115,116]

## E. Mechanisms Underlying the Effects of Caloric Restriction

Several hypotheses have been proposed to explain the mechanisms by which caloric restriction prolongs mean and maximum life span as well as what we are calling the "health span" (period of life without disease and disability). To what extent these hypotheses overlap or intersect is not known. None has yet adequately accounted for the diversity of effects reported, their specificity in terms of animal species or strain, and the interactions between genetic and environmental influences. Early theories suggested that *neuroendocrine signals* that regulate development, metabolism, reproduction, adaptation to the environment, and senescence (regarded as failure or exhaustion of adaptive capability) (Chapter 10) may be modified by caloric restriction.[93,117,118]

*Evolutionary theories* (Chapter 6) suggested that caloric restriction could be viewed as part of a spectrum of neuroendocrine responses to environments with variable food levels that may have adaptive value during periods of short–term food shortage in the wild.[119–121] According to current theories, caloric restriction is thought to exert its effects by protecting *against free radical accumulation* (Chapter 5), by *upregulating apoptosis* (Chapter 4) or by *changing gene expression* and *suppressing DNA damage*.

Studies in yeast have shed some light on interactions between the genetics and the physiology of aging and reaffirmed the usefulness of model organisms.[122] Yeast undergo only a finite number of divisions before they die. Caloric restriction can be induced in yeast by limiting glucose availability or by genetically impairing their ability to sense or respond to glucose. Under caloric restriction, yeast longevity is extended by 20 to 40%, comparable to life span extension in mammals.[122,123] For this extension to take place, two yeast genes are required: NPTI which encodes one of two enzymes that produce nicotinamide adenine dinucleotide, a key intermediate in energy metabolism, and SIR2, which encodes a protein that by promoting a compact chromatin structure prevents (silences) gene transcription at selected loci. Caloric restriction may facilitate repair of the damage induced by reactive oxygen species and restore energy production; it may also reduce the loss of chromatin silencing and thereby prevent alterations in gene expression with restoration of cell function to withstand stressful stimuli.[123,124] Experiments in worms, flies, and yeast suggest a common process that controls life spans and that can be modified by caloric restriction by synergizing energy metabolism

through gene expression of specific proteins.[125] However, more research is needed before these observations can be extrapolated to mammals (and particularly to humans).

## F. Human Studies

Of the many caloric restriction studies conducted in humans, some use historical anecdotes or draw on social or religious customs for data, whereas others rely on more rigorously conducted experiments which, however, are hampered by the lack of a sufficient number of subjects and a sufficiently long study period; as a result, no definite conclusions about the impact of caloric restriction on aging in humans can be put forth. Various other research projects designed to explain the potential effects of caloric restriction in humans are in the planning stage.[126–128]

In the first category is a 3-year study of Spanish nursing home residents (average age: 72 years) in whom morbidity was reduced by limiting food intake, although mortality rates remained unaffected.[79,128] Another study was conducted among Islamic individuals who, during the month of Ramadan, fasted during the daylight and then consumed a meal after sundown; their HDL plasma levels were found to have increased by 30%.[79,129] In a study of Japanese living in Okinawa, it was reported that Okinawans have a longer life span (81 years) and six times as many centenarians per 100,000 people than those living in the United States, and 97% of their lives were spent free from disabilities.[79,130] Their longevity has been correlated with a diet relatively low in calories and high in fruits and vegetables. However, in addition to differences in the amount and composition of the diet, other factors (e.g., strong social, cultural, and religious ties, insular isolation) distinguished the Okinawan group.

Among the controlled experiments is an early study conducted by Keys in a small number of subjects placed on a diet that was not only restricted in quantity but also was deficient in critical micronutrients. Two recent examples are the Biosphere 2 (in Arizona)[131] and the Toxicology and Nutrition Institute study (in the Netherlands).[132] In Biosphere 2, although some changes in body weight and in blood biochemistry (reduced levels of fasting glucose, cholesterol, triglyceride) were reported, the small number of subjects and the lack of appropriate controls made it impossible to draw definitive conclusions. Similarly, for the Toxicology and Nutrition Institute study, a reduction in body mass and an increase in HDL levels were reported, but again, definitive conclusions could not be drawn.

More comprehensive studies have been proposed,[133] but so far, none have been launched. Research in this area is currently focused on identifying appropriate criteria that will allow us to extrapolate the findings demonstrated in calorie-restricted animals to humans. Thus far, criteria include (1) identification, development, and

**Prevention and Rehabilitation**

confirmation of biomarkers indicative of vitality, disease, and longevity in humans; (2) criteria for selection of subjects; (3) strategy(ies) for caloric restriction, including identification of substitute interventions (e.g., chemicals, herbs, hormones) mimicking the effects of dietary restriction, and (4) assessment of risks versus benefits. These are only a few of the factors to be considered. Social and cultural aspects of human populations compound our physiologic complexity and make research in this area extremely difficult.

## V.   FUTURE STRATEGIES IN BIOGERONTOLOGY

The major purpose of this book has been to present an overall, systematic view of physiologic aging in humans and for comparison and elucidation, in experimental animals and cultured cell models. Given the increased incidence of disease associated with aging, a few diseases have been selected for brief discussion to serve as examples of how pathology influences physiologic adjustments with advancing age. Maintaining or restoring a state of "wellness" in old age demands that physiologic responses be strengthened and disease eliminated. While we have seen some success in preventing and eliminating a number of diseases in recent years (primarily infectious diseases), much less has been achieved when it comes to strengthening basic functions and preventing or repairing functional decline in old age. As presented in this chapter, attention to an appropriate diet and regular physical activity are two interventions that, in combination, promote functional fitness at all ages.

Other types of strategies for combating disease and prolonging "health span" and "life span" come from daily biomedical advances and their technologic applications. Some of them have been discussed throughout the book. In this 21st century, bioscience is benefiting from the previous century's discoveries, especially those involving the role of antibiotics in reducing infectious diseases, as mentioned above and discussed in Chapter 2, and the identification of DNA as the carrier molecule of genetic information in cells. At the beginning of this century, the sequencing of the genome in several animal species, including humans, has opened exciting avenues for basic and applied research.

The recognition that cells such as neuron and muscle and cardiac cells, previously considered incapable of proliferating in old age, retain the potential for plasticity and regeneration has opened the way to the study of intrinsic and extrinsic factors that may facilitate tissue repair in neurodegenerative diseases and in gene therapy. Likewise, the identification of pluripotent cells, such as stem cells, potentially capable of acquiring the structure and function of a variety of specialized cells promises to let us effectively replace lost tissues and functions.

As suggested in Chapter 1, continuing advances in human genetics will lead to the identification not only of the genetic etiology of several diseases but also of the genetic variations that, combined with environmental risks, make people vulnerable to numerous diseases. Progress in this area will open the way for genetic therapy to restore the normal genome, and will provide the basis for individualizing therapeutic treatments.

### LIFE EXTENSION SCIENCES

Research in this area purports to extend as well as to improve life in old age. Some of the areas being explored, under the rubric of "life extension sciences," are still controversial. They include, among many others:

- Use of dietary caloric restriction discussed above
- Tissue and organ preservation for transplants (e.g., by cryopreservation) and prevention of tissue and organ rejection
- Uses of deep hypothermia (for special surgical procedures) and of cryonics (for subsequent resuscitation)
- Bioengineered improvements for artificial internal organs (e.g., artificial kidney, heart) and prostheses for replacing malfunctioning body parts
- Progress in medical visualization (imaging) technology to improve our understanding of function and the diagnosis and treatment of diseases
- Gene therapy and technology for artificial reproduction
- Development of new, improved, and genetically individualized pharmacological products for therapeutic purposes
- Cloning techniques to produce "human cloned bodies" to be used as a source of young organs capable of being back-transplanted to replace failing organs in the older individual (from whom the original nucleus had been obtained). After removal of part of the brain (to prevent the development of a sentient human being), and rapid hormone-stimulated growth, the clone could serve as a cell, tissue, or organ donor to supply healthy youthful parts. Nuclear transfer techniques, proven successful for cloning goats, cattle, mice, and pigs, may be applied to human clones, despite the potential risks to the clone and strong ethical opposition.

While some of the predictions for life extension may sound like science fiction, given the monumental changes taking place every day, these predictions are not that outrageous. Seeking immortality in the arduous old-fashioned way, doing good deeds, and taking care of [one's] children remains a goal worthy of pursuit; but it needs not halt efforts aimed at improving and lengthening life. Aging and death remain, indeed, the last sacred enemies, a fact that is particularly frustrating to humans who have now harnessed nuclear energy, circled the moon, decoded the human genome, artificially reproduced DNA, and significantly extended life expectancy. Such intrepid individuals can be expected to continue striving to improve the quality of life at all stages, as well as to extend the duration of life.

## ACKNOWLEDGMENT

The authors wish to thank Dr. L. Ritchie from the Department of Nutritional Sciences and Toxicology, College of Natural Resources, University of California at Berkeley, and Ms. M.L. Zernicke, from the Alameda County Area Agency on Aging, California, for their valuable comments and updates, especially with respect to nutrition.

## REFERENCES

1. Roth, G.S., Ingram, D.K., and Lane, M.A., Calorie restriction in primates: Will it work and how will we know?, *J. Am. Geriatr. Soc.*, 47, 896, 1999.

2. Mattson, M.P., Neuroprotective signaling and the aging of the brain: Take away my food and let me run, *Brain Res.*, 886, 47, 2000.

3. Miller, H., in *The Oxford Book of Ages,* Sampson, A. and Sampson S., Eds., Oxford University Press, Oxford, 1985, p. 151.

4. Picasso, P., in *The Oxford Book of Ages,* Sampson A. and Sampson S., Eds., Oxford University Press, Oxford, 1985, p. 118.

5. Minkler, M. and Pasick, R.J., Health promotion in the elderly: A critical perspective on the past and future, in *Wellness and Health Promotion in the Elderly*, Dychtwald, K., Ed., Aspen Systems Corp., Rockville, MD, 1986.

6. Minkler, M., Schauffer, H., and Clemens-Nolle, K., Health promotion for older Americans in the 21st century, *Am. J. Health Promot.*,14, 371, 2000.

7. Henretta, J.C., The future of age integration in employment, *The Gerontologist,* 40, 286, 2000.

8. Loscocco, K., Age integration as a solution to work-family conflict, *The Gerontologist,* 40, 292, 2000.

9. Damnefer, D., Paradox of opportunity, education, work and age integration in the United States and Germany, *The Gerontologist*, 40, 282, 2000.

10. Tolstoy, L., in *The Oxford Book of Ages*, Sampson A. and Sampson S., Eds., Oxford University Press, Oxford, 1985, p. 148.

11. American Dietetic Association, Nutrition, Aging and the Continuum of Health Care, *J. Am. Diet. Assoc.*, 93, 81, 1993.

12. Reuben, B.D., Greendale, A.G., and Harrison, G.G., Nutrition screening in older persons, *J. Am. Geriatr. Soc.*, 43, 415, 1995.

13. Dwyer, J.T., Dietary assessment, in *Modern Nutrition in Health and Disease*, 8th ed., Shils, M.E., Olson, J.A., Shike, M., Eds., Lea & Febiger, Philadelphia, 1994.

14. Chumlea, W.C., Roche, A.F., and Muklerjee, D., Some anthropometric indices of body composition for elderly adults, *J. Gerontol.*, 41, 36, 1986.

15. Ferguson, R.P. et al., Serum albumin and pre-albumin as predictors of clinical outcomes of hospitalized nursing home residents, *J. Am. Geriatr. Soc.,* 41, 545, 1993.

16. Cusak, B.J. and Parker, B.M., Pharmacology and appropriate prescribing, in *Geriatric Review Syllabus: A Core Curriculum in Geriatric Medicine,* 3rd ed., Reuben, D.B., Yoshikawa, T.T., and Besdine, R.W., Eds., The American Geriatric Society, New York, 1996.

17. Beers, M.H. and Berkow, R., Obesity, in *The Merck Manual of Geriatrics*, 3rd ed., Merck Research Laboratories, Whitehouse Station, New Jersey, 2000.

18. Corpas, E., Harman, S.M., and Blackman, M.R., Human growth hormone and human aging, *Endocrinol. Rev.*, 14, 20, 1993.

19. Forse, R.A. and Shizgal, H.M., Serum albumin and nutritional status, *J. Parenter. Enter. Nutr. (JPEN)*, 4, 450, 1980.

20. Food and Nutrition Board, Committee on Dietary Allowances, *Recommended Dietary Allowances*, 9th ed., National Academy of Sciences, National Research Council, Washington, DC,1989.

21. Food and Nutrition Board, How should the recommended dietary allowances be revised? *Nutr. Rev.*, 52, 216, 1994.

22. Keys, A., Mediterranean diet and public health: Personal reflections, *Am. J. Clin. Nutr.*, 61, 1321S, 1995.

23. United States Department of Agriculture, *Dietary Guidelines for Americans*, 4th ed., U.S. Departments of Agriculture, Health, and Human Services, Washington, DC, 1995.

24. Summary of the Surgeon General's Report addressing Physical Activity and Health, *Nutr. Rev.*, 54, 280, 1996.

25. Premuda, L., *Storia della Medicina*, Milani, Padova, 1975.

26. Malizia, E., *Il ricettario delle Streghe*, Edizioni Mediterranee, Roma, 1992.

27. Gersovitz, M. et al., Human protein requirement: Assessment of the adequacy of the current recommended dietary allowance for the dietary protein in elderly men and women, *Am. J. Clin. Nutr.*, 35, 6, 1982.

28. Millward, D.J. et al., Aging, protein requirement and protein turnover, *Am. J. Clin. Nutr.*, 66, 774, 1997.

29. Schneeman, B.O. and Tietyen, J., Dietary fiber in *Modern Nutrition in Health and Disease*, Shils, M.E., Olson, J.A., and Shike, M., Eds., Lea & Febiger, Philadelphia, PA, 1994.

30. Apple, L.J. et al., Does supplementation of diet with fish oils reduce blood pressure? A meta-analysis of controlled clinical trials, *Arch. Int. Med.*, 153, 1429, 1993.

31. Albert, M.C. et al., Blood levels of long-chain n-3 fatty acids and the risk of sudden death, *N. Engl. J. Med.,* 346, 1113, 2002.

32. Ernst, N.D., Consistency between U.S. fat dietary intake and serum total cholesterol concentration: The National Health and Nutrition Examination Surveys, *Am. J. Clin. Nutr.*, 66, 965, 1997.

33. Bettie, L.B. and Louie, V.Y., Nutrition and Aging, in *Reichel's Care of the Elderly, 5th ed.,* Gallo, J.J. et al., Eds., Williams and Wilkins, New York, 1999.

34. Nowak, R., Beta-carotene: Helpful or harmful?, *Science*, 264, 500, 1994.

35. Bikle, D.D., Agents that affect bone mineral homeostasis, in *Basic and Clinical Pharmacology*, 7th ed., Katzung, B.G., Ed., McGraw-Hill, New York, 1998.

36. Vitamins and trace mineral disorders, in *Merck's Manual of Geriatrics*, 3rd ed., Beers, M.H. and Berkow, R., Eds., Merck Research Laboratories, Whitehouse Station, NJ, 2000.

37. Myelin, S.N., et al., Vitamin $B_6$ deficiency impairs interleukin-2 production and lymphocyte proliferation in elderly adults, *Am. J. Clin. Nutr.*, 53, 1275, 1991.

38. McNulty, H., Folate requirements for health in different population groups, *Br. J. Biomed. Sci.*, 52, 110, 1995.

39. Alpert, J.E. and Fava, M., Nutrition and depression: The role of folate, *Nutr. Rev.*, 55, 145, 1997.

40. Bernard, M.A., Nakonezny, P.A., and Kashner, T.M., The effect of Vitamin $B_{12}$ deficiency on older veterans and its relationship to health, *J. Am. Geriatr. Soc.*, 46, 1199, 1998.

41. Norman, E.J., Urinary methylmalonic acid/creatinine ratio defines true tissue cobalamin deficiency, *Br. J. Haematol.*, 100, 614, 1998.

42. Crane, M.G., Studies on two total vegetarian (vegan) families, *Veg. Nutr.* 2, 87, 1998.

43. Burns, J.J. et al., Third conference on Vitamin C., *Ann. N.Y. Acad. Sci.*, 498, 534, 1987.

44. Wood, R.J., Suter, P.M., and Russell, R.M., Mineral requirements of elderly people, *Am. J. Clin. Nutr.*, 62, 493, 1995.

45. Harris, S.S. and Dawson-Hughes, B., Caffeine and bone loss in healthy post-menopausal women, *Am. J. Clin. Nutr.*, 60, 573, 1994.

46. Salonen, J.T. et al., High stored iron levels are associated with excess risk of myocardial infarction in Eastern Finnish men, *Circulation*, 86, 803, 1992.

47. Ascherio, A. and Willett, C.W., Are body iron stores related to the risk of coronary artery disease? *N. Engl. J. Med.*, 330, 1152, 1994.

48. Berr, C. et al., Selenium and oxygen-metabolizing enzymes in elderly community residents: A pilot epidemiological study, *J. Am. Geriatr. Soc.*, 41, 143, 1993.

49. Boosalis, M., Stuart, M.A., and McClain, C.J., Zinc metabolism in the elderly, in *Geriatric Nutrition: A Comprehensive Review*, Morley, J.E., Glick, Z., and Rubenstein, L.Z., Eds., Raven Press, New York, 1995.

50. Taubes, G., A DASH of data in the salt debate, *Science*, 288, 1319, 2000.

51. McGinnis, G.M. and Foege, W.H., Actual causes of death in the United States, *J. Am. Med. Assoc.*, 270, 2207, 1993.

52. Hines, M.L. et al., Genetic variation in alcohol dehydrogenase and the beneficial effect of moderate alcohol consumption on myocardial infarction, *N. Engl. J. Med.*, 344, 549, 2001.

53. Orthostatic Hypotension, in *Merck's Manual of Geriatrics*, 3rd ed., Beers, M.H. and Berkow, R., Eds., Merck Research Laboratories, Whitehouse Station, New Jersey, 2000, p. 845.

54. Pendergast, M., *Uncommon Grounds, The History of Coffee and How It Transformed Our World*, Basic Book Ed., New York, 1999.

55. *Dietary Reference Intakes for Vitamin C, E, Selenium and Carotenoids*, National Academy Press, Washington, D.C., 2001.

56. Jameson, J.L., Principles of endocrinology, in *Harrison's Principles of Internal Medicine*, 15th ed., Braunwald, E. et al., Eds., McGraw-Hill, New York, 2001.

57. Beck, D., Casper, R., and Andersen, A., Truly late onset of eating disorders: A study of 11 cases averaging 60 years of age at presentation, *Int. J. Eating Disord.*, 20, 389, 1996.

58. Surgeon General's Report on the Physical Activity and Health, U.S. Department of Health and Human Services, National Center for Disease Prevention and Health Promotion, Atlanta, GA, 1996.

59. Brooks, G.A. et al., *Exercise Physiology*, 3rd ed., Mayfield Publishing Co., Mountain View, 2000.

60. Andersen, J.L., Schierling, P., and Saltin, B., Muscles, genes, and athletic performance, *Sci. Am.*, 11, 48, 2000.

61. Kiens, et al., Skeletal muscle substrate utilization during submaximal exercise in man: Effects of endurance training, *J. Physiol. (Camb.)*, 469, 459, 1993

62. Svedenhag, J., Henrickson, J., and Sylven, C., Dissociation of training effects on skeletal muscle mitochondrial enzymes and myoglobin in man, *Acta. Physiol. Scand.*, 117, 213, 1983.

63. Davies, K.J.A. et al., Muscle mitochondria bioenergetics, oxygen supply, and work capacity during dietary iron deficiency and repletion, *Am. J. Physiol.*, 242, E418, 1982.

64. Coggan, A.R., Spina, R.J., and King, D.S., Skeletal muscles adaptation to endurance training in 60 to 70 year old men and women, *J. Appl. Physiol.*, 72, 1780, 1992.

65. Cress, M.E. et al., Effects of training in $VO_2$ thigh strength, muscle morphology in septuagenarian women, *Med. Sci. Sports Exerc.*, 23, 752, 1991.

66. Pyka, G. et al., Muscle strength and fiber adaptation to a year-long resistance training program in elderly men and women, *J. Gerontol.*, 49, 22, 1994.

67. Fiatarone, M. et al., Exercise training and nutritional supplementation for physical frailty in very elderly people, *N. Engl. J. Med.*, 330, 1769, 1994.

68. Chestnut, C.H., Bone mass and exercise, *Am. J. Med.*, 959 Suppl 5A, 34, 1993.

69. Drinkwater, B.L., Physical activity, fitness, and osteoporosis, in *Physical Activity, Fitness and Health: International Proceedings and Consensus Statement*, Boucher, C., Sheperd, R.J., and Stephens, T., Eds., Human Kinetics Publishers, Champaign, IL, 1994.

70. Tipton, C.M. and Vailas, A.C., Bone and connective tissue adaptation to physical activity, in *Exercise, Fitness and Health: A Consensus of Current Knowledge*, Bouchard, C. et al., Eds., Human Kinetics Books, Champaign, IL, 1994.

71. Cheitlin, M.D. et al., ACC/AHA guidelines for the clinical application of echocardiography: A report of the ACC/AHA Task Force on the practice guidelines, *Circulation*, 95, 1686, 1997.

72. Pina, I.L. et al., Guidelines for clinical exercise testing laboratories: A statement from the committee on exercise testing and cardiac rehabilitation, *Circulation*, 91, 912, 1995.

73. Lakatta, E.G., Gerstenblith, G., and Weisfeld, M.L., The aging heart, structure, function, and disease, in *Heart Disease, A Textbook of Cardiovascular Medicine*, 5th ed., Braunwald, E., Ed., W.B. Saunders Co., Philadelphia, PA, 2000.

74. O'Brian Cousins, S., *Exercise, Aging and Health: Overcoming Barriers to an Active Old age,* Taylor and Francis, Eds., Philadelphia, PA, 1997.

75. Balady, J.G., Recommendations for cardiovascualar screening, staffing, and emergency policies at health/fitness facilities, *Circulation*, 97, 2283, 1998.

76. Busby-Whitehead, J., Exercise: The right prescription, *J. Am. Geriatr. Soc.*, 43, 308, 1995.

77. Swanson, E.A., Tripp-Reimer, T., and Buckwalter, K., *Health Promotion and Disease Prevention in the Older Adults*, Springer, New York, 2001.

78. McCay, C.M., Crowell, M.F., and Maynard, L.A., The effect of retarded growth upon the length of the lifespan and ultimate body size, *J. Nutr.*, 10, 63, 1935.

79. Weindruch, R. and Walford, R.L., *The Retardation of Aging and Disease by Dietary Restriction*, Charles C. Thomas, Ed., Springfield, 1988.

80. Sprott, R.L. and Austad, S.N., Animal models for aging research, in *Handbook of the Biology of Aging*, Schneider, E.L. and Rowe, J.W., Eds., Academic Press, London, England, 1996.

81. Lane, M.A., Ingram, D.K., and Roth, G.S., Beyond the rodent model: Caloric restriction in rhesus monkeys, *Age*, 20, 45, 1997.

82. Roth, G.S., Ingram, D.K., and Lane, M.A., Calorie Restriction in Primates: Will it work and how will we know?, *Am. J. Geriatr. Soc.*, 47, 896, 1999.

83. Nakamura, E. et al., A strategy for identifying biomarkers of aging: Further evaluation of hematology and blood chemistry data from a calorie restriction study in rhesus monkeys, *Exp. Gerontol.*, 33, 421, 1998.

84. Hursting, S.D., Perkins, S.N., and Phang, J.N., Caloric restriction delays spontaneous tumorigenesis in p53-knockout transgenic mice, *Proc. Natl. Acad. Sci., USA*, 91, 7036, 1994.

85. Merry, B.J., Dietary restriction in aging, in *Studies of Aging*, Sternberg, H. and Timiras, P.S., Eds., Springer, Berlin, 1999.

86. Segall, P.E. and Timiras, P.S., Patho-physiologic findings after chronic tryptophan deficiency in rats: A model for delayed growth and aging, *Mech. Ageing Dev.*, 5, 109, 1976.

87. Masoro, E.J. et al., Temporal pattern of food intake not a factor in the retardation of aging processes by dietary restriction, *J. Gerontol.*, 50A, B48, 1995.

88. Merry, B.J., A radical way to age, *Biologist*, 46, 114, 1999.

89. McCay, C.M. et al., Retarded growth, lifespan, ultimate body size and age changes in the albino rat after feeding diets restricted in calories, *J. Nutr.*, 18, 1, 1939.

90. Holehan, A.M. and Merry, B.J., The control of puberty in the dietary restricted rat, *Mech. Ageing Dev.*, 32, 179, 1985.

91. Koizumi, A. et al., Effects of energy restriction on mouse mammary tumor virus mRNA levels in mammary glands and uterus and on uterine endometrial hyperplasia and pituitary histology in C3H/SHN F1 mice, *J. Nutr.*, 120, 1401, 1990.

92. Xu, X. and Sonntag, W.E., Moderate caloric restriction prevents the age-related decline in growth hormone receptor signal transduction, *J. Gerontol.*, 51, B167, 1996.

93. Sonntag, W.E. et al., Pleiotropic effects of growth hormone and insulin-like growth factor (IGF)-1 on biological aging: Inferences from moderate caloric-restricted animals, *J. Gerontol.*, 54A, B521, 1999.

94. Roecker, E.B. et al., Reduced immune responses in rhesus monkeys subjected to dietary restriction, *J. Gerontol.*, 51, B276, 1996.

95. Venkatraman, J. and Fernandes, G., Modulation of age-related alterations in membrane composition and receptor-associated immune functions by food restriction in Fischer-344 rats, *Mech. Ageing Dev.*, 63, 27, 1992.

96. De Lany, J.P. et al., Long-term caloric restriction reduces energy expenditure in aging monkeys, *J. Gerontol.*, 54, B5, 1999.

97. Colman, R.J. et al., Body fat distribution with long-term dietary restriction in adult male rhesus macaques, *J. Gerontol.*, 54, B283, 1999.

98. Barzilai, N. and Gupta, G., Revisiting the role of fat mass in the life extension induced by caloric restriction, *J. Gerontol.*, 54, B89, 1999.

99. Ramsey, J.J., Harper, M.E., and Weindruch, R., Restriction of energy intake, energy expenditure, and aging, *Free Radic. Biol. Med.*, 29, 946, 2000.

100. Gazdag, A.C. et al., Effect of long-term caloric restriction on GLUT4, phosphatidylinositol-3kinase p85 subunit, and insulin receptor substrate-1 protein levels in rhesus monkey skeletal muscle, *J. Gerontol.*, 55, B44, 2000.

101. Cefalu, W.T. et al., Chronic caloric restriction alters muscle membrane fatty acid content, *Exp. Gerontol.*, 35, 331, 2000.

102. Wang, Z.Q. et al., Effect of age and caloric restriction on insulin receptor binding and glucose transporter levels in aging rats, *Exp. Gerontol.*, 32(6), 671, 1997.

103. Weindruch, R. et al., Caloric restriction mimetics: Metabolic interventions, *J. Gerontol.*, 56A, 20, 2001.

104. Poehlman, E.T. et al., Caloric restriction mimetics: Physical activity and body composition changes, *J. Gerontol.*, 56A, 45, 2001.

105. Lee, J.M. et al., Epidemiologic data on the relationship of caloric intake, energy balance and weight gain over the life span with longevity and morbidity, *J. Gerontol.*, 56A, 7, 2001.

106. Haas, B.S. et al., Dietary restriction in humans: Report on the Little Rock conference on the value, feasibility and parameters of a proposed study, *Mech. Ageing Dev.*, 91, 79, 1996.

107. Lane, M.A. et al., Dehydroepiandrosterone sulfate: A biomarker of primate aging slowed by caloric restriction, *J. Clin. Endocrinol. Metab.*, 82, 2093, 1997.

**Prevention and Rehabilitation**

108. Miller, R.A., Biomarkers of aging: Prediction of longevity by using age-sensitive T-Cell subset determinations in a middle-aged, genetically heterogeneous mouse population, *J. Gerontol.*, 56A, 180, 2001.

109. Ding, H. and Gray, S.D., Senescent expression of genes coding collagens, collagen-degrading metalloproteinases, and tissue inhibitors of metalloproteinases in rat vocal folds: Comparison with skin and lungs, *J. Gerontol.*, 56, B145, 2001.

110. Reiser, K. et al., Effects of aging and caloric restriction on extracellular matrix biosynthesis in a model of injury repair in rats, *J. Gerontol.*, 50A, 40, 1995.

111. Roth, G.S. et al., Effect of age and caloric restriction on cutaneous wound closure in rats and monkeys, *J. Gerontol.*, 52, B98, 1997.

112. Lee, C.K. et al., Gene expression profile of aging and its retardation by caloric restriction, *Science*, 285, 1390, 1999.

113. Tatar, M. et al., A mutant *Drosophila* insulin receptor homolog that extends life-span and impairs neuroendocrine function, *Science*, 292, 107, 2001.

114. Allison, D.B. et al., Genetic variability in responses to caloric restriction in animals and in regulation of metabolism and obesity in humans, *J. Gerontol.*, 56A, 55, 2001.

115. Warner, H.R., Fernandes, G., and Wang, E., A unifying hypothesis to explain the retardation of aging and tumorigenesis by caloric restriction, *J. Gerontol.*, 50A, B107, 1995.

116. Lipman, R.D., Dallal, G.E., and Bronson, R.T., Effects of genotype and diet on age-related lesions in ad libitum fed and calorie-restricted F344, BN, and BNF3F1 rats, *J. Gerontol.*, 54, B478, 1999.

117. Timiras, P.S., Neuroendocrinology of aging: Retrospective, current, and prospective views, in *Neuroendocrinology of Aging*, Meites, J., Ed., Plenum Press, New York, 1983.

118. Mobbs, C.V. et al., Neuroendocrine and pharmacological manipulations to assess how caloric restriction increases life span, *J. Gerontol.*, 56A, 34, 2001.

119. Walford, R.L. and Spindler, S.R., The response to calorie restriction in mammals shows features also common to hibernation: A cross-adaptation hypothesis, *J. Gerontol.*, 52A, B179, 1997.

120. Masoro, E.J. and Austad, S.N., The evolution of the antiaging action of dietary restriction: A hypothesis, *J. Gerontol.*, 51, B387, 1996.

121. Kirk, K.L., Dietary restriction and aging: Comparative tests of evolutionary hypotheses, *J. Gerontol.*, 56, B123, 2001.

122. Guarente, L. and Kenyon, C., Genetic pathways that regulate ageing in model organisms, *Nature*, 408, 255, 2000.

123. Lin, S.J., Defossez, P.A., and Guarente, L., Requirement of NAD and SIR2 for life-span extension by caloric restriction in Saccharomyces cerevisiae, *Science*, 289, 2126, 2000.

124. Imai, S. et al., Transcriptional silencing and longevity protein Sir2 is an NAD-dependent histone deacetylase, *Nature*, 403, 795, 2000.

125. Fabrizio, P. et al., Regulation of longevity and stress resistance by Sch9 in yeast, *Science*, 292, 288, 2001.

126. Strauss, E., Growing old together, *Science*, 292, 41, 2001.

127. Hadley, E.C. et al., Human implications of caloric restriction's effects on aging in laboratory animals: An overview of opportunities for research, *J. Gerontol.*, 56A, 5, 2001.

128. Vallejo, E.A., La dieta de hombre a dias alternos en la alimentacion de los viejos, *Rev. Clin. Exp.*, 63, 25, 1957.

129. Maislos, M. et al., Gorging and plasma HDL-cholesterol-The Ramadan model, *Eur. J. Clin. Nutr.*, 52, 127, 1998.

130. Kagawa, Y., Impact of westernization on the nutrients of Japanese: Changes in physique, cancer, longevity and centenarians, *Prev. Med.*, 7, 205, 1978.

131. Walford, R.L., Harris, S.B., and Gunion, M.W., The calorically restricted low-fat nutrient dense diet in Biosphere 2 significantly lowers blood glucose, total leukocyte count, cholesterol and blood pressure in humans, *Proc. Natl. Acad. Sci. USA*, 98, 11533, 1992.

132. Velthuis, T.E. et al., Energy restriction, a useful intervention to retard human aging? Results of a feasibility study, *Eur. J. Clin. Nutr.*, 48, 138, 1994.

133. Roberts, S.B. et al., Physiologic effects of lowering caloric intake in nonhuman primates and nonobese humans, *J. Gerontol.*, 56A, 66, 2001.

# Index